Natural Product Biosynthesis

Chemical Logic and Enzymatic Machinery

Natural Product Biosynthesis

Chemical Logic and Enzymatic Machinery

Christopher T. Walsh

Chem-H, Stanford University, USA
Email: cwalsh2@stanford.edu

and

Yi Tang

Chemical and Biomolecular Engineering,
Chemistry and Biochemistry, UCLA, USA
Email: yitang@ucla.edu

ROYAL SOCIETY OF CHEMISTRY

THE QUEEN'S AWARDS FOR ENTERPRISE: INTERNATIONAL TRADE 2013

Print ISBN: 978-1-78801-076-4
EPUB eISBN: 978-1-78801-131-0

A catalogue record for this book is available from the British Library

The Royal Society of Chemistry is a charity, registered in England and Wales, Number 207890, and a company incorporated in England by Royal Charter (Registered No. RC000524), registered office: Burlington House, Piccadilly, London W1J 0BA, UK, Telephone: +44 (0) 207 4378 6556.

Visit our website at www.rsc.org/books

Printed in the United Kingdom by CPI Group (UK) Ltd, Croydon, CR0 4YY, UK

Preface

The two words natural products have different impacts on distinct audiences. On the one hand there is the sustained consumer interest in organic foods, supplements, and nutraceuticals from nature, built around the simple and abiding premise that natural is good. To chemists and pharmacognosists (specialists who study medicinal substances from nature) natural products are jointly about the molecules themselves, their isolation, structural, and initial functional characterization (the chemists) and the effects they have in acute and chronic therapeutic settings in humans (the pharmacognosists). In recent decades specific natural products, such as wortmannin, brefeldin, staurosporin, trapoxin and many others, have been powerful tools for cell biologists to study almost any facet of signal transduction, cell growth and division, protein secretion, and apoptosis.

Historically, natural products have been used in folk medicine for at least 4500 years, stretching back to early Egyptian and Sumerian written records. Starting in the first decades of the 19th century, chemists began to isolate alkaloids in pure form, beginning with morphine in 1817, marking the first major turning point from millennia of use of plant extracts containing multiple mixtures of compounds to purified single molecule preparations. Over the intervening two centuries the characterization of the inventory of natural products, from plants, fungi and bacteria, has spurred many of the conceptual and practical advances in organic chemistry. These have included analytical tools such as mass spectrometry and sophisticated two dimensional high field nuclear magnetic resonance (NMR) and have encompassed the golden age of total synthesis of natural products. For medicinal chemists the scaffolds of different classes of

Natural Product Biosynthesis: Chemical Logic and Enzymatic Machinery
By Christopher T. Walsh and Yi Tang

Published by the Royal Society of Chemistry, www.rsc.org

natural products have been starting points and/or inspiration for both semisynthetic and fully synthetic approaches to new drugs (*e.g.* podophyllotoxin to etoposide, staurosporine to the plethora of synthetic heterocyclic protein kinase inhibitors).

The second major turning point, and one that has sparked a renaissance in the natural products arena, has been the availability of thousands of microbial genomes and the beginnings of a few plant genomes over the past two decades. Bioinformatic analysis of natural product biosynthetic capacity, known as genome mining, has become a core discipline at the intersections of chemistry, life sciences, and medicine in all its manifestations. In addition to generalizing the premise that microbial biosynthetic genes are clustered (and so easier to find, interrogate, and move *en bloc*), bioinformatic analyses show that many actinomycetes, myxobacteria, and fungi have some 30–50 predicted biosynthetic gene clusters for polyketides, nonribosomal peptides, and terpenes. Such strains typically produce one to five natural products under standard laboratory culturing techniques: the universe of natural products to be explored has thus instantly gone up by an order of magnitude. Given an expected 20 000 microbial genome sequences to be available in the next decade or so, that would give about 10^6 biosynthetic gene clusters for the indicated three classes of natural products, with >99% of those as yet unexamined. In addition, there are large classes of plant and fungal alkaloids and plant-derived phenylpropanoids not counted in that set of predictions.

In this period of natural product renaissance, there are many strands of investigation being brought to bear on determining what new molecules are left to find in nature. In many therapeutic arenas, from infectious disease (beta lactam antibiotics, erythromycins, tetracyclines, aminoglycosides), to cancer (vincristine, taxol), to immunosuppression (cyclosporine, rapamycin, FK506), to cholesterol lowering agents (lovastatin), natural products have been key contributors to new and effective medicines. To what extent as yet unknown scaffolds of natural products will continue to illuminate new therapeutic modalities is one of the challenges and new opportunities in the postgenomic era.

This volume takes up the biosynthesis of major classes of natural products – polyketides, peptides, nucleosides, isoprenoids/terpenoids, alkaloids, phenylpropanoids, and glycosides. The approach is not to be encyclopedic, nor to illuminate every subclass or intriguing chemical functional group. Instead the approach is to codify the chemical logic that underlies each natural product structural class as they are assembled from building blocks of primary metabolism. A few simple reaction types are used in each natural product class, some that are common in primary metabolism and some that are much more frequent in secondary pathways than primary pathways.

Of the seven natural product cases noted above, two of them – polyketides (PK) and nonribosomal peptides (NRP) – but none of the others, are built on enzymatic assembly lines where the growing chains are covalently tethered as thioesters to carrier protein domains. Hybrid nonribosomal peptide–polyketide scaffolds comprise some of the most therapeutically interesting molecules from nature – rapamycin, FK506, bleomycin, epothilones – and they are built on convergent PK and NRP assembly lines. The dominant reaction in enzymatic polyketide chemistry is an iterated decarboxylative thioclaisen condensation for C–C bond formation, essentially carbanion chemistry.

The isoprenoid/terpenoid class of natural products, currently the largest natural product set, at over 50 000 known molecules, instead are assembled by enzymes that use *diffusible* substrates, intermediates, and products. The chemical logic in this class is overwhelmingly carbocation chemistry. This modality starts with initial allylic carbocation formation from the Δ^2-prenyl-PP building blocks, through alkylative condensations to scaffolds where cations induce some dramatic scaffold rearrangements. The cyclizations of triterpene squalene and 2,3-oxidosqualene to tetracyclic and pentacyclic product scaffolds continue the cation-driven reaction manifolds and emphasize the nucleophilic role of the π electrons of double bonds in C–C forming reactions.

In contrast to the polyketides, isoprenoids, and phenylpropanoids where the product scaffolds are essentially devoid of nitrogen atoms, the peptides, nucleosides, and alkaloids contain nitrogens that are central to the chemistry of these classes. The largest and most structurally diverse class comprises the alkaloids, in part because they were defined experimentally by the minimal criterion of at least one basic nitrogen in a heterocycle, driven by isolation schemes that went back and forth between free base and salt forms. The alkaloids range from simple monocyclic amines to amazingly complex frameworks in opioids, the poison strychnine, the antimalarial quinine, and the dimeric antitumor vinca alkaloids vinblastine and vincristine.

One of the pervasive features of the maturation of natural product scaffolds across all these classes is the tailoring of nascent product frameworks by oxygenase enzymes. Oxygenases are used sparingly in primary metabolic pathways, steroids being the major exception (and given the plethora of plant steroidal scaffolds one might put steroids in a bridging category between primary and secondary metabolism). By contrast, oxygenases proliferate in plant and microbial maturations of secondary natural products. A particularly clear example is the introduction of eight oxygen atoms around the periphery of the C_{20} isoprenoid hydrocarbon taxadiene to get to the tubulin inhibitor and anti-ovarian-cancer drug taxol.

We take up oxygenases in detail as key elements of natural product enzymatic machinery but also because they bring carbon radicals and homolytic C–C bond formation manifolds into play. We note many cases where molecular oxygen reduction generates an enzyme-bound high valent oxoiron reagent that initiates substrate C–H bond homolysis. In several contexts intramolecular C–C or C–X bond formation competes with net hydroxylation. In those cases the oxygen input is cryptic, but the reaction manifold has been radical-based, in contrast to the heterolytic C–C bond-forming flux in polyketides and isoprenoids. The phenylpropanoid lignan scaffolds are also built by phenoxy radical dimerization enzymology.

Two additional chapters are devoted to enzyme classes typically not broken out as features in natural product biosynthesis in other treatments of the subject. One of them deals with *S*-adenosylmethionine (SAM), not in its familiar role as a donor of [CH_3^+] equivalents to cosubstrate nucleophiles but as a radical initiator, continuing the theme of C–C bond formation *via* radical intermediates. At two extremes, aerobically with O_2 and anaerobically with SAM and four iron/four sulfur cluster enzymes, radical chemistry is in play for C–C bond formation in a diverse set of natural product frameworks.

The second focused chapter deals with natural product glycosides. Often treated as an afterthought in natural product maturation, the sugars serve many purposes in the localization, solubility, and functional activity of essentially every natural product class. The biosynthesis of dedicated deoxy- and aminodeoxyhexoses by enzymes encoded within biosynthetic gene clusters testifies to the importance of these sugar moieties in those pathways. The chemical logic of activation of glucose or ribose as electrophiles at $C_{1'}$ reveals why *all* glycosylated natural products are linked though $C_{1'}$.

The final section of the book takes up strategies for isolation of natural products, from the arc of medicinally important products historically, to the challenges in the postgenomic era of how to prioritize hundreds of thousands of unexamined gene clusters for novel frameworks and new activities. Given the intense coalescence of new methodologies, from genetics, genomics, synthetic biology, heterologous expression, and gene cluster activation, we do not delve deeply into any one set of current technologies, on the assumption they may be rapidly outdated. Instead we set an overview perspective and raise the strategic questions of what criteria could be assembled to increase the probability of finding members of the natural product inventory with novel structural and functional properties.

Acknowledgements

We are deeply indebted to John Billingsley at UCLA for his original artwork throughout the book, especially the chapter opener illustrations. We also thank Dr Yang Hai at UCLA for rendering the crystal structures of proteins and peptides in various chapters. Yi Tang thanks other members of his lab, including John, Nicholas Liu, Leibniz Hang, Sunny Hung and Yan Yan for proofreading the chapters.

Natural Product Biosynthesis: Chemical Logic and Enzymatic Machinery
By Christopher T. Walsh and Yi Tang

Published by the Royal Society of Chemistry, www.rsc.org

About the Authors

Prof. Walsh was on the faculty of MIT and Harvard Medical School and is now affiliated with the ChEM-H Institute at Stanford University. Yi Tang is a professor of Chemical and Biochemical Engineering, and Chemistry, at UCLA. Between them they have published more than 300 research papers on the biosynthesis of the major classes of natural products. Their research groups have deciphered chemical principles and novel enzymes for assembly of polyketides, nonribosomal peptides and posttranslationally modified nascent proteins, oxygenated isoprene scaffolds, peptidyl nucleosides, and fungal alkaloids. Through sequencing of fungal genomes to identify novel biosynthetic gene clusters, heterologous expression in engineered yeast cells, overproduction, isolation and structural assignment, they have identified genes, encoded enzymes, and the structure of novel secondary metabolites in several natural product classes. This broad set of research experience and expertise makes them an ideal pair of authors for *Natural Product Biosynthesis: Chemical Logic and Enzymatic Machinery.*

Natural Product Biosynthesis: Chemical Logic and Enzymatic Machinery
By Christopher T. Walsh and Yi Tang

Published by the Royal Society of Chemistry, www.rsc.org

Dedicated by CTW to Diana, Allison, Thomas, and Sean

Dedicated by YT to Minlei, Melody, Connor, and Justin

Contents

Section I: Introduction to Natural Products

Natural Product Biosynthesis: Chemical Logic and Enzymatic Machinery
By Christopher T. Walsh and Yi Tang

Published by the Royal Society of Chemistry, www.rsc.org

Section II: Six Natural Product Classes

Section I

Introduction to Natural Products

This section constitutes the introduction to the main structural groups of natural products, divided into six major classes, many with tens of thousands of molecular variants. These major structural types are polyketides, peptides that have been morphed into stable small molecule frameworks, isoprenoid/terpenoid scaffolds, alkaloids, nucleosides, and phenylpropanoid metabolites.

The intent of this volume is not to try to provide exhaustive coverage of structural subtypes and molecular variants (an impossible task given the >300 000 known natural product molecules). Instead the intent is to illustrate some of the rules of chemical logic that have evolved to deal with particular building blocks, the nature of the C–C and C–N bonds that build up the particular classes of natural product scaffolds, and the limited inventory of enzyme types that catalyze characteristic transformations. This should allow prediction of patterns of assembly for molecules not directly covered in the text and for discerning likely patterns of reactivity in molecules yet to be characterized.

Each class follows different but comprehensible chemical rules for building up structural and functional group complexity in the scaffolds of advanced secondary metabolites, consistent with distinct building blocks siphoned from primary metabolic pathways. Analogously there are distinct gate keeper enzymes that commit flux of one or more primary metabolites into the conditional pathways that lead to the natural product end point metabolites.

The introductory chapter brings into focus types of enzymatic reactions which occur in secondary metabolism but are rare in primary metabolism. Most important in sculpting the final architectures, functional group content and polarity in all six of the natural product

Natural Product Biosynthesis: Chemical Logic and Enzymatic Machinery
By Christopher T. Walsh and Yi Tang

Published by the Royal Society of Chemistry, www.rsc.org

classes are the actions of many oxygenases, many of them iron-based catalysts but some utilizing the coenzyme forms of vitamin B2 riboflavin. Next in natural product pervasiveness are glycosylations, often occurring as late steps in metabolite maturation.

The third molecule introduced for subsequent examination of its mechanism in detail is *S*-adenosylmethionine (SAM). SAM is the premier biological methyl donor. It turns out to be able to transfer other $[CH_3^+]$ or $[CH_3^\bullet]$ equivalents to cosubstrates, depending on substrate demand and the presence of an iron–sulfur cluster that can act as one electron initiator of radical mechanistic pathways. Biosynthesis in large part is about C–C bond constructions and both heterolytic and hemolytic pathways are introduced for detailed discussion in later sections.

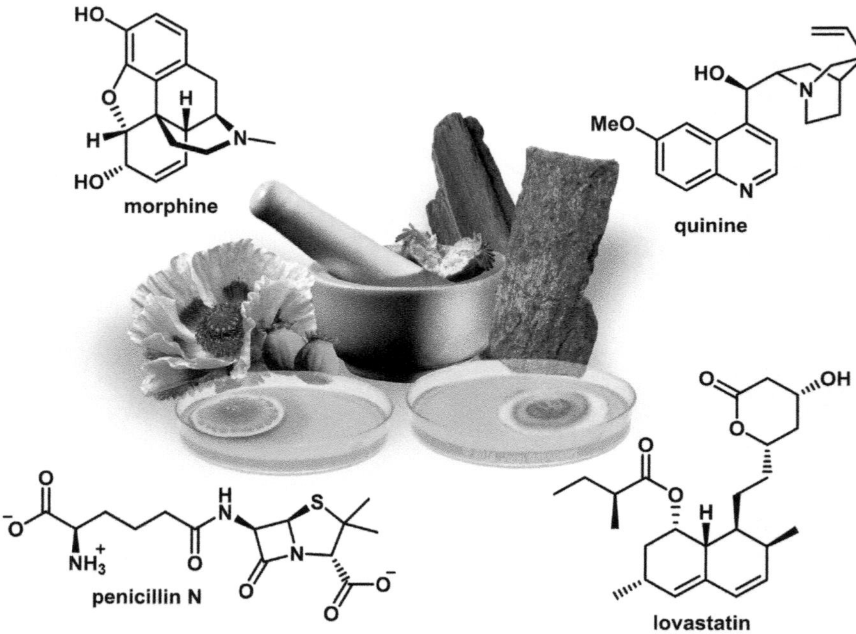

Notable natural products that have significant therapeutic values. Copyright (2016) John Billingsley.

1 Major Classes of Natural Product Scaffolds and Enzymatic Biosynthetic Machinery

1.1 Introduction

Natural products could be defined broadly as any molecules found in Nature. More traditionally in organic and medicinal chemistry communities natural products are defined as small organic molecules (molecular weight (MW) <1500 daltons) generated from conditional metabolic pathways (but see Vignette 1.1 in this chapter). That is the definition used here. Conditional metabolic pathways are also known as secondary pathways, not present in all organisms and not essential for life. Producer organisms include microbes such as bacteria, algae, fungi and also plants of every variety.

The natural products they generate from conditional pathways presumably confer some form of advantage or protection to the producers. The physiological functions may differ and are often not clear to the chemists who have done the isolation. On the other hand, many of the natural product classes isolated historically have either useful pharmacologic activities in human medicine or the reverse, showing mammalian toxicity through diverse mechanisms.

The adjective "natural" has a strong positive resonance in this era with consumers of food, cosmetics, medicines, nutraceuticals, even clothing and furnishings. In part that may be a reaction to the synthetic and abiotic materials that pervade our environments and in

Natural Product Biosynthesis: Chemical Logic and Enzymatic Machinery
By Christopher T. Walsh and Yi Tang

Published by the Royal Society of Chemistry, www.rsc.org

part is probably a connection back to humanity's past and a time when there was a closer dependence and harmony on what the natural world provided for carving out a simpler existence. One (incompletely examined) assumption is that humans have evolved with the plants and microbes that generate the natural materials and small molecules and have coadapted. This has led, over millennia, to a learned avoidance of toxic substances and conversely the utilization of natural extracts for treatment of health problems.

Starting some 200 years ago and continuing into the present, chemists have focused on isolating biologically and pharmacologically active substances, first from plants and then from fungi and bacteria, characterizing them molecularly, and producing useful molecules as pure compounds. At this point there are some 32 000 compounds tabulated from Chinese traditional medicine sources, including the antimalarial drug artemisinin (for which the 2015 Nobel Prize in chemistry was awarded). In parallel, the Dictionary of Natural Products database, which records information on purified natural molecules, contains some 210 000 compounds (Rodrigues, Reker *et al.* 2016). Natural products have been a continuing source of architectural and synthetic inspiration (Jurjens, Kirschning *et al.* 2015) to eight generations of chemists, since the first decades of the 19th century. It is estimated that 50% of natural products still have no synthetic counterparts and up to 80% of the natural product ring systems, which generate the constrained molecular architectures, are not mimicked by synthetic molecules (Rodrigues, Reker *et al.* 2016).

Figure 1.1 shows the structures of eight natural products isolated and characterized by their pharmacologic activities. Ergotamine, rebeccamycin, tubocurarine, and morphine have diverse biologic roles as foreign substances in humans. All four of these molecules have amino acid-derived scaffolds and can be broadly classified in the realm of alkaloid natural products, by virtue of one or more basic nitrogens embedded in a ring system. In structural terms morphine and the lysergic acid tetracyclic moiety of ergotamine are clearly related but the indolecarbazole framework of the antitumor rebeccamycin and the arrow poison tubocurarine bear no obvious overlap.

Ergotamine, in addition to the tetracyclic lysergic acid starter unit, is also built from the three amino acids L-alanine, L-proline, and L-phenylalanine on a nonribosomal peptide synthetase assembly line (Chapter 3). Similarly, the nitrogen atoms in the bicyclic antibiotic penicillin derive from a nonribosomally generated tripeptide aminoadipyl-cysteinyl-D-valine (Walsh and Wencewicz 2016).

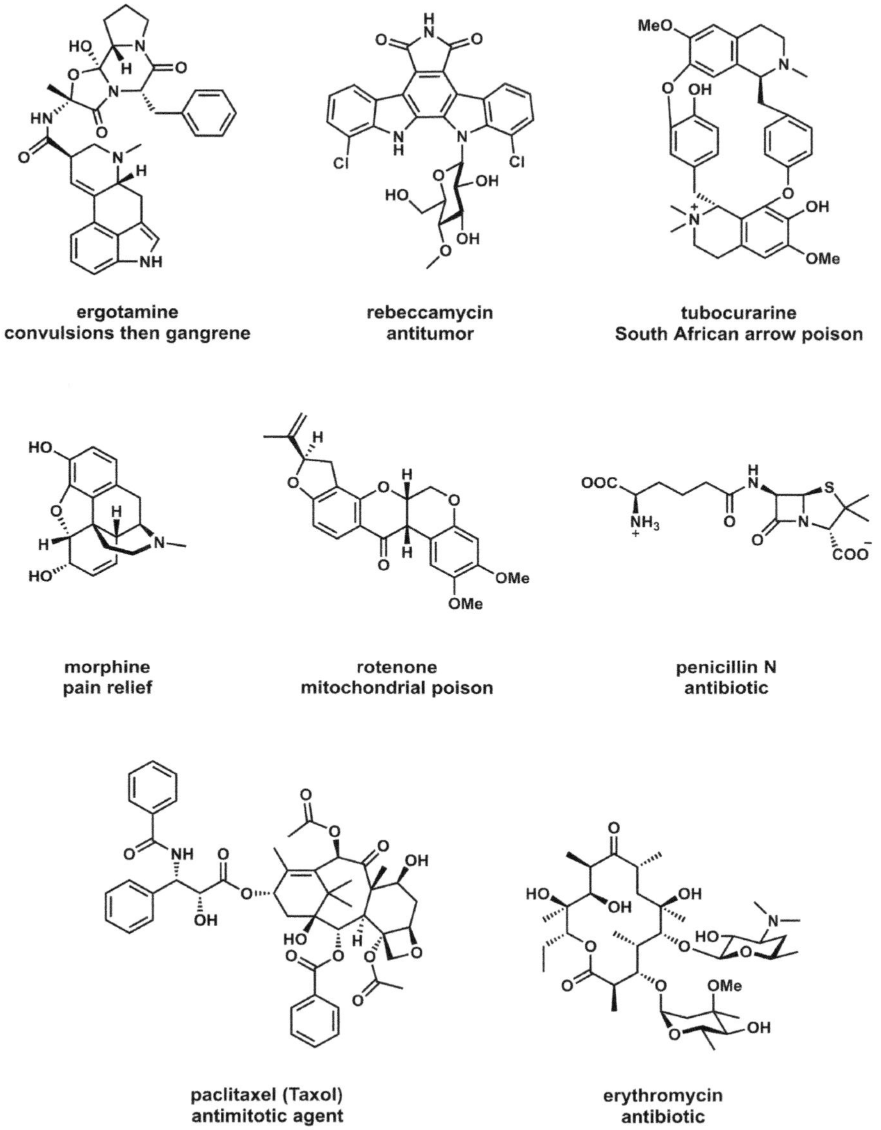

Figure 1.1 Eight natural products purified and identified by their pharmacologic activities: ergotamine (convulsant alkaloid); rebeccamycin (antitumor indolecarbazole); tubocurarine (alkaloid arrow poison); morphine (alkaloid analgesic); rotenone (phenylpropanoid respiratory chain inhibitor); penicillin N (nonribosomal peptide antibiotic); taxol (isoprenoid antimitotic agent); erythromycin (polyketide antibiotic).

The remaining three molecules in Figure 1.1 come from three additional distinct natural product classes. The anticancer microtubule blocking agent taxol (paclitaxel) is of diterpene origin. A late stage hydrocarbon intermediate taxadiene (Chapter 4) is subsequently heavily oxygenated and multiply acylated to yield taxol. Erythromycin is a venerable antibiotic with a 14-membered macrolactone core and a pair of deoxysugars. The substitution pattern on the macrolactone arises from a polyketide synthase assembly line (Chapter 2). The eighth natural product shown is rotenone, a mitochondrial respiratory blocker that is a member of the plant phenylpropanoid class of natural products (Chapter 7).

Taxol and rotenone lack any nitrogen atoms in their scaffolds, reflecting distinct building blocks and assembly logic from the alkaloids and penicillin, respectively. The presence or absence of nitrogens, particularly basic nitrogen atoms in natural product frameworks, affects physical and functional properties of the metabolite classes and is key factor in subclass definitions.

Natural products in general and dozens of particular compounds that have become therapeutic agents or inspired design of structural mimics have come to the attention of human investigators over the past 150–200 years on the basis of their diverse biologic activities. Figure 1.2 summarizes a gamut of pharmacologic activity of just 11 of the hundreds of thousands of known natural products. Among contemporary natural products of therapeutic interest, lovastatin, which lowers cholesterol by targeting the rate-determining enzyme in cholesterol biosynthesis, and the immunosuppressives rapamycin and cyclosporine have probably been the most significant human therapeutic leads. Lovastatin biogenesis is examined in Chapter 2 and cyclosporine and rapamycin in Chapter 3.

Natural Products Run the Gamut of Pharmacological Activity	
toxins	rotenone
antibiotics	penicillins, erythromycin
cholesterol lowering	lovastatin
tremorgenic	fumitremorgens
immunosuppressive	rapamycin, cyclosporine
anticancer	vincristine
analgesic	morphine, cannabinols

Figure 1.2 Natural products run the gamut of pharmacologic activity.

Estimates of the natural product inventory, defined as above, are in the range of 300 000 to 600 000 compounds. Three quarters have been isolated from plants, indicating their prodigious commitment to secondary metabolites; the remainder are microbial metabolites. There are no good estimates on the inventory yet to be discovered and whether many new molecular classes will be found. In the future as more plant genomes are sequenced better estimates may become available.

1.2 Primary Metabolites *vs.* Secondary Metabolites

Primary metabolites are the molecules that populate the pathways essential for life. At one limit they comprise the molecules in both the biosynthesis and degradation of the classes of biopolymers: nucleic acids, proteins, polysaccharides, and lipids. They also populate the pathways for generation and storage of energy, including glycolysis, the citrate cycle, aromatic biosyntheses, amino acid metabolism, the pentose phosphate pathways and others.

Secondary metabolites instead populate pathways that may only occur in some cells or in some organisms (Demain and Fang 2000) in some circumstances, for example when plants respond to predators by synthesis of defensive small molecules (phytoalexins and phytoanticipins) (Schenk, Kazan *et al.* 2000, War, Paulraj *et al.* 2012). They may represent specialized molecular scaffolds that are not found in primary metabolism. Often the natural products that sit as the end metabolites of secondary pathways have substantially more complex scaffolds than found in primary metabolites, reflecting C–C bond-forming reactions in their biosynthesis.

The boundaries between primary and secondary metabolic pathways often have a gate keeper enzyme which acts to shuttle some of the flux of a primary metabolite into the secondary pathways. For example, lignan (Chapter 7) is a key structural polymer in woody plants. After cellulose it is the most abundant form of plant biomass. The proteinogenic amino acid phenylalanine provides all the carbon framework for lignan polymers. The gate keeper enzyme, the first one committed to moving L-phenylalanine into phenylpropanoid metabolites, is phenylalanine deaminase. We will note in Chapter 7 that this enzyme has an unusual covalently attached cofactor that allows a low energy mechanistic path for elimination of the elements of NH_3 across Cα and Cβ to produce cinnamate.

Analogously, acetyl-CoA carboxylase and propionyl-CoA carboxylase, generating malonyl-CoA and 2*S*-propionyl-CoA, respectively, are on the border between primary metabolism and secondary pathways that lead to polyketide natural products. Malonyl-CoA can go either way in producer organisms, to fatty acids (primary pathway) or to polyketides (conditional pathway). 2*S*-Methylmalonyl-CoA is not used in fatty acid synthesis but is a key elongation substrate in erythromycin assembly.

Figure 1.3 tabulates a set of primary metabolites that are building blocks for many of the structural classes of natural products discussed in Chapters 2–7. Glucose is the most common sugar in cells and the glucose-1-phosphate derivative is the entry point for commitment of glucose flux to glycosylated natural products: this is the subject of Chapter 11.

The isomeric pair of isopentenyl diphosphates, the Δ^2- and Δ^3-isomers, in head to tail alkylative couplings are progenitors to >50 000 isoprenoid natural products. When such molecules are isolated from plants they have been known historically and even today as terpenoid molecules (Pichersky, Noel *et al.* 2006). The C_{30} isoprenoid squalene-2,3-oxide is a borderline primary/secondary metabolite (triterpene) that on directed enzymatic cyclizations gives rise to hundreds to thousands of sterol type natural products, as we shall note in Chapter 4.

The two aromatic amino acids L-tryptophan (Trp) and L-phenylalanine (Phe) are important building blocks for the thousands of proteins made in every free-living cell and organism. They are also utilized in nonribosomal peptide assemblies. As shown in Figure 1.3 they are also the building blocks for D-(+)-lysergic acid and the dimeric lignin (+)-pinoresinol, respectively (Chapters 5 and 7).

As noted above, the two carbon acetyl-CoA and its three carbon enzymatic carboxylation product malonyl-CoA are key acyl thioesters in primary metabolism but also in the genesis of the large and various natural product class of polyketides. Shown in Figure 1.3 is the antifungal ionophore monensin which is distinguished from other polyketide subclasses by the presence of furan and pyran cyclic ethers embedded in the molecular backbone.

Figure 1.3 contains two additional molecules in the primary metabolite column: molecular oxygen (O_2) and *S*-adenosylmethionine (SAM). O_2 is such a pervasive cosubstrate in the tailoring of all the major natural product classes of Chapters 2–8 that a separate chapter, Chapter 9, is devoted to the chemical logic and enzymatic catalysts that have evolved for its selective reductive activation.

S-Adenosylmethionine, with its trigonal sulfonium cation interspersed between a methionine residue and an adenosyl residue, is a

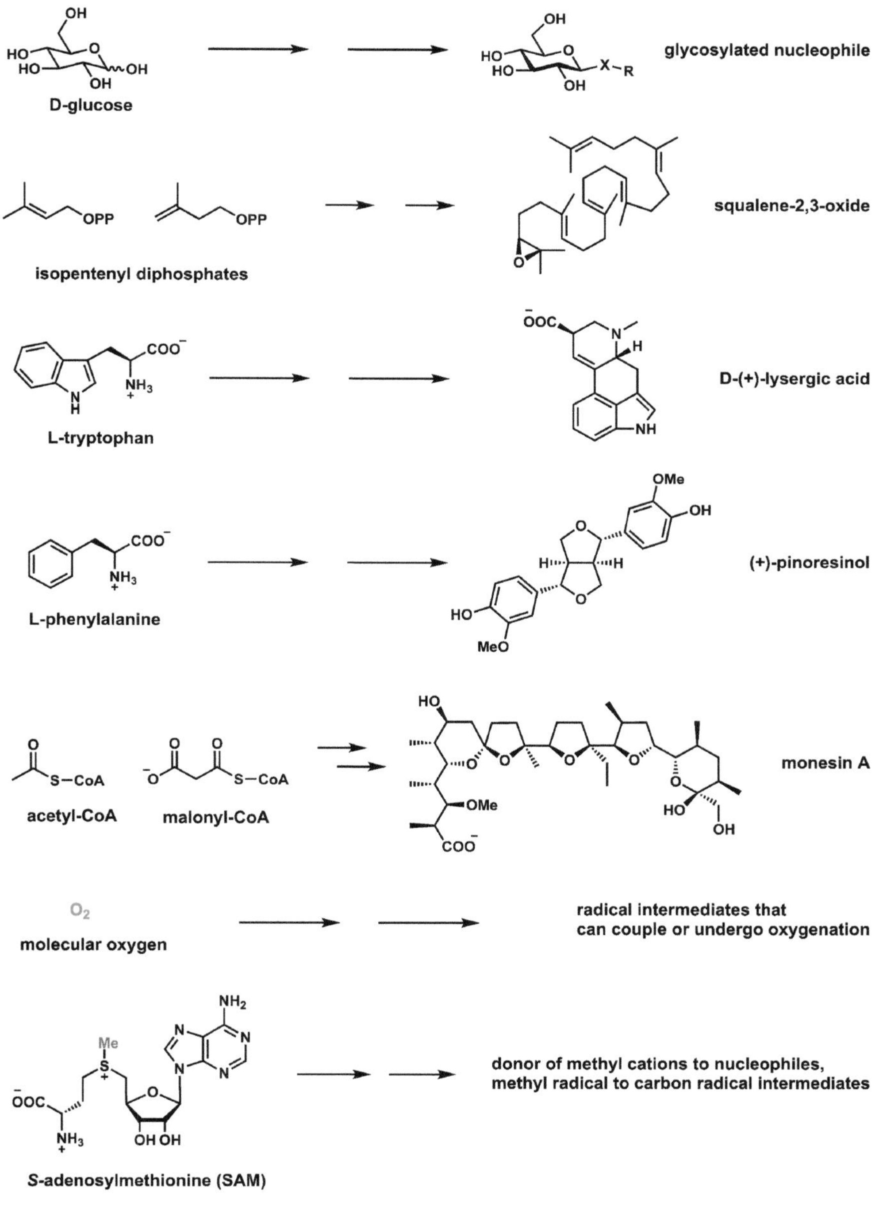

Figure 1.3 Primary metabolites serve as building blocks for specific classes of natural products.

crucial reactant in both primary and secondary metabolic pathways. We will note the iterative use of SAM as methyl donor to a diverse array of cosubstrate oxygen, nitrogen, and carbon nucleophiles of isoprenoid, polyketide, alkaloid, peptide, and phenylpropanoid frameworks. Most of these involve transfer of a [CH_3^+] equivalent. A significant set of methyl transfers go to substrates at unactivated carbon centers and these are dealt with in Chapter 10 where radical intermediates, including [$CH_3^\bullet$] equivalents, are emphasized.

1.3 Polyketide Natural Products

Figure 1.4 shows five structurally distinct subclasses of polyketide natural products: polycyclic aromatics, macrolactones, decalin-containing scaffolds, polyenes, and polyethers containing furans and pyran cyclic ether rings. All of these are built with equivalent logic, as detailed in Chapter 2, that borrows the chemical and protein precepts from fatty acid biosyntheses. In this sense it is a good place to start, at the boundary of logic between a primary and a set of secondary metabolic pathways.

The three aromatic metabolites, oxytetracycline, xanthones, and urdamycin, represent the large subgroup of polycyclic fused aromatic ring polyketides. They are all made as polyketone-containing acyl chains (hence the class name of polyketides) covalently tethered to acyl carrier proteins. Such chains are reactive, a ready source of enolates acting as carbanions and intramolecular aldol condensations, followed by aromatizing dehydrations that lead to the characteristic aromatic polycyclic frameworks.

The antibiotic erythromycin and the antiparasitic drug ivermectin fall in the subclass of polyketide macrolactones, with deoxysugar moieties attached to oxygens of the macrocyclic core. Members of this class typically span 12-atom to 22-atom macrolactone rings that form a platform for display of the various substituents, including the sugar moieties, to biological targets.

Lovastatin is featured because of the presence of the bicyclic decalin ring system. This is a molecular signature in a small number of polyketide metabolites that suggest Diels–Alder [4 + 2] cyclization chemistry at some stage in the biosynthetic pathway.

Nystatin represents a number of polyketides with polyene moieties, in this case six olefins. The polyene rigidifies the macrocycle and contributes to its ability to insert into the membranes of fungal pathogens. The final structure in Figure 1.4 is the potassium ionophore lasalocid which has insecticidal properties. The distinguishing features of this subclass of polyketides are the cyclic ethers, in this instance a five ring furan and six ring pyran.

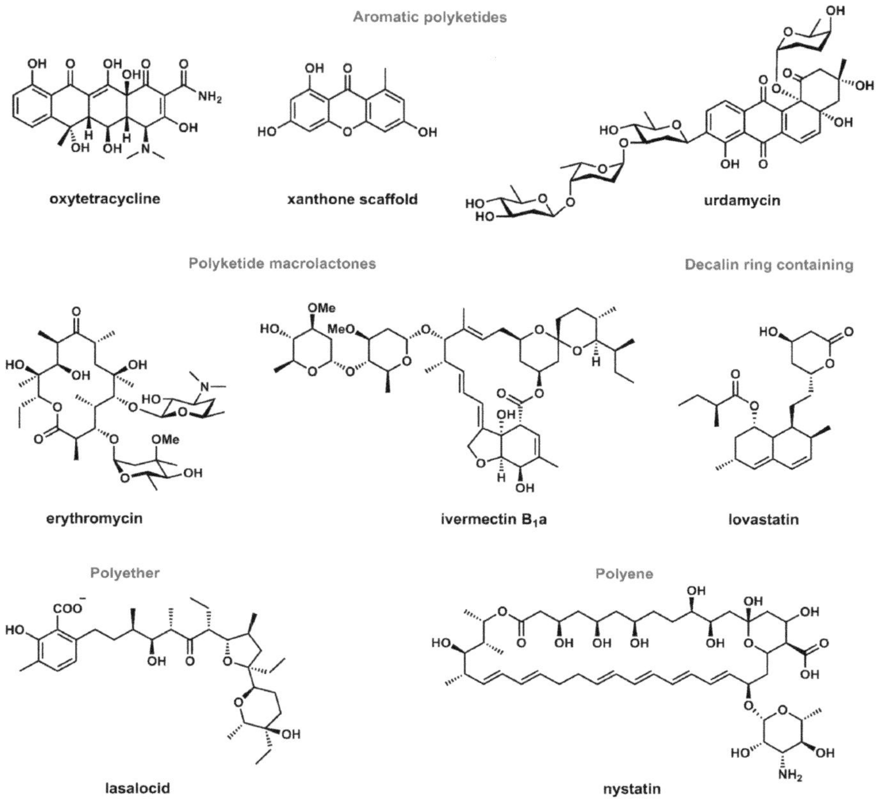

Figure 1.4 Five distinct subclasses of polyketide natural products: aromatic polyketides; macrolactones; decalin-containing polyketides; polyenes; polyethers.

We will delve into the common strategy for carbon–carbon bond formation that constitutes the chain elongations that generate the core scaffold constructions in all these polyketide subclasses. The carbon nucleophile arises by thioclaisen decarboxylative condensation of malonyl thioesters, in all polyketide synthases and all fatty acid synthases. The carbonyl group of acyl thioesters are the electrophilic partners in those C–C bond formations. We will also observe how tailoring of the initial β-ketoacyl-*S*-carrier proteins from such condensations lead to the array of –OH, CH_2 and CH=CH functional groups in various types of mature polyketide frameworks.

Figure 1.5 indicates that two of the founding members of the most widely used types of polyketide antibiotics, the macrolides,

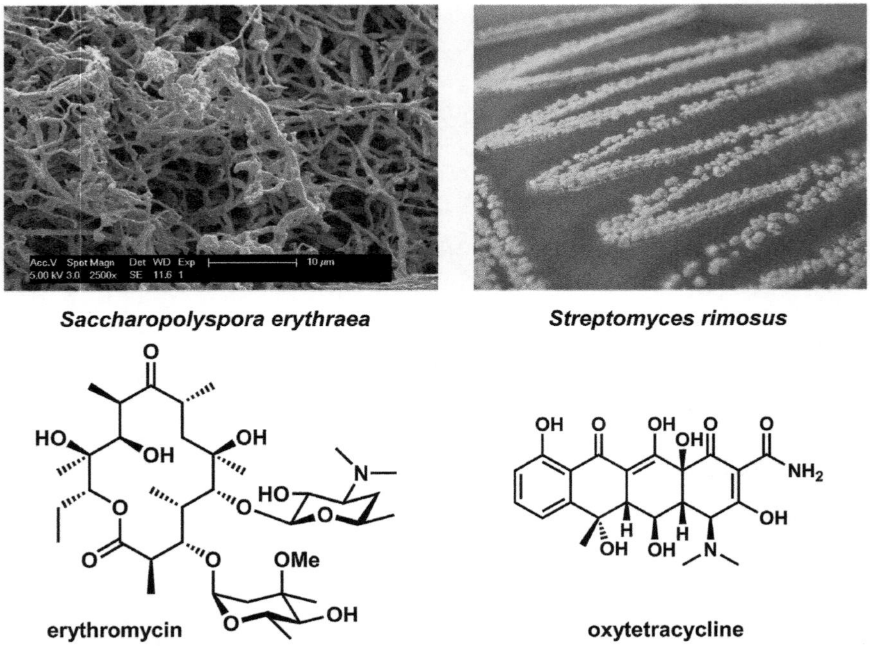

Figure 1.5 Filamentous streptomycete soil bacteria are prolific polyketide producers. *Saccharopolyspora erythraea* is the source of erythromycin. *Streptomyces rimosus* produces oxytetracycline. The electron micrograph of *S. erythraea* mycelium is courtesy of Jeremy Skepper, University of Cambridge. The *S. rimosus* photo is courtesy of Dr O. Gromyko, Ivan Franko National University of Lviv, Ukraine.

represented by erythromycin, and the aromatics, represented by oxytetracycline, are conditional metabolites from soil bacteria. *Saccharopolyspora erythraea* gives its name to the macrolide antibiotic while *Streptomyces aureofaciens* and *rimosus* make the tetracycline scaffold.

1.4 Peptide Based Natural Products

The peptide bonds that form the backbone of proteins and smaller peptides are chemically stable in physiologic aqueous media. However, they are susceptible to the diverse set of proteases that recycle proteins and peptides back to constituent amino acids. A number of strategies are in play for producer organisms to turn proteins or peptides into long-lived low molecular weight natural products.

These strategies include the generation of cyclic peptides to thwart amino- and carboxypeptidase action. Modification of side chains and/or *de novo* utilization of nonproteinogenic amino acids can produce side chains that resist proteolytic hydrolysis of peptide bonds. We will note in Chapter 3 the combined heterocyclization of side chains to oxazoles and thiazole rings in cyanobactins, and the macrocyclization to create highly morphed cyanobactin scaffolds that behave as stable small molecules, despite their protein origin (Figure 1.6). The heterocyclizations in particular have converted protease-susceptible amide bonds into heterocyclic backbones that are protease-resistant linkages.

A complementary route to peptide-derived natural products involves molecules built on nonribosomal peptide synthetase assembly lines (Walsh and Wencewicz 2016). The logic is parallel to that of polyketide synthases. Growing peptidyl chains are covalently tethered as thioesters on peptidyl carrier proteins. In this universe the common chain elongation chemical step is C–N rather than C–C bond formation. The attacking nucleophile is the amine group of aminoacyl carrier proteins onto upstream peptidyl thioester carbonyl groups. Release of the full length chains is often *via* macrocyclization, in full analogy to the polyketide synthase release logic to give compact macrolactams and macrolactones. The third aspect of NRPS assembly lines that build in structural and functional diversity is the use of many dozens of nonproteinogenic amino acids as building blocks for the enzymatic assembly lines. In the nonribosomal natural product kutzneride A all six building blocks are nonproteinogenic amino acids (Figure 1.7) and the scaffold is a cyclic macrolactam. Both features impart protease resistance to this antifungal natural product produced by root-associated bacteria.

Two other examples of nonribosomal peptide metabolites are displayed in Figures 1.8 and 1.9. The topical antibiotic cream polysporin contains three nonribosomal peptides: polymyxin B, bacitracin, and gramicidin A (Figure 1.8). Figure 1.9 emphasizes that the best selling class of antibiotics world wide, the fused 4,5-bicyclic ring penicillins and the downstream metabolites, the 4,6-fused cephalopsorins, are fungal metabolites that arise from nonribosomal peptide synthetase assembly lines.

Polyketides and nonribosomal peptides are the only classes of natural products that are formed by assembly line chemical logic and

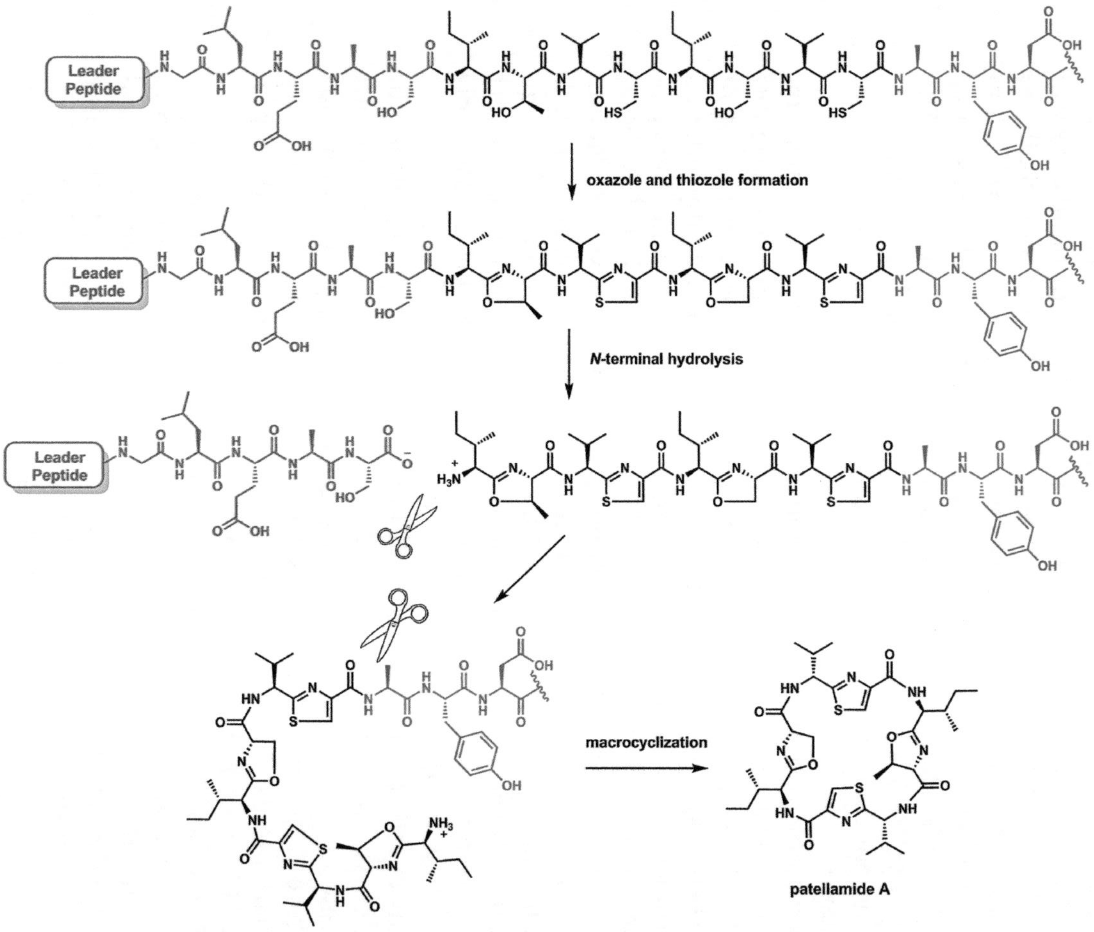

Figure 1.6 Cyanobactin biosynthesis involves heterocyclization of Ser, Thr, and Cys side chains and protease action to cut out the modified scaffold and circularize it as a macrolactam.

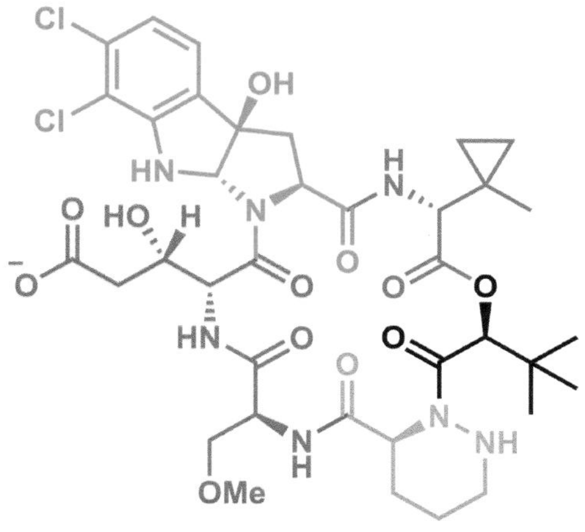

Figure 1.7 Kutzneride A, a nonribosomal hexapeptidolactam, is built entirely from six nonproteinogenic amino acid building blocks.

protein machinery. We will note that the other major classes use different assembly logic.

On the other hand, because PKS and NRPS assembly lines both use carrier proteins as common chain tethering elements it may not be surprising that producer microbes have learned to practice evolutionary convergence and build hybrid assembly lines. Figure 1.10 notes five such hybrid natural products, reflecting different combinations from polyketide and nonribosomal peptide portions. The most famous of these in human therapeutics may be the immunosuppressive drugs rapamycin and FK506. Bleomycin and epothilone D both have anticancer activity although they are not front line agents. The fifth molecule, yersiniabactin, is a virulence factor of the plague bacterium *Yersinia pestis* that functions as an iron chelating agent. Hybrid NRP–PK and PK–NRP assembly lines will be examined in Chapter 3.

1.5 Isoprenoid/Terpenoid Natural Products

The family of natural products built from one or both of the Δ^2 and Δ^3 isomers of isopentenyl diphosphate is thought to comprise the largest set of natural products of a single structural origin. More than 50 000 isoprenoid scaffolds have been isolated to date (Lange 2015). There are a small number of bacterial isoprenoid metabolites, many more in

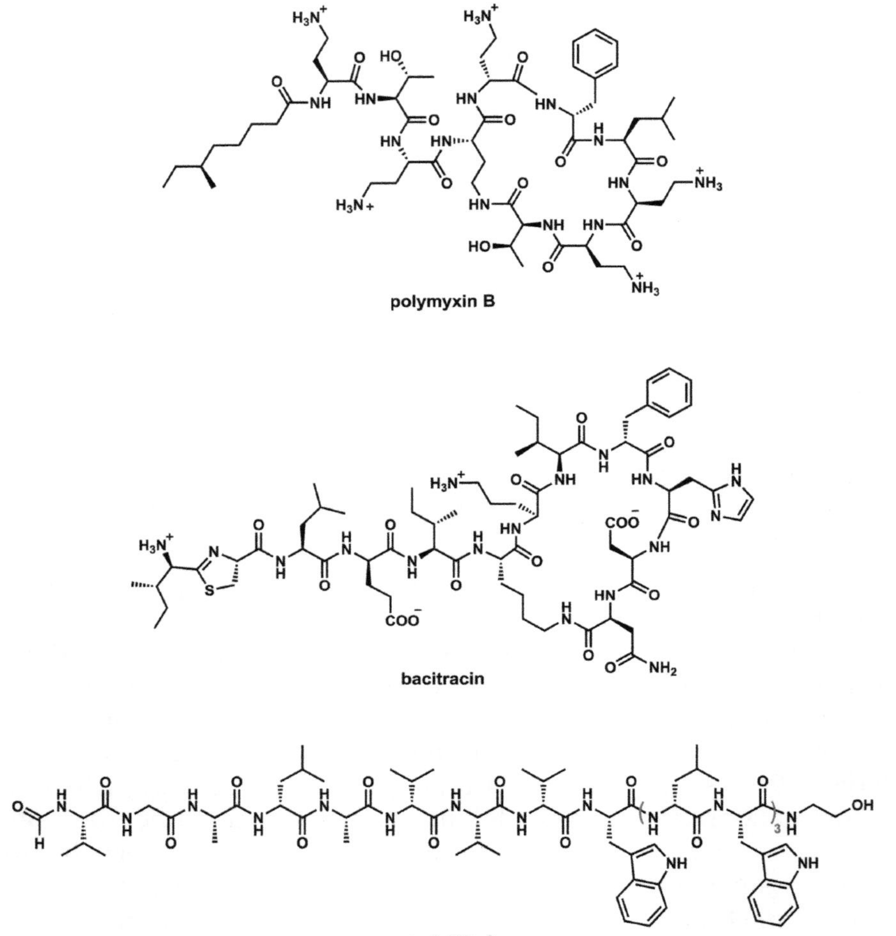

Figure 1.8 The topical antibiotic ointment polysporin contains three active nonribosomal peptide antibiotics: polymyxin B, bacitracin, and gramicidin A, all working to block bacterial cell wall biosynthesis and/or membrane integrity.

the fungal realm, but most have been isolated from plants. The plant-centric history led to the dominant nomenclature of terpenes for isoprenoid metabolites. One central set of chain-building steps occur as Δ^2-prenyl-PP dissociate to allyl cations and are captured by the π-electrons of the terminal olefin in the Δ^3-IPP partner (Figure 1.11). In this head to tail alkylation mode the prenyl chains grow by a five carbon unit each time. In the terpene nomenclature, C_{10} molecules

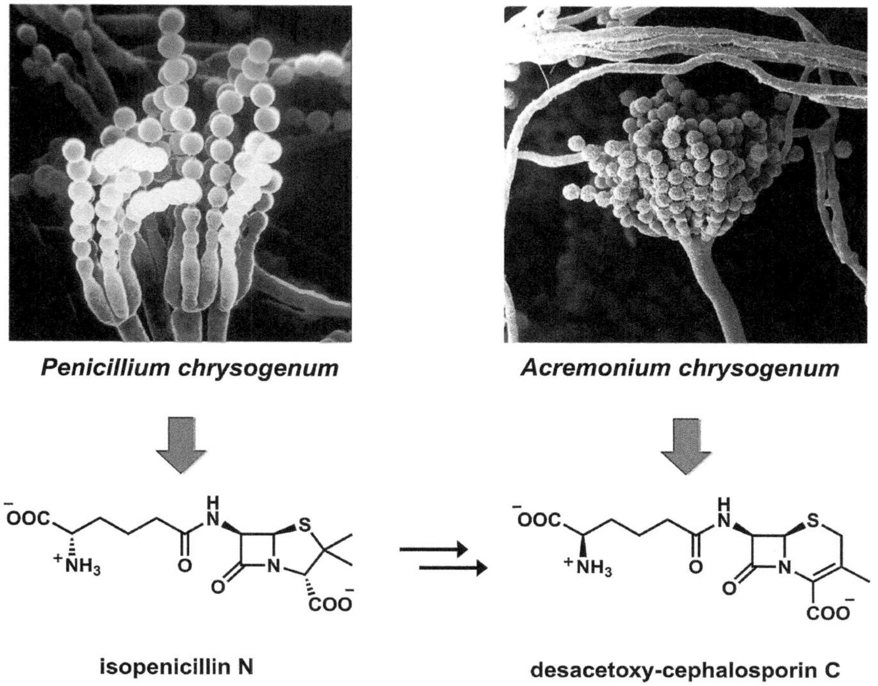

Figure 1.9 The fungus *Penicillium chrysogenum* is the source of isopenicillin N, the first biosynthetic metabolite containing the 4-5-fused β-lactam ring system. *Acremonium chrysogenum* produces the ring-expanded 4–6 bicyclic cephalosporin antibiotic desacetoxy-cephalosporin C. Both antibiotics derive from the acyclic nonribosomal tripeptide aminoadipyl-L-Cys-D-Val. The scanning electron micrograph of conidiophores of *P. chrysogenum* is courtesy of Dr Robert A. Samson at CBS-KNAW Fungal Diversity Centre, The Netherlands. Electron micrograph of *A. chrysogenum* conidia copyright (2009) of Hans von Döhren, https://www.flickr.com/photos/36881808@N06/4195269877/.

are monoterpenes, C_{15} are sesquiterpenes, C_{20} are diterpenes, C_{30} are triterpenes and so on.

Terpene/isoprenoid biochemistry is dominated by the reactions of the allyl cations that arise from the Δ^2-prenyl substrates. In many cases these involve intramolecular rearrangements of the initial allyl cations before the positive charge is quenched (by adjacent proton removal to form a double bond, by water addition to give an alcohol, and/or by single bond migrations. Figure 1.12 shows a small set of

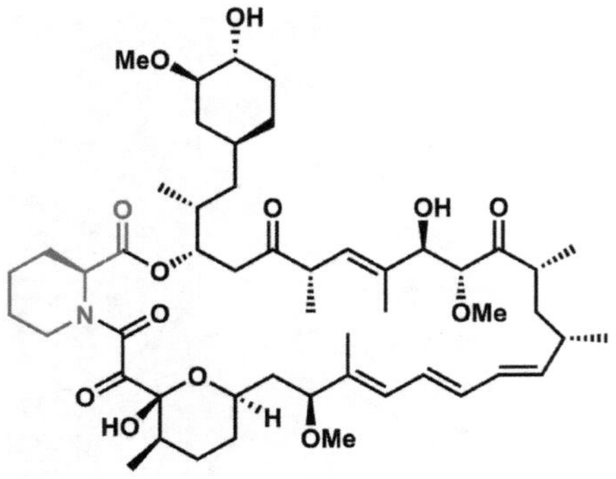

rapamycin

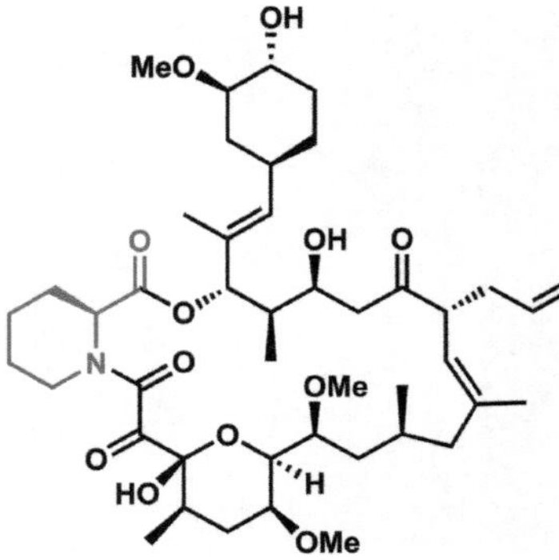

FK506

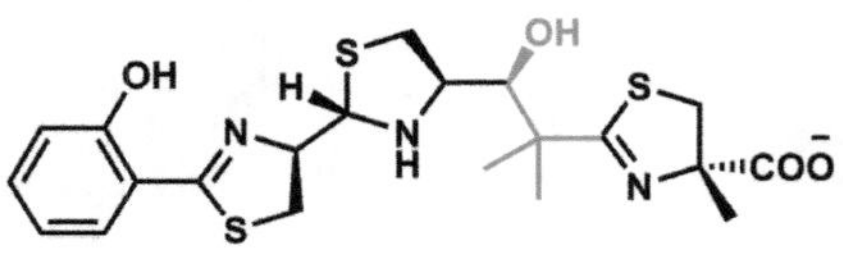

yersiniabactin

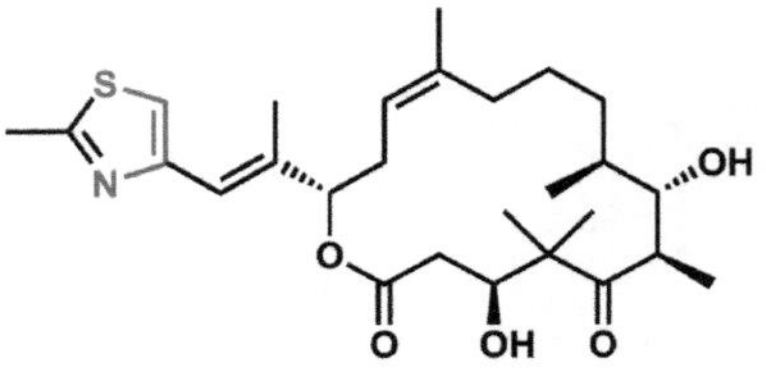

epothilone D

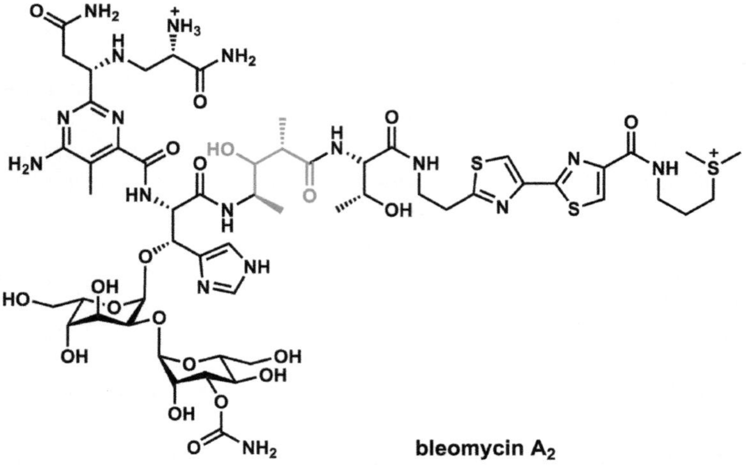

bleomycin A_2

Figure 1.10 Hybrid nonribosomal peptide–polyketide scaffolds: the peptidic moieties in rapamycin, FK506, and epothilone D are shown in blue, emphasizing that these are mostly polyketide in origin. The polyketide moieties in bleomycin A2 and the *Yersinia pestis* siderophore yersiniabactin are shown in red, indicating that these two molecules are mostly peptidic in origin.

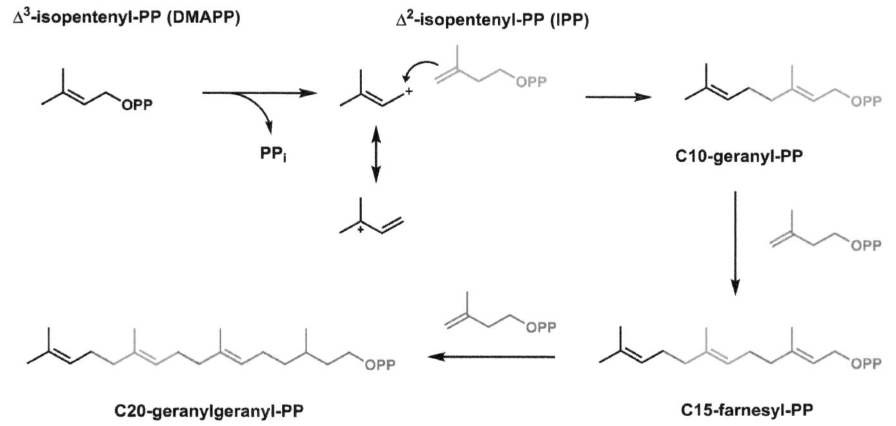

Figure 1.11 Isoprenoid chain extension by five carbons at a time. Addition of Δ^3-IPP units to growing Δ^2-prenyl chains.

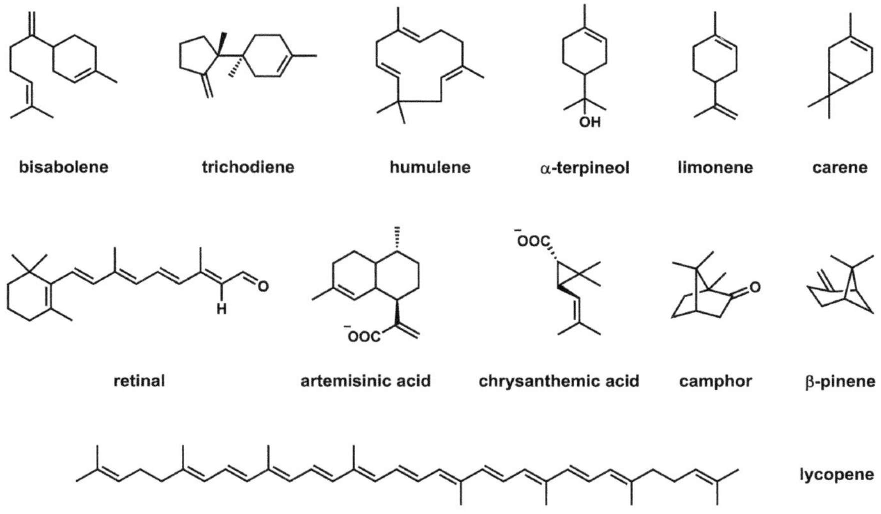

Figure 1.12 More than 50 000 terpenoid/isoprenoid natural products make this the largest natural product class. C_{10} molecules are monoterpenes, C_{15} are sesquiterpenes, retinal is a C_{20} diterpene aldehyde, and lycopene is a C_{40} polyene in the carotenoid family.

monoterpene and sesquiterpene natural products that have formed *via* such enzyme-directed cation rearrangements. Vitamin A aldehyde (retinal), artemisinic acid as the precursor to the antimalarial drug artemisinin, and the carotenoid lycopene are also derived from the isoprenyl-PP isomers biosynthetically. The active final metabolite artemisinin is shown in Figure 1.13, as is the plant of origin, *Artemisia annua.* The conversion from artemisinic acid to artemisinin clearly involves multiple rounds of oxygenase action, including insertion of the transannular endoperoxide, a reminder of the key role O_2 plays in natural product scaffold maturations.

The widely used anticancer drug taxol comes from the Pacific yew tree and more recently has been harvested from needles of the English yew. The 6-8-6 tricyclic core of taxol is a C_{20} diterpene framework that has been heavily morphed by cation rearrangements (Figure 1.14). The initial diterpene product is the C_{20} hydrocarbon taxadiene. In the late stages of the pathway taxadiene gets oxygenated at eight of the peripheral carbons by a set of tailoring oxygenases. Four hydroxyls out of those eight functional groups then get acylated to yield the final molecule taxol.

One last example of the diversity of structure and function of isoprenoid molecules is the molecule β-carotene, the precursor to vitamin A aldehyde (see Figure 1.12). This is one of a set of carotene isomers. As noted in Figure 1.15, rice plants are deficient in carotene because of inability to convert a C_{40} precursor phytoene to the carotenoid scaffold. Golden rice represents a genetically modified rice

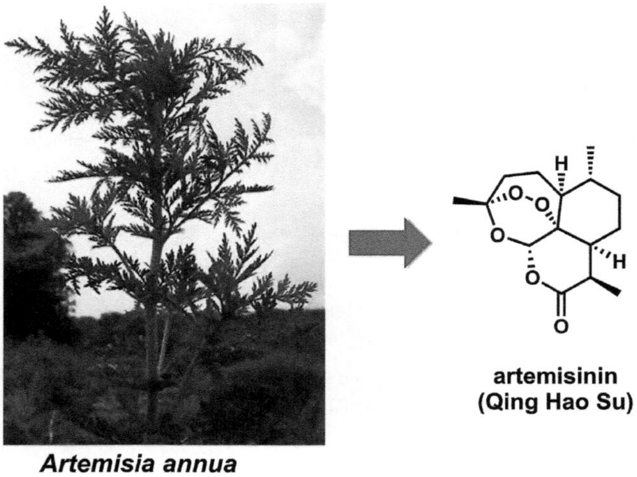

Figure 1.13 *Artemisia annua* is the source of the antimalarial agent artemisinin, also known as Qing Hao Su. The highly oxygenated scaffold with cyclic endoperoxide is an unusual structural feature.

***Taxus* (Yew) tree**

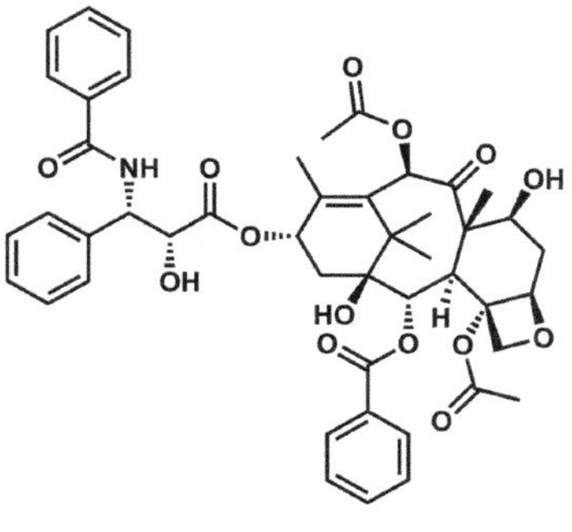

Paclitaxel (Taxol)

Figure 1.14 Taxol is a C_{20} diterpene that has undergone intramolecular rearrangements to a 6-8-6 tricyclic core, then tailored by extensive oxygenation and acylation. Taxol's clinically relevant anticancer activity stems from its activity as an antimitotic agent. Photo credit to Jason Hollnger, accessed *via* flickr (https://creativecommons.org/licenses/by/2.0/legalcode).

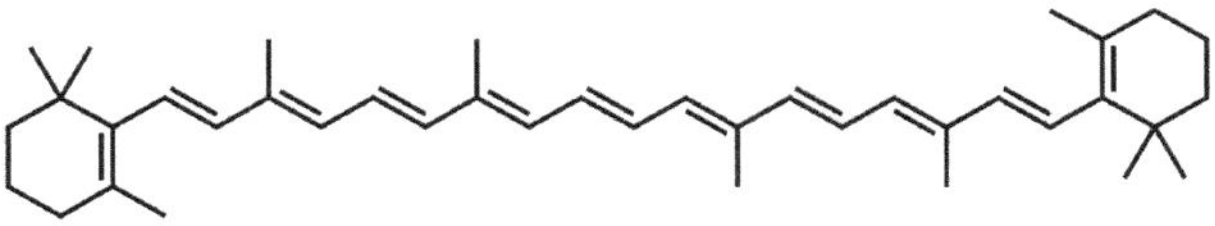

β-carotene

Figure 1.15 Golden rice contains two transgenes that encode phytoene synthase and dehydratase, enabling the plants to make β-carotene (yellow) that is the precursor to retinal, required as the light absorbing chromophore in visual pigments.
Image credit: International Rice Research Institute in the Philippines. License: CC-BY-2.0 (https://creativecommons.org/licenses/by/2.0/).

that has sparked controversy over the genome manipulations that bring in two gene replacements.

1.6 Alkaloids

In contrast to the polyketide and isoprenoid families of natural products, which are deficient in nitrogen atoms, the alkaloid families are defined by the presence of (at least) one basic nitrogen atom embedded within a heterocyclic ring. Peptide-based natural products, either from morphing of protein precursors or from NRPS assembly lines, do contain nitrogen atoms, but many are tied up as nonbasic atoms in peptide or amide linkages. We will note in Chapters 5 and 12 that the nitrogen basicity in alkaloids allows isolation by partitioning back and forth between organic and aqueous phases, depending on the pH and protonation state of the amine.

Figure 1.16 shows three distinct types of simple alkaloid scaffolds from three distinct amino acid primary metabolites. Although the five carbon dibasic amino acid L-ornithine is not incorporated into proteins, it is a central metabolite in arginine metabolism and is the

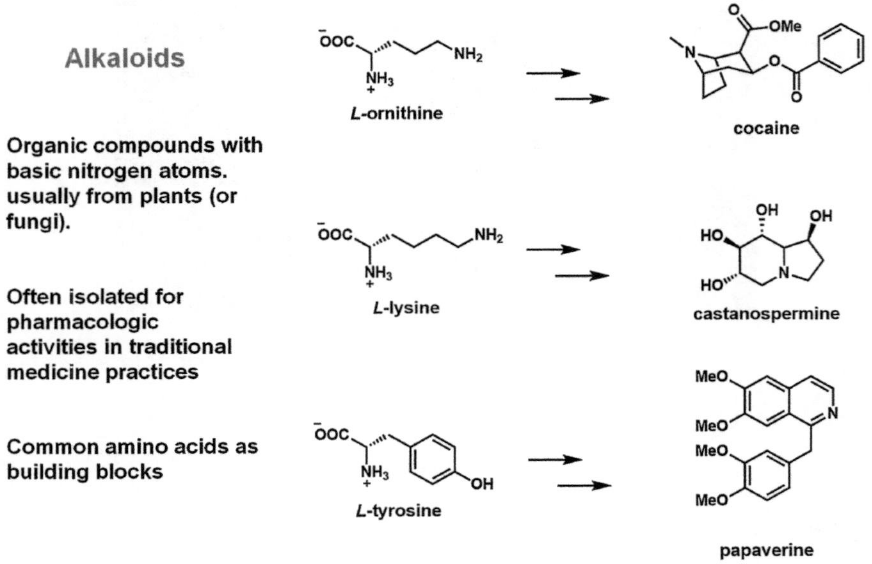

Figure 1.16 Alkaloids utilize primary amino acids as metabolic building blocks. Ornithine is the source for castanospermine, lysine for cocaine, and tryptophan for papaverine.

precursor to the bicyclic castanospermine. Lysine is the precursor to the addictive stimulatory cocaine from the South American coca plant. Tyrosine is precursor to the opium antispasmodic, vasodilatory molecule papaverine. The three nitrogen ring systems are quite distinct as are additional alkaloids in Figure 1.17, which makes the general point that Nature builds many different versions of fused five- and six-membered ring systems in its secondary metabolite inventories. Figure 1.17 also illustrates that cyclic scaffolds are built from the other classes of natural products that lack nitrogen atoms.

Alkaloids depicted in Figure 1.17 include the dimeric bicyclic coumarin dicumarol, the tetracyclic antiarrhythmic agent sparteine from Scotch broom, and a dramatically different tetracyclic framework in D-(+)-lysergic acid. The most complex alkaloid framework in Figure 1.17 is the dimeric tubocurarine, a poisonous ingredient in South American arrow poisons. Indeed, many of the alkaloids are known for their vertebrate toxicity, among them the highly complex hexacyclic strychnine, from *Strychnos toxifera* (Figure 1.18).

A large subfamily of alkaloids is built by the convergence of two natural product biosynthetic pathways: tryptophan-based alkaloids and the terpene/prenyl group routes. The so-called indole terpenes comprise more than a thousand metabolites, four of which are shown in Figure 1.19. Note that lysergic acid is included in this alkaloid subgroup along with the plant products vindoline, catharanthine, and cytochalasins (Figure 1.17), as well as the spiro framework of the fungal alkaloid spirotryprostatin A.

Much of the chemistry in this complex alkaloid-terpene class derives from the behavior of the indole ring of tryptophan (and its decarboxylation product tryptamine) as a nucleophile at multiple carbon sites. Among the outcomes are net annulations of the indole ring at C_2 and C_3 to produce tricyclic β-carbolines that get further modified. The condensation of tryptamine and the glycosylated terpene aldehyde secologanin yields strictosidine (Figure 1.20) which is a central precursor to more than a thousand downstream indole terpene natural products in different plant systems. The indole terpene biosynthetic pathways are treated in Chapter 8.

1.7 Purine and Pyrimidine Natural Products

The purine and pyrimidine nitrogen heterocycles are the key building blocks and informational units in both RNA and DNA.

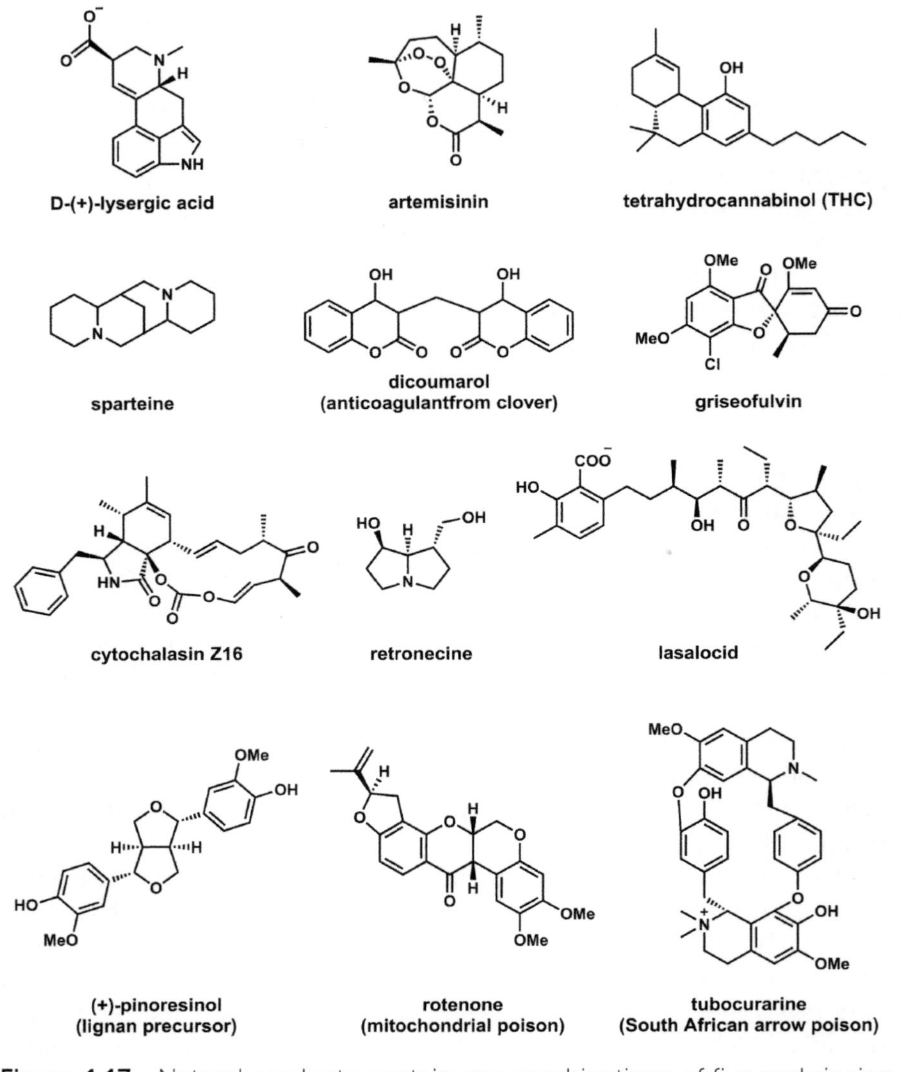

Figure 1.17 Natural products contain any combinations of five and six ring carbacycles, and heterocycles, during creation of constrained architectures and functional groups to interact with biological targets.

They are anchors of primary metabolism. Additionally, there are a number of nucleosides that function independently of their roles as nucleic acid building blocks. These can be modified in the nucleo bases, in the sugar backbones, or at the 5′ position to function in

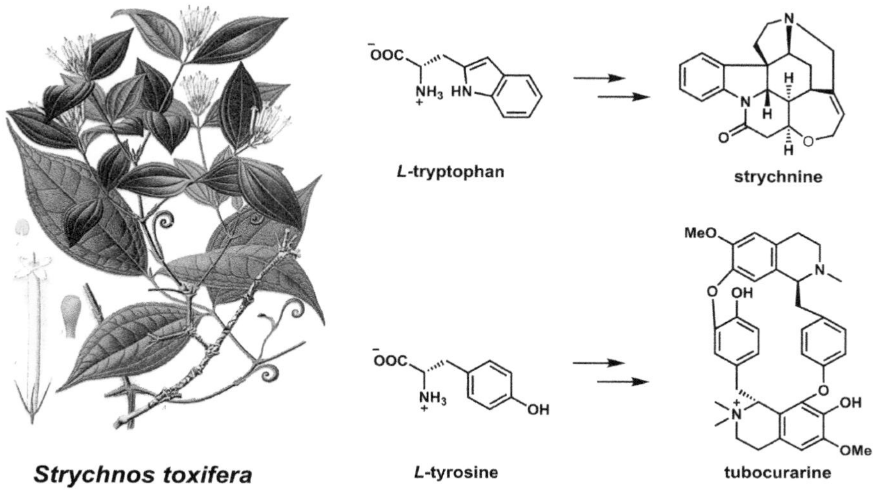

Figure 1.18 *Strychnos toxifera* is the source for several poisonous alkaloids including strychnine and tubocurarine.

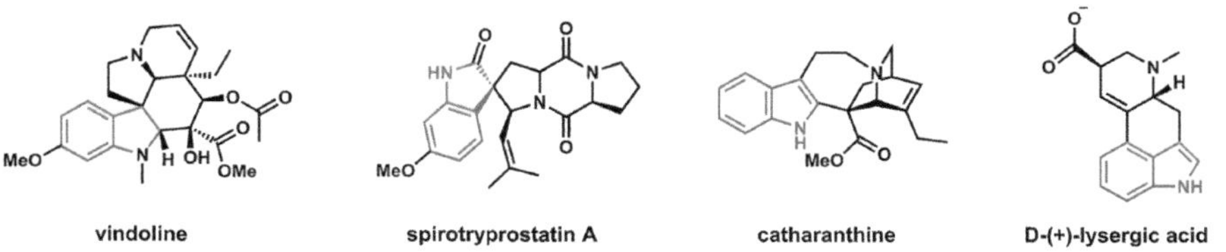

Figure 1.19 Indole mono- and diterpenes have highly diversified scaffolds.

different biological niches. The logic of their assembly is taken up in Chapter 6.

1.8 Phenylpropanoid Scaffolds

In some starting analogy to alkaloids, where the entry points to secondary pathways are from amino acid primary metabolites, all plant phenylpropanoids derive from rerouting some of the metabolic flux of L-phenylalanine from protein synthesis. However, in contradistinction to alkaloids, where the basic nitrogens are central to assembly and subsequent functions, the nitrogen in phenylalanine

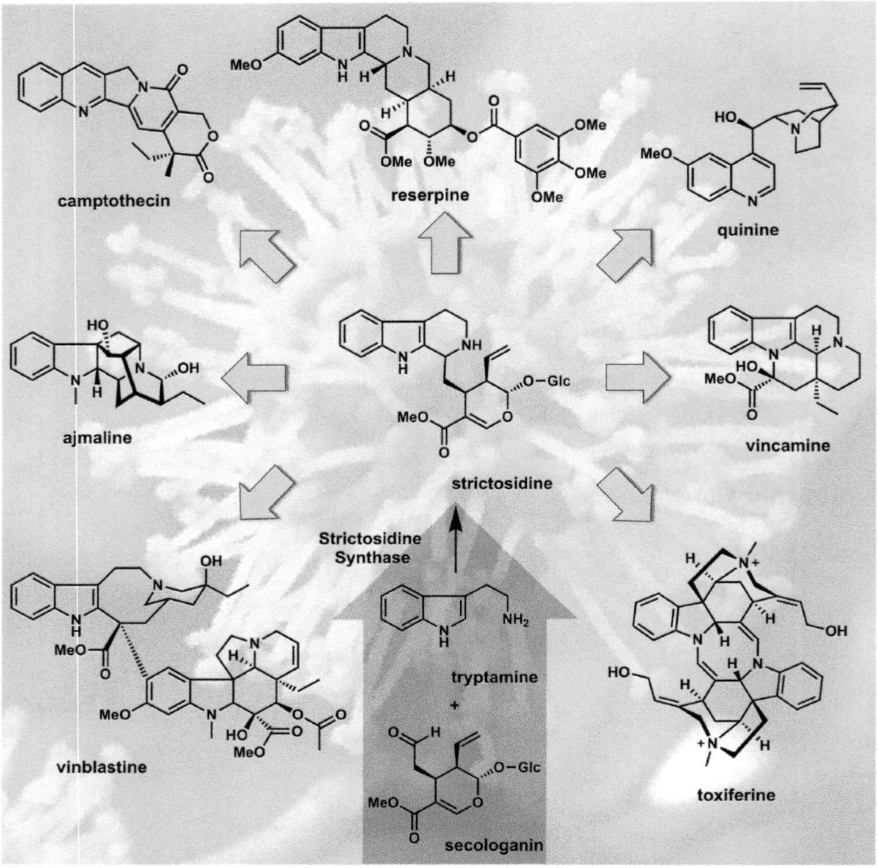

Figure 1.20 Strictosidine synthase condenses tryptamine and secologanin in a variant of a Mannich condensation known as a Pictet–Spengler reaction. The resultant tricyclic β-carboline strictosidine is precursor to more than 1000 indole terpene metabolites in plants. Shown in the background is the flower of *Camptotheca acuminata*, the producer of camptothecin.

is jettisoned in the first committed step to phenylpropanoids. Figure 1.21 shows how phenylalanine is converted to the anti-asthmatic alkaloid ephedrine with retention of the amine group. By contrast, the gate keeper enzyme to all phenylpropanoids, phenylalanine deaminase, eliminates the element of ammonia to get to the nitrogen-free cinnamate skeleton. This is the key nine carbon unit in this natural product family, illustrated by the flavonone scaffold, typically deficient in nitrogen content.

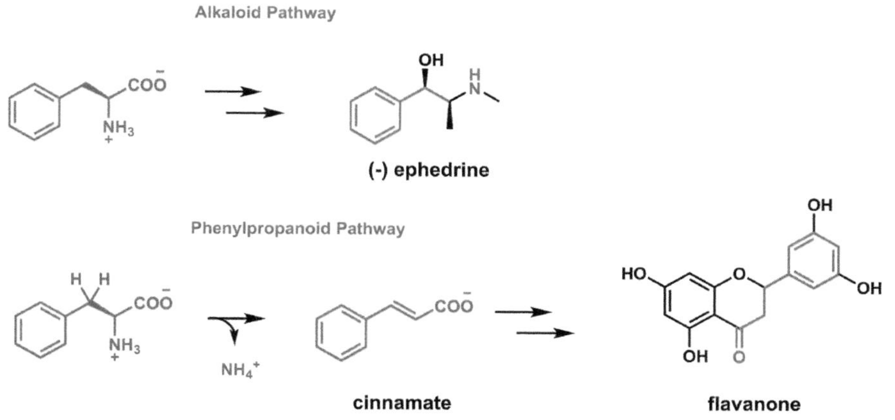

Figure 1.21 L-Phenylalanine as building block for both alkaloids and phenylpropanoid metabolites. The alkaloids retain the basic nitrogen as exemplified in ephedrine formation. In phenylpropanoid metabolites the amine group is ejected by the gate keeper enzyme phenylalanine deaminase and cinnamate is the proximal building block for all downstream C_9 metabolites in the phenylpropanoid family.

We will note in Chapter 7 that phenoxy radical chemistry is abundant in the transformations of cinnamate into many dimeric lignans, exemplified by pinoresinol in Figure 1.17. These lignan dimers can in turn be peroxidatively oligomerized into lignin, the major structural polymer in woody plants. Cinnamate flux can also be directed to chalcones, stilbenes and flavanones, one of which is shown in Figure 1.21. The nine carbon scaffold of cinnamate can also be used as a starting point for a variety of oxygen heterocycles, including the dimeric dicumarol, an anticoagulant agent in clover (see Figure 1.17).

Once the flavanones have been biosynthesized they play a number of roles in plant physiology including defense against herbivores. They also can be oxidized and/or rearranged to anthocyanin glycosides, such as pelargonidin-3-*O*-glucoside or to aureusin, depicted in Figure 1.22 as providing the red and yellow colors, respectively, to certain fruits and flowers. The colors are important in attracting pollinating insects. Figure 1.22 also illustrates that color in natural products can arise from other types of chemical scaffolds, including the red and yellow of the carotenoids lycopene and neoxanthin. The cationic portulacaxanthins and betanins illustrate yet another chemical logic for creating visible chromophores, this time from natural product alkaloids.

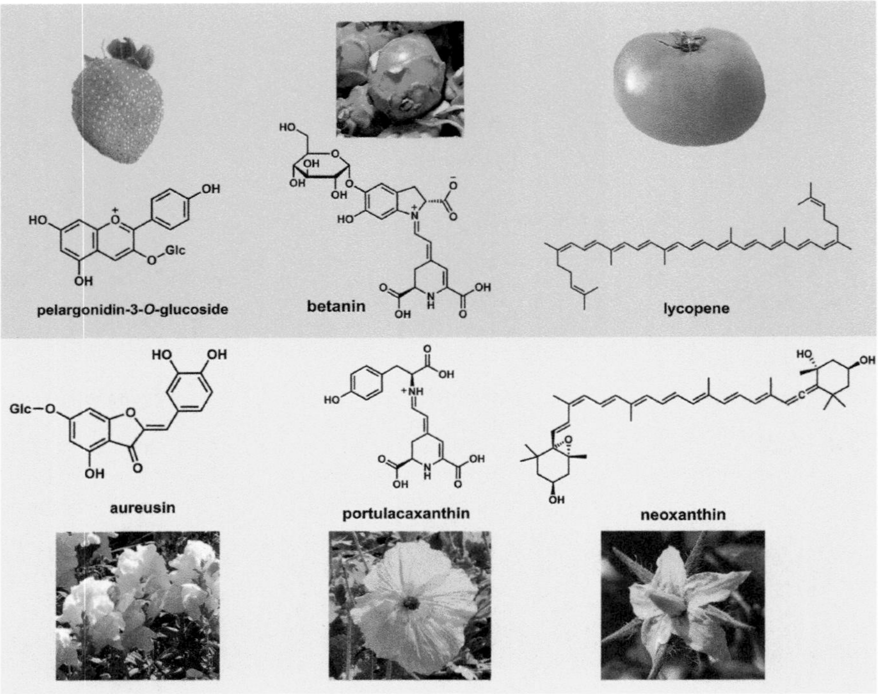

Figure 1.22 Colors in fruit and flowers can arise from different types of chemical chromophore, including anthocyanin glycosides, aurones, carotenoids, and cationic conjugated betaines. Adapted from Pichersky, E. and E. Lewinsohn (2011). "Convergent Evolution in Plant Specialized Metabolism". *Annu. Rev. Plant Biol.* **62**.

1.9 Glycosylated Natural Products

Thousands of natural products, in all the classes noted so far in this chapter, occur in glycosylated forms. The sugars are often treated as an afterthought with most attention on the non-sugar, aglycone portion of the natural product. However, given the pervasiveness of the glycosylations and the requirement that the sugars can provide for activity, proper localization, solubility and transport, we devote Chapter 11 to the logic and machinery for forming specialized hexoses found in natural products and the enzymes that carry out the glycosyl transfers. Figure 1.23 includes the major class of natural products that are constituted only from sugars. These are the

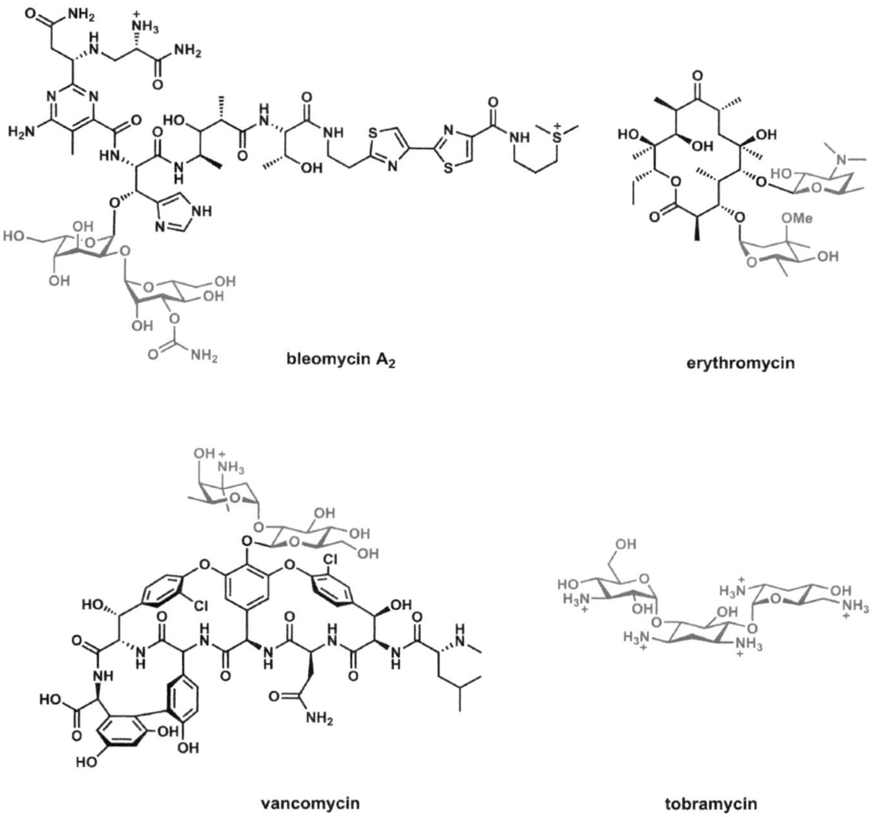

Figure 1.23 Glycosylated natural product scaffolds include the trisaccharides of the aminoglycoside antibiotic family, the glycosylated polyketide erythromycin, the glycosylated peptide antibiotic vancomycin, and the NRP–PK hybrid antitumor agent bleomycin A2.

trisaccharide aminoglycosides (tobramycin shown) which have been useful antibiotics for treating bacterial infections in humans since their discovery as streptomycete metabolites almost 70 years ago. Erythromycin is active as an antibiotic, blocking bacterial protein biosynthesis in the 50S ribosomal subunit, *only* when the two deoxyhexoses are attached. Both vancomycin and bleomycin, antibiotic and antitumor agent, respectively, are glycopeptides. In all four molecules the sugars are not the abundant, primary hexoses glucose, mannose, or galactose. Rather, they are specialized deoxy- or deoxyamino-sugars that play significant recognition roles with targets and have a distinct hydrophobic/hydrophilic balance from glucose and its congeners.

Vignette 1.1 Natural Products as Small Molecules.

Metabolites from secondary (conditional) biosynthetic pathways are typically in the category of small molecules. Small is a subjective criterion. For most organic chemists molecules in the molecular weight range of 300–500 daltons would qualify. Also, for medicinal chemists, that is a common upper limit for molecules that are orally available as human therapeutic agents. In this volume we exclude biopolymers including RNAs, DNAs, proteins, and polysaccharides from the definition of secondary metabolites and they are not by any definition small molecule natural products.

Nonetheless, biologically active natural products span an enormous range of molecular composition and molecular weight. A variety of bacteria evolve H_2 gas from two protons and two electrons. Those hydrogenases can serve as conduits to dump electrons during primary metabolism and H_2 can diffuse away. It is typically captured in mixed microbial communities by organisms that run hydrogenases in the opposite direction, using H_2 as an energy source. By this criterion H_2, with a molecular weight of 2 daltons, would be the smallest natural product.

There are multiple varieties of hydrogenases, all of them with iron atoms as part of the catalytic centers, but some with nickel and others with an Fe–Fe core. One of the Fe atoms is in a typical Fe_4/S_4 cluster but the other is part of a two-iron cluster (Figure 1.V1) that contains three molecules of carbon monoxide (CO) and two cyanide ions as ligands to iron. In addition, there is an as yet unidentified ligand. The source of the carbon monoxide and cyanide is a molecule of tyrosine, explained in the vignette for Chapter 10. Counting cyanide as a natural product, both in this context, in ethylene production, and when released from cyanogenic glycosides as part of plant defenses (Chapter 10), cyanide ion would count as another very low molecular weight natural product, with a mass of 26 daltons, while CO comes in at 28 daltons. Another secondary metabolite of mass 26 daltons is the fruit-ripening hormone ethylene (C_2H_2), which is produced from *S*-adenosylmethionine in plants under a variety of conditions (elaborated in Chapter 9). Anaerobic bacteria can use H_2 as an energy source and produce natural gas, methane (CH_4, MW = 16) as perhaps the second lowest molecular weight natural product. Mammalian cells can convert one of the nitrogen atoms of the amino acid arginine into the gas NO *via* action of NO synthase as a signaling agent that acts to affect smooth muscle tone by binding to the heme cofactor of guyanylate cyclases.

Many of the polyketide classes discussed in Chapter 2 fall in the <500 daltons molecular weight range but the marine polyether toxins

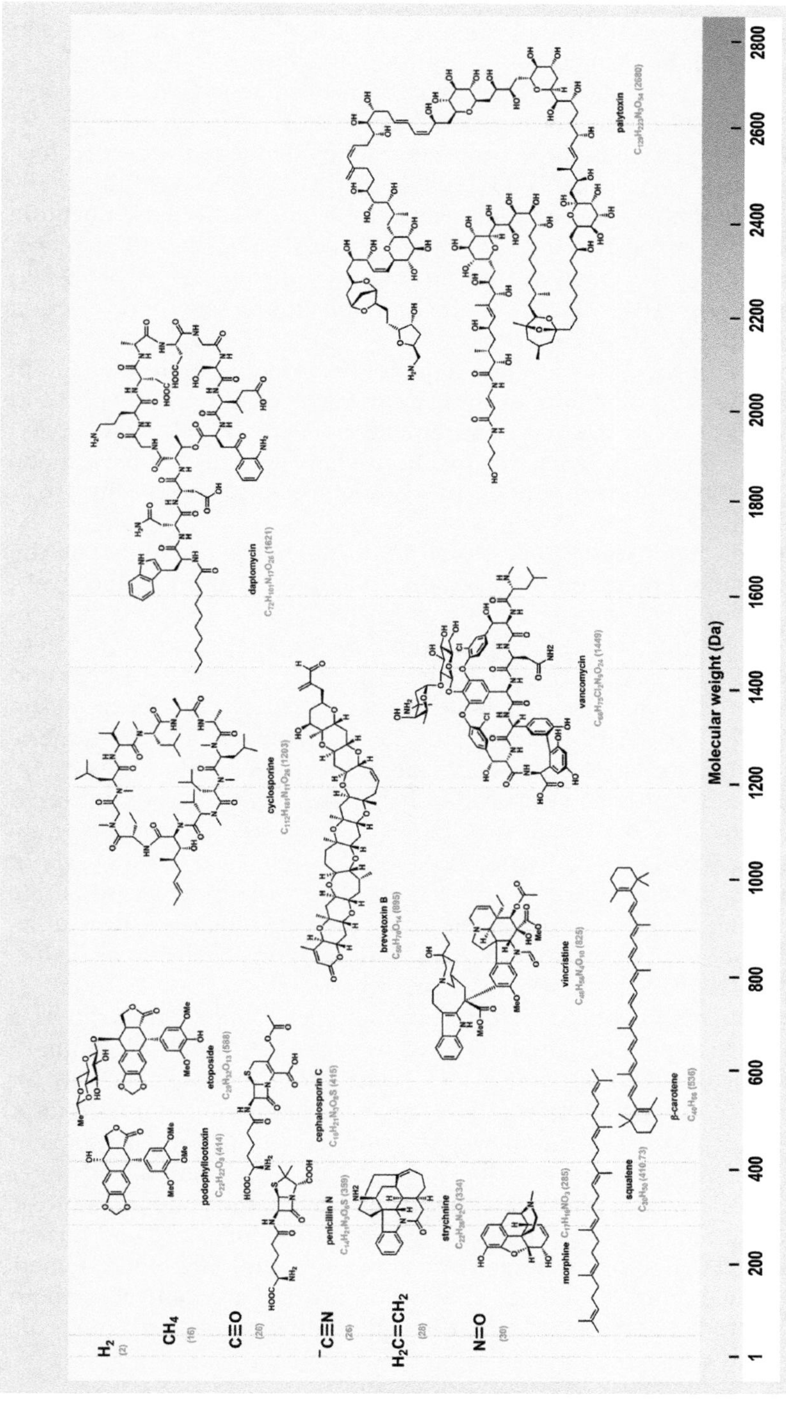

Figure 1.V1 The range of sizes of natural products.

can exceed that range, with brevetoxin at a MW of 895 and palytoxin ($C_{129}H_{223}N_3O_5$) at 2680 daltons, straining the limit of definition of a natural small molecule. Peptides that have been morphed into stable natural products often are macrocyclized and/or heterocyclized (as detailed in Chapter 3) to build in protease stability. Some molecules such as isopenicillin N (MW = 359) and cephalosporin C (MW = 415) are orally active (with suitable side chain derivatization), while other larger peptide frameworks of therapeutic agents are not orally absorbable and are given intravenously. These include the antibiotics vancomycin (MW = 1449) and daptomycin (MW = 1621) and the immunosuppressant, anti-rejection drug cyclosporine A (MW = 1203).

Almost all of the 50 000 known terpenoid/isoprenoid molecules fall in the <500 MW range, including simple monoterpenes, sesquiterpenes, and diterpenes as well as the acyclic triterpene squalene ($C_{30}H_{50}$, MW = 411) that is the immediate precursor to hundreds of tetracyclic sterol metabolites. On the other hand, the C_{40} polyisoprene beta-carotene comes in at 536 daltons.

Two other major classes of natural products are the alkaloids and the phenylpropanoid metabolites, taken up in Chapters 5 and 7, respectively. Thousands of alkaloids, containing at least one basic nitrogen atom in a heterocyclic ring, come in under the 500 Da MW parameter. These include the analgesic morphine, a ligand for the μ-opioid receptors and the poisonous strychnine that is an antagonist at glycine receptors on neurons. The largest of the alkaloids in common use are the dimeric anticancer agents vinblastine and vincristine (MW = 825 and 849, respectively) targeted against microtubules; these agents are administered intravenously. We shall note in Chapter 11 that steroids can be glycosylated, increasing their aqueous solubility but also adding about 180 daltons for every hexose unit added. Thus, digoxin, a cardiotonic glycoside, has a MW of 780, bulked up by the three digitose sugar units (~420 daltons). It is used to treat heart failure and is orally bioavailable, despite its relatively high MW for a small molecule.

In the phenylpropanoid natural product class, podophyllotoxin is a prominent example of a lignan, created by regioselective dimerization of coumaryl alcohol phenoxy radicals. Podophyllotoxin provides the core for the anticancer drug etoposide and has required harvesting of hundreds of thousands of roots from the mayapple producers to provide a sufficient therapeutic supply. Etoposide, at MW = 589, is poorly soluble in water and is formulated as a phosphate salt. In contrast to the pododphyllotoxic lignin, the coumaryl alcohol building blocks can be dimerized and further oligomerized to lignins (Chapter 6), meshworks of building block isomers that can reach many thousands in MW and provide mechanical strength layers in woody plants. Lignins fall outside the realm of small molecule natural products.

1.10 Natural Product Scaffold Diversity From a Limited Set of Building Blocks and a Limited Set of Enzyme Families

From the discussion so far of the major classes of natural products built in secondary metabolic pathways, and shown in schematic form back in Figure 1.3, it is clear that producer organisms use a limited set of primary metabolic building blocks to create an amazing diversity of end product architectures and functional group arrays. In turn there is a comparable finite, limited set of enzyme families that create these architectures. Figure 1.24 lists eight catalytic protein families that function in many conditional metabolic pathways as they do in primary pathways, reflecting the adaptation and evolution of specific functions.

In almost every class of natural products there are redox adjustments of functional groups during scaffold assemblies and maturations. The polyketide conversions of β-ketone to β-OH and of the α/β-enoyl double bond to the saturated CH_2–CH_2 in any catalytic cycle are classic two electron redox reactions catalyzed by NADH- and $FADH_2$-dependent oxidoreductases. These are the two central redox coenzymes in primary metabolism. Given the mimicry of fatty acid maturation enzymology, it is not surprising that NADH and $FADH_2$ are also serving secondary metabolic redox interconversions.

A Limited Set of Enzymes and Chemical Reaction Types Employed to Establish Scaffold Diversity	
oxidoreductases	alcohol to ketone to acids
decarboxylases	removal of carboxylate groups
Aldol and Claisen Condensation	C-C bond formation
C-N condensation catalysts	amide bond formation, *N*-heterocycles
transferases	addition of methyl, acyl or glycosyl groups
isoprenoid synthases	head-to-tail and head-to-head alkylation
epoxide hydrolases	furan and pyran formation
Mannich reactions	amino alkylation
In addition: • a plethora of types of oxygenation reactions • augmented by some rare enzyme classes	

Figure 1.24 A limited set of enzyme families and chemical reaction types are employed in natural product scaffold construction.

In alkaloid biosynthetic pathways the amino acid building blocks are often decarboxylated to the corresponding amines. These reactions are catalyzed by standard amino acid decarboxylases which use pyridoxal-phosphate as bound coenzymes. In a sense the decarboxylases act at the borders of primary and secondary metabolism.

The C–C bond-forming Claisen condensations in every PKS-mediated chain elongation are also identical in mechanism to the Fab enzymes in fatty acid biosynthesis so they are not unique to secondary pathways but are used iteratively therein. The intramolecular cyclodehydrations by the aromatizing aldolases in the formation of the aromatic fused tri- and tetracyclic frameworks also have precedents in other aldolase chemistry but create distinct scaffolds in secondary metabolic pathways.

In nonribosomal peptide synthetase assembly lines the sole chain elongation mechanism is amide bond synthesis. The responsible condensation domains are homologous to acyl transferases such as chloramphenicol acetyl transferase. The glycosyltransferases that act on natural product aglycones have evolved to recognize those specific cosubstrates but fall into the structural folds of glycosyltransferases in primary metabolism.

Depending on definitions of the boundaries of primary and secondary metabolism, one can argue that the prenyltransferases span such a boundary. The head to tail alkylations that grow isoprenoid/terpenoid chains to C_{10} geranyl-PP, C_{15} farnesyl-PP, and C_{20} geranylgeranyl-PP are key steps in generation of the 50 000 isoprenoid natural products. But farnesyl-PP and geranylgeranyl-PP also act as the prenyl donors for posttranslational protein farnesylations and geranylgeranylations (Walsh 2005). Similarly, the head to head condensations of farnesyl-PP to squalene are at the interface of primary and secondary metabolic pathways. One set of oxidosqualene cyclases subsequently yield lanosterol on the way to cholesterol, a primary constituent of eukaryotic cell membranes. A distinct subset of squalene/squalene-2,3-oxide cyclases create the sterol backbones for hundreds to thousands of conditional metabolites, as we will note in Chapter 4.

1.11 Some Notable and Unusual Transformations in Secondary Pathways

So, are there types of enzymes catalyzing chemical transformations that are not seen in primary metabolism? One might argue for the paired epoxidases and epoxide hydrolases that lead to the cascade of furan and pyran rings in polyketide polyether metabolites (Figure 1.25). Also of note are the presumptive Diels–Alderases that produce the decalin ring system during lovastatin assembly and a fused 5-6-5 tricyclic system during spinosyn maturation. The Pictet Spengler versions of Mannich reactions to convert indoles into tricyclic β-carbolines (*e.g.* strictosidine synthase) have no obvious homologs in primary metabolism.

Phenylalanine deaminase, the gate keeper enzyme for all molecules in phenylpropanoid pathways, is unusual in self-assembly of a tethered methylidene-imidazolone (MIO) cofactor from three active site residues to carry out the conversion of L-Phe to cinnamate.

The conversions of serine and threonine residues to oxazoles and methyloxazoles and, correspondingly, of cysteines to thiazoles are remarkable examples of chemistry confined so far to secondary pathways. We will note that these five ring heterocycles, which have not only the side chains morphed but also the peptide backbone, can be accessed either from posttranslational catalytic machinery on nascent proteins or *via* nonribosomal peptide synthetase assembly lines. Finally, the conversion of the acyclic tripeptide aminoadipyl-L-Cys-D-Val into the bicyclic β-lactam scaffold is a remarkable achievement (see Figure 1.25 for all these examples) and one that consumes a cosubstrate molecule of molecular oxygen.

We will argue in Chapter 10 that *S*-adenosylmethionine, when cleaved to methionine and 5′-deoxydenosine, signals transfer of [$CH_3^{\bullet}$] to unactivated carbon sites. This is a reactivity pattern distinct from the canonical transfer of [CH_3^{+}] to nucleophilic cosubstrates during cleavage to *S*-adenosylhomocysteine (see Figure 1.29 below).

1.12 Oxygenases are Pervasive in Natural Product Biosynthetic Pathways

A special place has to be accorded to oxygenases in secondary metabolism. They are dramatically more abundant in secondary metabolic pathways than in primary pathways.

Although higher eukaryotes are aerobic organisms, most of the reductive flux of oxygen metabolism occurs at the active site of cytochrome oxidase, the terminal enzyme in mitochondrial respiratory chains. At that terminal oxidase, O_2 is reduced by four electrons to two molecules of water, releasing a large amount of energy that ultimately is stored in adenosine triphosphate (ATP) molecules to

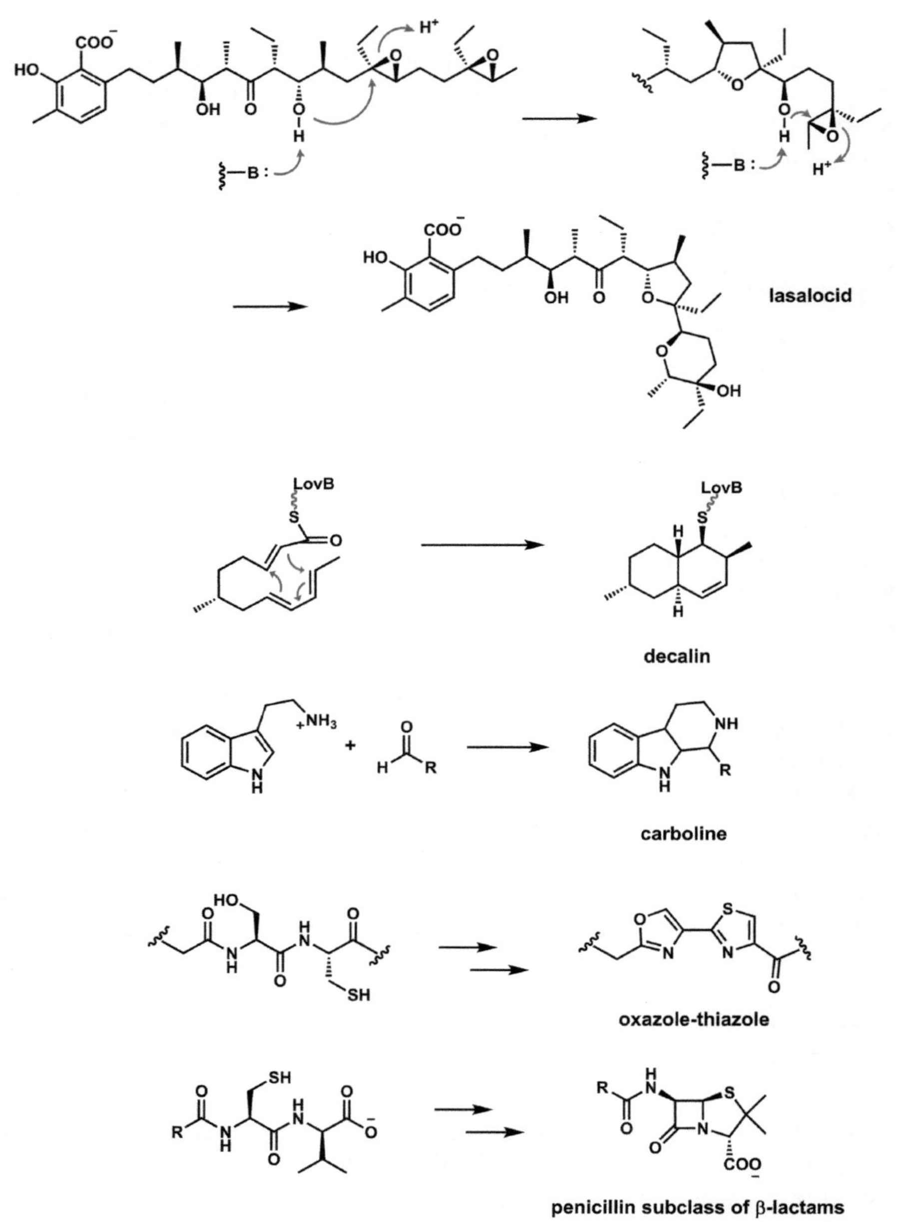

Figure 1.25 Five notable transformations during natural product biosyntheses.

drive many cellular chemical transformations. There are some substrate-oxidizing enzymes that reduce O_2 by two electrons to H_2O_2 but those are often located in the peroxisome compartments where catalase is also present to detoxify the hydrogen peroxide coproduct. The relative dearth of oxygenases in primary metabolism may reflect a history of primary pathways evolving in anaerobic organisms before the atmosphere was substantially oxygenated.

In contrast, we will note the key features of oxygenases in the maturation of each of the major natural product classes described in Chapters 2–8 and integrate the separate examples in Chapter 9, where the unusual constraints of O_2 reactivity are described. Briefly, O_2 is a triplet ground state molecule and reacts very sluggishly with spin-paired organic molecules found in cells and tissues.

Nature has evolved two parallel strategies to activate O_2 for reduction by one electron transfer pathways. One route involves redox active transition metals, predominantly iron in the 2^+ or 3^+ oxidation state, in the resting form of oxygenases (Figure 1.26). The iron atom in oxygenases is most commonly found in the equatorial plane of the heme cofactor in a family of cytochromes designated cytochromes P450 (450 refers to the diagnostic absorption maximum of the Fe(II)–CO complex). An alternative arrangement is an iron atom in a nonheme active site environment where side chains of two histidines and one Glu/Asp carboxylate provide the three ligands from the enzyme active site (Figure 1.26). Most often a cosubstrate molecule of α-ketoglutarate occupies the fourth and fifth ligand positions to the active site iron atom.

The second solution to reductive activation of O_2 on demand is to use the vitamin B2-based flavin coenzymes as one-electron conduits to O_2. The tricyclic isoalloxazine ring system of flavins (*e.g.* FAD) can

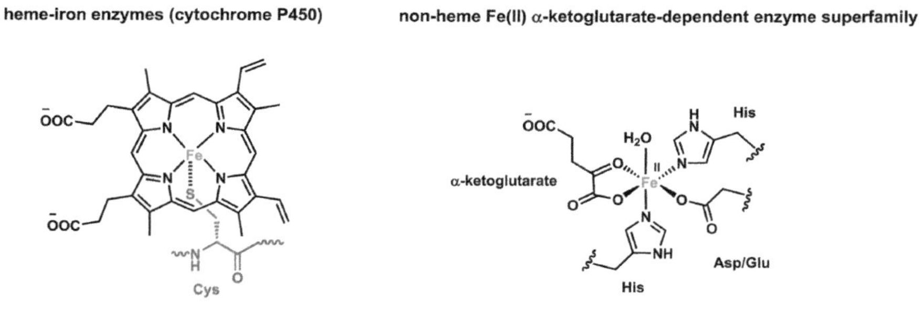

Figure 1.26 Two distinct enzyme microenvironments for iron-based oxygenases that generate high valent oxo-iron intermediates.

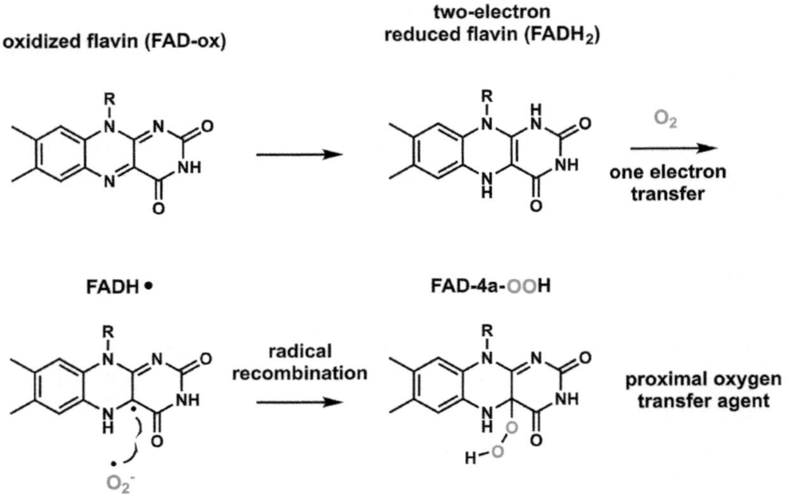

Figure 1.27 The $FADH_2$ oxidation state of the FAD coenzyme reacts rapidly with O_2 by one electron transfer, and the resultant $FADH^•$ and superoxide anion ($O_2^{-•}$) recombine to give the FAD-4a-OOH as proximal oxygen transfer agent.

accept either two electrons at a time to give $FADH_2$ directly or one electron at a time *via* the flavin semiquinone $FADH^•$ on the way to $FADH_2$. The $FADH_2$ form is reactive with oxygen, giving a superoxide radical and an $FADH^•$ which recombine to the FAD4a–OOH (Figure 1.27). The flavin hydroperoxide is the oxygen donor to co-substrates. We will note in Chapter 9 that the iron-based oxygenases are more robust oxygen transfer agents than the flavoenzyme oxygenases. Thus, oxygenation at chemically unactivated carbon sites is usually undertaken by the iron enzymes.

1.13 Carbon–Carbon Bonds in Natural Product Biosynthesis

Much of natural product scaffold assembly from monomeric building blocks involves construction of carbon–carbon bonds as complexity is built into the secondary metabolite frameworks. As always there are two limiting options: C–C bonds can be built or taken apart by heterolytic or homolytic routes (Figure 1.28). The heterolytic, two electron routes require carbanions and carbocations as cosubstrates. The homolytic routes are one electron transfers and utilize a pair of

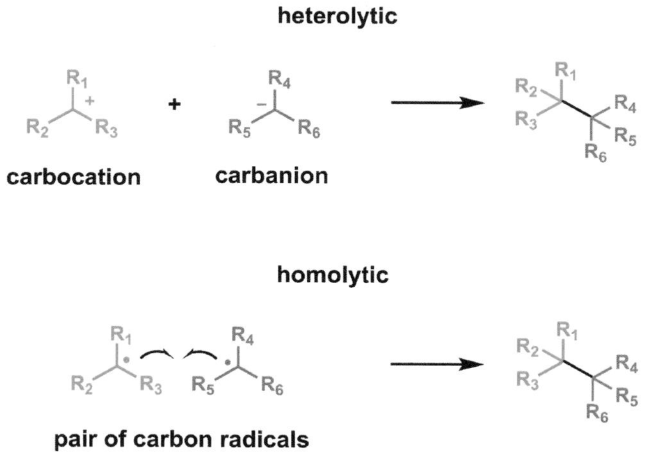

Figure 1.28 Two limiting options in enzymatic formation of C–C bonds. Heterolytic mechanisms involve carbanions and carbocations. Homolytic pathways involve pairs of carbon radicals.

carbon radicals. The heterolytic routes to C–C bond formation predominate in primary metabolism and also in secondary metabolism.

However, we will note two circumstances where carbon-centered radicals are enzymatically generated. The first is in oxygenase catalysis, and those catalysts are pervasive in natural product maturations (Chapter 9). The second is at the other end of the oxygenation spectrum: *S*-adenosylmethionine in anaerobic microenvironments can act as a radical generator (Chapter 10).

Figure 1.29 illustrates two such examples. The aerobic one involves the O_2-consuming enzyme in the morphine biosynthetic pathway that converts *R*-reticuline to the tetracyclic product salutaridine. The new C–C bond, joining the A and C rings as the B ring is formed, arises from intramolecular coupling of two carbon radicals. The anaerobic example involves enzymatic conversion of guanosine triphosphate (GTP) to 3′, 8-cyclo-GTP. This, too, is a coupling of two carbon radicals but the radical generator is not oxygen but rather *S*-adenosymethionine, *via* cleavage to methionine and the 5′-deoxyadenosyl radical which is the proximal radical generator (Chapter 10).

Readily accessible carbanion equivalents include enolates and most notably thioester enolates for fatty acid and polyketide chain elongation chemistry (Figure 1.30). Phenolate anions represent a variant where the carbanion resonance forms can act as nucleophiles, for example in C-methylations of *ortho* to the phenol. The π-electrons

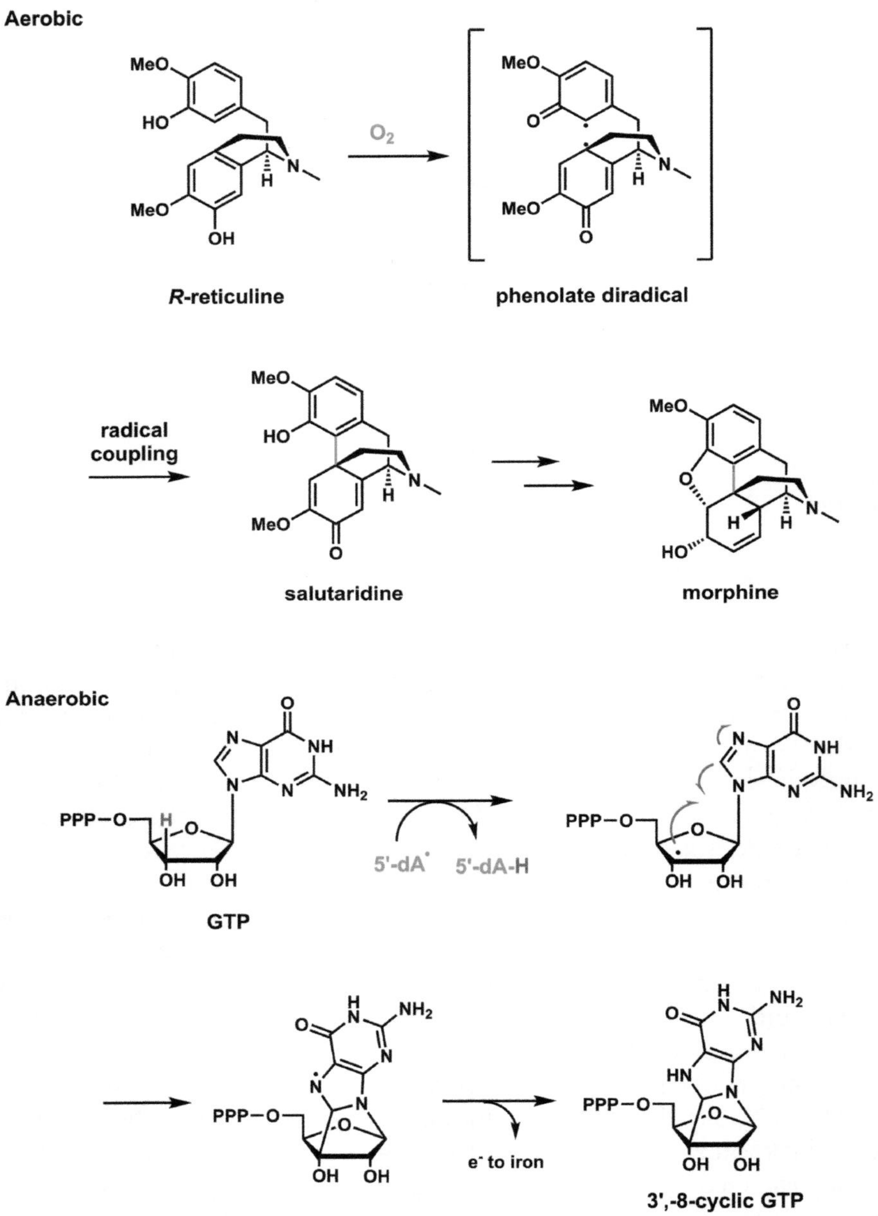

Figure 1.29 C–C bonds from radical coupling: aerobic *vs*. anaerobic reaction manifolds. O_2-dependent conversion of *R*-reticuline to salutaridine during morphine biosynthesis. Formation of the B ring involves C–C bond formation by a (phenoxy) radical pair. Generation of 3′,8-cyclo GTP during molybdopterin coenzyme biogenesis involves radical chemistry initiated by hemolytic cleavage of *S*-adenosylmethionine.

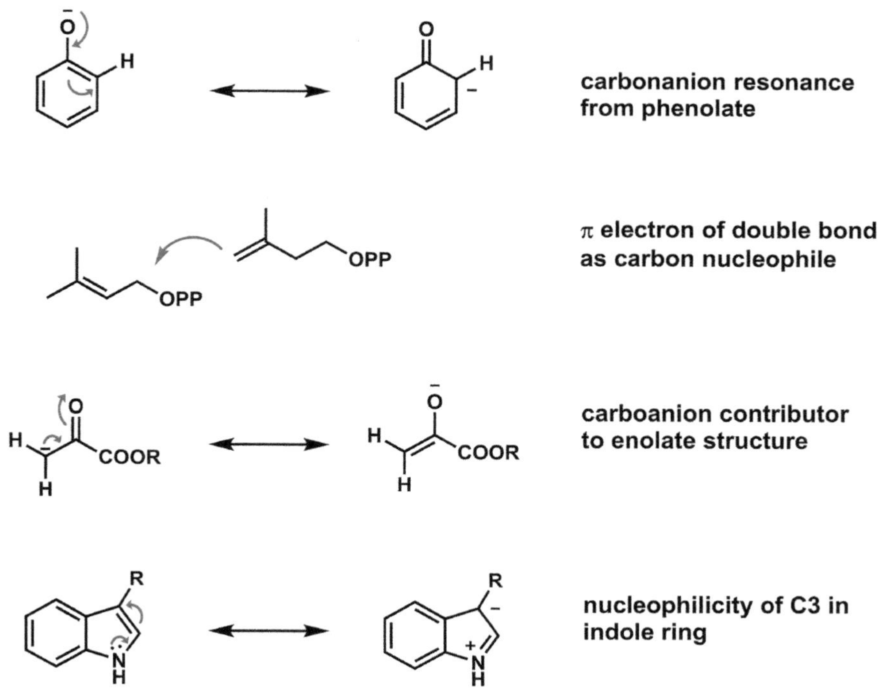

Figure 1.30 Kinetically accessible carbanion equivalents in secondary metabolic pathways for heterolytic C–C bond formation.

of the Δ^3-regioisomer of isoprenyl diphosphate are the carbon nucleophile in the head to tail chain elongations in the early steps of isoprenoid scaffold growth. A fourth example, discussed in detail in Chapter 8, is the ability of the indole ring of tryptophan and tryptamine to function as a carbanion equivalent, shown as the C_3 carbanion in Figure 1.31.

The most abundant sources of carbon electrophiles in both primary and secondary metabolism are carbons in the aldehyde and ketone oxidation states, and also carboxyl groups activated as thioesters. Aldehydes in general are rarer than ketones but they are featured cosubstrates in the Pictet Spengler reactions that convert indoles into tricyclic carbolines in alkaloid biosynthesis. Ketones are the work-horse electrophiles in the polyketonic acyl chains that undergo intramolecular aldol condensation in aromatic metabolites such as tetracyclines and daunomycin.

Members of the most common natural product class, the iso-prenoid/terpenoid family, use allylic cations from Δ^2-prenyl

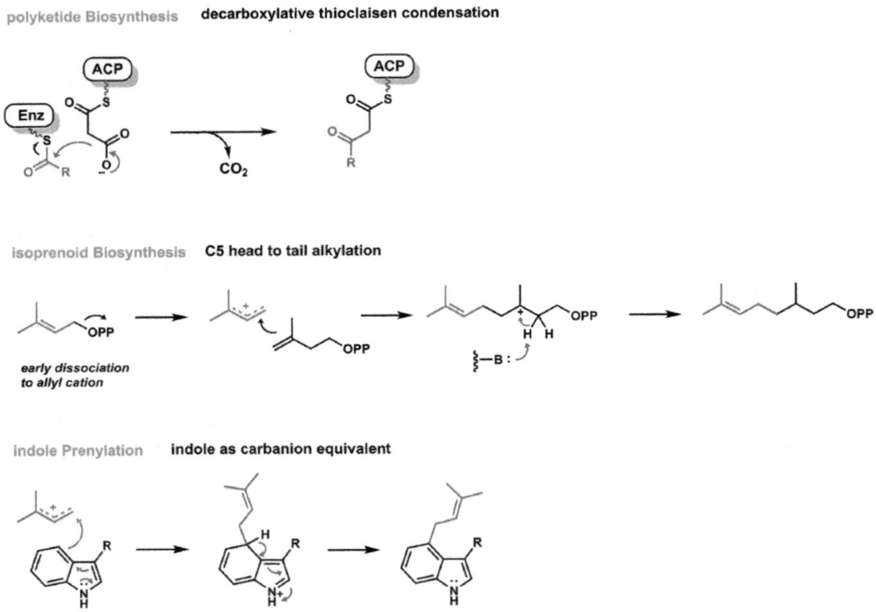

Figure 1.31 Three types of C–C bond forming reactions in natural product assembly.

diphosphates as the initiating chemical step in thousands of intermolecular and intramolecular coupling reactions (Figure 1.31). The initial allyl cations can rearrange to drive structural changes before they are quenched regio- and stereospecifically by particular enzyme catalysts. The baruol synthase of *Arabidopsis thaliana* generates a suite of tetracyclic steroid frameworks from squalene-2,3-oxide that indicate that 14 distinct carbocations have been generated during the electrocyclization process (Chapter 4). Comparable cation-driven skeletal rearrangements are not common in any of the other natural product families.

1.14 Testing Biosynthetic Hypotheses by Feeding Isotopically Labeled Building Blocks

During the second half of the 20th century the chemical logic for assembly of the major classes of natural products, described in Chapters 2–8, came into view as the details of primary metabolic pathways became essentially complete, and the building blocks for conditional pathways became clear. For those secondary pathways the availability of building blocks containing stable heavy atom isotopes provided central reagents for feeding studies, often in microorganisms, to determine the connectivity of such fragments in mature, often complex natural product scaffolds.

Two classic kinds of starting substrate for feeding studies were glucose molecules labeled synthetically at specific carbons with ^{13}C and acetic acid, doubly labeled with ^{13}C at both carbons 1 and 2 (to enable determination of retained connectivity or biosynthetic separation). The natural 1% abundance of ^{13}C in a sea of 99% ^{12}C molecules was sufficient for ^{13}C-NMR analyses of the natural product metabolites (and higher enrichment fractions could be generated by synthetic methods). Also of utility were samples of trideutero CD_3COOH and oxygen-18-labeled $CH_3C^{18}O^{18}OH$ (Figure 1.32). The deuterium and oxygen-18 atoms create diagnostic splitting patterns on adjacent carbon atoms which allow deduction of biosynthetic logic and determine whether the C–H and C–O bonds in acetate units remained intact during incorporation into end metabolites.

Isotopomer-guided feeding studies have become less central in biosynthetic projects over the past few decades as pathway hypotheses have been filled out for the major natural product classes. This is particularly true in the post genome era where bioinformatic predictions of open reading frames in gene clusters give high information content about both operant chemical logic and enzymatic machinery. Nevertheless, there are contemporary occasions where isotopomer

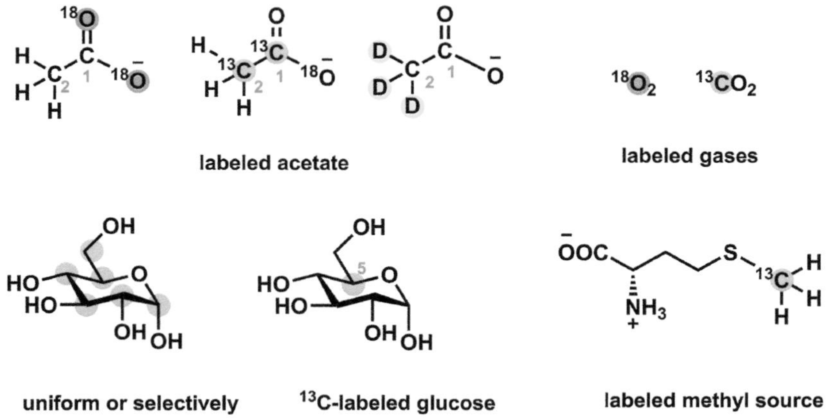

Figure 1.32 Commonly used isotopomers in natural product labeling studies.

labeling studies may still be essential. A recent example utilized [^{13}C-methyl]-methionine to probe the biosynthetic pathway to polyketide toxins from dinoflagellate blooms. Genome mining has been difficult in dinoflagellates given that some have genome sizes 100 fold those of human genomes. Transcriptome and proteome studies have also been difficult owing to the physical states of dinoflagellate chromosomes, so isotopomer labeling has been the preferred route to pathway deconvolution in polyketide spiroimine toxins (Anttila, Strangman *et al.* 2016).

We will illustrate the role of isotopomer-guided thinking in examples in three subsequent chapters to emphasize how the results constrain and inspire biosynthetic hypotheses. Readers with special interest in the topic may wish to look at those examples, even in this brief summary context. Indeed, a recent review (Bacher, Chen *et al.* 2016) describes how advances in methodology, particularly around pulse chase studies with [$^{13}CO_2$] can still be useful in whole organism feeding efforts.

A polyketide example, noting the use of both ^{13}C and ^{18}O labels, is provided in Chapter 2 as the experimental backdrop to formulation of the Cane–Celmer–Westley rules (Cane, Celmer *et al.* 1983) for understanding the chemical logic of creation of the furan and pyran ring scaffold of monensin and related polyether metabolites. In Chapter 4, which deals with the largest family of natural products, the terpenoid–isoprenoid class, we note how isotopomer labeling from ^{13}C-acetate allows ready distinction of formation of the isopentenyl diphosphate isomers by the classical mevalonate pathway *vs.* the nonclassical methylerythritol phosphate (MEP) pathway. The third example, in Chapter 8, illustrates the pattern of isotopomer labeling in the mixed polyketide/phenylpropanoid family member kaemphenol, a plant flavanone diglycoside.

One further example of the potentially exquisite utility of isotopomer approaches to assess a mechanistic question in biosynthesis is in the use of 5-[^{13}C]-glucose to address a regiochemistry question during biogenesis of the *Bacillus subtilis*-generated dipeptide antibiotic bacilysin (L-Ala-L-anticapsin) (Figure 1.33A) (Parker and Walsh 2012). The anticapsin (2,3-epoxy-4-keto-cyclohexylalanine) arises from glucose *via* the shikimate pathway intermediates chorismate and prephenate, as diagrammed. Prephenate has two prochiral olefins. To determine whether the enzyme BacA acted regiospecifically on only one of them during decarboxylation, 5-[^{13}C]-glucose was fed to a strain of *Aerobacter aerogenes* that accumulated chorismate (Parker and Walsh 2012). The chorismate was isolated and shown by

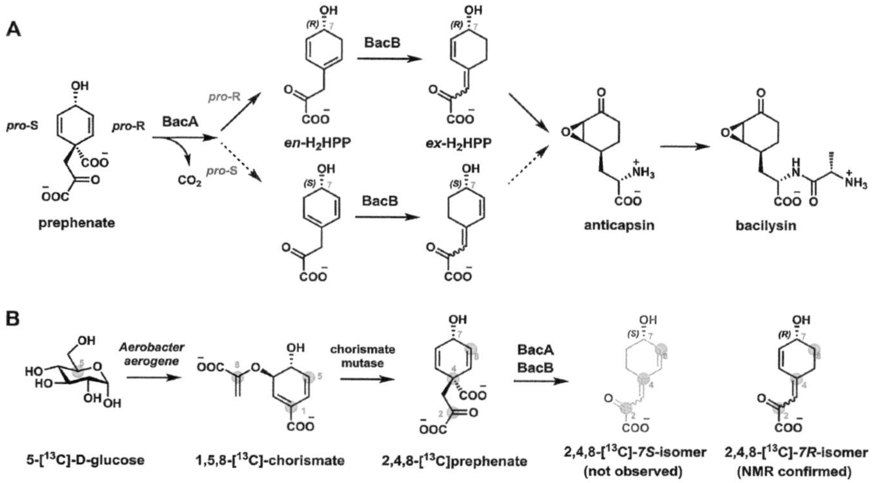

Figure 1.33 Use of a labeled precursor to study olefin isomerization regiochemistries in bacilysin biosynthesis.

[^{13}C]-NMR analysis to have the anticipated 1,5,8-[^{13}C]-labeling pattern (Figure 1.33B). The isolated chorismate was then incubated with purified chorismate mutase to yield 2,4,6-[^{13}C]-prephenate (Figure 1.33B). At this juncture, submission of purified prephenate to the pure BacA and BacB enzymes gave only 2,4,8-[^{13}C]-7*R*-exocyclic H_2-hydroxyphenylpyruvate, on the way to bacilysin. These data conclusively show that BacA isomerizes only the 7*R* double bond of prephenate while avoiding the aromatization fate that is the default mode of the BacA family prephenate decarboxylase homologs, a result that would be difficult to ascertain by any other method.

1.15 Historical and Contemporary Approaches to the Detection and Characterization of Natural Products

The scientific enterprise of characterizing natural products can be split into three eras. The first reached from at least as far back at the Sumerians in 2600 BC to the beginning of the 19th century. In that ~4400 year span, natural products with biological activity and therapeutic utility were used largely as crude extracts and/or mixtures of plants. This was an era of ethnobotany (Soejarto, Fong *et al.* 2005), hard won experience of medicinal plants, and ethnopharmacology

(Acharya and Shrivastava 2008), as different cultures around the globe found plants in their local environments that had real or perceived medicinal activity.

The second era spanned the roughly 200 years, from 1800 to the end of the 20th century. In that period the science of natural products evolved from mixtures of crude botanic extracts to the isolation of pure compounds. The first six pure alkaloids were isolated within the decade 1816–1826 (see Chapter 5), changing the standards from incompletely characterized and variable mixtures of molecules in an extract to purified molecular entities. The determination of the molecular structures lagged behind the isolation protocols by decades as the analytical and structural tools for small molecule structure determination developed over a century and a half. The inventory of pure natural products broadened explosively from plant sources to microbial sources in the second half of the 20th century as microbially sourced antibiotics were discovered in the two decades between 1940 and 1960.

The third era is the current postgenomic era when microbial genomes of 5–10 megabases can be completed within one day of genome sequence time (Chapter 13). Thousands of fungal and bacterial genomes have been sequenced in the past two decades and tens of thousands more are likely in the near term. Given 30–50 biosynthetic gene clusters just for the three natural product classes of polyketides, nonribosomal peptides, and terpenes in many of those genomes, it is reasonable to think that millions of bioinformatically predicted gene clusters will be available for investigation. Most of these clusters are silent under standard laboratory growth conditions and their activation, expression, and the identification of encoded small molecule natural products are the focus of many initiatives in synthetic biology, and heterologous host optimizations (Chapter 13).

Figure 1.34 schematizes four stages of inquiry that illustrate a contemporary approach to the logic for detection and activation of the tens of thousands of silent microbial biosynthetic gene clusters (BGCs). **Genomics** is at the front end to choose microorganisms of interest (perhaps an emerging pathogen, perhaps occupying an unusual and underexamined biological niche) and determine the genome sequence. The next activity is **genome mining**: a bioinformatic exercise to catalog predicted gene clusters for major classes of natural products based on prior knowledge of the kinds of open reading frames in such pathways. The third stage is the **molecular biology** and, increasingly, **synthetic biology** effort to place the gene clusters to be interrogated into contexts for activation and expression

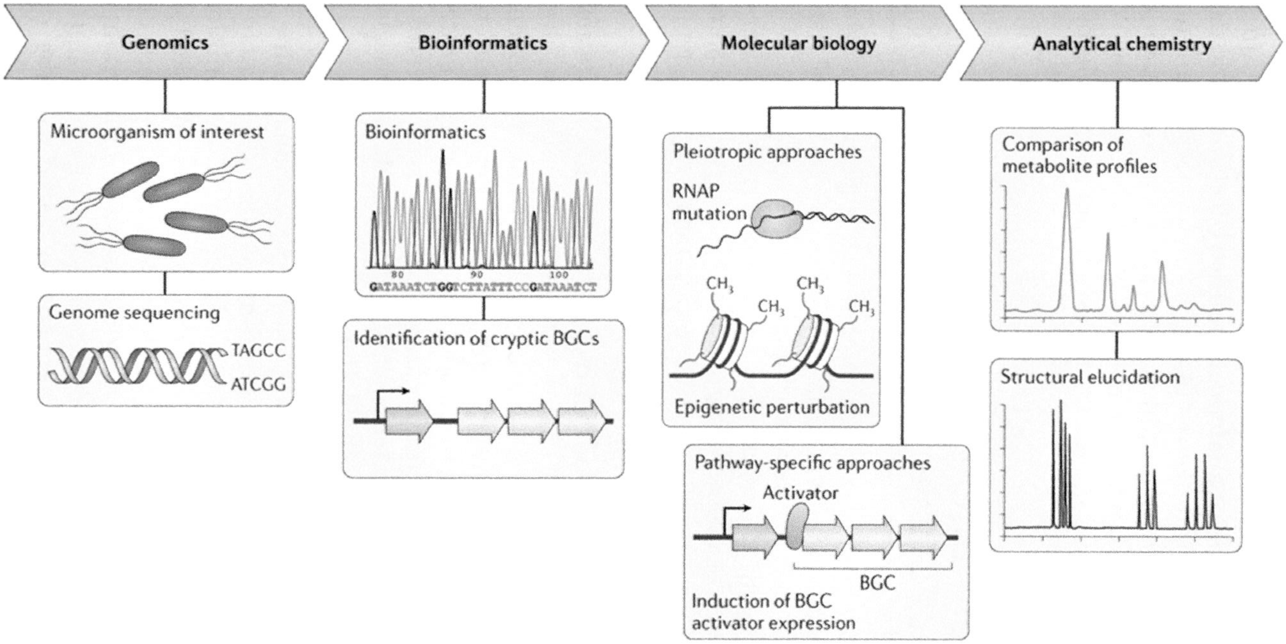

Figure 1.34 A contemporary approach to the logic for detection and activation of the tens of thousands of silent microbial biosynthetic gene clusters.
Reprinted by permission from Macmillan Publishers Ltd: Nature Reviews Microbiology (Rutledge, P. J. and G. L. Challis. "Discovery of Microbial Natural Products by Activation of Silent Biosynthetic Gene Clusters". *Nat. Rev. Microbiol.* **13**: 509), copyright (2015).

of end products, including any suite of scaffold tailoring maturation steps. The parameters for prioritization of gene clusters to be examined can vary with the specific goals of the investigating team. The fourth activity is the suite of **analytical chemistry techniques to detect any products specific to the gene cluster.** If any molecules are novel, their structures must be determined, typically by a combination of mass spectrometry and two-dimensional NMR, but also by X-ray analysis of single crystals if the structures are not resolvable by MS and NMR alone.

1.16 Summary: Distinct Assembly Logic for Different Classes of Natural Products

As we embark on the analysis of the chemical logic and enzymatic machinery operant for biosynthesis of the major classes of natural products, we note that distinct assembly logic is in play.

For both the polyketides and nonribosomal peptides, iterative addition of monomers occurs using covalently tethered thioesters on enzymatic assembly lines. The nascent products released, typically hydrolytically or *via* macrocyclization, are then subjected to maturation by sets of dedicated post assembly line tailoring enzymes.

The characteristic feature of isoprenoids/terpenoids is the use of allylic cations as electrophilic partners in intermolecular alkylations. The cations are also subject to isomerizations that can drive extensive intramolecular skeletal rearrangements. A predominant pattern is head to tail self-condensation to C_{15} and C_{20} lengths, then a switch to head to head condensation, preparatory to cyclization to sterol and hopane frameworks. In general prenylation enzymes do not covalently tether any of the reaction intermediates.

In the alkaloid family the basic nitrogens are key reactants as nucleophiles in a variety of C–N bond forming reactions. Imine and enamine chemistry facilitates C–C bond formation as well. Mannich and Pictet Spengler versions of the Mannich reaction are complexity-generating transformations that allow construction of fused ring scaffolds. Central to the biosynthetic chemistry of prenylated indole alkaloids is the protean reactivity of indole as carbon nucleophile at six-carbon positions around the bicyclic ring. The C_5 and C_{10} prenyl-PPs are favored electrophilic partners in C–C bond forming decorations of the indole scaffold.

The purine and pyrimidine heterocycles, strictly speaking, are categorizable as alkaloids. However, they have such a central place in primary metabolism that they are almost invariably treated as primary rather than secondary metabolites. Nonetheless the logic for modification of the heterocycles, the sugar and the 5′ substituents is revealing of distinct niches for modified nucleosides in secondary metabolism.

The logic in phenylpropanoid assembly seems the reverse of the alkaloid reactivity. Both start with amino acid building blocks but L-Phe is deaminated at the outset and the basic nitrogen atom discarded. Indeed, the vast range of phenylpropanoids lack nitrogen in their frameworks.

Oxygenases play key roles in this set of natural products, from creating phenolic and catecholic versions of cinnamate (*e.g.* 4-OH-coumarate), to subsequent generation of phenoxy radicals on the way to lignans and lignins. Oxygenases are also key to oxidation of flavanones to the colored pigments in various flowers and fruits.

1.17 Approach of This Volume

Given the enormous number of natural products already known (>300 000 molecules), with many more to come from bioinformatic predictions in fungal and bacterial genomes, there is no way that coverage of biosynthesis can be exhaustive or comprehensive.

Instead, we have endeavored to present selected examples of major natural product classes to illustrate chemical logic in bond constructions, and in tailoring and maturation of small molecule scaffolds, particularly in complexity generating reactions. We have focused on major classes of enzyme catalysts that operate in secondary pathways, with attention on enzymes that both make and break carbon–carbon and carbon–nitrogen bonds. Oxidation/reduction reactions feature prominently in natural product scaffold maturations: accordingly these features, and the enzymes that accelerate these reactions, are highlighted in the specific natural product classes.

While chemical efforts – isolations, structural characterizations, total and semi-syntheses – have appropriately dominated the field of natural product biosynthesis historically, the current and likely future efforts to find and reprogram the biosynthetic pathways to the molecules of nature use a broader swath of scientific competence. The genomic era has brought bioinformatics experts, synthetic biologists, metabolic pathway reconstruction, and gene expression specialists into full partnerships in multidisciplinary teams with chemists.

Among the questions going forward are how many new molecular frameworks and classes are yet to be discovered, given estimates that we have detected <10% of Nature's small molecule inventory. What novel biologic activities will they possess and what physiologic roles do they play in producer organisms? To what extent will they continue to serve as probes of specific biologic events and as inspiration for new therapeutic agents?

Additional reading: In shaping this volume, we have built on the foundational book by P. Dewick, *Medicinal Natural Products, a Biosynthetic Approach, 3rd Edition*, 2009 (John Wiley and Sons) which provides an extensive coverage of molecular classes and proposed biosynthetic chemical mechanisms, as well as scholarly commentary particularly on plant metabolites.

Many of the references that are provided in the ensuing chapters of this volume are to recent review articles, many in the journal *Natural Product Reports*. They are intended to be entry points into relevant research literature for readers who want to delve more deeply into specific topics.

Some readers may also wish to consult the ten volume series *Comprehensive Natural Products II* (Mander and Liu 2010), which appeared in 2010. It spans 7388 pages but its current price of ~$6000 (as of June, 2016) makes it likely to be of restricted distribution in personal libraries.

References

Acharya, D. and A. Shrivastava (2008). *Indigenous Herbal Medicines: Tribal Formulations and Traditional Herbal Practices*. Jaipur, India, Avishkar Publishers.

Anttila, M., W. Strangman, R. York, C. Tomas and J. L. Wright (2016). "Biosynthetic Studies of 13-Desmethylspirolide C Produced by Alexandrium ostenfeldii (= A. peruvianum): Rationalization of the Biosynthetic Pathway Following Incorporation of (13)C-Labeled Methionine and Application of the Odd-Even Rule of Methylation". *J. Nat. Prod.* **79**(3): 484–489.

Bacher, A., F. Chen and W. Eisenreich (2016). "Decoding Biosynthetic Pathways in Plants by Pulse-Chase Strategies Using (13)CO(2) as a Universal Tracer dagger". *Metabolites* **6**(3): 21–45.

Cane, D. E., W. D. Celmer and J. W. Westley (1983). "Unified stereochemical model of polyether antibiotic structure and biogenesis". *J. Am. Chem. Soc.* **105**: 3594–3600.

Demain, A. L. and A. Fang (2000). "The natural functions of secondary metabolites". *Adv. Biochem. Eng./Biotechnol.* **69**: 1–39.

Jurjens, G., A. Kirschning and D. A. Candito (2015). "Lessons from the synthetic chemist nature". *Nat. Prod. Rep.* **32**(5): 723–737.

Lange, B. M. (2015). "The evolution of plant secretory structures and emergence of terpenoid chemical diversity". *Annu. Rev. Plant Biol.* **66**: 139–159.

Mander, L. and H. W. Liu, eds. (2010). *Comprehensive Natural Products II: Chemistry and Biology*, Elsevier Science.

Parker, J. B. and C. T. Walsh (2012). "Olefin isomerization regiochemistries during tandem action of BacA and BacB on prephenate in bacilysin biosynthesis". *Biochemistry* **51**(15): 3241–3251.

Pichersky, E., J. P. Noel and N. Dudareva (2006). "Biosynthesis of plant volatiles: nature's diversity and ingenuity". *Science* **311**(5762): 808–811.

Rodrigues, T., D. Reker, P. Schneider and G. Schneider (2016). "Counting on natural products for drug design". *Nat. Chem.* **8**: 531–541.

Schenk, P. M., K. Kazan, I. Wilson, J. P. Anderson, T. Richmond, S. C. Somerville and J. M. Manners (2000). "Coordinated plant defense responses in Arabidopsis revealed by microarray analysis". *Proc. Natl. Acad. Sci. U. S. A.* **97**(21): 11655–11660.

Soejarto, D. D., H. H. Fong, G. T. Tan, H. J. Zhang, C. Y. Ma, S. G. Franzblau, C. Gyllenhaal, M. C. Riley, M. R. Kadushin, J. M. Pezzuto, L. T. Xuan, N. T. Hiep, N. V. Hung, B. M. Vu, P. K. Loc, L. X. Dac, L. T. Binh, N. Q. Chien, N. V. Hai, T. Q. Bich, N. M. Cuong, B. Southavong, K. Sydara, S. Bouamanivong, H. M. Ly, T. V. Thuy, W. C. Rose and G. R. Dietzman (2005). "Ethnobotany/ethnopharmacology and mass bioprospecting: issues on intellectual property and benefit-sharing". *J. Ethnopharmacol.* **100**(1–2): 15–22.

Walsh, C. T. (2005). *Posttranslational Modification of Proteins: Expanding Nature's Inventory*. Englewood, Colorado, Roberts and Company.

Walsh, C. T. and T. Wencewicz (2016). *Antibiotics Challenges, Mechanisms, Opportunities*. Washington DC, ASM Press.

War, A. R., M. G. Paulraj, T. Ahmad, A. A. Buhroo, B. Hussain, S. Ignacimuthu and H. C. Sharma (2012). "Mechanisms of plant defense against insect herbivores". *Plant Signaling Behav.* **7**: 1306–1320.

Section II

Six Natural Product Classes

This section takes up each of the six major natural product structural types, in Chapters 2–8, respectively.

The starting points are polyketides (Chapter 2) and peptides (Chapter 3) because of similarities in the chemical logic and enzymatic machinery of biosynthetic assembly lines. On such assembly lines the monomer units to be added and the growing ketidyl/peptidyl chains are tethered as covalent thioester intermediates to carrier protein domains. The other categories of natural products (isoprenes/terpenes, alkaloids, phenylpropanoids) do not employ assembly line logic, but instead collect and assemble freely diffusible intermediates.

Polyketides encompass a remarkable range of structural variation and emphasis is placed on how the common chain-extending thioclaisen condensations are routed into such distinct molecular frameworks. Thioester enolate chemistry dominates the chain extension processes in all the polyketide frameworks.

Subclasses include the fused tetracyclic aromatic scaffolds of such molecules as tetracyclines and daunomycin, and the macrolactones exemplified by erythromycin, fidaxomicin, and ivermectin. The biosynthesis of the decalin-containing lovastatin and the tricyclic 5-6-5 ring system of spinosyn represent likely examples of biological Diels–Alder enzyme chemistry. Polyethers in which epoxide intermediates are opened enzymatically to furans and pyrans comprise a substantial subclass of polyketides. The remarkable scaffolds of the enediyne anticancer polyketides reflect a convergence of four distinct biosynthetic lines.

Nonribosomal peptide natural products such as vancomycin, cyclosporine, and penicillins and cephalosporins arise by assembly line thioester logic and enzymatic machinery. Their assembly lines

Natural Product Biosynthesis: Chemical Logic and Enzymatic Machinery
By Christopher T. Walsh and Yi Tang

Published by the Royal Society of Chemistry, www.rsc.org

can use dozens of nonproteinogenic amino acid building blocks to build in side chain functional group diversity. Many of the NRPS assembly lines and the PKS assembly lines of Chapter 2 release products by macrocyclization, and the cyclic frameworks are essential for biologic activity. Post assembly line maturation by dedicated tailoring enzymes is a common strategy for modification of the core scaffold in polyketides, nonribosomal peptides and hybrid PK–NRP and NRP–PK metabolites. A complementary strategy to generate peptide-derived stable small molecular frameworks involves limited proteolysis of ribosomally produced precursor proteins. Such peptide fragments are again often macrocyclized and contain other rigidifying chemical links. These include the thioether bonds in lantipeptides and the oxazole and thiazole heterocycles in cyanobactins and thiazolyl peptide antibiotics.

The largest class of natural products is thought to be the isoprene-based molecules, also known as terpenes, reflecting the importance of plant science in the characterization of this natural product class. Two types of C–C bond formations dominate the early steps in terpenoid biosynthetic pathways. The first is head to tail alkylative condensations between Δ^2-prenyl-PPs and Δ^3-isopentenyl-PP, extending chains five carbons at a time. The Δ^2-partners are facile sources of allyl carbocations and these can drive many kinds of intramolecular skeletal rearrangements, notably at the C_{10}, C_{15}, and C_{20} levels.

The second type of C–C bond formation is head to head alkylative condensation between pairs of C_{15}-farnesyl-PPs or C_{20}-geranylgeranyl-PPs to generate squalene and phytoene, respectively. Squalene cyclization then leads to many hundreds of cyclic triterpenes, including the steroid family. Phytoene is not comparably cyclized but gives rise to carotenoids and vitamin A.

While both polyketide and isoprenoid metabolites are essentially devoid of nitrogen atoms, the alkaloids contain at least one basic nitrogen atom, typically with a heterocycle. The nitrogens reflect the fact that the starter units are half a dozen of the common amino acids of primary metabolism. Imine and enamine chemistry is in play for C–C bond constructions. Many of the more complex transformations in alkaloid biosynthesis involve oxygenase-mediated radical chemistry. The indole-terpene class of alkaloids is treated in a separate chapter (Chapter 8), in part to focus on the reactivity of the indole ring as a carbon nucleophile at six positions around the bicyclic ring system and to explore the morphing of the indole scaffold into scaffolds with up to seven fused rings.

The purines and pyrimidines constitute a relatively small class of natural products (Chapter 6). These nitrogen heterocycles are the building blocks for RNA and DNA and are thus central for the informational macromolecules of life. The biosynthetic routes in primary metabolism lead directly to nucleosides so the free base forms of the heterocycles, as in the famous alkaloid caffeine, reflect enzymatic disconnection of the nucleobase from the sugar. A variety of secondary purine and pyrimidine metabolites have biosynthetic alterations in one or more of the three parts of the scaffold: the nucleo base, the sugar, or the substituents attached to the 5′ end of the sugar (typically a D-ribose sugar). Some of these molecules are antiviral, some antiparasitic, some antibacterial.

The phenylpropanoid class of secondary metabolites (Chapter 7) is widely distributed in plant metabolism. L-Phenylalanine is the exclusive primary metabolite siphoned into these pathways. Unlike the alkaloids, which use the amino acid amine as a key reactant in construction of complex scaffolds, the amino group of phenylalanine is jettisoned in the first committed step to phenylpropanoids. It is the C_6–C_3 framework that the plants utilize in combination with *ad seriatim* oxygenase chemistry to create phenol-containing coumaryl, coniferyl, and sinapyl alcohols, the monolignols. These are acted on for phenoxy radical dimerization to lignans and oligomerization to lignins, a major structural support polymer in woody plants.

A separate set of pathways funnel cinnamoyl-CoA flux *via* the action of type III polyketide synthases to generate stilbenes and chalcones. The chalcones are cyclized to flavonoids which are entry points to many scaffold variants, including defensive substances such as phytoanticipins and phytoalexins, as well as anthocyanin pigments.

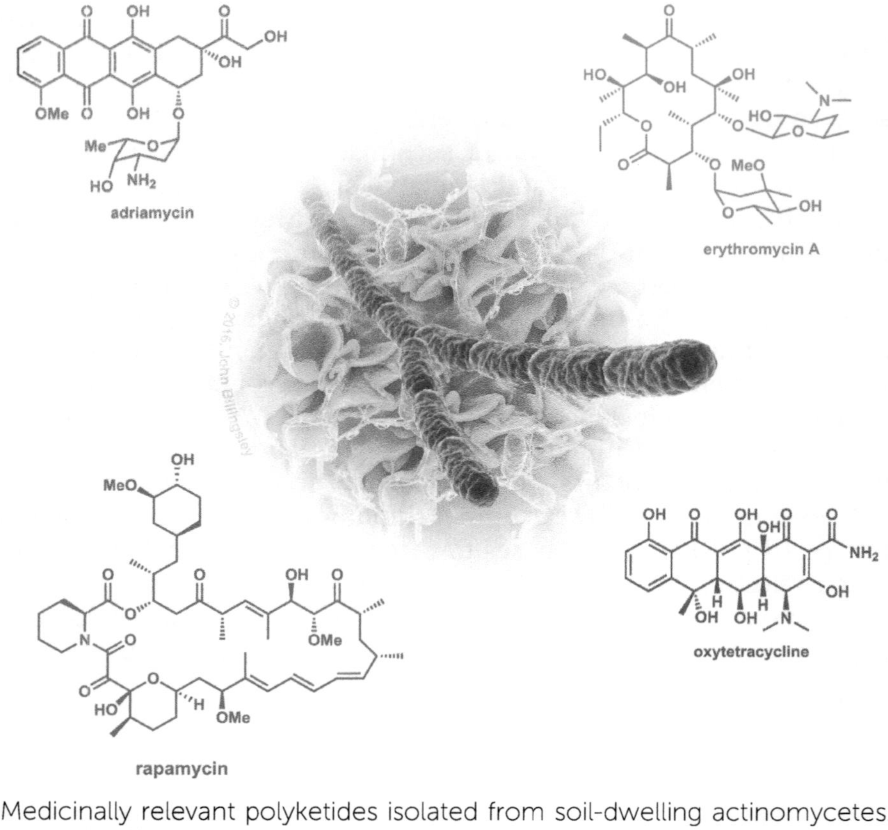

Medicinally relevant polyketides isolated from soil-dwelling actinomycetes. Copyright (2016) John Billingsley.

2 Polyketide Natural Products

2.1 Introduction

The first major class of natural products to be analyzed is the diverse set of molecules that comprise the polyketides (Dewick 2009, Mander and Liu 2010). The first molecule synthesized and structurally characterized by Collie in 1893 was the simple 5-methyl-1,3-catechol resorcinol (Staunton and Weissman 2001). The hypothesis that polyketides originated from head to tail condensation of acetate-derived units was worked out for 6-methylsalicylate by Birch in 1955 using radioactive tracer studies (Figure 2.1). Nuclear magnetic resonance (NMR) analyses of feeding studies with doubly ^{13}C-labeled acetates, trideuteromethyl acetate, and ^{18}O-labeled acetate isotopomers subsequently allowed detailed hypotheses about biosynthetic pathways for many scaffold variants in this natural product class.

Among several possible definitions of polyketides as complex natural products is a biosynthetic one, that the final products arise from *polyketonic intermediates*. This is valid for aromatic polyketides, including the bicyclic tetrahydroxynapthalene core in napyradiomycin A (Figure 2.2) which features a tetraketonyl thioester intermediate. It also holds for biosynthesis of the antitumor agent daunomycin (Figure 2.3) where a linear nonaketonyl thioester enzyme intermediate undergoes three aldol cyclizations and one thioclaisen condensation to release the characteristic fused aromatic tetracyclic framework. Unfortunately this is not true for other structural types

Natural Product Biosynthesis: Chemical Logic and Enzymatic Machinery
By Christopher T. Walsh and Yi Tang

Published by the Royal Society of Chemistry, www.rsc.org

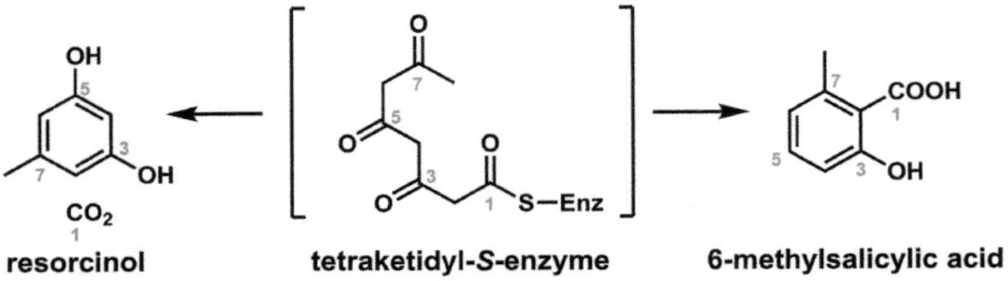

Figure 2.1 Resorcinol and 6-methylsalicylic acid: two phenolic polyketides of historical significance.

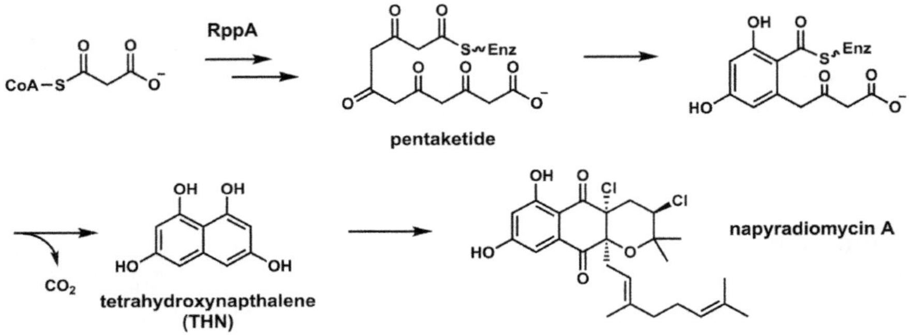

Figure 2.2 The tetrahydroxynaphthalene core of napyradiomycin A arises from a tetraketonyl thioester enzyme intermediate.

of polyketides, among them erythromycin, ivermectin, lovastatin, rapamycin, lasalocid, and nystatin, shown in Figure 2.4. Monoketonic but not polyketonic intermediates are generated during assembly.

Each of the molecules in Figure 2.4 except doxycycline is biosynthesized by iterative condensations involving β-ketoacyl thioester enzyme intermediates, like the aromatic polyketides. However, the initial β-keto group can undergo up to three kinds of successive enzymatic transformations, reduction to the β-OH-acyl thioester, dehydration of the alcohol to the α,β-enoyl thioester, and then saturation to the β-CH_2 group, before the next iterative chain extension. Polyketonic intermediates are thus avoided. We shall examine these two extremes of catalytic processing in subsequent sections of this chapter. Clearly nystatin and erythromycin (Figure 2.4) have undergone some but not all three tailoring modifications of the β-keto groups, as various hydroxyl and olefinic functionalities persist in the mature scaffolds.

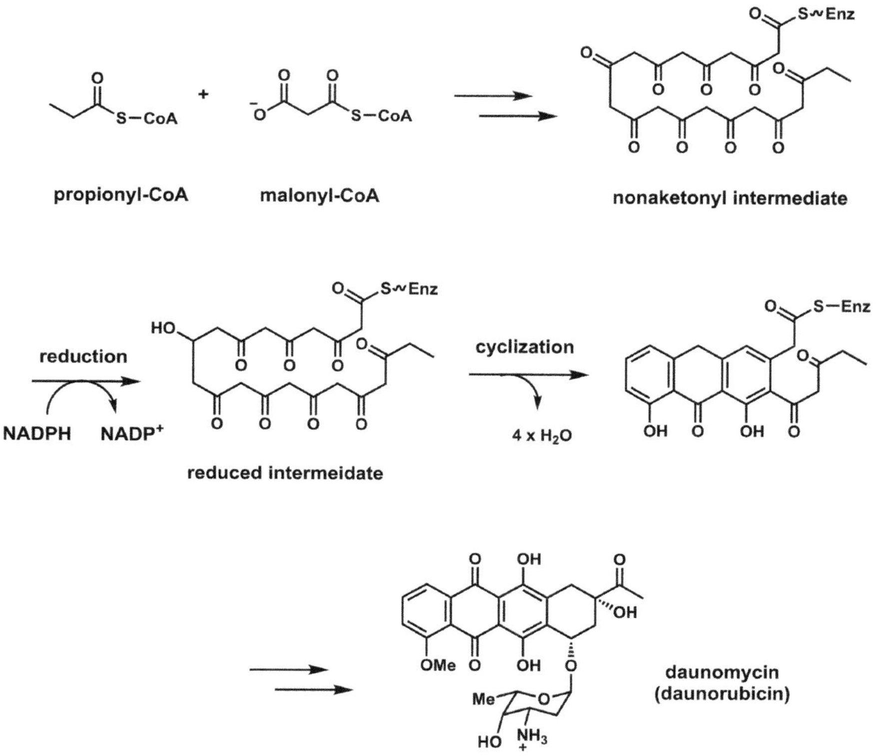

Figure 2.3 The fused tetracyclic aromatic scaffold of the anticancer agent daunomycin arises by intramolecular aldol condensations from a covalent decaketidyl-*S*-enzyme, with nine keto groups.

2.2 Polyketides Have Diverse Scaffolds and Therapeutic Utilities

Figure 2.5 notes several distinct molecular subclasses of polyketides. The macrolactone class is represented by the antibiotic erythromycin as well as the anthelminthic ivermectin (and the insecticide spinosyn; see Figure 2.29). Nystatin, with its six double bonds, is an example of the polyene antifungal subgroup, while doxycycline represents the tetracyclic aromatic class of antibiotics. The decalin-containing lovastatin, inspiration for the best selling cholesterol lowering drugs, is formed by a [4 + 2] cyclization, likely *via* a biological Diels–Alder reaction (as is spinosyn). The veterinary drug lasalocid contains a furan and a pyran ring that are the prototypic

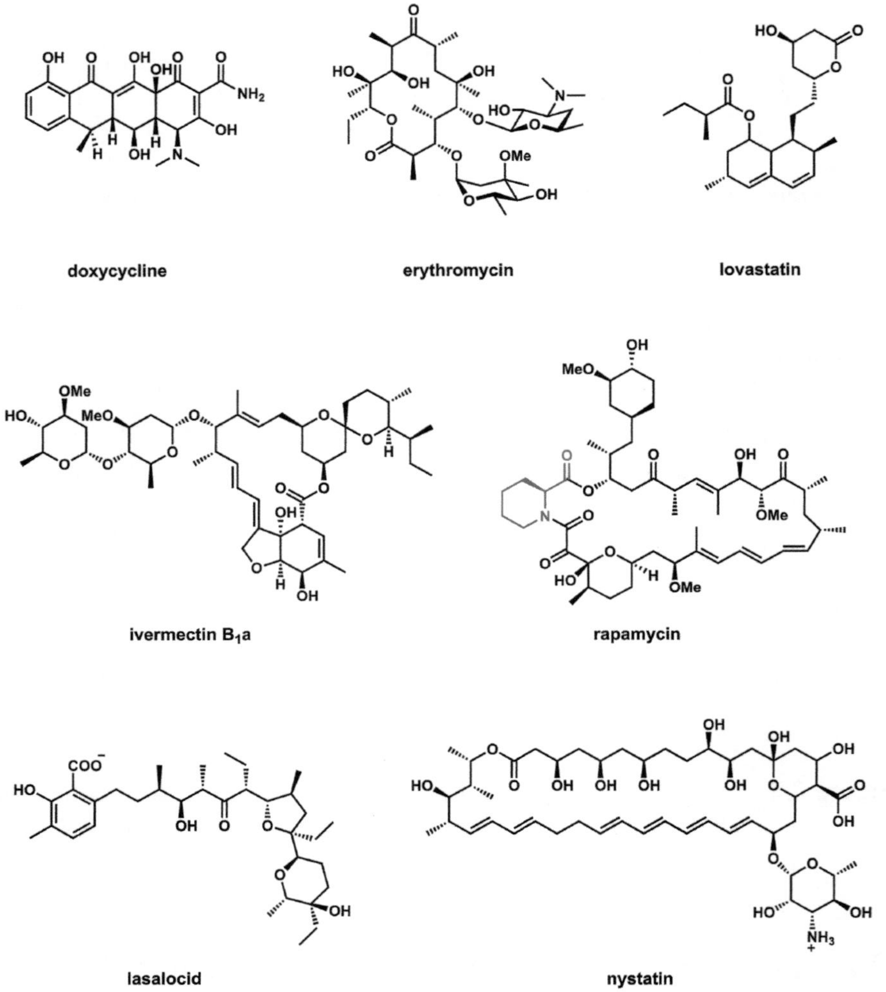

Figure 2.4 Seven diverse polyketide scaffolds of therapeutic utility.

ring structures that dominate the marine toxin polyether subclass of polyketides.

Rapamycin is a widely used immunosuppressive drug that is a macrolactone but also contains the amino acid building block pipecolate inserted into the polyketide backbone. This is an example of a hybrid polyketide–nonribosomal peptide structure, a subclass examined in the next chapter after the logic of nonribosomal peptide synthetase assembly lines has been detailed. The final subclass noted in Figure 2.5 also represents an intersection of two

Notable Scaffold Variants in Polyketides: A Diverse Range of Chemical Matter	
macrolactones	erythromycin class
polyenes	nystatin class
aromatics	tetracycline, daunorubicin
[4+2] cyclized	lovastatin, spinosyn
polyether	lasalocid, brevetoxins
PK-NRP hybrids	rapamycin, epothilones
PK-terpene hybrids	hyperforin, fumagillin

Figure 2.5 Subclasses of polyketides.

different types of natural product biosynthetic pathways, isoprenyl groups (Chapter 4) and polyketide scaffolds. The biosynthesis of hyperforin and fumagillin is taken up in Chapters 4 and 9, respectively, when the prenylation and oxygenative tailoring strategies are discussed.

From this brief and decidedly partial enumeration of the biological and pharmacological activities of these distinct subgroups of polyketides it is clear that these molecules span a huge range of therapeutic utility. A quick scan of the structures of Figure 2.4 also reveals them to be rich in oxygen substituents but almost totally lacking in nitrogen atoms, particularly in the polyketide core frameworks. These asymmetries reflect the use of acetyl and malonyl thioester building blocks, which lack nitrogen atoms. In the next chapter we will examine how the use, instead, of amino acid building blocks, but similar assembly line logic, builds nitrogen rich peptidic scaffolds. As a final introductory comment we note that polyketide biosynthetic pathways are the domain of carbanion chemistry for providing the nucleophiles and thioesters as electrophilic carbon partners required for C–C bond assembly as acetyl and malonyl groups are incorporated and morphed into complex molecular scaffolds.

2.3 Acetyl-CoA, Malonyl-CoA and Malonyl-*S*-Acyl Carrier Proteins as Building Blocks for Fatty Acids and Polyketides

Figure 2.6 is a diagram that indicates that the two carbon acid acetate is the source of both fatty acids such as the C_{16} palmitic acid and the

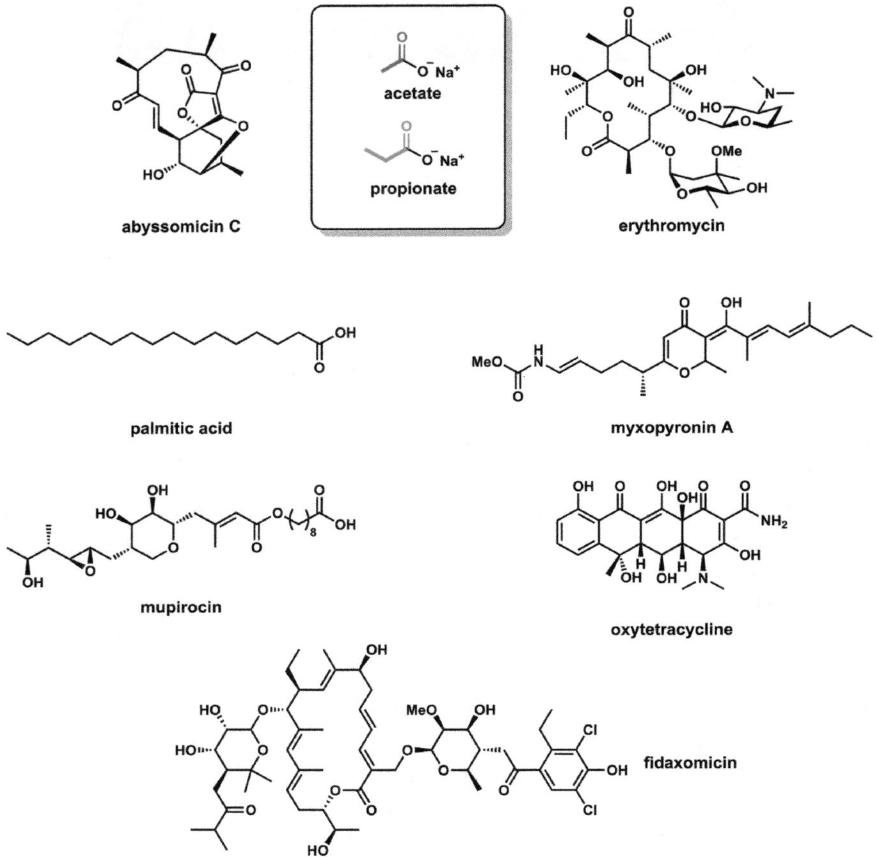

Figure 2.6 The two carbon primary metabolite acetate, activated as acetyl-CoA, is the building block for all fatty acids and polyketide classes.

six varieties of diverse polyketide scaffolds from myxopyronin A to the topical antibiotic mupirocin. This is a dramatic oversimplification of the actual enzyme chemistry involved but makes the central point that the two carbons of acetate building blocks stay connected during iterative chain elongations of the C_2 unit.

In fact, it is with the thioester acetyl-CoA that the great bulk of acetyl group metabolism occurs in both primary and secondary pathways. As the thioester, the acetyl group is electrophilic at C_1 and potentially nucleophilic at C_2 by virtue of stabilization of the C_2 carbanion as the thioester enolate (Figure 2.7). For both fatty acid and

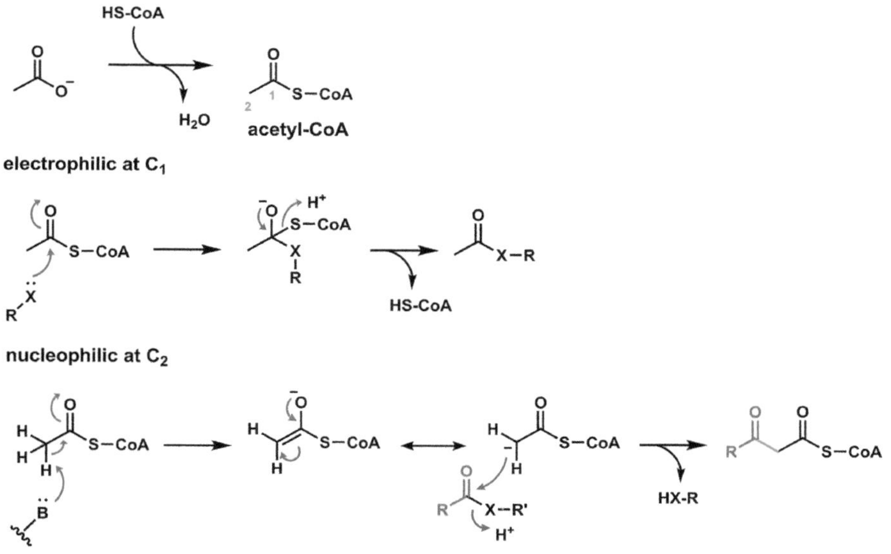

Figure 2.7 Acetyl-CoA is doubly activated in the acetyl moiety: as a nucleophilic thioester enolate-stabilized carbanion at C_2 and as an electrophilic carbonyl at C_1 by virtue of the thioester linkage.

polyketide biosynthesis, while acetyl-CoA is often the starter unit the elongation unit is instead malonyl-CoA.

The malonyl thioester can undergo facile enzyme-mediated decarboxylation to lose CO_2 irreversibly and generate the acetyl-CoA C_2 carbanion directly. The loss of CO_2 drives carbanion formation in the biosynthetic direction. This decarboxylative thioclaisen condensation is the *sole reaction type* for fatty acid and all subclasses of polyketide chain elongations. Figure 2.8 also shows how malonyl-CoA is generated by ATP-dependent carboxylation of acetyl-CoA by acetyl-CoA carboxylase, the first committed enzyme for both fatty acid and polyketide biosynthesis (Broussard, Price *et al.* 2013). While acetyl-CoA carboxylase generates malonyl-CoA, the malonyl group is transferred to a thiol on a posttranslationally modified side chain of an 8–10 kDa protein, termed an acyl carrier protein (Majerus, Alberts *et al.* 1965, Majerus, Alberts *et al.* 1965, Vagelos, Majerus *et al.* 1966) before utilization by fatty acid synthesis (FAS) and PKS enzymatic machineries. This will be delineated explicitly in the next sections.

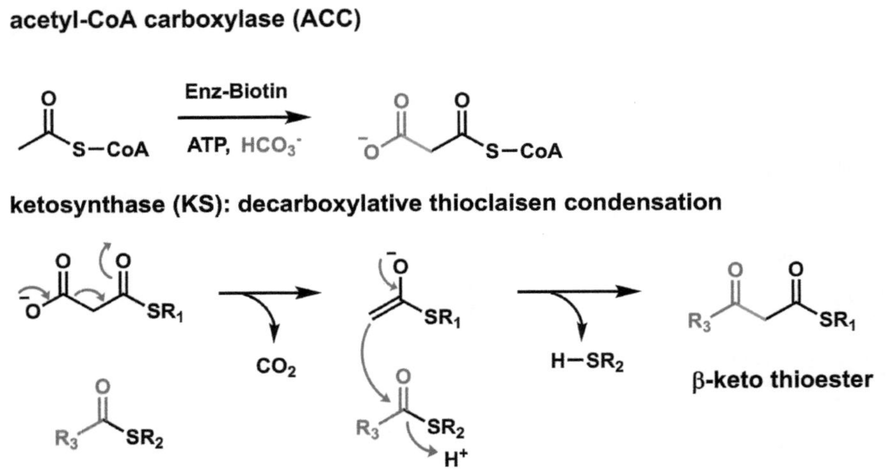

Figure 2.8 Malonyl-CoA is generated by ATP-dependent carboxylation of acetyl-CoA by the biotin-containing acetyl-CoA carboxylase, the gate keeper enzyme for both fatty acid and polyketide biosyntheses. Malonyl-*S*-ACP is a source of the C_2 carbanion utilized in chain elongation steps for fatty acids and polyketides.

2.4 The Logic and Enzymatic Machinery of Fatty Acid Synthesis is Adapted by Polyketide Synthases

Fatty acid biosynthesis is a primary metabolic pathway in all free living organisms. The building blocks, chemical logic, and enzymatic machinery appear to have been conscripted for the conditional, secondary pathways of polyketide assembly in producer microorganisms and plants (Hopwood and Sherman 1990).

Central features of fatty acid synthesis include the covalent tethering of elongating acyl chains in thioester linkage to 80–100 residue, four helix bundle protein domains known as acyl carrier proteins (ACPs) (Figure 2.9). The thiol to which the acyl chains are appended and on which they grow is the terminus of a post-translationally introduced phosphopantetheinyl moiety (Figure 2.9) at a particular Ser side chain within the ACP domain/subunit. In turn, the phosphopantetheinyl arm derives from the readily available primary metabolite coenzyme A *via* the action of dedicated phosphopantetheinyl transferases (Lambalot, Gehring *et al.* 1996, Beld, Sonnenschein *et al.* 2014) (Figure 2.10). This priming of *apo* carrier proteins to *holo* carrier proteins installs the –SH group on which the

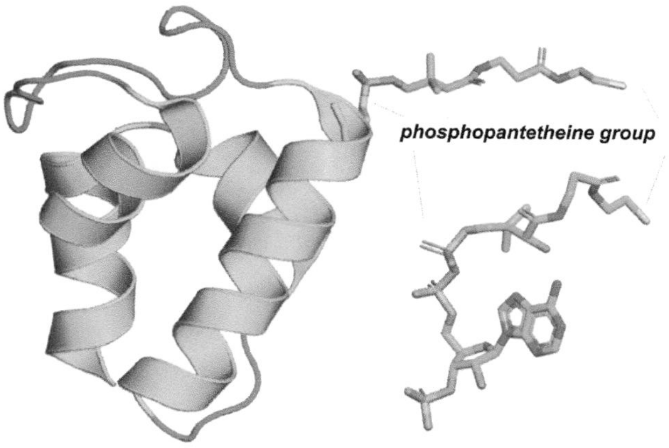

Figure 2.9 Acyl carrier protein (ACP): an autonomously folding 80–100 residue (8–10 kDa) domain/subunit that must be posttranslationally phosphopantetheinylated on a particular serine side chain to become the active *holo* form in fatty acid and polyketide synthase action.

acyl chains are built and is an essential prequel to the functioning of any FAS, PKS or nonribosomal peptide synthetase (Chapter 3). Every *apo*-ACP in an assembly line must be converted to the thiol-terminated *holo* forms for assembly lines to run.

Two other features of FAS catalysis bear comparison to PKS logic. Chain elongation occurs by malonyl S-ACP decarboxylative thioclaisen chemistry in every condensation, with loss of CO_2 as a chemical device to draw the C–C bond formation in the condensation direction. Each chain elongation adds a C_2 acetyl unit to the end of the growing acyl chain. Typically, seven elongation cycles generate the C_{16} palmityl group while eight cycles yield the C_{18} stearyl chain.

The initial product in each FAS-mediated condensation cycle is a two carbon extended β-ketoacyl-*S*-ACP. A three step enzymatic conversion of the β-C=O to the β-CH_2 occurs in every cycle before the next chain elongation event (Figure 2.11). Thus, every ketosynthase (KS) domain/subunit is accompanied by a ketoreductase (KR), a dehydratase (DH), and an enoyl reductase (ER) domain/subunit, which act consecutively: the β-keto group is reduced with cosubstrate NAD(P)H to the β-OH-acyl-*S*-ACP by KR domains. The dehydratase domain abstracts one of the acidic α-hydrogens and eliminates that β-OH to yield the α,β-olefin, conjugated to the thioester carbonyl. Because

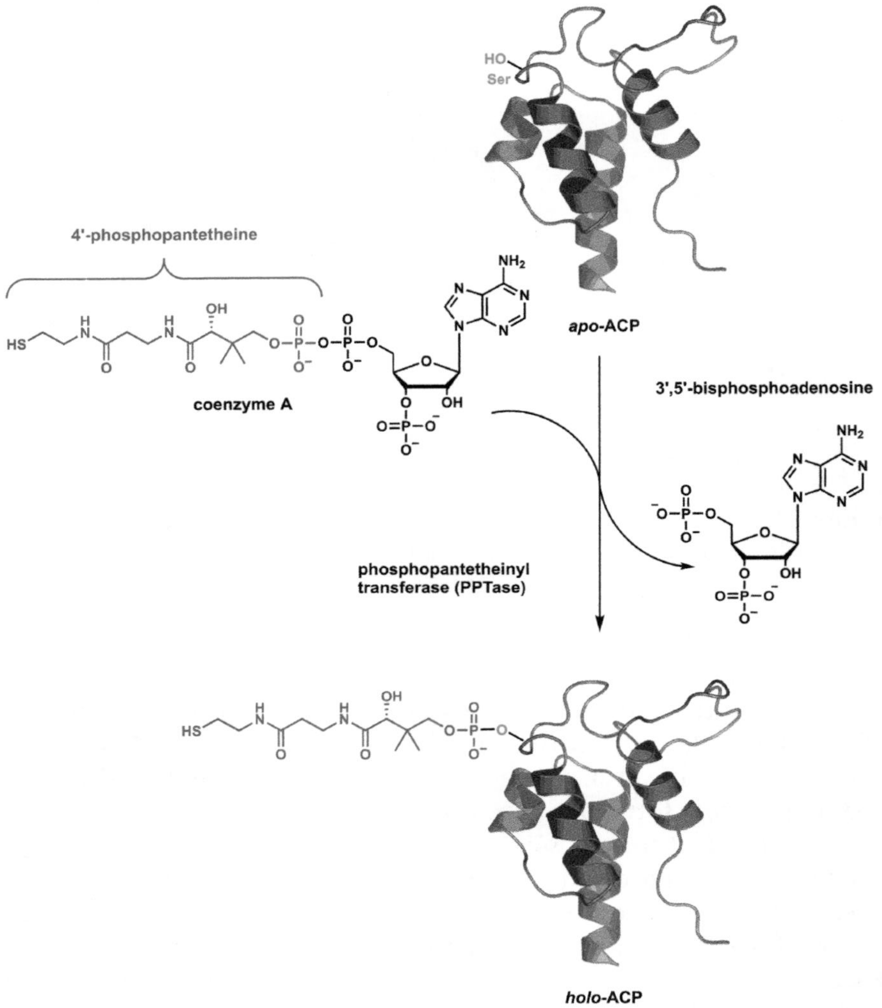

Figure 2.10 Phosphopantetheinyl transferases prime the inactive *apo*-ACP forms to the active *holo*-ACP forms by transfer of the phosphopantetheinyl moiety from the substrate coenzyme-A to the nucleophilic –OH side chain of a specific serine residue.

of this conjugation the terminus of the olefin is electrophilic and can accept a hydride from $FADH_2$ in the active site of the ER enzyme, resulting in the saturated acyl-*S*-ACP. The KR and ER each add two electrons to the just added C_2 unit.

The net result is conversion of the initial β-C=O to the β-CH_2 group for a net elongation of the growing chain by a CH_2–CH_2 unit. After

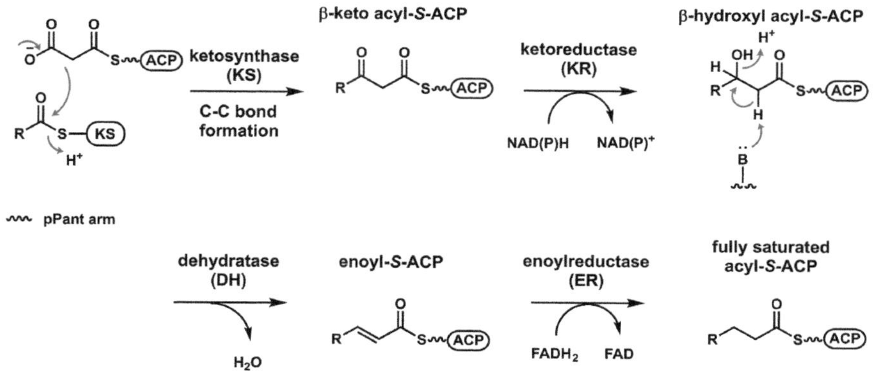

Figure 2.11 Three step enzymatic tailoring of the nascent β-ketoacyl-*S*-ACP occurs in every chain extension cycle of fatty acid synthases. The ketoreductase (KR) acts first, then the dehydratase (DH) and, lastly, the enoyl reductase (ER) to generate the completely saturated methylene chain.

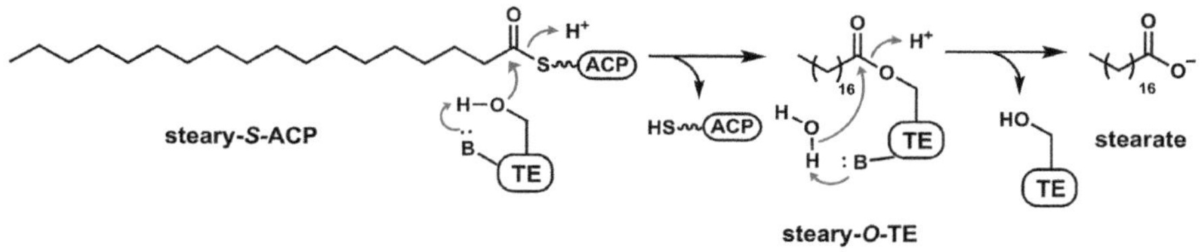

Figure 2.12 The fatty acid synthase thioesterases (TEs), as embedded domains or separate subunits, act hydrolytically to release the saturated full length C16 and C18 acyl chains as free fatty acids.

eight elongation cycles the product would be the fully saturated acyl-*S*-ACP (stearyl-*S*-ACP). This is fully consistent with the physiologic role of fatty acid synthesis, to store reducing equivalents until the cell has a need to recover the reduced carbon and its attendant electrons for some other form of cellular work.

The long chain saturated fatty acids are released from their thioester linkage by a hydrolytic thioesterase (TE) domain, typically situated at the C-terminal end of the FAS proteins. The full length acyl chain is attacked by an active site Ser-OH side chain in the TE domain/subunit, converting acyl thioester to acyl oxoester. The TEs are part of the serine hydrolome, an enzyme superfamily that generates acyl-*O*-Ser-enzyme intermediates and typically hydrolyzes them to the free acids as the release step (Figure 2.12). The fatty acids can subsequently be converted to various storage forms, including triglycerides or phosphoglyceryl-based phospholipids.

2.5 Polyketide Synthases (PKS)

While most of the chemical logic and enzymatic machinery that has evolved into functional FAS catalysts is mimicked by PKS biosynthetic enzymes, a major distinction is in the *incomplete reductive processing* of the chain elongated β-ketoacyl-*S*-ACPs that are produced in each chain elongation event (Walsh and Wencewicz 2016, chapter 16)

At one end of the chemical spectrum, none of the three processing enzymes (KR, DH, ER) is present. Each chain elongation cycle introduces another ketone into the growing chain. As shown above for the tetrahydroxynapthalene and daunomycin biosynthetic PKS enzymes, four ketones and nine ketones respectively decorate the growing acyl-*S*-ACP moieties. As we will note in a subsequent section these polyketonic chains are highly susceptible to condensations, dehydration, and aromatization to morph the linear polyketonic chains into polycyclic scaffolds.

At the other end of the chemical spectrum, all three of the KR, DH, and ER tailoring enzymes may be present and function in a catalytic cycle. In that event, the two carbon unit that has just been added will be at the CH_2–CH_2 oxidation state, akin to a fatty acid unit.

More interesting from the perspective of retention of functional groups will be if one or more of the tailoring enzymes is defective or absent. A hypothetical illustration in Figure 2.13 shows that if the KR domain/subunit is nonfunctional, the β-ketone will persist in the elongating acyl chain. Analogously, if the second enzyme, the dehydratase (DH), is nonfunctional/absent the β-OH functional group will persist through subsequent chain elongations. Finally, if the third tailoring enzyme (ER) is absent, then the olefin will persist.

In the 2-enoyl-5-OH-7-keto-octanoate example of Figure 2.13 all three types of functional group are present. This example shows the value of incomplete reduction in PKS chain elongation cycles to create any of three functional groups that can carry out chemistry, interact with targets, and generally provide versatility to the polyketide scaffolds.

The PK frameworks have not evolved to be energy storage molecules, but instead to display an array of the three functional groups for other biological purposes. A scan of the polyketide structures in Figures 2.2, 2.3, and 2.5 quickly reveals the presence of

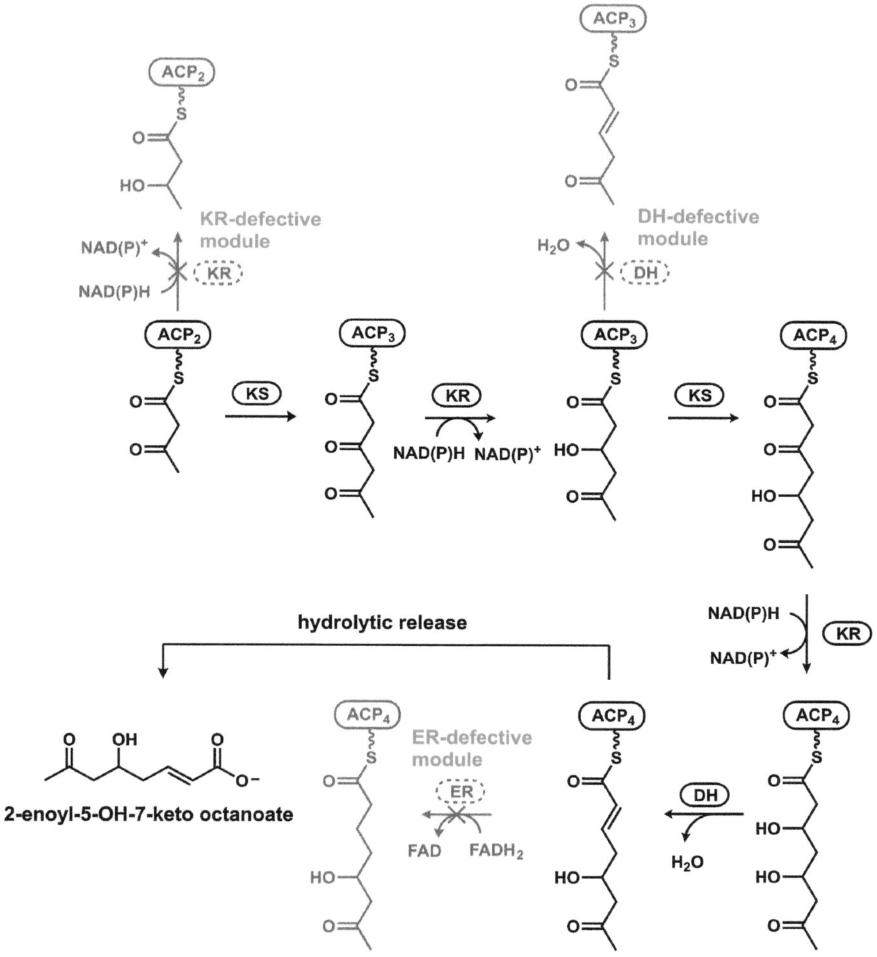

Figure 2.13 Incomplete tailoring of the β-ketoacyl chains in polyketide synthase chain elongation cycles can lead to retention of ketone, hydroxyl, or olefinic substituents in the growing chains.

ketones, alcohols, and olefins, distributed through the polyketide molecules.

Three additional points are to be made before taking up the biosynthesis of major subclasses of polyketides.

2.5.1 Alternate Acyl-CoA Substrates for PKS

The first point deals with substrates, in particular alternate chain initiation and chain-extending building blocks, for PKS enzymes *vs.* FAS enzymes. For chain initiation, a variety of acyl-*S*-CoA substrates can stand in for acetyl-CoA for particular PKS assembly lines. For example erythromycin begins with a propionyl-CoA starter unit (see Figure 2.22). Tetracycline assembly (see Figure 2.17) begins with a malonamyl-*S*-ACP and rapamycin with a cyclohexenyl-*S*-ACP. For chain extension the most common substitute for malonyl-*S*-CoA is 2S-methylmalonyl-*S*-CoA, as will be evident from inspection of the deoxyerythronolide B synthase assembly line in Figure 2.22. This adds a C3 rather than a C2 extender unit in each such catalytic cycle and introduces a chiral center. Additional variant malonyl-CoA derivatives, such as hydroxymalonyl- and aminomalonyl-, are less frequently utilized (Khosla, Gokhale *et al.* 1999).

2.5.2 Organization of Multi-protein Assemblages

The second feature deals with the organization of the PKS proteins and the congeneric FAS proteins. Type I PKS and FAS have all the constituent domains (KS-ACP-KR-DH-ER, TE, and acyltransferases (AT)) strung together in large proteins with multiple autonomously folding domains (Schweizer and Hofmann 2004) (Figure 2.14*)*.

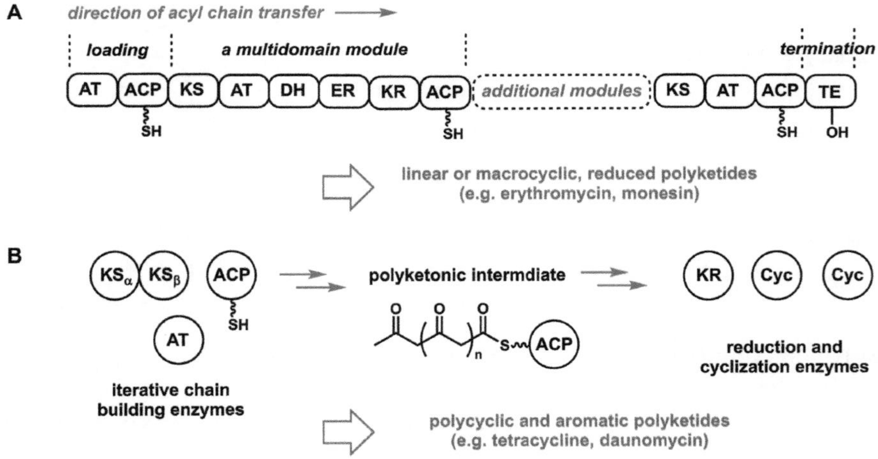

Figure 2.14 (A) Type I PKSs have modular organization: each domain is used once; (B) iterative PKS: one copy of each catalytic and carrier protein domain, used iteratively in each catalytic cycle of chain elongation.

Type II FAS (White, Zheng *et al.* 2005) and PKS assemblages have the constituent proteins required for chain initiation, elongation, and termination as separate proteins that purify separately and interact only transiently. Presumably the separate Type II constituent proteins were the early evolutionary solution before gene fusions occurred for the proteins to interact *in cis* (Type I) rather than *in trans* (Type II), which may offer kinetic efficiencies for chain assembly.

There is a third organizational subtype, Type III PKS, that operate in plants to make chalcones and stilbenes. We take these up in Chapter 7 in the biosynthesis of phenylpropanoid metabolites. The Type III PKS are single proteins, use the soluble malonyl-CoAs as extender units rather than malonyl-*S*-ACP protein bound extenders, and do not extend beyond the chalcone/stilbene frameworks, although some 900 members of this catalyst subclass are predicted.

2.5.3 PKS Chain Termination Modes

The TE domains at the most downstream end of Type I PKS assemblies and the free-standing TEs in Type II PKS organizational arrays can release acyl-*O*-TE chains hydrolytically, in full analogy to the TE catalysts working in the FAS system. Figure 2.15 shows a representative set of linear polyketide scaffolds. Monensin shows the free carboxylate but many of the others have undergone extensive post assembly line enzymatic tailoring steps that transform the nascent carboxylate product into the mature natural products. However, the PK core is still linear, not macrocyclic.

A large subset of PKS TEs are, on the other hand, macrocyclizing. Water is excluded from the active site or is not located in a kinetically competent locus to attack the acyl-*O*-TE in competition with intramolecular attack by a nucleophile within the acyl chain. We have noted above the dearth of nitrogen atoms, especially basic amines, in PK chains, so the intramolecular nucleophiles are almost always alcohol groups or carbanions that are part of stabilized enolates.

The *alcohols* involved in cyclization of macrolides to give 14-member (erythromycin), 16-member (tylosin), or 18-member macrolactones (fidaxomicin) (Figure 2.16) are alcohol groups that were left during chain elongation in cycles lacking functional DH catalysts. By contrast, the release step in the Type II PKS for oxytetracycline

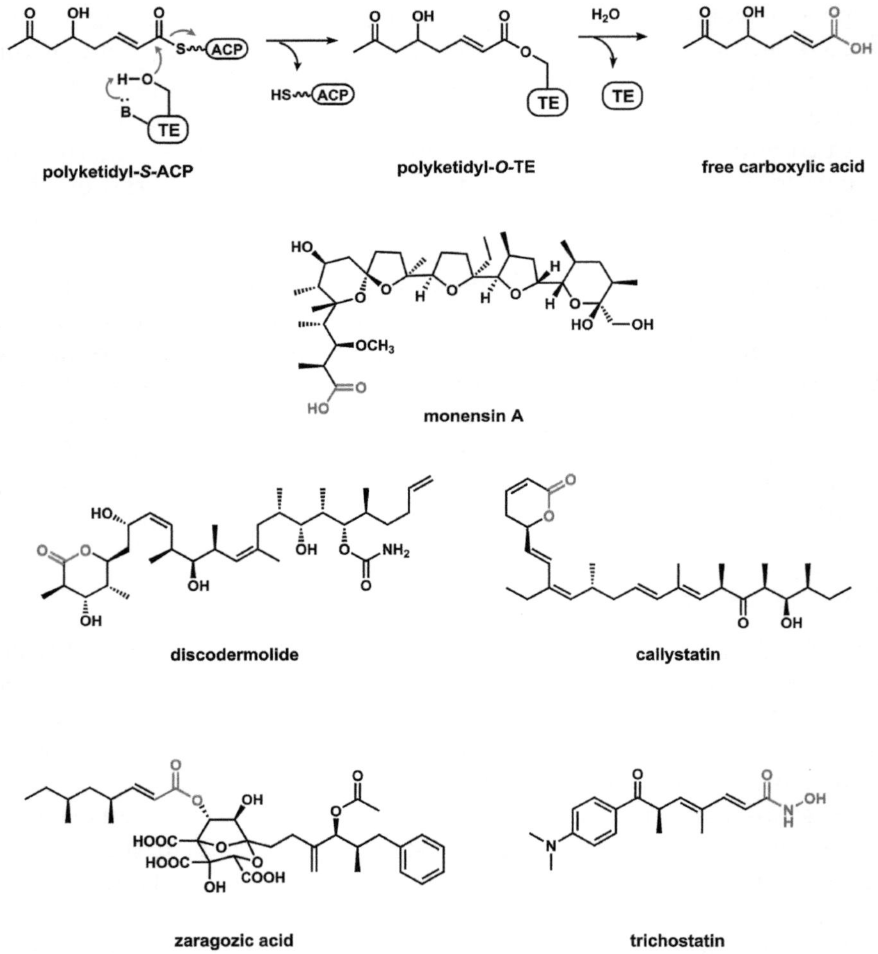

Figure 2.15 PKS termination modes: One main mechanism is hydrolysis of the full length polyketidyl-*O*-TE acyl enzyme to release the free carboxylic acids (monensin). These can be subsequently modified in post assembly line coupling reactions to generate lactones (discodermolide, callystatin), esters (zaragozic acid), and hydroxamic acids (trichostatin).

(Figure 2.17) is proposed to undergo a two-step process in which the tricyclic product is first released *via* spontaneous hydrolysis. The carboxylate is then activated as an acyl-*O*-adenylate, and can undergo a Claisen condensation where one of the two C_{18}-hydrogens is abstracted as a proton, yielding the stabilized *enediolate carbanion* as

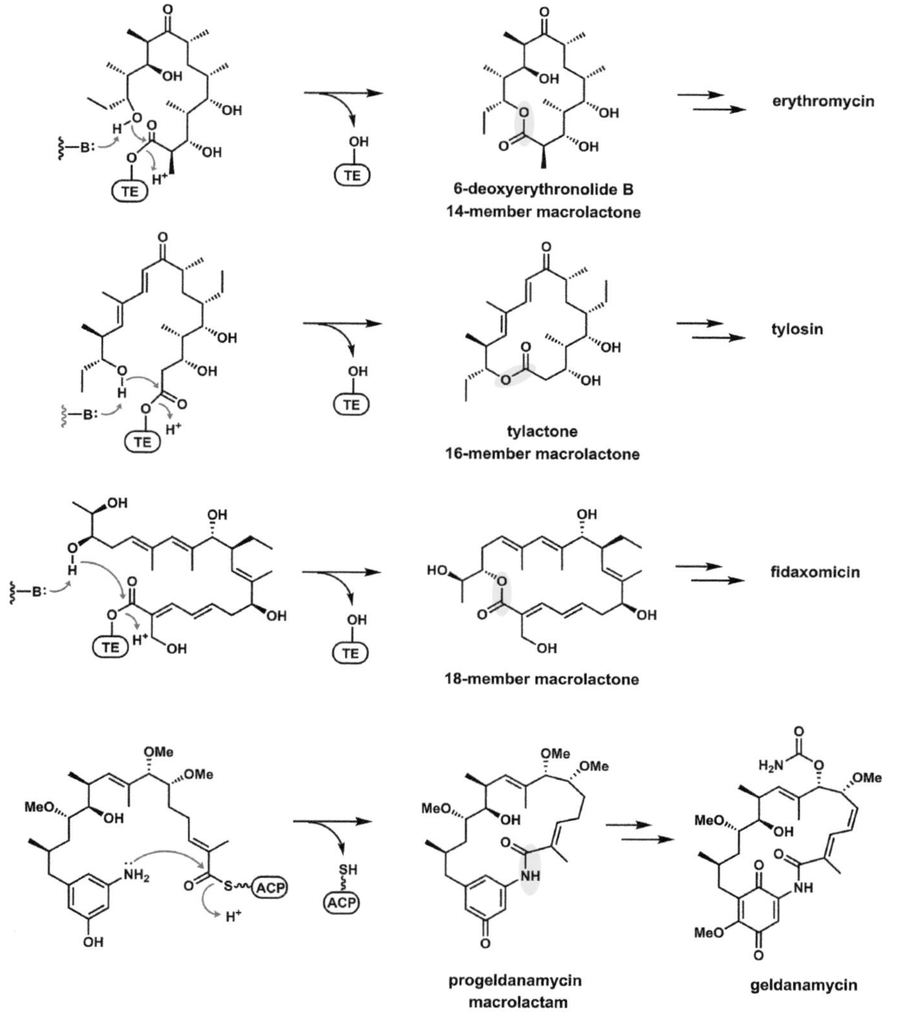

Figure 2.16 Polyketide synthase TE domains as macrocyclization catalysts. Chain release *via* internal capture of the acyl-*O*-TE by an intramolecular –OH group yields a 14-member macrolactone for 6-deoxyerythronolide B, a 16-member macrolactone for tylactone, and an 18-member lactone in fidaxomicin formation. When the competent intramolecular nucleophile is an amine, a macrolactam results as shown for geldanamycin.

nucleophile for attack on the C_1-carbonyl. This forms the tetracyclic pretetramid as the first free intermediate *en route* to the tetracycline products.

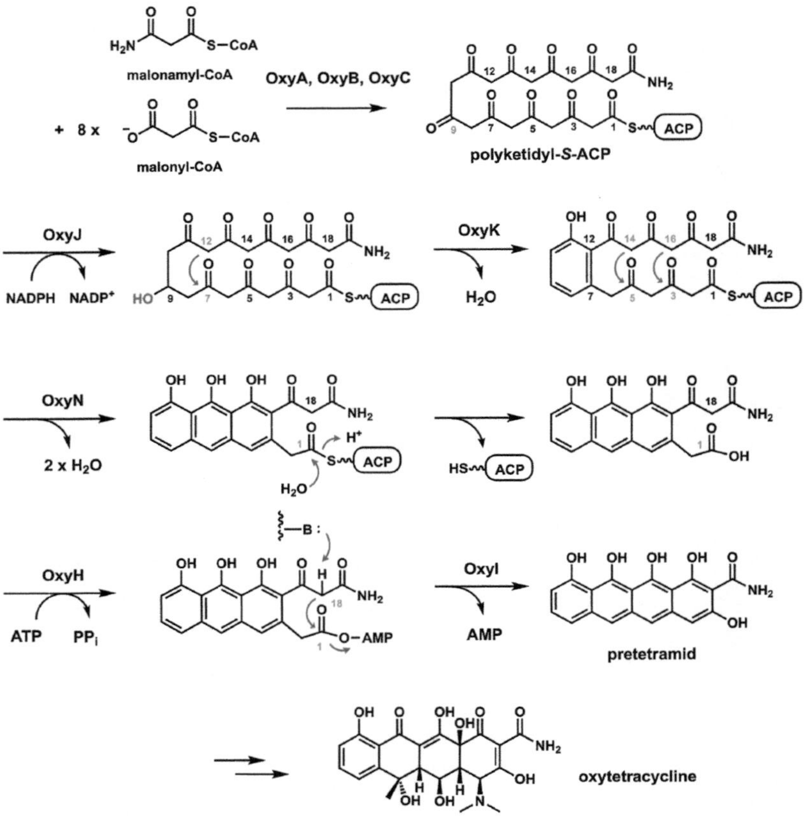

Figure 2.17 Prior reduction of the C_9 ketone in the polyketonic acyl-SACP of the tetracycline type II PKS sets up C_7–C_{12} aldol condensation and dehydration to form the first aromatic ring. C_5–C_{14}, C_3–C_{16}, and C_1–C_{18} C–C bonds follow to release pretetramid. The release step involves C–C bond formation rather than O–C or N–C formation. This occurs by a Claisen condensation with a C_{13}-enolate attacking the C_1 carbonyl.

2.6 Biosynthesis of Major Polyketide Structural Classes

2.6.1 Oxytetracycline Biosynthesis: Aromatic Polyketides that Initiate Cyclization at C_7–C_{12} or C_9–C_{14}

Bicyclic, tricyclic, and polycyclic aromatic ketide scaffolds, including the tetrahydroxynapthalene core of napyradiomycin (see Figure 2.2), daunomycin (Figure 2.3), and oxytetracycline (Figure 2.17), are built from minimal PKS enzymatic components with accumulating polyketonic chains (Zhan 2009, Pickens and Tang 2010, Zhou, Li *et al.* 2010, Zhang, Pang *et al.* 2013). There are no tailoring enzymes to reduce the ketone, dehydrate the resultant β-OH, or reduce any αβ-olefinic species.

Such polyketonic species are potentially highly reactive to intramolecular aldol condensations and could undergo off-pathway derailments if not protected both physically and kinetically.

The fate of these enzyme tethered polyketonic thioesters involves directed, regiospecific intramolecular aldol condensations and thioclaisen cyclizing release steps, typically mediated by partner proteins that act on the bound acyl chains. In bacterial aromatic PKS synthases, there are two modes of cyclization seen with decaketidyl-*S*-enzymes that generate the predominant tetracyclic frameworks (Zhou, Li *et al.* 2010).

Most common is formation of the first aromatic ring *via* C_7–C_{12} bond formation, as shown for both the tetracycline (Figure 2.17) and the jadomycin PKSs (Figure 2.18). In both cases an oxidoreductase intervenes to reduce the C_9-ketone to the alcohol (maroon color) before the first intramolecular aldol. Among other things this prevents an aldol to C_9 as the electrophilic partner and may set up the linear foldamer conformation that is productive for aromatizing cyclodehydration. The next ring to form occurs by another aldol, with a C_{14} carbanion attacking the ketone at C_5 under the agency of an aromatase partner enzyme. The third ring arises from a third aldol, C_{16} onto C_3. This aldol and the chain-releasing thioclaisen, C_{18} onto C_1, may be spontaneous. The released product is the tetracyclic pretetramid as noted above in Figure 2.17. A similar sequence occurs for creation of the aromatic scaffold in daunomycin.

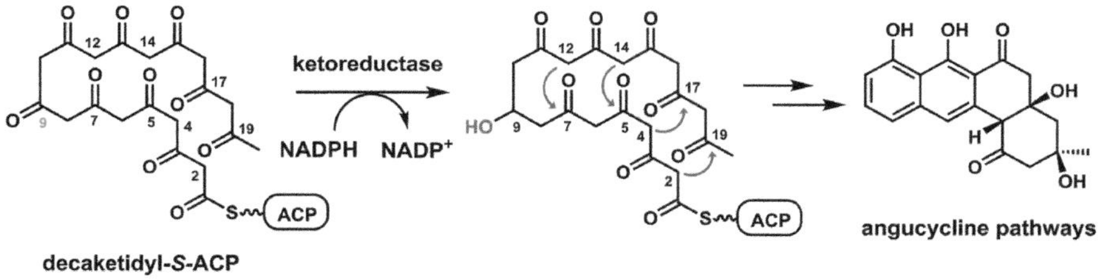

Figure 2.18 The cyclizations to the angucycline jadomycin follow the same regiospecificity for the first two rings as tetracyclines, then deviate to form C_4–C_{17} and C_2–C_{19} C–C bonds to create the angular tetracyclic aromatic scaffold.

For the angular tetracyclic frameworks (angucycline) the jadomycin PKS is representative. It folds the decaketidyl-*S*-enzyme in a different conformation, as indicated in Figure 2.18. As with the tetracycline pathway, the first two rings arise by aldols between C_{12} and C_7 and C_{14} and C_5. The third ring occurs with a unique connectivity as a carbanion at C_4 attacks the C_{17} ketone. This C–C bond creates the angular connectivity of the angucycline. The fourth ring is formed by an aldol, C_2–C_{19}, rather than a Claisen, so a separate release step must occur.

The second mode of connectivity of the first ring to form in aromatic polyketides is C_9–C_{14}, found in the action of the TcmN enzyme in tetracenomycin biosynthesis (Figure 2.19) (Zhou, Li *et al.* 2010). As shown, the polyketonic acyl-*S*-PKS in this case does not have the C_9 ketone reduced and it must be in opposition to the C_{14} methylene locus. Aldol condensation and dehydrative aromatization yield the A ring (C_{14}–C_9). The next ring forms by attack of C_{16} on the C_7 ketone, then C_5 onto C_{18} and C_2 onto C_{19}. The chain release step is hydrolytic, liberating the tetracyclic carboxylic acid nascent product Tcm F2 which undergoes a series of post assembly line tailorings to the final end product tetracenomycin (not shown).

The bottom portion of Figure 2.19 shows in detail how these A ring-forming aldol condensations lead to aromatic rings by subsequent dehydration and tautomerization of the cyclohexenediones to the aromatic diphenols.

2.6.2 Fungal Aromatic Polyketides: Cyclizations that Start at C_2–C_7, C_4–C_9 or C_6–C_{11}

Fungal polyketide synthases tend to be more complex proteins, less obvious in prediction of catalytic and noncatalytic domain boundaries, but the logic underlying formation of the tricyclic norsolorinic acid, an intermediate in the aflatoxin pathways, has been well studied (Korman, Crawford *et al.* 2010). Cyclization specificity depends on a product template (PT) domain within the PKS (Crawford, Korman *et al.* 2009). A hexanoyl-CoA starter is subjected to seven rounds of chain elongation to yield a tethered octaketidyl-*S*-enzyme. The regiospecificity and timing of ring formation involve two successive aldol condensations, between C_4–C_9 and then C_2–C_{11}. The TE domain then mediates the C_{14}–C_1 thioclaisen-type of chain cyclization and release to give tricyclic norsolorinic acid (Figure 2.20A).

Conversely, the tricyclic fungal anthraquinones, represented by emodin, undergo a first ring aromatizing aldol cyclodehydration

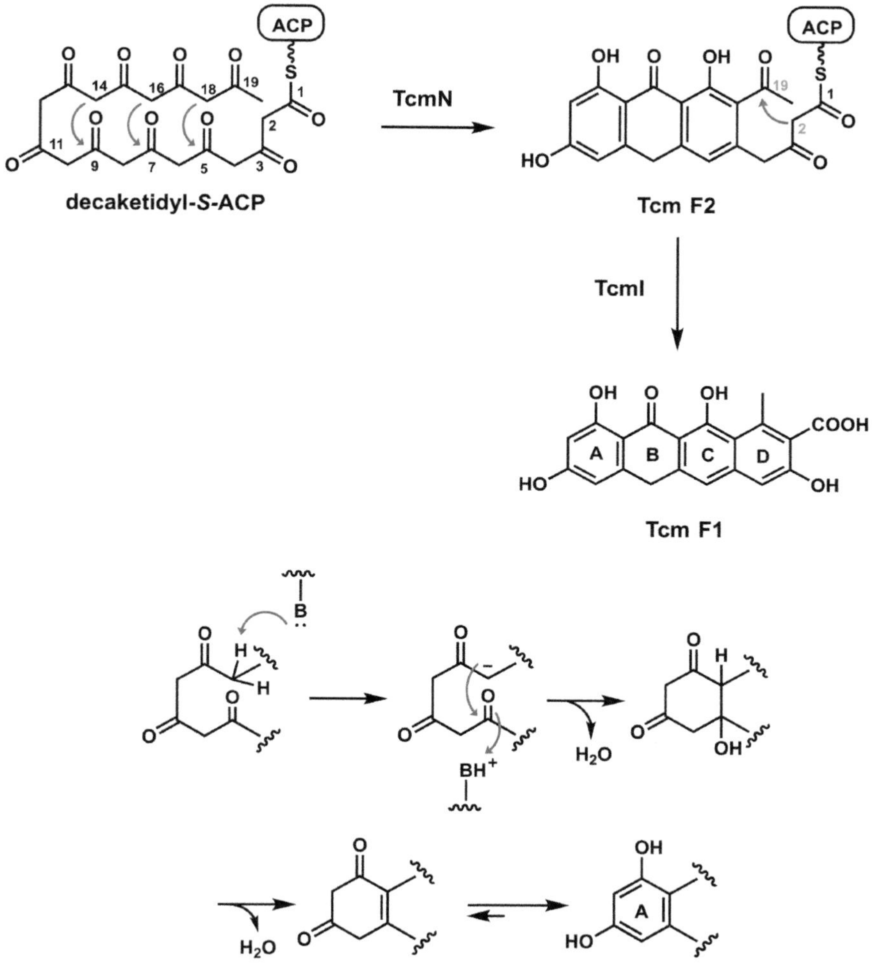

Figure 2.19 The polyketidyl-*S*-ACP in the related tetracyclic aromatic polyketide tetracenomycin is folded in a different linear conformation in the active site of the TcmN cyclase. The first C–C bond formed by intramolecular aldol condensation is between C_9 and C_{14} (C_9 has not been reduced to the –OH). The connectivity in the next three rings is C_7–C_{16}, C_5–C_{18}, and C_2–C_{19}. A linear tetracyclic scaffold is released but the substituents are in distinct positions from those in pretetramid.

sequence that starts with a C_6–C_{11} bond formation instead of the C_4–C_9 bond, reflecting a distinct acyclic foldamer in the enzyme active site (Zhou, Li *et al.* 2010). This sets the register for the following two

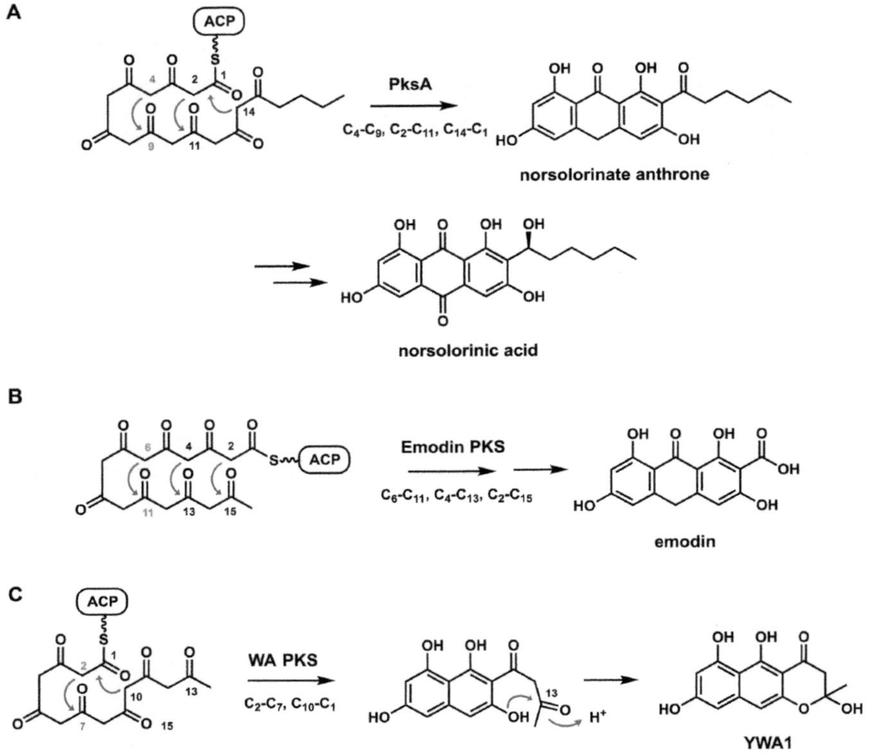

Figure 2.20 Three distinct regiospecificities for first ring formation in fungal aromatic polyketides. (A) The fungal iterative norsolorinic acid synthase folds the heptaketonyl-acyl-*S*-ACP chain in yet a distinct conformer such that the first aldol creates the C_4–C_9 C–C bond, then C_2–C_{11}, and C_{14}–C_1 are connected to release the tricyclic product. (B) In the generation of the tricyclic fungal anthraquinones represented by emodin, the cyclization begins with a C_6–C_{11} C–C bond, followed by C_4–C_{13}, and C_2–C_{15}. (C) In the formation of the naphtha-pyran YWA1, cyclization begins with a C_2–C_7 aldol condensation, followed by product release through a C_{10}–C_1 Claisen cyclization. The released product can then be spontaneously cyclized into the tricyclic hemiacetal pyrone.

aromatic rings and gives a different connectivity from the norsolorinate framework (Figure 2.20B).

In another variation of cyclization regioselectivity observed during formation of the linear naphthopyrone YWA1 (Figure 2.20C), formation of the first ring is through the C_2–C_7 aldol cyclization catalyzed by the PT domain in the WA PKS. This is followed by C_{10}–C_1 Claisen-like cyclization to forge the second ring and release the naphthalene intermediate, which can undergo spontaneous pyrone formation in solution to afford YWA1.

The structures of several of the bacterial aromatase/cyclase enzymes have been solved by X-ray crystallography (Figure 2.21A) for both C_7–C_{12} and C_9–C_{14} initiating aldol condensations (Tsai and Ames 2009, Ames, Lee *et al.* 2011, Lee, Ames *et al.* 2012, Caldara-Festin, Jackson *et al.* 2015). Analogously, the cyclization cavity of the PT domain of PksA (Figure 2.21B) that initiates C_6–C_{11} and C_4–C_{13} cyclizations has also been examined to gain insights into

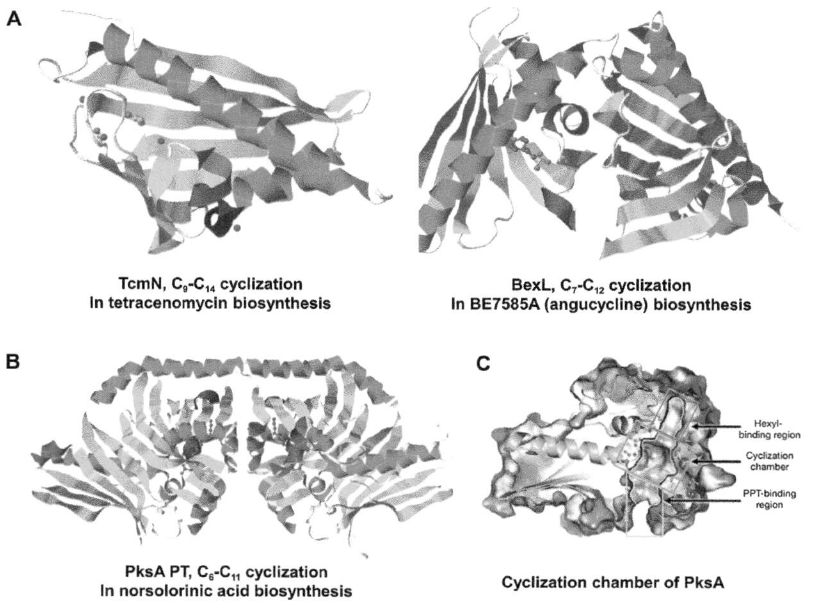

Figure 2.21 (A) Structures of bacterial aromatase/cyclase enzymes that catalyze distinct cyclization modes. Structures generated from PDB entries 2RER and 4XRW for TcmN and BexL, respectively. (B) Structure of a fungal PksA PT domain cavity that catalyzes fungal-specific cyclization modes, generated from PDB entry 3HRQ. (C) Cyclization chamber of PksA PT, reprinted by permission from Macmillan Publishers Ltd: Nature (Crawford J. M., T. P. Korman, J. W. Labonte, A. L. Vagstad, E. A. Hill, O. Kamari-Bidkorpeh, S.-C. Tsai and C. A. Townsend, "Structural basis for biosynthetic programming of fungal aromatic polyketide cyclization." *Nature*, **461**, 1139), copyright (2009).

selectivity rules and as guideposts for reengineering specificities (Crawford and Townsend 2010).

All subsequent subclasses of polyketides discussed in the next sections differ from these aromatic PKS machineries in that at least some of the catalytic cycles involve further processing of the β-ketones that arise *via* each ketosynthase action. While occasional ketone functionalities are carried out through chain elongations, inspection of the various nonaromatic scaffolds of polyketides in this chapter shows a preponderance of hydroxy and olefinic functional groups decorating the mature PK acyl chains.

2.6.3 Polyketide Macrolactones: Type I Assembly Line Logic and Machinery

Figure 2.16 above noted macrolactonizing release of C_{14}, C_{16}, and C_{18} PK chains from the respective TE domains in erythromycin, tylosin, and fidaxomycin biosynthetic pathways. In each case the acyclic acyl chain in the acyl-*O*-TE enzyme active site must be folded such that only one side chain-OH group becomes the catalytically competent nucleophile. Each of these three PKS assembly lines uses multiple methylmalonyl-CoA extender units to introduce methyl branches at what were β-carbons in those chain elongation cycles. The starter unit for the erythromycin and tylosin assembly lines is propionyl-CoA rather than acetyl-CoA but it is less clear for fidaxomycin, whose biosynthesis has been less well characterized.

The prototypic Type I PKS assembly line is comprised of the three protein subunits, DEBS1, 2, 3, of deoxyerythronolide B synthase (DEBS) (Figure 2.22) (Donadio, Staver *et al.* 1991, Haydock, Dowson *et al.* 1991, Cane 2010). DEBS1 has the starter module and two extender modules. DEBS2 has two extender modules, while DEBS3 has the fifth extender module and the termination module, ending with the macrocyclizing TE domain. The seven modules each have the obligate ACP for covalent tethering of the growing acyl chain [here the interchangeable single letter T domain (thiolation) abbreviation is shown in place of the three letter ACP] and also a KS domain to make a new C–C bond in each elongation cycle. Each module also has an acyltransferase (AT) domain for moving the methylmalonyl group from methylmalonyl-CoA to become methylmalonyl-*S*-ACP. Because methylmalonyl-CoA is the extender unit, chiral methyl groups are generated in each chain elongation where it is used. One can also see an array of the three tailoring enzyme domains (KR, DH, ER) distributed differentially across the last six modules.

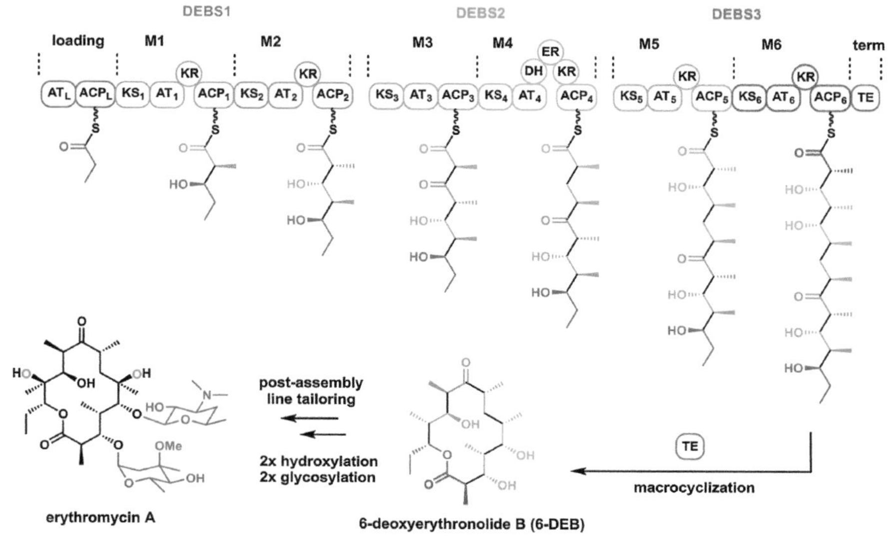

Figure 2.22 The DEBS three protein, seven module Type I PKS assembly line.

Indeed, assuming all the KR, DH, and ER domains are functional (as they turn out to be), one can read the nature of the functional groups in the growing polyketidyl chain from their presence or absence. Thus module 2 (the first extender module) has only a KR tailoring domain so the β-OH functional group will persist through all subsequent elongations. Module 3 also has only a KR domain so a second OH group is retained in the growing chain.

Of the two modules in DEBS2, the first one has no tailoring enzyme domains so the ketone goes through, while the second module has all three tailoring enzymes so the nascent ketone gets reduced by four electrons to the CH_2 group. Modules 6 and 7 are in DEBS3 and both contain only the KR tailoring enzymes, sending β-OH groups on. By the time the full length polyketidyl chain reaches the termination module before transfer to the final macrolactonizing TE domain, the chain bears 4-OH functional groups, one ketone and 6-stereogenic methyl groups. Thirteen of the fourteen carbon atoms bear functionality. Chain release of the acyl-*O*-TE involves directed attack of the C_{13}-OH on the C_1 thioester carbonyl group.

The released product is 6-deoxyeryrythronolide B. It does not possess antibiotic activity. Instead it must undergo three types of post assembly line chemical maturation event, summarized in Figure 2.23. Of the five tailoring enzymatic steps shown, two are carried out by cytochrome P450 type monooxygenases, discussed in detail in

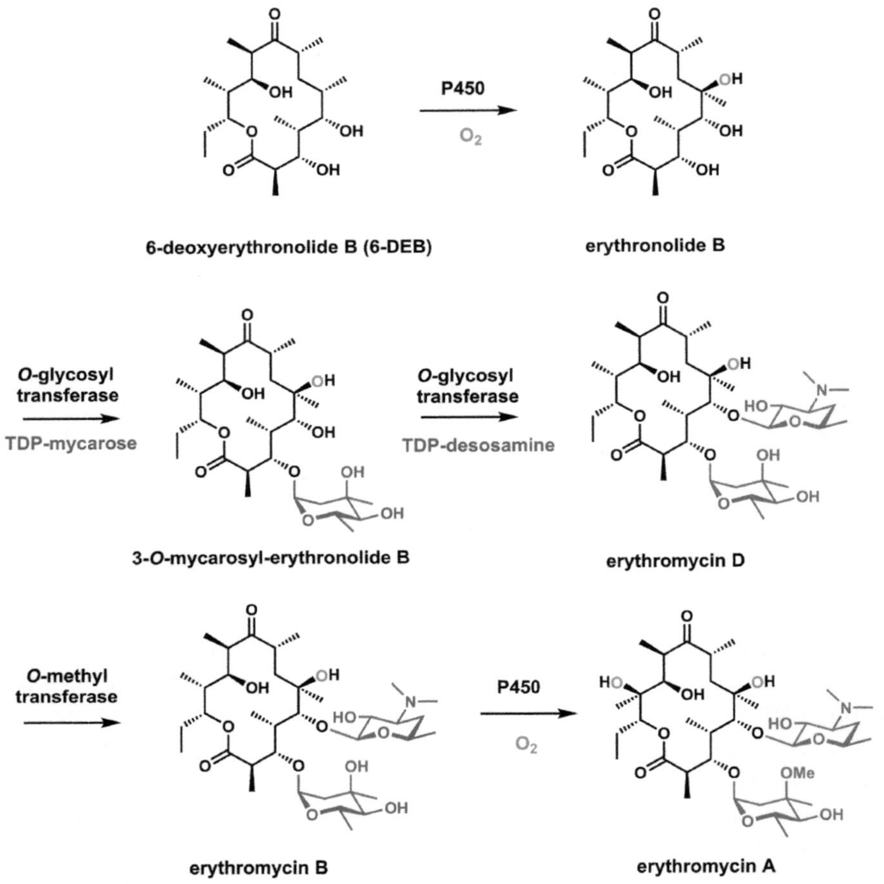

Figure 2.23 Five post assembly line tailoring enzymes convert inactive deoxyerythronolide B into the active antibiotic erythromycin A.

Chapter 9. Two are glycosyltransferase steps, adding deoxysugars at the 3-OH and 5-OH substituents. Glycosylations are addressed separately in Chapter 11. The fifth step is an *O*-alkylation of the mycarosyl sugar in an *S*-adenosylmethionine-dependent transfer. The final end product metabolite, erythromycin A, is a potent inhibitor of the peptidyl transferase center on the 50S subunit of bacterial ribosomes (Walsh and Wencewicz 2016). The macrolactone is the core scaffold and the deoxysugar residues make key high affinity interactions with 23S ribosomal (r)RNA.

The DEBS assembly line logic is generalizable to the biogenesis of many other macrolides by comparable Type I assembly lines. As noted earlier, macrolactonization is not the only route for disconnection of the covalent polyketidyl-*O*-TE species, but is a particularly useful one for building cyclic, compact scaffolds likely to have higher affinity for many biological targets. The high density of functional groups exemplified in the erythromycin, tylosin, and fidaxomicin frameworks also imparts conformational constraints to and functional group interactions with different types of biological targets.

2.7 Polyketides with Ring-forming [4 + 2] Cyclizations on or After PKS Assembly Lines: Concerted or Stepwise?

Figure 2.24 shows seven natural products for which ring-forming [4 + 2] cyclizations, between a conjugated diene and another double or triple bond to create a six membered ring, have been proposed. All have polyketide frameworks, including chaetoglobosin which is a hybrid alkaloid–polyketide molecule. Macrophomate has a single aromatized six membered ring while lovastatin and chlorothricin have bicyclic decalin rings. Decalin rings may be a signature for [4 + 2] cyclizations. Evidence has accrued over the past decade that at least some of these are concerted, reflecting acceleration by enzymes acting as Diels–Alder catalysts. The cautiously skeptical view from 15 years ago (Stocking and Williams 2003) about whether Diels–Alderases exist has morphed into evidence backing such assertions for SpnF in spinosyn construction (Kim, Ruszczycky *et al.* 2011) and in abyssomycin biosynthesis (Byrne, Lees *et al.* 2016).

The enzymes SpnF and SpnL in spinosyn assembly set up the tricyclic 5-6-5 ring system in a pair of consecutive reactions. While SpnF is of particular interest for accelerating the nonenzymatic [4 + 2] cyclization to the 6-5 system, SpnL is also unusual in building the third ring (5-6-5) by a Rauhut–Currier type mechanism (Aroyan, Dermenci *et al.* 2010, Kim, Ruszczycky *et al.* 2011).

2.7.1 Lovastatin

Lovastatin, along with erythromycin and tetracycline, may be one of the most famous polyketide natural scaffolds, largely for its activity at blocking cholesterol biosynthesis at the rate-determining HMG-CoA (3-hydroxy-3-methyl-glutaryl-CoA) reductase step. This inspired the whole field of cardiovascular therapeutics with synthetic statins that mimic

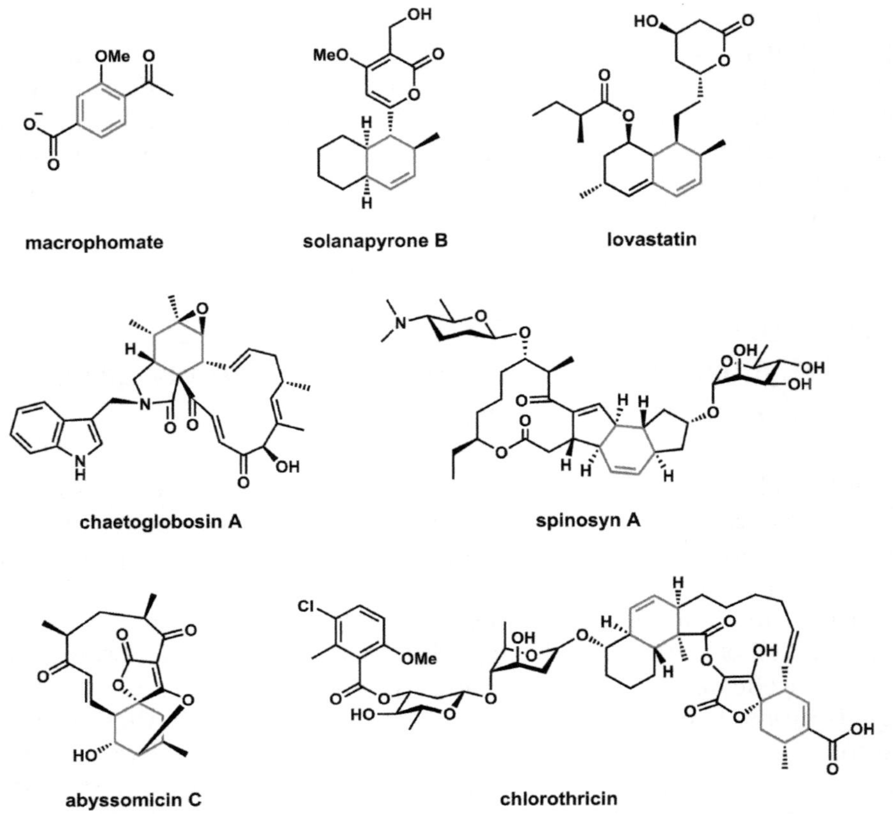

Figure 2.24 Seven natural products containing scaffold elements that suggest Diels–Alderase catalysts in their assembly.

the actions of lovastatin (Tobert 2003). As depicted in Figure 2.25, two PKS proteins work in parallel to make two pieces of the lovastatin molecule (Xie, Watanabe *et al.* 2006, Campbell and Vederas 2010, Xu, Chooi *et al.* 2013). LovF uses acetyl-CoA as starter unit, adds one two-carbon unit from malonyl-CoA, methylates adjacent to the newly installed ketone and then reduces the nascent β-keto all the way to β-CH_2 to release the branched chain methylbutyryl-CoA. This will become esterified to the monacolin J framework in the last step by an acyltransferase LovD (Campbell and Vederas 2010, Xu, Chooi *et al.* 2013).

The main scaffold construction is carried out by the single module PKS protein LovB, which acts iteratively *via* eight cycles of malonyl-CoA addition to build the tethered nonaketidyl-*S*-thioester framework on the ACP domain. The ER domain in LovB is present but inactive. The assembly line needs the participation of a separate ER protein,

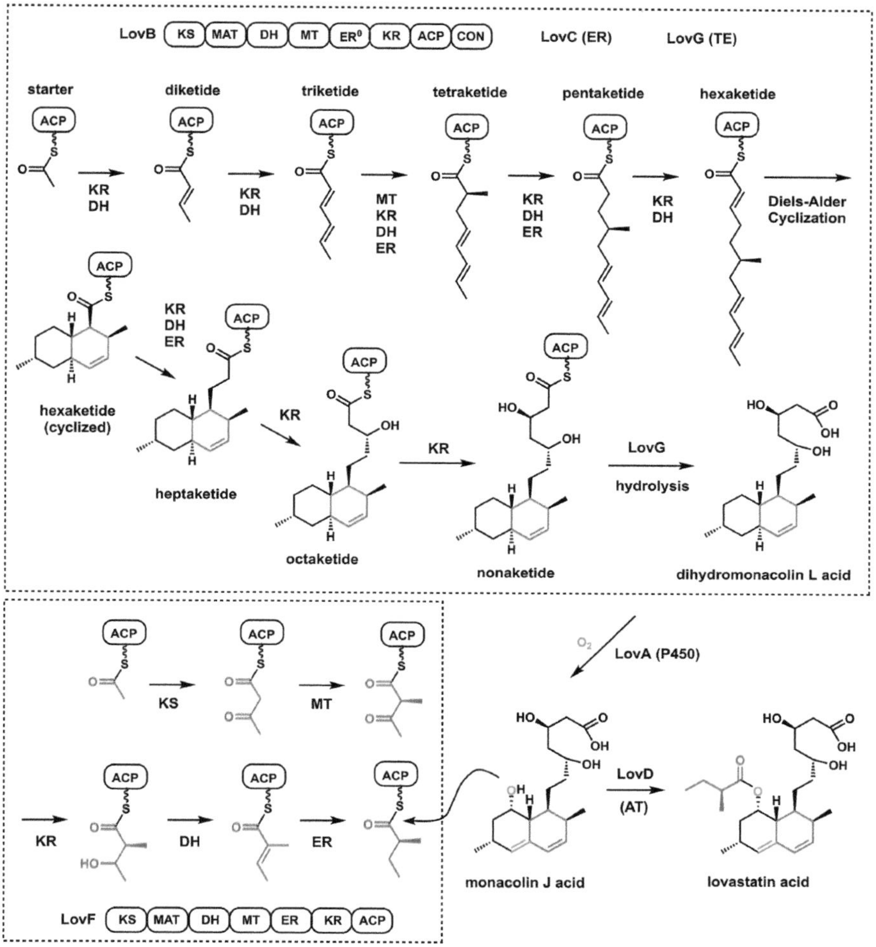

Figure 2.25 The lovastatin PKS assembly line, showing cooperation between two PKS enzymes, LovB and LovF. The decalin formation occurs at the hexaketidyl-*S*-enzyme stage.

LovC (Ames, Nguyen *et al.* 2012), which can work *in trans* with LovB but does so only at particular stages of polyketidyl chain growth.

As schematized in Figure 2.25 the KR and DH domains of LovB work in the first and second elongation cycles to give a triketidyl-*S*-LovB that carries a diene forward. In the next two elongation cycles LovC participates and generates saturated units, so the diene persists at the pentaketide level. In the next elongation cycle LovC does not work on the nascent hexaketidyl chain, allowing a third olefin, isolated from the conjugated diene, to persist. The figure shows the subsequent

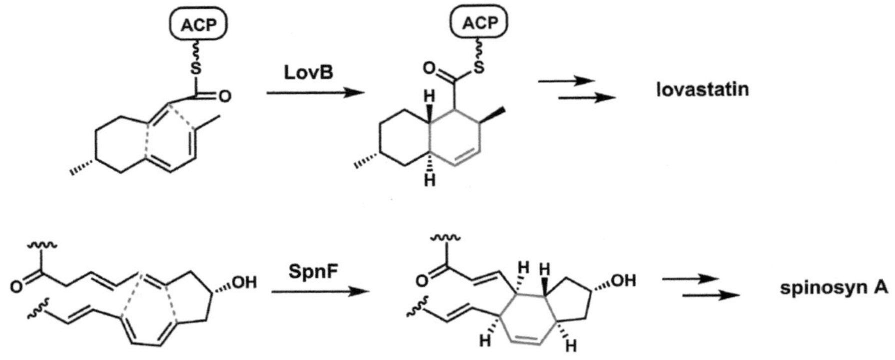

Figure 2.26 Enzyme-mediated Diels–Alder cyclizations in the lovastatin and spinosyn pathways.

domain participations in heptaketide to nonaketidyl enzyme formation with hydrolytic release by the TE LovG to yield dihydromonacolin L (Xu, Chooi *et al.* 2013). After introduction of a hydroxyl group and another double bond by the P450 LovA, monacolin J is formed and coupled with the LovF product to give lovastatin.

It is at the hexaketidyl-*S*-enzyme that that decalin formation occurs *via* cyclization between the 2-enoyl and 8,10-dienyl olefinic systems. A single [4 + 2] cyclization transition state – a Diels–Alder reaction – is one proposed mechanism (Figure 2.26) (Witter and Vederas 1996).

Figure 2.26 also shows the proposed Diels–Alder cyclization during maturation of the spinosyn insecticidal framework (Kim, Ruszczycky *et al.* 2011). It remains to be determined by detailed mechanistic analysis whether the ring-forming reactions truly occur in a single concerted transition state for these and other related transformations, although the action of the abyssomycin enzyme AbyU that creates the cyclohexene ring in the spirotetronate moiety of an abyssomycin intermediate (Figure 2.27) is argued to be a "natural Diels–Alderase" (Byrne, Lees *et al.* 2016).

2.7.2 Sch210972: Imposition of Stereochemical Control by CghA

The biosynthesis of the decalin- and tetramate-containing natural product Sch210972 (Figure 2.28) from the fungus *Chaetomium globosum* has been investigated recently. Sch210972 has drawn interest as a molecule that blocks the binding of human immunodeficiency virus (HIV) to the CCR5 receptor on human T cells. The biosynthetic

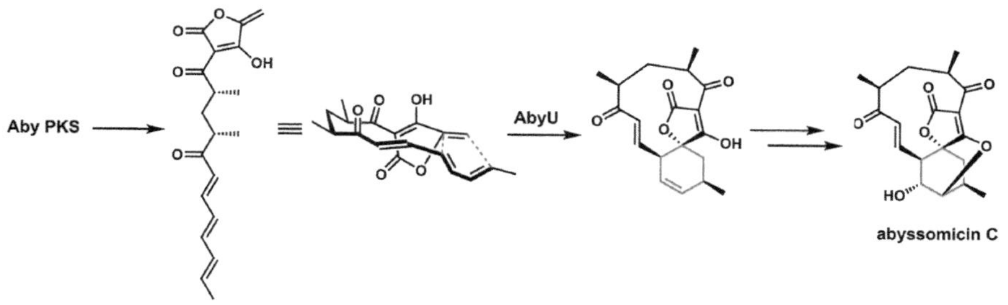

Figure 2.27 The "Diels–Alderase" AbyU catalyzes formation of the cyclohexene ring in the spirotetronate moiety found in abyssomycin.

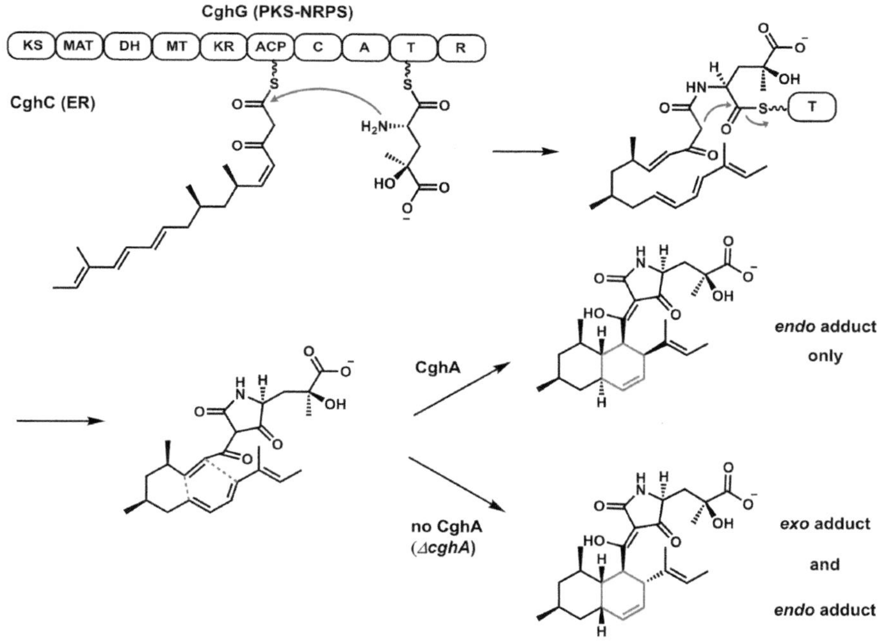

Figure 2.28 A hybrid PKS–NRPS assembly line for Sch210972 features a [4 + 2] cyclization and a Dieckmann condensation release step. The enzyme CghA in Sch21092 formation directs all the reaction flux through the *endo* transition state. The *exo*-adduct product forms along with *endo* adduct in *ΔcghA* strains.

gene cluster was identified, the structure of the product confirmed by spectroscopic analysis, and, importantly, the absolute stereochemistry of the decalin established by X-ray analysis. CghA was established as the catalyst for the [4+2] cycloaddition reaction. Disruption of the *cghA* gene led to a 30-fold reduction in product formation but also the appearance of a new product that turned out to be the alternative stereoisomer from the [4+2] cyclization (Sato, Yagishita *et al.* 2015). As diagrammed in Figure 2.28, CghA catalyzes formation of the *endo* adduct stereospecifically with no detectable amount of the *exo* adduct. When the enzyme is not produced, the *endo* and *exo* products form in a 2:1 ratio (presumably reflecting the energetics of the two transition states) at one twentieth the total amount.

While these results do not unambiguously speak to whether the decalin is formed in a single Diels–Alder transition state (however synchronous or asynchronous), they do suggest that the enzyme imposes a clear stereochemical preference. In this and the spinosyn case below where SpnF accelerates another slow, nonenzymatic [4+2] cycloaddition, it may be that the enzymes "merely" accelerate the rate of the nonenzymatic cycloadditions. The decalin-forming energetics in all these natural products may lie along a continuum where the reaction proceeds slowly in the absence of protein catalyst but can be speeded up by enzymes and stereochemical control imposed to conserve the useful mass to take forward in subsequent steps. To what extent the imposition of stereochemical control also alters the mechanism of the bond-forming steps, *e.g.* from stepwise to concerted or *vice versa*, is not established for the general class of these cycloadditions.

2.7.3 Spinosyn: SpnF as Diels–Alderase?

Figure 2.29 indicates the structure of the most downstream polyketidyl chain in thioester linkage to the phosphopantetheinyl arm of ACP_{10} of the spinosyn synthase. Macrolactonizing release catalyzed by the TE domain generates the 22-membered lactone macrocycle as the first soluble product, containing a diene and two isolated olefins. A set of nine further post assembly line enzymes act to generate the mature spinosyn A structure. SpnF has been isolated, its structure determined (Fage, Isiorho *et al.* 2015), and shown to accelerate a slow nonenzymatic cyclization (Kim, Ruszczycky *et al.* 2011) (Figure 2.26, Figure 2.30) by some 60-fold. Note that the particular configuration of diene and dienophile generates a bicyclic 6-5 system rather than the 6-6 bicyclic decalin system in the lovastatin scaffold.

The substrate for SpnF (Figure 2.29) bound to the enzyme is presumed to be converted to a higher fraction of the *s-cis* conformer

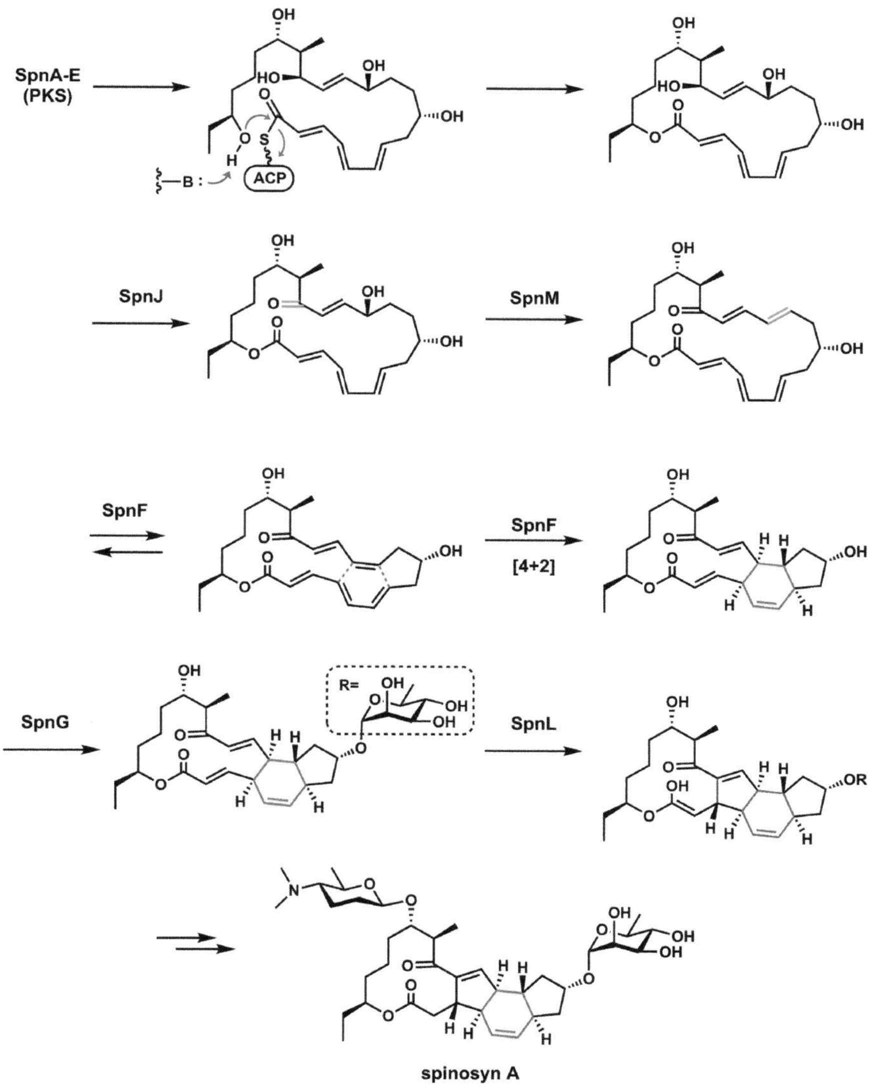

Figure 2.29 Intermediates in the spinosyn biosynthetic pathway. SpnF accelerates the [4 + 2] cycloaddition reaction that creates the 6,5 ring system while SpnL creates the cyclopentene ring.

required for the transannular [4 + 2] cyclization. This must also occur on solution given that the reaction occurs nonenzymatically at 1/500th the rate catalyzed by SpnF. SpnF has the fold of an SAM-dependent methyltransferase but does not catalyze any methylations, although it does copurify with a molecule of *S*-adenosylhomocysteine

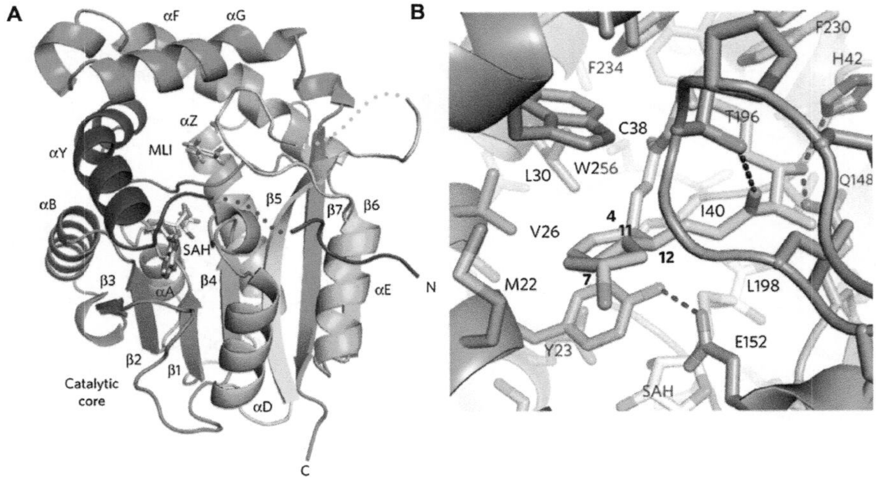

Figure 2.30 Structure of SpnF with (A) bound SAH and (B) the product modeled into the active site region.
Reprinted by permission from Macmillan Publishers Ltd: Nature Chemical Biology (Fage C. D., E. A. Isiorho, Y. Liu, D. T. Wagner, H.-W. Liu. and A. T. Keatinge-Clay., "The structure of SpnF, a standalone enzyme that catalyzes [4+2] cycloaddition". *Nat. Chem. Biol.*, **11**, 256), copyright (2015).

(SAH) that appears in the enzyme X-ray structure (Figure 2.30). Although crystals with the SpnF product were not reported, a model for how it could bind in the SpnF active site is shown in Figure 2.30B. The structural and mechanistic studies set the stage for determining whether the SpnF catalyzed cycloaddition is truly a Diels–Alder single step process or is multistep. Fage *et al.* (2015) note that:

> "SpnF may facilitate cycloaddition through a combination of (i) removing water molecules surrounding the substrate; (ii) stabilizing the reactive geometry, perhaps by lowering the energy of the C5–C6 s-*cis* conformation relative to those of other conformations; and (iii) enhancing the reactivity of the dienophile; T196 is in position to form a hydrogen bond with the C15 carbonyl and could facilitate withdrawing of electron density through the C_{11}–C_{15} π-system."

Computational analysis of the SpnF reaction, by quantum mechanical calculations and dynamic simulations, suggests that the reaction does not clearly compute to either a concerted or a stepwise process (Patel, Chen *et al.* 2016). A Diels–Alder route is computed

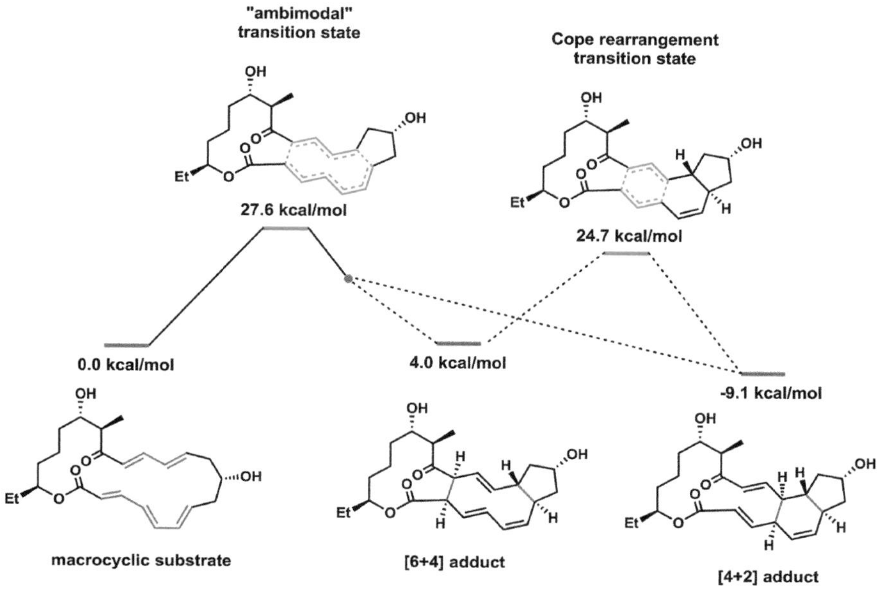

Figure 2.31 Energetic computations of SpnF transition states and a [6 + 4] cycloadduct intermediate on the way to the observed [4 + 2] product. Gibbs free energies in kcal mol^{-1}.
Data from Patel A., Z. Chen, Z. Yang, O. Gutiérrez, H.-W. Liu, K. N. Houk and D. A. Singleton (2016). "Dynamically Complex [6 + 4] and [4 + 2] Cycloadditions in the Biosynthesis of Spinosyn A". *J. Am. Chem. Soc.*, **138**(11): 3631.

to be reasonable but so is a [6 + 4] cycloadduct. Such a product is not observed, but a rapid Cope rearrangement (Figure 2.31) would give the observed [4 + 2] product. Combining this analysis with the ikarugamycin route, two sections below, to a comparable 5-6-5 tricyclic system by a completely different chemical strategy indicates that much is still to be learned about the chemical logic Nature uses to accelerate carbocyclic ring formations in natural product assembly.

2.7.4 Solanapyrone

Solanapyrones are polyketides elaborated by the fungus *Alternaria solani*, an organism that causes forms of potato and tomato blight. The embedded decalin suggests a possible Diels–Alder cyclase origin [see (Walsh and Wencewicz 2013) for a summary]. It turns out that

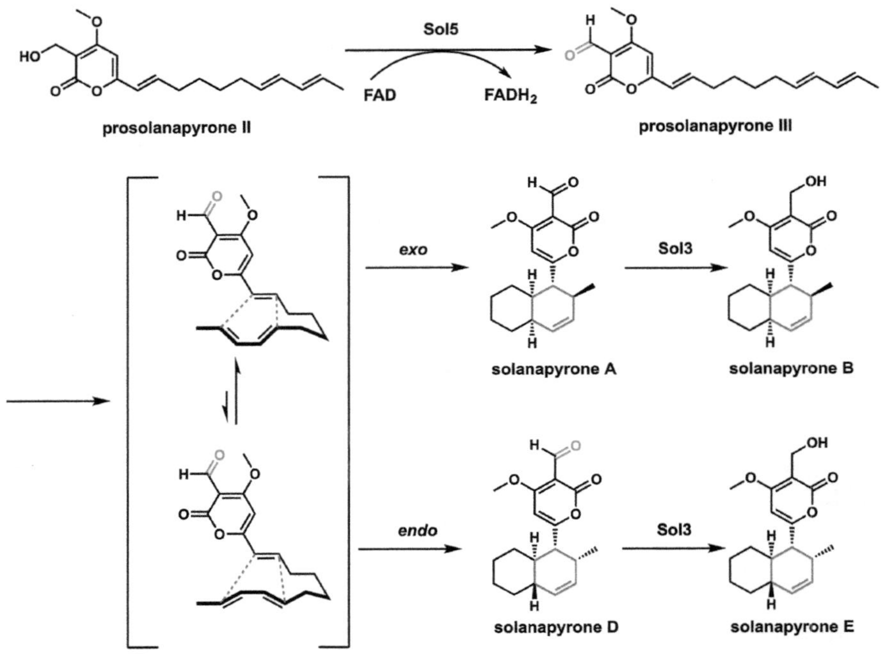

Figure 2.32 Diels–Alder cyclization of prosolanapyrone II following oxidation of a remote alcohol to an aldehyde.

oxidation of an alcohol to a ketone remote from the diene moiety, but in electronic contact with the lowest unoccupied molecular orbital of the olefin that becomes the dienophile *via* the intervening aromatic ring system, is sufficient to set off the decalin-forming reaction. A proposed Diels–Alder route is shown in Figure 2.32, but is not yet proven.

Vignette 2.1 Two Types of [4+2] Cyclizations in One Biosynthetic Pathway

The pentacyclic core of pyrridomycins contains both a decalin system and a 6,5-spiro tetramate system (Figure 2.V1). Remarkably, both of these bicyclic systems have recently been shown to arise by a cascade of [4+2] cyclizations (Tian, Sun *et al.* 2015). The spiro tetramate is reminiscent of a comparable spirotetronate assembled by AbyU in abyssomycin formation, while the decalin resembles that in lovastatin and other metabolites.

These are, to date, the two types of canonical ring system assembled enzymatically by [4+2] cyclizations. Recently, they have been found to occur sequentially in the same pathway, catalyzed by PyrE3 (decalin formation) and PyrI4 (spiro tetronate) respectively, as

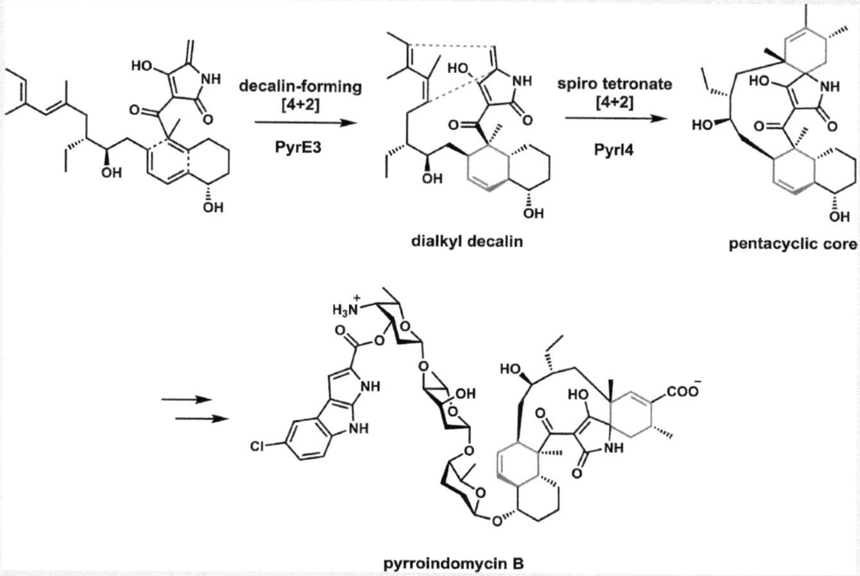

Figure 2.V1 A cascade of [4 + 2] cycloadditions leading to the formation of pyrridomycins.

schematized in the bottom part of Figure 2.V1. PyrE3 is proposed to act first on the nascent product released from a hybrid PKS–NRPS assembly line as shown. Then PyrI4 acts on the dialkyl decalin-containing scaffold.

PyrE3 contains FAD but no redox role is assigned to the cofactor in the decalin formation. PyrI4 has no detectable cofactor requirement. Together they build the pentacyclic core of pyrroindomycin as a post assembly line tailoring sequence. Whether either or both PyrE3 and PyrI4 act as Diels–Alderases, *e.g.* as entropic traps, or carry out stepwise core construction remains to be determined, but the authors note that if Diels–Alder routes are involved the dialkyl decalin would be *endo* selective while the spirotetronate cyclization would be *exo* selective. PyrI4 homologs are detected in several microbial genomes, sometimes coupled with PyrE3 homologs, suggesting a family of such spirotetronate and decalin-forming enzymes.

2.7.5 A Stepwise Alternative to the Same Tricyclic Ring System Found in Spinosyn

Stepwise construction occurs in the fused 5-6-5 tricyclic ring system in ikarugamycin and related polycyclic tetramate macrolactams. Despite having the same 5-6-5 fused tricyclic core as in spinosyn, the cyclizations to this core are clearly reductive in mechanism (Antosch,

Schaefers *et al.* 2014, Zhang, Zhang *et al.* 2014). The first five–six ring forming step and the second, forming the third of the rings (the right hand cyclopentane of Figure 2.33) require NADPH. The proposed

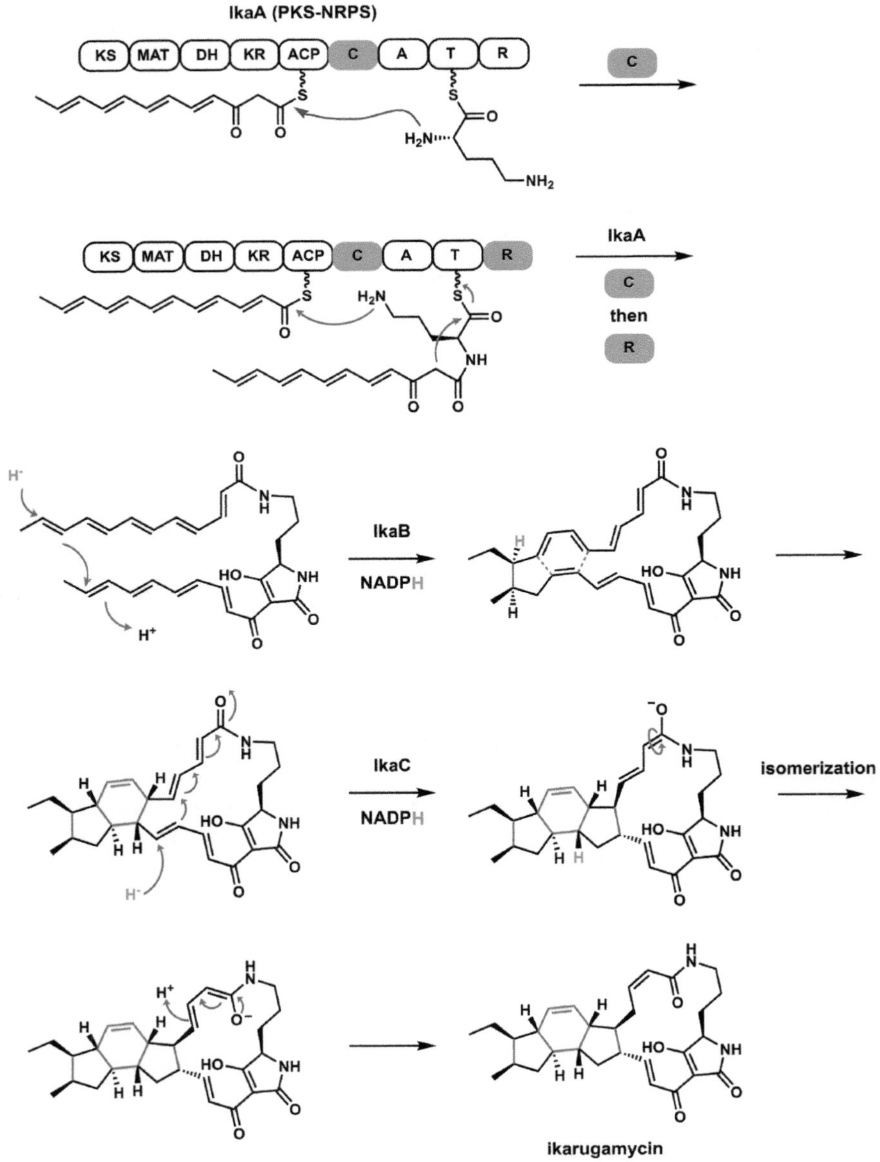

Figure 2.33 Stepwise construction of the 5-6-5 tricyclic ring system in ikarugamycin.

mechanism shows where the hydrides from each of the two NADPH coenzymes would end up in ikarugamycin.

A three gene cassette from a marine streptomycete is sufficient to produce ikarugamycin in the heterologous host *S. lividans* TK64 (see Chapter 13 for an overview of heterologous gene cluster expression) (Zhang, Zhang *et al.* 2014). IkaA is a hybrid polyketide synthase–nonribosomal peptide synthease (PKS–NRPS), one module each, proposed to generate the linear product shown in Figure 2.33. Then, IkaB and IkaC are both dihydronicotinamide-dependent reductases which are proposed to effect the indicated Michael type cyclizations.

The first cyclization is proposed to initiate by hydride addition to the terminus of the conjugated polyene of one chain with C–C bond formation to the polyene terminus of the adjacent polyene. An analogous hydride-initiated cyclization would build the third ring (cyclopentane) fused to the first two. Obviously, this is a distinct chemical logic and set of enzymes from the spinosyn system and reflects the versatility of carbocyclic ring construction. In passing we note that IkaB, in generating the 5,6-bicyclic framework, also closes the macrolactam by a surprising route. This has analogies to the timing of the trithiazolylpyridine core formation in the thiazole peptide antibiotics whose biosynthesis is summarized in the next chapter.

Ikarugamycin is one of a family of polycyclic tetramate macrolactams, as illustrated in Figure 2.34. They have either two or three fused rings within the macrolactam in different configurations of 5,5-bicyclic and 5,5,6- and 5,6,5-tricyclic moieties. It is likely that the reductive cyclization mode, with one enzyme (bicyclic) or two reductases working sequentially (three rings), is operant in these biosyntheses, in analogy with the logic and machinery for ikarugamycin.

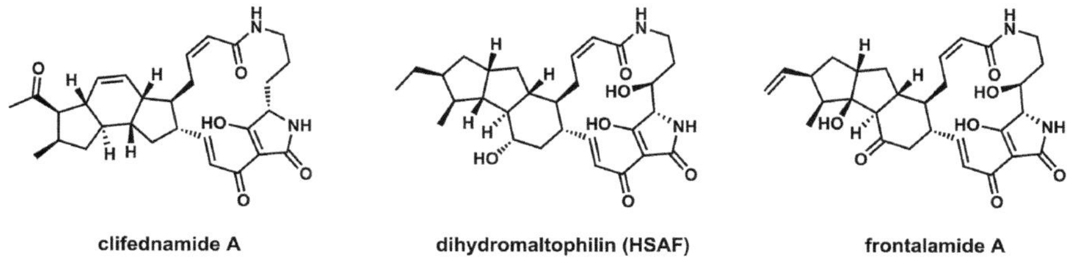

Figure 2.34 Other tetramate macrolactams with fused multicyclic ring systems.

2.8 The Polyene Subclass of Polyketides

A Type I PKS assembly line that has 19 modules is responsible for construction of the antifungal molecule nystatin (Brautaset, Sekurova *et al.* 2000)(Figure 2.35). A related PKS array creates the congeneric antifungal agent amphotericin B (not shown) (Caffrey, Lynch *et al.* 2001). The 19 modules are spread over six proteins and that fact emphasizes that pairwise interactions of the PKS subunits with each other must occur for the chain to be processed in the order indicated. That is, NysB must interact with NysA as upstream partner and NysB as downstream partner for growing chain acceptance and donation, respectively. Similar protein–protein discrimination and selective affinity must hold for NysC, I, and J, with their upstream and downstream partners in this divided assembly line. These recognition properties are requisite features of all such multiprotein assembly lines, both for the PKS complexes in this chapter and the NRPS in the next chapter (Weissman and Muller 2008, Weissman 2015).

This is an amazingly complex biosynthetic assembly line with 91 functional domains distributed over the 19 modules in the six proteins, encompassing more than 3 million daltons of protein mass, and involving >3000 amino acid residues. The producer organism *Streptomyces noursei* spends a lot of ATP to build the nystatin assembly line, so presumably the end product is of great metabolic value to it.

Modules 4, 5, 7, 8, 9, 10 lack an ER domain so six olefins accumulate in the growing chain and impose conformational rigidity on the chain. (These are not subject to Diels–Alderase type cyclizations.) The run of olefins gives rise to the class name of nystatin and amphotericin as polyene antibiotics. They insert into fungal membranes and disrupt membrane integrity, causing cell death, but have toxicities associated with insertion into membranes of higher eukaryotes. The sugar residue is essential for biological activity and is a post assembly line tailoring event, as shown.

Different PKS organizational logic is involved in the microbial production of polyunsaturated fatty acids, including those that have the double bonds in deconjugated 1,4-*cis* arrays or in conjugated 1,3-*trans* arrays (Jiang, Zirkle *et al.* 2008). These are carried out by iterative use of Type II single PKS modules (Okuyama, Orikasa *et al.* 2007). For example, myxobacterial PuFA PKS enzymes make $\Delta^{5,8,11,14,17}$ C20 pentaenoic acid and also the C22 homolog $\Delta^{4,7,10,13,16,19}$ docosahexaenoic acid, the homolog of mammalian polyunsaturated fatty acids (Gemperlein, Garcia *et al.* 2014).

Of particular interest are the iterative PKS that make polyolefinic precursors for enediyne antibiotics (Figure 2.36A). Formation of a

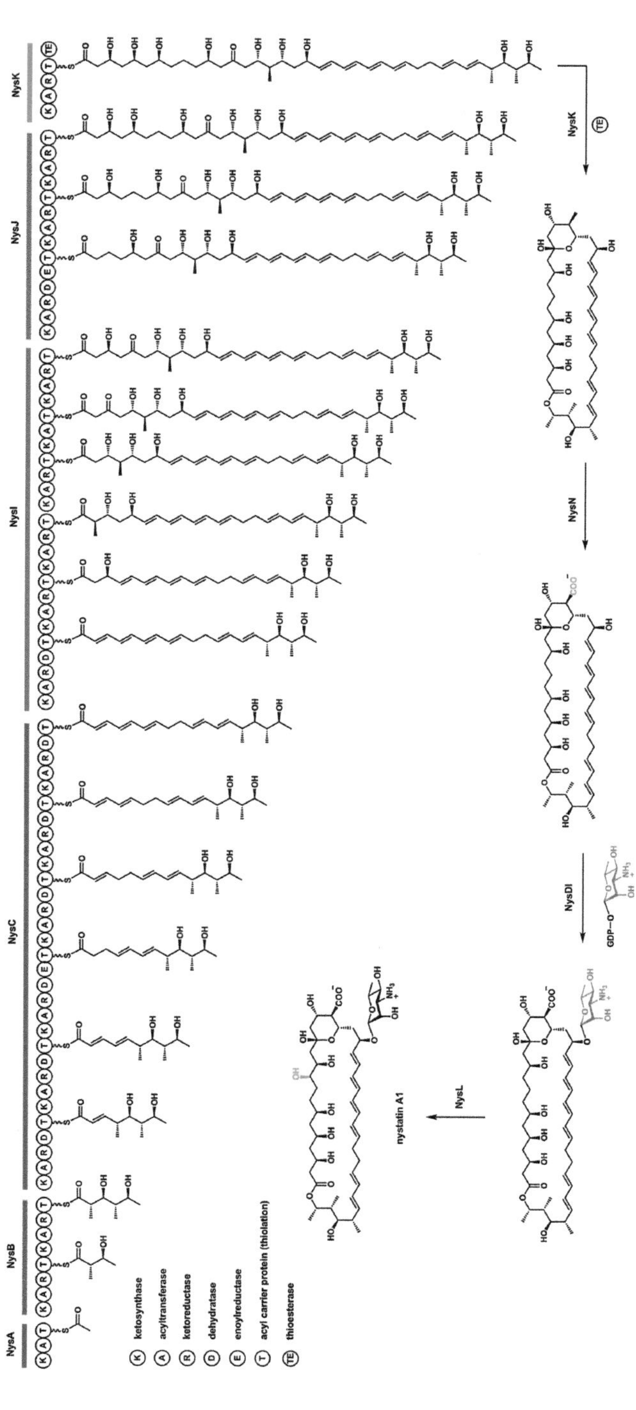

Figure 2.35 A six protein 19 module assembly line with 91 domains builds the polyene macrolactone antifungal agent nystatin.

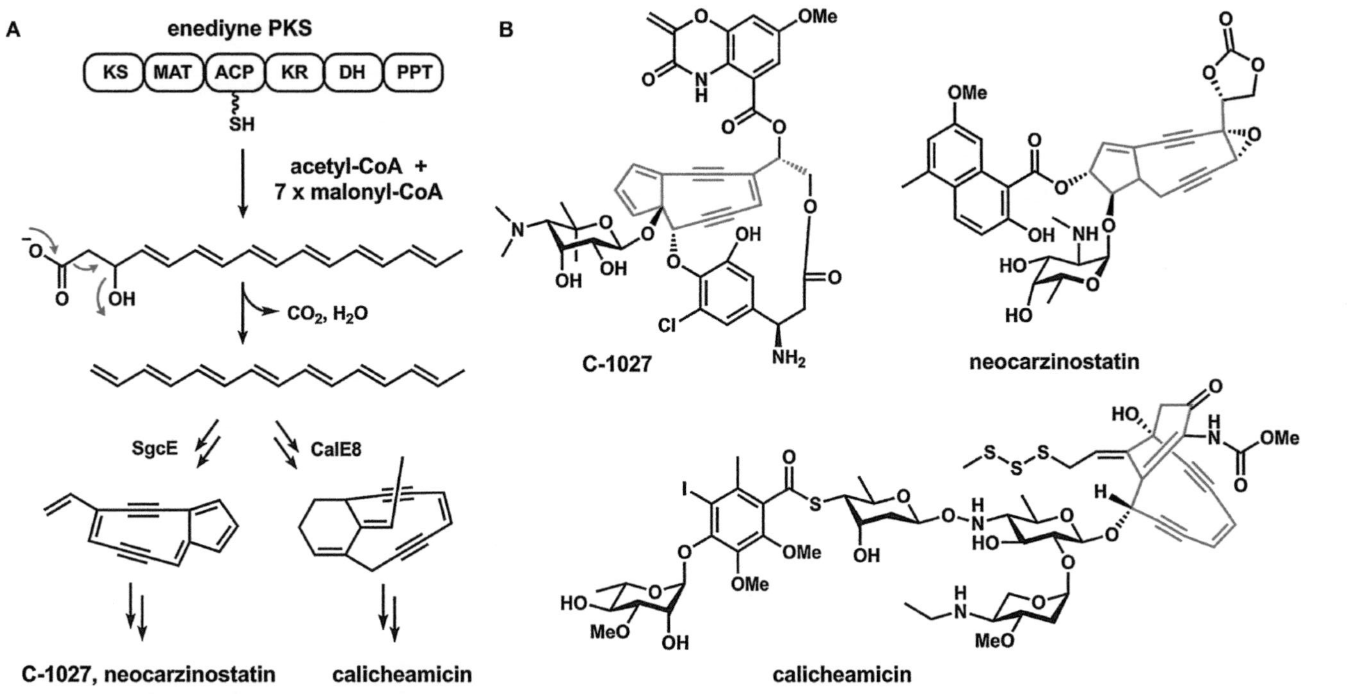

Figure 2.36 Enediynes from morphing of polyunsaturated fatty acids. (A) A single PKS module from the enediyne cluster releases the conjugated heptaene 1, 3, 5, 7, 9, 11, 13 pentadecaheptaene as proposed precursor to the nine-member enediyne ring core. (B) Representative nine-membered enediyne natural products.

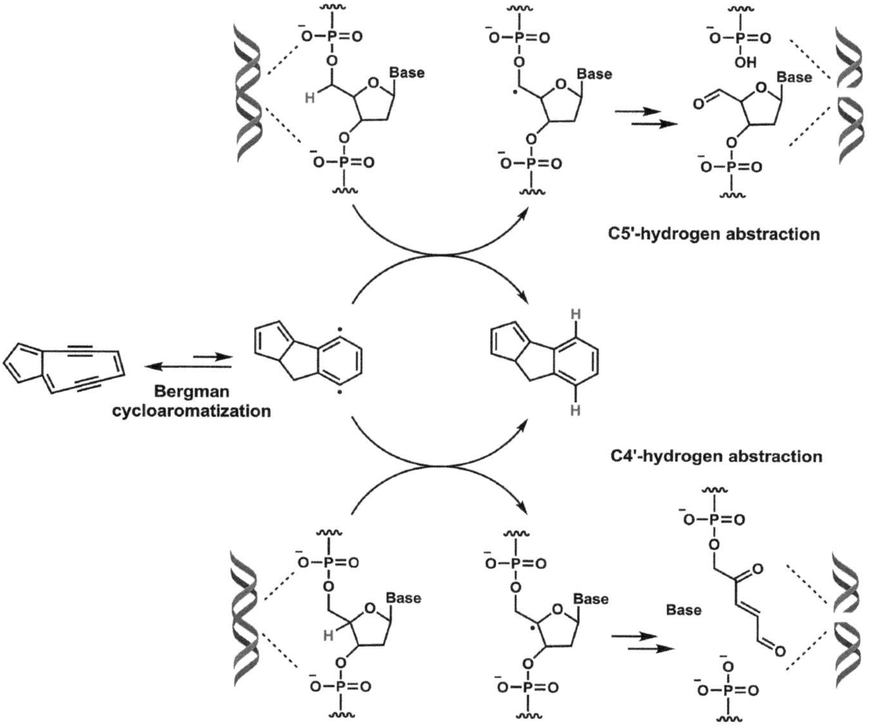

Figure 2.37 Bergman cycloaromatization of enediyne radicals to benzene diradicals in molecules bound to DNA leads to strand cleavages.

3-OH, 4, 6, 8, 10, 12,14 hexadecenoyl-ACP bound intermediate that is released and decarboxylates to 1, 3, 5, 7, 9, 11, 13 pentadecaheptaene (Zhang, Van Lanen *et al.* 2008) is the basis for the proposal that such a polyolefin is proximal precursor to the bicyclic nine-membered enediyne class of natural products (Liu, Christenson *et al.* 2002) (Figure 2.36B). Molecules such as esperamicin, neocarzinostatin, C-1027, and calicheamycin are of medicinal interest because the enediynes bind to DNA and are precursors to aromatic diradical chemistry (Figure 2.37) (Van Lanen and Shen 2008).

To assemble a fully mature enediyne natural product scaffold requires input from four converging natural product pathways. As shown in Figure 2.38, while the unsaturated enediyne core of C-1027 arises *via* PKS-generated polyene intermediates, the benzoxazolinate derives from cyclization of chorismate, the chloro-β-tyrosine from

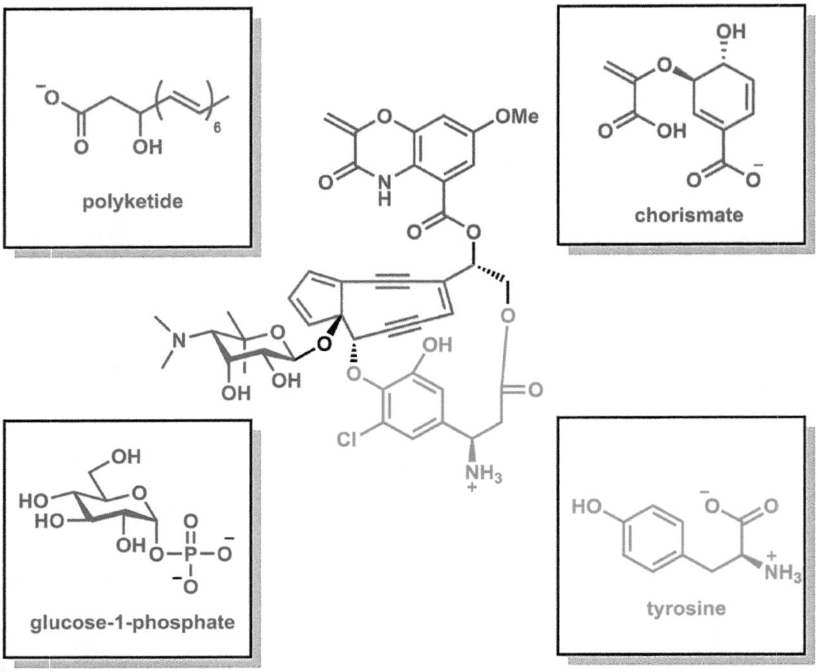

Figure 2.38 Convergence of several pathways from distinct primary building blocks is required to produce the enediyne C-1027.

amino acid metabolism, and the deoxyhexose from glycosyltransfer enzymology (discussed in Chapter 11) (Liu, Christenson *et al.* 2002).

2.9 Polyketide to Polyether Metabolites

A substantial number of polyketide metabolites contain cyclic ethers (Gallimore 2009, Liu, Cane *et al.* 2009). The rings can range from five to nine members but by far the most common are five-membered ring furan and six-membered ring pyran rings, as seen in the molecules in Figure 2.39. We will note ring-forming reactions in lasalocid and isolasalocid, the related potassium ionophore monensin and a speculative route to the fused ether rings in ciguatoxin. Figure 2.40 indicates six common ether ring formation mechanisms which apply to these polyketidyl ether chains, and introduces cyclization nomenclature (Baldwin 1976, Baldwin, Thomas *et al.* 1977, Van Wagoner, Satake *et al.* 2014). The first route (A) forms a six-member pyran by

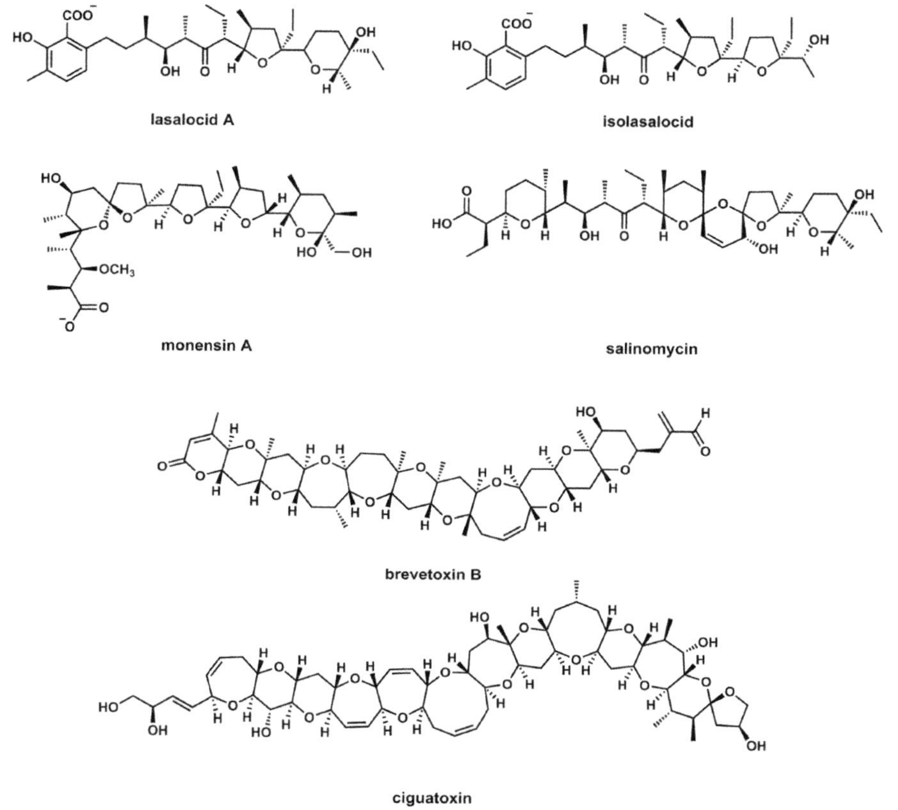

Figure 2.39 Polyketides containing cyclic ethers (polyether class) include lasalocid, monensin, salinomycin, and ciguatoxin, which has 13 fused cyclic ethers.

exocyclic attack of an OH on a carbonyl group (6-*exo*-trig). The second route (B) is a formal 6-*endo*-trig pathway since the olefinic carbon attacked is an sp^2 carbon, and so on. Routes E and F involve prior conversion of an olefin to an epoxide and then its opening by a neighboring –OH group in either a 6-*endo*-tet or 5-*exo*-tet mode, giving pyran or furan, respectively. Many of these routes are relevant to the enzymology noted below.

The A ring in salinomycin is formed by a 6-*endo*-trig pathway. The enzyme dehydrates the β-OH-acylthioester to generate the α,β-conjugated enoyl thioester that serves as the electrophile in pyran ring formation as the C_7-OH attacks the C_3 terminus of the conjugated system (Luhavaya, Dias *et al.* 2015). The protein

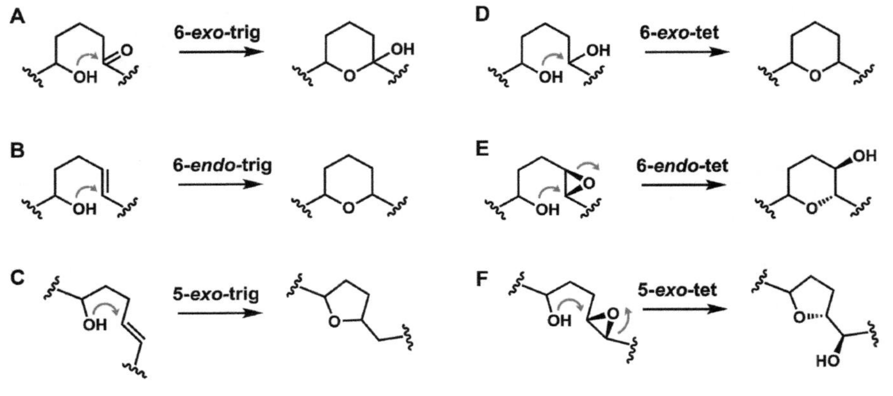

Figure 2.40 Ether ring-forming mechanisms that can apply to the polyketide class.
Adapted from Van Wagnoer, R. M., M. Satake and J. L. C. Wright (2014). "Polyketide biosynthesis in dinoflagellates: what makes it different?" *Nat. Prod. Rep.*, **31**: 1101 with permission from The Royal Society of Chemistry.

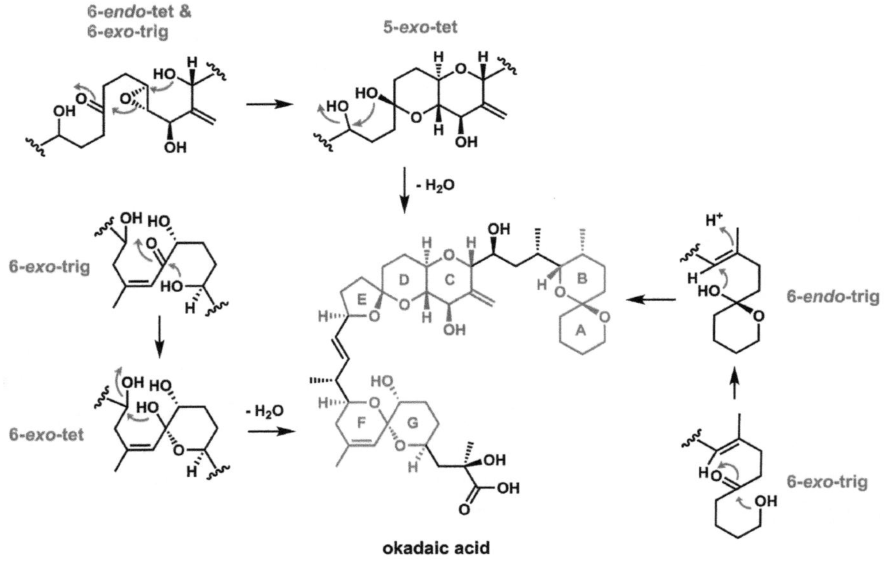

Figure 2.41 Three routes to the seven cyclic ethers in okadaic acid.

phosphatase inhibitory toxin okadaic acid has seven cyclic ethers in its scaffold (Van Wagoner, Satake *et al.* 2014). As shown in Figure 2.41, three route variations are utilized. Formation of the first spiro pair of pyrans involves attack on a ketone and olefin, respectively. The fused 5-6-6 tricyclic ether system in rings CDE involves one epoxide intermediate (from prior olefin enzymatic epoxidation) and one displacement of –OH as a leaving group as the furan forms. The second spiro pyran pair involves the 6-*exo*-trig attack in a ketone and then a 6-*exo*-tet displacement of an –OH substituent.

The monensin and lasalocid ionophores are processed by the epoxidation, epoxide opening cascade route as noted in Figures 2.42 and 2.43, respectively (Migita, Watanabe *et al.* 2009) (Oliynyk, Stark *et al.* 2003). This two-step process was first suggested by Cane, Celmer and Westley in 1983 as biosynthetic logic that would underlie all the polyethers known at the time (Cane, Celmer *et al.* 1983). Those authors drew on isotopomer (see Figure 1.32 in Chapter 1) labeling patterns from feeding studies of triply labeled acetate CH_3-[^{13}C-$^{18}O_2H$] and propionate CH_3CH_2-[^{13}C-$^{18}O_2H$] to the triene precursor of monensin and then to the final product, monensin, with its three furan and two pyran rings (Figure 2.42). In complementary studies with $^{18}O_2$ it was shown that those three oxygens ended up in furans 2 and 3 and the terminal pyran of monensin. This pattern suggested the triene–triepoxide chemistry, which was fully substantiated in later studies showing the triene as nascent product and the finishing role of the epoxidases. It has also been robust in predicting

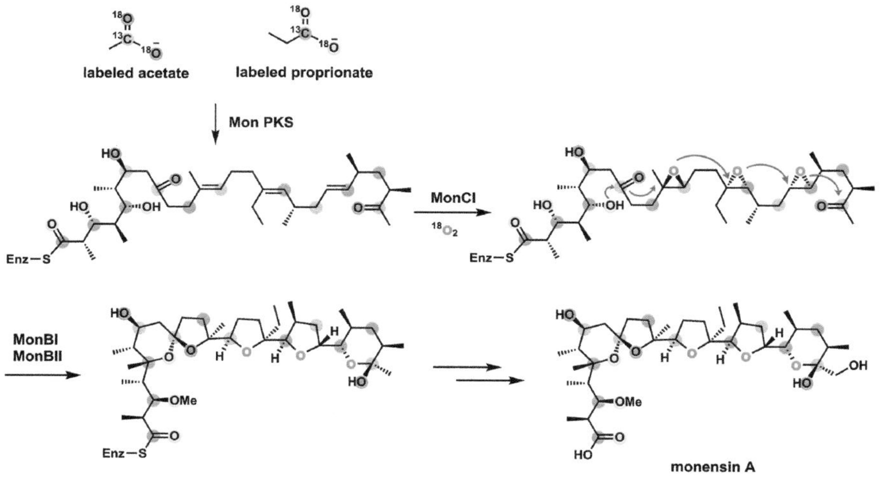

Figure 2.42 Olefins to epoxides to pyrans and furans in monensin A biosynthesis. Labeling results from the Cane studies are shown.

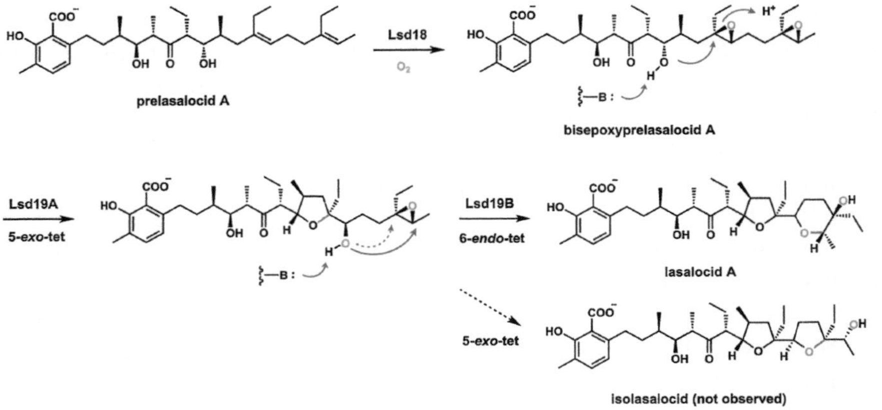

Figure 2.43 Olefins to epoxides to pyrans and furans in lasalocid A biosynthesis.

and accounting for subsequently discovered polyether scaffolds (Vilotijevic and Jamison 2009).

Two tandem chemical steps occur: tandem double bond epoxidations by flavoenzymes (reminiscent of squalene epoxidase in Chapter 4) are followed by cofactor-independent epoxide hydrolases which are the actual cyclic ether-forming catalysts.

For lasalocid, the Lsd18 and Lsd19 enzymes fit this two step catalytic logic. Opening of a bis epoxide by an intramolecular, neighboring –OH under the aegis of Lsd19 sets both the furan and pyran rings (Minami, Shimaya *et al.* 2012). Analogously, in monensin maturation, the tris-epoxide, generated by action of a flavoenzyme oxygenase/epoxidase on the three indicated double bonds, leads to the pentacyclic ether array by action of a partner epoxide hydrolase.

The lasalocid epoxide hydrolase/ether forming enzyme Lsd19 has been examined for mechanism and structure to evaluate how it binds substrate to promote the thermodynamically disfavored 6-*endo*-tet pathway over the 5-*exo*-tet route (to isolasalocid) (Minami, Shimaya *et al.* 2012) (Hotta, Chen *et al.* 2012) (Figure 2.43).

The structure of lasalocid and monensin, with two and five cyclic ethers, respectively, looks quite simple in comparison to the 11 fused ethers in brevetoxins (Baden 1989, Turner, Higgins *et al.* 2015) or ciguatoxins (Nicholson and Lewis 2006) from marine dinoflagellate PKS assembly lines (Figure 2.39, Figure 2.44). (Note the eight- and nine-membered cyclic ethers embedded within the ciguatoxin

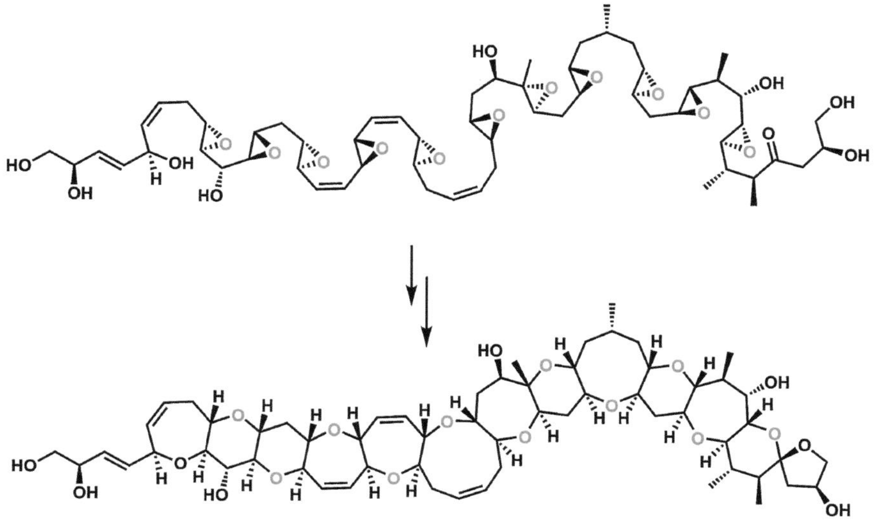

Figure 2.44 A proposal for an epoxide hydrolase cyclization cascade to form the fused polyether scaffold in ciguatoxin by unraveling of an undecaepoxy intermediate.

molecule.) None of these complex marine-derived, fused polyether scaffolds, ion channel blockers (Baden 1989), has been examined in any significant mechanistic or structural biosynthetic detail because of barriers to genetics and protein biochemistry in the host producers.

How concerted such cascades are, where they start (*e.g.* with the proposed initiator epoxide ring), how the selectivity of epoxidation of some double bonds and not others are determined, and how the cascade directionality proceeds remain to be determined when purified enzymes eventually become available.

2.10 Convergence of Polyketide and Other Natural Product Pathways

As we shall note with other classes of natural products in subsequent chapters, biosynthetic pathways can and do converge to create additional variation and functionality. This can occur at an early stage of assembly, *via* different types of building blocks, exemplified with the origins of the starter units of the enediyne C1027 shown in Figure 2.38. In addition to the polyketide core, the amino acid tyrosine, the sugar phosphate glucose-1-P, and the central precursor to aromatics, chorismate, are conscripted for C-1027 (Van Lanen and Shen 2008).

Pathways can also converge at later stages. Figure 2.45 displays the structures of hyperforin and Δ^9 tetrahydrocannabinol, two molecules that are hybrids of polyketides and prenylation (Chapter 4) biosynthetic machinery. The polyketide origin of hyperforin is drawn in blue and the four isoprenyl units are shown red. The biosynthetic logic for hyperforin is considered in detail in Chapter 4. Tetrahydrocannibinol, with a distinct human pharmacology from hyperforin, is also a PK–prenyl hybrid with the same color scheme and is likewise dissected for biosynthetic machinery in Chapter 4. In both cases the PKS enzymes build the aromatic core for later prenyl decoration.

The third molecule shown in Figure 2.45 is the antibiotic andrimid (Feredenhagn, Tamura *et al.* 1987), a hybrid of polyketide assembly (blue) and nonribosomal collection of amino acid units (red). Nonribosomal peptide synthetases are the subject of the next chapter.

Andrimid is a potent inhibitor of bacterial acetyl-CoA carboxylase (Clardy, Fischbach *et al.* 2006), the first committed enzyme in both fatty acid and polyketide biosynthesis noted at the beginning of this chapter, so it seems a fitting example with which to close. Two elements of the android scaffold are provided by PKS enzymatic machinery. One is the polyunsaturated 2, 4, 6-octatrienoate as a starter unit and then another malonyl piece in between some amino acid fragments. The phenylalanine residue is actually a β-Phe, generated by action of a phenylalanine mutase (Heberling, Masman *et al.* 2015) that has an unusual cofactor, found also in phenylalanine deaminase which is the gate keeper enzyme to the large

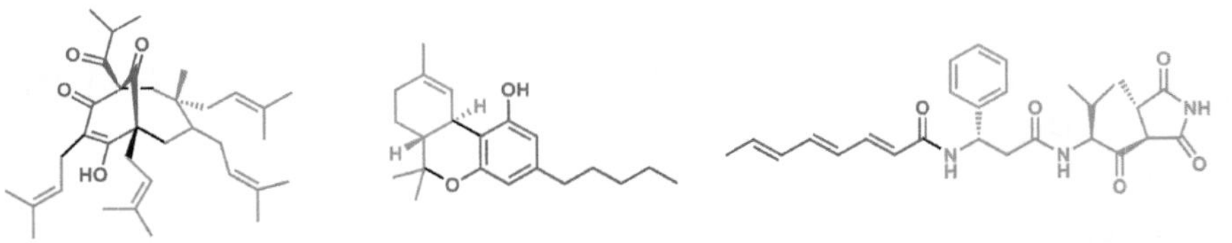

hyperforin tetrahydrocannabinol andrimid

Figure 2.45 Convergence of different natural product biosynthetic pathways leads to hyperforin, tetrahydrocannabinol, and andrimid.

phenylpropanoid class of natural products (Chapter 7). The methylsuccinamide tail in andrimid is the pharmacophore that mimics the biotin cofactor of the target enzyme (Liu, Fortin *et al.* 2008).

Vignette 2.2 Ivermectin, Lovastatin, Erythromycin, Tetracycline, Adriamycin

The polyketide family of natural products has yielded many molecules that have found utility as human therapeutic agents. Among them are the five molecules shown in Figure 2.V2.

Erythromycin and tetracycline, isolated during the golden age of antibiotic discovery in the 1940s and 1950s, are broad spectrum antibiotics. The glycosylated macrolactone ivermectin has become the premier agent for the treatment of river blindness. Adriamycin is one of a related series of aromatic glycosides that targets DNA as part of anticancer regimens. Lovastatin has given rise to the semisynthetic class of statin drugs that became the best selling drugs in the cardiovascular space for lowering cholesterol levels.

Erythromycin was isolated as a related complex of four molecules from the soil bacterium *Saccharopolyspora erythraea* in 1952. Erythromycin A is the most active form of the natural product family and the one used in treatment of human respiratory infections. It contains a 14-member macrolactone ring with two deoxysugar substituents which are key for antibiotic activity. It targets the peptidyltransferase center in the 50S subunit of bacterial ribosomes and thereby blocks bacterial protein synthesis. Over its decades of use, bacteria have evolved various resistance mechanisms, including efflux pumps and mutation of rRNA in the 50s ribosomal subunit. To combat the decreased efficacy of the natural product medicinal chemists have made semisynthetic or synthetic second generation (azithromycin, clarithromycin), third generation (telithromycin), and fourth generation (solithromycin – in late clinical development) versions of this macrolide antibiotic class (Walsh and Wencewicz 2016).

The tetracyclines were isolated in the mid-1940s and chlortetracycline and oxytetracycline commercialized by the late 1940s and early 1950s. They are also of polyketide origin but, as fused aromatic tetracyclic ring structures, have radically different scaffolds from the macrolactone subclass of polyketides. The main text of Chapter 2 details how the polyketonic thioesters covalently tethered to the acyl carrier proteins undergo cyclodehydrative aromatization to the released fused aromatic ring framework. Tetracyclines also target bacterial ribosomes, but act on regions of 16S rRNA at particular sites in the small 30S ribosomal subunit. They have been on the World Health Organization (WHO) list of essential medicines and were historically consequential in cholera epidemic treatments. As with other antibiotics, widespread use has selected for

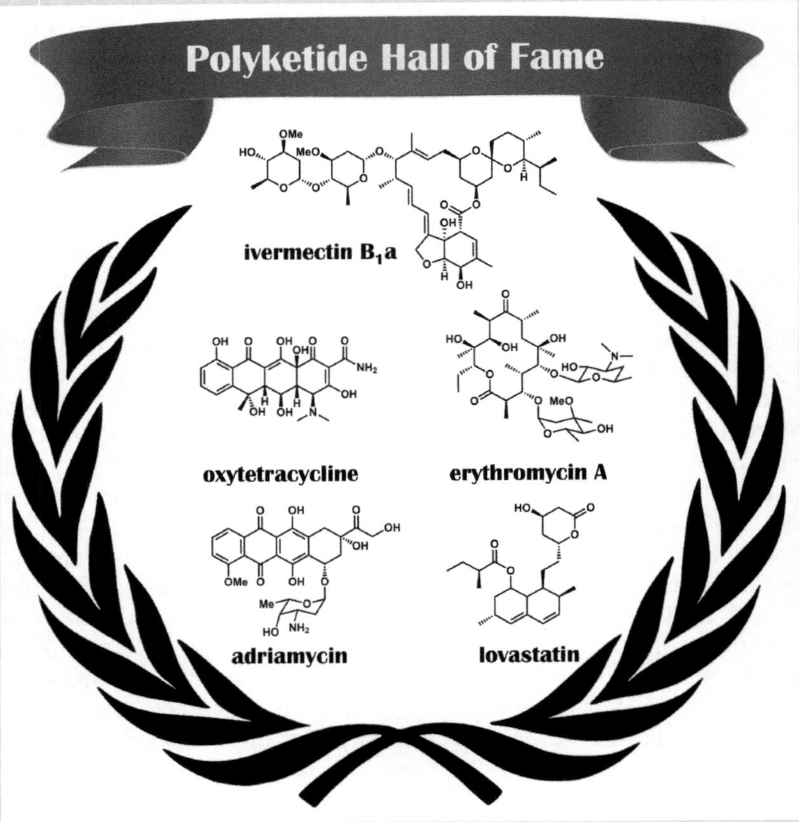

Figure 2.V2 Five polyketide metabolites that changed the world.

tetracycline resistance, leading chemists to counter with second and third generation molecules.

Avermectin is another example of a polyketide macrolactone. It is produced by *Streptomyces avermitilis* and was in a collection of samples, from a Japanese golf course, sent from the Kitasato Institute in Japan to Merck antiparasitic scientists in 1975 (Omura and Crump 2004). In the course of structure–activity work to optimize the properties of avermectin, reduction of the 22, 23-double bond gave a more stable compound and this became ivermectin. The Merck group recognized ivermectin's potential to treat river blindness by killing the causative parasite *Onchocerca volvulus* (Ottesen and Campbell 1994). Later the drug was observed to be effective in lymphatic filariasis (elephantiasis). Merck donates the compound to more than 40 countries and an estimated 60 million people per year are on extended treatment regimens. Avermectin has not been isolated from any other microorganism, suggesting the rarity of this biosynthetic assembly line. Satoshi Ōmura, the head of the Kitasato Institute and William Campbell who headed the parasitology team at Merck shared the 2015 Nobel Prize in Medicine for the discovery and development of ivermectin.

Adriamycin, also known clinically as doxorubicin, is the 14-hydroxy derivative of the natural anthracycline daunomycin that blocks topoisomerase 2 in human cells (Yang, Teves *et al.* 2014). It does not have selectivity for cancer cells over normal cells but relies on the more rapid growth rates of tumor cells for any selective toxicity/therapeutic index. Doxorubicin is proposed to intercalate *via* the planar anthracycline core between DNA bases and also to undergo free radical redox cycling to damage DNA. Doxorubicin is part of chemotherapy regimens in a broad range of tumor types, including in combinations with vincristine and dexamethasone for multiple myeloma.

The anthracyclines are glycosylated aromatics that follow the biosynthetic logic for tetracycline assembly. The acyclic polyketonic thioester intermediates are folded in the assembly line thioesterase domain active site to set up regiospecific cyclizations which account for the observed substituent patterns. Daunomycin was isolated in the 1950s by the Italian company Farmitalia from *Streptomyces peucetius*. Mutagenesis of the strain in an effort to improve fermentation yields also led to a new derivative, the 14-hydroxy daunomycin, which was named adriamycin as homage to the Adriatic Sea. The name was later changed to doxorubicin. Adriamycin has a better therapeutic index than daunomycin, but both have dose-limiting cardiotoxicity which may derive from redox cycling and radical generation in that tissue.

Lovastatin: By the 1970s, inhibition of the rate-limiting enzyme hydroxymethylglutaryl-CoA reductase in cholesterol biosynthesis had become a goal for new strategies to prevent coronary disease. Akira Endo had discovered the microbial metabolite compactin from *Penicillium citrinum* and shown that it could block activity of HMG CoA reductase. In 1978 a group at Merck found the related microbial metabolite from *Aspergillus terreus*, at first named mevinolin, but later renamed lovastatin. It was taken through clinical development, and shown to be highly effective in lowering cholesterol levels *via* blockade of synthesis (Tobert 2003). Lovastatin was approved by the US Food and Drug Administration (FDA) in 1987 and became the first in class of what would become a burgeoning family of semisynthetic statins that have become the most widely prescribed cardiovascular medicines, with global sales of $29 billion in 2014.

2.11 Post-assembly Line Tailoring Enzymes

In the course of examining the chemical logic and enzymatic reactions that constitute the biosynthetic machinery for the several classes of polyketides in this chapter we have noted several examples of tailoring reactions that occur while the growing polyketidyl chains are still tethered as thioesters on ACP domains of the particular PKS. These have been termed "on assembly line" tailorings.

The most notable of the "on assembly line" tailorings are the reactions catalyzed by the KR, DH, and ER domains in any given elongation cycle. The aromatic PKS catalysts lack all three of these and in that sense constitute the simplest pure polyketide synthases. All others carry out differential amounts of tailoring as they engage ketoreductases, dehydratases, and enoyl reductases differentially in any elongation cycle, *e.g.* the DEBS seven module assembly line or the nystatin 19 module assembly line.

A second set of on line tailorings is found in the lovastatin assembly line where the ER domain *in cis* is nonfunctional and the growing chain on LovB interacts with an ER *in trans* which leaves a diene and an isolated olefin to persist and undergo the decalin-forming [4 + 2] reaction at the hexaketidyl stage of chain elongation, noted earlier in this chapter.

We have also noted the different mechanisms of the chain-releasing TE domains, giving linear acids, cyclic lactones or lactams, or reduced aldehydes or alcohols.

In many, if not most, the nascent products of the PKS assemblies are not the metabolic end products. They often do not possess the biologic activities of the mature products. The maturation steps are usually carried out by dedicated ***post-assembly line*** tailoring enzymes typically encoded among the biosynthetic genes for the PKS domains and subunits. This allows coordinated regulation to turn on all the genes needed to make the active end product with just in time inventory control.

The post-assembly line modifications utilize reactions common to primary metabolism. They include acylations, alkylations, glycosylations, prenylations, oxidoreductions, and oxygenations. The maturation of deoxyerythronolide B to the active antibiotic erythromycin A, shown in Figure 2.23, depicts three of these six types of post assembly line modification. Oxygenation occurs at C_6 and C_{12}, glycosylation occurs at C_3 and C_5, and methylation at the 3′OH of the mycarosyl sugar. Analogously, although they are not called out in Figure 2.17, there are post assembly line methylations and oxygenations that occur as the nascent pretetramid (inactive as an antibiotic) is converted to oxytetracycline. The polyene antifungal nystatin, after 91 domains of the 19 module PKS assembly line have had a turn, still requires post assembly line glycosylation to gain activity (Figure 2.35).

We take up the topic of oxygenases in Chapter 9, *S*-adenosylmethionine-dependent methylations in Chapter 10, and glycosylation machinery in Chapter 11, as a measure of their biosynthetically central roles.

Methylation is formally an alkylation with a one carbon unit. Prenylation, as just noted in hyperforin and THC, occurs in units of five carbons or multiples thereof. These are the subject of Chapter 4. Most of the acylation events in both primary and secondary metabolism involve acetylation from acetyl-CoA as the activated acetyl donor. However, there are long chain acylations that we will note in the next chapter for teicoplanin maturation, and in Chapter 4 for the tetra-acylation of taxol.

A notable set of "post-assembly line" tailorings that are consequential are the olefin epoxidations shown for lasalocid and monensin chains after release from the ACP domain thioester linkages (Figures 2.42 and 2.43). The subsequent enzyme-catalyzed epoxide openings generate the characteristic cyclic ether moieties of those ionophores and are predictive precedents for the more complex fused cyclic ether marine toxins.

There are "one off" kinds of post assembly line chemical changes mediated by tailoring enzymes worth attention. The Diels–Alderase activity of SpnF (Figure 2.29) in creating the 5-6-5 tricyclic ring system in the spinosyn core is such a noteworthy example.

As we will note in Chapter 3, the nonribosomal peptide scaffolds are the (only) other major class of natural products built on multimodular enzymatic assembly lines. Post assembly line tailoring enzymes play a similarly crucial maturative role in yielding biologically active end products.

All the other classes of natural products to be considered, isoprenoids (Chapter 4), alkaloids (Chapters 5 and 8), purines and pyrimidines (Chapter 6), and phenylpropanoids (Chapter 7), are not built on comparable enzymatic assembly lines. They all involve soluble pathway intermediates, so the concept of post assembly line enzymatic tailoring is not operant for those classes.

References

Ames, B. D., M. Y. Lee, C. Moody, W. Zhang, Y. Tang and S. C. Tsai (2011). "Structural and biochemical characterization of ZhuI aromatase/cyclase from the R1128 polyketide pathway". *Biochemistry* **50**(39): 8392–8406.

Ames, B. D., C. Nguyen, J. Bruegger, P. Smith, W. Xu, S. Ma, E. Wong, S. Wong, X. Xie, J. W. Li, J. C. Vederas, Y. Tang and S. C. Tsai (2012). "Crystal structure and biochemical studies of the trans-acting polyketide enoyl reductase LovC from lovastatin biosynthesis". *Proc. Natl. Acad. Sci. U. S. A.* **109**(28): 11144–11149.

Antosch, J., F. Schaefers and T. A. Gulder (2014). "Heterologous reconstitution of ikarugamycin biosynthesis in E. coli". *Angew. Chem. Int. Ed.* **53**(11): 3011–3014.

Aroyan, C. E., A. Dermenci and S. J. Miller (2010). "Development of a cysteine-catalyzed enantioselective Rauhut-Currier reaction". *J. Org. Chem.* **75**(17): 5784–5796.

Baden, D. G. (1989). "Brevetoxins: unique polyether dinoflagellate toxins". *FASEB J.* **3**(7): 1807–1817.

Baldwin, J. E. (1976). "Rues for Ring Closure". *J. Chem. Soc., Chem. Commun.* .

Baldwin, J. E., R. Thomas, L. Jruse and L. Silberman (1977). "Rules for Ring Closure: Ring Formation by Conjugate Addition of Oxygen Nucleophiles". *J. Org. Chem.* **42**: 3846.

Beld, J., E. C. Sonnenschein, C. R. Vickery, J. P. Noel and M. D. Burkart (2014). "The phosphopantetheinyl transferases: catalysis of a post-translational modification crucial for life". *Nat. Prod. Rep.* **31**(1): 61–108.

Brautaset, T., O. N. Sekurova, H. Sletta, T. E. Ellingsen, A. R. StrLm, S. Valla and S. B. Zotchev (2000). "Biosynthesis of the polyene antifungal antibiotic nystatin in Streptomyces noursei ATCC 11455: analysis of the gene cluster and deduction of the biosynthetic pathway". *Chem. Biol.* **7**(6): 395–403.

Broussard, T. C., A. E. Price, S. M. Laborde and G. L. Waldrop (2013). "Complex formation and regulation of Escherichia coli acetyl-CoA carboxylase". *Biochemistry* **52**(19): 3346–3357.

Byrne, M. J., N. R. Lees, L. C. Han, M. W. van der Kamp, A. J. Mulholland, J. E. Stach, C. L. Willis and P. R. Race (2016). "The Catalytic Mechanism of a Natural Diels-Alderase Revealed in Molecular Detail". *J. Am. Chem. Soc.* **138**(19): 6095–6098.

Caffrey, P., S. Lynch, E. Flood, S. Finnan and M. Oliynyk (2001). "Amphotericin biosynthesis in Streptomyces nodosus: deductions from analysis of polyketide synthase and late genes". *Chem. Biol.* **8**(7): 713–723.

Caldara-Festin, G., D. R. Jackson, J. F. Barajas, T. R. Valentic, A. B. Patel, S. Aguilar, M. Nguyen, M. Vo, A. Khanna, E. Sasaki, H. W. Liu and S. C. Tsai (2015). "Structural and functional analysis

of two di-domain aromatase/cyclases from type II polyketide synthases". *Proc. Natl. Acad. Sci. U. S. A.* **112**(50): E6844–6851.
Campbell, C. D. and J. C. Vederas (2010). "Biosynthesis of lovastatin and related metabolites formed by fungal iterative PKS enzymes". *Biopolymers* **93**(9): 755–763.
Cane, D. E. (2010). "Programming of erythromycin biosynthesis by a modular polyketide synthase". *J. Biol. Chem.* **285**(36): 27517–27523.
Cane, D. E., W. D. Celmer and J. W. Westley (1983). "Unified stereochemical model of polyether antibiotic structure and biogenesis". *J. Am. Chem. Soc.* **105**: 3594–3600.
Clardy, J., M. A. Fischbach and C. T. Walsh (2006). "New antibiotics from bacterial natural products". *Nat. Biotechnol.* **24**(12): 1541–1550.
Crawford, J. M., T. P. Korman, J. W. Labonte, A. L. Vagstad, E. A. Hill, O. Kamari-Bidkorpeh, S. C. Tsai and C. A. Townsend (2009). "Structural basis for biosynthetic programming of fungal aromatic polyketide cyclization". *Nature* **461**(7267): 1139–1143.
Crawford, J. M. and C. A. Townsend (2010). "New insights into the formation of fungal aromatic polyketides". *Nat. Rev. Microbiol.* **8**(12): 879–889.
Dewick, P. (2009). *Medicinal Natural Products, a Biosynthetic Approach.* UK, Wiley.
Donadio, S., M. J. Staver, J. B. McAlpine, S. J. Swanson and L. Katz (1991). "Modular organization of genes required for complex polyketide biosynthesis". *Science* **252**(5006): 675–679.
Fage, C. D., E. A. Isiorho, Y. Liu, D. T. Wagner, H. W. Liu and A. T. Keatinge-Clay (2015). "The structure of SpnF, a standalone enzyme that catalyzes [4 + 2] cycloaddition". *Nat. Chem. Biol.* **11**(4): 256–258.
Feredenhagn, A., S. Tamura, P. Kenny, H. Komura, Y. Naya, K. Nakanishi, K. Nishiyama, M. Sugiura and H. Kita (1987). "Andrimid, a new peptide antibiotic produced by an intracellular bacterial symbiont isolated from a brown planthopper". *J. Am. Chem. Soc.* **109**: 4409–4411.
Gallimore, A. R. (2009). "The biosynthesis of polyketide-derived polycyclic ethers". *Nat. Prod. Rep.* **26**(2): 266–280.
Gemperlein, K. R. S., R. Garcia, S. Wenzel and R. Muller (2014). "Polyunsaturated fatty acid biosynthesis in myxobacteria: different PUFA synthases and their product diversity". *Chem. Sci.* **5**: 1733–1741.

Haydock, S. F., J. A. Dowson, N. Dhillon, G. A. Roberts, J. Cortes and P. F. Leadlay (1991). "Cloning and sequence analysis of genes involved in erythromycin biosynthesis in Saccharopolyspora erythraea: sequence similarities between EryG and a family of S-adenosylmethionine-dependent methyltransferases". *Mol. Gen. Genet.* **230**(1–2): 120–128.

Heberling, M. M., M. F. Masman, S. Bartsch, G. G. Wybenga, B. W. Dijkstra, S. J. Marrink and D. B. Janssen (2015). "Ironing out their differences: dissecting the structural determinants of a phenylalanine aminomutase and ammonia lyase". *ACS Chem. Biol.* **10**(4): 989–997.

Hopwood, D. A. and D. H. Sherman (1990). "Molecular genetics of polyketides and its comparison to fatty acid biosynthesis". *Annu. Rev. Genet.* **24**: 37–66.

Hotta, K., X. Chen, R. S. Paton, A. Minami, H. Li, K. Swaminathan, Mathews II, H. Watanabe, K. N. Oikawa, Houk and C. Y. Kim (2012). "Enzymatic catalysis of anti-Baldwin ring closure in polyether biosynthesis". *Nature* **483**(7389): 355–358.

Jiang, H., R. Zirkle, J. G. Metz, L. Braun, L. Richter, S. G. Van Lanen and B. Shen (2008). "The role of tandem acyl carrier protein domains in polyunsaturated fatty acid biosynthesis". *J. Am. Chem. Soc.* **130**(20): 6336–6337.

Khosla, C., R. S. Gokhale, J. R. Jacobsen and D. E. Cane (1999). "Tolerance and specificity of polyketide synthases". *Annu. Rev. Biochem.* **68**: 219–253.

Kim, H. J., M. W. Ruszczycky, S. H. Choi, Y. N. Liu and H. W. Liu (2011). "Enzyme-catalysed [4 + 2] cycloaddition is a key step in the biosynthesis of spinosyn A". *Nature* **473**(7345): 109–112.

Korman, T. P., J. M. Crawford, J. W. Labonte, A. G. Newman, J. Wong, C. A. Townsend and S. C. Tsai (2010). "Structure and function of an iterative polyketide synthase thioesterase domain catalyzing Claisen cyclization in aflatoxin biosynthesis". *Proc. Natl. Acad. Sci. U. S. A.* **107**(14): 6246–6251.

Lambalot, R. H., A. M. Gehring, R. S. Flugel, P. Zuber, M. LaCelle, M. A. Marahiel, R. Reid, C. Khosla and C. T. Walsh (1996). "A new enzyme superfamily - the phosphopantetheinyl transferases". *Chem. Biol.* **3**(11): 923–936.

Lee, M. Y., B. D. Ames and S. C. Tsai (2012). "Insight into the molecular basis of aromatic polyketide cyclization: crystal structure and in vitro characterization of WhiE-ORFVI". *Biochemistry* **51**(14): 3079–3091.

Liu, T., D. E. Cane and Z. Deng (2009). "The enzymology of polyether biosynthesis". *Methods Enzymol* **459**: 187–214.
Liu, W., S. D. Christenson, S. Standage and B. Shen (2002). "Biosynthesis of the enediyne antitumor antibiotic C-1027". *Science* **297**(5584): 1170–1173.
Liu, X., P. D. Fortin and C. T. Walsh (2008). "Andrimid producers encode an acetyl-CoA carboxyltransferase subunit resistant to the action of the antibiotic". *Proc. Natl. Acad. Sci. U. S. A.* **105**(36): 13321–13326.
Luhavaya, H., M. V. Dias, S. R. Williams, H. Hong, L. G. de Oliveira and P. F. Leadlay (2015). "Enzymology of Pyran Ring A Formation in Salinomycin Biosynthesis". *Angew Chem., Int. Ed.* **54**(46): 13622–13625.
Majerus, P. W., A. W. Alberts and P. R. Vagelos (1965). "Acyl Carrier Protein. 3. An Enoyl Hydrase Specific for Acyl Carrier Protein Thioesters". *J. Biol. Chem.* **240**: 618–621.
Majerus, P. W., A. W. Alberts and P. R. Vagelos (1965). "Acyl Carrier Protein. Iv. The Identification of 4'-Phosphopantetheine as the Prosthetic Group of the Acyl Carrier Protein". *Proc. Natl. Acad. Sci. U. S. A.* **53**: 410–417.
Mander L. and H. W. Liu, eds. (2010). *Comprehensive Natural Products II: Chemistry and Biology*. Elsevier Science.
Migita, A., M. Watanabe, Y. Hirose, K. Watanabe, T. Tokiwano, H. Kinashi and H. Oikawa (2009). "Identification of a gene cluster of polyether antibiotic lasalocid from Streptomyces lasaliensis". *Biosci. Biotechnol. Biochem.* 73(1): 169–176.
Minami, A., M. Shimaya, G. Suzuki, A. Migita, S. S. Shinde, K. Sato, K. Watanabe, T. Tamura, H. Oguri and H. Oikawa (2012). "Sequential enzymatic epoxidation involved in polyether lasalocid biosynthesis". *J. Am. Chem. Soc.* **134**(17): 7246–7249.
Nicholson, G. and R. Lewis (2006). "Ciguatoxins: Cyclic Polyether Modulators of Voltage-gated Ion Channel Function". *Mar. Drugs* **4**: 82–118.
Okuyama, H., Y. Orikasa, T. Nishida, K. Watanabe and N. Morita (2007). "Bacterial genes responsible for the biosynthesis of eicosapentaenoic and docosahexaenoic acids and their heterologous expression". *Appl. Environ. Microbiol.* **73**(3): 665–670.
Oliynyk, M., C. B. Stark, A. Bhatt, M. A. Jones, Z. A. Hughes-Thomas, C. Wilkinson, Z. Oliynyk, Y. Demydchuk, J. Staunton and P. F. Leadlay (2003). "Analysis of the biosynthetic gene cluster for the polyether antibiotic monensin in Streptomyces cinnamonensis

and evidence for the role of monB and monC genes in oxidative cyclization''. *Mol. Microbiol.* **49**(5): 1179–1190.

Omura, S. and A. Crump (2004). "The life and times of ivermectin - a success story''. *Nat. Rev. Microbiol.* **2**(12): 984–989.

Ottesen, E. A. and W. C. Campbell (1994). ''Ivermectin in human medicine''. *J. Antimicrob. Chemother.* **34**(2): 195–203.

Patel, A., Z. Chen, Z. Yang, O. Gutierrez, H. W. Liu, K. N. Houk and D. A. Singleton (2016). "Dynamically Complex [6 + 4] and [4 + 2] Cycloadditions in the Biosynthesis of Spinosyn A''. *J. Am. Chem. Soc.* **138**(11): 3631–3634.

Pickens, L. B. and Y. Tang (2010). "Oxytetracycline biosynthesis''. *J. Biol. Chem.* **285**(36): 27509–27515.

Sato, M., F. Yagishita, T. Mino, N. Uchiyama, A. Patel, Y. H. Chooi, Y. Goda, W. Xu, H. Noguchi, T. Yamamoto, K. Hotta, K. N. Houk, Y. Tang and K. Watanabe (2015). "Involvement of Lipocalin-like CghA in Decalin-Forming Stereoselective Intramolecular [4 + 2] Cycloaddition''. *ChemBioChem* **16**(16): 2294–2298.

Schweizer, E. and J. Hofmann (2004). "Microbial type I fatty acid synthases (FAS): major players in a network of cellular FAS systems''. *Microbiol. Mol. Biol. Rev.* **68**(3): 501–517.

Staunton, J. and K. J. Weissman (2001). "Polyketide biosynthesis: a millennium review''. *Nat. Prod. Rep.* **18**(4): 380–416.

Stocking, E. M. and R. M. Williams (2003). "Chemistry and biology of biosynthetic Diels-Alder reactions''. *Angew. Chem., Int. Ed.* **42**(27): 3078–3115.

Tian, Z., P. Sun, Y. Yan, Z. Wu, Q. Zheng, S. Zhou, H. Zhang, F. Yu, X. Jia, D. Chen, A. Mandi, T. Kurtan and W. Liu (2015). "An enzymatic [4 + 2] cyclization cascade creates the pentacyclic core of pyrroindomycins''. *Nat. Chem. Biol.* **11**(4): 259–265.

Tobert, J. A. (2003). ''Lovastatin and beyond: the history of the HMG-CoA reductase inhibitors''. *Nat. Rev. Drug Discovery* **2**(7): 517–526.

Tsai, S. and B. Ames (2009). ''Structural Enzymology of Polyketide Synthases''. *Methods Enzymol.* **459**: 17–47.

Turner, A. D., C. Higgins, K. Davidson, A. Veszelovszki, D. Payne, J. Hungerford and W. Higman (2015). "Potential threats posed by new or emerging marine biotoxins in UK waters and examination of detection methodology used in their control: brevetoxins''. *Mar. Drugs* **13**(3): 1224–1254.

Vagelos, P. R., P. W. Majerus, A. W. Alberts, A. R. Larrabee and G. P. Ailhaud (1966). "Structure and function of the acyl carrier protein''. *Fed. Proc.* **25**(5): 1485–1494.

Van Lanen, S. and B. Shen (2008). "Biosynthesis of enediyne antibioitcs". *Curr. Top. Med. Chem.* **8**: 448–459.
Van Lanen, S. G. and B. Shen (2008). "Biosynthesis of enediyne antitumor antibiotics". *Curr. Top. Med. Chem.* **8**(6): 448–459.
Van Wagoner, R. M., M. Satake and J. L. Wright (2014). "Polyketide biosynthesis in dinoflagellates: what makes it different?". *Nat. Prod. Rep.* **31**(9): 1101–1137.
Vilotijevic, I. and T. F. Jamison (2009). "Epoxide-opening cascades in the synthesis of polycyclic polyether natural products". *Angew. Chem., Int. Ed.* **48**(29): 5250–5281.
Walsh, C. T. and T. Wencewicz (2016). *Antibiotics Challeneges, Mechanisms, Opportunities*. Washington DC, ASM Press.
Walsh, C. T. and T. Wencewicz (2016). *Antibiotics Challeneges, Mechanisms, Opportunities*. Washington DC, ASM Press, ch. 16.
Walsh, C. T. and T. A. Wencewicz (2013). "Flavoenzymes: versatile catalysts in biosynthetic pathways". *Nat. Prod. Rep.* **30**(1): 175–200.
Weissman, K. J. (2015). "The structural biology of biosynthetic megaenzymes". *Nat. Chem. Biol.* **11**(9): 660–670.
Weissman, K. J. and R. Muller (2008). "Protein-protein interactions in multienzyme megasynthetases". *ChemBioChem* **9**(6): 826–848.
White, S. W., J. Zheng, Y. M. Zhang and Rock (2005). "The structural biology of type II fatty acid biosynthesis". *Annu. Rev. Biochem.* **74**: 791–831.
Witter, D. J. and J. C. Vederas (1996). "Putative Diels-Alder-Catalyzed Cyclization during the Biosynthesis of Lovastatin". *J. Org. Chem.* **61**(8): 2613–2623.
Xie, X., K. Watanabe, W. A. Wojcicki, C. C. Wang and Y. Tang (2006). "Biosynthesis of lovastatin analogs with a broadly specific acyltransferase". *Chem. Biol.* **13**(11): 1161–1169.
Xu, W., Y. H. Chooi, J. W. Choi, S. Li, J. C. Vederas, N. A. Da Silva and Y. Tang (2013). "LovG: the thioesterase required for dihydromonacolin L release and lovastatin nonaketide synthase turnover in lovastatin biosynthesis". *Angew. Chem., Int. Ed.* **52**(25): 6472–6475.
Yang, F., S. S. Teves, C. J. Kemp and S. Henikoff (2014). "Doxorubicin, DNA torsion, and chromatin dynamics". *Biochim. Biophys. Acta* **1845**(1): 84–89.
Zhan, J. (2009). "Biosynthesis of bacterial aromatic polyketides". *Curr. Top. Med. Chem.* **9**(17): 1958–1610.
Zhang, G., W. Zhang, Q. Zhang, T. Shi, L. Ma, Y. Zhu, S. Li, H. Zhang, Y. L. Zhao, R. Shi and C. Zhang (2014). "Mechanistic insights into

polycycle formation by reductive cyclization in ikarugamycin biosynthesis''. *Angew. Chem., Int. Ed.* **53**(19): 4840–4844.
Zhang, J., S. G. Van Lanen, J. Ju, W. Liu, P. C. Dorrestein, W. Li, N. L. Kelleher and B. Shen (2008). ''A phosphopantetheinylating polyketide synthase producing a linear polyene to initiate enediyne antitumor antibiotic biosynthesis''. *Proc. Natl. Acad. Sci. U. S. A.* **105**(5): 1460–1465.
Zhang, Q., B. Pang, W. Ding and W. Liu (2013). ''Aromatic Polyketides Produced by Bacterial Iterative Type I Polyketide Synthases''. *ACS Catal.* **3**: 1439–1447.
Zhou, H., Y. Li and Y. Tang (2010). ''Cyclization of aromatic polyketides from bacteria and fungi''. *Nat. Prod. Rep.* **27**(6): 839–868.

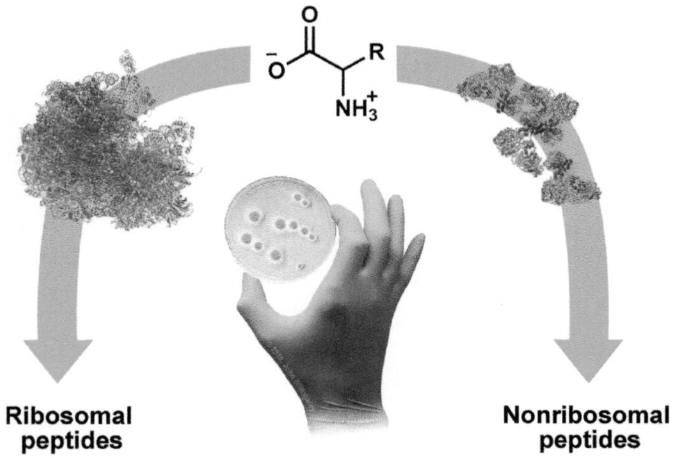

Peptide-derived natural products can be synthesized from both ribosomal and nonribosomal pathways (PDB IDs: 4ZXH and 4V88).
Copyright (2016) John Billingsley.

3 Peptide Derived Natural Products

3.1 Introduction

Both prokaryotic and eukaryotic organisms biosynthesize small peptides that have diverse biological activities, from signaling molecules to antibiotics, from immunosuppressives to agents of innate immunity. Some of the peptides are proteolytic fragments of protein precursors. Others, almost exclusively from bacterial and fungal sources, are products of nonribosomal peptide synthetase assembly line machinery, mimicking the logic of polyketide synthase assembly lines detailed in Chapter 2 (Figure 3.1). In this figure, SapB, a developmental morphogen in *Streptomyces* bacteria, and the antibiotic thiocillin (both shown in right of the figure) are products of post-translational modification of precursor protein; while daptomycin, colistin, vancomycin, and isopenicillin N derive from nonribosomal peptide synthetase assembly line enzymes (see Chapter 15 in Walsh and Wencewicz 2016). Clearly, both strategies can generate architectural and functional group complexity (Nolan and Walsh 2009).

A major set of biosynthetic strategies for extending the lifetime of peptide scaffolds involves cyclization of the typical acyclic, linear, conformationally flexible and floppy peptide chains to compact cyclic structures as illustrated by the five peptide natural products that have undergone different modes of macrocyclization (Walsh 2004) (Figure 3.2). We will note a variety of cyclization modes and enzymatic machinery but the typical consequence is reduction in protease

Natural Product Biosynthesis: Chemical Logic and Enzymatic Machinery
By Christopher T. Walsh and Yi Tang

Published by the Royal Society of Chemistry, www.rsc.org

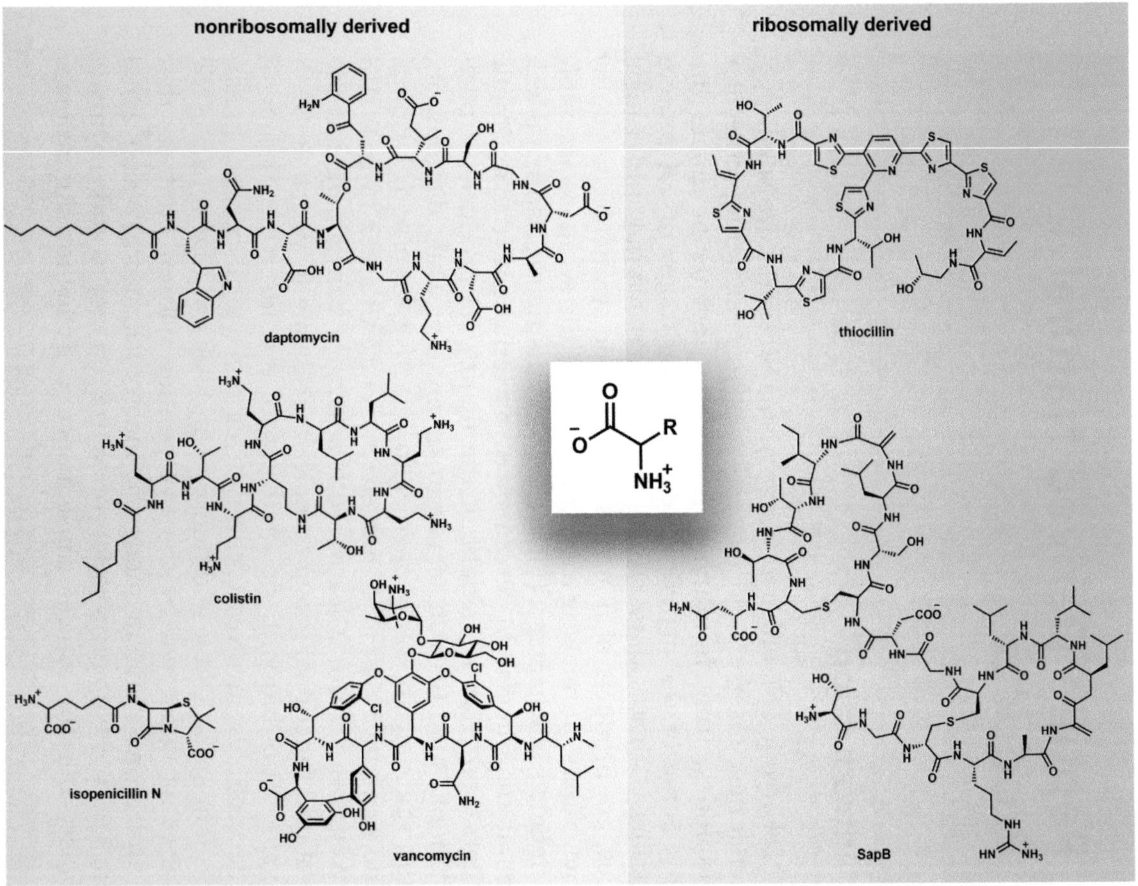

Figure 3.1 Amino acid building blocks in peptide linkages. The scaffolds in red (SapB, thiocillin) arise from posttranslational modification of protein precursors. The structures in blue are produced on nonribosomal peptide synthetase assembly lines (colistin, vancomycin, isopenicillin N and daptomycin).

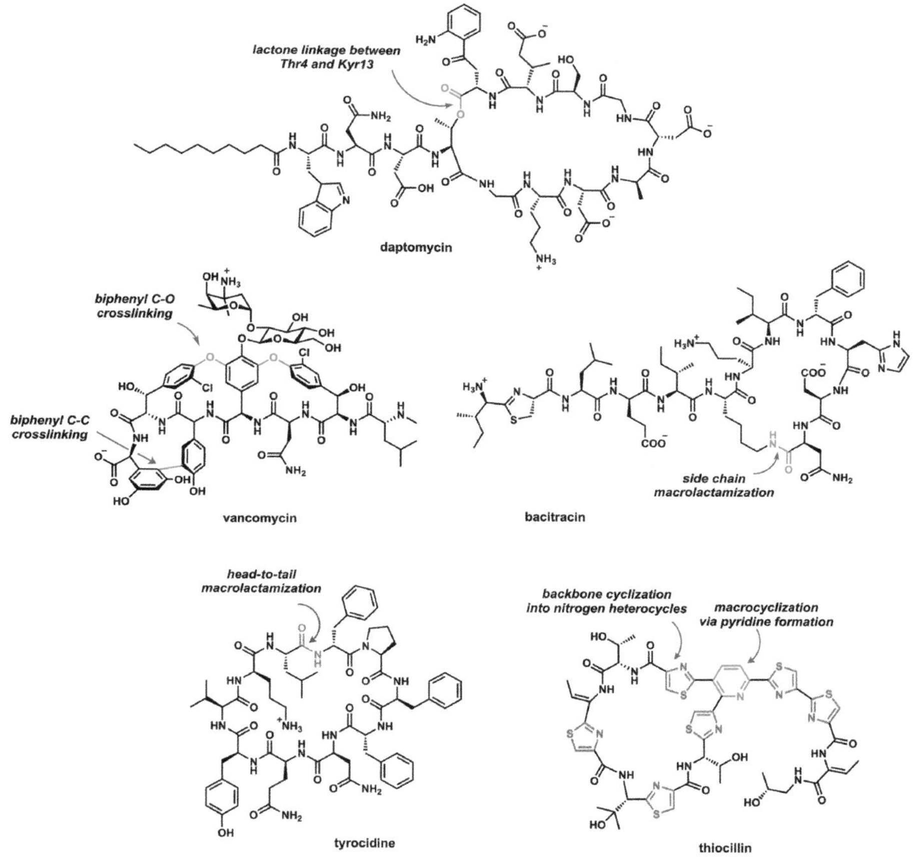

Figure 3.2 Heterocyclic and macrocyclic peptide scaffolds can arise from either ribosomal proteins or nonribosomal peptides. Multiple modes of macrocyclization can impose peptide scaffold rigidity.

hydrolytic lability, and gain in affinity for biological targets, including proteins, nucleic acids, and lipids. Tyrocidine has undergone a head to tail macrolactamization as the free amine of Phe_1 has captured the carbonyl of Val_{10}. The antibiotic daptomycin has a macrolactone substructure between the side chain -OH of Thr_4 and the carbonyl group of $kyneurinine_{13}$. The heptapeptide scaffold of the antibiotic vancomycin has instead been crosslinked by two aryl ethers and one direct C–C bond, dramatically reducing conformational flexibility and imposing a cup-shaped architecture on the aglycone portion, required for high affinity interaction with its bacterial cell wall target. Thiocillins have a central trithiazolyl pyridine core. The last step in biosynthesis is creation of the pyridine ring, which also creates the macrocycle. The fifth example is the topical antibiotic bacitracin which has engaged in side chain macrolactamization, *via* attack of N_6 of Lys_6 on the carbonyl of Asn_{12}.

A complementary strategy involves heterocyclization of Ser, Thr, or Cys side chains to oxazoles, methyloxazoles, and thiazoles, respectively, which converts hydrolytically labile peptide bonds into hydrolytically stable five-membered ring heterocycles (Figure 3.3). These five-membered ring heterocycles can arise from protein post-translational modification (Schmidt, Nelson *et al.* 2005) or from NRPS machinery, as exemplified by the siderophores vibriobactin (*Vibrio cholerae*) and yersiniabactn (*Yersinia pestis*) (Crosa and Walsh 2002). The mechanisms of these reactions will be discussed in detail in Figure 3.10.

A third strategy available to nonribosomal peptide synthetase assembly lines but not to ribosomal factories for protein synthesis is the selection, activation, and incorporation of hundreds of non-proteinogenic amino acid monomers (Walsh, O'Brien *et al.* 2013). We will examine all three of these strategies in detail.

3.2 Ribosomal *vs.* Nonribosomal Amino Acid Oligomerization Characteristics

Ribosomes are effectively RNA machines, polymerizing amino acid monomers at rates of from 10 to 30 peptide bonds per second. The order of amino acids selected and incorporated into elongating chains of peptide bonds is dictated by a particular messenger RNA (mRNA) serving as template. There are 20 canonical amino acids that are proteinogenic, although in some organisms selenocysteine and pyrrolysine can constitute 21st and 22nd building blocks. The amino acids are selected by cognate aminoacyl transfer (t)RNA synthetase enzymes and activated as mixed carboxylic–phosphoric anhydrides in the form of aminoacyl-AMPs. The thermodynamically activated aminoacyl moieties are transferred to cognate tRNAs in the active site of the aminoacyl-tRNA synthetases. These loaded aminoacyl-tRNAs (oxoesters) are kinetically stable enough to be chaperoned to the ribosome yet thermodynamically activated enough to form chain-elongating peptide bonds in the peptidyl transferase centers of the 50S ribosomal subunit.

By contrast, the nonribosomal peptide synthetase assembly lines are independent of any mRNA or other RNA template (Figure 3.4). The specificity of the adenylation domains in each NRPS module

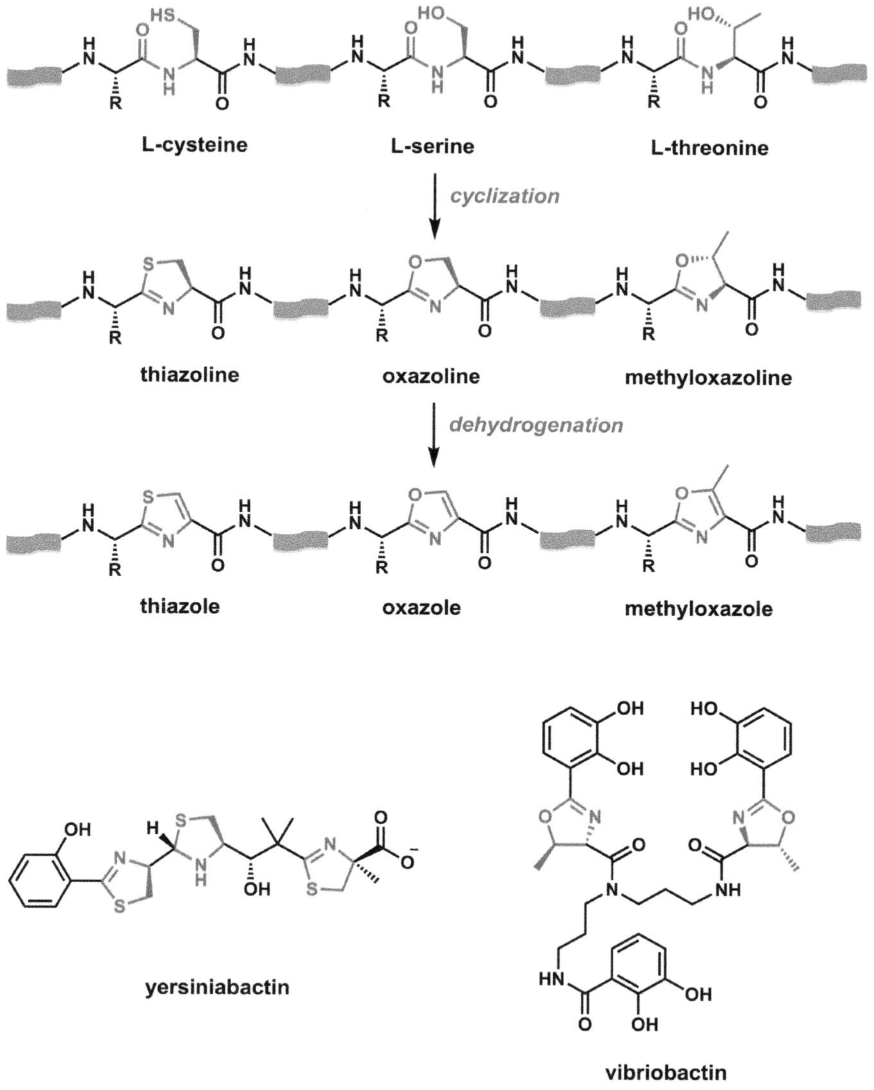

Figure 3.3 The side chains of Ser, Thr, and Cys residues can be heterocyclized to oxazolines and thiazolines and oxidatively aromatized to oxazoles and thiazoles. These heterocycles can be formed by both ribosomal and nonribosomal routes.

dictates what amino acid will be selected and activated. The same logic of cleaving an ATP molecule into PPi as each amino acid is activated as an aminoacyl-AMP is utilized by the adenylation domains, which are thought to represent convergent evolution with respect to

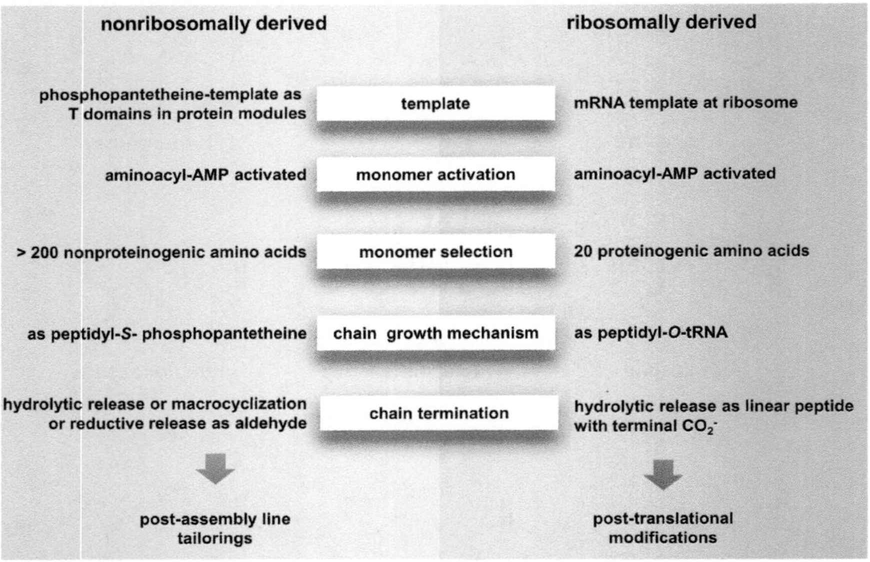

Figure 3.4 Characteristics of ribosomal *vs.* nonribosomal pathways, exemplified by peptide antibiotics.

aminoacyl-tRNA synthetases. The amino acid units are then transferred to the thiol groups at the end of phosphopantetheinyl arms that have been installed on each carrier protein domain in each NRPS module. Thus, two distinctions apply. The aminoacyl groups to be incorporated into growing peptidyl chains are thioesters not oxoesters. Also, they are covalently tethered, one to each peptidyl carrier protein. As the chain grows the elongating peptidyl chain is transferred to the next downstream PCP domain where the incoming aminoacyl unit acts as nucleophile. The analogy to macroscopic assembly lines comes from the concept of each PCP domain filled by an elongating peptidyl thioester chain. The further downstream the PCP domain in the NRPS assembly line, the more "finished" the peptidyl chain. When a full length peptidyl chain reaches the most downstream module, it undergoes catalyzed disconnection from its thioester linkage. We will note a variety of intramolecular (such as macrocyclization) and intermolecular (such as hydrolysis) routes that break the covalent thioester attachment to release either cyclic or linear peptide products.

Because the NRPS assembly lines do not use aminoacyl-tRNA synthetases and their aminoacyl-specifying adenylation domains have evolved convergently, NRPS enzymes can select up to 200 different, nonproteinogenic amino acids as well as the 20 proteinogenic amino acids (Walsh, O'Brien *et al.* 2013). These include β-alanine, γ-aminobutyrate, and notably the δ-aminoadipate, as will be noted in the section on aminoadipyl-cysteinyl-valine (ACV) synthetase. Finally, although ribosomes release the nascent peptide chains with a free carboxylate at the C-terminal residue, we shall note several chemically distinct routes of release of the full length peptidyl chains from NRPS assembly lines. Hydrolysis is one such route but direct macrocyclization by internal capture of a nucleophilic amine or hydroxyl group is the route that leads directly to cyclic scaffolds essential for biological activity. We shall note that biosynthetic gene clusters for nonribosomal products, like those for polyketides (Chapter 2), typically include genes encoding dedicated tailoring enzymes for maturation of the nascent NRP scaffolds. They may also contain genes encoding non-proteinogenic amino acid building blocks (Hubbard and Walsh 2003).

We take up examples of posttranslational morphing of ribosomally generated proteins into stable small molecule ligand frameworks before turning to the nonribosomal peptide assembly logic.

3.3 Posttranslational Modifications That Convert Nascent Proteins into Morphed, Compact Scaffolds: RIPPs

A variety of ribosomally produced proteins turn out to be processed by dedicated enzymatic machinery for sets of particular posttranslational modifications, followed by another set of specific proteolytic cleavages and export *via* dedicated protein pumps from the producing cells into extracellular spaces and/or specialized compartments for secretion (Arnison, Bibb *et al.* 2013). The mature product can be highly morphed in sidechain, backbone, and/or end to end connectivity to behave more like a stable small molecule scaffold than a protease sensitive peptide scaffold. Most of these are microbial protein morphings but there is also a class of cone snail toxins that fit into this category.

A recent review has codified what had been viewed as diverse and idiosyncratic processing of nascent proteins into a class of natural products termed RIPPs (*ri*bosomal *p*osttranslational *p*eptides) (Arnison, Bibb *et al.* 2013). As denoted in Figure 3.5, several segments of the precursor peptide serve distinct features for the posttranslational enzymatic machinery. Essentially all the proteins to be morphed in the six examples that follow have N-terminal signal sequences that in turn are upstream of leader peptides that often serve a recognition function

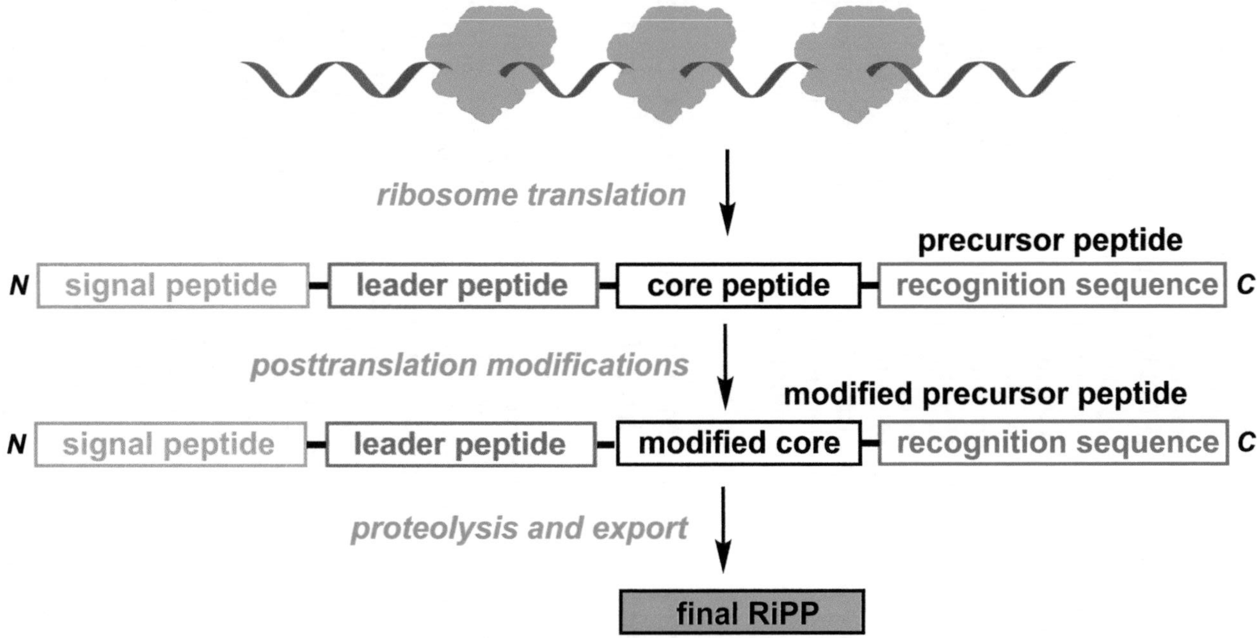

Figure 3.5 Schematic of the maturation of ribosomal posttranslational protein (RIPP) with signal, leader, and recognition peptide elements around a core region to be modified.
Data from van der Donk, W. A. *et al.* (2013). "Ribosomally synthesized and post-translationally modified peptide natural products: overview and recommendations for a universal nomenclature". *Nat. Prod. Rep.* **30**: 108.

for the PTM enzymes that will modify the core peptide region. In some cases, the recognition element for the tailoring enzymes may be downstream rather than upstream of the core sequence, which may be ~5–20 residues. The tailoring enzymes typically contain proteases or protease domains that make cuts to excise and sometimes circularize the modified core that is being cut out hydrolytically.

Skinnider *et al.* (Skinnider, Johnston *et al.* 2016) have recently reported bioinformatic analysis on 65 000 prokaryotic genomes and predict some 30 000 RIPP peptide gene clusters containing the genes for precursor proteins and the tailoring enzymes to morph nascent proteins into stable peptide products. They posit that 2200 unique mature peptide scaffolds are likely encoded in those 30 000 gene clusters, which would greatly expand this class of microbial natural products. This argues that genome mining in the RIPP class will be productive for novel scaffolds and perhaps novel biologic activities, over and above the categories taken up in the next sections.

3.3.1 Lantipeptides

The most extensively studied of RIPP classes are the lantipeptides, named for the bifunctional, thioether-containing amino acid lanthionine and its congener methyl lanthionine. The thioether arises during posttranslational processing from a Ser or Thr residue which undergoes net α,β-dehydration to dehydro-Ala (Dha)/dehydrobutyrine (Dhb) resides. The olefinic terminus is electron deficient and can be attacked by the side chain thiolate of a Cys residue under enzymatic catalysis to yield the thioether bridged residues (Figure 3.6). Lantipeptides, such as nisin, were originally purified on the basis of their antibiotic activity, but SapB (see Figure 3.1) is a developmental signaling molecule in *Streptomyces*. The five thioether bridges in nisin, introduced with particular topology and directionality, constrain nisin to a folded conformation that has nanomolar affinity for lipid II, the limiting carrier molecule in bacterial cell wall peptidoglycan biosynthesis (Hsu, Breukink *et al.* 2004). Bacterial genomics indicates hundreds of lantipeptides sequenced to date, as detected by AntiSMASH algorithms versions 2 and 3, suggesting a broad involvement of this class of RIPP in bacterial physiology.

3.3.2 Cyanobactins

The cyanobactins are defined as small ribosomally derived cyclic peptides from species of cyanobacteria (McIntosh, Donia *et al.*

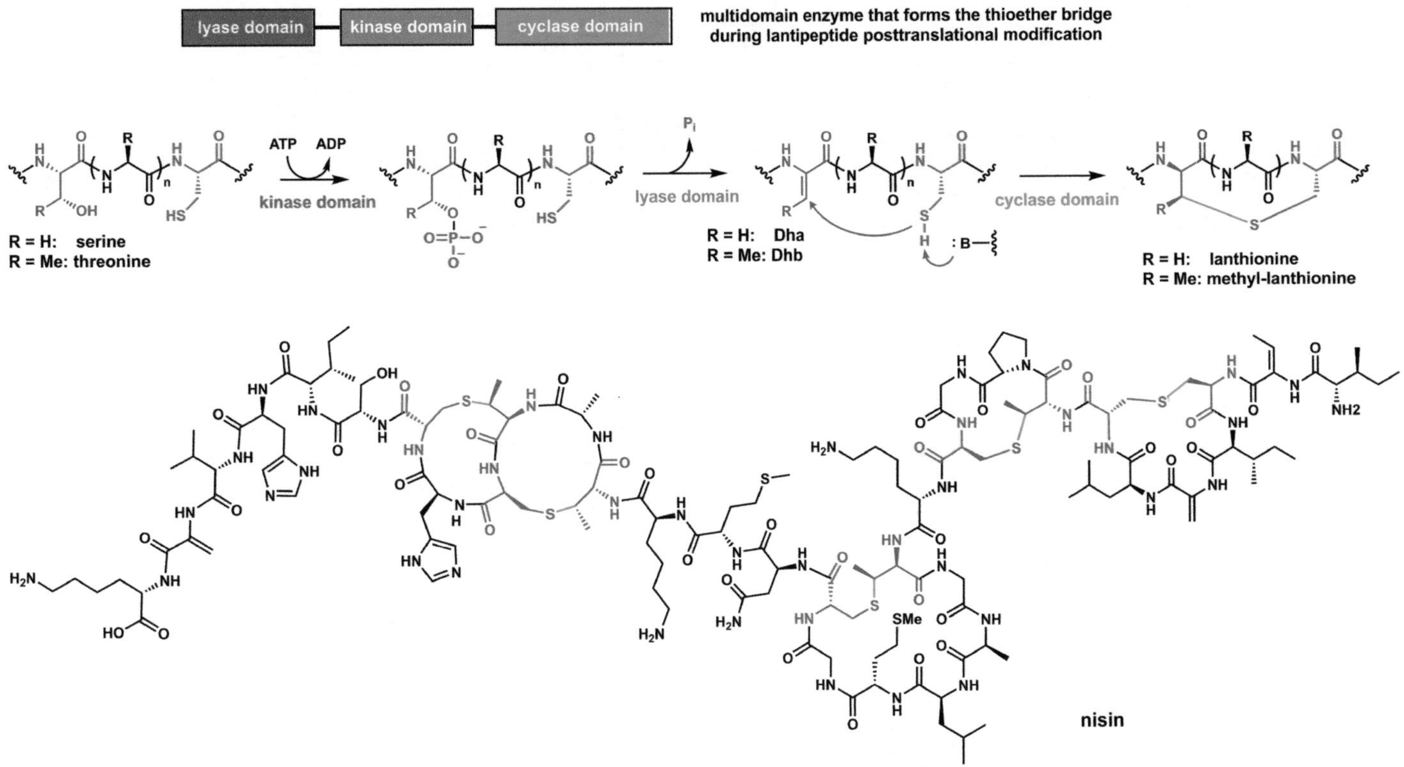

Figure 3.6 Lantipeptides undergo posttranslational addition of the thiolate of Cys residues into dehydro-Ala (Dha) or dehydrobutyrine (Dhb) residues arising from prior conversion of Ser and Thr to the dehydro-amino acid residues, respectively. The resulting bifunctional amino acids are lanthionines (from Ser) and methyllanthionines (from Thr), respectively, and contain a stable thioether linkage.

2009, Jaspars 2014, Martins and Vasconcelos 2015). They are post-translationally excised from precursor proteins after undergoing heterocyclizations, oxidations, and/or prenylations (see Chapter 4) by tailoring enzymes. Among the first set of cyanobactins to be characterized chemically and then biosynthetically are the patellamides, A and C (Figure 3.7). They are produced by *Prochloron didemni*, bacterial symbionts in marine sponge colonies (Schmidt, Nelson *et al.* 2005).

The *PatE* gene encodes a 71-residue protein that contains the eight-residue sequence VTACITFC (in black in Figure 3.7) as well as the eight-residue sequence ITVCISVC. The first is precursor to patellamide C, the second to patellamide A, demonstrating that one protein precursor gives rise to both patellamides in a combinatorial biosynthetic process. Notably, the four cysteines (two in each patellamide) have been morphed into five-membered ring heteroaromatic thiazoles, while the three Thr residues and the one Ser residue are at the dihydroaromatic oxidation state of methyloxazoline and oxazolines, respectively, as though they have been halted halfway along to the oxazoles. In each product four of the eight residues (Ser, Thr, Cys) have become heterocycles, alternating with the unchanged side chains in the PatE precursor protein.

The second notable feature in addition to the heterocycles themselves is that they are embedded within a macrocyclic lactam ring in both patellamide A and C. The sequence of events turns out to be heterocyclization first and then proteolytic excision and cyclization, as diagrammed in Figure 3.7, mediated by a protease PatG encoded in the biosynthetic cluster. Many other cyanobactins (a few are depicted in Figure 3.8) are known, shown with characteristic heterocyclic and macrocyclic features. Some are prenylated on Ser, Thr, and Tyr residues as another posttranslational modification which affects the hydrophilicity of these morphed peptide scaffolds (McIntosh, Donia *et al.* 2011). The leader and recognition sequences for these cyanobactins and for the linear heterocyclic-containing peptides noted below are determinants for recognition by the tailoring/processing enzymes.

3.3.3 Linear Heterocyclic Peptides

Figure 3.9 shows three additional representatives of peptides that, while linear rather than macrocyclic, show elements of both the cyanobactin and lantipeptide posttranslational processing logic. Microcin B17 (Li, Milne *et al.* 1996) is a bacterial DNA gyrase inhibitor,

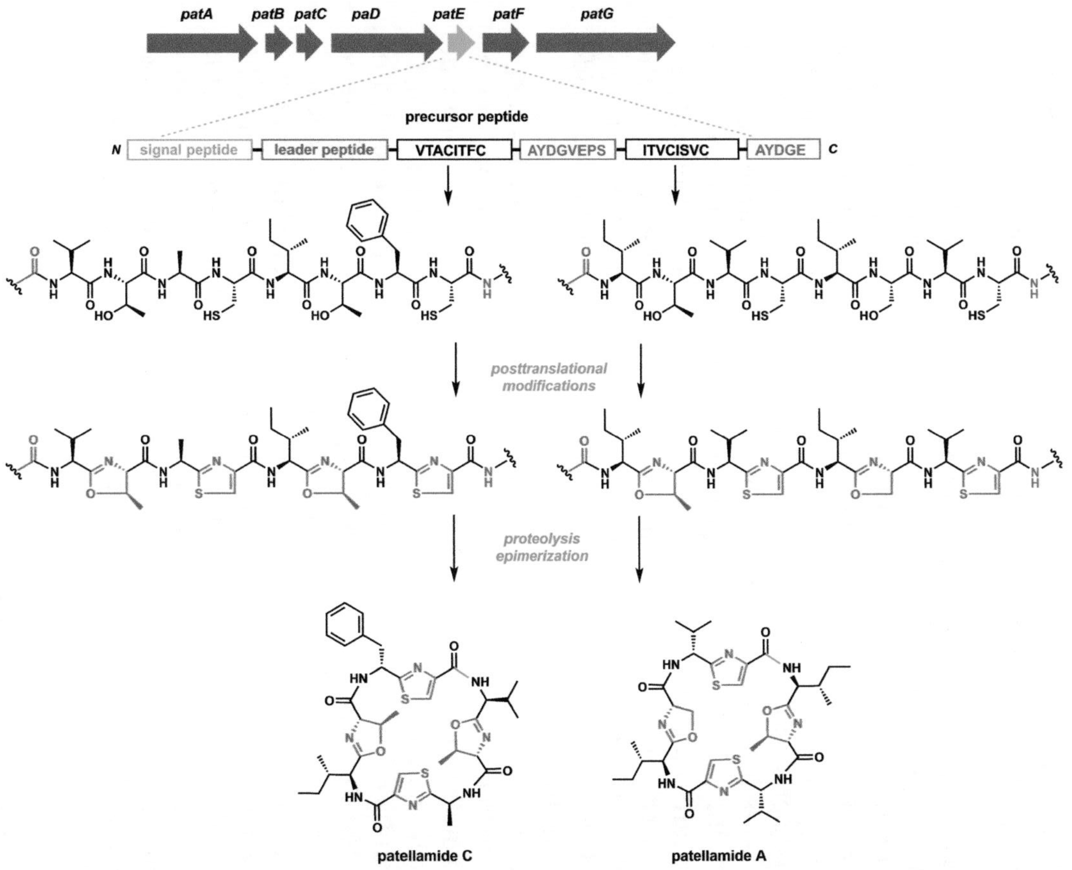

Figure 3.7 The structures of cyanobactin patellamides A and C. The octapeptide sequences that give rise to both patellamides A and C are embedded within the sequence of the 71-residue precursor protein PatE.

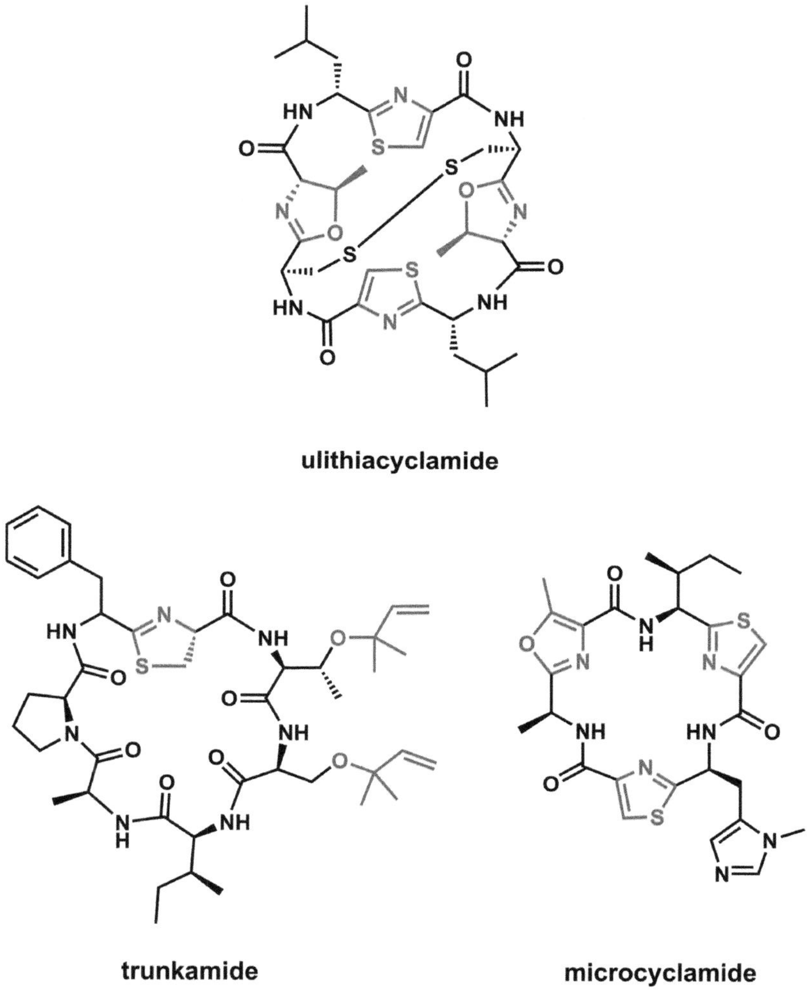

Figure 3.8 A representative diversity set of additional cyanobactin heterocyclic/macrocyclic peptide scaffolds.

goadsporin is a developmental morphogen for some *Streptomyces* strains (Onaka, Nakaho *et al.* 2005), while plantazolicin is a narrow spectrum *Bacillus anthracis* inhibitor (Scholz, Molohon *et al.* 2011).

Studies on the tailoring enzymes in the microcins and related systems have shed light on the heterocyclization catalysts (Dunbar, Melby *et al.* 2012). As shown in Figure 3.10, a Ser side chain can be added into the upstream carbonyl to form a transient tetrahedral adduct. The oxyanion of that adduct can attack cosubstrate ATP and become

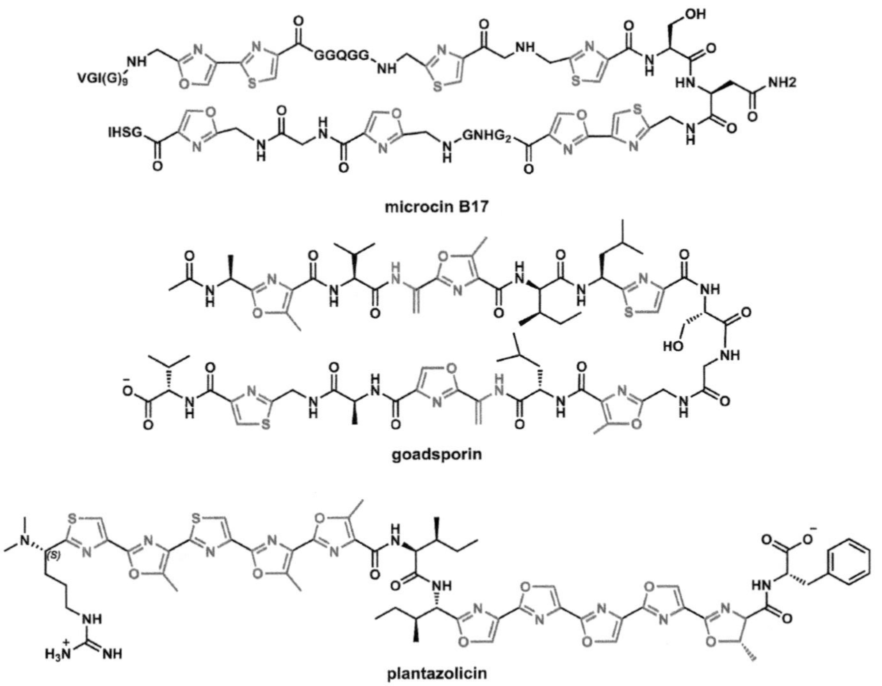

Figure 3.9 Linear peptides arising from posttranslational processing of Ser, Thr, and Cys side chains.

phosphorylated. Now, elimination of the elements of phosphate produces the oxazoline. Similar logic would lead to methyloxazoline from Thr and thiazoline from Cys side chains. A companion enzyme is the thiazoline to thiazole and oxazoline to oxazole oxidoreductase, using bound flavin mononucleotide (FMN) as hydride acceptor to yield the heteroaromatic ring systems (Li, Milne *et al.* 1996, Milne, Eliot *et al.* 1998). Thiazole formation was about 100-fold faster than oxazole formation, perhaps accounting for the observations in patellamides where the thiazoles have been formed but the Thr and Ser residues have become stuck at the oxazoline oxidation state. Plantazolicin has five heterocycles in a row, surely generating restricted conformation, while goadsporin shows two dehydroalanine residues to go along with the thiaozles and oxazole. The dehydroalanine probably arises by the lantipeptide type of posttranslational logic on Ser side chains: phosphorylation then α-H, β-PO_3 elimination in competition with the intramolecular closure and net pyrophosphate elimination that leads to oxazoline ring formation (see Figure 3.6).

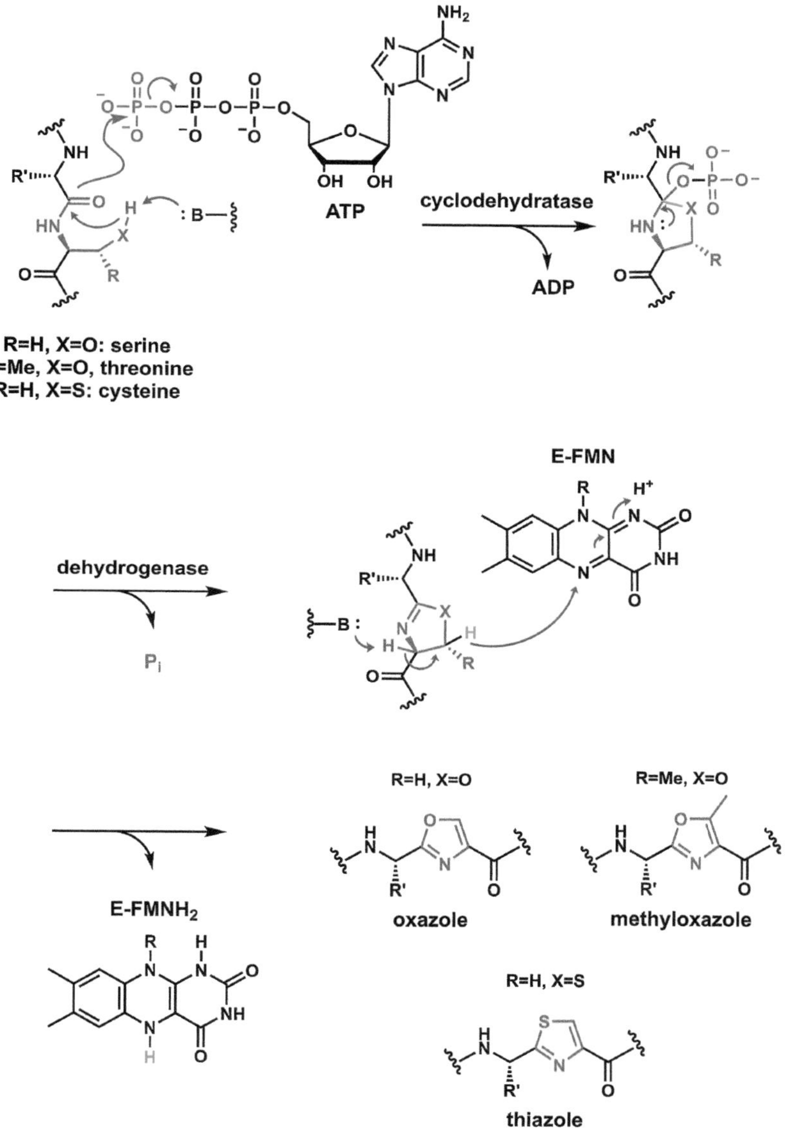

Figure 3.10 Mechanism for cyclodehydration and oxidative aromatization of Ser/Thr and Cys residues.

3.3.4 Thiazolyl-, Oxazolyl-, and Pyridine Ring-containing Peptide Scaffolds

Over 80 peptide natural products have been isolated that contain a central pyridine or dihydropyridine, decorated with between two and three thiazole/oxazole substituents in a characteristic 2,3 or 2,3,6 pattern (Bagley, Dale *et al.* 2005) as noted in Figure 3.11. The pyridine and the thiazoles/oxazoles in the 2- and 3-positions are embedded

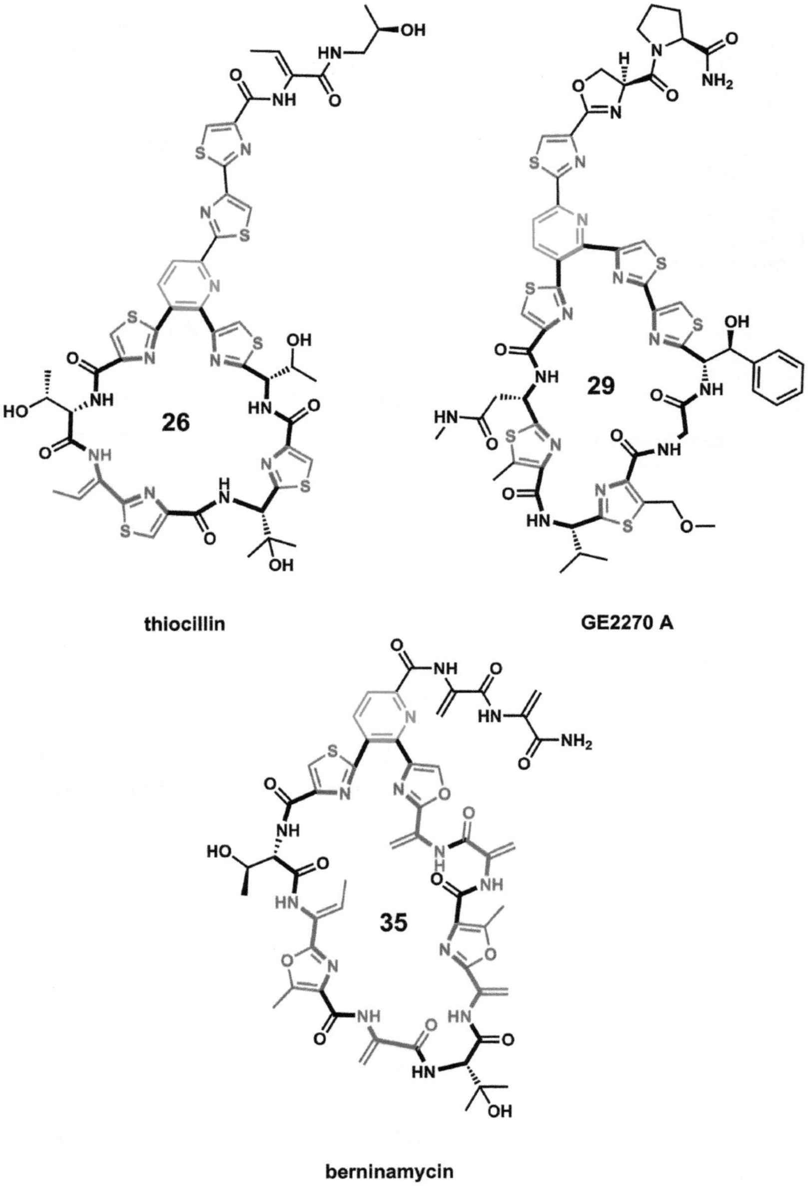

Figure 3.11 Thiazolyl peptides containing a central pyridine ring, decorated in a 2,3-bicyclic or 2,36-tricyclic pattern at the core of 26, 29, or 35 atom macrocycles.

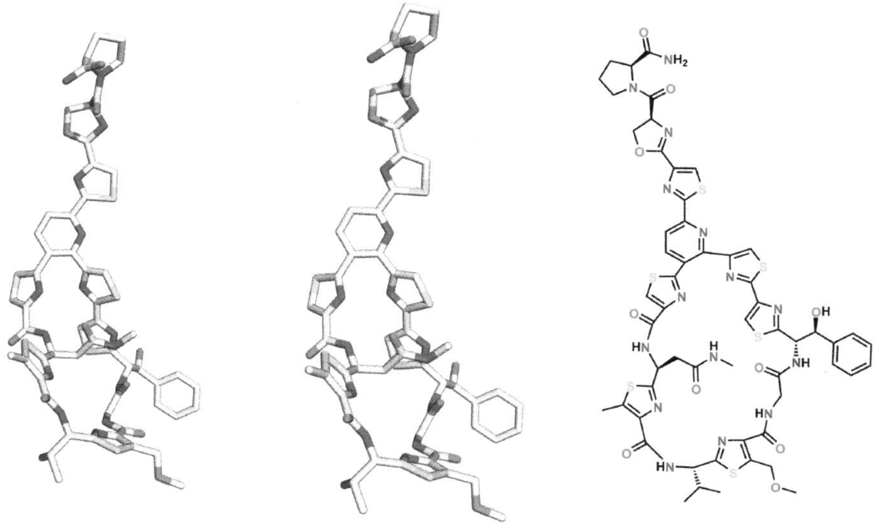

Figure 3.12 The trithiazolylpyridine core has a nonplanar geometry and sets the architecture of the macrocycles within which the core is embedded. The three-dimensional structure of GE2270 A bound to EF-Tu was generated from PDB entry 2C77 by Dr Yang Hai.

within a macrocycle that can range from 26 atoms (thiocillins), to 29 atoms (GE2270 A) to 35 atoms (berninamycin). The thiocillins and berninamycin target a specific site on bacterial 50S ribosomes as the basis of antibiotic action (Kelly, Pan *et al.* 2009). Remarkably, the 29-membered macrocyclic GE2270 A and congeners instead block protein synthesis by blockade of the elongation factor EF-Tu and its aminoacyl-tRNA chaperone activity as a GTP analog (Liao, Duan *et al.* 2009) (Walsh, Acker *et al.* 2010). The three dimensional structure of GE2270 A crystalized with EF-Tu is shown in Figure 3.12 (Parmeggiani, Krab *et al.* 2006); note that the trithiazole rings connected to the central pyridine ring are not planar, but are rather in a propeller architecture that confers structural constraints in the overall structure that are essential for biological functions.

Genetic and genomic analysis of these thiazolyl peptide antibiotics shows that the mature scaffolds derive from the C-terminus of a ribosomally produced protein, as depicted in Figure 3.13 for berninamycin (Malcolmson, Young *et al.* 2013). The mature berninamycin shows both the thiazole/oxazole rings along with dehydroalanines and dehydrobutyrine, reflective of both modes of posttranslational processing of serine residues (Figure 3.13) in pre-berninamycin.

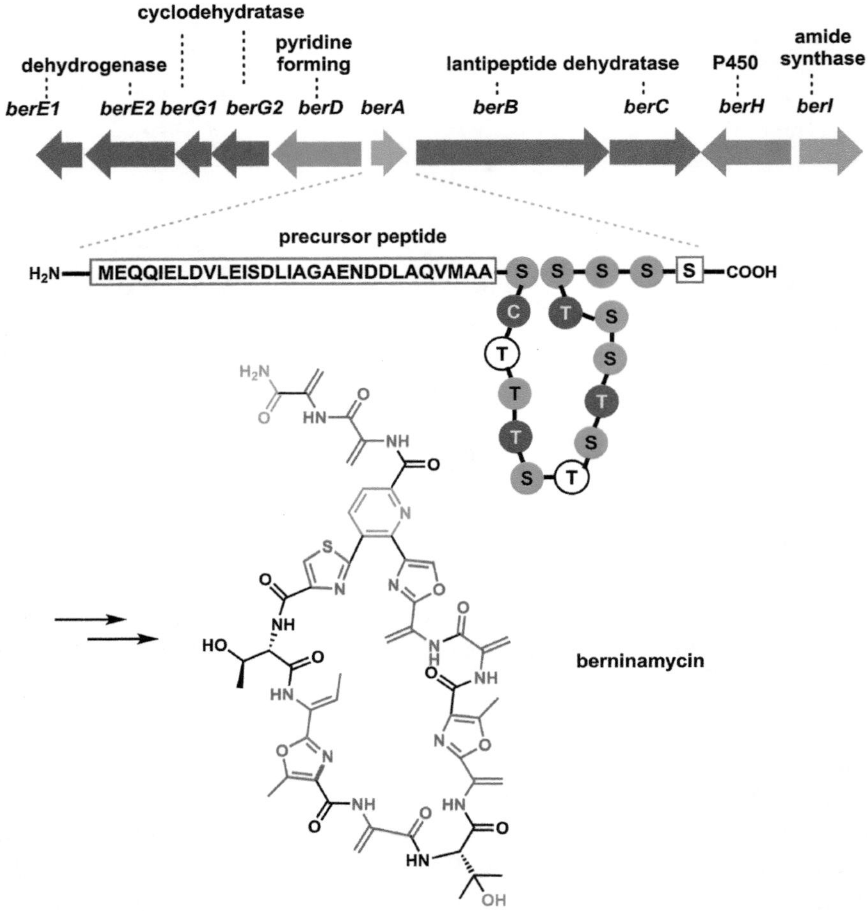

Figure 3.13 Schematic of the gene cluster for berninamycin and post-translational processing of the nascent 47-residue protein to the mature antibiotic framework.
Data from Malcolmson, S. J., T. S. Young, J. G. Ruby, P. Skewes-Cox and C. T. Walsh (2013). "The posttranslational modification cascade to the thiopeptide berninamycin generates linear forms and altered macrocyclic scaffolds". *Proc. Natl. Acad. Sci. U. S. A.* **110**(21): 8483.

Analysis of the thiocillin scaffold maturation has established that pyridine ring formation is a late step, arising from head to tail coupling of two dehydroalanine residues, loss of water and then aromatization by expulsion of the leader peptide. The formation of the pyridine ring is

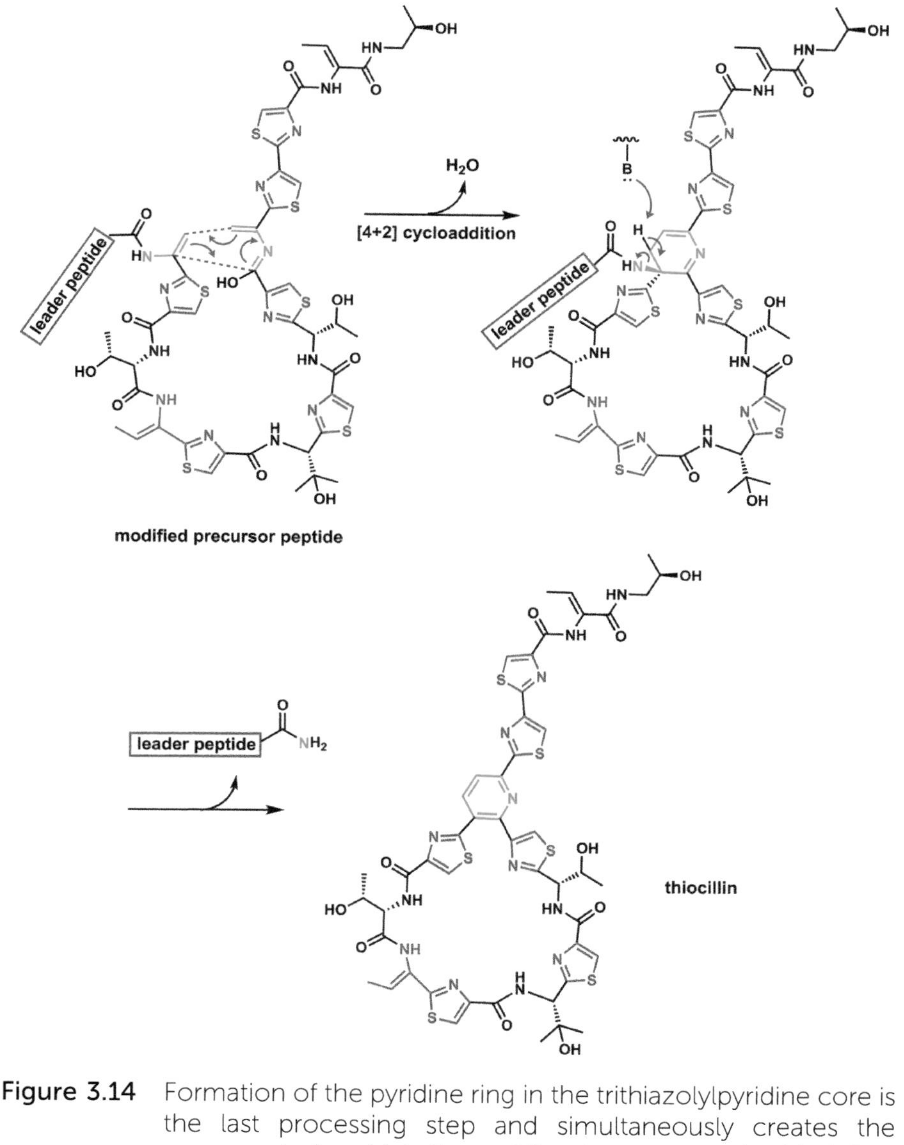

Figure 3.14 Formation of the pyridine ring in the trithiazolylpyridine core is the last processing step and simultaneously creates the macrocycle in which the pyridine becomes embedded. The leader peptide is proposed to be cleaved in the aromatization of the pyridine ring.

featured as a [4 + 2] cycloaddition (Wever, Bogart *et al.* 2015), coincident with generation of the macrocycle (Figure 3.14).

Thiostrepton is a more complex structural representative of this 50S ribosome targeting antibiotic class (Figure 3.15). The central

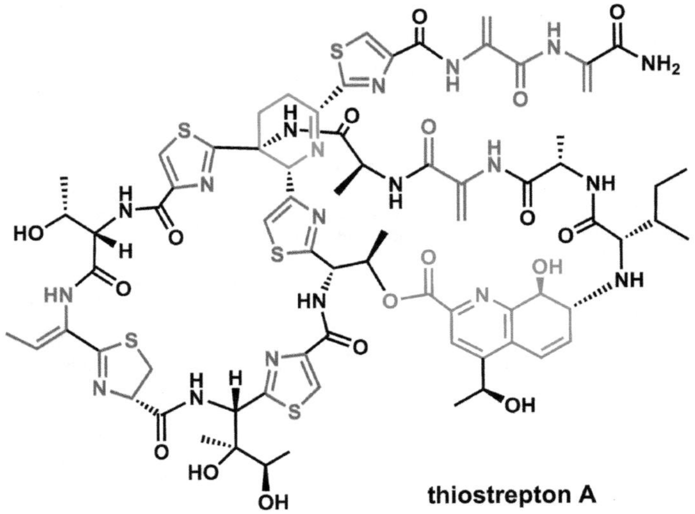

Figure 3.15 Thiostrepton has a tetrahydropyridine (red) at its trithiazolyl core (maroon) and still contains a portion of the upstream leader peptide as a second macrocyclic loop.

ring (red) still has the 2,3,6 trithiazolyl substituent pattern but is at the oxidation state of a tetrahydropyridine rather than the fully aromatized pyridine. Part of the upstream leader peptide is still attached as a second macrocyclic loop, which contains a bicyclic quinaldic acid (pink) that has been morphed from a tryptophan residue in the pre-thiostrepton nascent protein.

3.3.5 Bottromycin and Polytheonamide

Figure 3.16 shows the structure of polytheonamide A, a 48 residue peptide with 18 D-amino acid residues, alternating with L-residues through the core of the scaffold to create a helix that can function as a transmembrane pore (Hamada, Matsunaga *et al.* 2005) (Figure 3.17). Figure 3.16 also contains the structure of the microbial metabolite bottromycin. At first glance it is not obvious that polytheonamide A and bottromycin A deserve joint consideration but they both are products of posttranslational modifications, notably a set of C-methylations at unactivated carbon centers.

As anticipated for this class of natural products, peptidase action removes an upstream sequence. In the case of bottromycin, this is only the N-terminal methionine, preparatory to catalyzed imine

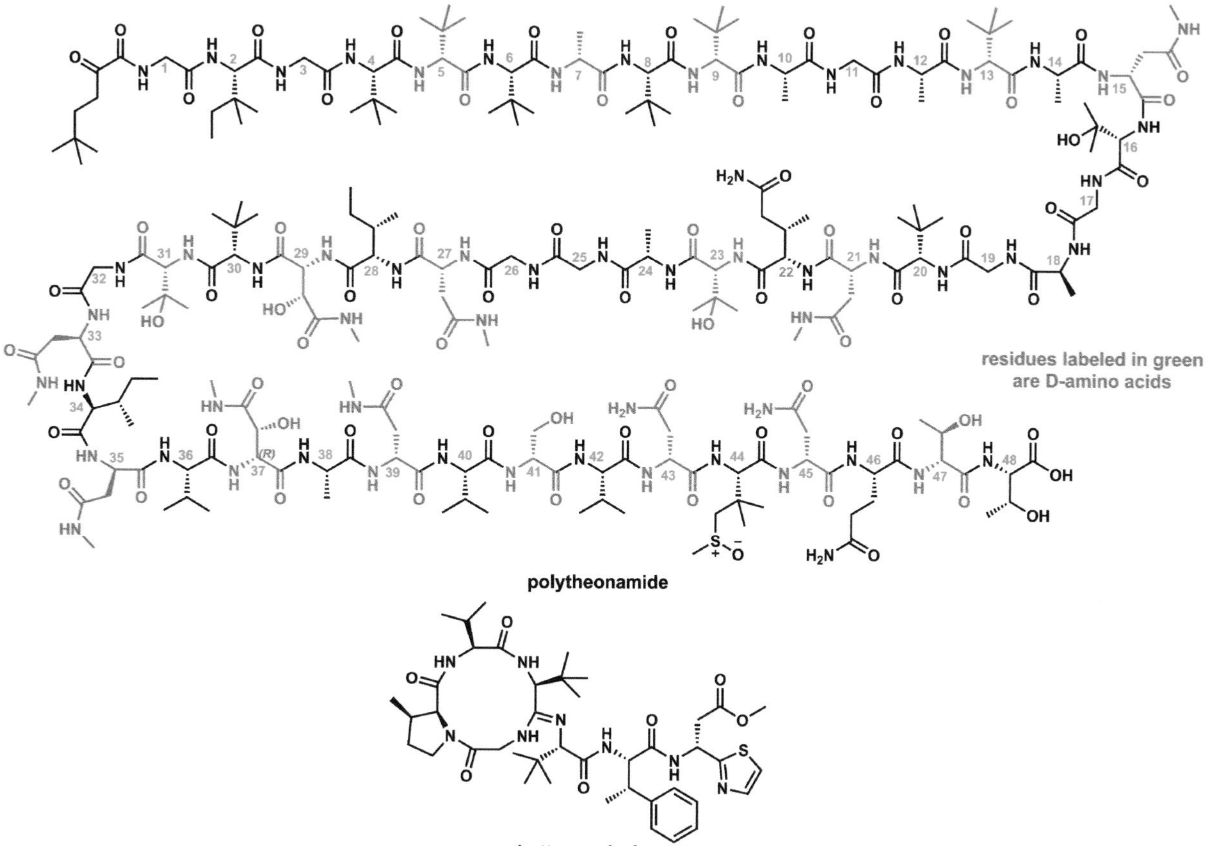

bottromycin A_2

Figure 3.16 Polytheonamide A, a 48-residue peptide ion channel, has 17-D-residues introduced during posttranslational processing of its precursor protein. Bottromycin is likewise generated by a set of unusual posttranslational modifications including five C-methylations at unactivated carbon sites.

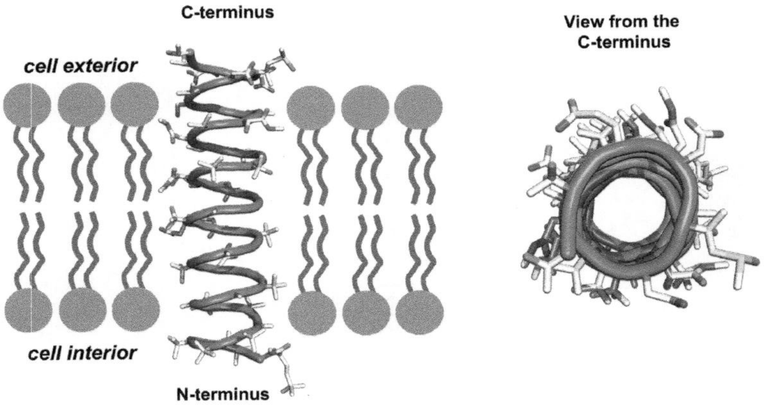

Figure 3.17 Polytheonamide adopts a helical conformation (thanks to the alternating stretch of D- and L-residues) that can span cell membranes and act as a pore for ion flux. Images generated from PDB entry 2RQO by Dr Yang Hai.

formation between the amine of newly revealed Gly_1 and the carbonyl of Val_4 to yield the 12-membered macrocyclic imine (Crone, Leeper *et al.* 2012, Gomez-Escribano, Song *et al.* 2012, Hou, Tianero *et al.* 2012). Four C-methylations, denoted in purple in Figure 3.18, ensue. Each methylation consumes two molecules of cosubstrate *S*-adenosylmethionine and involves a radical intermediate at the carbon atom being methylated, to pair with a [$CH_3^{\bullet}$] equivalent being transferred from the second SAM that is consumed. These mechanisms are covered in detail in Chapter 10. The methyl ester group at the Asp side chain of bottromycin, on the other hand, is presumed to arise by the normal route of [CH_3^{+}] transfer from SAM to the carboxylate oxygen as nucleophile. Formation of the thiazole is proposed to go by the internal formation of tetrahedral adduct and subsequent dehydration and dehydrogenation as described above.

Polytheonamide posttranslational processing is even more irregular. After proteolytic cleavage between residues 96 and 97 of the precursor protein, 18 of the remaining 48 residues undergo *in situ* epimerization from L- to D-configuration (Figure 3.19) (Freeman, Gurgui *et al.* 2012). The racemization or epimerization mechanism is not well defined at present. The process must be catalyzed, presumably *via* isoamide contributors to planar carbanionic transition states with stereorandom return of the α-H, since L-residues in proteins are almost indefinitely stable towards configurational

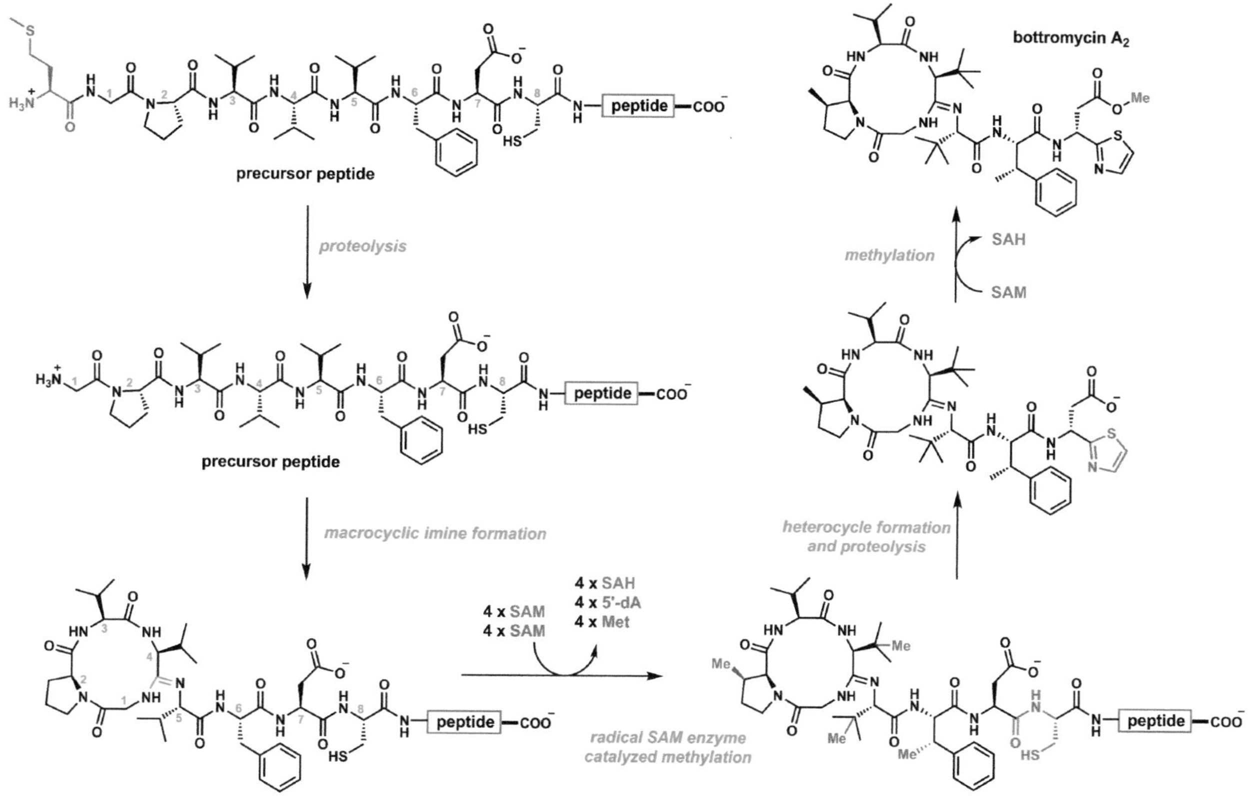

Figure 3.18 The biosynthesis of bottromycin A2 from the precursor peptide involves significant posttranslational modifications. Most notably, radical SAM methyltransferases consume eight molecules of *S*-adenosylmethionine (SAM) to methylate four unactivated carbons during the biosynthesis. See Chapter 10 for radical SAM enzyme mechanisms. An additional molecule of SAM is used as a conventional [CH_3^+] source in the final step of the pathway to yield the methyl ester.

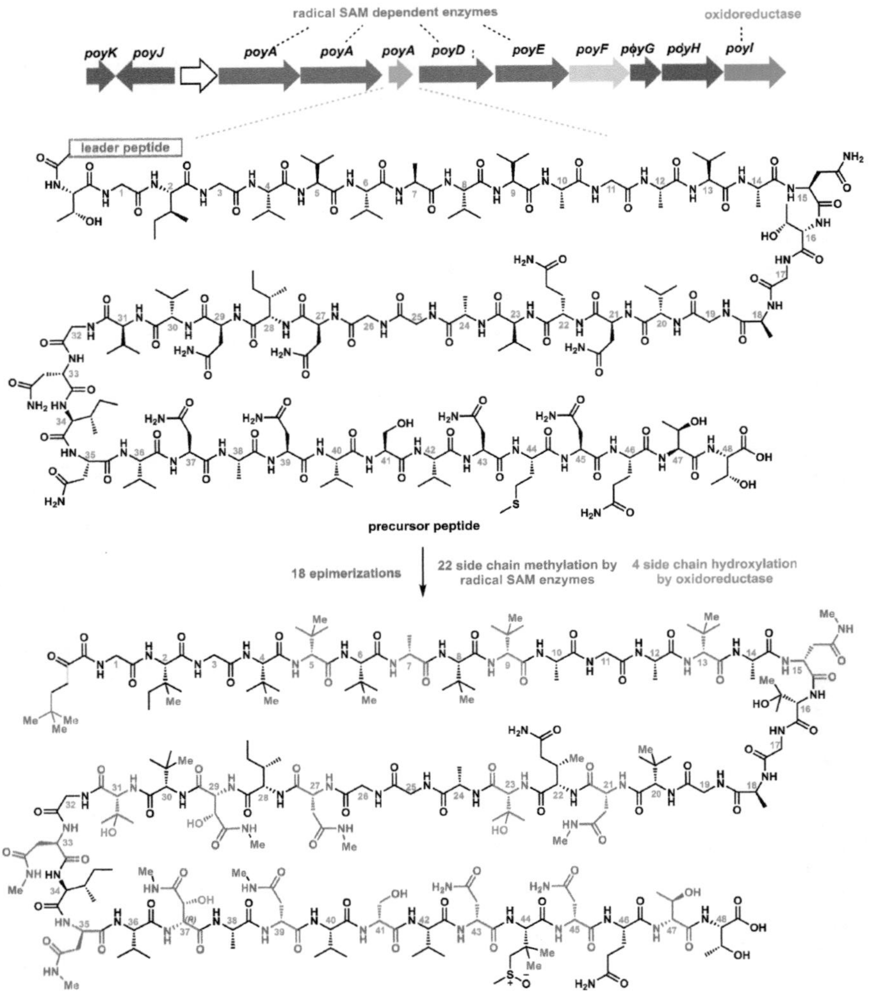

Figure 3.19 Racemization/epimerization of 18 L-amino acid residues to D-residues occurs in the peptide backbone of the precursor during posttranslational morphing of the 48-residue polytheonamide core.

equilibration. In addition to the 18 C_{α} epimerizations, there are 22 C-methylations, mostly at the β-carbons of Val and Ile residues, unactivated sites, that raise the radical SAM mechanism prospect as postulated for bottromycin (and discussed in Chapter 10).

3.3.6 Conotoxins

Cone snails are efficient predators of fish, which they harpoon and into which they inject a mix of peptide-based toxins. Three such toxins, out of many (Akondi, Muttenthaler *et al.* 2014), are shown in Figure 3.20. Contulakin-G and conopeptide-MrlA have a blocked N-terminus due to cyclized pyroglutamate (5-oxoproline) residues. Conantokin-G is a representative of a subset of conotoxins that have had glutamate side chains converted to γ-carboxy-glutamate side chains. These malonic acid derivatives chelate calcium ions and undergo conformational re-arrangements thereby, typically, yielding membrane-active forms.

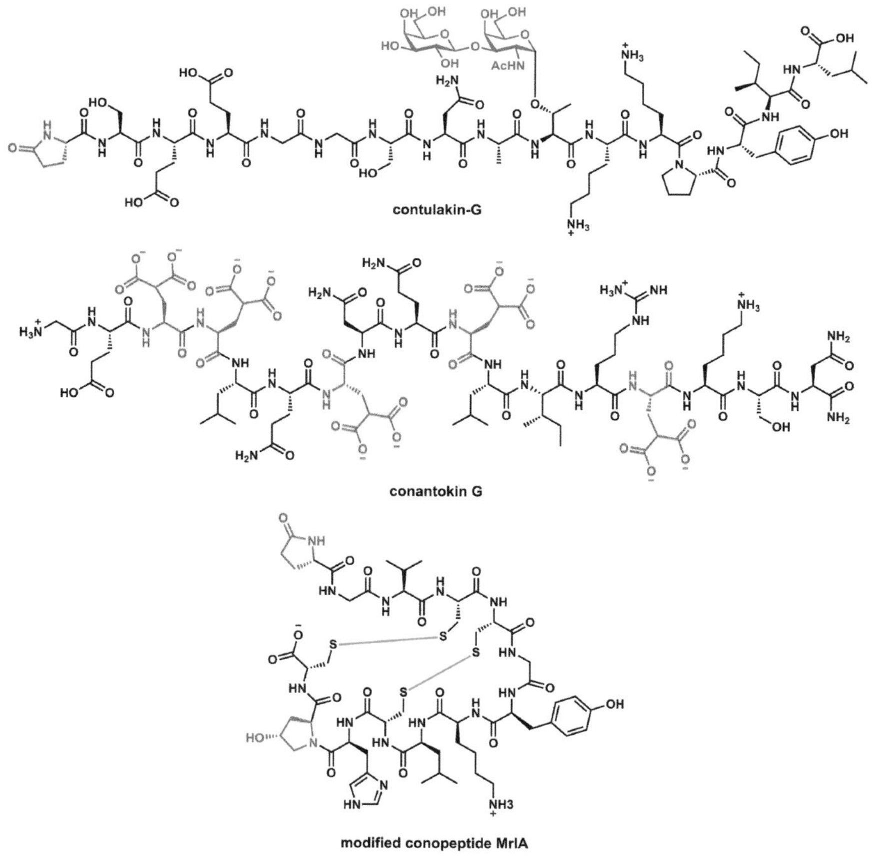

Figure 3.20 Predator cone snails make a set of modified peptide toxins from posttranslational modifications of precursor proteins.

3.3.7 Amanitin and Phalloidin

The final examples of modified peptide scaffolds deriving from selective proteolysis of protein precursors are the pair of mushroom toxins (Vetter 1998) amanitin (Hallen, Luo *et al.* 2007) and phalloidin (Lynen and Wieland 1937) (Figure 3.21). Amanitin is a cyclic octapeptide macrolactam, arising by excision/transpeptidation from a 35-residue protein precursor in *Amanita bisporigera* by action of a prolyloligopeptidase. Amanitin is a potent inhibitor of RNA polymerase in both initiation and elongation phases. A notable feature of amanitin is the crosslink between an oxidized cysteine and the indole-C_2 position of a 6-hydroxtryptophan, which actually creates a bicyclic framework. There is other evidence that oxygenation is at work in posttranslational maturation, including a hydroxy-Pro residue and a bis-hydroxylated Ile moiety. The timing of Cys-*O*-Trp crosslinking is not yet established, nor whether this involves radical intermediates and what the oxidation state of the cysteine side chain sulfur is during attachment.

The structurally related phalloidin from the death cap mushroom is toxic by binding to F-actin fibers and preventing their depolymerization. The liver is the organ most sensitive to phalloidin destruction

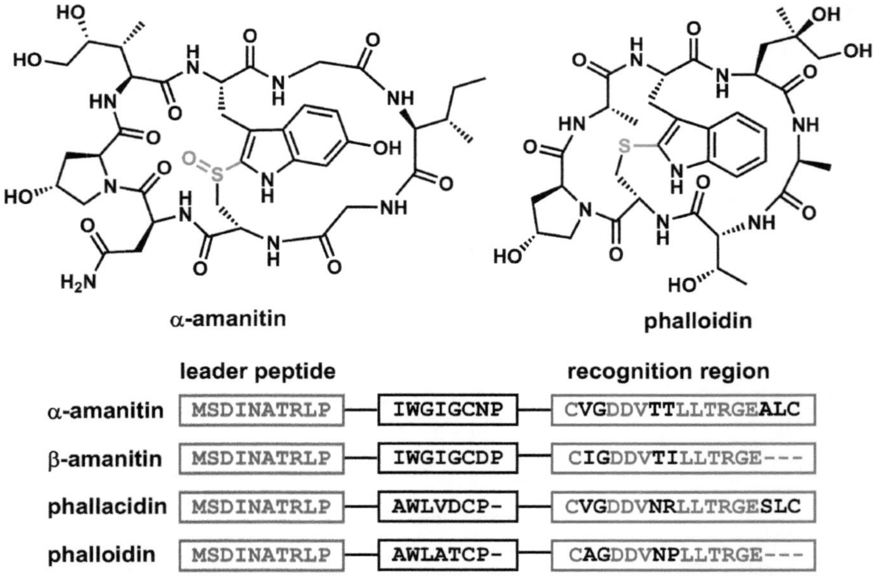

Figure 3.21 Amanitin and phalloidin: intramolecular oxidative coupling generates rigid peptide frameworks in mushroom toxins.

due to selective transport of the toxin. It is a bicyclic heptapeptide with a similar transannular Cys–Trp crosslink but without the hydroxylation at C_6 of Trp or the oxidation of the Cys sulfur atom. On the other hand, the threonine residue at some point has been isomerized to the D-configuration. Prolyl endopeptidases are also implicated in excision and cyclization of the heptapeptide from the phalloidin precursors.

Vignette 3.1 Ustiloxins: Fungal Cyclic Peptides from Ribosomal Precursors.

Genome mining of *Aspergillus flavus* and other *Aspergillus* fungal species has focused on the bioinformatic predictions of the capacity to make three major natural product classes: biosynthetic gene clusters for polyketides, nonribosomal peptides (and NRP–PK hybrids), and terpenes. A shift in focus to examine open reading frames of fungal ribosomal proteins led Umemura and colleagues to note a series of fungal proteins with N-terminal signal sequences and sixteen fold repeats of the tetrapeptide sequence Tyr-Ala-Ile-Gly at the C-termini (Umemura, Nagano *et al.* 2014). The natural product ustiloxin (Figure 3.V1) contains this tetrapeptide core that has undergone oxidation and two crosslinking reactions.

Processing of the precursor peptide to release the 16 copies of the YAIG peptide is proposed to be catalyzed by the endoproteinase Kex2 during transport from the endoplasmic reticulum to the Golgi apparatus. Based on a series of knockout and biochemical characterization steps, Oikawa and coworkers elucidated the maturation steps as shown in Figure 3.V1 (Ye, Minami *et al.* 2016). The tetrapeptide is first crosslinked between the phenolic oxygen of the Tyr and the β-carbon of the Ile residue by a set of oxidases. It is likely that the phenol ring of Tyr is hydroxylated to the catechol by the tyrosinase homolog UstQ. One electron oxidation to the catecholate radical and the Ile β-radical could couple to give the ether crosslink. Following methylation by UstM to give ustiloxin F, oxidation of the catechol ring to the orthoquinone by the P450 UstC could set up capture by L-cysteine to build the S–C crosslink. Subsequent oxidations of the sulfur and nitrogen atoms of the coupled cysteine by two flavin-dependent monooxygenases (Chapter 9) set up the oxime intermediate, which is proposed to undergo further decomposition to the aldehyde (not shown). In the last step of the pathway a PLP-dependent (Chapter 5) enzyme decarboxylates the β-carboxylate of aspartate regiospecifically and then uses the alanyl-β-carbon-PLP enamine equivalent as carbanion source to build the side chain C–C bond in ustiloxin B.

Additional *Aspergillus* genome mining, spurred on by the ustiloxin findings, indicates additional morphed peptide scaffolds derived from maturation of protein precursors, suggesting even more natural product

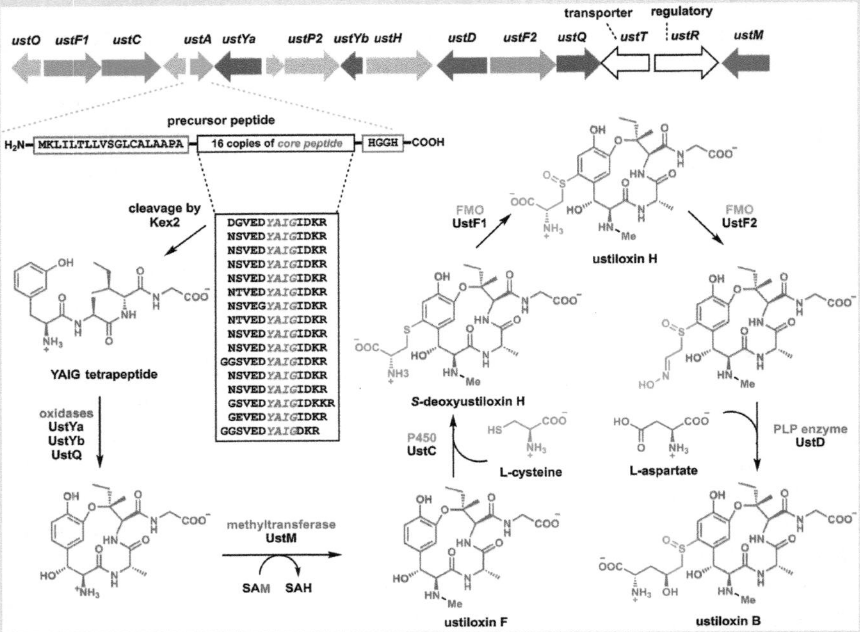

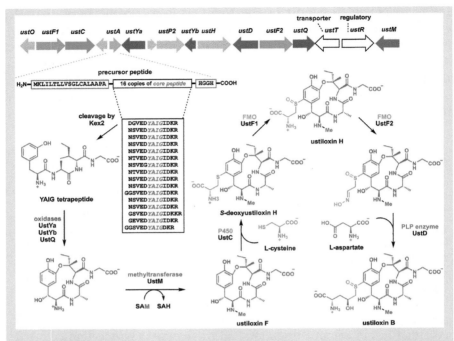

Figure 3.V1 Proposed biosynthetic pathway for posttranslational cleavage and oxidative conversion of 16 tetrapeptide repeats to ustiloxins.

diversity than usually uncovered by the algorithms in general use. One such novel framework is found in the bicyclic architecture of asperipin-2a and many others may remain to be discovered and characterized (Nagano, Umemura *et al.* 2016).

3.4 Nonribosomal Peptide Synthetase Assembly Lines: Alternative Routes to Highly Morphed Peptide Scaffolds

As noted in the introduction to this chapter, a complementary route to stable small molecule scaffolds of peptidic origin is to bypass mRNAs and the ribosomal factories altogether (see Figure 3.4). Instead, nonribosomal peptide synthetase machinery occurs in bacteria and fungi and serves to build peptides up to 21 residues in length without any RNA template (Walsh and Wencewicz 2016). Figure 3.22

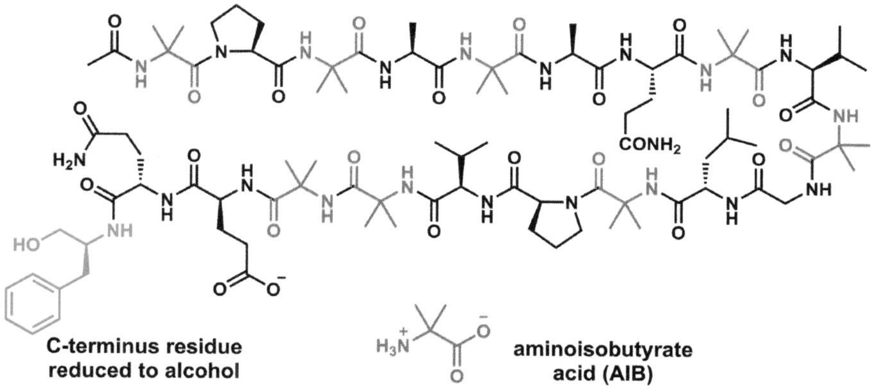

Figure 3.22 Alamethicin from *T. viride* is made by an NRPS pathway. These linear metabolites from fungi are also referred to as peptaibols, which stands for *pept*ide, *AIB* and amino alcoh*ol*. AIB is the unnatural amino acid aminoisobutyrate shown in green in the structure.

shows the structure of alamethicin, a 20-residue non-ribosomally derived peptide from *Trichoderma viride* that exhibits a wide spectrum of biological activities through formation of an ion channel.

3.5 Nonproteinogenic Amino Acid Building Blocks

Features of NRPS assembly lines include the use of nonproteinogenic amino acid building blocks, as well as "on assembly line" epimerizations and *N*-methylations. Figure 3.23 shows structures of four nonribosomal peptides that highlight nonproteinogenic amino acid building blocks (Walsh, O'Brien *et al.* 2013). The immunosuppressant cyclosporine A is a cyclic undecapeptidyl lactam that contains one D-alanine residue and two nonproteinogenic residues: L-aminobutyrate and butenyl-methyl-L-threonine. Miraziridine A has, within its pentapeptide framework, aziridine dicarboxylate, statine, aminobutyrate, and vinyl-arginine as four out of five nonproteinogenic units. Didemnin B is a cyclic depsipeptide (peptidolactione) from a marine tunicate. It has antitumor properties by virtue of inhibition of the action of palmitoylated protein thioesterases. In addition to *N*,*O*-dimethyltyrosine, didemnin has an isostatine residue, a homolog of the statine residue in miraziridine A. Lastly, the antifungal hexapeptide macrolactones of the kutzneride family from soil *Kutzneria* spp. bacteria

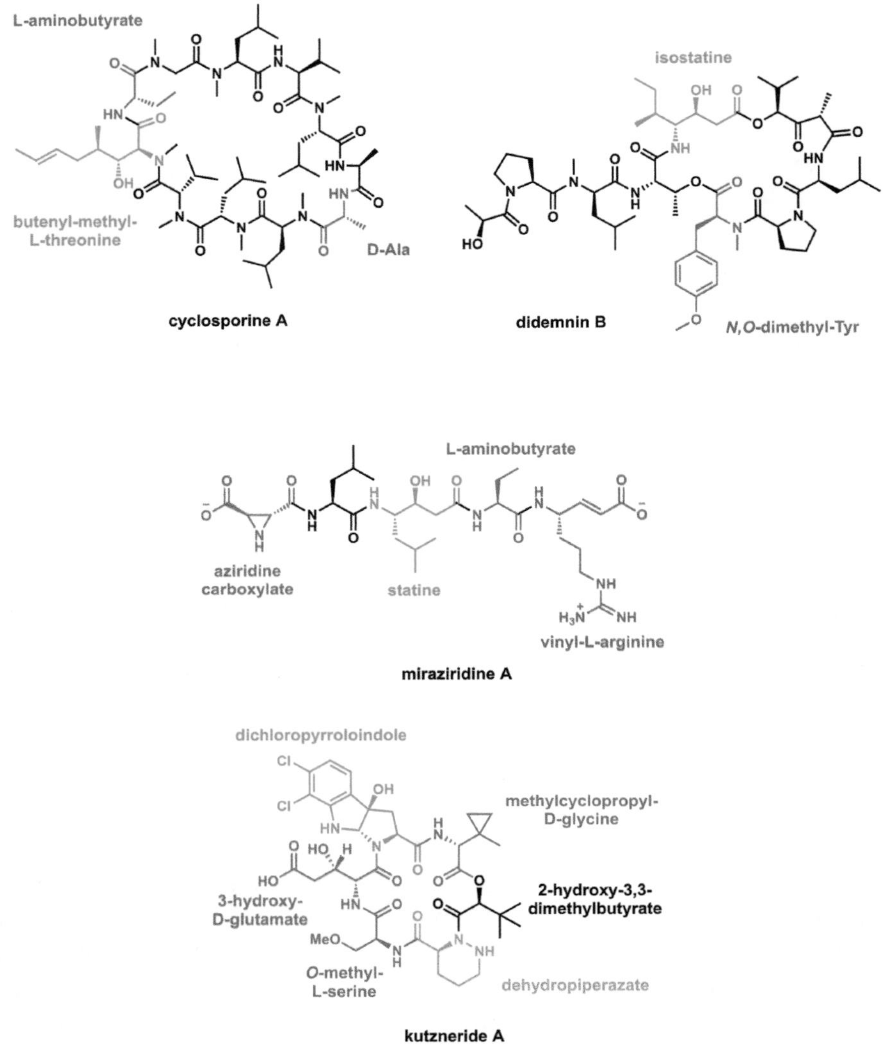

Figure 3.23 Examples of nonribosomally synthesized peptides that contain nonproteinogenic amino acids.

have all six constituent amino acids that are nonproteinogenic, including piperazate, a dichloropyrroloindole unit, and a methylcyclopropylglycine (Broberg, Menkis *et al.* 2006, Fujimori, Hrvatin *et al.* 2007).

Statine, isostatine, and vinyl-arginine are examples of building blocks that are hybrids of polyketide and amino acid units that are combined on a two module assembly line (Figure 3.24)

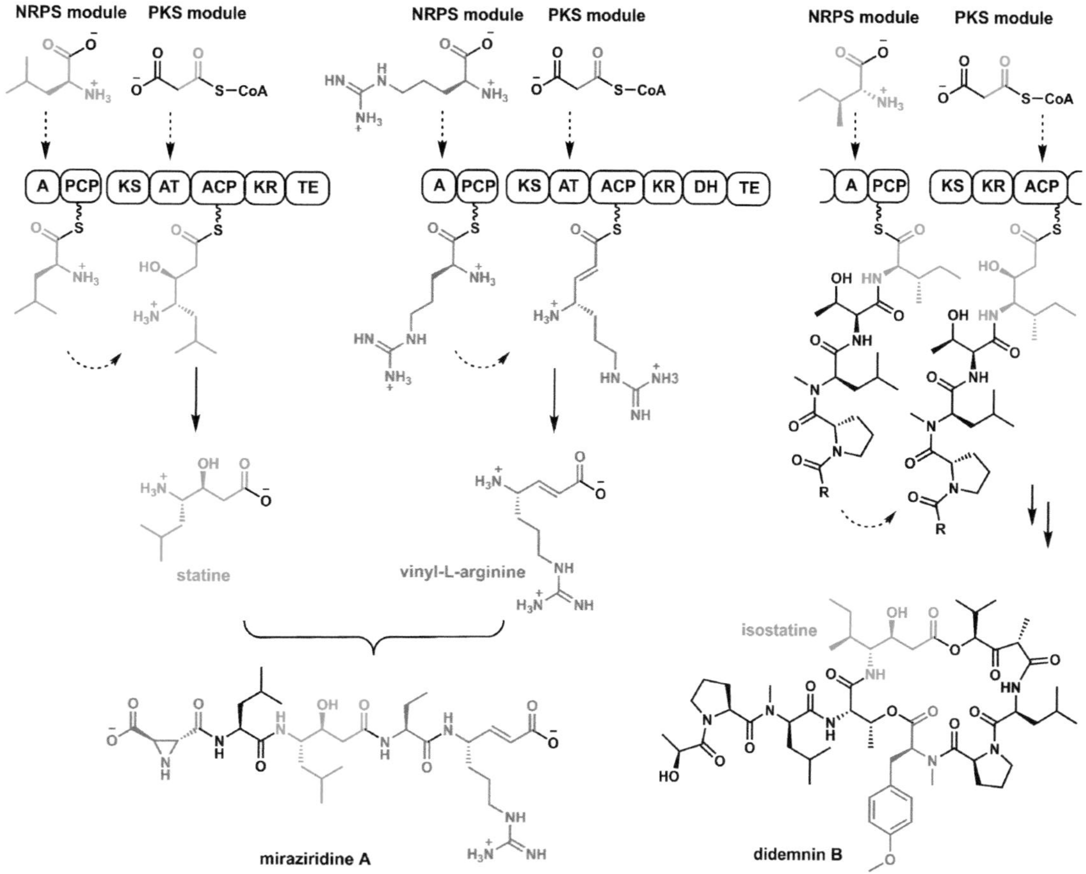

Figure 3.24 The assemblies of the statine, isostatine, and vinyl-arginine occur on hybrid assembly lines. The NRPS assembly line components are discussed below. The domain abbreviations of A and PCP stand for adenylation and peptidyl carrier protein, respectively. For details see Figure 3.29.

(Walsh, O'Brien *et al.* 2013). The amino acid is loaded first and then condensed with the malonyl-*S*-ACP to yield a β-keto-δ-amino-*S*-ACP nascent product. Action of the ketoreductase domain creates the β-OH, δ-NH_2-substitution pattern of statine (and isostatine) on thioesterase-mediated release from the assembly line. Subsequent loading of such a substrate on a PKS module containing a functional dehydratase domain would lead to the α,β-vinyl-δ-amino pattern of vinyl-Arg (miraziridine A) and also vinyl-Tyr and vinyl-Val in other metabolites.

Vancomycin and teicoplanin, shown in Figure 3.25, have two kinds of hydroxyphenylglycine units. There are 4-OH-phenylglycines at residues 4 and 5 of vancomycin and also at residue 1 of teicoplanin, while 3,5-dihydroxyphenylglycine residues are found at residue 7 of vancomycin and residues 3 and 7 of teicoplanin (Hubbard and Walsh 2003). Residues 2 and 6 in vancomycin are diastereomers of chloro-β-hydroxy-tyrosines. All these electron rich phenolic rings are susceptible to enzyme-mediated phenoxy radical generation (discussed in detail in Chapter 9), leading to three and four crosslinks, respectively, between the side chains of vancomycin and teicoplanin. In turn these oxidative crosslinks restrict the heptapeptide scaffolds to a cup-shaped architecture that enables hydrogen bonding to *N*-acyl-D-Ala-D-Ala units in uncrosslinked peptidoglycan termini of bacterial cell walls.

The vancomycin congener chloroeremomycin has biosynthetic genes for the NRPS assembly line clustered together with the genes encoding three cytochrome P450 crosslinking oxidases, three glycosyltransferases, and the genes that direct biosynthesis of the two types of hydroxyphenylglycine building blocks (Figure 3.26).

As noted in Figure 3.27, the primary metabolite prephenate is converted to 4-hydroxyphenylpyruvate and then three dedicated enzymes are deployed to oxidatively shorten the three-carbon ketoacid chain to a two-carbon ketoacid chain before transaminase action generates the 4-hydroxyphenylglycine building block.

By contrast, the 3,5-dihydroxyphenylglycine comes by a completely different route (Figure 3.28). Four malonyl-CoAs are condensed in chain-elongating thioclaisen reactions to the eight-carbon aryl-CoA. An oxygenase-mediated conversion of the acetyl-CoA side chain to the glyoxalate side chain is the preamble to transaminative conversion to the 3,5-dihydroxyphenylglycine. Cotranscription of the two sets of hydroxyphenylglycine biosynthetic genes with the NRPS

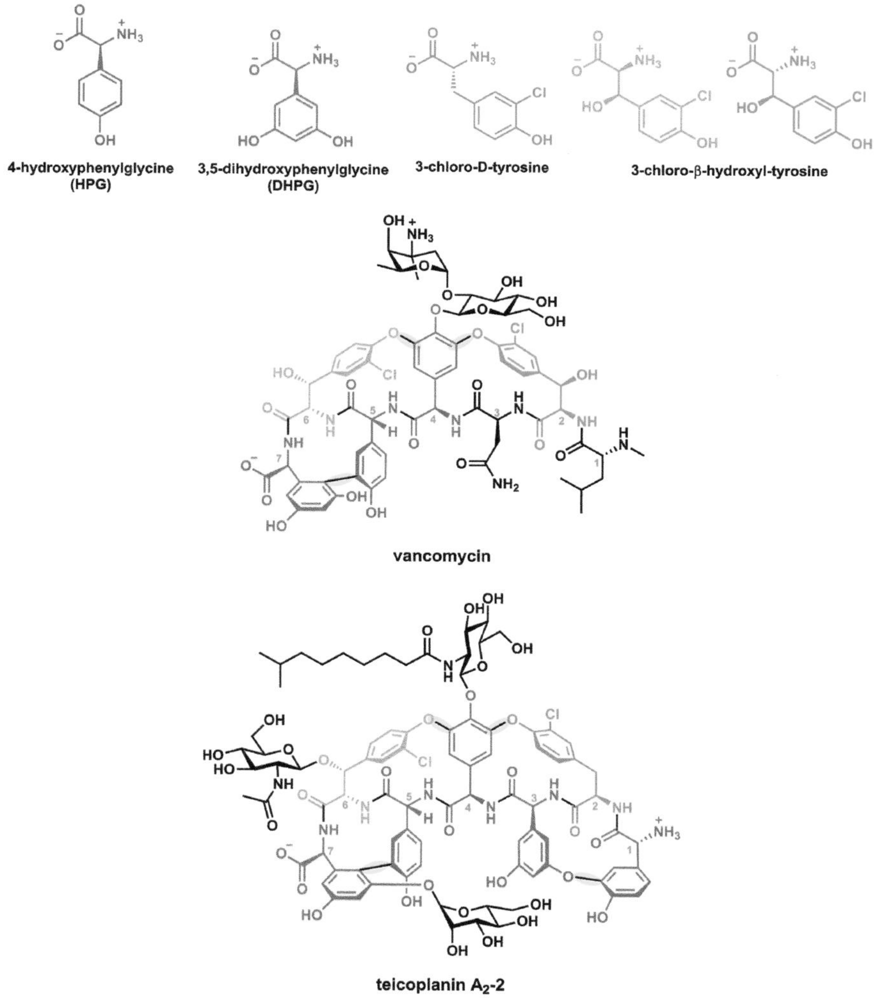

Figure 3.25 The glycopeptide antibiotics vancomycin and teicoplanin contain nonproteinogenic hydroxyphenylglycine regioisomers at several residues. They also contain side chain aryl ether and direct C–C crosslinks (yellow), catalyzed by dedicated P450 monooxygenases.

assembly line subunits assures a "just in time" inventory control to be able to run the vancomycin, teicoplanin, and chloroeremomycin assembly lines.

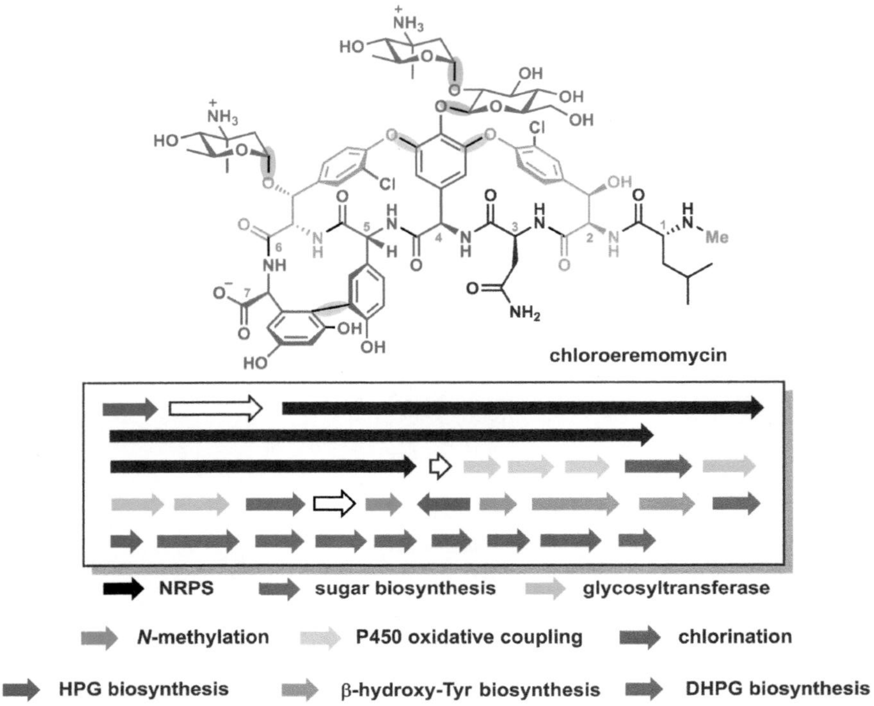

Figure 3.26 The chloroeremomycin biosynthetic gene cluster encodes enzymes required for synthesis of the nonproteinogenic aromatic amino acids, as well as P450 enzymes that catalyze phenyl couplings.

3.6 NRPS Assembly Line Logic: Priming, Initiation, Elongation, Termination

In analogy to the polyketide synthase assembly line logic described in Chapter 2, NRPS assembly lines follow parallel chemical logic (Liu and Walsh 2009, Walsh and Wencewicz, 2016). There are initiation modules, elongation modules, and termination modules in canonical NRPS assembly lines. The modules may be contained in a single protein, as is the case for the ACV tripeptide synthetase assembly line involved in providing the scaffold for penicillin and cephalosporin antibiotics (Byford, Baldwin *et al.* 1997), or they can be distributed over multiple separate proteins. The vancomycin/chloroeremomycin assembly line has seven modules, one for each amino acid selected and incorporated into the heptapeptide backbone.

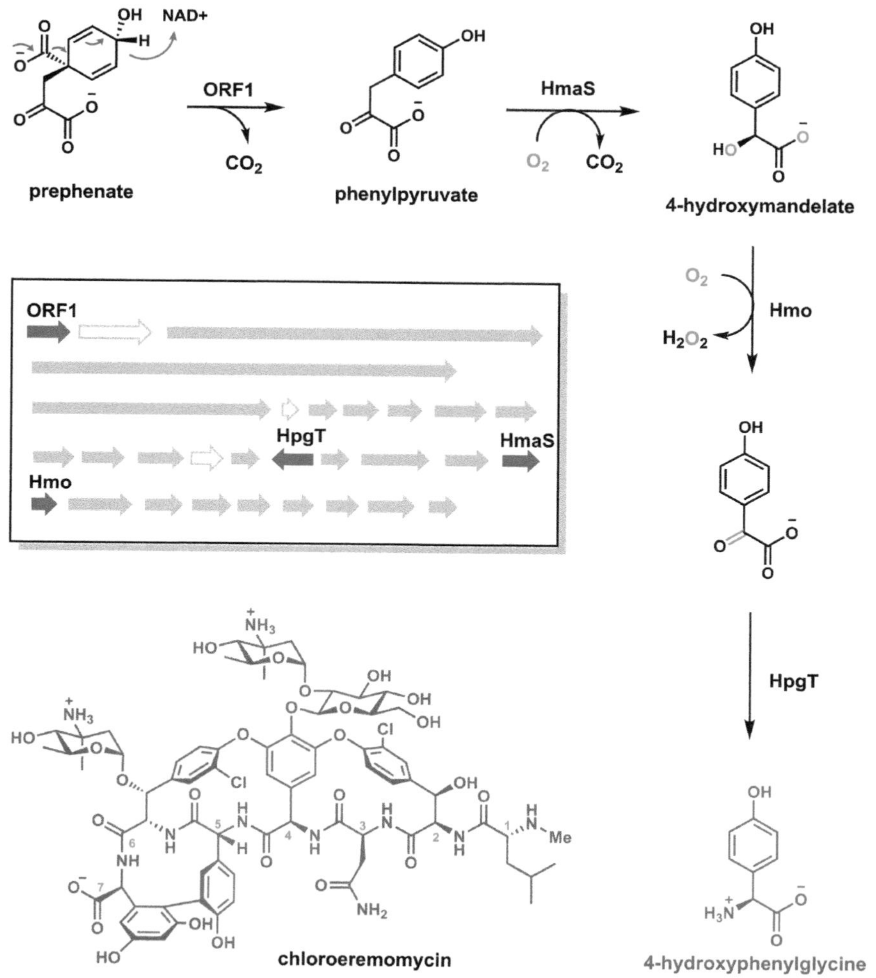

Figure 3.27 Conversion of prephenate to 4-hydroxyphenylglycine by three enzymes encoded in the biosynthetic gene cluster.

Before NRPS assembly lines can be functional every peptidyl carrier protein (PCP) domain of 80–100 amino acid residues must be posttranslationally primed with a phosphopantetheinyl prosthetic group (Figure 3.29) (Walsh and Wencewicz 2016). Coenzyme A is the donor substrate and a specific serine side chain is the attacking nucleophile. Failure to convert even one *apo*-PCP domain to the pantetheinylated *holo*-form will render an NRPS assembly line nonfunctional. This is identical logic to the posttranslational

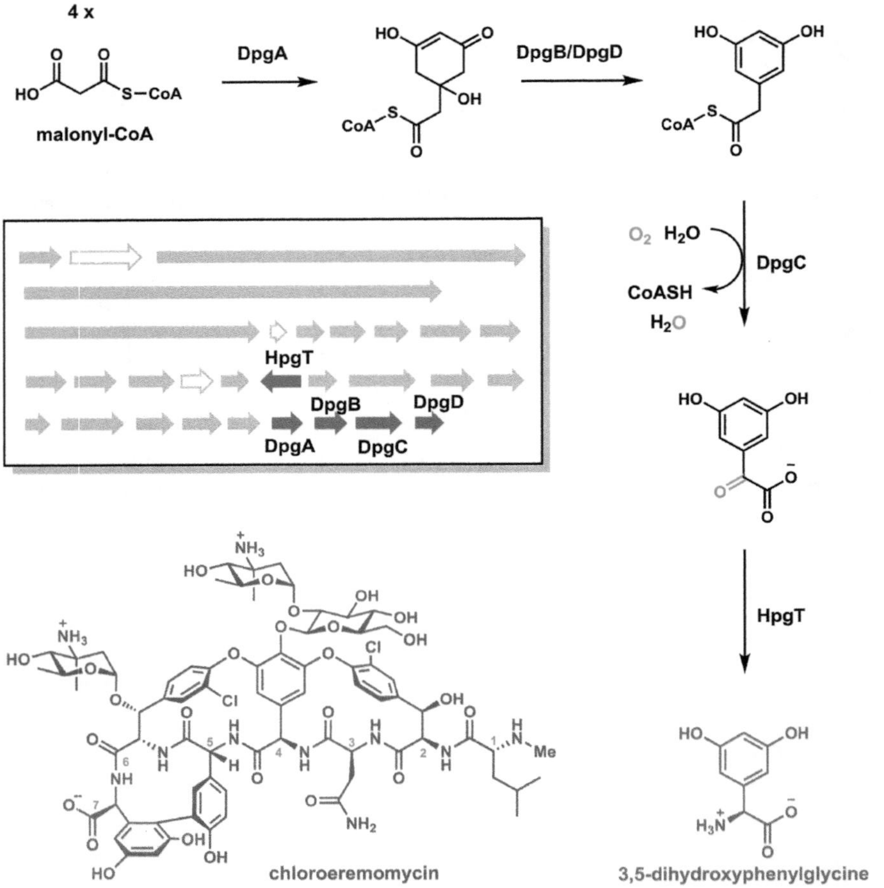

Figure 3.28 The 3,5-dihydroxyphenylglycine regioisomer is fashioned by PKS logic from four molecules of malonyl-CoA to release 3,5-dihydroxyphenylacetyl-CoA. The C_2 acyl-CoA side chain is oxidized to the glycolic acid side chain and transaminated to give the amino acid.

modification of acyl carrier proteins (ACPs) in polyketide biosynthesis (see Figures 2.9 and 2.10).

The *initiation module* in an NRPS assembly line typically has a 50 kDa adenylation (A) domain, noted earlier in this chapter, and a carrier protein domain, abbreviated either as PCP or T (thiolation domain, to emphasize the thiol group from the pantetheinyl arm). The A domain selects the amino acid, including any nonproteinogenic ones, activates it as the aminoacyl-AMP and then tethers it on thioester linkages to the P: pantetheinyl thiol.

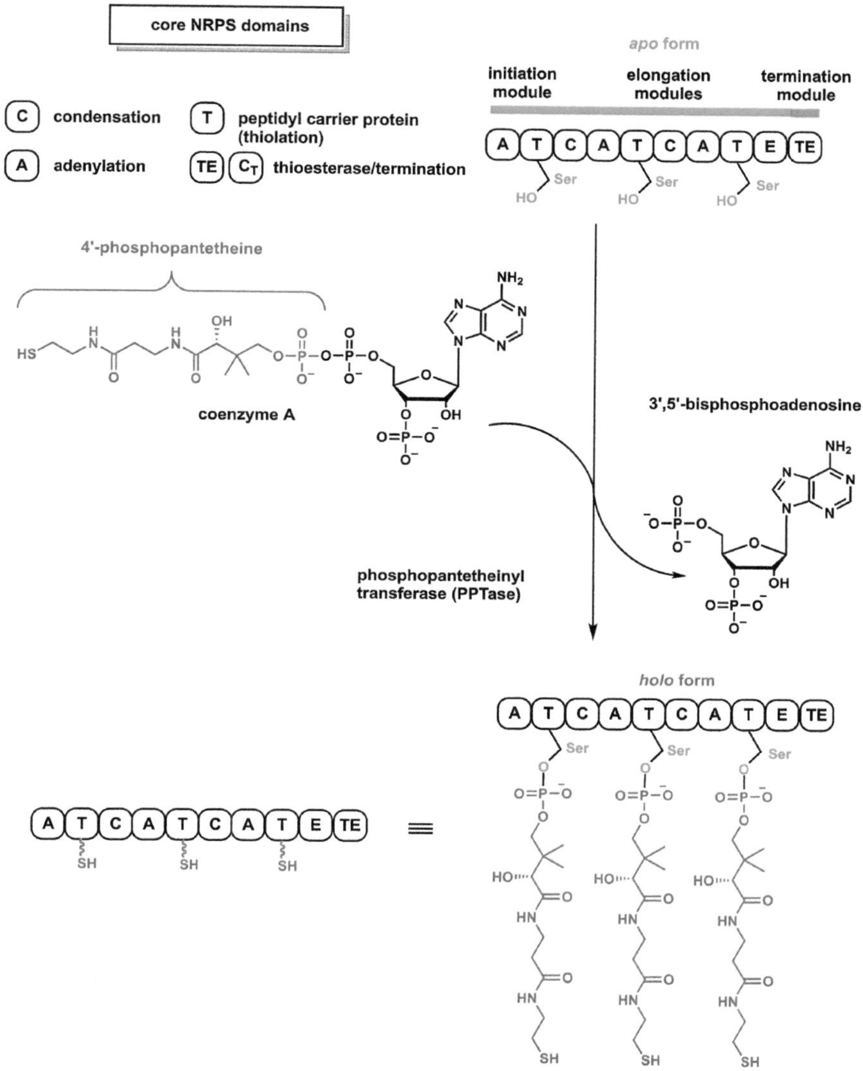

Figure 3.29 Conversion of *apo* forms of peptidyl carrier proteins (PCPs) to *holo* forms by posttranslational phosphopantetheinyl donation from coenzyme A to a seryl side chain in each PCP domain. The PCP and T abbreviations are used interchangeably as the T domain takes up less space and implies function.

Elongation modules have the same A–T (A-PCP) pair of modules and also a 50 kDa condensation domain (C = condensation) that catalyzes the peptide bond forming step between modules (C–A–T/C–A–PCP).

The upstream module has the elongating peptidyl chain whose thioester carbonyl will function as electrophile. The downstream module has the incoming aminoacyl thioester where the amine acts as the nucleophile. Peptide bond formation accompanies chain transfer to the downstream module. Thus, chain growth is from upstream to downstream, from N- to C-terminus of the (multimodule/multiprotein) NRPS assembly line.

When the full length peptidyl chain reaches the most downstream module of an NRPS assembly line it has reached the chain *termination module*. In bacterial systems, but not necessarily in fungal NRPS ones, the termination module has a fourth domain, a 35 kDa thioesterase domain as the last domain (C–A–T–TE/C–A–PCP–TE). In fungal systems the arrangement may be C–A–T–C, where the terminal C domain functions in chain release (Gao, Haynes *et al.* 2012). Peptidyl chain transfer in C–A–T–TE termination modules is effected from the thiolation domain to a nucleophilic Ser-side chain-OH, switching at this juncture from peptidyl thioester to peptidyl oxoester chemistry. The TE domains are part of the serine hydrolome superfamily that uses nucleophilic serines in enzymes to carry out acyl/peptidyl transfers to various cosubstrates.

The prototypic NRPS assembly line is the one protein 450 kDa fungal tripeptide synthetase for aminoadipoyl-cysteinyl-D-valine (ACV). ACV is the immediate progenitor to isopenicillin N, in turn a precursor to cephalosporin C, the two classes of β-lactam antibiotics. As depicted in Figure 3.30, there are three modules, encompassing 10 autonomously folding domains in ACV synthetase. The order of amino acids selected, activated, and incorporated is determined by the order of the three A domains.

The first domain recognizes the nonproteinogenic aminoadipate, specifically activating the C_6 carboxylate not the C_1 carboxylate. A_2 activates Cys, A_3 activates L-Val. The aminoacyl-AMPs then get installed as aminoacyl thioesters on their neighboring T (thiolation) domains by way of the phosphopantetheinyl arms. The C domain of module 2 transfers the aminoadipyl group as it builds the aminoadipyl–cysteinyl–dipeptidyl thioester on T_2. Analogously, the C domain on module 3 uses the Val-NH_2 group as nucleophile, and creates A–C–V, in thioester linkage to T_3. The presence of an epimerase (E) domain in module 3 indicates that, either before condensation or after tripeptidyl-*S*-pantetheinyl thioester formation, the α carbon of the Val residue is epimerized from L- to D-configuration. Released ACV is indeed aminoadipyl-L-Cys-D-Val and that is the chirality required for subsequent processing to penicillins.

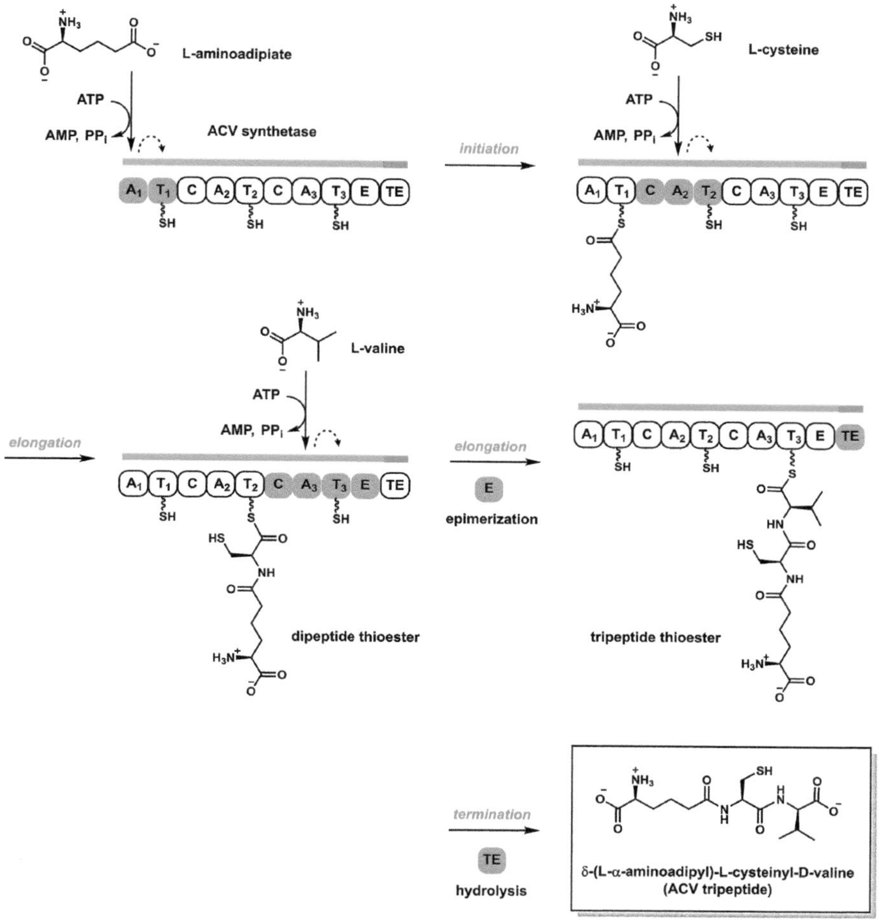

Figure 3.30 Aminoadipyl-L-cysteinyl-D-valine (ACV) synthetase is a three module, 10 domain, single chain 450 kDa enzyme. In addition to adenylation (A) and thiolation (T) domains in each module, it has condensation (C) domains in modules two and three, an epimerase (E) domain in module 3 and a chain-releasing thioesterase (TE) domain, also in module 3, as the most downstream domain.

3.7 Different Chain Release Fates in the NRPS Termination Step

Figure 3.31 shows action of the thioesterase as a hydrolase, releasing the tripeptide as the free acid. Similarly, the vancomycin assembly

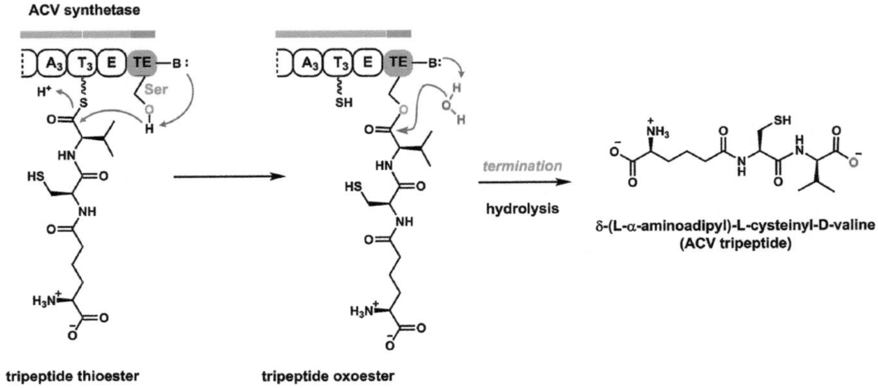

Figure 3.31 The TE domain of ACV synthetase acts as a hydrolase.

line TE also acts as an oxoester hydrolase, converting the crosslinked heptapeptidyl-*O*-TE to the aglycone free acid (see Figure 3.29). However, hydrolysis is only one of several known modes of release of full length chains from NRPS TE domains.

A second, predominant, mode is macrocyclization, where an intramolecular nucleophile within the chain outcompetes water attack (Walsh and Wencewicz 2016). Typical internal nucleophiles include amines, as in Figure 3.32, which shows amine capture in tyrocidine (head to tail) and in bacitracin (side chain on C-terminus, leading to lariat structure). Kinetically competent side chain –OH groups are illustrated by the 4–13 macrolactone in the antibiotic daptomycin. Figure 3.32 also shows two additional cyclic peptide scaffolds from NRPS release. Head to tail dimerization of identical pentapeptidyl chains leads to the cyclic decapeptidyl lactam gramicidin S, while cyclotrimerization of dihydroxybenzoyl–seryl moieties produces the *Escherichia coli* iron chelator enterobactin.

Almost all nonribosomal lipopeptides, such as daptomycin (shown in Figure 3.32), are acylated on the first residue of the peptide chain and carried through the elongation and termination modules. Additional lipopeptide structures are shown in Figure 3.33. Like daptomycin, the lariat structures in the macrolactone lipopeptides syringomycin, fengycin, and ADEP1 arise through the attack of a side chain hydroxyl nucleophile (serine or tyrosine) on the carbonyl of the terminal residue. Attack of side chain amines to form macrolactam lipopeptides is observed in polymxin B2 and fruilimicin, where the side chains of lysine and the unusual 2,3-diaminobutyrate are

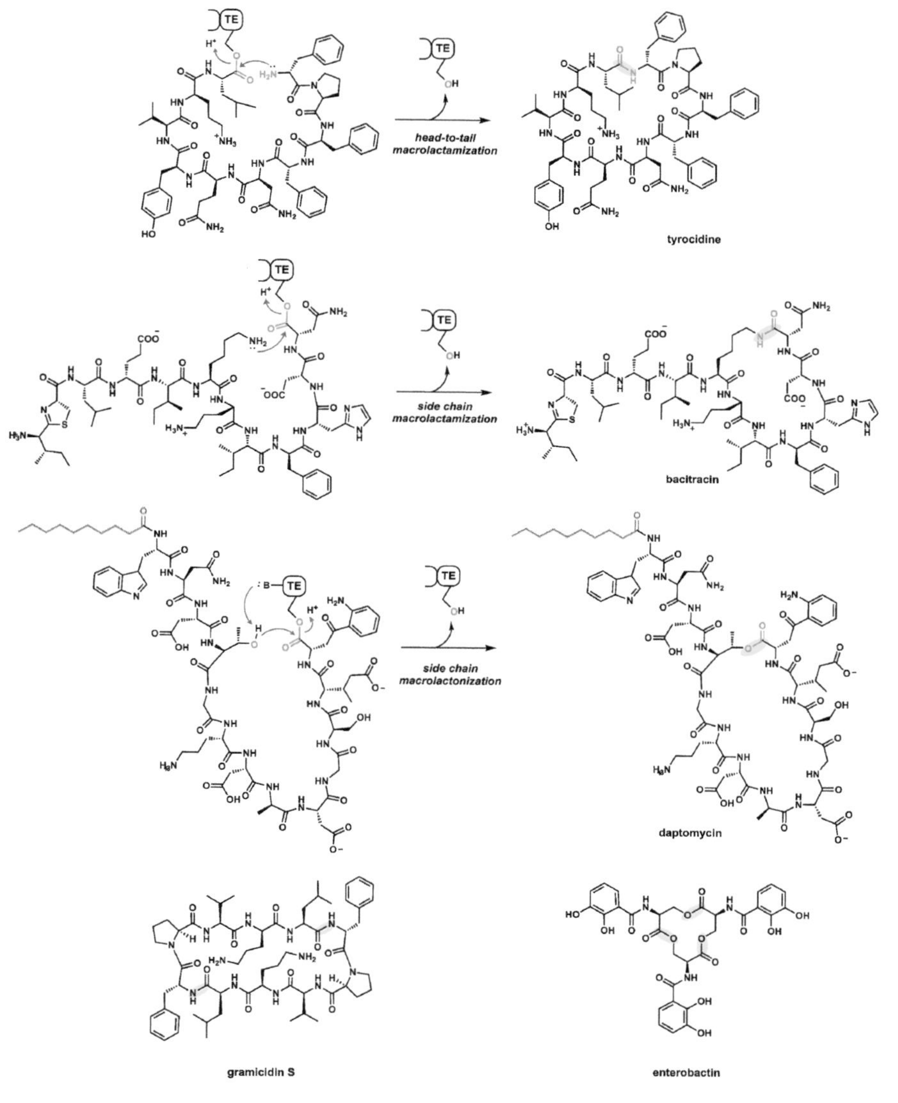

Figure 3.32 Macrocycle formation by the thioesterase domain in NRPS. Shown are macrolactamization by head to tail capture of the peptidyl-*O*-TE in tyrocidine, and through the side chain to C-terminus cyclization in bacitracin. Macrolactone formation can occur *via* side chain hydroxyl attack once (daptomycin) or even three times (enterobactin), while two pentapeptidyl intermediates dimerize to gramicidin S.

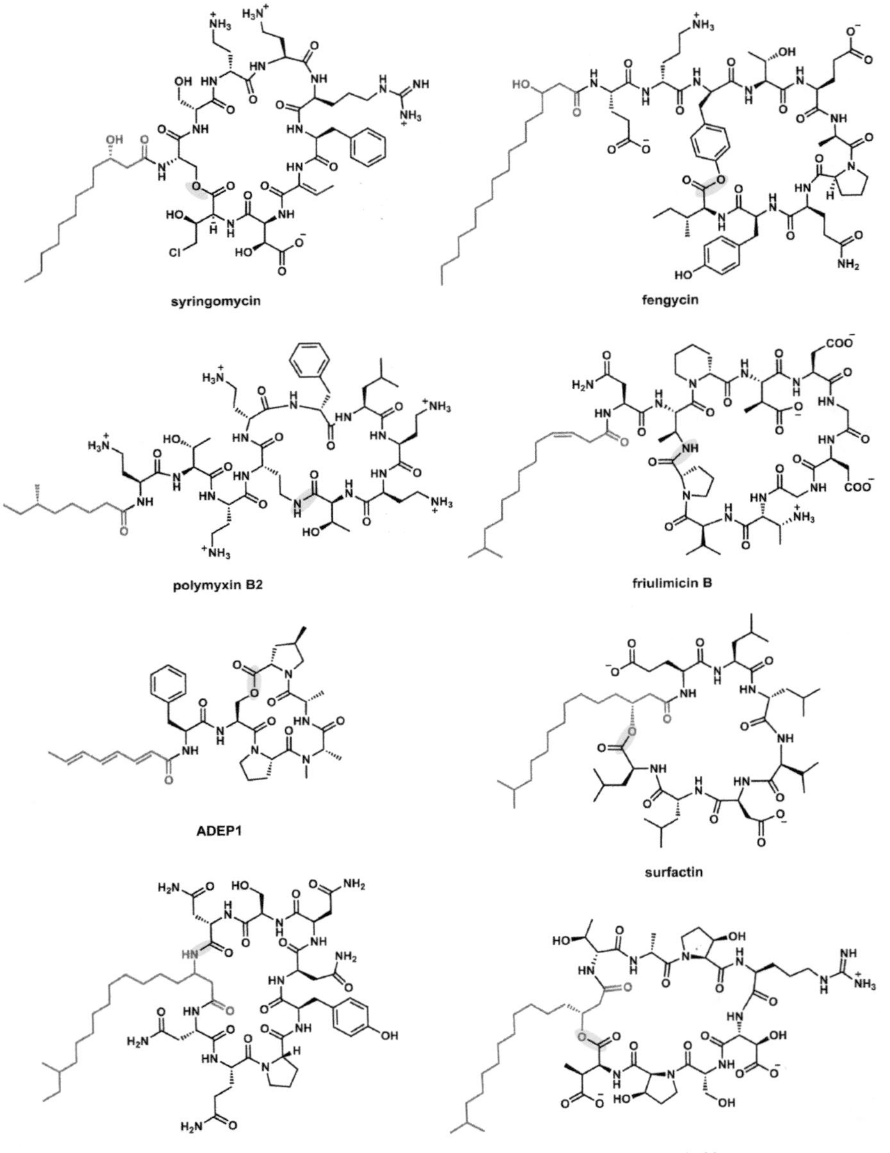

Figure 3.33 Lipopeptides produced from NRPS assembly lines. The lipid acyl chains are installed at the first NRPS module and are shown in maroon. The bond that forms the macrocycle is highlighted in yellow.

involved. Plusbacin A3, surfactin, and mycosubtilin (also shown in Figure 3.34), are formed when the attacking –OH or $-NH_2$ in the macrocyclization comes from the β-hydroxy or β-amine of the fatty acyl chain on residue 1 of the peptide rather than from any amino acid residue side chain.

The third release outcome is mediated by a C-terminal reductase domain, the mechanism of which is elaborated on in Figure 3.34. The TE domains in the lyngbyatoxin, nostocyclopeptide, myxochelin, and myxalamide NRPS assembly lines have been replaced by an NAD(P)H-utilizing reductase (R) domain (Barajas, Phelan *et al.* 2015). One cycle of reduction converts the covalently tethered peptidyl thioester to the thiohemiacetal, which can unravel to the noncovalently connected product aldehyde. This is the case for nostocyclopeptide. The free aldehyde cyclizes to the imine. In the lyngbyatoxin and myxochelin cases the nascent aldehyde is reduced a second time to the alcohol, before product diffuses out of the active site. The four electron reduction of a thioester to an alcohol also gives the signature terminal amino alcohol found in the peptaibol natural products, as shown in alamethicin in Figure 3.22.

The fourth outcome, depicted in Figure 3.35, is a Dieckmann condensation during formation of cycloacetoacetyl-Trp in the cyclopiazonic acid pathway (also see Chapter 8). A catalytically dead reductase domain (R*) mediates enolate formation and internal cyclization to produce the tetramate ring system in the released cycloacetoacetyl-Trp (Liu and Walsh 2009). Similar Dieckmann cyclization release modes have been found in other assembly lines, including those for equisetin, aspyridone, and tenellin. The distinct release modes, arising from fusion of alternative terminal domains in place of the canonical TE domain, allow scaffold variation combinatorially.

3.8 Structural Considerations of NRPS Assembly Lines

Structures of the constituent condensation, adenylation, thiolation = peptidyl carrier protein (see Figure 3.29), and thioesterase domains of NRPS modules have been available for some two decades (Koglin and Walsh 2009). However, the orientation of the domains within a module has been more difficult to determine structurally, in part because of the mobility that occurs within a catalytic cycle. For

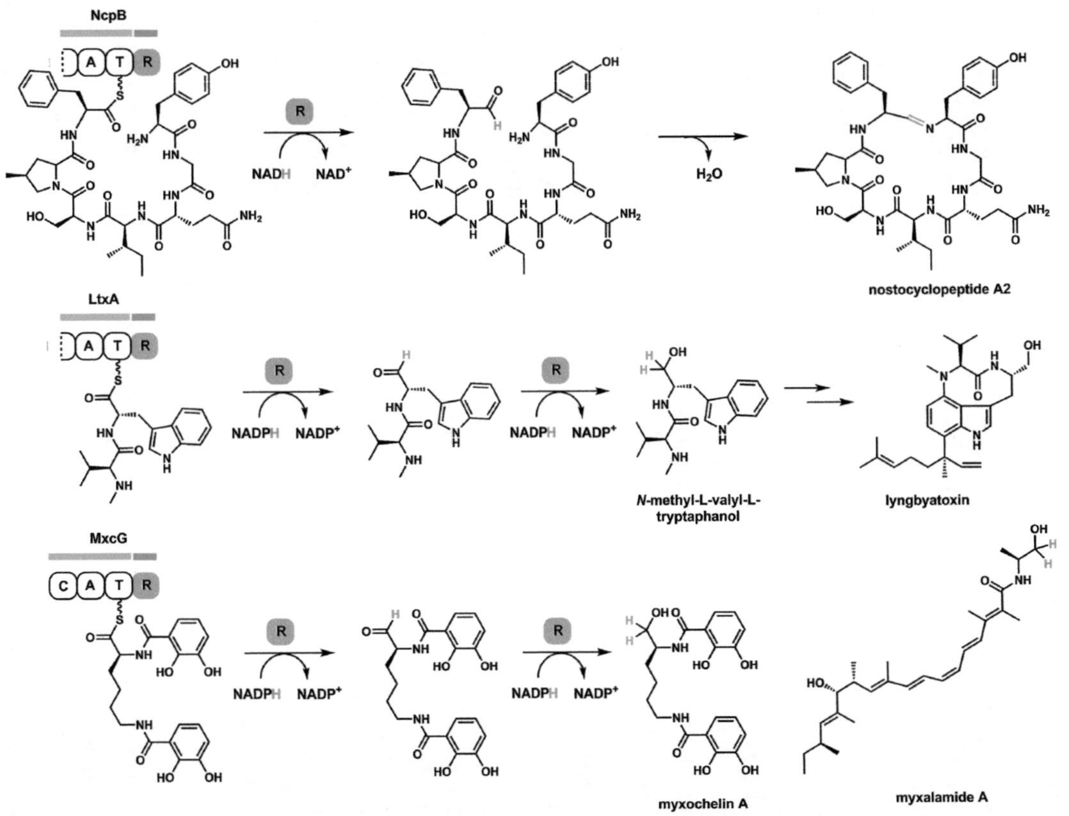

Figure 3.34 Peptidyl chain release by reduction. Two-electron reduction releases aldehydes. Four-electron reduction releases the peptidyl alcohol.

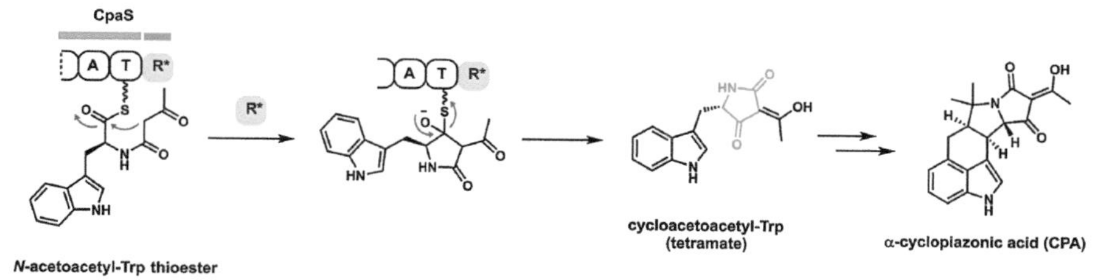

Figure 3.35 Mechanism of Dieckmann condensation mediated product release. The R* domain is a defective reductase fold and cannot carry out electron transfer. The C–C bond Dieckmann tetramate-forming condensation is the default catalytic mode for these R* domains.

example, as schematized in Figure 3.36A, in a termination module the peptidyl carrier protein has to visit the active site of the condensation domain, then the adenylation domain, and finally the thioesterase domain (Drake, Miller *et al.* 2016).

Through a series of protein fusions and immobilizing substrate analogs, the structures of three termination modules have been determined and are displayed in Figure 3.36B (Strieker, Tanovic *et al.* 2010, Drake, Miller *et al.* 2016). They are the termination module of surfactin synthase, of a functionally unassigned NRPS from *Acinetobacter baumanii*, and the macrocyclizing terminal module for the *E. coli* siderophore enterobactin. There are differences in domain orientations towards each other but the overall placement of the four domains is constant, indicating the blueprint for module assembly. To that point Marahiel (Marahiel 2016) has proposed a three dimensional model for how the seven modules of surfactin synthetase could be organized (Figure 3.36C), to indicate how the heptapeptidyl chain could be assembled, elongated, and released.

3.9 Pre-assembly Line *vs.* On-assembly Line *vs.* Post-assembly Line Tailoring of Peptidyl Chains

There are three stages where additional enzymatic tailoring of nonribosomal peptide scaffolds could occur. One can be categorized as pre-assembly line. In a sense all the nonproteinogenic amino acids qualify in this category (Hubbard and Walsh 2003, Walsh, O'Brien

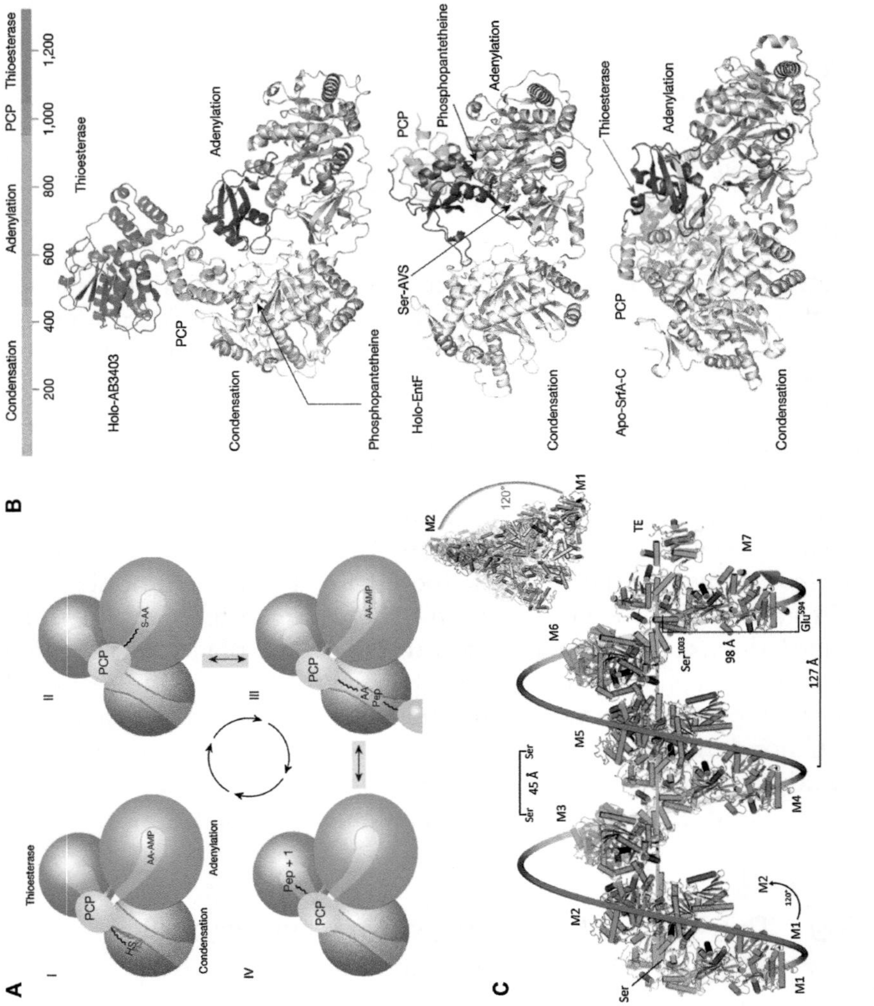
A
I
II
III
IV
Thioesterase
Condensation
Adenylation
PCP
AA-AMP
S-AA
HS
Pep + 1
AA
Pep
B
Condensation
Adenylation
PCP
Thioesterase
200
400
600
800
1,000
1,200
Holo-AB3403
Phosphopantetheine
Holo-EntF
Ser-AVS
Apo-SrfA-C
C
M1
M2
M3
M4
M5
M6
M7
TE
Ser
45 Å
98 Å
127 Å
Ser1003
Glu594
120°

et al. 2013). This is particularly relevant in the 4-OH-PheGly and 3,5-$(OH)_2$-PheGly cases we noted above as specialized building blocks for the vancomycin/teicoplanin class of antibiotics under "just in time" inventory control.

We noted above that epimerizations typically happen on the assembly lines. In part this may be due to the limited availability of D-amino acids in metabolite pools, compared to L-amino acids. The ACV synthetase is a prime example of a built-in epimerase domain (in module 3). Polytheonamide undergoes 18 epimerization steps on alternating L-amino acid residues during posttranslational maturation. On the other hand, cyclosporine synthetase utilizes D-alanine directly, perhaps because it is relatively abundant from bacterial cell wall metabolism. Vancomycin and teicoplanin have D-amino acids at residues 2, 4, and 5. Inspection of the assembly line reveals epimerase domains in modules 2, 4, and 5, consistent with selection of L-tyrosine and L-4-OH-PheGly monomers and epimerization on the assembly line.

N-Methylation is a fairly common feature of some classes of nonribosomal peptides, including cyclosporine A (see Figure 3.23). Seven of the 11 residues are *N*-methylated, altering both the resonance stability and planarity of this undecapeptidyl lactam such that it can pass through biological membranes diffusively. *N*-Methylation domains, using SAM as electrophilic methyl donor, are inserted into each of these seven modules of cyclosporine synthetase.

Figure 3.36 (A) Dynamics of NRPS catalytic cycles: the PCP domain bearing the growing peptidyl chain visits the condensation, adenylation and thioesterase domains in the final module. (B) Architecture of three NRPS termination modules: The seventh module of surfactin synthetase, the termination module of an unassigned *A. baumanii* module, the EntF protein involved in enterobactin chain elongation and trimerization. (C) Model for the seven module surfactin synthetase. Parts (A) and (B) reprinted by permission from Macmillan Publishers Ltd: Nature (Drake, E. J., B. R. Miller, C. Shi, J. T. Tarrasch, J. A. Sundlov, C. L. Allen, G. Skiniotis, C. C. Aldrich and A. M. Gulick. "Structures of Two Distinct Conformations of holo-Nonribosomal Peptide Synthetases". *Nature*, **529**: 235) copyright (2016). Part (C) reproduced from Marahiel, M. A. (2016). "A Structural Model for Multimodular NRPS Assembly Lines". *Nat. Prod. Rep.* **33**: 136 with permission from The Royal Society of Chemistry.

One additional notable set of on-assembly line modifications are the oxidative crosslinking events connecting chains 2–4, 4–6, 5–7 in vancomycin and 1–3, 2–4, 4–6, 5–7 in teicoplanin. The mechanisms are discussed in Chapter 9 under oxygenase chemistry but are noted here in terms of timing.

Post-assembly line tailoring events are catalyzed by dedicated enzymes that can carry out acylations (teicoplanin), *N*-methylations (vancomycin), glycosylations (vancomycin and teicoplanin), and side chain hydroxylations (echinocandins) (Walsh and Fischbach 2010). Chapter 11 takes up natural product glycosylations in detail, including the vancomycin and teicoplanin aglycone scaffolds. Of special note are the consecutive enzymes processing acyclic ACV to bicyclic 4,5-fused isopenicillin N and then on to fused bicyclic 4,6-fused cephalosporin antibiotics (Figure 3.37), which will be discussed in Chapter 9.

3.10 NRP–PK Hybrids: Machinery and Examples

Given that many polyketides discussed in Chapter 2 are built on assembly lines that tether monomers and elongating chains as phosphopantetheinyl-thioesters covalently attached to carrier protein domains (ACPs), it is not surprising that NRPS assembly lines, which use parallel chemical logic and PCPs (thiolation domain) as covalent way stations, could evolve to interface with PKS assembly lines and *vice versa* (Richter, Nietlispach *et al.* 2007, Weissman 2015). Productive transfer of elongating polyketide chains onto aminoacyl-*S*-PCPs and, inversely, of peptidyl chains onto downstream malonyl-*S*-ACPs as carbanion equivalents, highlights two sets of recognition issues, as depicted in Figure 3.38.

When an NRPS module is upstream of a PKS module and the growing chain is transferred, a new C–C (rather than an N–C amide) bond is made. This requires the ketosynthase domain of the downstream PKS module to recognize the thioester carbonyl of the growing peptidyl chain rather than the canonical polyketidyl chain. Correspondingly, when the PKS module is upstream of an NRPS module, the condensation domain must recognize the polyketidyl thioester as an acceptable electrophilic partner for N–C amide bond formation. In addition to the permissivity with regard to heterologous growing peptidyl and polyketidyl chains, if the modules are on two distinct proteins there must be enough protein–protein interaction to guide specificity between the heterologous subunits. Presumably it is less of

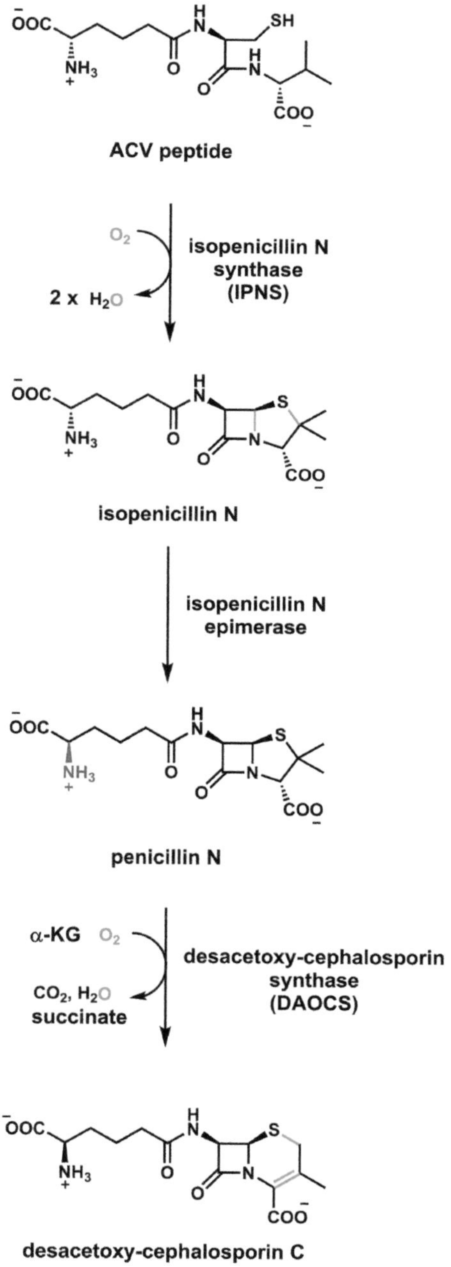

Figure 3.37 Post-NRPS modification of the linear ACV peptide into the β-lactam antibiotics penicillin and cephalosporin requires iron dependent oxygenases.

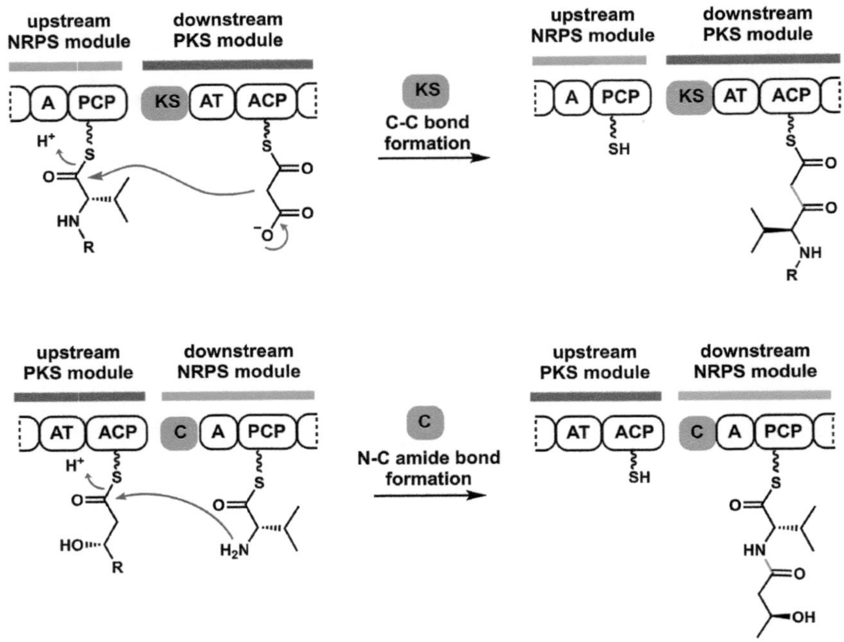

Figure 3.38 Recognition issues at PKS–NRPS and NRPS–PKS module interfaces.

a problem if the PKS and NRPS modules have become fused in a single protein.

Figure 3.24 details the reaction at the PKS/NRPS interface in formation of the β-OH, δ-NH_2 amino acid statine, itself a hybrid building block noted earlier in the example of didemnin B (Xu, Kersten *et al.* 2012). This is a stripped down case of recognition where the ketosynthase decarboxylates malonyl-*S*-ACP to generate the carbon nucleophile to attack the upstream valyl-*S*-PCP. The C–C bond formation occurs with formation of the hybrid scaffold and transfer to the ACP way station. Subsequent ketoreduction and thioesterase-mediated hydrolytic cleavage in the PKS module release the hybrid building block.

Figure 3.39 shows a Venn diagram of predicted NRP, PK, and hybrid NRP–PK molecules from a genome mining survey of ∼2700 genomes (Wang, Fewer *et al.* 2014). About 30% of the ∼3300 biosynthetic gene clusters are predicted to encode enzymes that make hybrid scaffolds. This suggests a huge opportunity for discovery since only a few dozen NRP–PK hybrids have been characterized to date. Further, the mix of PK and NRP structural and functional group

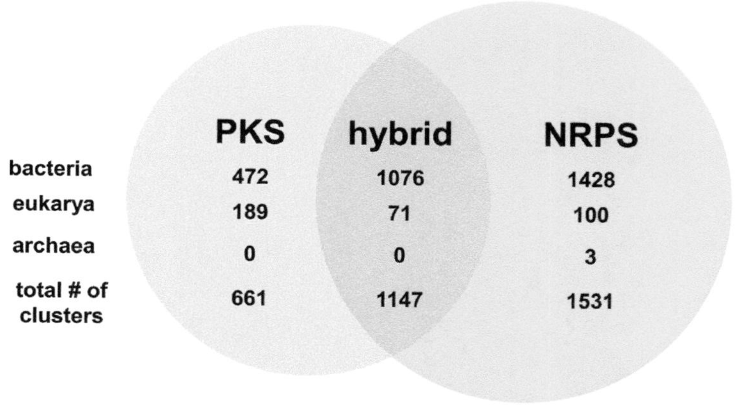

Figure 3.39 Venn diagram of PKS, NRPS, and NRPS–PKS/PKS–NRPS hybrid open reading frames (ORFs) from genome mining. Modified from Wang, H., D. P. Fewer, L. Holm, L. Rouhianinen and K. Sivonen (2014). "Atlas of Nonribosomal Peptide and Polyketide Biosynthetic Pathways Reveals Common Occurrence of Nonmodular Enzymes". *Proc. Natl. Acad. Sci. U. S. A.* **111**(25).

elements offers total synthesis challenges and novel molecular architectures.

The integrated PKS/NRPS assembly line for the cyanobacterial hybrid molecule nostophycin contains 10 modules distributed over three proteins (Fewer, Osterholm *et al.* 2011) (Figure 3.40). The NpnB protein has one PKS module and two NRPS modules, presumably reflecting an evolutionary gene fusion. The nostophycin lariat scaffold is mostly peptidic with one PK-derived building block that is actually constructed on the hybrid assembly line. The first three are PKS modules, the last six are NRPS modules. The three PKS modules build a 2,5-dihydroxy-3-oxo-8-phenyl octanoyl scaffold tethered to ACP_3. Module 3 has an embedded transaminase (AMT) domain that acts to generate the 3-amino group in what is still a polyketidyl background. Then the six NRPS modules build a hexapeptidyl chain and engineer a macrolactamizing release with that amino group from the polyketide moiety attacking the Pro_6-O-TE acyl enzyme. The released nostophycin has the macrolactam cyclic core with the polyketidyl unit as the tail of the lariat structure.

Two additional examples of building blocks that contain an amino group but are built on PKS assembly lines are shown in Figure 3.41. The 3-amino-9-methoxy-2,6,8 trimethyl-10-phenyl-deca-4,6-dienoic acid (Adda) is found in cyanobacterial peptide toxins

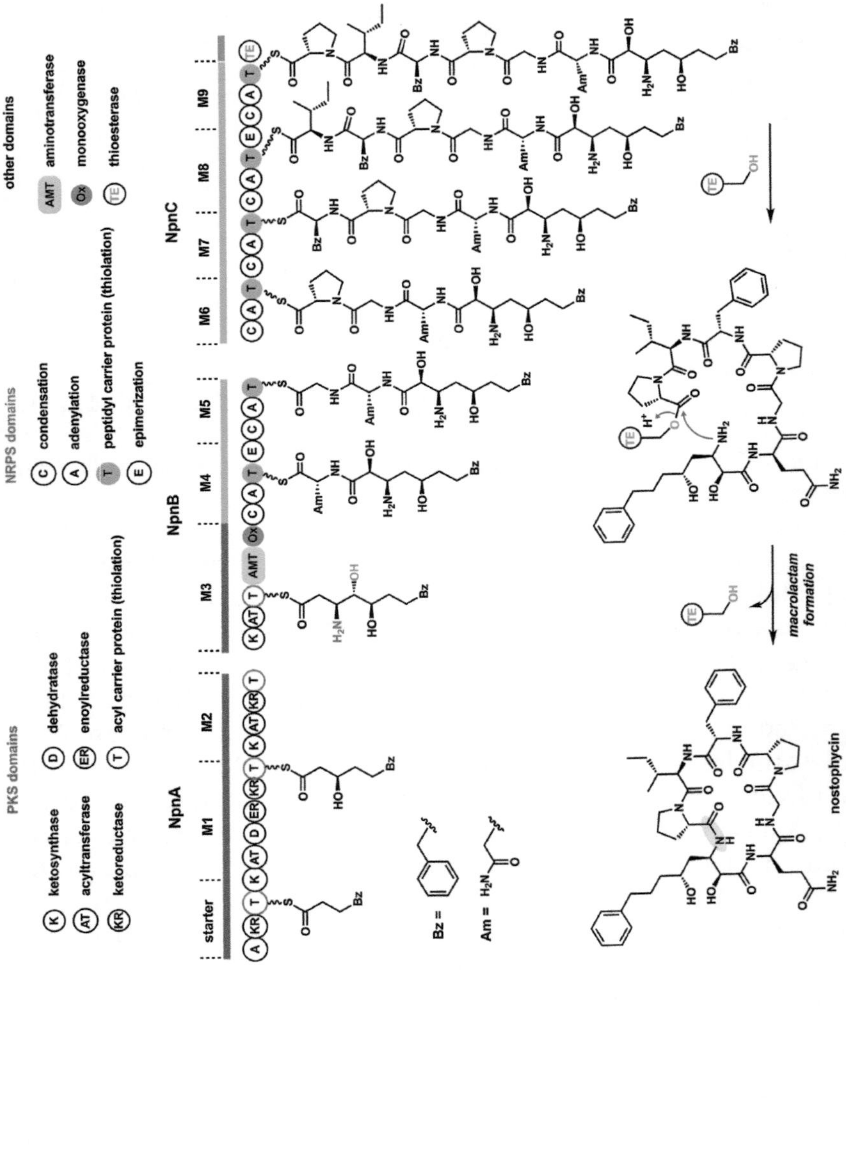

Figure 3.40 Nostophycin is built on a hybrid PKS/NRPS integrated assembly line.

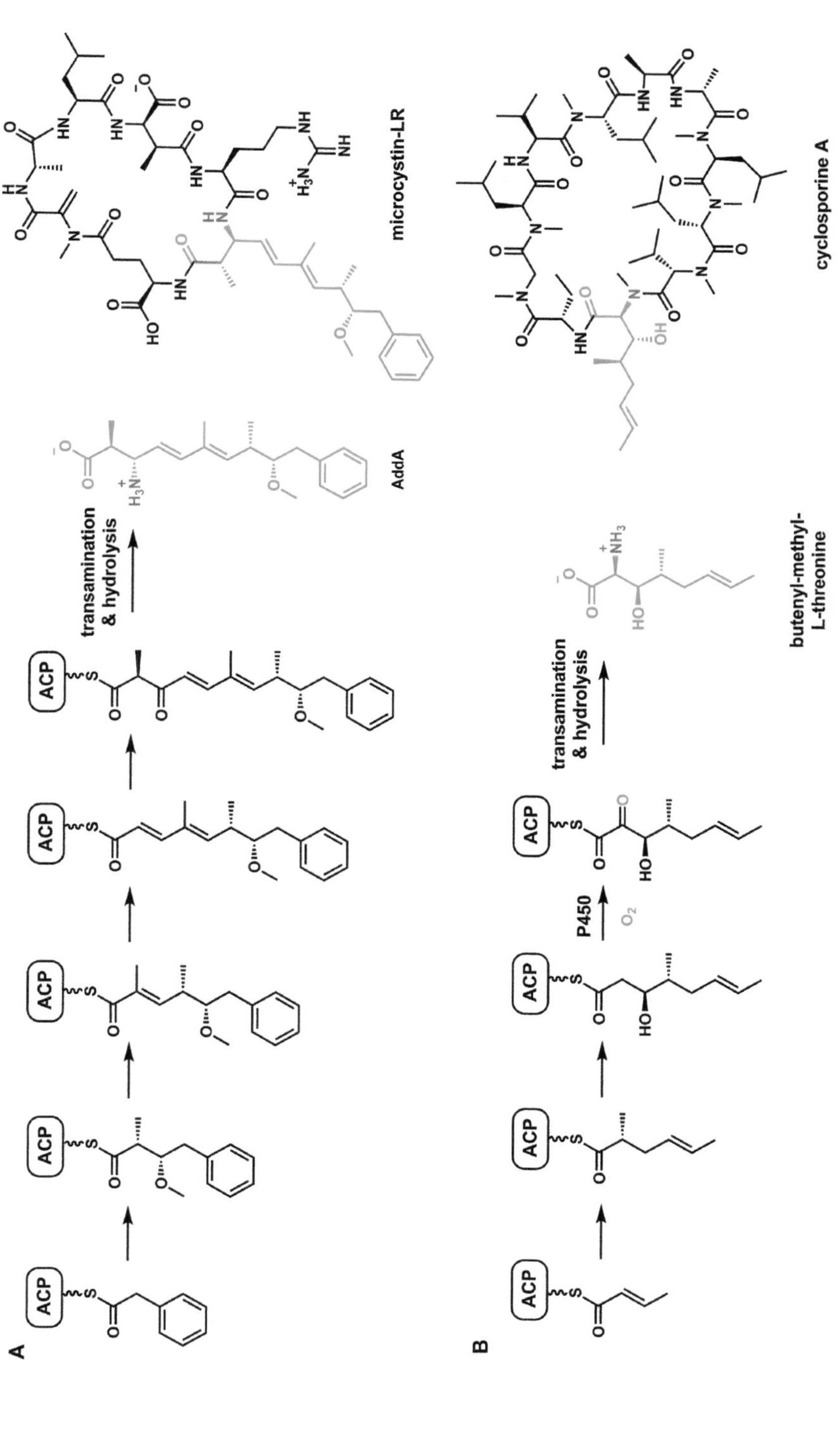

Figure 3.41 Unusual amino acids built on polyketide synthase assembly lines and transaminated at late stages: Adda and butenyl-methyl-L-threonine.

such as microcystin-LR. Late stage transamination of the 3-keto group yields the β-amino rather than a conventional α-amino acid framework (Walsh, O'Brien *et al.* 2013). Analogously, the butenyl-methyl-L-threonine residue in cyclosporine A is built on a PKS assembly line and subjected to enzymatic reductive transamination late in its assembly. The α-position in the polyketide chain is first subjected to P450-mediated oxidation to an α-keto-acyl chain, which can be transaminated and hydrolyzed off the PKS to yield the α-amino acid.

Additional NRP–PK hybrids of note include the epothilone anticancer class of agents (tubulin binders), the bleomycin class of DNA-targeting drugs, the sangliferin class of immunosuppressive agents, and the iron-chelating siderophore yersiniabactin, a virulence factor from the plague bacterium *Yersinia pestis* (Figure 3.42). The epothilone assembly line has only one NRPS module, as protein EpoB, in a sea of PKS modules (Julien, Shah *et al.* 2000, Chen, O'Connor *et al.* 2001, O'Connor, Chen *et al.* 2002). The peptide portion is shown in blue. The bleomycin assembly line has the opposite characteristics, with NRPS modules dominating the elongation phase (Shen, Du *et al.* 2002). The polyketide portions derived from

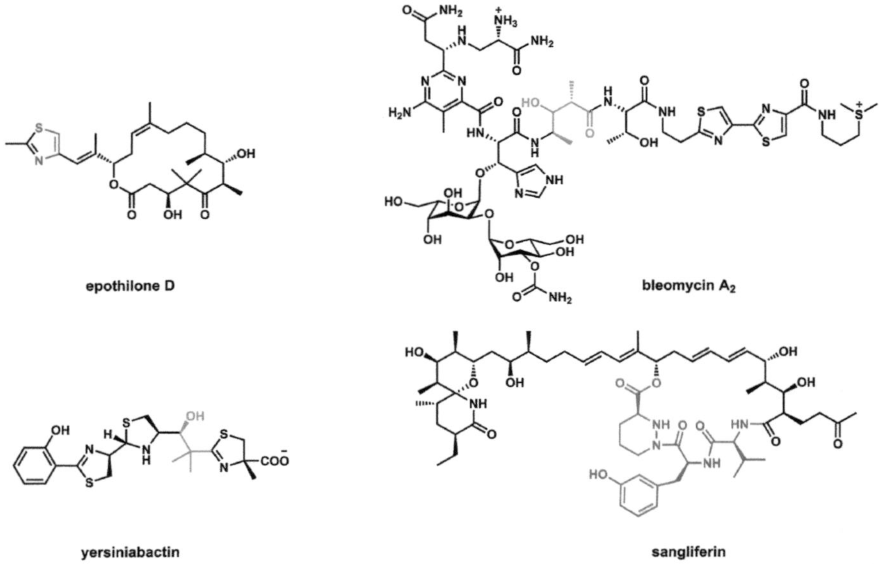

Figure 3.42 Four additional NRP–PK/PK–NRP hybrid natural products: epothilone, bleomycin, sangliferin, and yersiniabactin.

the single PKS module in both bleomycin and yersiniabactin are shown in red.

The assembly line for the immunosuppressant drug rapamycin contains 15 modules: the first 14 are PKS modules, the 15th and last is the sole NRPS module (Figure 3.43) (Schwecke, Aparicio *et al.* 1995, Aparicio, Molnar *et al.* 1996). The NRPS module has four domains: C_1-A-PCP-C_2. The A domain is specific for pipecolic acid,

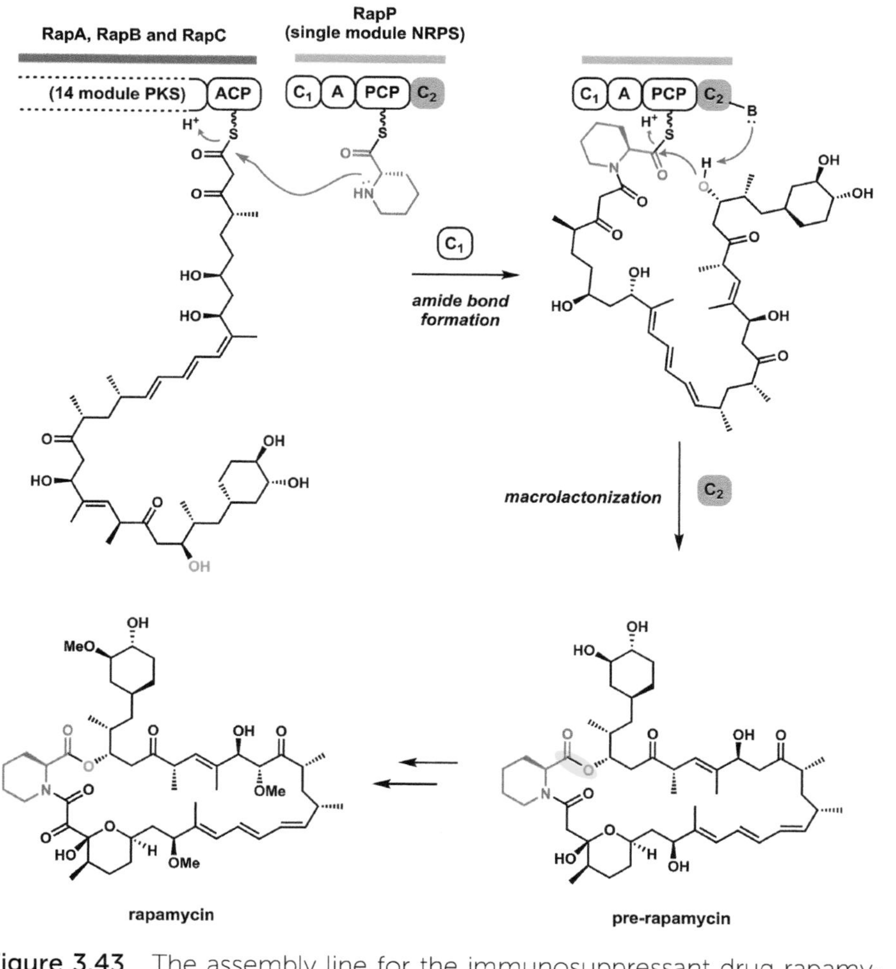

Figure 3.43 The assembly line for the immunosuppressant drug rapamycin consists of 14 PKS modules and then a 15th and final NRPS module. The NRPS module selects, activates, and inserts pipecolate into the both ends of the polyketide chain to make a hybrid macrolactone.

the higher homolog of proline (Konig, Schwecke *et al.* 1997, Gatto, McLoughlin *et al.* 2005). The C_1 domain is presumed to use the secondary nitrogen of the pipecolyl-*S*-PCP to transfer the full length polyketidyl chain resting on ACP_{14} to generate the amide bond and move the chain onto the PCP domain. Then the C_2 domain of the NRPS module is thought to assist in intramolecular attack of the indicated side chain-OH of the polyketidyl moiety onto the pipecolyl-thioester carbonyl. The product is the macrocyclic pre-rapamycin. The net effect of action of the NRPS module is insertion of the lone amino acid into the cyclic polyketide framework (Gatto, McLoughlin *et al.* 2005).

Vignette 3.2 β-Lactam Antibiotics and Immunosuppressant Cyclic Peptides.

The β-lactam antibiotics and the immunosuppressant cyclic peptides cyclosporine and the rapamycin/FK506 pair are two sets of nonribosomal peptide natural products, derived from fungi or bacteria (rapamycin), that have had enormous impact in two different areas of medical therapy. The lactam antibiotics were discovered in the 1920s (penicillin) and the 1940s (cephalosporins), while the immunosuppressant peptides were isolated and characterized in the 1970s (cyclosporine, rapamycin) and 1980s (FK506) (Figure 3.V2).

The two major classes of β-lactams, the 4,5-bicyclic penicillins and the 4,6-ring expanded cephalosporins, have a precursor to product relationship metabolically. Fungal NRPS assembly lines release the acyclic tripeptide aminoadipyl-cysteinyl-D-valine. This is converted to isopenicillin N by a nonheme iron enzyme, isopenicillin N synthase (see Figure 3.37 and Chapter 9) that reduces a cosubstrate O_2 by four electrons to two molecules of H_2O, as it builds the lactam and then the thiane rings by radical routes. After epimerization of the aminoadipyl side chain, the next enzyme also reduces O_2 while expanding the five ring thiane to the six ring cephem as desacetoxycephalosporin is fashioned.

The β-lactams target the crosslinking bacterial cell wall transpeptidases, forming long-lived covalent acyl enzyme intermediates (Walsh and Wencewicz 2016). Interdiction of peptidoglycan strand crosslinking leaves the cell walls mechanically weak, and susceptible to osmotic lysis as the cause of bacterial cell death. An enormous range of semisynthetic penicillins and cephalosporins, with the 4,5- and 4,6-warheads intact but with variant acyl chains, have been designed and synthesized over the decades to combat waves of pathogenic bacteria, and together this antibiotic class accounts for over 50% of the annual global $40 billion antibiotic market. If people around the globe were asked to name one antibiotic, it would probably be penicillins.

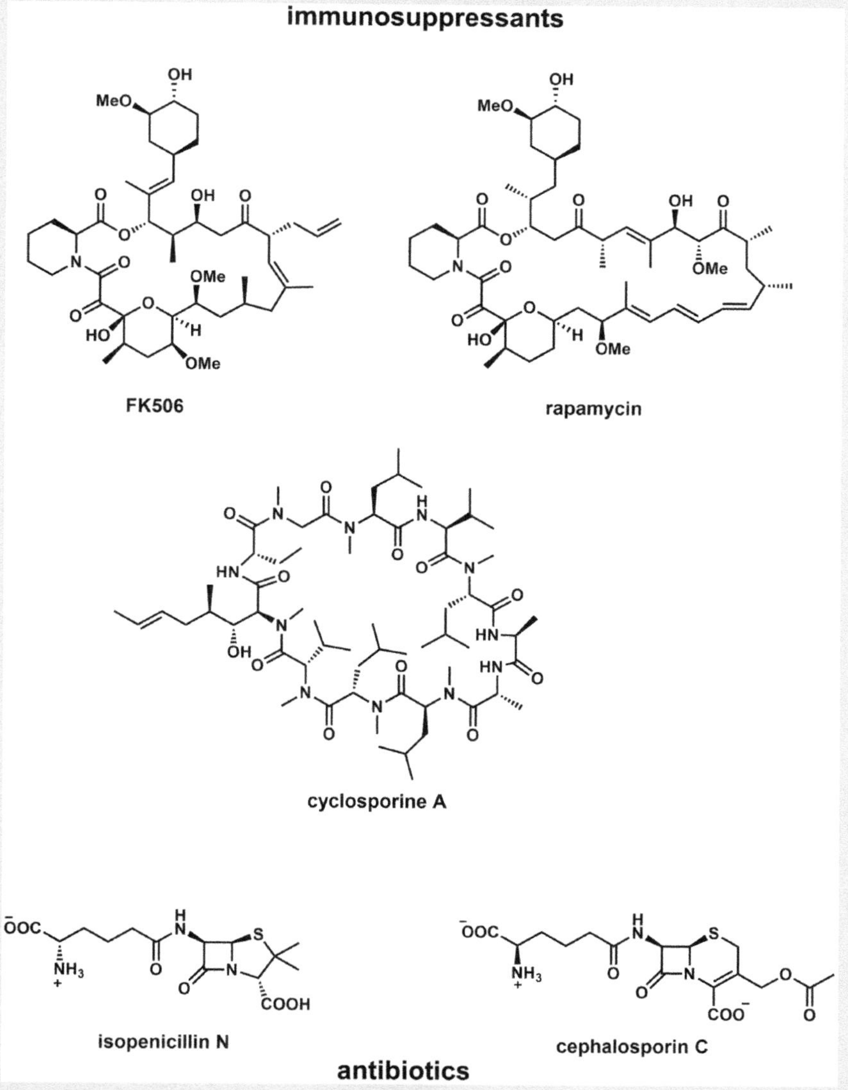

Figure 3.V2 Five NRPS-derived natural products that changed the world.

Cyclosporine (CsA) was isolated from fungi collected by Sandoz employees in both Wisconsin and Norway in 1971. Its immunosuppressive effects on T cells led to its approval as a drug to stop allograft rejection in organ transplants, allowing explosive growth in the medicine of replaceable human parts. Cyclosporine was more selective than previous combinations of immunosuppressive drug regimens and less toxic. The

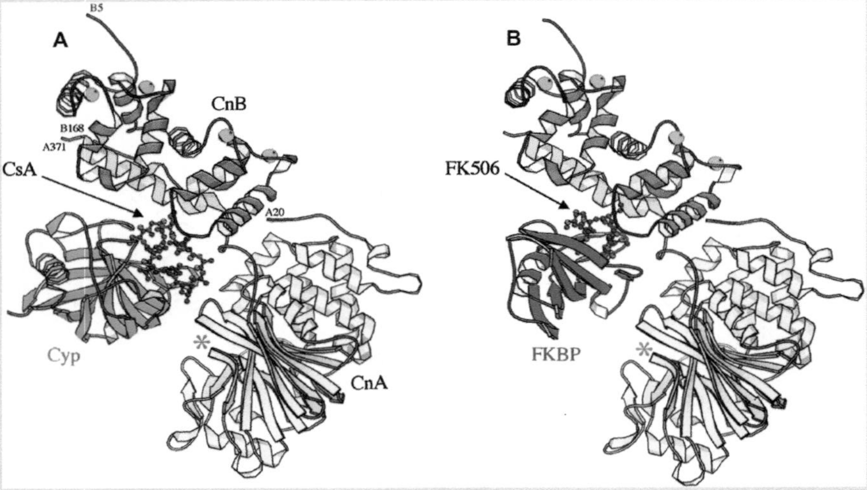

Figure 3.V3 Reprinted with permission from Jin, L. and S. C. Harrison (2002). "Crystal Structure of Human Calcineurin Complexed with Cyclosporine A and Human Cyclophilin". *Proc. Natl. Acad. Sci. U. S. A.* **99**(21): 13522. Copyright (2002) National Academy of Sciences. Abbreviations: CsA: cyclosporine; Cyp: cyclophilin; CnA: calcineurin subunit A; CnB: calcineurin subunit B; FKBP: FK506-binding protein.

mechanism of action was sorted out by Schreiber and colleagues (Liu, Farmer *et al.* 1991) when they found that cyclosporine bound to cyclophilin protein and was thereby presented to and inhibited the protein phosphatase activity of calcineurin (Figure 3.V3) (Jin and Harrison 2002).

Rapamycin, a 23 membered macrolactone, is 94% polyketide and 6% peptide, as estimated from the number of PK synthase to NRPS modules in the biosynthetic assembly line. The molecule is produced by *Streptomyces hygroscopicus* and was originally discovered in a soil sample from Rapa Nui (Easter Island) in 1972. It targets a mammalian protein mTOR, a protein complex with protein kinase activity that integrates several cell signaling pathways. The clinical result is blockade of antigen-induced T cell and B cell proliferation. Rapamycin was then developed as Sirolimus and approved for prevention of kidney rejection. A semisynthetic *O*-hydroxyethyl derivative of FK506, everolimus, has proven useful as an anticancer agent in renal carcinoma patients.

FK506 was isolated from *Streptomyces tsukubaensis* in 1987. It has been commercialized as tacrolimus and approved as an antirejection agent, initially for liver transplants and then more broadly for solid organ and bone marrow transplants. FK506 is bound by a series of FK-binding proteins (FKBP) and in that complex, in analogy to cyclosporine, is presented to and inhibits the phosphatase activity of calcineurin (Figure 3.V3). A key protein substrate for calcineurin is NFAT, nuclear factor of activated T cells. When dephosphorylated it moves from the T cell cytoplasm to the nucleus and turns on the transcription of genes, including that encoding interleukin 2, which is a proliferative molecule. Thus, blockade of calcineurin, *via* immobilization of phospho-NFAT out of the nucleus, prevents IL-2 gene expression and T-cell proliferation. That is the molecular basis of immunosuppression for both FK506 and cyclosporine.

With any of the three immunosuppressive regimens, the reduced immunocompetency of patients leads to increased vulnerability to infections and some forms of cancers such as skin cancers and non-Hodgkin's lymphomas.

One last example of the kind of scaffold complexity generation that can arise in a hybrid assembly line is provided by myceliothermophin E, isolated from the fungus *Myceliophthora thermophila* (Yang, Lu *et al.* 2007) (Figure 3.44). An iterative PKS builds the 20 carbon

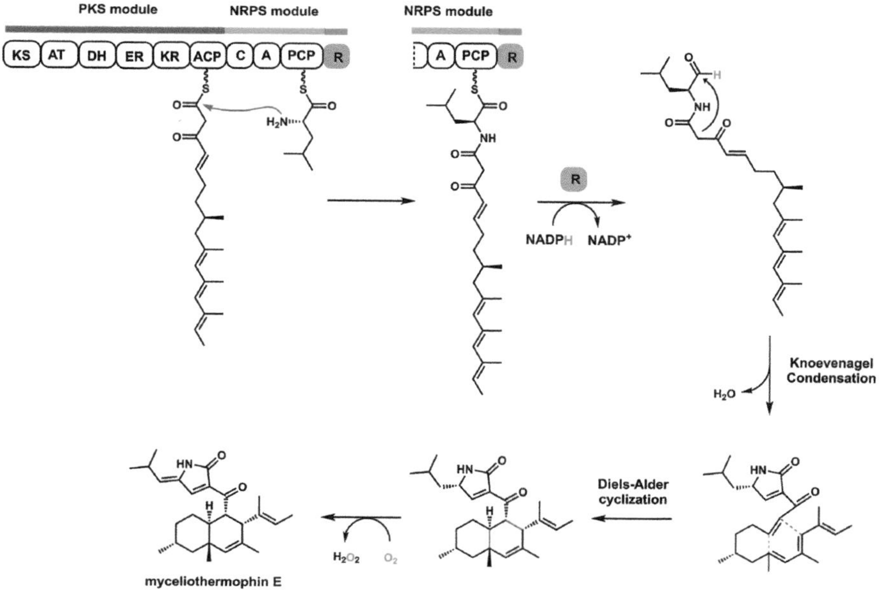

Figure 3.44 Myceliothermophin E is a hybrid PK–NRP molecule. It contains a decalin bicyclic system, most probably arising from a Diels–Alder cyclization (PK moiety) and a leucyl pyrrolidone ring system.

tetra-olefinic polyketide acyl framework that gets transferred to the NRPS portion of the integrated assembly line which bears a single leucine-specific A domain. Reductive release of the leucyl-polyketide chain followed by Knoevenagel Condensation yields the five-membered pyrrolidone. Subsequent Diels–Alderase-mediated cyclization of the triene (recall Diels–Alderase chemistry in Chapter 2) is proposed to create the bicyclic decalin ring system found in the mature myceliothermophin molecule, which exhibits submicromolar toxicity towards various cancer cell lines (Li, Yu *et al.* 2016).

3.11 Summary

Peptide-derived small molecular frameworks with the stability and conformational constraints that place them in the universe of natural small molecules can arise by two biosynthetic strategies.

One is posttranslational modification of nascent chains by dedicated tailoring enzymes, including proteases that excise stretches of modified peptides from their precursor protein backbones. Macrocyclization and heterocyclization of Ser, Thr, and Cys side chains create the architectural constraints and resistance to protease cleavage that allow the morphed products to have long lifetimes and conformations that function as high affinity ligands for a diverse array of biological targets.

The second strategy employs nonribosomal peptide synthetase assembly lines. Perhaps the most useful feature of NRPS machinery is the utilization of dozens of nonproteinogenic amino acids that bring in otherwise rare functional groups. Post assembly line tailoring by dedicated companion enzymes decorates and morphs the nascent products by acylation, alkylation, glycosylation, and most notably oxidation/oxygenation to fashion the mature, biologically active natural product framework.

Both strategies involve the transformation of linear, conformationally mobile, protease-sensitive peptide backbones (and side chains) into compact, stable, often macrocyclic frameworks. The evolution of hybrid NRP–PK assembly lines further broadens the scope of molecular architectures and functional group density that can be built into NRP–PK and PK–NRP hybrids, as exemplified in the preceding sections.

References

Akondi, K. B., M. Muttenthaler, S. Dutertre, Q. Kaas, D. J. Craik, R. J. Lewis and P. F. Alewood (2014). "Discovery, synthesis, and structure-activity relationships of conotoxins". *Chem. Rev.* **114**(11): 5815–5847.

Aparicio, J. F., I. Molnar, T. Schwecke, A. Konig, S. F. Haydock, L. E. Khaw, J. Staunton and P. F. Leadlay (1996). "Organization of the biosynthetic gene cluster for rapamycin in Streptomyces hygroscopicus: analysis of the enzymatic domains in the modular polyketide synthase". *Gene* **169**(1): 9–16.

Arnison, P. G., M. J. Bibb, G. Bierbaum, A. A. Bowers, T. S. Bugni, G. Bulaj, J. A. Camarero, D. J. Campopiano, G. L. Challis, J. Clardy, P. D. Cotter, D. J. Craik, M. Dawson, E. Dittmann, S. Donadio, P. C. Dorrestein, K. D. Entian, M. A. Fischbach, J. S. Garavelli, U. Goransson, C. W. Gruber, D. H. Haft, T. K. Hemscheidt, C. Hertweck, C. Hill, A. R. Horswill, M. Jaspars, W. L. Kelly, J. P. Klinman, O. P. Kuipers, A. J. Link, W. Liu, M. A. Marahiel, D. A. Mitchell, G. N. Moll, B. S. Moore, R. Muller, S. K. Nair, I. F. Nes, G. E. Norris, B. M. Olivera, H. Onaka, M. L. Patchett, J. Piel, M. J. Reaney, S. Rebuffat, R. P. Ross, H. G. Sahl, E. W. Schmidt, M. E. Selsted, K. Severinov, B. Shen, K. Sivonen, L. Smith, T. Stein, R. D. Sussmuth, J. R. Tagg, G. L. Tang, A. W. Truman, J. C. Vederas, C. T. Walsh, J. D. Walton, S. C. Wenzel, J. M. Willey and W. A. van der Donk (2013). "Ribosomally synthesized and post-translationally modified peptide natural products: overview and recommendations for a universal nomenclature". *Nat. Prod. Rep.* **30**(1): 108–160.

Bagley, M. C., J. W. Dale, E. A. Merritt and X. Xiong (2005). "Thiopeptide antibiotics". *Chem. Rev.* **105**(2): 685–714.

Barajas, J. F., R. M. Phelan, A. J. Schaub, J. T. Kliewer, P. J. Kelly, D. R. Jackson, R. Luo, J. D. Keasling and S. C. Tsai (2015). "Comprehensive Structural and Biochemical Analysis of the Terminal Myxalamid Reductase Domain for the Engineered Production of Primary Alcohols". *Chem. Biol.* **22**(8): 1018–1029.

Broberg, A., A. Menkis and R. Vasiliauskas (2006). "Kutznerides 1–4, Depsipeptides from the Actinomycete Kutzneria sp. 744 Inhabiting Mycorrhizal Roots of Picea abies Seedlings". *J. Nat. Prod.* **69**: 97–102.

Byford, M. F., J. E. Baldwin, C. Y. Shiau and C. J. Schofield (1997). "The Mechanism of ACV Synthetase". *Chem. Rev.* **97**(7): 2631–2650.

Chen, H., S. O'Connor, D. E. Cane and C. T. Walsh (2001). "Epothilone biosynthesis: assembly of the methylthiazolylcarboxy starter unit on the EpoB subunit". *Chem. Biol.* **8**(9): 899–912.

Crone, W., F. Leeper and A. Truman (2012). "Identification and characterisation of the gene cluster for the anti-MRSA antibiotic bottromycin: expanding the biosynthetic diversity of ribosomal peptides". *Chem. Sci.* **3**: 3516–3521.

Crosa, J. H. and C. T. Walsh (2002). "Genetics and assembly line enzymology of siderophore biosynthesis in bacteria". *Microbiol. Mol. Biol. Rev.* **66**(2): 223–249.

Drake, E. J., B. R. Miller, C. Shi, J. T. Tarrasch, J. A. Sundlov, C. L. Allen, G. Skiniotis, C. C. Aldrich and A. M. Gulick (2016). "Structures of two distinct conformations of holo-non-ribosomal peptide synthetases". *Nature* **529**(7585): 235–238.

Dunbar, K. L., J. O. Melby and D. A. Mitchell (2012). YcaO domains use ATP to activate amide backbones during peptide cyclodehydrations". *Nat. Chem. Biol.* **8**(6): 569–575.

Fewer, D. P., J. Osterholm, L. Rouhiainen, J. Jokela, M. Wahlsten and K. Sivonen (2011). "Nostophycin biosynthesis is directed by a hybrid polyketide synthase-nonribosomal peptide synthetase in the toxic cyanobacterium Nostoc sp. strain 152". *Appl. Environ. Microbiol.* 77(22): 8034–8040.

Freeman, M. F., C. Gurgui, M. J. Helf, B. I. Morinaka, A. R. Uria, N. J. Oldham, H. G. Sahl, S. Matsunaga and J. Piel (2012). "Metagenome mining reveals polytheonamides as posttranslationally modified ribosomal peptides". *Science* **338**(6105): 387–390.

Fujimori, D. G., S. Hrvatin, C. S. Neumann, M. Strieker, M. A. Marahiel and C. T. Walsh (2007). "Cloning and characterization of the biosynthetic gene cluster for kutznerides". *Proc. Natl. Acad. Sci. U. S. A.* **104**(42): 16498–16503.

Gao, X., S. W. Haynes, B. D. Ames, P. Wang, L. P. Vien, C. T. Walsh and Y. Tang (2012). "Cyclization of fungal nonribosomal peptides by a terminal condensation-like domain". *Nat. Chem. Biol.* **8**(10): 823–830.

Gatto Jr., G. J., S. M. McLoughlin, N. L. Kelleher and C. T. Walsh (2005). "Elucidating the substrate specificity and condensation domain activity of FkbP, the FK520 pipecolate-incorporating enzyme". *Biochemistry* **44**(16): 5993–6002.

Gomez-Escribano, J., L. Song, M. Bibb and G. L. Challis (2012). "Posttranslational β-methylation and macrolactamidination in the biosynthesis of the bottromycin complex of ribosomal peptide antibiotics". *Chem. Sci.* **3**: 3522–3525.

Hallen, H. E., H. Luo, J. S. Scott-Craig and J. D. Walton (2007). "Gene family encoding the major toxins of lethal Amanita mushrooms". *Proc. Natl. Acad. Sci. U. S. A.* **104**(48): 19097–19101.

Hamada, T., S. Matsunaga, G. Yano and N. Fusetani (2005). "Polytheonamides A and B, highly cytotoxic, linear polypeptides with unprecedented structural features, from the marine sponge, Theonella swinhoei". *J. Am. Chem. Soc.* **127**(1): 110–118.

Hou, Y., M. D. Tianero, J. C. Kwan, T. P. Wyche, C. R. Michel, G. A. Ellis, E. Vazquez-Rivera, D. R. Braun, W. E. Rose, E. W. Schmidt and T. S. Bugni (2012). "Structure and biosynthesis of the antibiotic bottromycin D". *Org. Lett.* **14**(19): 5050–5053.

Hsu, S. T., E. Breukink, E. Tischenko, M. A. Lutters, B. de Kruijff, R. Kaptein, A. M. Bonvin and N. A. van Nuland (2004). "The nisin-lipid II complex reveals a pyrophosphate cage that provides a blueprint for novel antibiotics". *Nat. Struct. Mol. Biol.* **11**(10): 963–967.

Hubbard, B. K. and C. T. Walsh (2003). "Vancomycin assembly: nature's way". *Angew. Chem., Int. Ed.* **42**(7): 730–765.

Jaspars, M. (2014). "The origins of cyanobactin chemistry and biology". *Chem. Commun.* **50**(71): 10174–10176.

Jin, L. and S. C. Harrison (2002). "Crystal structure of human calcineurin complexed with cyclosporin A and human cyclophilin". *Proc. Natl. Acad. Sci. U. S. A.* **99**(21): 13522–13526.

Julien, B., S. Shah, R. Ziermann, R. Goldman, L. Katz and C. Khosla (2000). "Isolation and characterization of the epothilone biosynthetic gene cluster from Sorangium cellulosum". *Gene* **249**(1–2): 153–160.

Kelly, W. L., L. Pan and C. Li (2009). "Thiostrepton biosynthesis: prototype for a new family of bacteriocins". *J. Am. Chem. Soc.* **131**(12): 4327–4334.

Koglin, A. and C. T. Walsh (2009). "Structural insights into nonribosomal peptide enzymatic assembly lines". *Nat. Prod. Rep.* **26**(8): 987–1000.

Konig, A., T. Schwecke, I. Molnar, G. A. Bohm, P. A. Lowden, J. Staunton and P. F. Leadlay (1997). "The pipecolate-incorporating enzyme for the biosynthesis of the immunosuppressant rapamycin–nucleotide sequence analysis, disruption and heterologous expression of rapP from Streptomyces hygroscopicus". *Eur. J. Biochem.* **247**(2): 526–534.

Li, L., P. Yu, M. Tang, Y. Zou, S. Gao, Y. Hung, M. Zhao, K. Watanabe, K. N. Houk and Y. Tang (2016). "Biochemical Characterization of

a Eukaryotic Decalin-Forming Diels–Alderase". *J. Am. Chem. Soc.* **138**(49): 15837–15840.
Li, Y. M., J. C. Milne, L. L. Madison, R. Kolter and C. T. Walsh (1996). "From peptide precursors to oxazole and thiazole-containing peptide antibiotics: microcin B17 synthase". *Science* **274**(5290): 1188–1193.
Liao, R., L. Duan, C. Lei, H. Pan, Y. Ding, Q. Zhang, D. Chen, B. Shen, Y. Yu and W. Liu (2009). "Thiopeptide biosynthesis featuring ribosomally synthesized precursor peptides and conserved posttranslational modifications". *Chem. Biol.* **16**(2): 141–147.
Liu, J., J. D. Farmer Jr., W. S. Lane, J. Friedman, I. Weissman and S. L. Schreiber (1991). "Calcineurin is a common target of cyclophilin-cyclosporin A and FKBP-FK506 complexes". *Cell* **66**(4): 807–815.
Liu, X. and C. T. Walsh (2009). "Cyclopiazonic acid biosynthesis in Aspergillus sp.: characterization of a reductase-like R* domain in cyclopiazonate synthetase that forms and releases cyclo-acetoacetyl-L-tryptophan". *Biochemistry* **48**(36): 8746–8757.
Lynen, F. and U. Wieland (1937). "Uber die Giftstoffe des Knollenblätterpilzes. IV". *Justus Liebigs Ann. Chem.* **533**: 93–117.
Malcolmson, S. J., T. S. Young, J. G. Ruby, P. Skewes-Cox and C. T. Walsh (2013). "The posttranslational modification cascade to the thiopeptide berninamycin generates linear forms and altered macrocyclic scaffolds". *Proc. Natl. Acad. Sci. U. S. A.* **110**(21): 8483–8488.
Marahiel, M. A. (2016). "A structural model for multimodular NRPS assembly lines". *Nat. Prod. Rep.* **33**(2): 136–140.
Martins, J. and V. Vasconcelos (2015). "Cyanobactins from Cyanobacteria: Current Genetic and Chemical State of Knowledge". *Mar. Drugs* **13**(11): 6910–6946.
McIntosh, J. A., M. S. Donia, S. K. Nair and E. W. Schmidt (2011). "Enzymatic basis of ribosomal peptide prenylation in cyanobacteria". *J. Am. Chem. Soc.* **133**(34): 13698–13705.
McIntosh, J. A., M. S. Donia and E. W. Schmidt (2009). "Ribosomal peptide natural products: bridging the ribosomal and nonribosomal worlds". *Nat. Prod. Rep.* **26**(4): 537–559.
Milne, J. C., A. C. Eliot, N. L. Kelleher and C. T. Walsh (1998). "ATP/GTP hydrolysis is required for oxazole and thiazole biosynthesis in the peptide antibiotic microcin B17". *Biochemistry* **37**(38): 13250–13261.
Nagano, N., M. Umemura, M. Izumikawa, J. Kawano, T. Ishii, M. Kikuchi, K. Tomii, T. Kumagai, A. Yoshimi, M. Machida, K. Abe,

K. Shin-ya and K. Asai (2016). "Class of cyclic ribosomal peptide synthetic genes in filamentous fungi". *Fungal Genet. Biol.* **86**: 58–70.

Nolan, E. M. and C. T. Walsh (2009). "How nature morphs peptide scaffolds into antibiotics". *ChemBioChem* **10**(1): 34–53.

O'Connor, S. E., H. Chen and C. T. Walsh (2002). "Enzymatic assembly of epothilones: the EpoC subunit and reconstitution of the EpoA-ACP/B/C polyketide and nonribosomal peptide interfaces". *Biochemistry* **41**(17): 5685–5694.

Onaka, H., M. Nakaho, K. Hayashi, Y. Igarashi and T. Furumai (2005). "Cloning and characterization of the goadsporin biosynthetic gene cluster from Streptomyces sp. TP-A0584". *Microbiology* **151**(Pt 12): 3923–3933.

Parmeggiani, A., I. M. Krab, S. Okamura, R. C. Nielsen, J. Nyborg and P. Nissen (2006). "Structural basis of the action of pulvomycin and GE2270 A on elongation factor Tu". *Biochemistry* **45**(22): 6846–6857.

Richter, C., D. Nietlispach, W. Broadhurst and K. Weissman (2007). "Multienzyme Docking in Hybrid Megasynthases". *Nat. Chem. Biol.* **4**: 75–81.

Schmidt, E. W., J. T. Nelson, D. A. Rasko, S. Sudek, J. A. Eisen, M. G. Haygood and J. Ravel (2005). "Patellamide A and C biosynthesis by a microcin-like pathway in Prochloron didemni, the cyanobacterial symbiont of Lissoclinum patella". *Proc. Natl. Acad. Sci. U. S. A.* **102**(20): 7315–7320.

Scholz, R., K. J. Molohon, J. Nachtigall, J. Vater, A. L. Markley, R. D. Sussmuth, D. A. Mitchell and R. Borriss (2011). "Plantazolicin, a novel microcin B17/streptolysin S-like natural product from Bacillus amyloliquefaciens FZB42". *J. Bacteriol.* **193**(1): 215–224.

Schwecke, T., J. F. Aparicio, I. Molnar, A. Konig, L. E. Khaw, S. F. Haydock, M. Oliynyk, P. Caffrey, J. Cortes, J. B. Lester *et al.* (1995). "The biosynthetic gene cluster for the polyketide immunosuppressant rapamycin". *Proc. Natl. Acad. Sci. U. S. A.* **92**(17): 7839–7843.

Shen, B., L. Du, C. Sanchez, D. J. Edwards, M. Chen and J. M. Murrell (2002). "Cloning and characterization of the bleomycin biosynthetic gene cluster from Streptomyces verticillus ATCC15003". *J. Nat. Prod.* **65**(3): 422–431.

Skinnider, M. A., C. W. Johnston, R. E. Edgar, C. A. Dejong, N. J. Merwin, P. N. Rees and N. A. Magarvey (2016). "Genomic charting of ribosomally synthesized natural product chemical space facilitates targeted mining". *Proc. Natl. Acad. Sci. U. S. A.* **113**: E6343–E6351.

Strieker, M., A. Tanovic and M. A. Marahiel (2010). "Nonribosomal peptide synthetases: structures and dynamics". *Curr. Opin. Struct. Biol.* **20**(2): 234–240.

Umemura, M., N. Nagano, H. Koike, J. Kawano, T. Ishii, Y. Miyamura, M. Kikuchi, K. Tamano, J. Yu, K. Shin-ya and M. Machida (2014). "Characterization of the biosynthetic gene cluster for the ribosomally synthesized cyclic peptide ustiloxin B in Aspergillus flavus". *Fungal Genet. Biol.* **68**: 23–30.

Vetter, J. (1998). "Toxins of Amanita phalloides". *Toxicon* **36**(1): 13–24.

Walsh, C. T. (2004). "Polyketide and nonribosomal peptide antibiotics: modularity and versatility". *Science* **303**(5665): 1805–1810.

Walsh, C. T. (2016). "Insights into the chemical logic and enzymatic machinery of NRPS assembly lines". *Nat. Prod. Rep.* **33**: 127–135.

Walsh, C. T., M. G. Acker and A. A. Bowers (2010). "Thiazolyl peptide antibiotic biosynthesis: a cascade of post-translational modifications on ribosomal nascent proteins". *J. Biol. Chem.* **285**(36): 27525–27531.

Walsh, C. T. and M. A. Fischbach (2010). "Natural products version 2.0: connecting genes to molecules". *J. Am. Chem. Soc.* **132**(8): 2469–2493.

Walsh, C. T., R. V. O'Brien and C. Khosla (2013). "Nonproteinogenic amino acid building blocks for nonribosomal peptide and hybrid polyketide scaffolds". *Angew. Chem., Int. Ed.* **52**(28): 7098–7124.

Walsh, C. T. and T. Wencewicz (2016). *Antibiotics Challeneges, Mechanisms, Opportunities*. Wshington DC, ASM Press.

Wang, H., D. P. Fewer, L. Holm, L. Rouhiainen and K. Sivonen (2014). "Atlas of nonribosomal peptide and polyketide biosynthetic pathways reveals common occurrence of nonmodular enzymes". *Proc. Natl. Acad. Sci. U. S. A.* **111**(25): 9259–9264.

Weissman, K. J. (2015). "The structural biology of biosynthetic megaenzymes". *Nat. Chem. Biol.* **11**(9): 660–670.

Wever, W. J., J. W. Bogart, J. A. Baccile, A. N. Chan, F. C. Schroeder and A. A. Bowers (2015). "Chemoenzymatic synthesis of thiazolyl peptide natural products featuring an enzyme-catalyzed formal [4 + 2] cycloaddition". *J. Am. Chem. Soc.* **137**(10): 3494–3497.

Xu, Y., R. D. Kersten, S. J. Nam, L. Lu, A. M. Al-Suwailem, H. Zheng, W. Fenical, P. C. Dorrestein, B. S. Moore and P. Y. Qian (2012). "Bacterial biosynthesis and maturation of the didemnin anticancer agents". *J. Am. Chem. Soc.* **134**(20): 8625–8632.

Yang, Y. L., C. P. Lu, M. Y. Chen, K. Y. Chen, Y. C. Wu and S. H. Wu (2007). "Cytotoxic polyketides containing tetramic acid moieties isolated from the fungus Myceliophthora thermophila: Elucidation of the relationship between cytotoxicity and stereoconfiguration''. *Chem. – Eur. J.* **13**(24): 6985–6991.

Ye, Y., A. Minami, Y. Igarashi, M. Izumikawa, M. Umemura, N. Nagano, M. Machida, T. Kawahara, K. Shin-Ya, K. Gomi and H. Oikawa (2016). ''Unveiling the Biosynthetic Pathway of the Ribosomally Synthesized and Post-translationally Modified Peptide Ustiloxin B in Filamentous Fungi''. *Angew. Chem., Int. Ed.* **55**: 8072–8075.

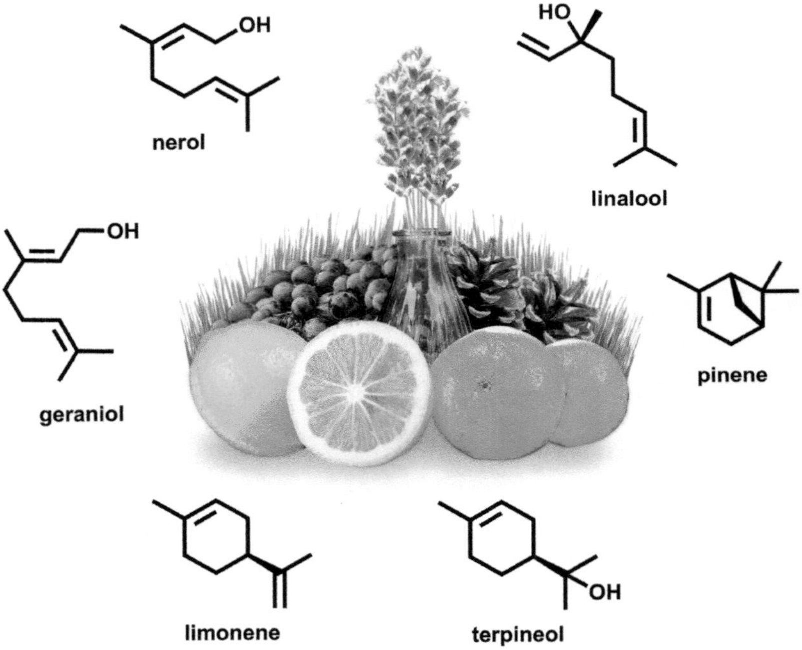

A collection of monoterpene natural products and their producing plant sources.

4 Isoprenoids/Terpenes

4.1 Isoprene-based Scaffolds Comprise the Most Abundant Class of Natural Products

This dominant class of natural products, largely from higher plants and fungi, derive from one or more units of the five carbon mono-olefinic isoprene. While the hydrocarbon isoprene (2-methyl-1,3-butadiene) itself is a natural product (*e.g.* in rubber trees), there are more than 50 000 molecules derived from biological isoprenyl donors that have been isolated and characterized structurally (Dewick 2009, Quin, Flynn *et al.* 2014).

Historically, the most abundant source of this natural product class has been from plants, where they have been known collectively as terpenes (Pichersky, Noel *et al.* 2006). We will use the terpenoid and isoprenoid descriptors interchangeably as the terpene formalisms dominate much of the research literature. Also, we will note that isoprenoid chain elongation enzymes add five carbon units at a time: the C_{10}, C_{15}, C_{20}, and C_{30} scaffolds are commonly termed monoterpenes, sesquiterpenes, diterpenes, and triterpenes, respectively, a nomenclature we shall use (Figure 4.1).

4.2 Δ^2- and Δ^3-Isopentenyl Diphosphates are the Biological Isoprenyl Building Blocks for Head to Tail Alkylative Chain Elongations

Two regioisomeric five carbon isoprenyl alcohol diphosphates are the biological isoprenyl (prenyl) donors that emerge from primary

Natural Product Biosynthesis: Chemical Logic and Enzymatic Machinery
By Christopher T. Walsh and Yi Tang

Published by the Royal Society of Chemistry, www.rsc.org

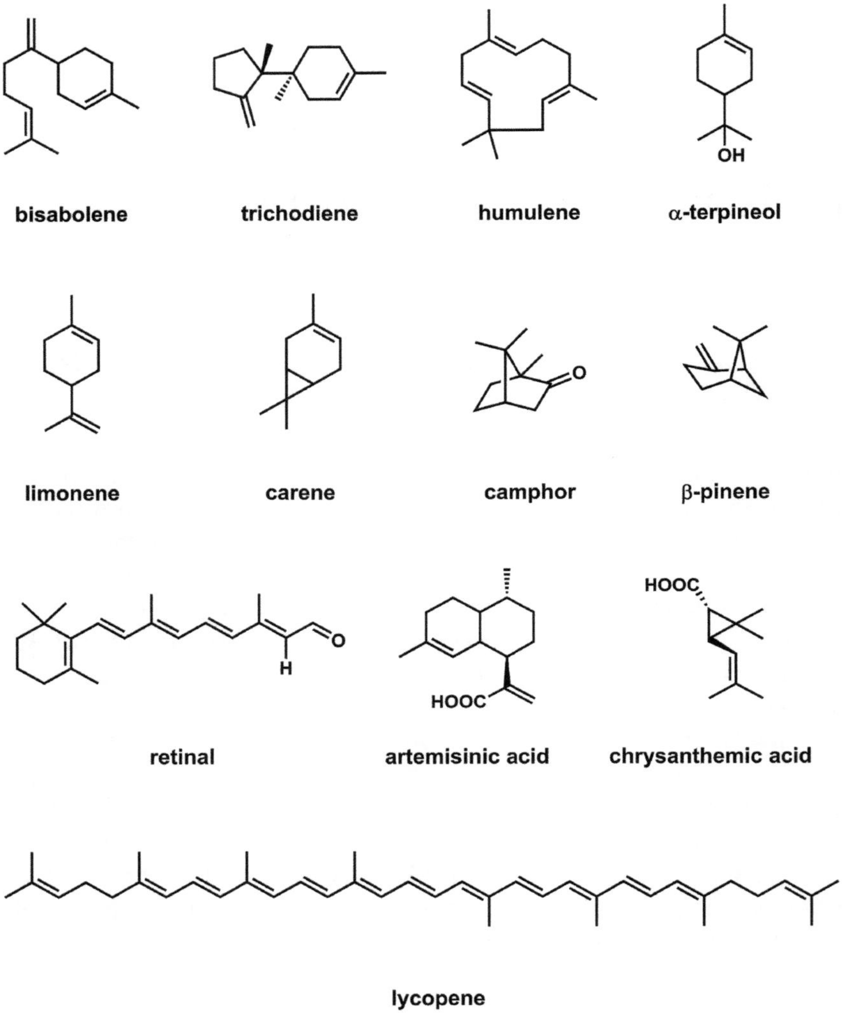

Figure 4.1 Terpenes, natural products with isoprenoid scaffolds, comprise the largest class of natural products. Shown are some monoterpenes (C_{10}), sesquiterpenes (C_{15} humulene, artemisinic acid), a diterpene (C_{20} retinal), and the C_{40} polyene lycopene.

metabolism and serve as building blocks for the frameworks of the secondary metabolites in this enormous molecular class. The Δ^2-isoprenyl diphosphate (Δ^2-IDP), also known historically as isoprenyl pyrophosphate (Δ^2-IPP), acts as the electrophilic partner in prenyl transferase catalyzed alkylation reactions (Figure 4.2).

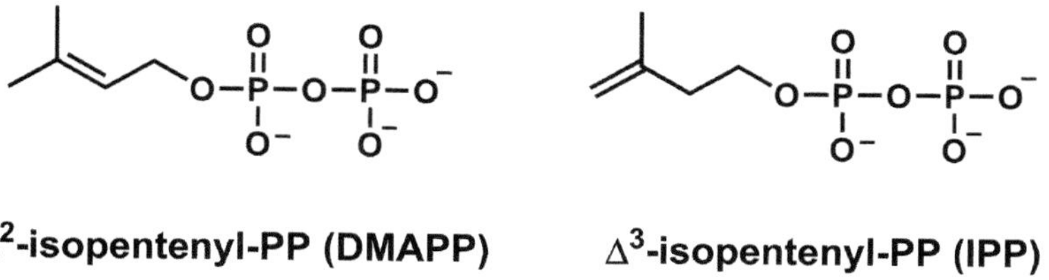

Figure 4.2 The Δ^2- and Δ^3-isopentneyl diphosphate regioisomers are the building blocks for all terpenoid/isoprenoid scaffolds.

The corresponding Δ^3-isomer acts as the nucleophilic partner in C–C bond formation in the first reaction of prenyl chain elongation to yield the monoterpene C_{10} geranyl-PP product (Figure 4.3) (Walsh and Wencewicz 2016).

The mechanism of this first committed prenyl transferase step has been scrutinized and involves an early dissociation of the C_1–OPP bond in the bound Δ^2-IPP substrate to yield a C_1 cationic transition state. This cation is kinetically accessible under physiologic conditions in the enzyme active site because it is thermodynamically stabilized by delocalization at C_3 as an allyl cation by virtue of the C_2 double bond. Indeed, we shall note in subsequent sections that this cation can be captured at C_3 as well as C_1 by cosubstrate nucleophiles, reflecting the delocalization of positive charge at these two carbon centers.

The nucleophilic partner in C–C bond formation to geranyl-PP is the Δ^3-IPP. Specifically, the π-electrons of the terminal olefin attack the allyl cation partner. This generates a C_{10} tertiary carbonium ion nascent product. Stereocontrolled abstraction of a proton from C_2 of this adduct recreates a Δ^2 double bond as geranyl-PP is formed and released from the prenyltransferase active site.

Because geranyl-PP has this Δ^2-olefin it can function as the electrophilic partner in the next prenyl transfer elongation cycle. Its partner will be another Δ^3-IPP five carbon unit as nucleophile. Thus, in every chain elongation cycle, the five carbon Δ^3-IPP is cosubstrate nucleophile by virtue of the π electrons in the terminal olefin. On the other hand, the Δ^2-partner in each elongation cycle is five carbons longer than the previous Δ^2-substrate. Figure 4.3 illustrates this pattern for $C_5 + C_5$ to geranyl-PP, for $C_5 + C_{10}$ to sesquiterpene farnesyl-PP, and for $C_5 + C_{15}$ to the C_{20} Δ^2-product geranylgeranyl-PP. Thus, a nested series of linear Δ^2-prenyl-PP scaffolds emerge, all with *E*-configuration (*trans*) in the double bonds, spaced three carbons apart.

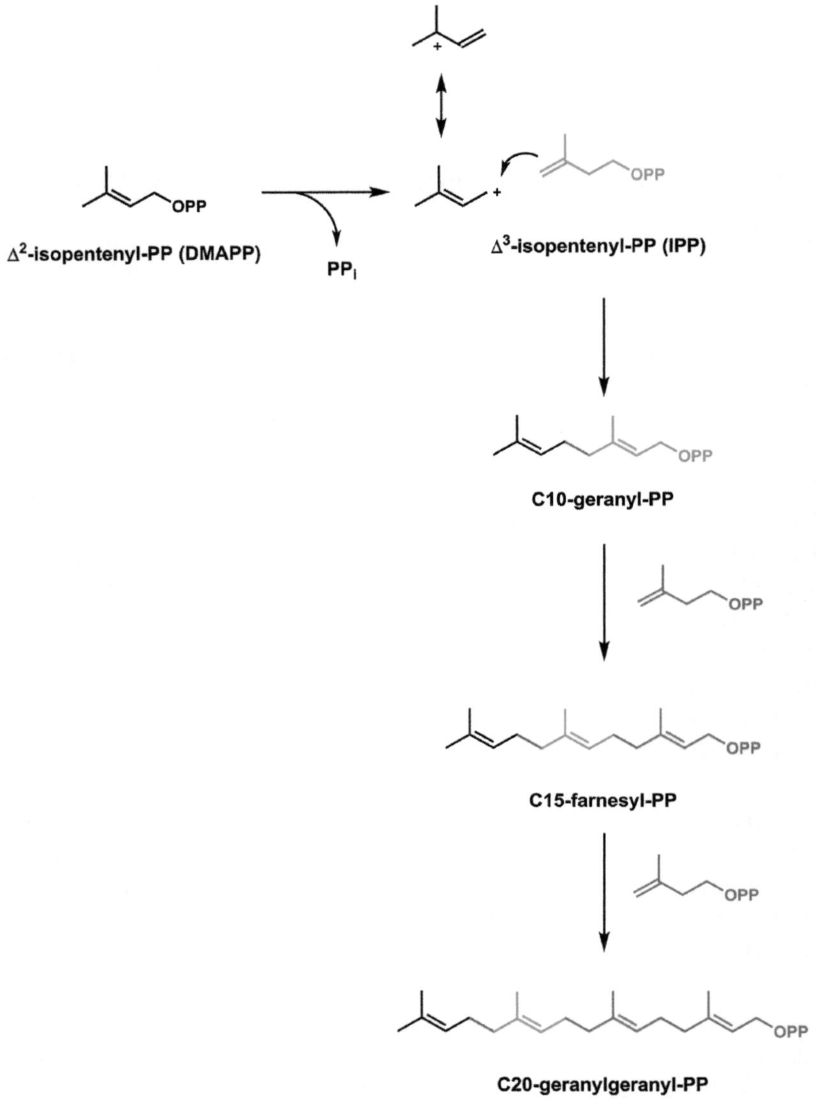

Figure 4.3 Chain growth of Δ^2-prenyl-PP chains by head to tail addition of the five-carbon Δ^3-IPP, iteratively to give geranyl-PP, farnesyl-PP, geranylgeranyl-PP.

4.3 Long Chain Prenyl-PP Scaffolds

The question arises of how long such linear Δ^2-prenyl chains get and what limits that size. Two of the best known long chain prenyl-PPs are

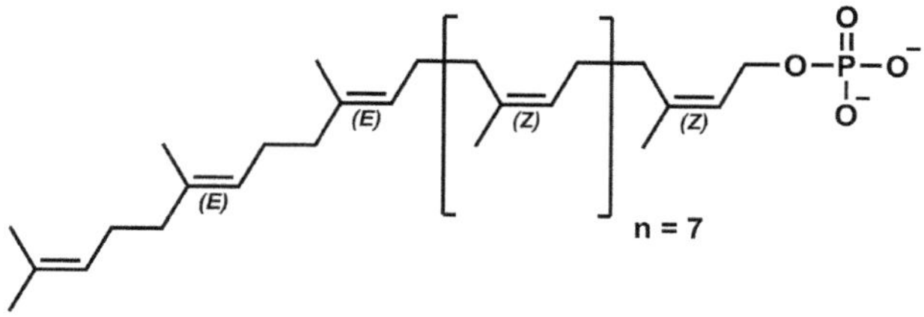

undecaprenyl (C_{55}) phosphate (*E. coli* and others)

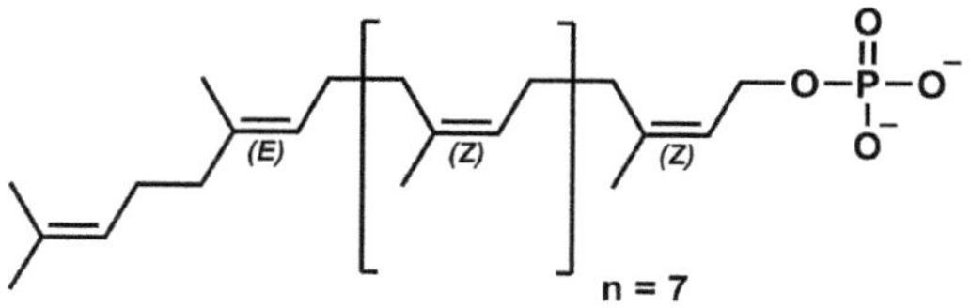

decaprenyl (C_{50}) phosphate (*M. smegmatis*)

Figure 4.4 C_{50} and C_{55} bactoprenyls function as carrier lipids in cell wall assembly and contain a number of *Z*-double bonds rather than the standard *E*-double bonds in the C_5–C_{20} prenyl-PP intermediates, indicating a distinct synthase.

the C_{55} undecaprenyl-PP, also known as bactoprenyl-PP, involved in bacterial peptidoglycan assembly (Walsh and Wencewicz 2016) and the C_{110}-dolichol-PP that is the essential carrier of a tetradecasaccharyl unit in eukaryotic *N*-glycoprotein formation in the endoplasmic reticulum of cells (Chojnacki and Dallner 1988). Both of these isoprenyl lipids act as membrane spanning carriers of their hydrophilic cargoes to deliver them across bacterial and ER membranes, respectively, to acceptors on the opposite face of the membranes (Figure 4.4).

Bactoprenyl-PP and its corresponding phosphate ester hydrolysis product bactoprenol are of note. Eight of the 11 olefins have the unusual *Z* (*cis*) geometry, while the last three are the standard *E* (*trans*) geometry typical of isoprenoid scaffolds. The *Z*-double bonds are installed by an octaprenyl synthase which generates a C_{40} product by seven cycles of addition of nucleophilic Δ^3-IPP to the growing Δ^2-chain (Guo, Kuo *et al.* 2004). The *Z*-configuration can arise in each elongation cycle, as suggested in Figure 4.5. Abstraction of one of the two prochiral C_2 hydrogens in the nascent C_3 product cation will give the *Z*-isomer, while abstraction of the other (shown here as H_a) gives the more common *E*-olefin geometry of the elongated product. The catalytic base performing the C_2-H abstraction must be on opposite

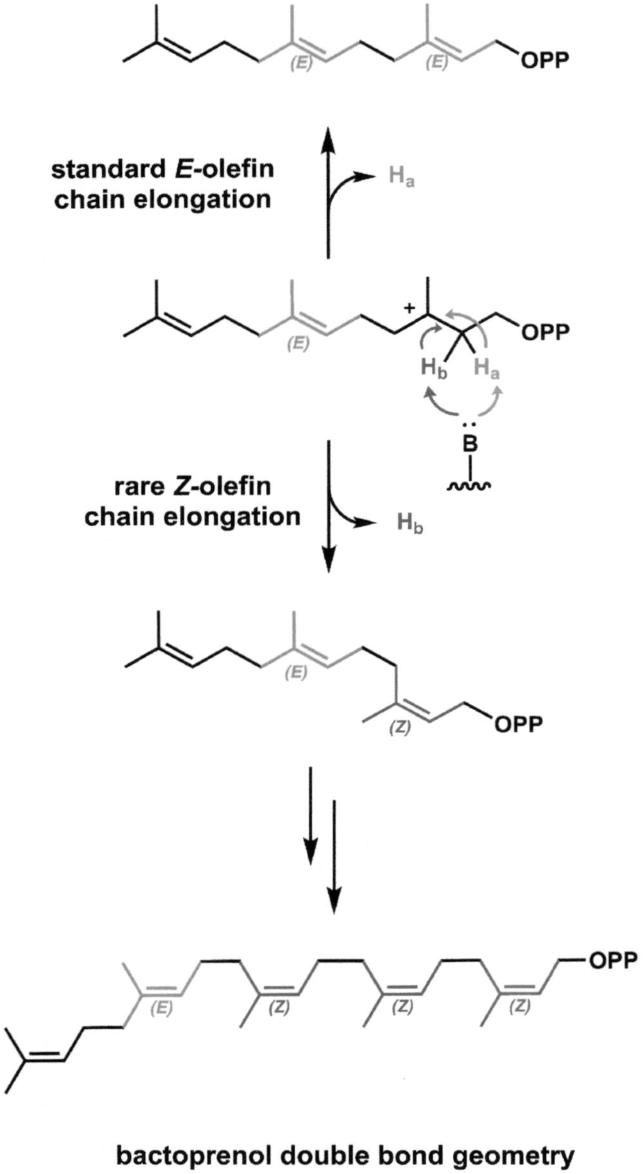

Figure 4.5 Control of *E*- *vs.* *Z*-Δ^2-olefin geometry during prenyl chain elongation by abstraction of one of the two prochiral hydrogens at C_2 in the last step.

faces of the active site in this rare *Z*-forming prenyl transferase compared to most of the *E*-forming prenyltransferases. X-ray analysis suggests a distinct fold for the *Z*- *vs.* *E*-selective prenyltransferases (Guo, Kuo *et al.* 2004). The released octaprenyl-PP is then subjected to three further cycles of prenylation, converting all the *Z*-octaprenyl-PP to bactoprenyl-PP with the last three olefin-containing units in the *E*-configuration. Presumably this generates a set of accessible conformations that allows bactoprenyl units to function effectively in carrying disaccharyl-pentapeptidyl units from inside to outside the bacterial membranes as peptidoglycan chains are elongated (Walsh and Wencewicz 2016).

4.4 Two Routes to the IPP Isomers: Classical and Nonclassical Pathways

The eukaryotic pathway was worked out in the mid 20th century, shortly after radioactive isotope methodologies were introduced to follow the path of radiolabeled carbons from one molecule to another. This is known as the classical pathway or also the mevalonate pathway, as shown in Figure 4.6. Tandem condensation of three molecules of the central metabolite acetyl-CoA yields 3-hydroxymethylglutary-CoA (HMG-CoA). Four electron reduction of the acyl thioester at C_1 to the alcohol generates mevalonate. Three consecutive phosphorylations then occur, two of which generate the pyrophosphate linkage at C_1. The third phosphorylation at the C_3 alcohol sets up decarboxylation and loss of Pi to yield the Δ^3-regioisomer of IPP. The required Δ^2-IPP isomer is created by the action of IPP isomerase which equilibrates the double bond position between the IPP regioisomers, by proton transfers (Zheng, Sun *et al.* 2007).

Some decades after establishment of the mevalonate pathway it became clear from labeling patterns that another biosynthetic source of IPP must exist in some organisms, including many bacterial pathogens that cause human disease. This is now termed the nonclassical pathway and also the MEP (methylerythritol-phosphate) pathway. As diagrammed in Figure 4.7A the primary metabolites pyruvate and glyceraldehyde-3-P condense to yield deoxyxylulose-5-P. This is subject to a reductoisomerase reaction, rearranging the skeletal connectivity and producing methylerythritol-P, for which this pathway is named. A cytidylyltransferase then goes to work, followed

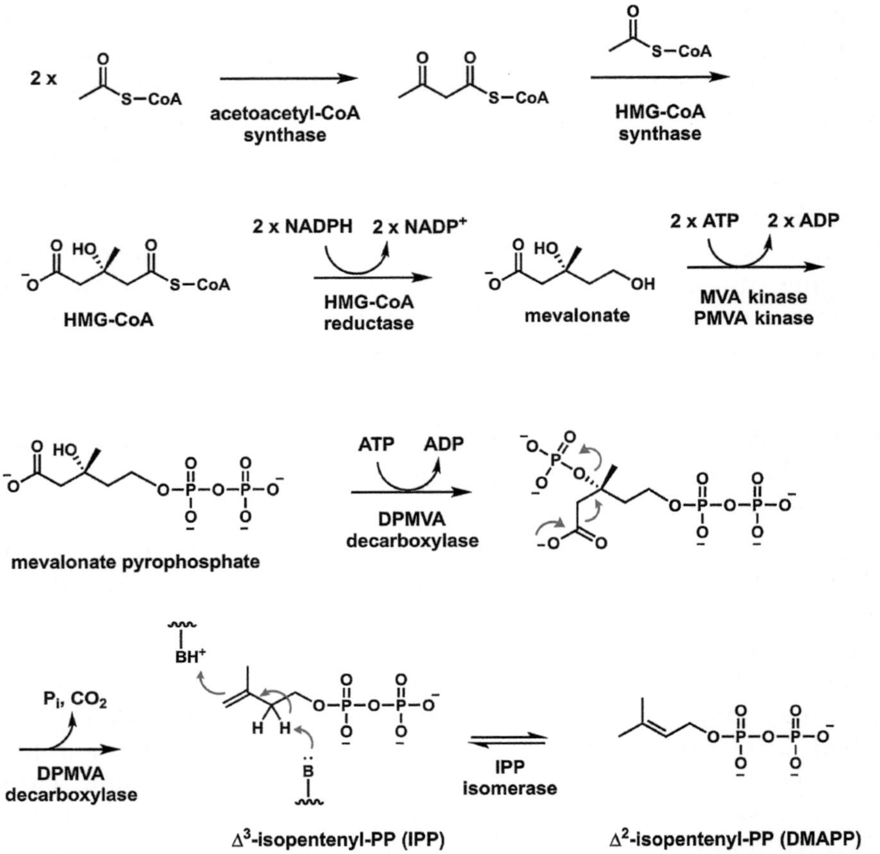

Figure 4.6 The mevalonate pathway leads first to Δ^3-IPP that is then isomerized to Δ^2-IPP, to provide equilibrium amounts of each.

by an alcohol kinase. Elimination of CMP occurs as a cyclic-pyrophosphorylester forms under the aegis of IspF catalysis. It is this cyclic eight-membered diphosphate that is the precursor to both the acyclic Δ^2- and Δ^3-IPP regioisomers. These conversions involve radical chemistry *via* iron/sulfur cluster one-electron transfers in and out of the fragmenting scaffold. One set of mechanistic possibilities is shown in Figure 4.7B.

Clearly, during evolution two convergent pathways to the IPP isomers have been built and retained. The two pathways could be distinguished by isotopomer pulse chase feeding studies with labeled [$^{13}CO_2$] rather than acetate samples (Bacher, Chen *et al.* 2016). As noted in Figure 4.6, the mevalonate pathway uses three molecules

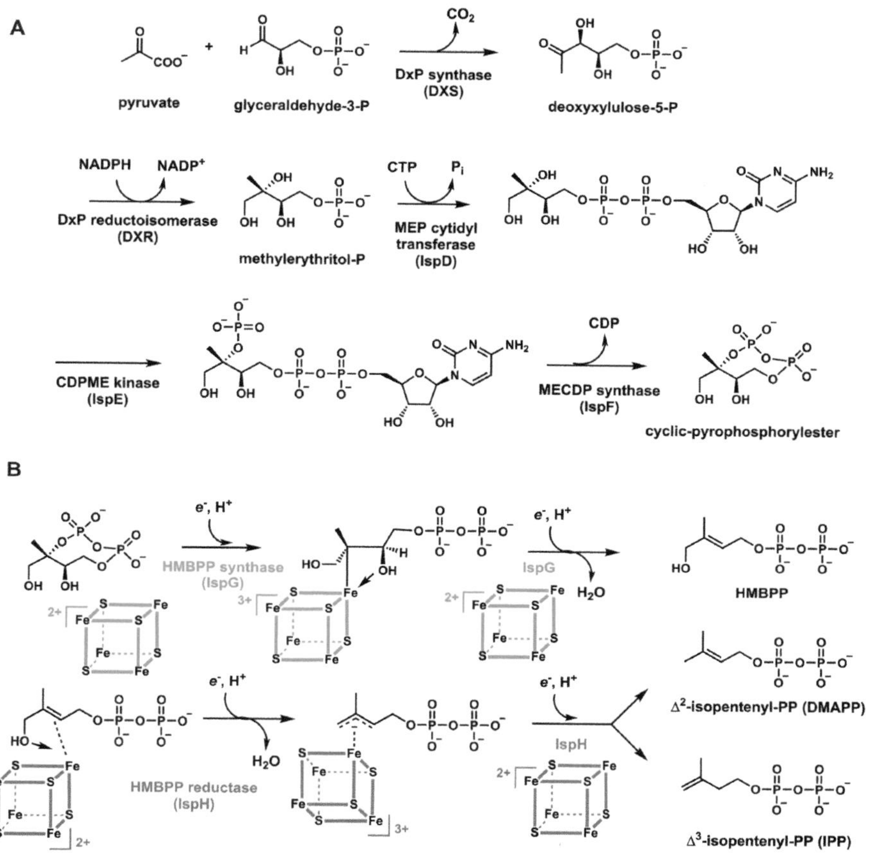

Figure 4.7 (A) The MEP pathway to the two isomeric IPPs. (B) Two mechanistic alternatives involving one-electron transfer to iron and coordinated radical intermediates.

of acetyl-CoA in two tandem reactions to build the six-carbon mevalonate. On decarboxylation and net dehydration of the tertiary alcohol, five [^{13}C]-labeled carbons exist in the indicated 2-2-1 pattern (yellow) for the Δ^2 and Δ^3-isopenetenyl diphosphate isomers (Figure 4.8). In particular, the mevalonate pathway can contribute connected pairs of [^{13}C] atoms. In contrast the MEP pathway involves enzymatic condensation of pyruvate and glyceraldehyde-3-P and as shown can yield *triply* connected 3–2 connections. The [^{13}C] pairs have the same locations but only the nonmevalonate pathway can contribute three [^{13}C] atoms from a precursor unit. This is the NMR pattern for terpene examples in *Arabidopsis thaliana*.

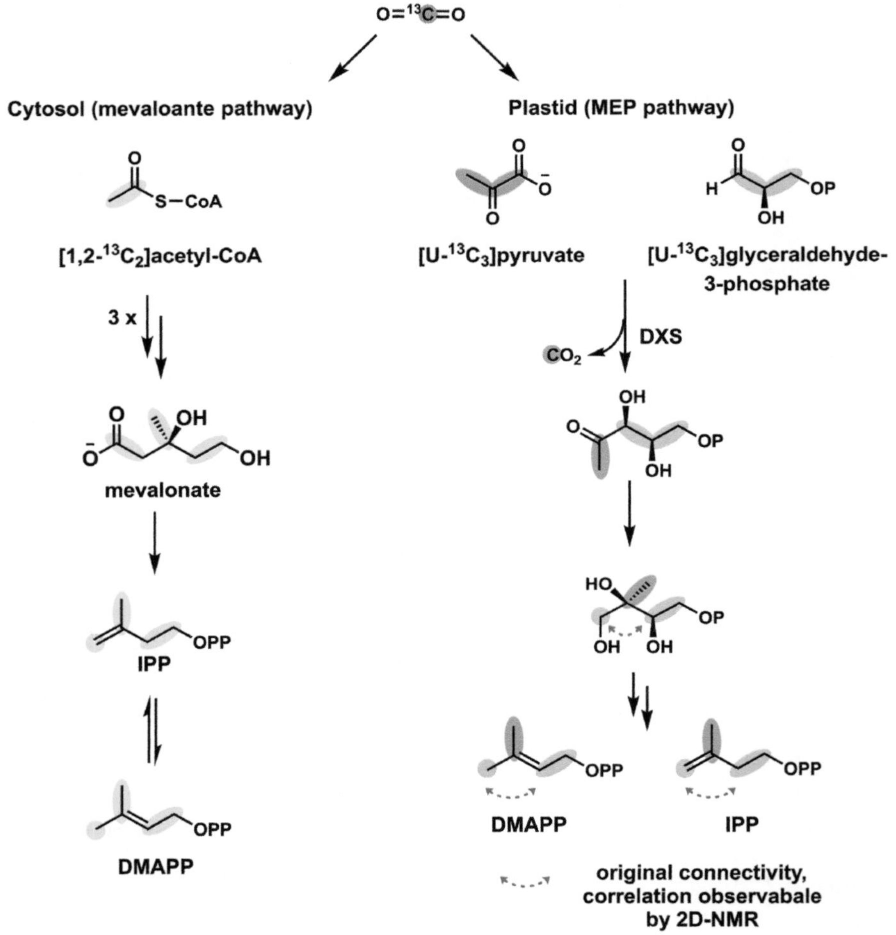

Figure 4.8 Use of ^{13}C labeled CO_2 in plants to distinguish origin of the isoprene building blocks.

The [$^{13}CO_2$] pulse labeling approach was also used to confirm a nonmevalonate route to the simple monoterpene alcohol thymol in *Thymus transcacucasius*. Perhaps more surprisingly this isotopomer approach to artemisinin biosynthesis indicated a mix of mevalonate and MEP pathway contributions of C_5 isoprenyl units, suggesting contributions from both cytoplasmic (mevalonate) and chloroplast (MEP) routes to the assembly of this front line antimalarial natural product (Bacher, Chen *et al.* 2016).

4.5 Self-condensation of Two Δ^2-IPPs to Chrysanthemyl Cyclopropyl Framework

The Δ^2-isomer of IPP is also referred to as a dimethylallyl pyrophosphate (DMAPP) to emphasize its allylic constitution. Although it has an internal double bond rather than the terminal, more unhindered olefin in the Δ^2–partner, that internal double bond can react in an alkylative condensation of two Δ^2-IPP molecules as shown in Figure 4.9, one as electrophilic allylic cation, the other as nucleophile. The corresponding branched tertiary cation can be quenched by proton abstraction, in some analogy to the more common alkylative condensation of a Δ^2- and a Δ^3-IPP. However, while the latter event leads to the linear geranyl-PP monoterpene framework, the migration of the C_3 carbon in the Δ^2-self alkylation pathway generates a cyclopropane ring.

The dimethylcyclopropane is the signature element of chrysanthemyl-PP, and undergoes alcohol oxidation to the eponymous chrysanthemic acid. We will return in the next section to cyclopropane formation as one of three routes that quench cation intermediates in terpene synthase reaction manifolds. We have already noted proton removal and olefin formation as the common quenching route in the chain elongation synthases that make geranyl-PP, farnesyl-PP, and geranylgeranyl-PP, and by extension longer linear isoprene chains.

4.6 Cation-driven Scaffold Rearrangements: and Quenching

4.6.1 Monoterpenes: Geranyl-PP to α-terpinyl Cation and Its Partitioning to Products

A variety of terpene synthases, acting at the monoterpene, sesquiterpene, and diterpene (C_{10}, C_{15}, C_{20}) levels, initiate catalysis by an S_N1-like dissociation of the C_1–OPP bond in those Δ^2-scaffold substrates. The initial allyl cation can then undergo a plethora of low energy barrier, unimolecular rearrangements to alternative carbocations, through a variety of mechanisms including hydride shifts and carbon bond shifts in 1,2 Wagner Meerwein type migrations. Any of the cations can be regioselectively quenched by one of three routes: (1) proton removal to create an olefin; (2) water addition to create an

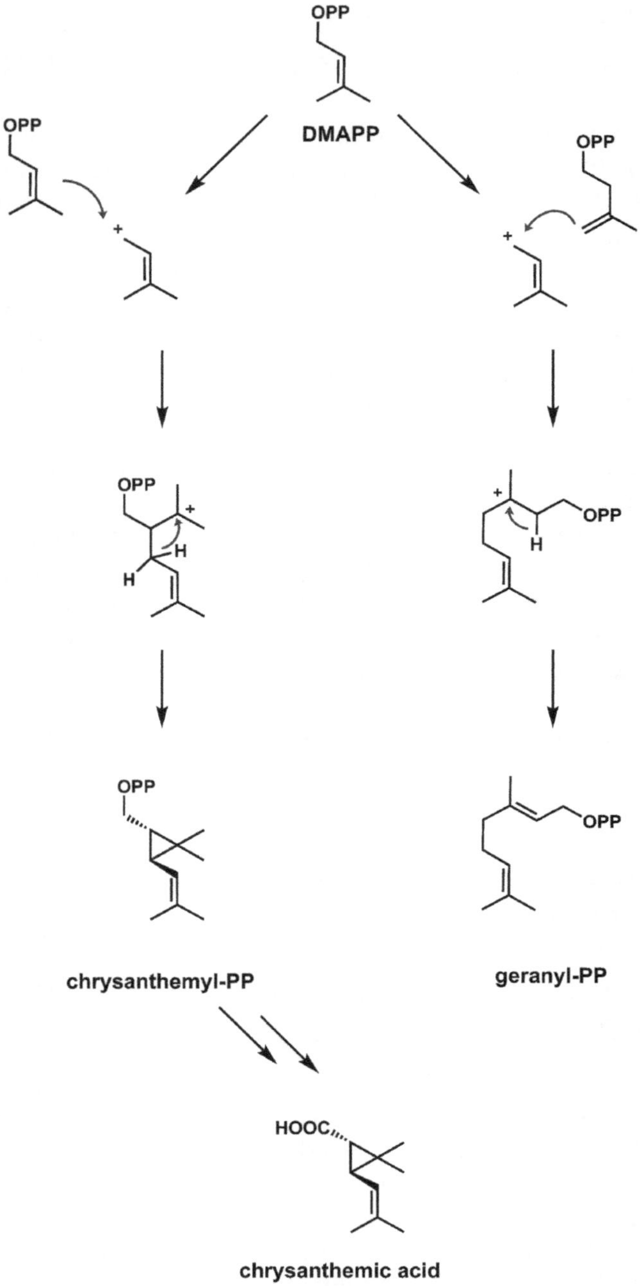

Figure 4.9 Self-alkylation of two molecules of Δ^2-IPP generates the cyclopropane ring of chrysanthemyl-PP.

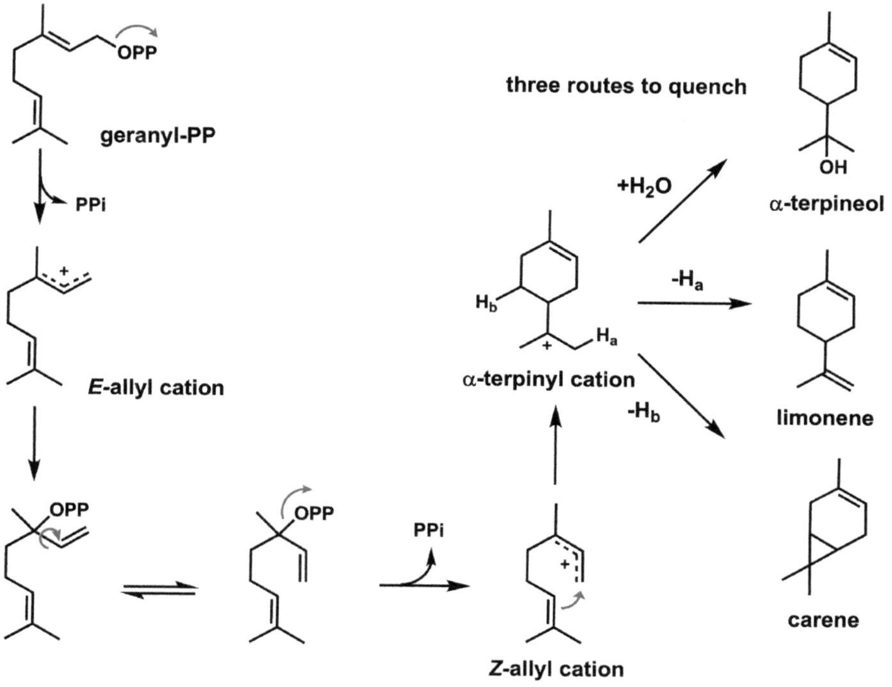

Figure 4.10 Three typical routes to quench rearranging prenyl cations in synthase active sites: conversion of geranyl-PP to α-terpineol, limonene, and carene.

alcohol; (3) carbon bond migration to create a ring structure (Gao, Honzatko *et al.* 2012, Dickschat 2016).

These fates are depicted in Figure 4.10 for a reaction manifold where geranyl-PP undergoes enzyme-mediated dissociation to the allyl cation and PP_i still in the active site. Rotation can occur at the E-C_2-olefin to the cisoid Z-isomer, with recapture of PP_i to give the *cisoid* nerolidol-PP transiently. That double bond configuration is required for the subsequently formed allyl cation to undergo facile intramolecular cyclization to the tertiary α-terpinyl carbocation. The three routes to quench, noted in the above paragraph, are water addition to give terpineol, proton abstraction from the adjacent methyl group to give limonene, or removal of one of the cyclohexene ring protons and capture of the carbocation with cyclopropane formation. This latter route converts the acyclic geranyl framework into the 3,6-bicyclic scaffold of carene.

Carene is by no means the only bicyclic alkane/alkene scaffold available from further rearrangement of the α-terpinyl cation. As shown in Figure 4.11, attack of either end of the cyclohexene

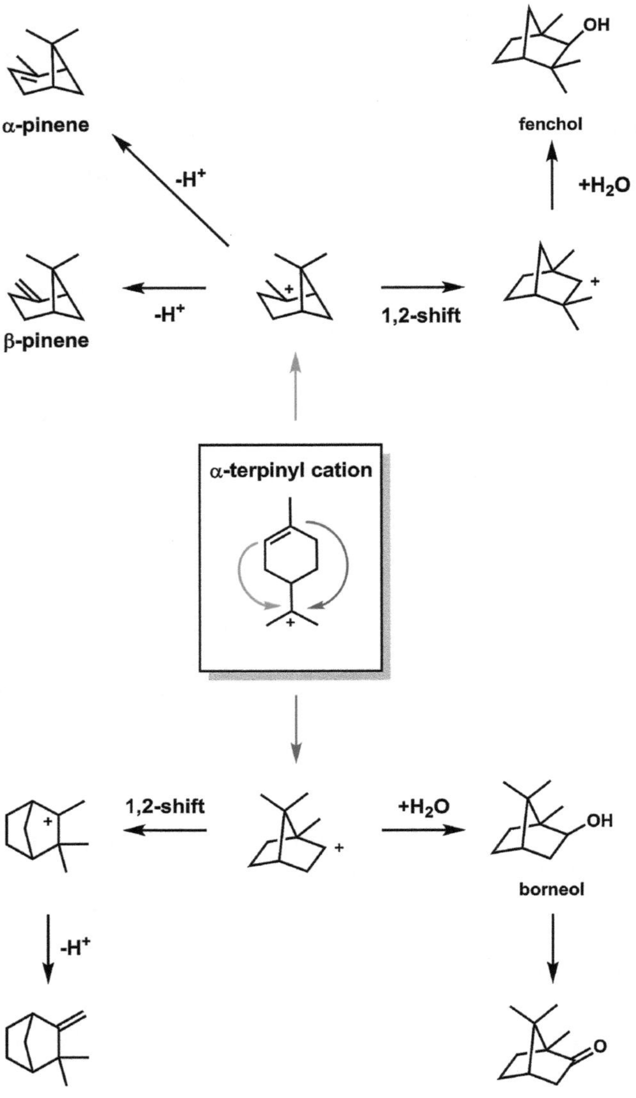

Figure 4.11 A range of bicyclic monoterpenes are also available from geranyl-PP *via* subsequent rearrangement of the α-terpinyl cation.

double bond on the exocyclic tertiary cation yields a pair of bicyclic cations. The α- and β-isomers of pinene are readily generated in cation quenching by proton loss. Also, a 1,2-shift of a C–C bond

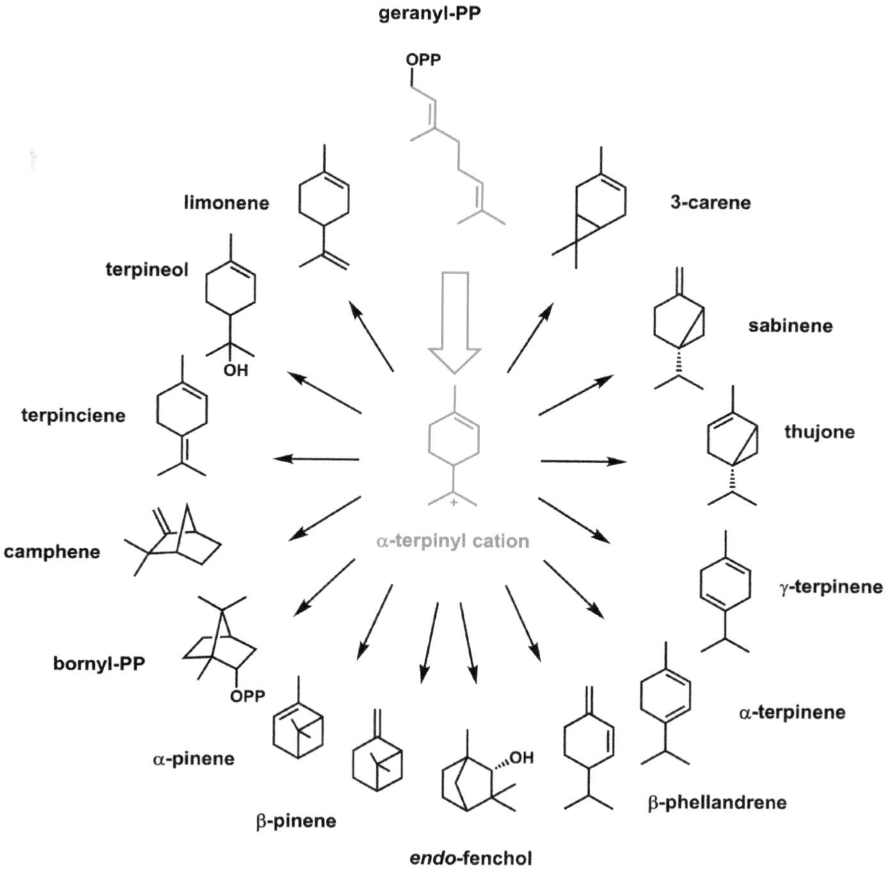

Figure 4.12 Cation rearrangement manifolds from the central terpinyl cation yield 14 distinct monoterpene products.

with subsequent H_2O quenching yields the fenchol framework. Analogously, the bicyclic borneol and camphor are also accessible. Broadening out the fate map a bit further in Figure 4.12 indicates a total of 14 monoterpene mono- and bicyclic frameworks in well-known plant volatile natural products from that same α-terpinyl cation.

4.6.2 Sesquiterpenes: Six Regiospecific Cyclizations from C_{15} Farnesyl-PP

An even more diverse set of rearranged scaffolds are available from enzymatic dissociation of the sesquiterpene farnesyl-PP given the presence of five more carbons in the starting substrate. Six

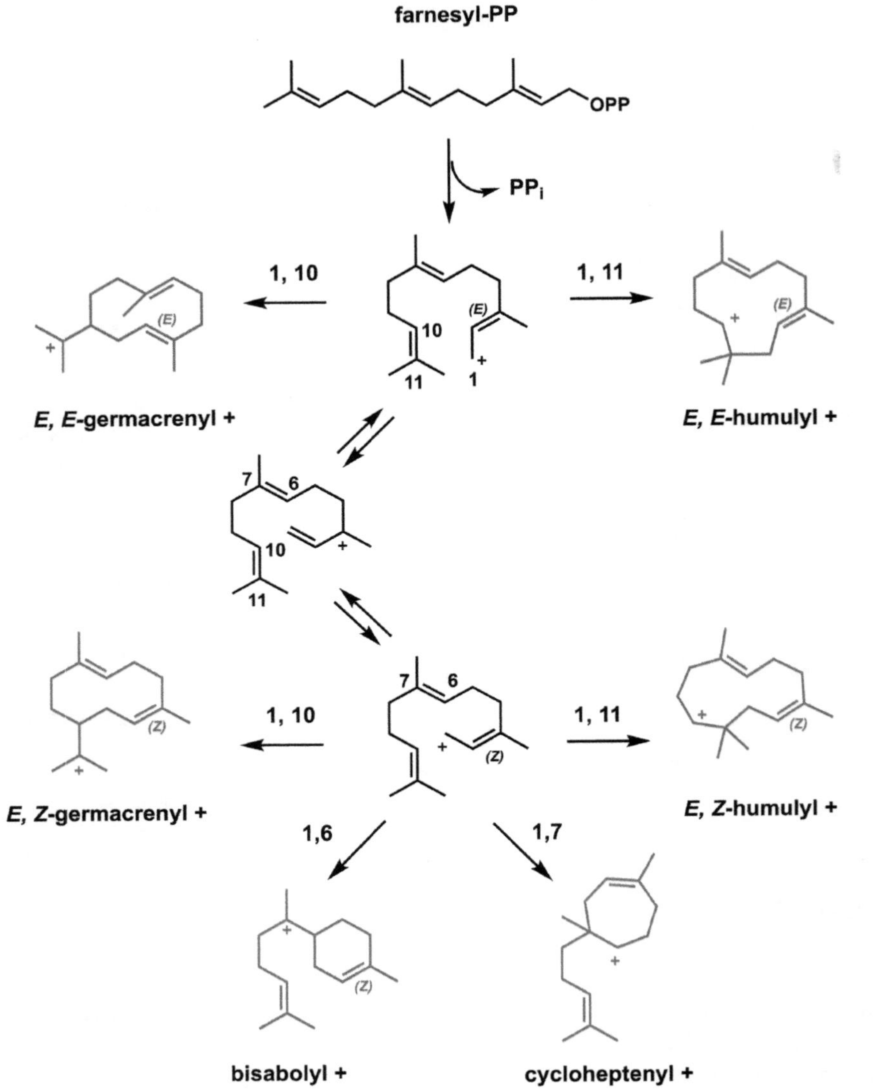

Figure 4.13 Six possible intramolecular rearrangements of the C_{15} farnesyl C_1 allylic cation. All quenched products are detected.

rearrangements, reflecting connection of the C_1 cation to carbons 6, 7, 10, and 11, are depicted in Figure 4.13. The 1,10-connection to generate the *E,E*-germacrenyl cation and the 1,11 connection to generate the *E,E*-humulyl cation are depicted as direct events. In competition with those C–C bond-forming steps, the initial farnesyl

E,E,E-C_1 cation can undergo isomerization, as depicted in Figure 4.5, to the *E,E,Z*-cation as shown. From this carbocation, a C_1–C_6 bond formation creates the bisabolyl cationic framework while the C_1–C_7 bond yields the seven-membered cycloheptenyl ring system. The fifth and sixth ring system variants are the C_1–C_{10} derived *E,Z*-germacrenyl cation and the C_1–C_{11} derived *E,Z*-humulyl cation.

To illustrate just one of many possible examples of further cation rearrangements that create scaffold complexity, Figure 4.14 shows how the *E,Z*-humulyl cation in the active site of pentalenene

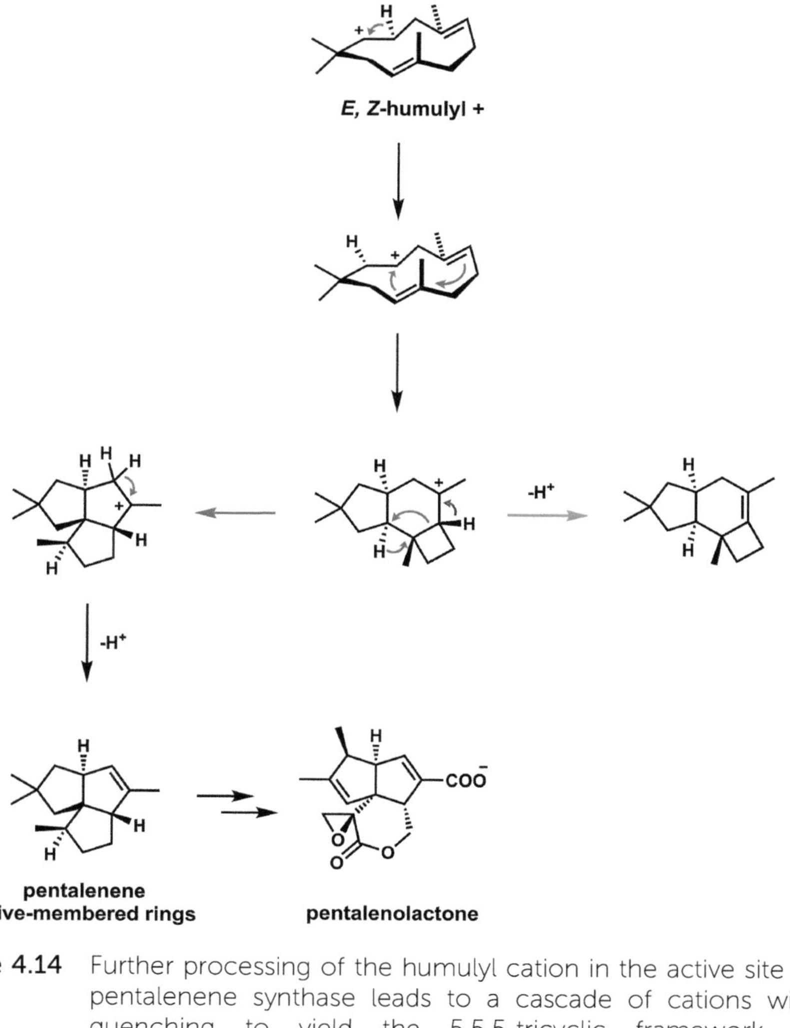

Figure 4.14 Further processing of the humulyl cation in the active site of pentalenene synthase leads to a cascade of cations with quenching to yield the 5,5,5-tricyclic framework of pentalenene.

synthase can undergo three more directed rearrangements before olefin-forming proton removal yields the fused 5,5,5-tricyclic frame of pentalenene. The hydrocarbon pentalenene is subsequently morphed by a series of oxygenases to the pentalenolactone with five oxygen atoms, creating a much more polar scaffold (Tetzlaff, You *et al.* 2006).

4.6.3 How Good are Terpenes Synthases at Directing Flux of the Sequential Series of Cations in their Active Sites?

Given the two examples noted for scaffold diversification at both the monoterpene and sesquiterpene levels, the general question arises as to how good are the synthase catalysts at directing flux down one manifold of cations to get a specific product. That can be framed experimentally by asking how promiscuous are terpene synthases (Steele, Crock *et al.* 1998). We will take this question up again in the triterpene section below but there is evidence that γ-humulene synthase shows some intrinsic promiscuity that can be increased/redirected by site-specific mutagenesis (Little and Croteau 2002). As noted in Figure 4.15, the native γ-humulene synthase makes predominantly the product for which it is named but also generates substantial amounts of siberene and longifolene. It even leaks smaller amounts of four other product scaffolds. Clearly it has a high degree of promiscuity. Another way of stating that outcome is that it may not be able to channel the reaction flux to stabilize only one or two carbocation intermediate structures. The energy barriers between the various cations that lead to the suite of seven sesquiterpene frameworks may be so low that the enzyme cannot prevent their population within its active site and the subsequent quenching routes to the observed hydrocarbons (Miller and Allemann 2012, Ueberbacher, Hall *et al.* 2012).

Mutations of from 2 to 5 residues in humulene synthase, guided by structural information, do alter the product profiles, to increase flux selectively to β-bisabolene, to α-longipinene, to longifolene, and to siberene, respectively (Yoshikuni, Ferrin *et al.* 2006). Minor leakage is still evident but this limited set of mutational results indicates how terpene synthases may have evolved to give an altered range of products. They also demonstrate that the terpene synthases may do rather little as catalysts. They may initiate C_1–OPP bond cleavage, hold on to the suite of cations that are then populated, based on bound substrate conformers and energy barriers of competing and/or

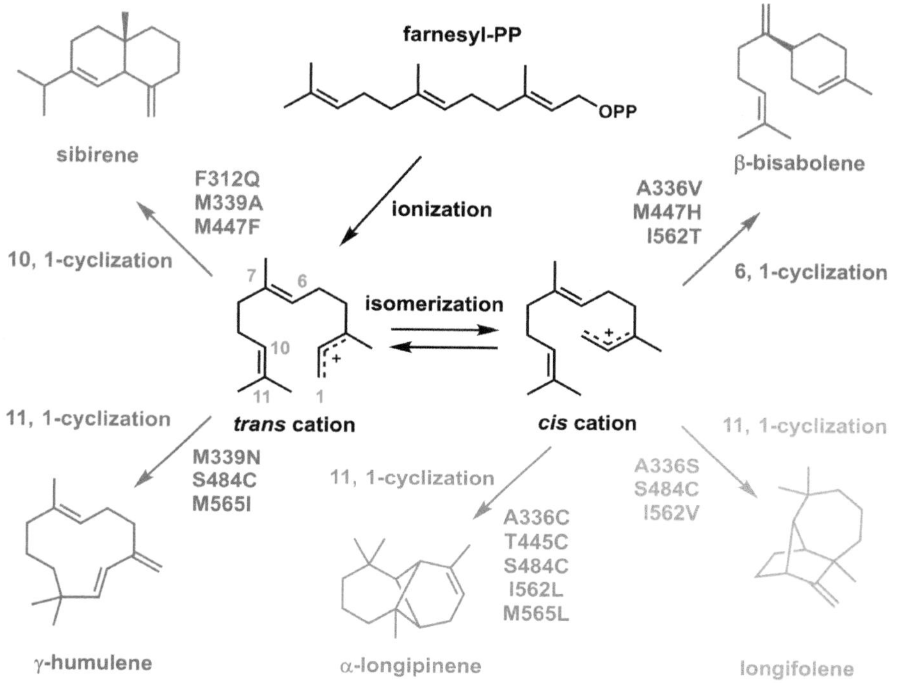

Figure 4.15 γ-Humulene synthase is a promiscuous catalyst, leaking flux to siberene and longifolene in addition to making humulene as major product. Mutagenesis causes redistribution of flux to products such as siberene, bisabolene, longifolene, and α-longipinene.
Data from Yoshikuni, Y., T. E. Ferrin and J. D. Keasling (2006). "Designed divergent evolution of enzyme function". *Nature* **440**: 1078–1082 and Cane, D. E. (2006). "How to evolve a silk purse from a sow's ear". *Nat. Chem. Biol.* **2**: 179–180.

sequential H and C migration aptitudes, and deliver quenching routes to get a range of product yields and frameworks (Cane 2006, Christianson 2008).

4.6.4 Diterpenes: Geranylgeranyl-PP to *ent*-Kaurene – Scaffold Complexity Generation

It is estimated that about 20% of the >50 000 terpenes are C_{20}-based diterpenes. Among them is the enantiomer of kaurene, termed *ent-kaurene*, a tricyclic hydrocarbon that is a pathway intermediate in the biosynthesis of the gibberellic acid class of plant hormones. Two

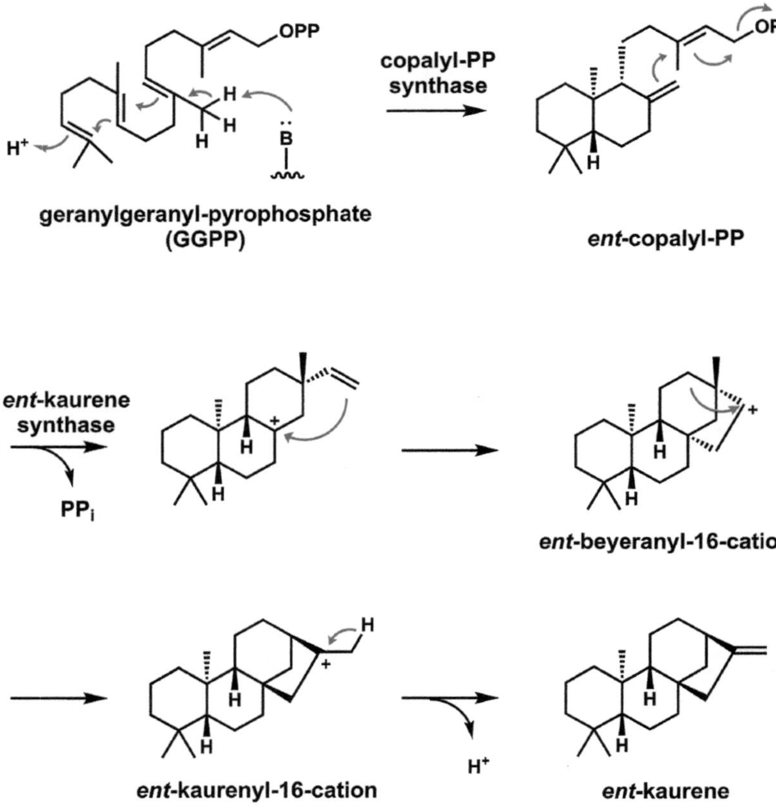

Figure 4.16 Conversion of linear C_{20} geranylgeranyl-PP to the tetracyclic framework of *ent*-kaurene, *via* a series of cationic rearrangements.

enzymes are involved in conversion of the major linear C_{20} prenyl donor geranylgeranyl-PP. The first is a copalyl-PP synthase, which converts the linear substrate into the bicyclic copalyl-PP. The mechanism is formulated as abstraction of the C_{10}-H as a proton to initiate formation of the two-ring decalin system (Figure 4.16) (Liu, Feng *et al.* 2014). This cyclization prefigures the kind of multicenter/multiring formation to be discussed in the context of triterpene squalene and oxidosqualene cyclizations. Copalyl-PP contains an exocyclic double bond, from the original abstraction of the C_{10}-H.

The next enzyme in the pathway, *ent*-kaurene synthase, is proposed to use that olefin in a doubly vinylogous, neighboring group assistance type of mechanism to promote cleavage of the C_1–OPP bond. (This could involve early loss of PP_i and capture of the C_1 allyl cation by the tandem movement of the pair of olefins) (Koksal, Potter *et al.* 2014). The initial product is a tricyclic ring structure with a tertiary carbocation. Intramolecular attack by the newly formed terminal olefin produces a tetracyclic cation; C–C bond migration converts this beyeranyl-16-cation to the *ent*-kaurenyl-16-cation which is finally quenched by proton removal to yield the exocyclic olefin. The net result of copalyl-PP synthase and *ent*-kaurene synthase is construction of a fused 6,6,6,5-tetracyclic ring system from a linear precursor.

4.6.5 Geranylgeranyl-PP to Taxadiene

Another example of enzymatic conversion of the linear C_{20}-isoprenyl-PP donor to a complex hydrocarbon scaffold is found in the action of taxadiene synthase (Figure 4.17) (Koksal, Jin *et al.* 2011). Because of the proven utility of the subsequent oxidation product taxol for treatment of a set of human cancers, the taxol biosynthetic pathway from the Pacific yew and related yew trees has been of great mechanistic and preparative interest (Chapter 12). As in all the other isoprenoid/terpenoid examples, carbocation reaction manifolds are in play. After early dissociation of the OPP leaving group in the taxadiene synthase active site, the bound linear C_1 allyl cation is folded such that the C_{14} olefinic terminus is the kinetically competent nucleophile to yield a 14-membered macrocyclic cation (C_1–C_{17} bond formation). This is proposed to undergo a series of four cation rearrangements, with C–C bond formations that convert the initial 14-membered macrocyclic cation to a 6,8,6-tricyclic tertiary cation, from which proton abstraction generates taxadiene. Two of the four original double bonds in geranylgeranyl-PP remain in taxadiene.

Note that isoprene/terpene biosynthesis leads to highly hydrophobic scaffolds in all the examples considered so far (humulene, *ent*-kaurene, taxadiene). Quenching with water generates mono-alcohols, but quenching by proton removal, which appears to be a default mode for many cyclizing terpene synthases, generates lipid-soluble, water-insoluble hydrocarbon frameworks.

To convert these water-insoluble scaffolds into molecular frameworks that show some water solubility, a second stage strategy comes into play. Producer plants, microorganisms, and even higher eukaryotes introduce oxygen functionality at different sites of the nascent hydrocarbon products. We will detail the nature of those oxygenase enzymes in Chapter 9 but note here that the maturation of the bis-olefinic C_{20} hydrocarbon taxadiene into the pharmacologically active taxol involves two kinds of tailoring enzyme strategies. First

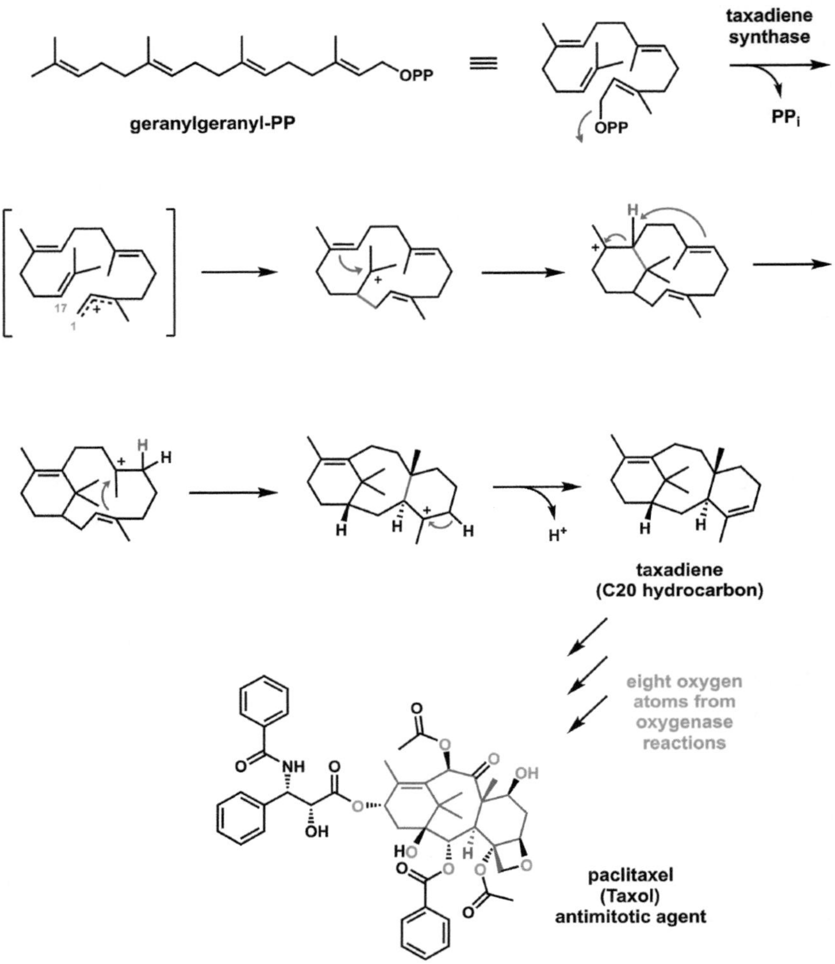

Figure 4.17 Taxadiene synthase converts a linear diterpene geranylgeranyl-PP into the 6-8-6 tricyclic framework of the hydrocarbon taxadiene. Subsequent oxygenative processing and acylation introduce eight oxygen atoms on the taxadiene periphery and four acyl groups to yield the pathway end product taxol.

there are eight oxygen atoms introduced, on eight of the 14 carbons on the periphery of the taxadiene framework. Four of those eight oxygen atoms are subsequently acylated to yield the taxol end product of the pathway. This example highlights a general oxygenation, acylation strategy for isoprene/terpene scaffold maturation. A third strategy, exemplified in Chapter 11, is glycosylation of the oxygenase-introduced functional groups to create glycosides with increased aqueous solubility.

4.7 Head to Head *vs.* Head to Tail Alkylative Couplings: C_{30} and C_{40} Terpene Compounds

A strategy for C–C bond formation occurs at the level of farnesyl-PP (C_{15}) and geranylgeranyl-PP (C_{20}) that is fundamentally different from the C_5 unit additions of Δ^3-IPP during chain elongation noted so far. Those chain elongation enzymes can all be characterized as catalyzing *head to tail* C_5 alkylation.

In contrast are the reactions of two C_{15} farnesyl-PP molecules to yield the C_{30} *linear hexaene* hydrocarbon squalene and the corresponding coupling of two C_{20} geranylgeranyl-PP molecules to yield the *linear nonaene* C_{40} phytoene (Figure 4.18). These can be characterized as *head to head* coupling alkylations as C_1 becomes joined to C_1. Squalene synthase is the most extensively studied and it was discovered early on that a cyclopropane-containing intermediate presqualene-PP was generated as a bound species (Figure 4.19A). This is subsequently opened by cation rearrangements. The reaction is completed by delivery of a hydride ion from NADPH which quenches the reaction (Tansey and Shechter 2001).

The early steps of the mechanism involve binding of the two molecules of farnesyl-PP in the squalene synthase active site, lined up head to head. The Δ^2-olefin of one FPP molecule attacks the incipient C_1 allyl cation of the second bound FPP to yield a C_2–C_1 carbon–carbon bond. At this point, abstraction of one of the C_1 prochiral hydrogens (from the electrophilic partner) as a proton leads to a C_3–C_1 bond as the cyclopropane forms. This is one of the three quenching routes noted in an earlier section for intervention in terminating the cation fluxes in terpene synthases. This is the structure of presqualene-PP. It is a discrete intermediate, but not released in mid-catalytic cycle.

In the second half reaction of squalene synthase (Figure 4.19B), S_N1 dissociation of the OPP group is featured, as usual, as the initiation step, yielding a primary cation but one that is adjacent to the cyclopropane such that rapid rearrangement to the more stable tertiary cation is anticipated, with formation of another cyclopropane (Blagg, Jarstfer *et al.* 2002). Termination of the reaction is proposed to occur by delivery of a hydride ion (shown in green) from cosubstrate NADPH. This fragments the new cyclopropane and generates the central

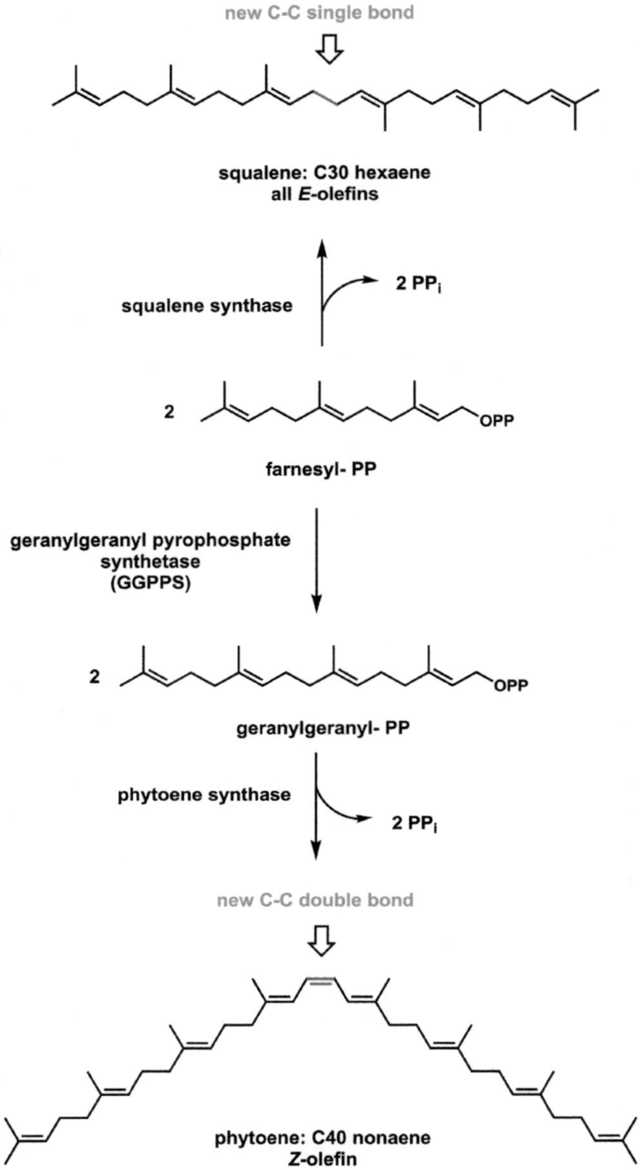

Figure 4.18 Head to head alkylations in squalene (C_{30}) and phytoene (C_{40}) formation are fundamentally distinct from the head to tail alkylations that generate the farnesyl-PP (C_{15}) and geranylgeranyl-PP (C_{20}) substrates for squalene synthase and phytoene synthase. The central double bond in phytoene has the *Z*-configuration.

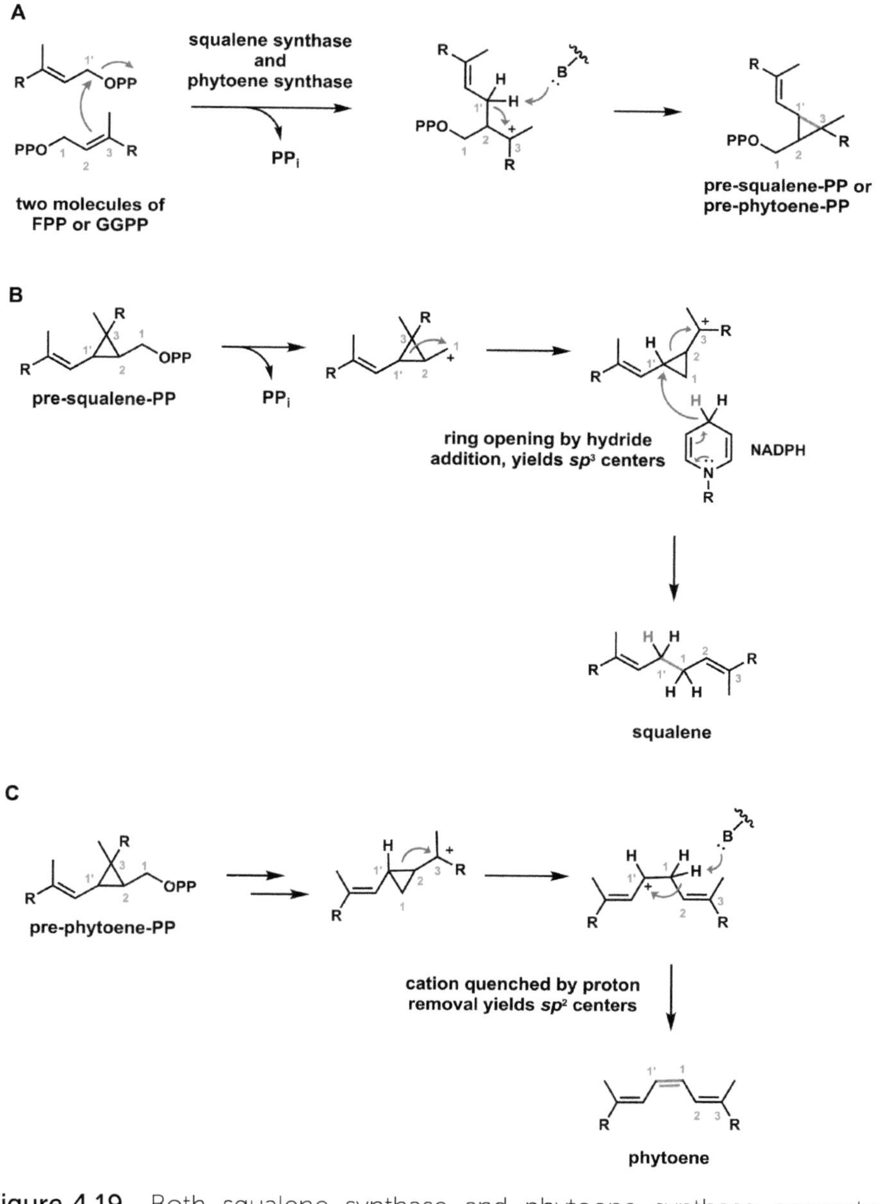

Figure 4.19 Both squalene synthase and phytoene synthase generate cyclopropane intermediates by 1,1-coupling, but then diverge in mechanism. Presqualene-PP is opened by hydride transfer, yielding a reduced CH_2–CH_2 linker. Prephytoene-PP is opened and quenched by proton removal to create a new olefin. In contrast to the squalene sp^3–sp^3 connection, the phytoene linkage is sp^2–sp^2.

CH_2–CH_2 functionality in squalene that represents the original C_1' and C_1 carbons of each of the two farnesyl-PP substrates. This emphasizes the *head-to-head nature* of the net coupling, albeit it began as a C_1–C_2 coupling. A global look at the squalene structure shows that the three double bonds from each farnesyl-PP are retained in the squalene product.

It is instructive to look briefly at the similarities and differences in the phytoene synthase reaction and product outcome (Figure 4.19C) (Misawa, Truesdale *et al.* 1994). Phytoene synthase analogously catalyzes a net head to head coupling, this time of two C_{20} geranylgeranyl-PP substrates. The first half reaction is essentially identical. As depicted in Figure 4.19A, a comparable cyclopropane-containing prephytoene-PP forms. The same mechanism is envisioned. Attack of the C_2 olefin from one bound GGPP on C_1 of the other bound GGPP generates the C_1–C_2 bond and cyclopropane formation generates the C_1–C_3 in the cyclopropane.

It is the deconvolution of prephytoene-PP to phytoene that differs in two related ways. First is that the rearranged cyclopropyl methylene cation is quenched by proton abstraction, not by a hydride addition from NADPH (Figure 4.19C). Thus, instead of reduction to a C_1H_2–$C_1'H_2$ link for the original C_1 carbons as seen in squalene, there is a CH=CH olefinic link in phytoene. So, in addition to the four double bonds contributed by each GGPP partner there is now a *ninth double bond in phytoene*, such that the three central olefins are conjugated rather than being out of conjugation. Second, the proton abstraction step has the stereochemistry noted in the previous *Z*-forming examples so the newly generated olefin is of the *Z*-configuration (Figure 4.18). Among other things this *E,Z,E*-triene system is thought to prevent subsequent enzymatic cyclization of the phytoene scaffold and is thus a very different metabolic fate from squalene, noted next (Moise, Al-Babili *et al.* 2014).

4.8 Squalene-2,3-Oxide and Cyclized Triterpenes

4.8.1 Formation of Squalene-2,3-Oxide

Squalene, once formed, is the substrate for a set of cyclases in microbes, plants, and animals that morph the linear hexaene into polycyclic six- and five-ring fused triterpene scaffolds. The cyclizations can occur from squalene itself but more commonly after squalene has undergone regiospecific epoxidation of one of its two terminal olefins to 3*S*-2,3-oxidosqualene with the epoxide above the plane (Figure 4.20) (Laden, Tang *et al.* 2000). The enzyme squalene epoxidase contains tightly bound FAD as the oxygen-activating coenzyme when in its reduced form. To that end, NADPH, along with molecular oxygen, is also a cosubstrate. The NADPH generates the $FADH_2$ form of the bound coenzyme that engages in one-electron transfer to O_2, generating a radical pair: the FADH$^{\bullet}$ semiquinone and superoxide anion ($O_2^{-\bullet}$). Radical recombination occurs specifically at the FAD-C_{4a} bridgehead position to yield the FAD-4a-OOH. This flavin peroxide is the proximal donor of one oxygen to the 2,3-olefin of bound squalene. As suggested in Figure 4.20 the FAD-OOH is a donor of an

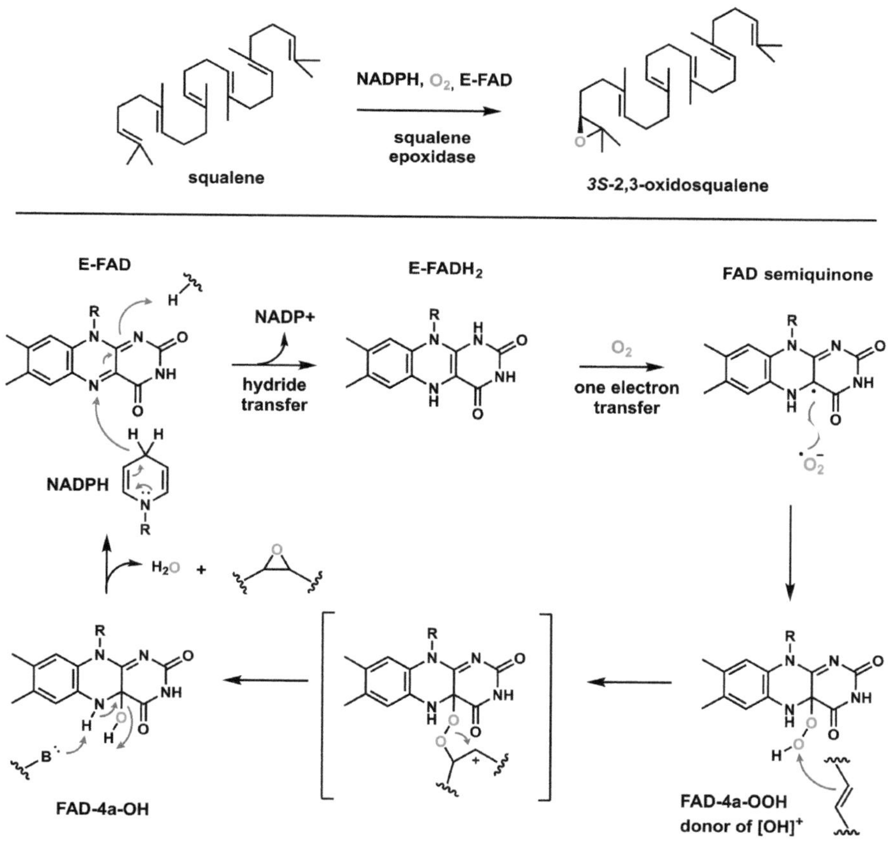

Figure 4.20 Squalene 2,3-epoxidase is a flavoprotein oxygenase. The bound FAD-4a-hydroperoxide is the proximal donor of an [OH^+] equivalent to the π-electrons of the 2,3-double bond of squalene.

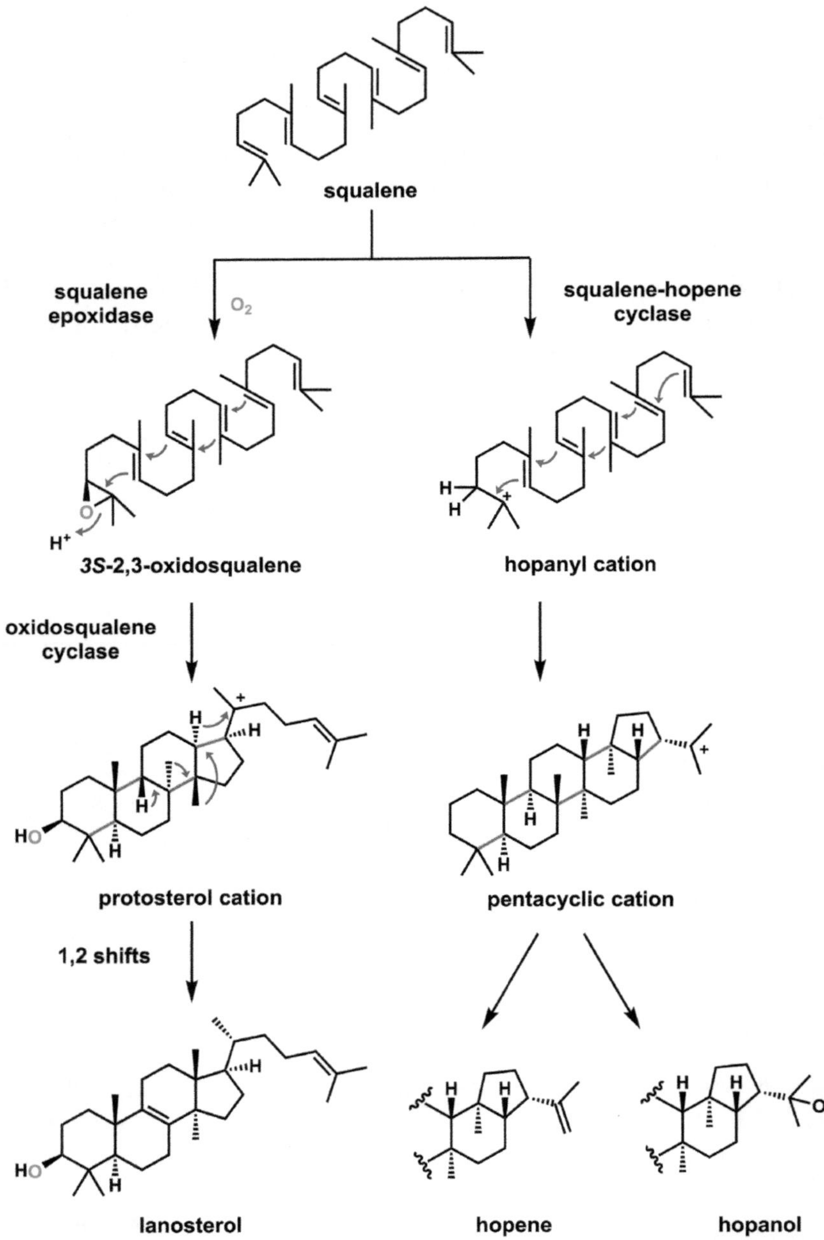

Figure 4.21 Squalene cyclizations. The 2,3-oxide goes to the indicated protosterol cation and this can be routed to lanosterol as released product. Squalene itself can undergo proton-assisted cyclization *via* a pentacyclic cation to hopene and hopanol, bacterial triterpene products.

$[OH^+]$ equivalent to the attacking double bond, with subsequent closure to the epoxide (Walsh and Wencewicz 2013).

The net effect of this action by squalene epoxidase is to polarize the squalene scaffold, with the epoxide at one end serving as an electrophilic functional group. That epoxide ring can be opened by *intramolecular attack of the π-electrons* of the neighboring olefins in a set of cascade reactions, mediated by a set of squalene cyclases described in the next section. As we will note in Chapter 9 it is relatively unusual for the flavin coenzyme FAD to serve as an epoxidation catalyst because the FAD-4a-OOH is a relatively weak oxygen transfer agent that is more typically used with electron rich substrates for oxygen transfer. Clearly, the 2,3-double bond in the active site of squalene-2,3-epoxidase is sufficiently active as a nucleophile to initiate oxygen transfer.

4.8.2 Squalene Cyclases: Squalene to Hopene Framework

The linear C_{30} hexaene squalene is precursor to hundreds to thousands of tricyclic to pentacyclic fused ring frameworks where the default scaffold in animal cells is the tetracyclic sterol nucleus (Figure 4.21). On the other hand, in higher plants β-amyrin is the most abundant product framework. Both of these scaffolds arise from cyclization of squalene-2,3-epoxide, discussed in the next section. However, cyclizations can be initiated from squalene itself by regiospecific protonation of the 2,3-double bond. This path, for example, is the route to bacterial hopenes and hopanols (Siedenburg and Jendrossek 2011). This resultant C_{22} pentacyclic cation from cyclization of squalene itself can be quenched by proton loss to the terminal olefin in hopene or by water addition to yield hopanol as shown in the top portion of Figure 4.21. Figure 4.22A indicates that the conformer of squalene bound in the squalene–hopene cyclase active site is likely to be in the all "pre-chair" conformation (Wendt 2005).

Vignette 4.1 Squalene Hopene Cyclases as Protonases in Brønsted Acid Catalysis.

We have noted that the leitmotif in enzymatic assembly of a plethora of terpene scaffolds is initial allylic carbocation formation, followed by series of cation rearrangements to alternative or rearranged frameworks. The organic cation flux gets terminated typically by one of three routes: adjacent proton abstraction to create an olefin, addition of water to create an alcohol, or migration of a carbon bond (*e.g.* in cyclopropane formation).

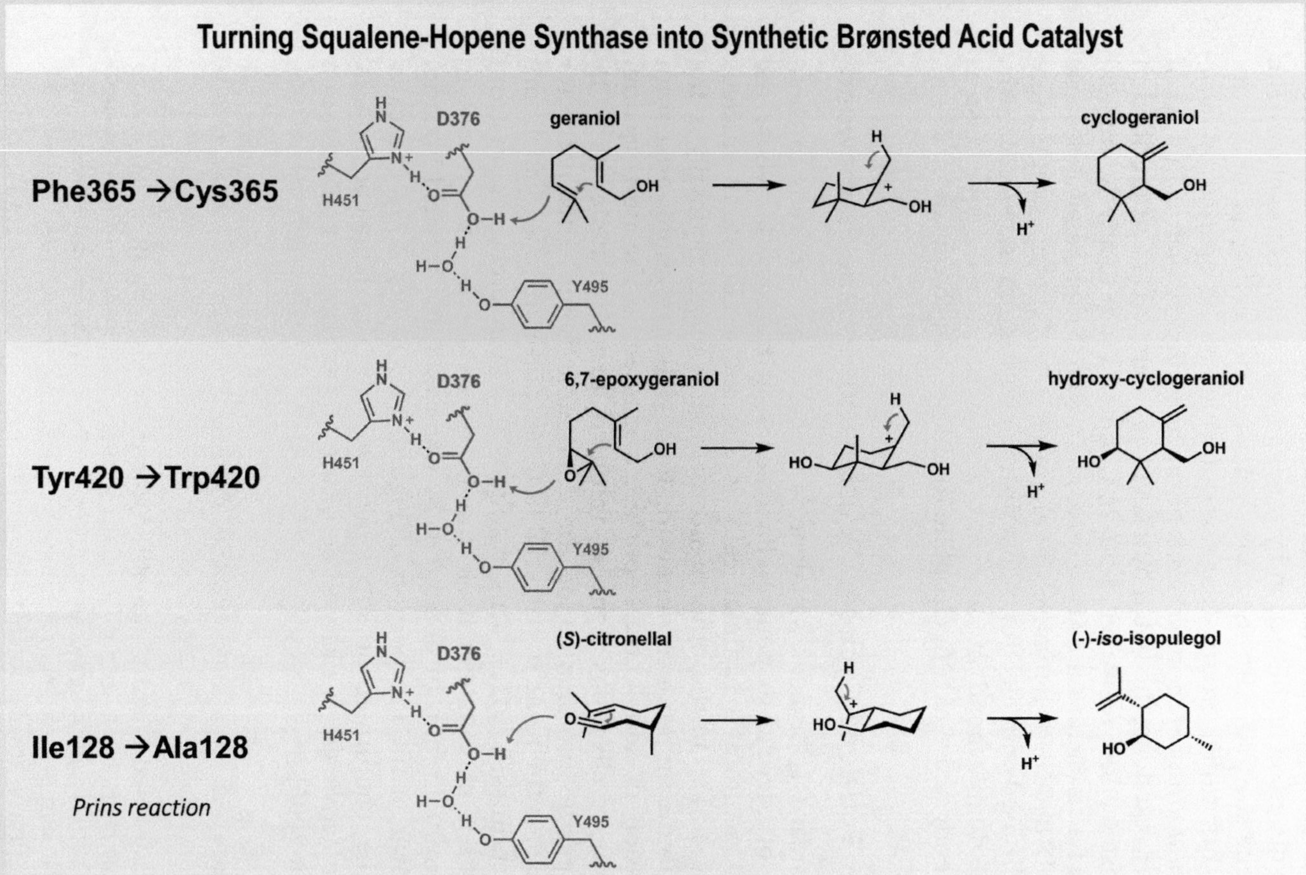

Figure 4.V1 Turning the squalene hopene synthase into a protonase for Brønsted-acid catalysis. Data from Hammer, S. C., A. Marjanovic, J. M. Dominicus, B. M. Nestl and B. Hauer (2015). "Squalene hopene cyclases are protonases for stereoselective Brønsted acid catalysis". *Nat. Chem. Biol.* **11**: 121–126.

Hauer and colleagues (Hammer, Marjanovic *et al.* 2015) have recently noted that the squalene hopene synthases are unique enzymes in catalysis of the cascade of C–C bond forming reactions to create the tetra and pentacyclic triterpene product scaffolds by virtue of protonating the 2,3-double bond in squalene. This is a nontrivial protonation event to produce the charge deficiency that sets off the cyclization cascade. They note that the substrate is hydrophobic and so is the active site, except for a side chain aspartate carboxylic acid as the incipient proton donor. Proper placement of the aspartate-β-COOH relative to bound squalene substrate is key to an effective and demanding proton donation to one face of one of the six double bonds in squalene.

They used the structure of squalene synthase to generate mutants that would be more general catalysts for Brønsted acid. Residues depicted in Figure 4.25 were targeted for mutagenesis to alter the substrate channel. A comparison of single mutants indicated that F365C cyclized geraniol to cyclogeraniol but Y420W was a better catalyst for cyclization of *S*-(6,7)-epoxygeraniol. The authors note that proton donation to the geraniol double bond as electrophilic initiation step is to the π-electrons (Figure 4.V1), while the epoxide gets protonated on the oxygen lone pair, and that difference may be exploited in the mutant active sites. A distinct point mutant of squalene synthase, I126A, carried out a regio- and stereospecific Prins reaction on citronellal to (−) iso-isopulegol. The creation of a chiral microenvironment where a proton can be transferred regio- and stereospecifically to a range of olefins, epoxides, and carbonyl substrates makes the squalene hopene cyclase family of protonation catalysts worth further exploration for evolution of their catalytic range.

4.8.3 Oxidosqualene to Lanosterol, Cycloartenol, β-Amyrin

When squalene-2,3-epoxide is instead the substrate (see next section), the cyclization pattern from the lanosterol-forming synthase differs and gives the protosterol cation shown in Figure 4.22B. Figure 4.22 suggests that the acyclic oxidosqualene conformer bound in the oxidosqualene–lanosterol synthase is folded in a pre-chair–boat-chair conformation, rather than the all-chair conformer proposed in the hopene cyclase active site. In contrast to the hopene cyclase, where no migrations of hydrogens or methyl groups occur before the C_{22} product precursor cation is quenched, the C_{20} protosterol cation in the lanosterol synthase undergoes a series of such migrations before proton abstraction quenches the process with introduction of the C_{11}–C_{12} double bond. The rearrangements may be some combination of altered energy surface that allows a long life of the protosterol

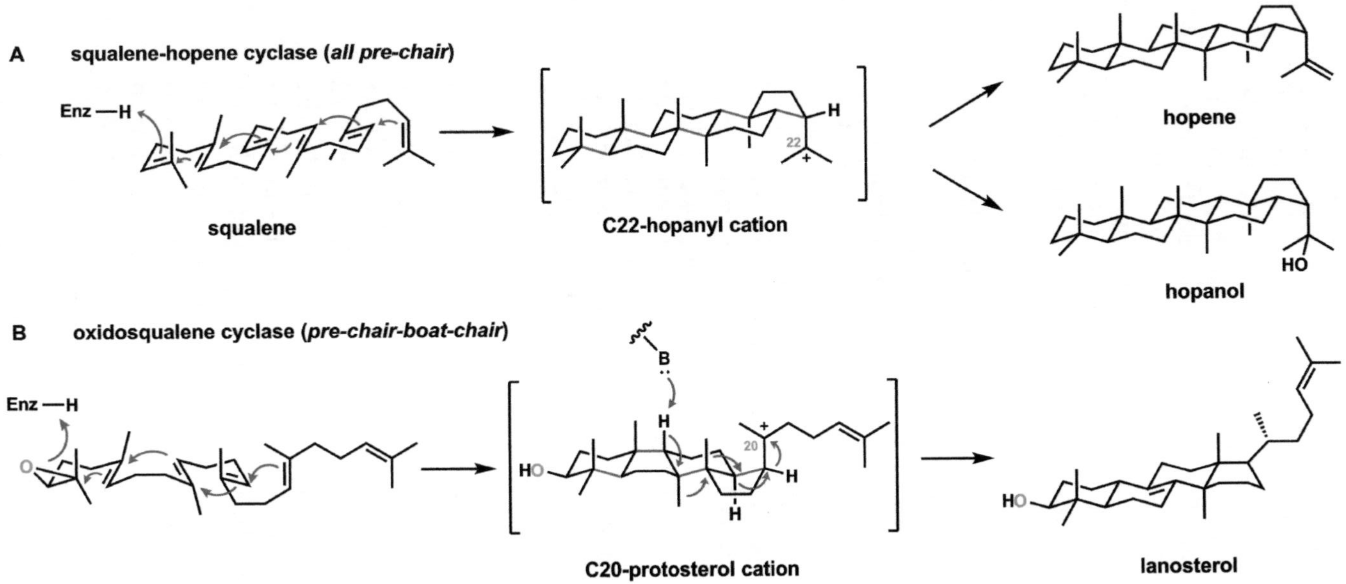

Figure 4.22 Chair *vs.* boat conformers to explain lanosterol *vs.* hopene product scaffolds.

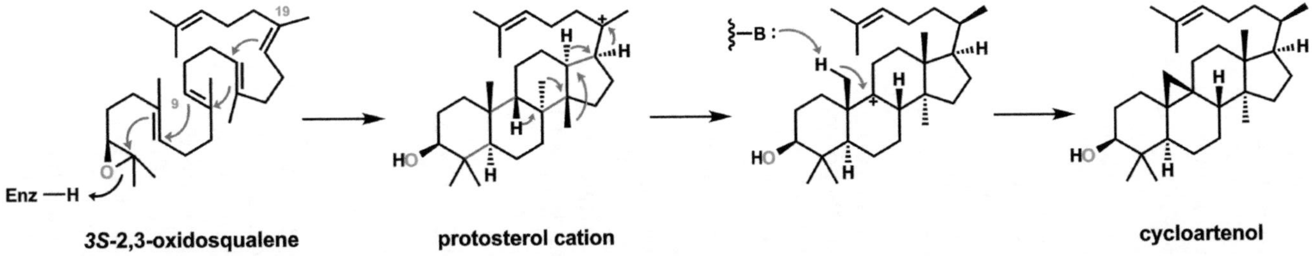

Figure 4.23 Squalene-2,3-oxide is converted to cycloartenol in *Cucurbita maxima* by cyclopropane formation as the cation-quenching step.

cation in the enzyme active site and/or lack of appropriately placed general acids to deliver a quenching proton.

In the related formation of cycloartenol in the plant *Cucurbita maxima* the same pattern of 1,2 rearrangements occurs up to the quenching step. In that cycloartenol synthase it is removal of one of the protons from the C_{19} angular methyl group that allows the resulting incipient C_{19} carbanion to attack the C_9 cation with formation of the cyclopropane ring (Figure 4.23), a process we have seen previously, both in the chrysanthemum and in presqualene-PP formation. The routing of protosterol cations down different quenching pathways suggests that the distinct synthases have active site bases disposed differentially with respect to the rearranging cation frameworks of intermediates and nascent products.

Figure 4.24 shows one more route variation for oxidosqualene cyclization, this time to the pentacyclic framework of β-amyrin. This is the most abundant cyclic triterpene in many higher plants (Thimmappa, Geisler *et al.* 2014). The proposed mechanism is drawn out as a stepwise process, rather than a single transition state, to reveal the sequential formation of cations that are named for the hydrocarbon frameworks observed in other triterpenes. This pathway involves the dammarenyl cation, lupenyl cation, and oleanyl cation on the way to the quenched beta amyrin pentacyclic (6,6,6,6,6) scaffold that is the released major product. Plant triterpenes are often found in the waxy coat of leaves, may function as signaling molecules such as brassinosteroids (Vert, Nemhauser *et al.* 2005), and when converted to glycosides may serve a defensive role against pests and plant pathogens.

4.8.4 Structural Biology Insights

The X-ray structures of both squalene–hopene cyclases and 2,3-oxidosqualene cyclases have been determined from several organisms and in complex with inhibitors. Figure 4.25 shows the active site region around a predicted bound substrate conformer of hopene cyclase (Hoshino and Sato 2002) with proposed functions for the indicated protein residues in sites of initial protonation, carbocation stabilization, stereochemistry, and substrate binding orientations. Figure 4.26 compares the placement and orientation of the same ligand, an anti-cholesterol drug candidate from Roche, R0-48-8071, in the bacterial hopene cyclase (Lenhart, Weihofen *et al.* 2002) and in the human lanosterol synthase which uses 2,3-oxidosqualene as natural substrate (Thoma, Schulz-Gasch *et al.* 2004).

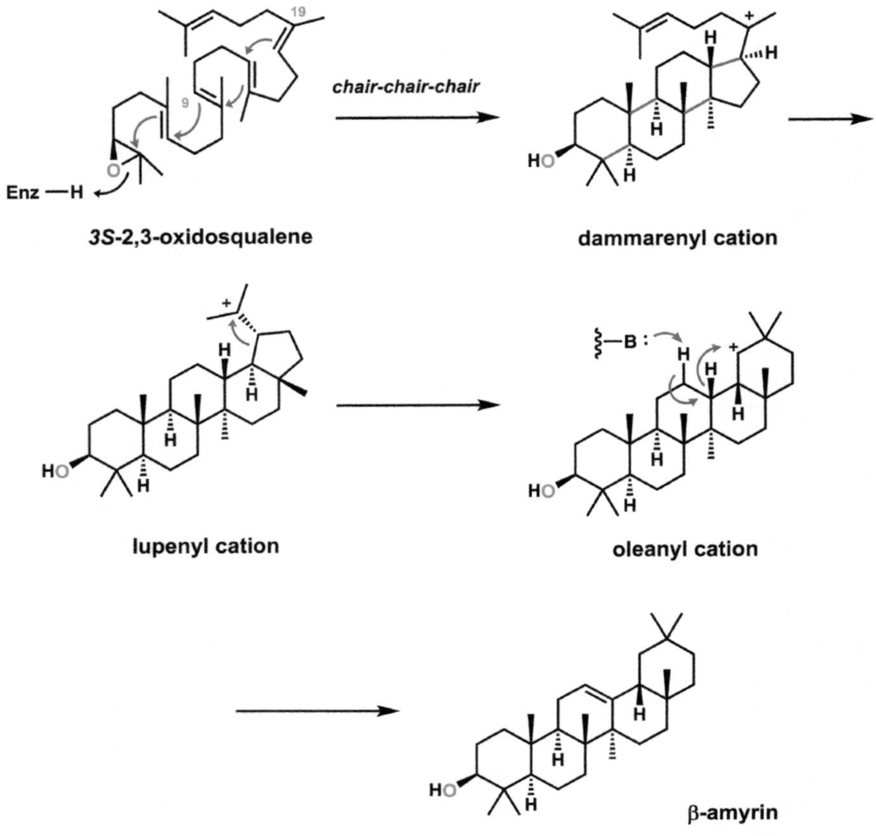

Figure 4.24 Cation rearrangements from squalene-2,3-oxide to generate the 6,6,6,6,6 pentacyclic scaffold of β-amyrin, the most abundant sterol in higher plants.

The structural findings have guided mutagenesis strategies to confirm/establish some of the predicted residue functions and have allowed evaluation of the production of sets of partially cyclized alternative products (Figure 4.27). As shown, mutations directed toward enzyme side chains involved in substrate binding, stereocontrol, or intermediate cation binding/stabilization yield a broad range of partially cyclized product scaffolds. In turn this set of derailed products helped establish the reaction pathways of native and mutant squalene hopene cyclases. These data and the ones discussed next also show the fine tuning and balance in routing cation-mediated flux to products in these triterpene cyclases.

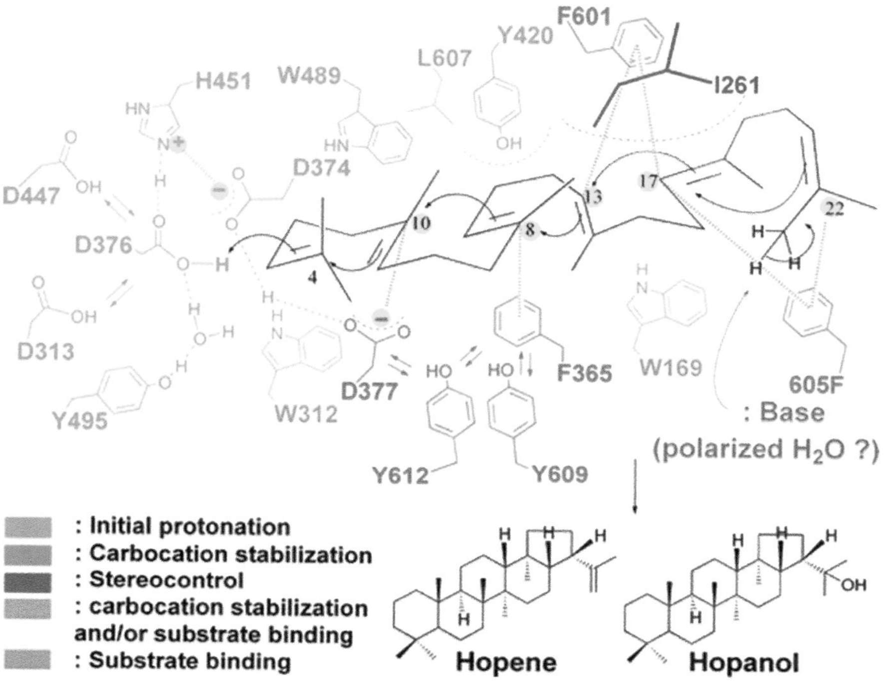

Figure 4.25 Active site residues in the *Alicyclobacillus acidocaldarius* squalene–hopene cyclase indicating proposed functions during catalysis.
Reproduced from Hoshino, T. and T. Sato (2002). "Squalene–hopene cyclase: catalytic mechanism and substrate recognition". *Chem. Commun.* 291–301 with permission from The Royal Society of Chemistry.

4.8.5 Promiscuity of Oxidosqualene Cyclases: How Good are the Oxidocyclases at Directing Flux Down One Reaction Manifold?

The reaction mechanism proposed in Figure 4.24 for conversion of squalene-2,3-oxide into the pentacyclic framework of β-amyrin suggested a population of cations which leads to other triterpenoids known in other plants, including the dammarene, lupene, and oleane structural types. In turn this raises the question of how efficient are the family of triterpene cyclases/oxidocyclases at directing reaction flux down only one pathway and corralling the different cationic species from going off pathway. The mutant data in Figure 4.27 for

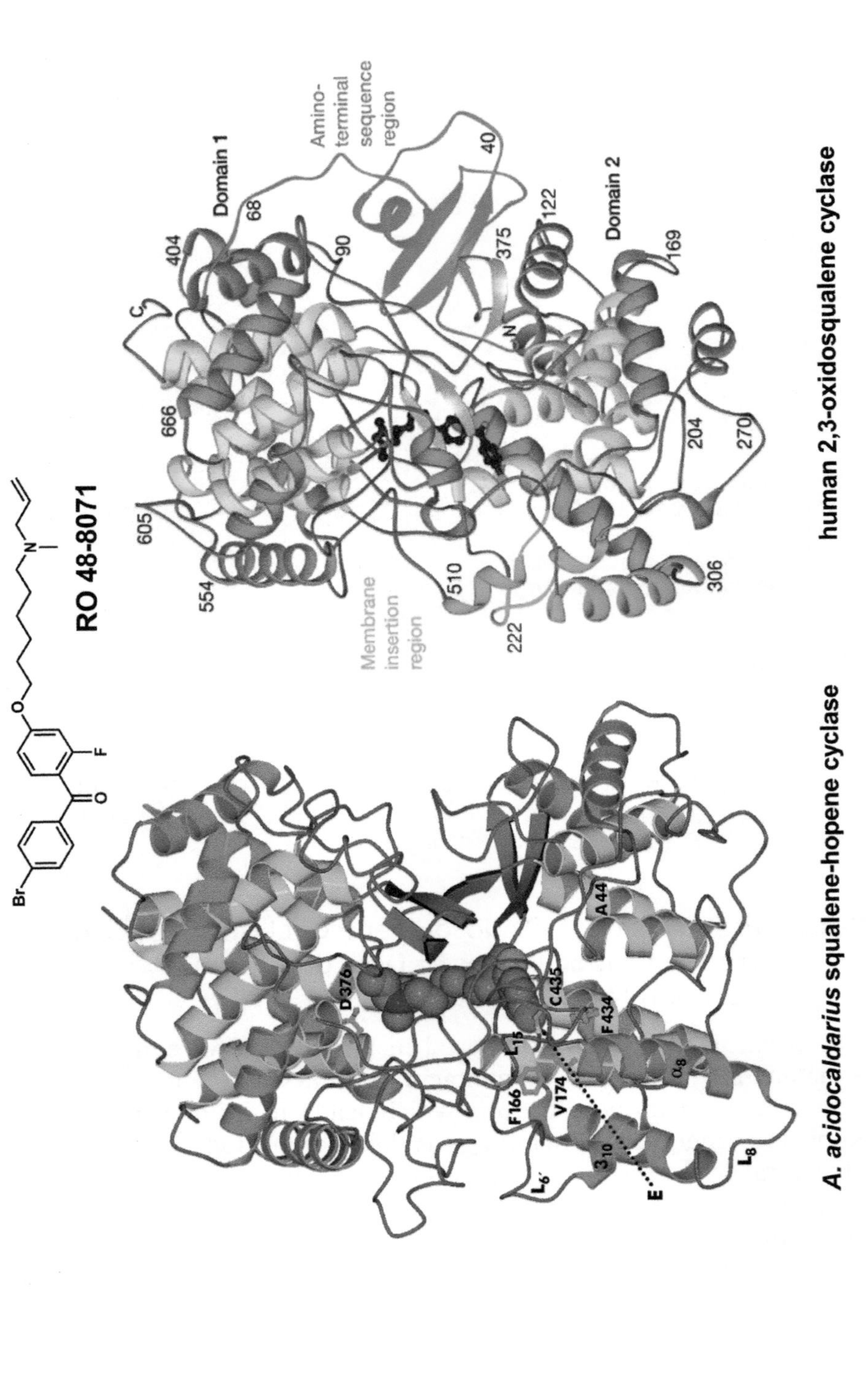
Br
O
F
N
RO 48-8071
A. acidocaldarius squalene-hopene cyclase
D376
C435
F434
A44
L15
F166
V174
α8
310
L6'
L8
E
human 2,3-oxidosqualene cyclase
Domain 1
Domain 2
Amino-terminal sequence region
Membrane insertion region
C
N
404
68
40
90
122
375
169
666
605
554
510
222
306
204
270

the squalene–hopene cyclases indicate that intermediates can clearly go off pathway in damaged enzymes.

This is in essence a measure of promiscuity on the part of the enzymes. We have noted earlier in the context of both C_{10} monoterpenes (Figures 4.11–4.13) and C_{15} sesquiterpenes (Figures 14–15) that essentially all possible combinations of products are formed in one setting or another. How do the triterpene synthases fare?

In fact there are reports of leakage of minor products in several studies but perhaps the most quantitative deals with baruol synthase from *Arabidopsis*. As noted in Figure 4.28, seven cationic species are proposed to form and account for the final product baruol (Lodeiro, Xiong *et al.* 2007). This is 89.7% of the product flux. The remaining 11.3% is distributed over 22 minor products, ranging in abundance from 2.7% down to 0.02%. Figure 4.28 shows only two of the minor products, the most abundant ones, columbiol and sasanqual, and how they branch off particular cationic intermediates. The original publication has a more complete scheme that counts 14 distinct cationic intermediates in the five A to E rings to give rise to the observed set of 23 triterpene products (raising the question of how concerted the cyclizations are, and whether there are multiple cationic transition states).

This set of findings reveals the complexity of the reaction manifolds of the oxidosqualene cyclases. Leakage of minor products may indicate a failure to be able to direct cation rearrangements along only one trajectory and the probability that an undesired quenching step intervenes during the multistep *ad seriatim* formation of distinct cationic species. The remarkable plasticity of the emergent triterpene

---

Figure 4.26 Structures of *A. acidocaldarius* squalene–hopene cyclase (left) and the human 2,3-oxidosqualene cyclase (lanosterol-forming) (right) with the same bound small molecule ligand RO 48-8071.
Left figure reprinted from Lenhart A., W. A. Weihofen, A. E. W. Pleschke and G. E. Schulz, "Crystal structure of a squalene cyclase in complex with the potential anticholesteremic drug Ro48-8071", *Chem. & Biol.* **9**: 639–645, Copyright (2002) Cell Press, with permission from Elsevier. Right figure reprinted by permission from Macmillan Publishers Ltd: Nature (Thoma, R., T. Schulz-Gasch, B. D'Arcy, J. Benz, J. Aebi, H. Dehmlow, M. Henning, M. Stihle and A. Ruf (2004). "Insight into steroid scaffold formation from the structure of human oxidosqualene cyclase". *Nature* **432**: 118–122), copyright (2004).

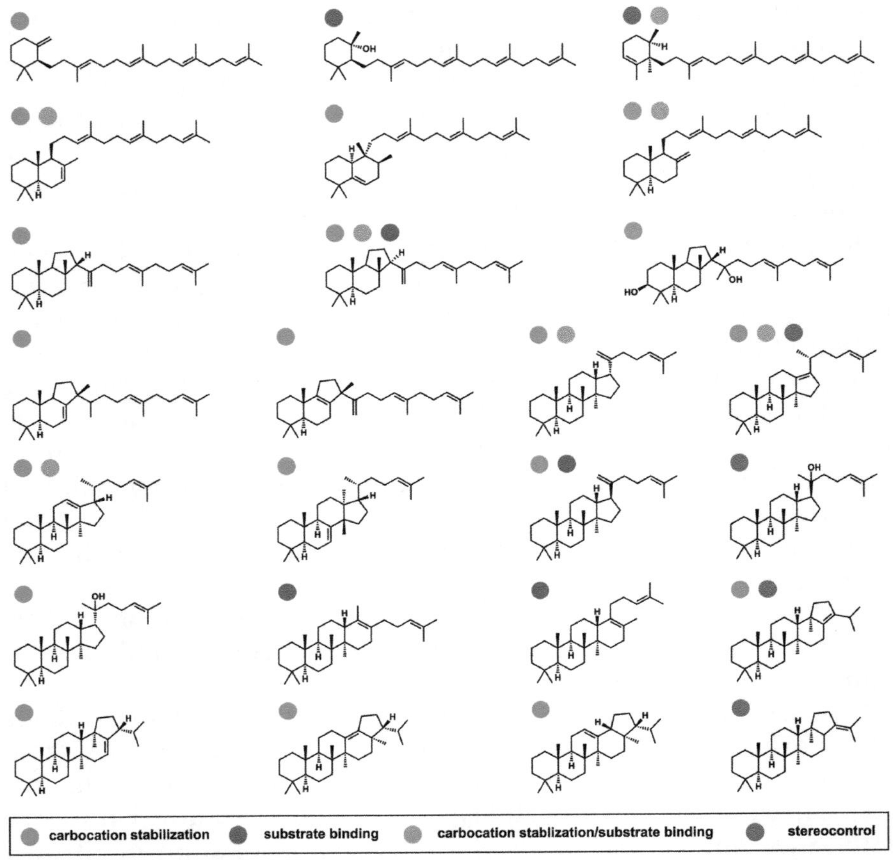

Figure 4.27 Structure-based mutagenesis of *A. acidocaldarius* squalene–hopene cyclase leads to disruption of the full cyclization cascade and results in a suite of partially cyclized products. Color coding is with reference to that in Figure 4.25.
Adapted from Hoshino, T. and T. Sato (2002). "Squalene–hopene cyclase: catalytic mechanism and substrate recognition". *Chem. Commun.* 291–301 with permission from The Royal Society of Chemistry.

cyclic scaffolds reflects the underlying low barriers to interconversion of the carbocations and the ease of 1,2 migrations of –H, $-CH_3$ and C–C single bonds within the nascent tri- to pentacyclic structures. The facile exploration of 14 cations in baruol synthase catalysis may also give some insight into how isoprenoid/terpenoid molecules have become so abundant among natural product classes.

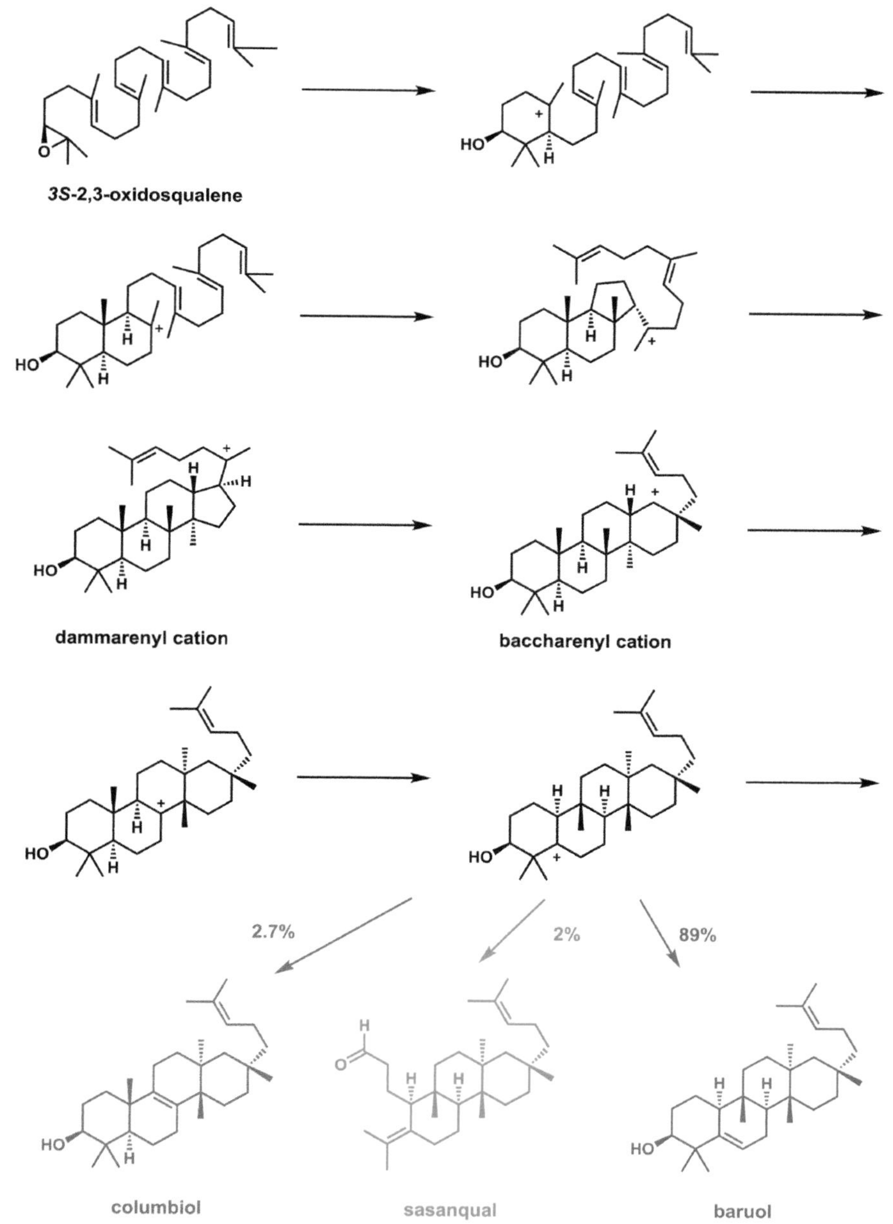

Figure 4.28 A range of minor products emerge along with baruol (89% of flux) during catalysis by the *A. thaliana* baruol synthase. Seven are shown of the 14 proposed cation manifolds in *Arabidopsis* baruol synthase that total 22 cyclized products.

4.8.6 Lanosterol to Cholesterol and Beyond: A Bevy of Oxygenases

From lanosterol to cholesterol, the predominant sterol in vertebrate metabolism, five discrete changes have to occur (Figure 4.29). Three of them involve the oxidative removal of the two methyl groups at C_4

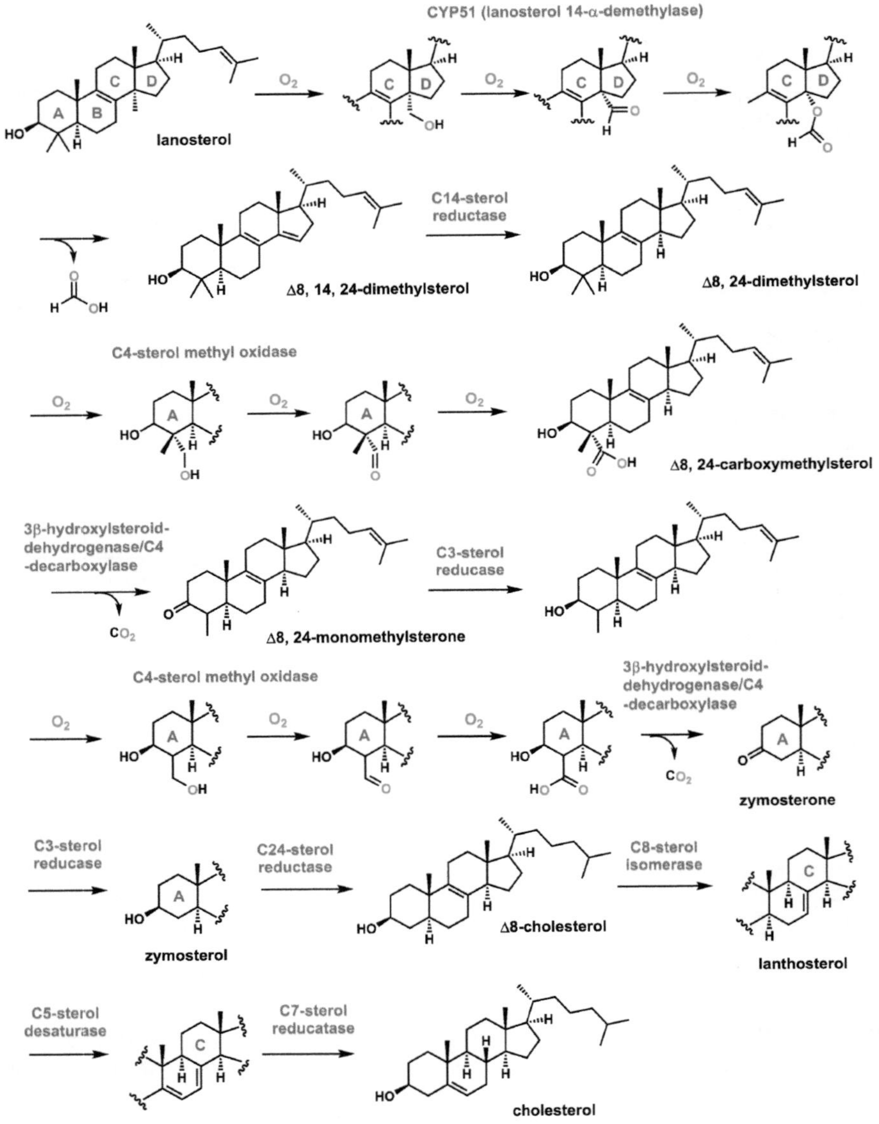

Figure 4.29 Nineteen steps are involved in enzymatic processing of lanosterol to cholesterol. These include consumption of nine molecules of O_2 as three C-methyl groups are carved out to two CO_2 and one formate as coproducts.

and the one at C_{14}. These involve the tandem action of cytochromes P450 in a series of nine oxygenation steps. At C_{14} the progression is methyl to alcohol to aldehyde to formyl ester with elimination of formate by CYP51. At C_4 the progression at each methyl is sequential oxygenation to alcohol, aldehyde, acid, and loss of CO_2. Saturation of the side chain olefinic terminus is the fourth change, and the last one involves formal migration of the double bond to the bridgehead position in the B ring of cholesterol. Overall, one can count 19 separate chemical steps (schematized in Figure 4.29). Because cholesterol is a primary rather than a secondary metabolite we do not spend more time on its formation here although extensive literature exists to validate the indicated pathway.

Also well documented are the subsequent conversions of cholesterol to adrenal and sex hormones. A suite of cytochrome P450 oxygenases are again central to those conversions, including the cholesterol side chain cleavage enzyme, the $C_{17,20}$ lyases, and the remarkable aromatase converting C_{19} androgens to C_{18} estrogens as the A ring is aromatized (Figure 4.30). Overall, the conversion of the C_{30} lanosterol, the released

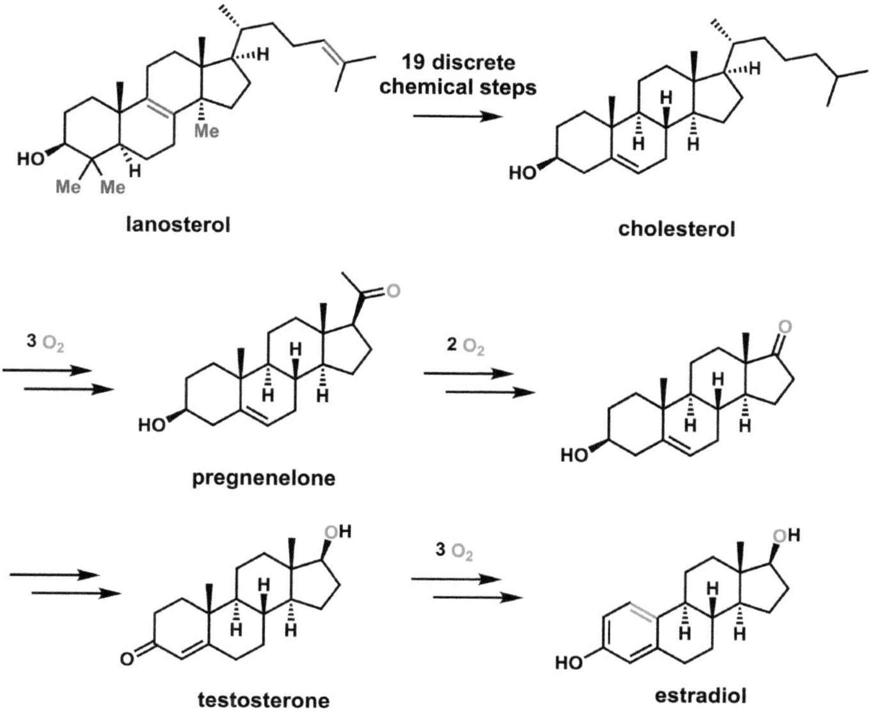

Figure 4.30 Lanosterol to cholesterol to steroid androgens and estrogens involves paring the C_{30} lanosterol down to the C_{19} testosterone and C_{18} estrogen by oxygenative removal of 11 and 12 carbons, respectively, from the initial lanosterol scaffold.

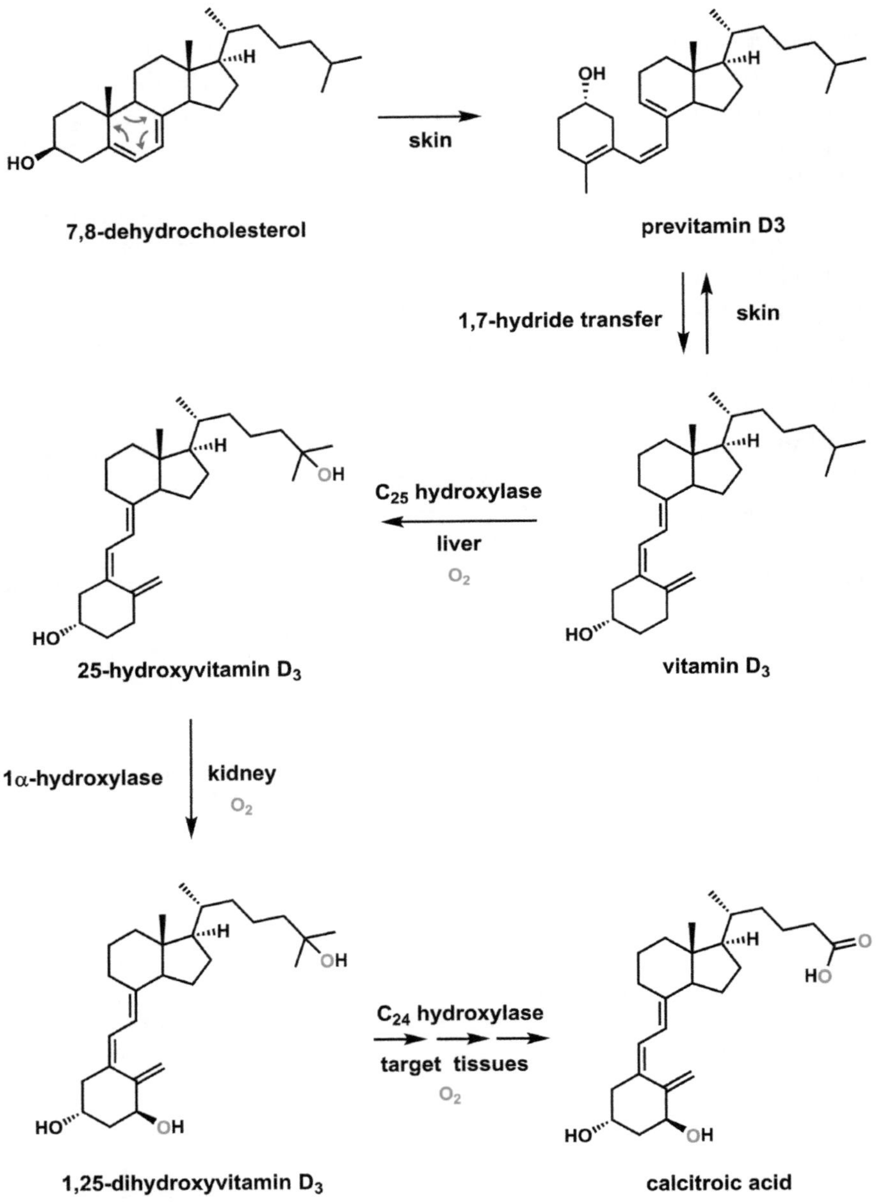

Figure 4.31 7,8-Dehydrocholesterol is converted to previtamin D in a nonenzymatic, light-induced conrotatory opening. Spontaneous rearrangement of pre-D3 to D3 occurs in the skin. The active form of vitamin D3 as a hormone requires three tissue-selective oxygenations, at C_1, C_{25}, and then C_{24}, to form the most active calcitroic acid.

product from oxidosqualene cyclase, to the 18-carbon estrogen represents oxygenative trimming away of four methyl groups and the eight-carbon side chain. It is the densest use of P450s in any set of vertebrate biosynthetic pathways.

One other remarkable chemical conversion in the cholesterol framework that is also then attended by three subsequent enzymatic oxygenations is the conversion of 7-dehydrocholesterol to previtamin D (Figure 4.31). This is a nonenzymatic reaction that happens in skin cells when ultraviolet (UV) light is adsorbed by 7-dehydrocholesterol. A six π-electron conrotatory opening of the B ring proceeds to yield the triene in previtamin D. Another nonenzymatic step happens subsequently, the 1,7-hydride shift to yield vitamin D3. While this is the vitamin form it is still a biological precursor to the active form of the hormone, calcitroic acid (1-α-hydroxy-23-carboxy-24,25,26,27 nortetravitamin D) that arises from action of three tissue-specific cytochromes P450.

Vignette 4.2 Wortmannin as Covalent Inactivator of PI-3-Kinase Family Members.

Penicillium funicolosum makes wortmannin, a highly oxygenated steroid metabolite, and the related steroidal furan viridin is another furan-containing fungal metabolite (Figure 4.V2). Both of these metabolites are inhibitors of phosphatidylinositol-3-kinase (PI-3-K) isozymes and structurally related kinases including mTor, polo like kinase-1, and, at higher concentrations, even myosin light chain kinase and mitogen activated protein kinases. Although the biosynthetic pathways have not been fully determined it is highly likely they arise from oxygenative maturation of nascent tetracyclic triterpenes.

Wortmannin received substantial attention because it works as a covalent inhibitor of its target kinases. After an initial reversible binding event in the ATP site (Figure 4.V2), the bound wortmannin undergoes a vinylogous Michael attack on the furan ring by the ε-NH_2 of the active site lysine 822 (Walker, Pacold *et al.* 2000). The resultant adduct, a vinylogous carbamate, is stable and irreversibly inactivates the catalytic p110 subunits.

Because wortmannin is cell permeable it has been a useful tool in deciphering aspects of PI3K family cell biology. On the other hand it is unstable, toxic, and isoform-nonselective. Another natural product, the phenylpropanoid quercetin (Chapter 7), is a reversible antagonist of ATP, less potent but more isoform-selective (Figure 4.V2). The X-ray structures of wortmannin and quercetin in the ATP site of porcine PI-3-Kγ show clearly that wortmannin is covalently connected to a Lys side chain of the enzyme while quercetin is the intact natural product. Both natural

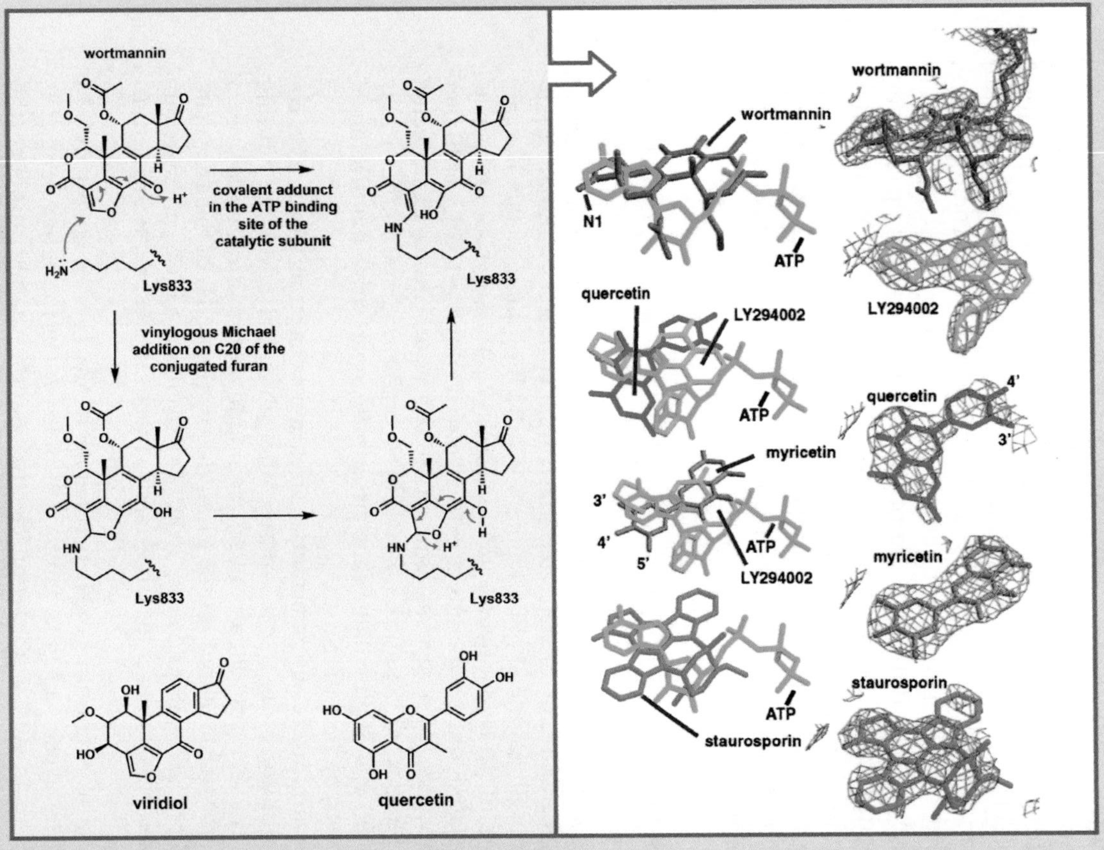

Figure 4.V2 Mechanism of inhibition of PI-3 kinase by wortmannin.
The crystal structure images in the right panel are reprinted from E. H. Walker, M. E. Pacold, O. Perisic, L. Stephens, P. T. Hawkins, M. P. Wymann and R. L. Williams. "Structural Determinants of Phosphoinositide 3-Kinase Inhibition by Wortmannin, LY294002, Quercetin, Myricetin, and Staurosporine", *Molecular Cell*, **6**: 909–919, Copyright (2000) Cell Press, with permission from Elsevier.

product scaffolds have inspired synthetic efforts to achieve clinical candidates that are either PI-3-K isoform specific or are dual PI3K and mTor inhibitors (Walker, Pacold *et al.* 2000).

The reactivity of wortmannin as a semiselective electrophilic ligand that covalently modifies its target protein is not a unique property of the furanosteroid natural product class. Many other natural product scaffolds contain latent electrophiles. Among them are the enones in the cyanobacterial peptides of the microcystin class that are captured by the active site cysteine thiolate side chain of protein tyrosine phosphatases. A recent estimate is that about one of six natural products have Michael acceptor functional groups that can become problematically reactive in covalent interaction with proteomes (Rodrigues, Reker *et al.* 2016).

Perhaps the most famous class are the penicillin and cephalosporin antibiotics that form long-lived acyl enzyme species when the beta-lactam ring is attacked by the active site serine side chain of target transpeptidases involved in bacterial cell wall crosslinking (Walsh and Wencewicz 2016).

4.9 Phytoene to Carotenes and Vitamin A

In Figure 4.18 above we noted that phytoene synthase, in contradistinction to squalene synthase, produced a head to head C_{40} self-alkylation product that had one more double bond than the starting C_{20} geranylgeranyl-PP pair of substrates, that it was uniquely in the *Z*-configuration, and was part of an *E,Z,E*-conjugated triene at the center of the released phytoene. A second distinction from squalene enzymatic logic is that phytoene is not subject to the cascade of cationic cyclizations to the tetra- and pentacyclic scaffolds so prevalent in triterpenes. Instead, phytoene is processed to linear carotenoids.

Two kinds of enzyme-mediated transformations occur on phytoene. The first is isomerization of the central *Z*-double bond to the *E*-configuration, producing the all *E*-configured linear nonaene. Figure 4.32 shows a proposed mechanism in which protonation of that central bond would generate an allylic cation which can rotate, in analogy to the *E*- to *Z*-isomerizations we have noted previously in this chapter (geranyl-PP to nerolidol-PP, and the *Z*-selective octaprenyl synthases).

At this juncture the all *E*-phytoene is substrate for a flavoenzyme that introduces another four olefins, presumably by a typical flavoenzyme proton/hydride desaturase mechanism (Schaub, Yu *et al.* 2012). The bacterial enzyme can function as an isomerase in the

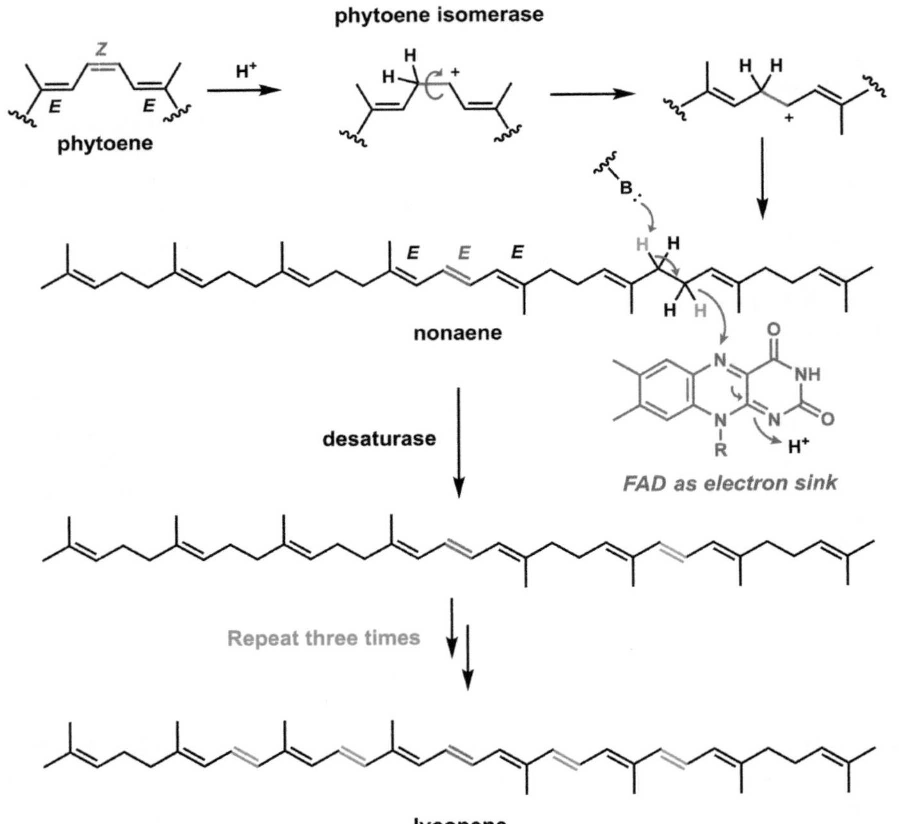

Figure 4.32 Isomerization of the central *Z*-double bond to *E*-configuration by phytoene isomerase to yield all the *E*-polyene products. Action by the flavoenzyme desaturase introduces four additional double bonds to produce the carotenoid lycopene.

absence of electron acceptor to carry out the *Z*- to *E*-olefin isomerization noted above. (It is believed that plants have two separate proteins for the isomerase and desaturase steps.) The product is lycopene, where 11 of its 13 olefins are conjugated, accounting for its red color and UV protection capabilities.

Lycopene can then be acted on by cyclases to turn the terminal nine carbons on each end of the molecule into cyclohexenes (Cunningham, Pogson *et al.* 1996, Moise, Al-Babili *et al.* 2014). As depicted in Figure 4.33, the cyclases take advantage one more time of the cation chemistry available to these poly-olefins. Cation quenching by removal of H_a or H_b can generate the epsilon or the beta dimethyl cyclohexene ring isomers. As shown, α-carotene has one beta and one epsilon cyclohexene at each terminus while β-carotene has beta cyclohexenes at both ends.

β-Carotene is then the precursor of the C_{20} aldehyde form of vitamin A, retinal, by action of carotene dioxygenase (Figure 4.34). Although mechanistic study has been limited by difficulties in purification and characterization of the membrane-associated oxygenase,

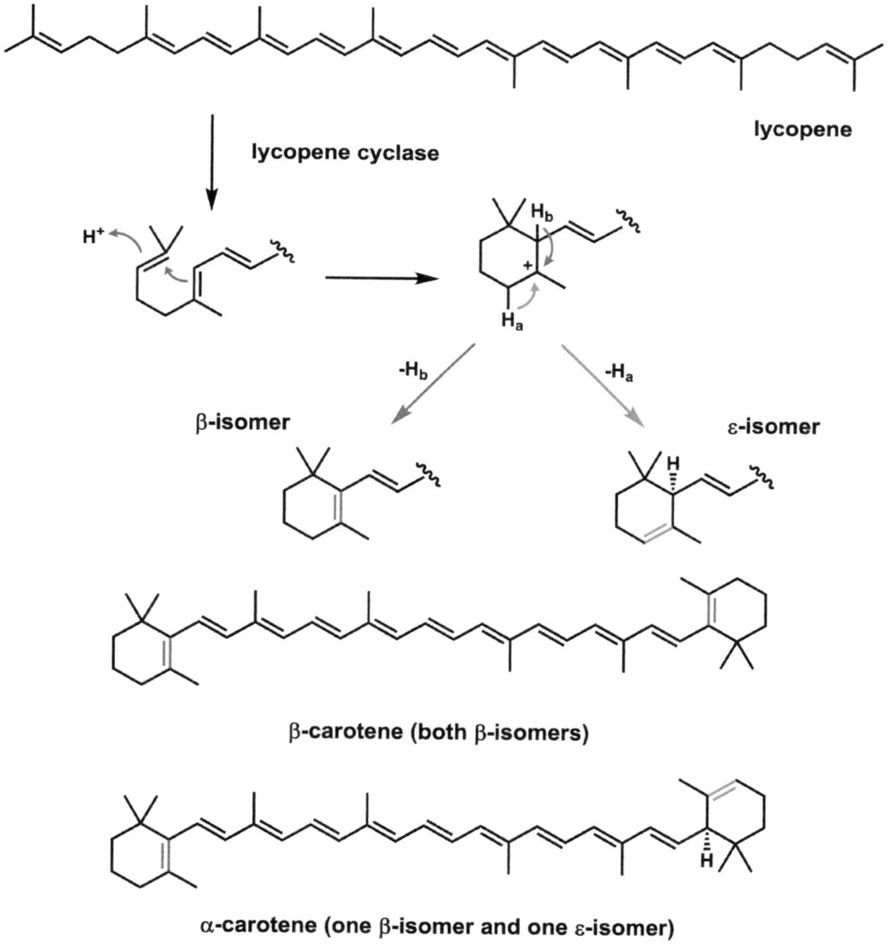

Figure 4.33 Enzymatic processing of lycopene to α-carotene or β-carotene by cation-based terminal cyclohexene ring formations.

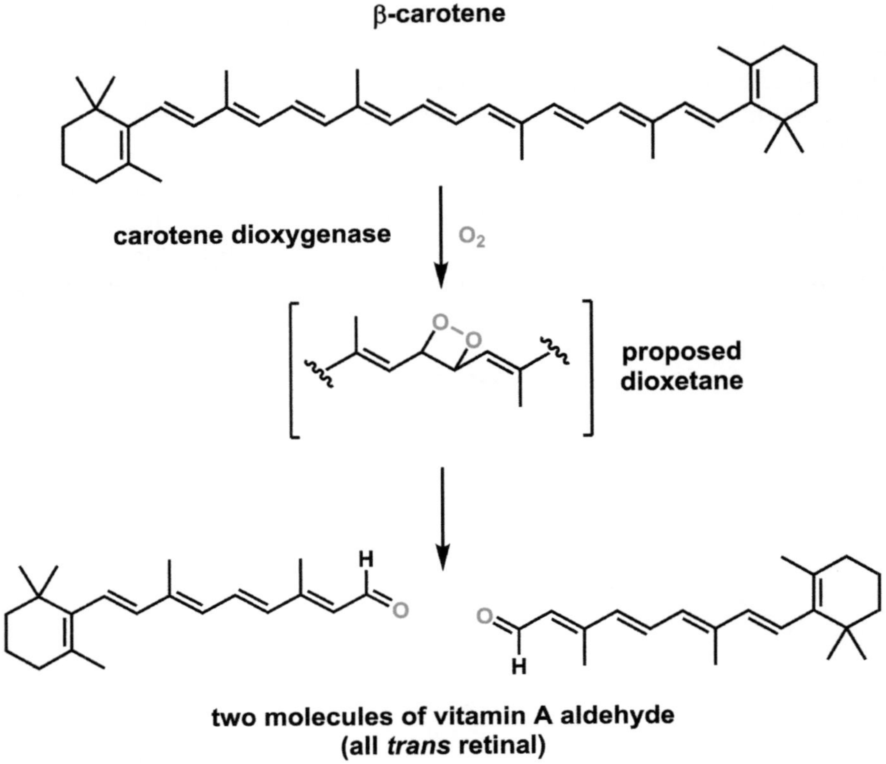

Figure 4.34 Carotene dioxygenase carries out a symmetric oxygenative cleavage of β-carotene to two molecules of retinal (vitamin A aldehyde), the essential chromophore in rhodopsin visual pigments.

evidence suggests that the aldehydic oxygens in the pair of product retinals derive from the same molecule of substrate O_2, indicating a dioxygenase logic. In turn that leads to formulation of the reaction as proceeding *via* a four-membered cyclic dioxetane that fragments to the two retinals (dela Sena, Narayanasamy *et al.* 2013, dela Sena, Riedl *et al.* 2014).

Vignette 4.3 Genetic Engineering and Golden Rice.

A significant part of the world population lives on rice as a major source of calories. Because rice is lacking functional phytoene synthase and phytoene desaturase genes and encoded proteins, many in these populations are severely deficient in vitamin A (Figure 4.V3), which arises from dietary phytoene by the pathway noted above. Genetic engineering of rice with

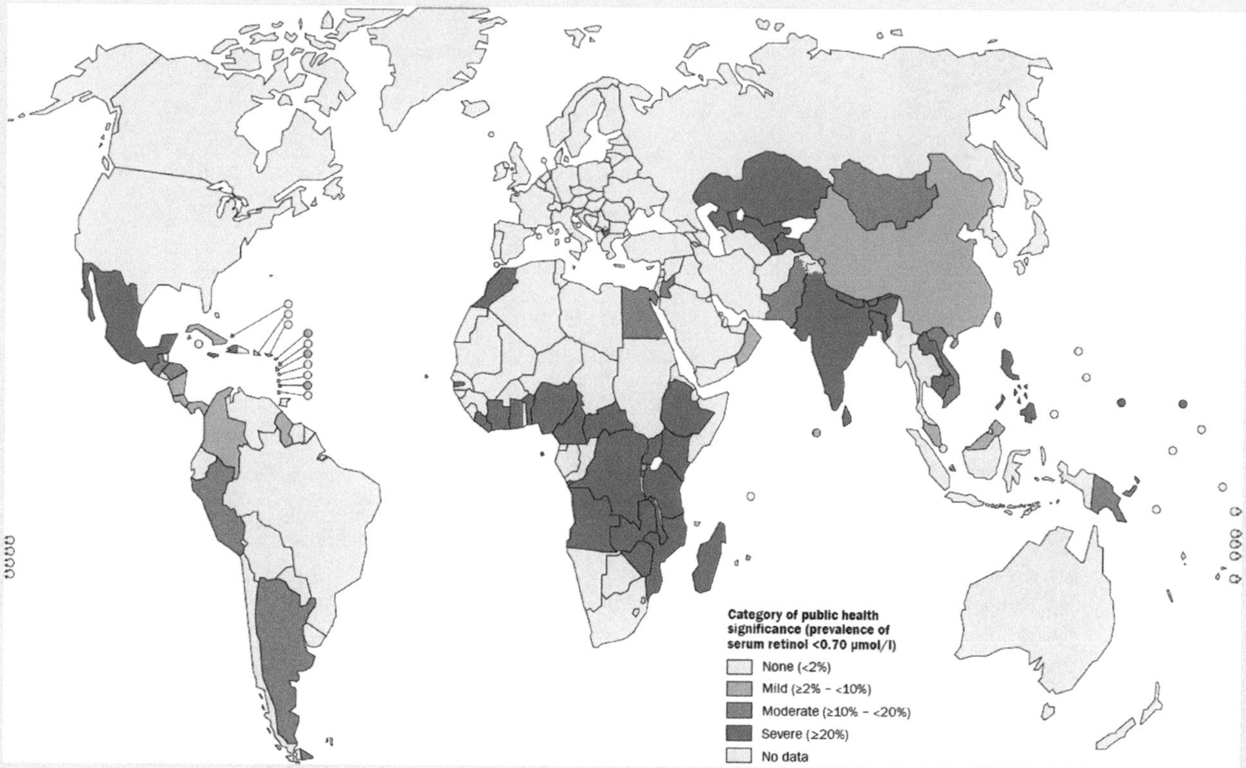

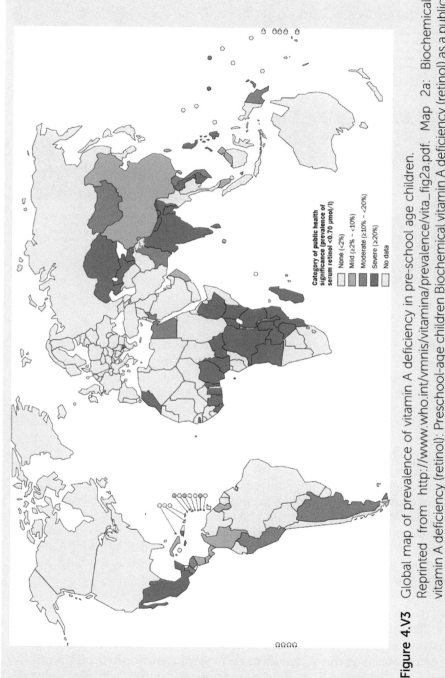

Figure 4.V3 Global map of prevalence of vitamin A deficiency in pre-school age children. Reprinted from http://www.who.int/vmnis/vitamina/prevalence/vita_fig2a.pdf. Map 2a: Biochemical vitamin A deficiency (retinol): Preschool-age children Biochemical vitamin A deficiency (retinol) as a public health problem 1995–2005: Countries and areas with survey data – http://www.who.int/vmnis/database/vitamina/status/en/. Copyright World Health Organization (2009).

Figure 4.V4 Golden rice and argument against GMO. Golden rice image credit: International Rice Research Institute in the Philippines. License: CC-BY-2.0 (https://creativecommons.org/licenses/by/2.0/).

transgenes for phytoene synthase and phytoene desaturase to correct the vitamin A deficiency has been achieved and the product developed as golden rice (Figure 4.V4). On the one hand this alleviates the severe vitamin deficiency and the attendant visual problems, but on the other hand it raises the paradigmatic concerns and controversies with genetically modified (GM) foods (some of which are noted in Figure 4.V4). The golden rice issue is just one of several around genetically modified crops and seeds. Although a US National Academy of Science report has been recently issued (The National Academies of Sciences 2016), concluding no detectable harm from GM crops, this has not satisfied the critics of purposeful genetic manipulation of food sources. With the advent of clustered regularly interspaced short palindromic repeats (CRISPR/CAS) technologies allowing modification of any specific target gene, it is likely that the ability to modify plant genomes will accelerate, so gaining a consensus on safety and food improvement will be ever more acute.

Additional carotene dioxygenases exist that cleave the polyene substrate asymmetrically. One such is involved in asymmetric cleavage of carotene at the 9′,10′ linkage to yield xanthophyll pigments in the retina (Li, Vachali *et al.* 2014). Another asymmetric dioxygenase functions in formation of strigolactones, plant hormones (first discovered in *Striga* weeds) that are important in root development and synergistic associations with microbes. Figure 4.35 shows that dioxygenative cleavage at the second olefin rather than the fifth olefin in the polyene linker of beta carotene substrate gives a pair of aldehydes. The left-hand aldehyde is further processed enzymatically to a set of strigolactone hormones (Alder, Jamil *et al.* 2012, Al-Babili and Bouwmeester 2015).

4.10 Reaction of Isoprenes with Other Natural Product Classes

Given the kinetic accessibility of the C_1 allyl carbocations from the Δ^2-prenyl diphosphates (DMAPP, geranyl-PP, farnesyl-PP, geranylgeranyl-PP), it is not surprising that they would be recruited to other natural product classes as electrophilic partners in C–C bond-forming, scaffold complexity generation enzymatic processes. This convergence is *a priori* particularly likely with natural product classes that have facile access to carbanion chemistry, because this is needed for carbon–carbon bond formations with the prenyl cations. The polyketide and indole natural product classes come to mind immediately since the enolate anions in polyketide assembly and the nucleophilicity of the indole ring as carbanion equivalents are hallmarks of those chemical frameworks. Examples of both such convergences are noted below.

4.10.1 Hyperforin is a Tetraprenylated Polyketide

Hyperforin (Figure 4.36) has been sold as a nutraceutical molecule from St John's wort. Biosynthesis of the polyketide core comes from a branched C_4 acyl-CoA starter unit and three chain elongations with malonyl-CoA extender units as noted in Chapter 2 (Figure 2.39). The released triphenol is activated at adjacent carbon atoms for nucleophilic capture of electrophilic prenyl groups. Two alkylations with the C_5 unit from DMAPP are then followed by alkylation by a C_{10} geranyl unit, which becomes additionally tethered by intramolecular attack on its C_2 olefin. The fourth and final prenylation is from the third Δ^2-IPP unit to build a remarkably complex final scaffold from only two types of biosynthetic reactions (decarboxylative Claisens in the PKS assembly line and electrophilic alkylations by the Δ^2-prenyl partners) (Adam, Arigoni *et al.* 2002).

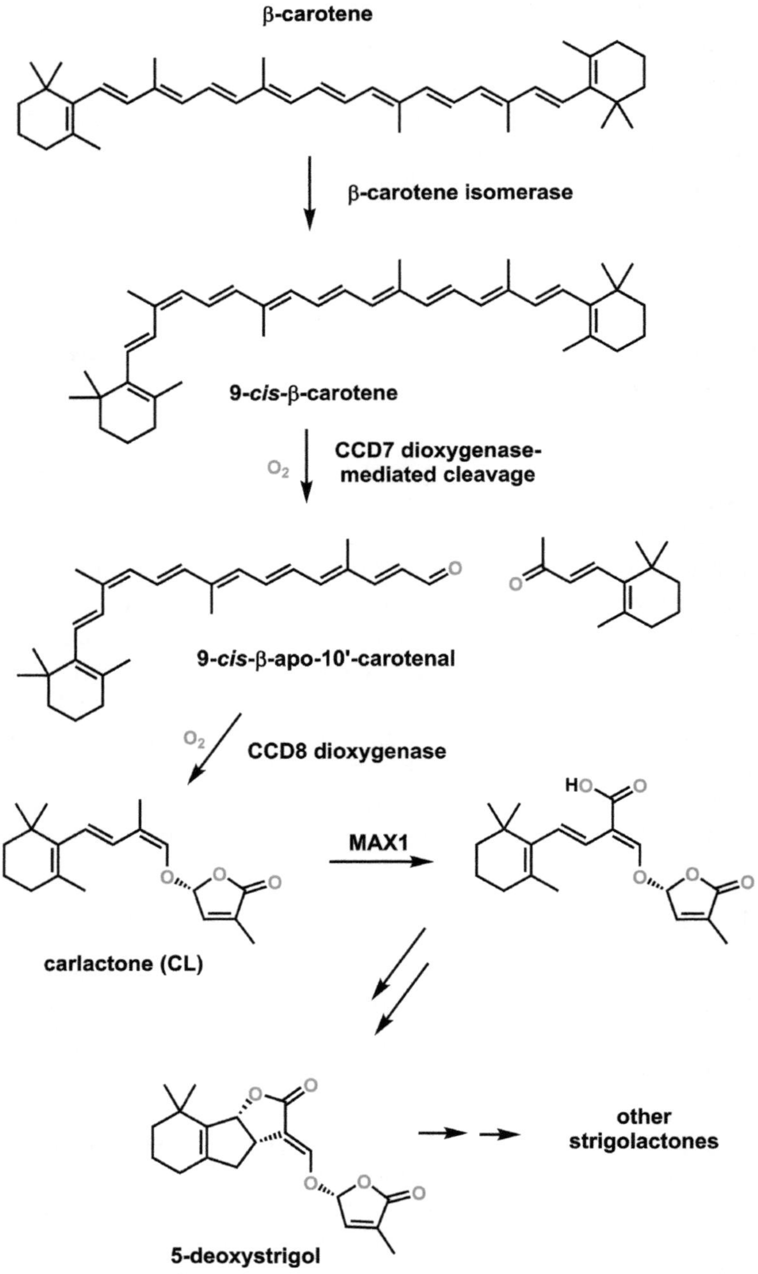

Figure 4.35 Asymmetric cleavage of β-carotene by a dioxygenase yields two aldehyde fragments, one of which is precursor to the strigolactone family of plant root hormones.

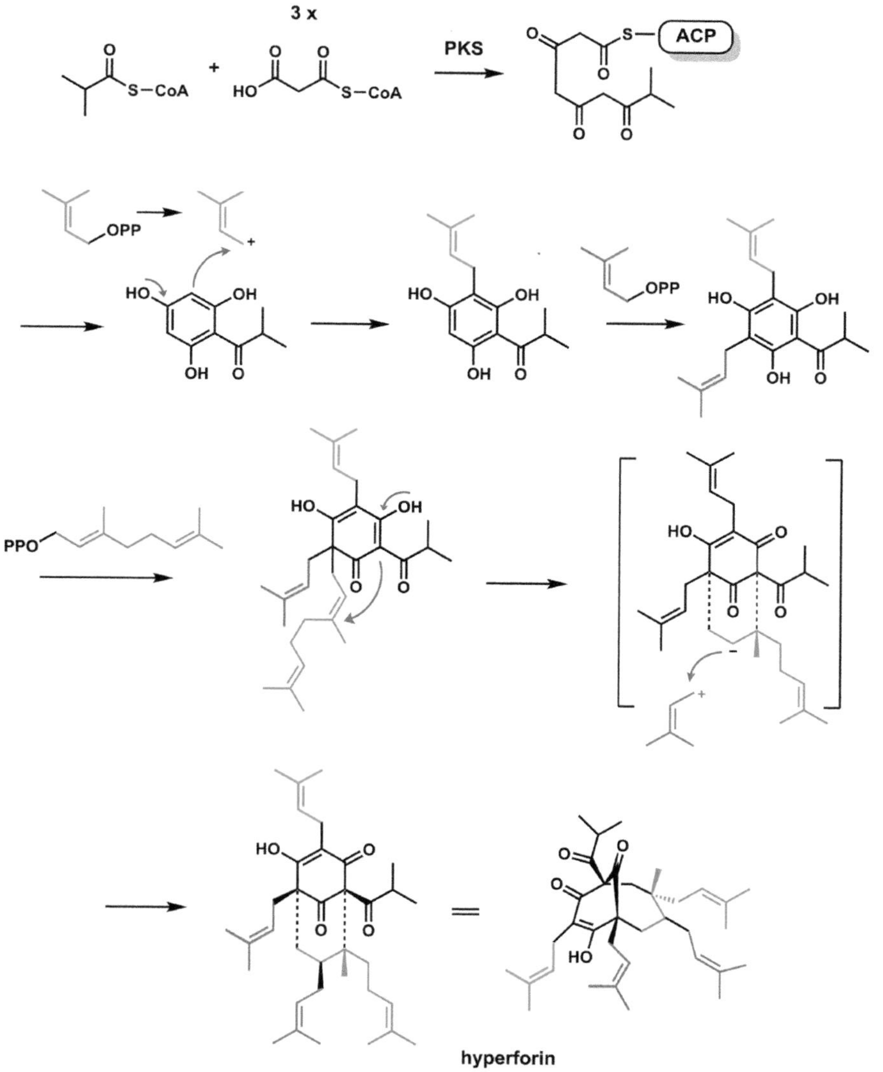

Figure 4.36 Hyperforin is biosynthesized by addition of one C_{10} geranyl unit and three C_5-dimethylallyl moieties to the polyketide core. The tetraprenylated polyketide is one of three active ingredients of St John's wort.

4.10.2 Tetrahydrocannabinol

A second example of the merging of PKS logic and isoprenoid logic occurs in the assembly of tetrahydrocannabinol (Sirikantaramas, Taura *et al.* 2005). In this case olivetolic acid is the product released

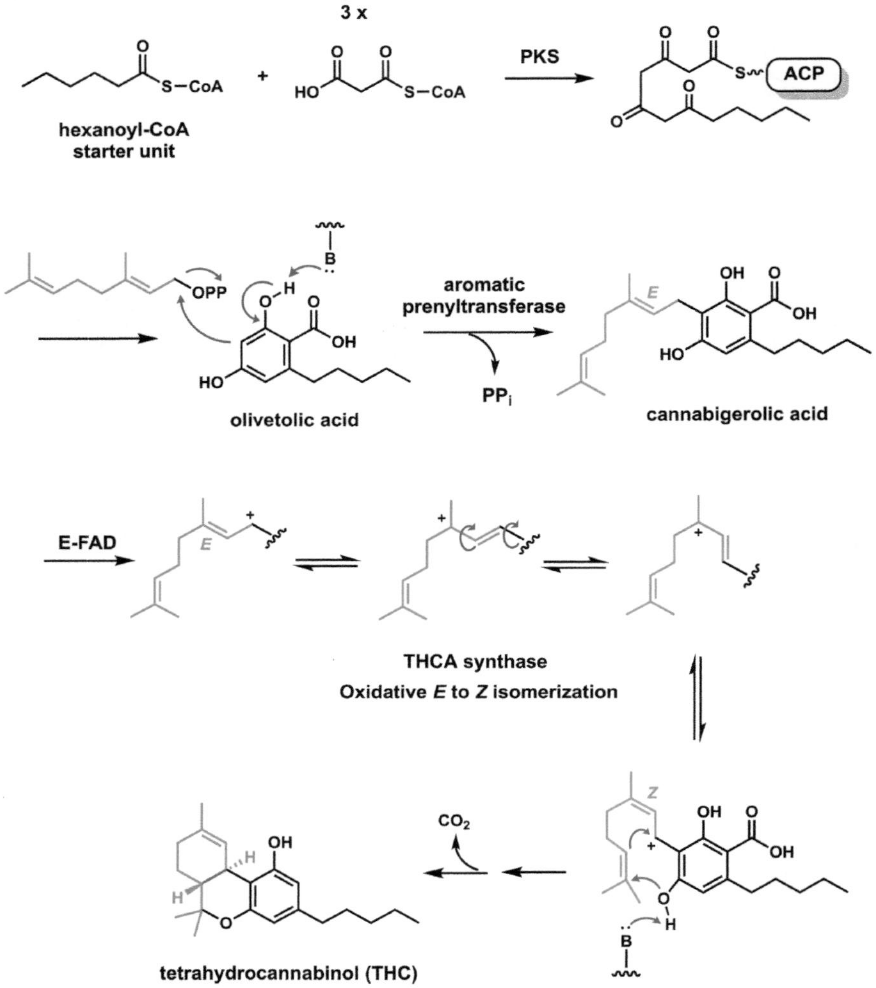

Figure 4.37 Tetrahydrocannabinol incorporates the C_{10} unit from geranyl-PP in two of the three rings of the final scaffold.

from the PKS assembly line (Figure 4.37). C-alkylation with a geranyl unit is followed by an *E*- to *Z*-olefin isomerization (as discussed in previous sections of this chapter) which sets up the geometry for closure to a cyclohexenyl cation that can be captured to form the tricyclic ring system in THC.

4.10.3 Paxilline

Paxilline is an indole diterpene alkaloid that is toxic to vertebrates by inhibition of voltage and calcium activated potassium channels.

It is assembled from the tryptophan metabolic precursor indoleglycerol-3-phosphate and the C_{20} diterpene donor geranylgeranyl-PP, as shown in Figure 4.38 (Saikia, Parker *et al.* 2007). In the active site of PaxC, indole glycerol-3-phosphate will undergo a reverse aldol to generate the primary metabolite glyceraldehyde-3-phosphate and free indole (Tagami, Liu *et al.* 2013). The nucleophilic C_3 of indole can attack the geranylgeranyl-PP-derived C_1 cation to yield 3-geranylgeranylindole. This is next subject to a cation mediated cascade by PaxM and tailoring by PaxB to produce the hexacyclic scaffold of paspaline in which the geranylgeranyl-derived unit is attached to both C_2 and C_3 of the indole nucleus. It is proposed that the $C_{10}=C_{11}$ double bond of geranylgeranylindole is epoxidized to set up the standard type of electrocyclization. Then, three oxygenations occur by P450 two cytochrome catalysts, PaxP and PaxQ, to install the ketone and 13-OH of paxilline.

Thus, three types of enzymatic chemistry suffice to build the hexacyclic framework of this ion channel toxin. One is indole acting as nucleophile at both C_2 and C_3. The C_{20} prenyl unit acts as C_1 allylic cation and then as a source of the three rings *via* a typical cyclization cascade. Finally, the ubiquitous oxygenases join the act, first to generate an oxido C_{20} moiety to set up the cyclization cascade, and finally to install two polar oxygen functionalities (ketone and alcohol).

4.11 Geranyl-PP to Secologanin: Entryway to Strictosidine and a Thousand Alkaloids

We will take up alkaloids and indole terpenes in subsequent chapters. The conversion of geranyl-PP *via* its hydrolysis product geraniol to loganin and secologanin provides a key isoprenoid-derived piece for thousands of alkaloid frameworks and is a bridge to those topics. The pathway is notable for its use of four consecutive oxygenases: oxygen carving the hydrocarbon into a highly oxygenated, water soluble reagent.

The nine steps in the pathway of geraniol to secologanin are summarized in Figure 4.39. First, geraniol is converted to the C10 dialdehyde which is reductively cyclized to the iridodial framework. This is then oxygenated to alcohol and then aldehyde (creating a reactive trialdehyde) which can cyclize to the 5,6-bicyclic hemiacetal. One more hydroxylation ensues, but most notable is the glucosylation of the hemiacetal *via* enzymatic transfer of a glucosyl moiety from

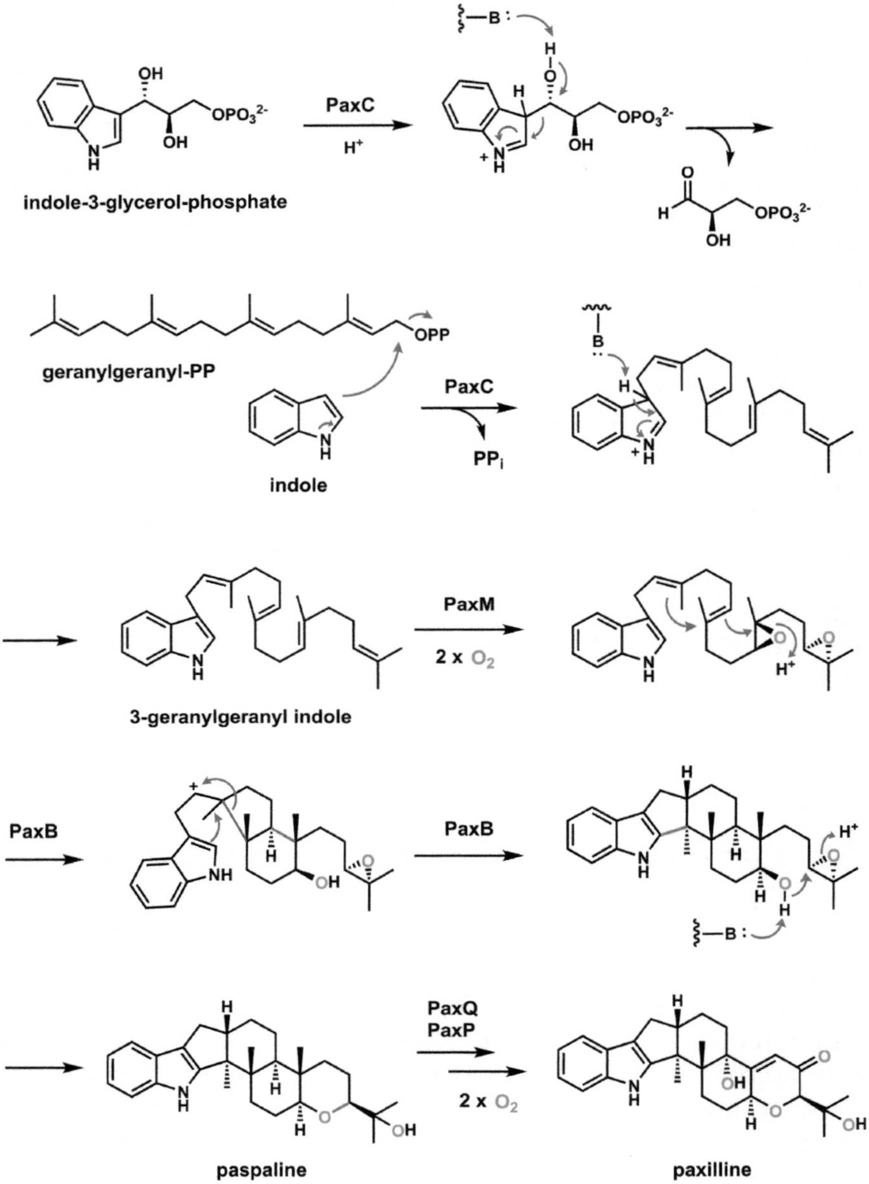

Figure 4.38 The indole diterpene paxilline arises from the C_{20} prenyl donor geranylgeranyl-PP and indoleglycerol-3-P. Indole–prenyl coupling is followed by cation rearrangement. A total of four molecules of dioxygen are consumed to reach the final end product paxilline.

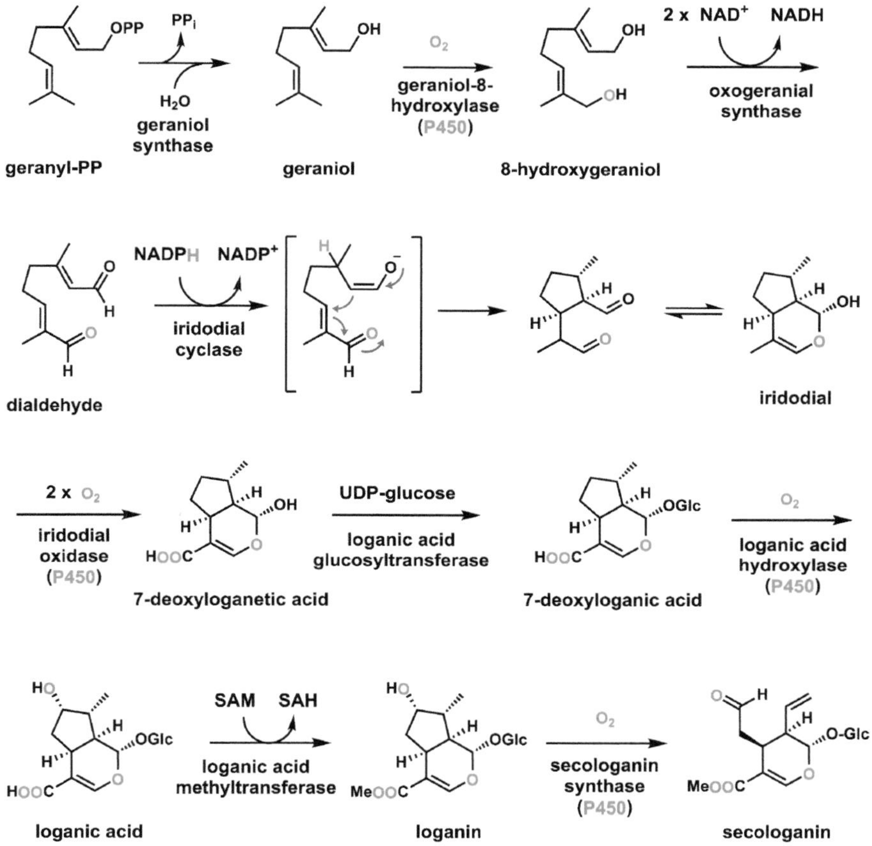

Figure 4.39 Geranyl-PP to secologanin, a key building block for the indole monoterpinome: four cytochromes P450 morph the geraniol scaffold oxidatively. Glucose creates a removable acetal protecting group.

UDP-glucose. This creates a stable acetal in place of the labile hemiacetal and protects the loganin/secologanin framework from unraveling in unwanted microenvironments. We will take up the subsequent enzymatic removal of the glycoside, after it has served its protective purpose, in alkaloid biosynthesis in Chapter 5. The last step in the pathway is oxygen-mediated fission, by the fourth cytochrome P450, of loganin into secologanin. This is carbon-radical chemistry, without net oxygen incorporation and is treated mechanistically in the oxygenase chapter (Chapter 9). The secologanin skeleton has five oxygen atoms, compared to the one in the starting geraniol.

Vignette 4.4 Artemisinin and Taxol.

Given the >50 000 isoprenoid/terpenoid molecules characterized to date, it is difficult to single out particular examples. However, two molecules that have had real impact in two areas of human medicine are the anti-malarial natural product artemisinin and taxol (Figure 4.V5), a mainstay in the treatment of ovarian cancers.

Artemisinin is a sesquiterpene (C_{15}) lactone that strikingly has an endoperoxide link within the lactone framework. The molecule is known as Qinghao Su in Chinese, reflecting a long history in Chinese herbal medicines, reputedly as far back as 340 AD. It is produced by *Artemisia annua* and is of terpene origin. Its biosynthesis up to the intermediate amorphadiene is well understood but the later stages involving lactone and endoperoxide formation are not worked out with regard to exact timing or mechanisms. This topic is examined in Chapter 13.

It has become a front line agent in treatment of malaria caused by *Plasmodium falciparum*, where it acts as a fast killer of the parasites. The killing mechanism is not fully elucidated but is thought to involve radical generation (Figure 4.V5) from the endoperoxide functionality on contact with heme and other sources of iron in infected cells. WHO guidelines argue against monotherapy, to mitigate against resistance development, but rather to use the drug in combination with other, earlier, synthetic antimalarials such as mefloquine amodiaquine or lumefantrine. As noted above for avermectin, half the Nobel Prize for medicine in 2015 went to the discoverers/developers of avermectin. The other half went to Dr Tu Youyou for her work on showing that artemisinin, despite its destabilizing endoperoxide, could be developed as an antimalarial drug.

Taxol and Other Natural Product Microtubule Ligands

Taxol is a diterpene (C_{20}), originally isolated from the bark of the Pacific yew tree, *Taxus brevifolia*, in 1962 under an NCI-funded plant screening program. Trials went on for some years before its approval as an anticancer agent in 1992, reaching front line agent status for ovarian cancer treatment. The availability of taxol for clinical supply was a limiting factor for some years before plant cell production technology from a *Taxus* cell line. The fermentation occurs in the presence of the endophytic fungus *Penicillium raistrickii*, and reports of taxol production from other endophytic fungi have been published but also disputed.

The target of taxol is tubulin, and binding to the β-tubulin subunits stabilizes microtubules against depolymerization. In turn, cell cycle progression is blocked, mitosis cannot be completed, and cells undergo checkpoint-induced apoptotic destruction. Biosynthetically, geranylgeranyl-PP is cyclized to the 6-8-6 tricyclic scaffold of taxadiene, a C_{20} hydrocarbon, as discussed in this chapter. This is then subjected to a set of enzymatic oxygenations that introduce eight oxygen substituents around

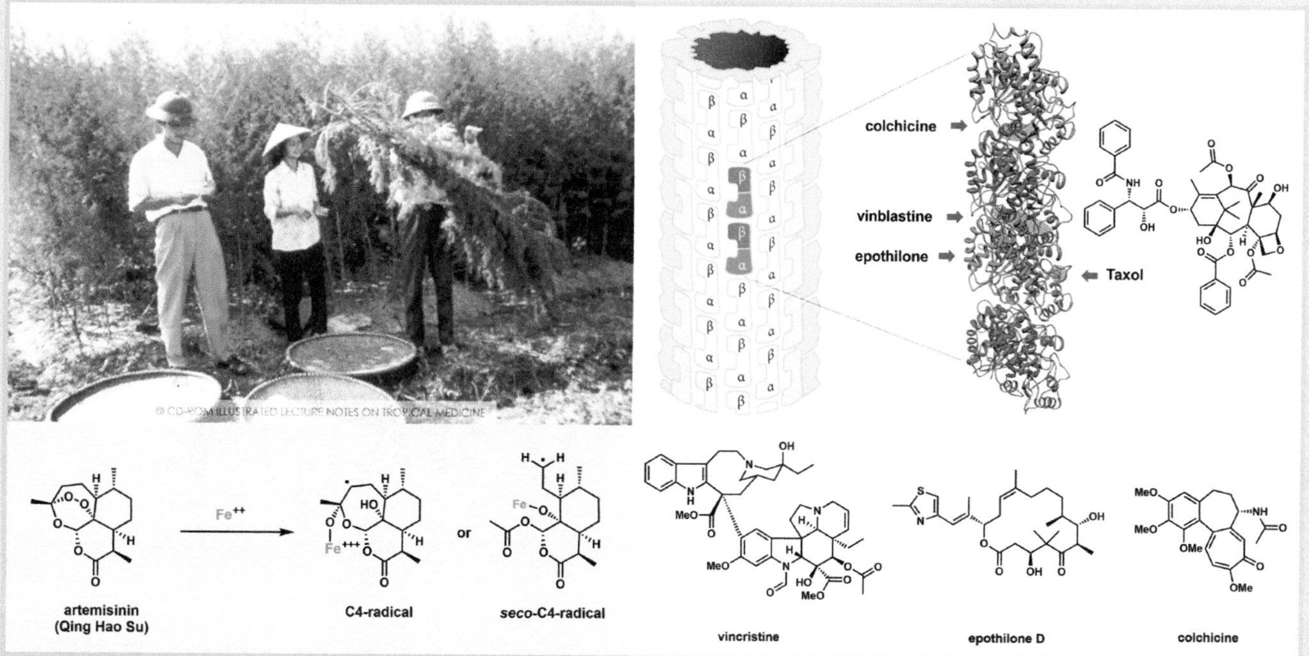

Figure 4.V5 The mechanisms of action of artemisinin and taxol. The figure on the left shows collection of *A. annua* leaves (copyright CD-ROM Illustrated Lecture Notes on Tropical Medicine). The figure on the right shows structure of the microtubule and sites of binding for various inhibitors, including taxol.
Reprinted and modified by permission from Macmillan Publishers Ltd: Nature Chemistry (Sackett, D. L. and D. Sept. "Protein-protein interactions: making drug design second nature". *Nat. Chem.* **1**: 596–597), copyright (2009).

the taxadiene periphery. Four of the just introduced hydroxyls are enzymatically acylated to yield taxol. Taxol is the front line drug in standard care for ovarian cancer.

Other natural products also target microtubules, either for stabilization or for destabilization (Figure 4.V5) (Sackett and Sept 2009). These include the epothilones (Chapter 3), which have a 16-member polyketide macrolactone core and a methylthiazole-containing side chain. The epothilone binding site partially overlaps the taxol site but some taxol-resistant mutants remain susceptible to epothilones. Additional natural product ligands are discodermolide, also overlapping the taxol site, and colchicines, which targets a distinct site as shown in Figure 4.V5 and acts as a tubulin destabilizer. The vinca alkaloids vinblastine and vincristine (Chapter 8) are also microtubule destabilizers and useful antitumor agents clinically. Most recently, eribulin, inspired by the polyether natural product halichondrin B, has also been approved as a microtubule-destabilizing agent in breast cancer treatment.

References

Adam, P., D. Arigoni, A. Bacher and W. Eisenreich (2002). "Biosynthesis of hyperforin in Hypericum perforatum". *J. Med. Chem.* **45**(21): 4786–4793.

Al-Babili, S. and H. J. Bouwmeester (2015). "Strigolactones, a novel carotenoid-derived plant hormone". *Annu. Rev. Plant Biol.* **66**: 161–186.

Alder, A., M. Jamil, M. Marzorati, M. Bruno, M. Vermathen, P. Bigler, S. Ghisla, H. Bouwmeester, P. Beyer and S. Al-Babili (2012). "The path from beta-carotene to carlactone, a strigolactone-like plant hormone". *Science* **335**(6074): 1348–1351.

Bacher, A., F. Chen and W. Eisenreich (2016). "Decoding Biosynthetic Pathways in Plants by Pulse-Chase Strategies Using (13)CO(2) as a Universal Tracer dagger". *Metabolites* **6**(3): 21–45.

Blagg, B. S., M. B. Jarstfer, D. H. Rogers and C. D. Poulter (2002). "Recombinant squalene synthase. A mechanism for the rearrangement of presqualene diphosphate to squalene". *J. Am. Chem. Soc.* **124**(30): 8846–8853.

Cane, D. E. (2006). "How to evolve a silk purse from a sow's ear". *Nat. Chem. Biol.* **2**(4): 179–180.

Chojnacki, T. and G. Dallner (1988). "The biological role of dolichol". *Biochem. J.* **251**(1): 1–9.

Christianson, D. W. (2008). "Unearthing the roots of the terpenome". *Curr. Opin. Chem. Biol.* **12**(2): 141–150.

Cunningham, F. X. Jr., B. Pogson, Z. Sun, K. A. McDonald, D. DellaPenna and E. Gantt (1996). "Functional analysis of the beta and epsilon lycopene cyclase enzymes of Arabidopsis reveals a mechanism for control of cyclic carotenoid formation". *Plant Cell* **8**(9): 1613–1626.

dela Sena, C., S. Narayanasamy, K. M. Riedl, R. W. Curley Jr., S. J. Schwartz and E. H. Harrison (2013). "Substrate specificity of purified recombinant human beta-carotene 15,15′-oxygenase (BCO1)". *J. Biol. Chem.* **288**(52): 37094–37103.

dela Sena, C., K. M. Riedl, S. Narayanasamy, R. W. Curley Jr., S. J. Schwartz and E. H. Harrison (2014). "The human enzyme that converts dietary provitamin A carotenoids to vitamin A is a dioxygenase". *J. Biol. Chem.* **289**(19): 13661–13666.

Dewick, P. (2009). *Medicinal Natural Products, a Biosynthetic Approach.* UK, Wiley.

Dickschat, J. S. (2016). "Bacterial terpene cyclases". *Nat. Prod. Rep.* **33**(1): 87–110.

Gao, Y., R. B. Honzatko and R. J. Peters (2012). "Terpenoid synthase structures: a so far incomplete view of complex catalysis". *Nat. Prod. Rep.* **29**(10): 1153–1175.

Guo, R. T., C. J. Kuo, C. C. Chou, T. P. Ko, H. L. Shr, P. H. Liang and A. H. Wang (2004). "Crystal structure of octaprenyl pyrophosphate synthase from hyperthermophilic Thermotoga maritima and mechanism of product chain length determination". *J Biol. Chem.* **279**(6): 4903–4912.

Hammer, S. C., A. Marjanovic, J. M. Dominicus, B. M. Nestl and B. Hauer (2015). "Squalene hopene cyclases are protonases for stereoselective Bronsted acid catalysis". *Nat. Chem. Biol.* **11**(2): 121–126.

Hoshino, T. and T. Sato (2002). "Squalene-hopene cyclase: catalytic mechanism and substrate recognition". *Chem. Commun.* **4**: 291–301.

Koksal, M., Y. Jin, R. M. Coates, R. Croteau and D. W. Christianson (2011). "Taxadiene synthase structure and evolution of modular architecture in terpene biosynthesis". *Nature* **469**(7328): 116–120.

Koksal, M., K. Potter, R. J. Peters and D. W. Christianson (2014). "1.55A-resolution structure of ent-copalyl diphosphate synthase and exploration of general acid function by site-directed mutagenesis". *Biochim. Biophys. Acta* **1840**(1): 184–190.

Laden, B. P., Y. Tang and T. D. Porter (2000). "Cloning, heterologous expression, and enzymological characterization of human squalene monooxygenase". *Arch. Biochem. Biophys.* **374**(2): 381–388.

Lenhart, A., W. A. Weihofen, A. E. Pleschke and G. E. Schulz (2002). "Crystal structure of a squalene cyclase in complex with the potential anticholesteremic drug Ro48-8071". *Chem. Biol.* **9**(5): 639–645.

Li, B., P. P. Vachali, A. Gorusupudi, Z. Shen, H. Sharifzadeh, B. M. Besch, K. Nelson, M. M. Horvath, J. M. Frederick, W. Baehr and P. S. Bernstein (2014). "Inactivity of human beta,beta-carotene-9′,10′-dioxygenase (BCO2) underlies retinal accumulation of the human macular carotenoid pigment". *Proc. Natl. Acad. Sci. U S A.* **111**(28): 10173–10178.

Little, D. B. and R. B. Croteau (2002). "Alteration of product formation by directed mutagenesis and truncation of the multiple-product sesquiterpene synthases delta-selinene synthase and gamma-humulene synthase". *Arch. Biochem. Biophys.* **402**(1): 120–135.

Liu, W., X. Feng, Y. Zheng, C. H. Huang, C. Nakano, T. Hoshino, S. Bogue, T. P. Ko, C. C. Chen, Y. Cui, J. Li, I. Wang, S. T. Hsu, E. Oldfield and R. T. Guo (2014). "Structure, function and inhibition of ent-kaurene synthase from Bradyrhizobium japonicum". *Sci. Rep.* **4**: 6214.

Lodeiro, S., Q. Xiong, W. K. Wilson, M. D. Kolesnikova, C. S. Onak and S. P. Matsuda (2007). "An oxidosqualene cyclase makes numerous products by diverse mechanisms: a challenge to prevailing concepts of triterpene biosynthesis". *J. Am. Chem. Soc.* **129**(36): 11213–11222.

Miller, D. J. and R. K. Allemann (2012). "Sesquiterpene synthases: passive catalysts or active players?". *Nat. Prod. Rep.* **29**(1): 60–71.

Misawa, N., M. R. Truesdale, G. Sandmann, P. D. Fraser, C. Bird, W. Schuch and P. M. Bramley (1994). "Expression of a tomato cDNA coding for phytoene synthase in Escherichia coli, phytoene formation in vivo and in vitro, and functional analysis of the various truncated gene products". *J. Biochem.* **116**(5): 980–985.

Moise, A. R., S. Al-Babili and E. T. Wurtzel (2014). "Mechanistic aspects of carotenoid biosynthesis". *Chem. Rev.* **114**(1): 164–193.

Pichersky, E., J. P. Noel and N. Dudareva (2006). "Biosynthesis of plant volatiles: nature's diversity and ingenuity". *Science* **311**(5762): 808–811.

Quin, M. B., C. M. Flynn and C. Schmidt-Dannert (2014). "Traversing the fungal terpenome". *Nat. Prod. Rep.* **31**(10): 1449–1473.

Rodrigues, T., D. Reker, P. Schneider and G. Schneider (2016). "Counting on natural products for drug design". *Nat. Chem.* **8**: 531–541.

Sackett, D. L. and D. Sept (2009). "Protein-protein interactions: making drug design second nature". *Nat. Chem.* **1**(8): 596–597.
Saikia, S., E. J. Parker, A. Koulman and B. Scott (2007). "Defining paxilline biosynthesis in Penicillium paxilli: functional characterization of two cytochrome P450 monooxygenases". *J. Biol. Chem.* **282**(23): 16829–16837.
Schaub, P., Q. Yu, S. Gemmecker, P. Poussin-Courmontagne, J. Mailliot, A. G. McEwen, S. Ghisla, S. Al-Babili, J. Cavarelli and P. Beyer (2012). "On the structure and function of the phytoene desaturase CRTI from Pantoea ananatis, a membrane-peripheral and FAD-dependent oxidase/isomerase". *PLoS One* 7(6): e39550.
Siedenburg, G. and D. Jendrossek (2011). "Squalene-hopene cyclases". *Appl. Environ. Microbiol.* 77(12): 3905–3915.
Sirikantaramas, S., F. Taura, Y. Tanaka, Y. Ishikawa, S. Morimoto and Y. Shoyama (2005). "Tetrahydrocannabinolic acid synthase, the enzyme controlling marijuana psychoactivity, is secreted into the storage cavity of the glandular trichomes". *Plant Cell Physiol.* **46**(9): 1578–1582.
Steele, C. L., J. Crock, J. Bohlmann and R. Croteau (1998). "Sesquiterpene synthases from grand fir (Abies grandis). Comparison of constitutive and wound-induced activities, and cDNA isolation, characterization, and bacterial expression of delta-selinene synthase and gamma-humulene synthase". *J. Biol. Chem.* **273**(4): 2078–2089.
Tagami, K., C. Liu, A. Minami, M. Noike, T. Isaka, S. Fueki, Y. Shichijo, H. Toshima, K. Gomi, T. Dairi and H. Oikawa (2013). "Reconstitution of biosynthetic machinery for indole-diterpene paxilline in Aspergillus oryzae". *J. Am. Chem. Soc.* **135**(4): 1260–1263.
Tansey, T. R. and I. Shechter (2001). "Squalene synthase: structure and regulation". *Prog. Nucleic Acid Res. Mol. Biol.* **65**: 157–195.
Tetzlaff, C. N., Z. You, D. E. Cane, S. Takamatsu, S. Omura and H. Ikeda (2006). "A gene cluster for biosynthesis of the sesquiterpenoid antibiotic pentalenolactone in Streptomyces avermitilis". *Biochemistry* **45**(19): 6179–6186.
The National Academies of Sciences (2016). *Genetically Engineered Crops: Experiences and Prospects*, The National Academies Press.
Thimmappa, R., K. Geisler, T. Louveau, P. O'Maille and A. Osbourn (2014). "Triterpene biosynthesis in plants". *Annu. Rev. Plant Biol.* **65**: 225–257.
Thoma, R., T. Schulz-Gasch, B. D'Arcy, J. Benz, J. Aebi, H. Dehmlow, M. Hennig, M. Stihle and A. Ruf (2004). "Insight into steroid

scaffold formation from the structure of human oxidosqualene cyclase". *Nature* **432**(7013): 118–122.

Ueberbacher, B. T., M. Hall and K. Faber (2012). "Electrophilic and nucleophilic enzymatic cascade reactions in biosynthesis". *Nat. Prod. Rep.* **29**(3): 337–350.

Vert, G., J. L. Nemhauser, N. Geldner, F. Hong and J. Chory (2005). "Molecular mechanisms of steroid hormone signaling in plants". *Annu. Rev. Cell Dev. Biol.* **21**: 177–201.

Walker, E. H., M. E. Pacold, O. Perisic, L. Stephens, P. T. Hawkins, M. P. Wymann and R. L. Williams (2000). "Structural determinants of phosphoinositide 3-kinase inhibition by wortmannin, LY294002, quercetin, myricetin, and staurosporine". *Mol. Cell.* **6**(4): 909–919.

Walsh, C. T. and T. Wencewicz (2016). *Antibiotics Challeneges, Mechanisms, Opportunities*. Washington DC, ASM Press.

Walsh, C. T. and T. A. Wencewicz (2013). "Flavoenzymes: versatile catalysts in biosynthetic pathways". *Nat. Prod. Rep.* **30**(1): 175–200.

Wendt, K. U. (2005). "Enzyme Mechanisms for Triterpene Cyclization: New Pieces of the Puzzle". *Angew. Chem., Int. Ed.* **44**: 3966–3971.

Yoshikuni, Y., T. E. Ferrin and J. D. Keasling (2006). "Designed divergent evolution of enzyme function". *Nature* **440**(7087): 1078–1082.

Zheng, W., F. Sun, M. Bartlam, X. Li, R. Li and Z. Rao (2007). "The crystal structure of human isopentenyl diphosphate isomerase at 1.7 A resolution reveals its catalytic mechanism in isoprenoid biosynthesis". *J. Mol. Biol.* **366**(5): 1447–1458.

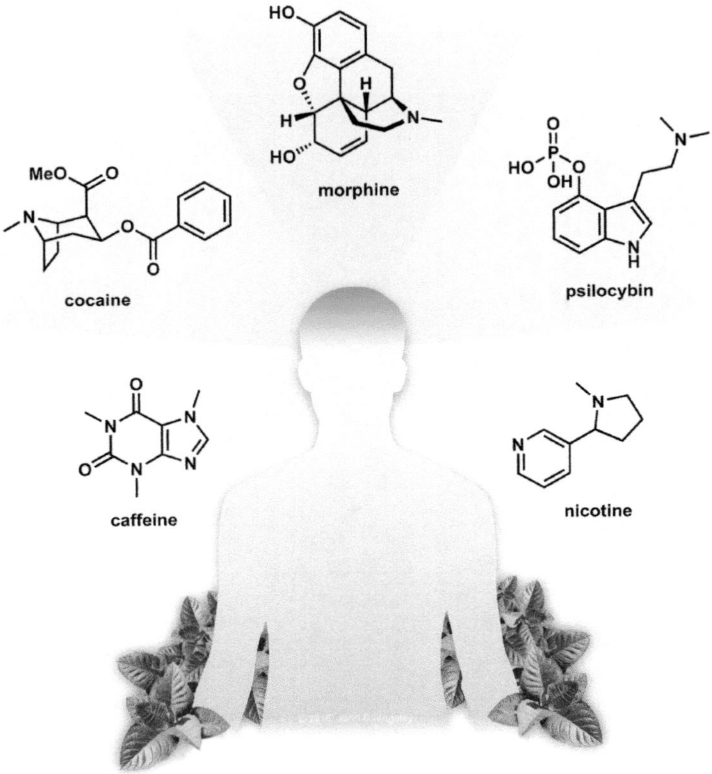

Psychoactive alkaloids that are derived from plants.

5 Alkaloids

5.1 Introduction

Alkaloids comprise a huge swath of natural product chemical space. Historically they have been united only by possession of a basic nitrogen, usually embedded in a heterocyclic ring (Dewick 2009). Most alkaloids, but not all, have been isolated by two-step extraction protocols, in which the free base forms are typically soluble in organic solvents. Then a wash with an aqueous acid layer extracts the amine salt into the aqueous phase. Basification then allows re-extraction into an organic phase for further separation from contaminants. Molecules with basic nitrogens, alkaloids = alkali like, partition between organic and aqueous phases for selective enrichment.

Some 27 000 alkaloids have been isolated and characterized from more than 2100 plant species (Dewick 2009) since the isolation of morphine from opium in 1816. As shown in Figure 5.1 the isolations of strychnine, piperine from black pepper, caffeine, quinine, and coniine all followed within the next decade but they remained mysterious at the molecular level. The first alkaloid structure to be determined was coniine, but not until 54 years later in 1870 (Hesse 2002).

In accord with the dramatic range of structural variants within the molecules grouped together as alkaloids there is a dramatically broad range of pharmacologic activity in humans, some of which is tabulated in Figure 5.2. From the antimalarial properties of quinine, the anticancer activity of camptothecin and vincristine, the hallucinogenic features of lysergic acid diethylamide, the analgesic properties of morphine, the antiasthma effects of ephedrine, to the stimulants

Natural Product Biosynthesis: Chemical Logic and Enzymatic Machinery
By Christopher T. Walsh and Yi Tang

Published by the Royal Society of Chemistry, www.rsc.org

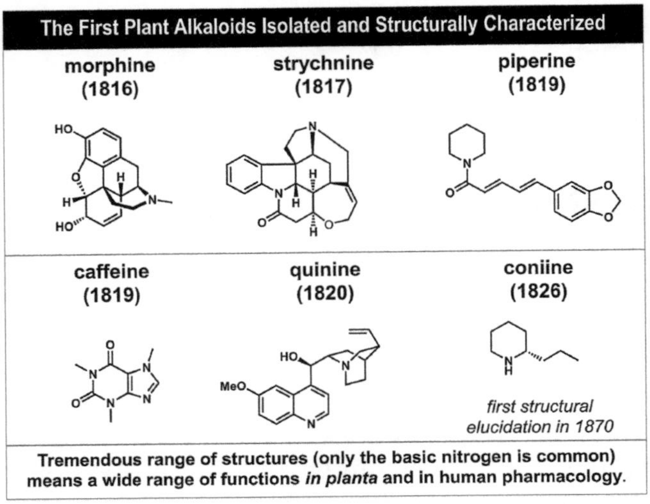

Figure 5.1 Alkaloids are signature natural product molecules from plants. Morphine, strychnine, piperine, caffeine, quinine, and coniine were all isolated within a 10 year period early in the 19th century, although determination of their molecular structures occurred many decades later. More than 27 000 alkaloids are known from ~2100 plant species.

Examples of Human Pharmacologic Activities of Alkaloids	
quinine	antimalarial
camptothecin	anticancer
lysergic acid	pre-hallucinogen
ephedrine	antiasthma
tubocurarine	arrow poison
morphine	analgesic
cocaine, caffeine	stimulant
atropine	poison
scopolamine	antinausea
vincristine	anticancer

Figure 5.2 Alkaloids show a dramatic range of pharmacologic activity in humans, in accord with a long history of isolation and human experimentation.

caffeine and cocaine, and finally the poisons atropine, strychnine, and tubocurarine, this alkaloid subset spans a huge therapeutic/toxic range.

5.2 Alkaloid Family Classifications

Several approaches to alkaloid classification can be useful. They vary from attention on the building blocks in primary metabolism, to the types of heterocycles in the final product, to the kinds of pharmacologic activity just noted.

5.2.1 Amino Acid Building Blocks

This building block perspective focuses on the six or seven amino acids from primary metabolism that give rise to distinct families of alkaloids. Four of the amino acids are proteinogenic. The fifth, ornithine, is involved in arginine and polyamine metabolism, not incorporated into proteins, but is the precursor to arginine. The sixth amino acid is anthranilate, a nonproteinogenic β-amino acid, but a key primary metabolic intermediate in tryptophan biosynthesis (Walsh, Haynes *et al.* 2012).

The indole ring of tryptophan is morphed into β-carbolines, quinolines, pyrroloindoles and to the more complex frameworks of ergot alkaloids. Lysine gives rise to piperidines, quinolizidines, and indolizidine alkaloids. Ornithine is progenitor to pyrrolidine rings, tropanes, and pyrrolizidine alkaloids. Phenylalanine is the source of half the carbons in the antinausea alkaloid scopolamine. Tyrosine yields phenylethylamines and tetrahydroquinolines. Histidine leads to imidazole-containing alkaloids. Finally, anthranilate is converted into quinazolines, quinolines, and acridine frameworks.

5.2.2 Additional Heterocyclic Frameworks

Many investigators would turn the classification scheme on its head and group alkaloids by those heterocyclic ring systems rather than by starting amino acid building block. Such a list might then be parsed into heterocyclic categories as pyrrolizidines (senecionine), tropanes (atropine), quinolines (quinine), isoquinolines (morphine), purines (caffeine), aminosteroids (solanidine), amino alkaloids (ephedrine), and diterpenes (aconitine). Figure 5.3 combines both of these

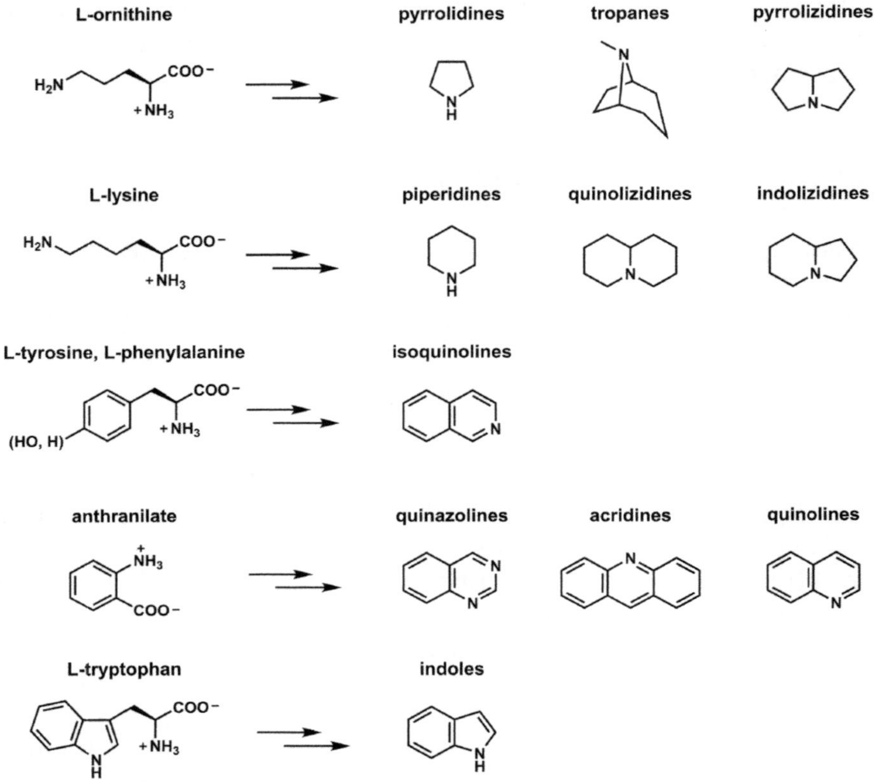

Figure 5.3 Alkaloids use amino acid building blocks and span a broad range of heterocyclic scaffolds.

classifications and the connection of starting amino acid to final product ring structures is crucial to deconvolution of biosynthetic chemical logic. By either criterion it is the alkaloids that represent the broad reach of organisms to build complex nitrogen heterocyclic scaffolds (Dewick 2009).

5.3 Common Enzymatic Reactions in Alkaloid Biosynthetic Pathways

5.3.1 Reactions Involving Amino Acid Building Blocks

Given that amino acids are the key building blocks of alkaloids, it is to be expected that amino acid decarboxylases would work at early stages to yield the corresponding amines for further elaboration.

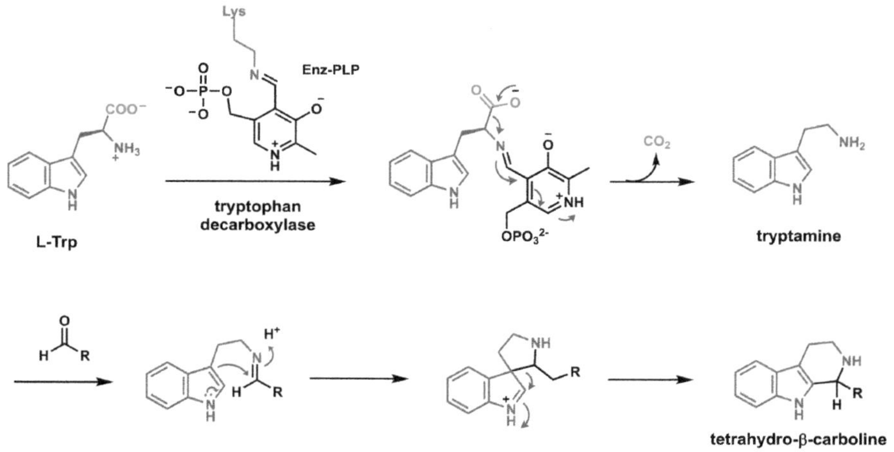

Figure 5.4 Decarboxylase and Mannich condensation *en route* from tryptophan to the tricyclic β-carboline (PLP: pyridoxal phosphate).

SAM-dependent methyltransferases convert alcohol and amine functionalities to the corresponding *N*-methyl and *O*-methyl derivatives that show up in many alkaloid pathways, as do acyl transfers to comparable *O*- and *N*-hetero atoms. Enzymatic oxidation of amine to imines and their hydrolysis to aldehydes, or imine reduction back to amines by oxidoreductase action, is also a common functional group manipulation strategy in alkaloid biosynthetic pathways, especially for the ornithine and lysine dibasic amino acid building blocks.

One of the hallmarks of alkaloid pathways is enzymatic Mannich type condensations, involving amine and carbonyl components that condense to imines that are then captured by carbanion equivalents to yield new C–C bonds at an amine bearing carbon center. The generation of a tricyclic β-carboline from such a Mannich reaction sequence is depicted in Figure 5.4.

5.3.2 Ornithine as Building Blocks for Cocaine and Retronecine

The five-carbon dibasic amino acid ornithine can undergo decarboxylation by a typical pyridoxal-phosphate (PLP) dependent amino acid decarboxylase to yield the C4 diamine putrescine. *N*-Methylation and oxidation of one of the amines to the imine and hydrolysis yield the *N*-methylbutanal, as shown in Figure 5.5. This is in equilibrium

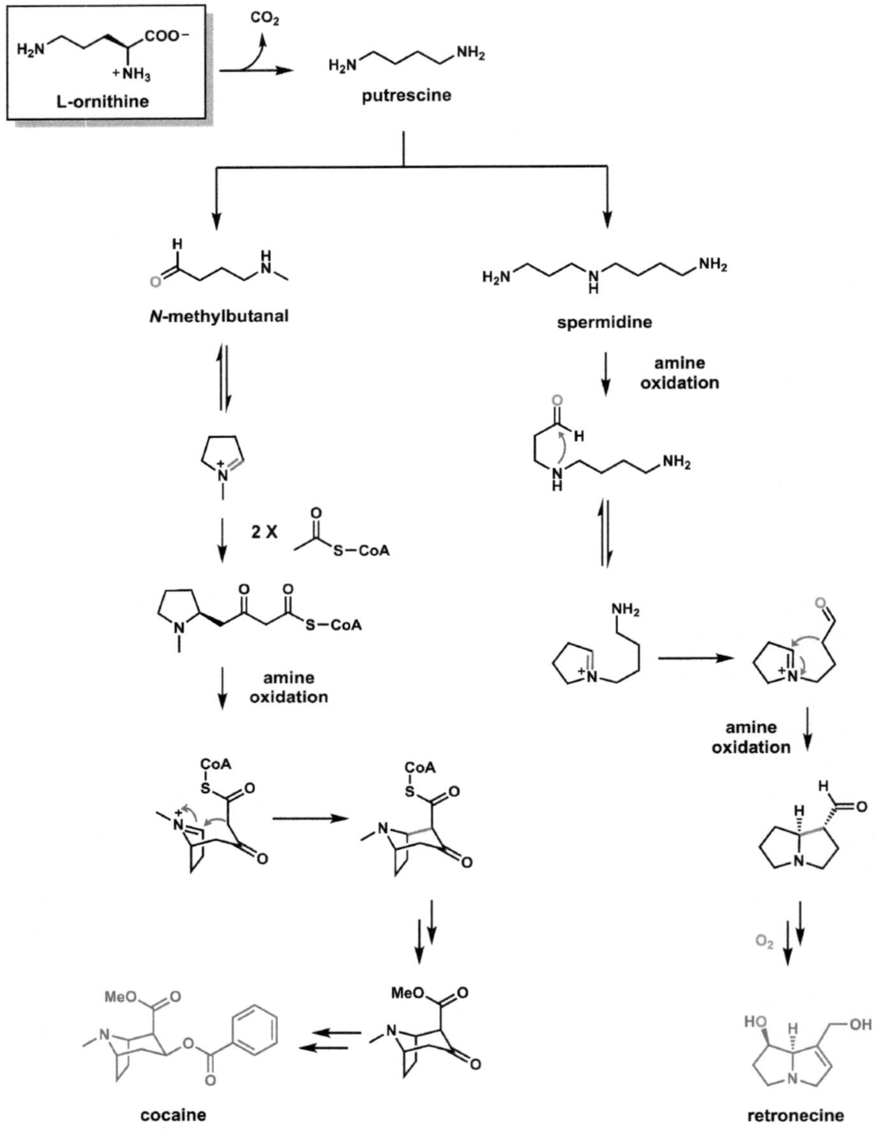

Figure 5.5 Ornithine to two different bis-heterocyclic scaffolds in cocaine and retronecine.

with the cyclic *N*-methyl pyrrolinium ion. This iminium can be attacked by the enolate anions of acetyl-CoA molecules in tandem, generating two C–C bonds as the C_2-substituted-acetoacetyl-CoA adduct to the *N*-methyl pyrrolidine is produced. The pyrrolidine must undergo a round of enzymatic dehydrogenation back to the iminium to set up an intramolecular Mannich condensation. The enolate at C_3 of the side chain attacks the imine, forming the new C–C bond of the bicyclic tropane system. The acyl-CoA thioester must undergo catalyzed hydrolysis and then *O*-methylation of the carboxylate *via* methyl transfer from SAM. NADPH-mediated reduction of the ketone to the β-alcohol provides the substrate for the last step in the cocaine biosynthetic pathway. Esterification *via* the cosubstrate benzoyl-CoA produces the benzoyl ester that is cocaine (Jirschitzka, Schmidt *et al.* 2012, Schmidt, Jirschitzka *et al.* 2015). The bicyclic 5,6-tropane ring system can be routed to more complex alkaloid scaffolds as will be noted two sections ahead in scopolamine assembly (Figure 5.6) (Humphrey and O'Hagan 2001).

In the biosynthesis of the pyrrolizidine alkaloid retronecine, ornithine can undergo a parallel process of enzymatic decarboxylation and then extension to the C7 triamine spermidine (Figure 5.5). Oxidation of one of the amines to the aldehyde sets up nonenzymatic cyclization to the first pyrrolinium ring. Dehydrogenation of the second, exocyclic amine sets up an aldol condensation to generate the fused 5,5-pyrrolizidine system on the way to retronecine (Hill and Yudin 2006).

Vignette 5.1 Pyridoxal-P: the Coenzyme that Mediates Amino Acid Metabolism.

The pyridine aldehyde form of vitamin B6, pyridoxal, undergoes enzymatic phosphorylation to pyridoxal-P which, in the active site of specific enzymes, mediates the suite of metabolic transformations that amino acids undergo at C_α, C_β, and C_γ, in both primary and secondary pathways.

The agency of PLP is particularly apparent in alkaloid biosynthesis, where several of the amino acid building blocks undergo early steps of decarboxylation to the corresponding amines, as noted in this chapter and in Chapter 8. PLP almost always functions as an electron sink, engaging the substrate amino group in aldimine linkage as shown in Figure 5.V1. In principle, any one of the four substituents at C_α can undergo bond cleavage, depending on the orbital alignments imposed by the particular enzyme active site geometry. Decarboxylation with release of CO_2 generates a transient C_α carbanion that is stabilized by delocalization, as shown by the indicated resonance structure. Directed protonation at C_α generates the product aldimine from which the product amine can dissociate by a transaldiminative attack by the side chain of the active site lysine residue which ensures that the PLP coenzyme is retained for the subsequent catalytic cycle.

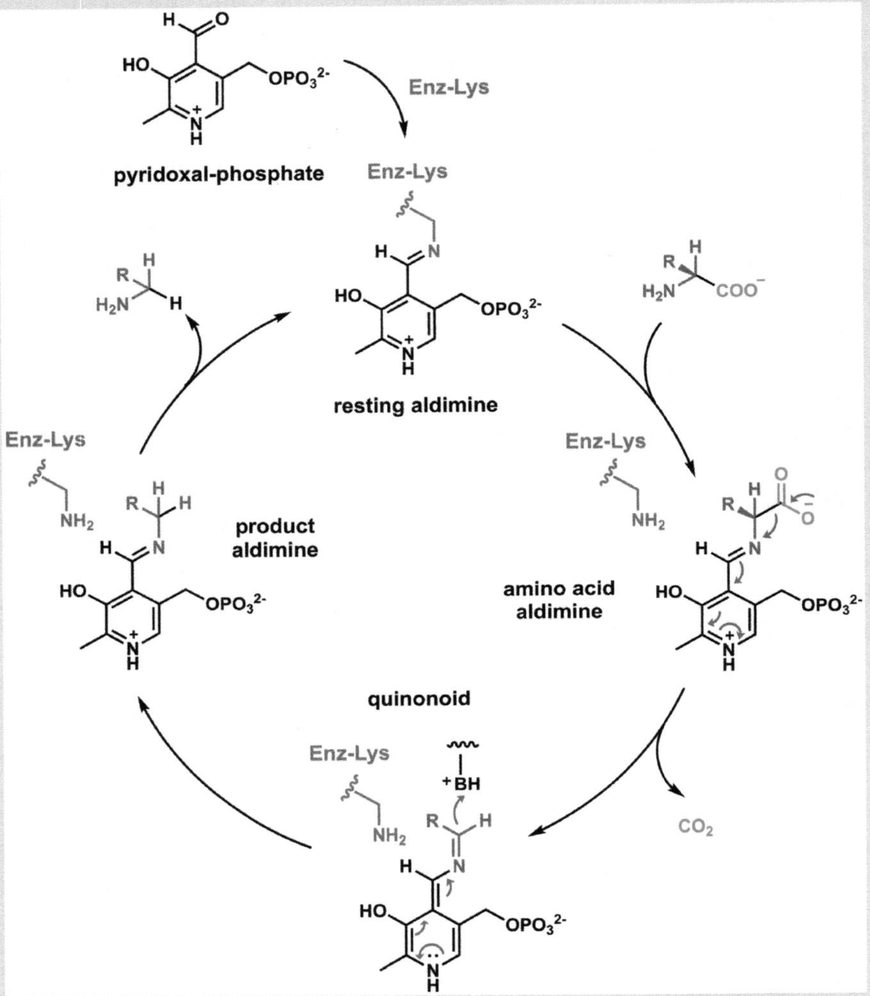

Figure 5.V1 Mechanism of PLP-dependent decarboxylation of amino acids.

5.3.3 Lysine to Pelletierine and Pseudopelletierine and Sparteine

In analogy to ornithine processing, the six-carbon dibasic amino acid lysine can be oxidized and cyclized to the corresponding six-membered imine ring (Figure 5.6) and serve as gateway to mono-, bi-, and tetracyclic alkaloids. The cyclic imine can be captured

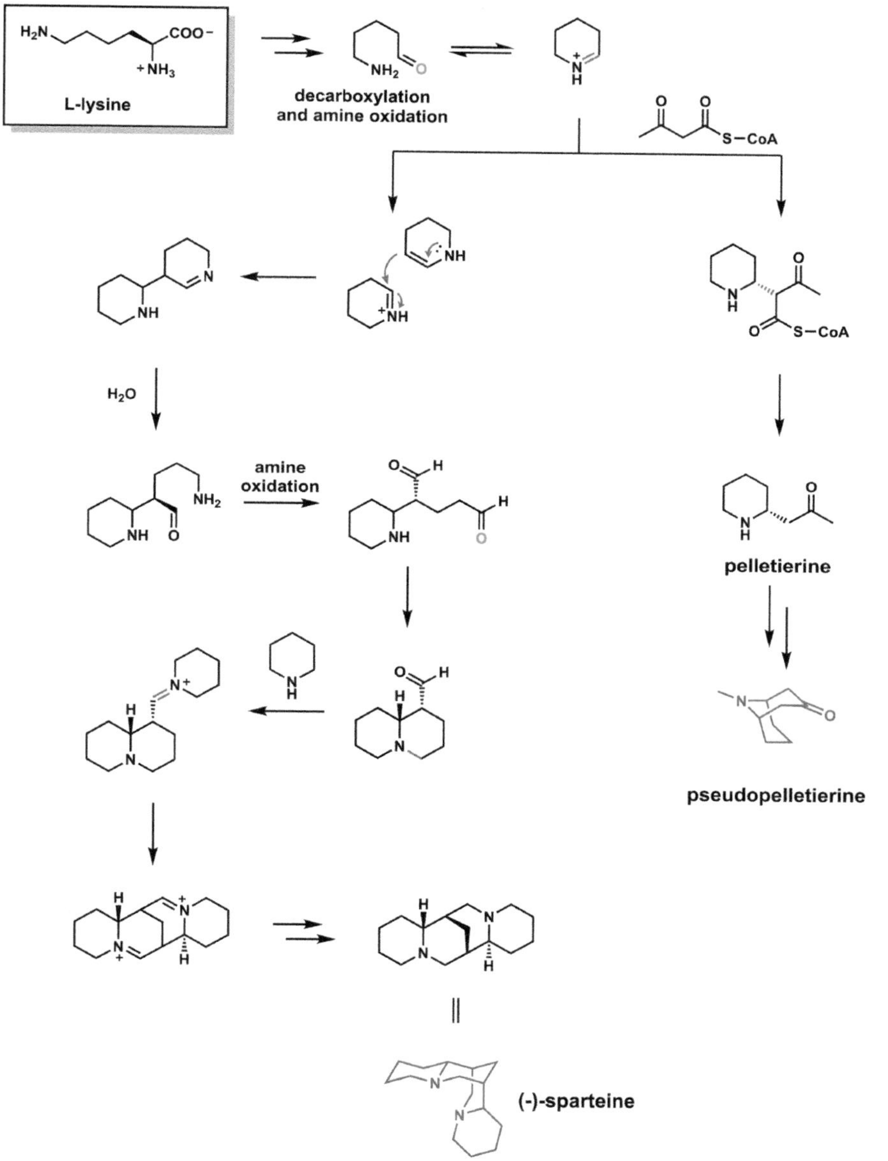

Figure 5.6 Conversion of lysine to bicyclic pseudopellitierine and tetracyclic (−)-sparteine.

in a Mannich reaction by acetoacetyl-CoA to yield the substituted piperidinyl-CoA. Hydrolysis of the thioester and decarboxylation of the β-keto acid yields pelletierine. Enzymatic methylation of the amine, oxidation to the imine away from the C_3 ketone, and an intramolecular Mannich condensation will yield the bicyclic pseudopelletierine alkaloid (Dewick 2009).

Further elaboration of this chemical logic is shown in the bottom part of Figure 5.6 where two of the cyclic imines condense, one as the electrophilic imine, the other as the nucleophilic enamine isomer to create a new C–C bond. Hydrolytic ring opening and enzymatic oxidation of the amine to the dialdehyde sets up condensation (and imine reduction) to the bicyclic system. Now a third version of the cyclic piperidine, as the basic amine, can add to the bicyclic aldehyde under enzymatic direction. Regioselective oxidation of that amine to the bis-imine, isomerization to the enamine, and Mannich condensation generates the corresponding tricyclic bis-imine on the way to the compact, nested tetracyclic diamine alkaloid sparteine (Herbert 1989).

5.3.4 Lysine to the 6,5-Indolizidine Bicyclic Framework

Lysine is also the progenitor for bicyclic indolizidine alkaloids, including castanospermine and swainsonine which are potent glycosidase enzyme inhibitors. As laid out in Figure 5.7, lysine can be enzymatically cyclized without prior decarboxylation to the cyclic pipecolate, activated, and chain extended with an acetyl-CoA carbanion equivalent as precursor to internal cyclization to the bicyclic 6,5-indolizidine scaffold. Castanosperime is a ketone reduction and three regio- and stereoselective hydroxylation steps away, emphasizing the late stage role oxygenases play in maturation of this alkaloid subfamily. It is the hydroxyl substituents which cause castanospermine to be an inhibitory sugar mimic for glycosyl transfer enzymes. To get

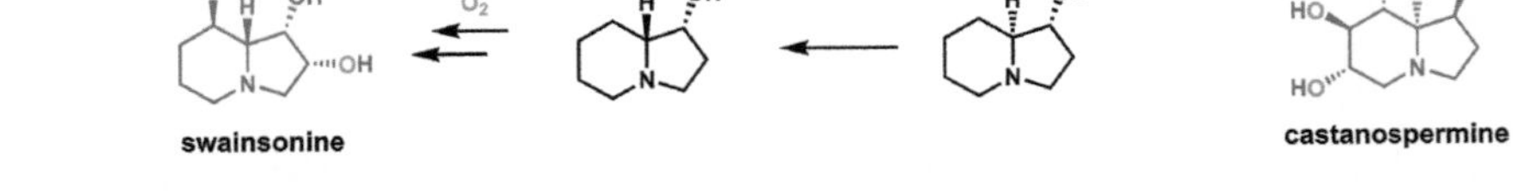

Figure 5.7 Lysine to the 6,5-bicyclic ring system in catsanospermine and swainsonine.

from the indolizidine precursor to swainsonine, the ketone is likewise enzymatically reduced (with opposite stereochemistry to the castanospermine reduction) but then the stereochemistry at the ring juncture is inverted *via* oxidation to the imine and re-reduction with opposite chirality. Two subsequent hydroxylations by distinct oxygenases generate the trihydroxy swainsonine alkaloid (Seigler 2012).

5.4 Three Aromatic Amino Acids as Alkaloid Building Blocks

In addition to the dibasic amino acids ornithine and lysine, the three proteinogenic amino acids phenylalanine, tyrosine, and tryptophan can serve as entry points from primary metabolism to alkaloid biosynthetic pathways (Figure 5.8). In turn this may mean increasing the flux from chorismate to Phe, Tyr, and Trp (Figure 5.9) to account for drawing off some of the aromatic amino acid pools from their use in protein biosynthesis. Anthranilate is an obligate intermediate between chorismate and tryptophan, and later in this chapter we will note it as a building block for fungal peptidyl alkaloids.

5.4.1 Phenylalanine to Hyoscyamine to Scopolamine: Radical Rearrangement

In the initial stages of this alkaloid biosynthetic pathway phenylalanine molecules are diverted from primary metabolic routes to undergo enzymatic oxidative deamination and then reduction of

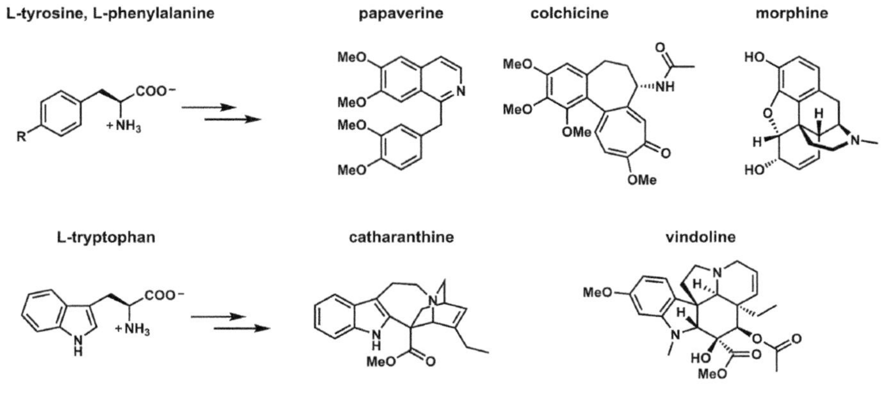

Figure 5.8 The three aromatic amino acids Phe, Tyr, and Trp are building blocks for a distinct set of alkaloids.

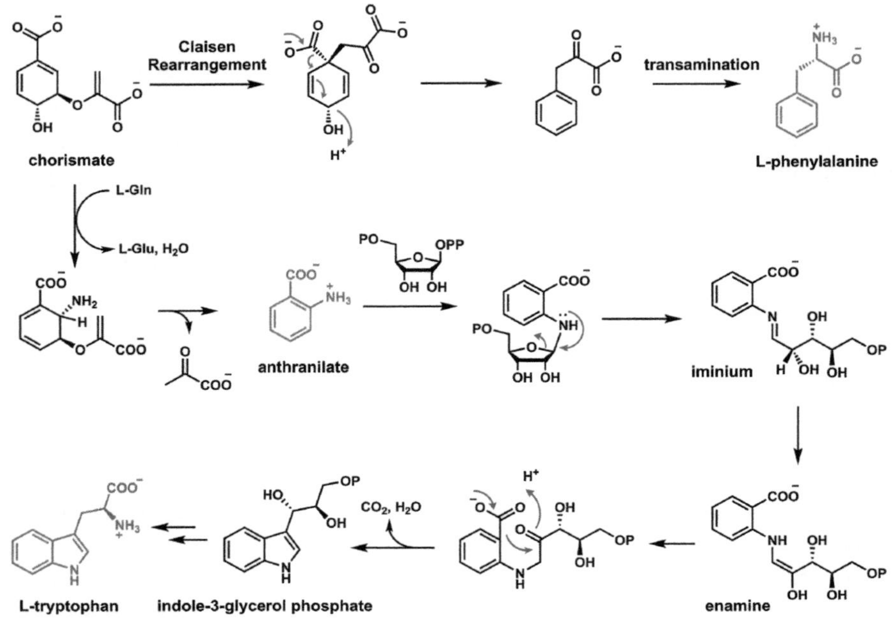

Figure 5.9 Chorismate is the common intermediate of primary metabolism, in alkaloid and nonalkaloid producers, that partitions between Phe, Tyr, and Trp. Anthranilate is an obligate intermediate on the way to tryptophan.

phenylpyruvate to phenyllactate, as noted in Figure 5.10. After activation as the CoA thioester the phenyllactyl acyl group is transferred by an ester synthase onto the alcohol oxygen of tropine to produce littorine. The tropine framework in turn arises from ornithine by a variant of the intramolecular Mannich reactions noted in Figure 5.4 (Humphrey and O'Hagan 2001).

At this juncture oxygenases again are key catalysts, in the subsequent conversion of littorine to hyoscyamine. Most notably the next enzyme catalyzes an oxygen-dependent skeletal rearrangement in the phenyl-lactyl moiety. This oxygenase is a member of the iron-containing enzyme family, elaborated in detail in Chapter 9, that carries out two half reactions. In the first half reaction a high valent oxo-iron species abstracts a hydrogen atom from a substrate C–H bond to produce the corresponding carbon radical. Typically, in a rebound reaction the oxo-iron center would then deliver an OH• equivalent, with radical coupling to form the nascent alcohol product.

In the littorine example the carbon radical species generated in the first half reaction undergoes a skeletal rearrangement *before* the

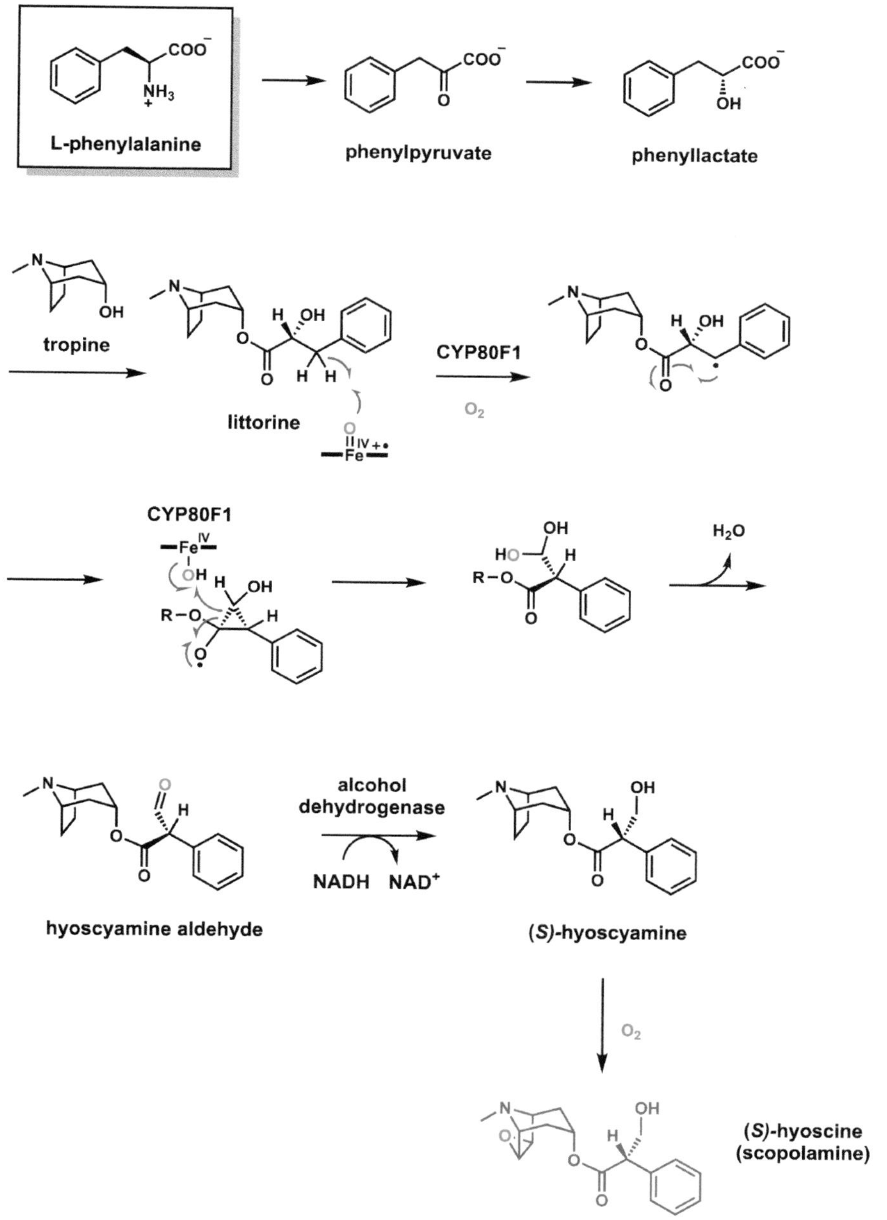

Figure 5.10 Phenylalanine to littorine to hyoscyamine to hyoscine (scopolamine) involves condensation with tropine. Note the radical rearrangement and intramolecular skeletal migration mediated by oxygenase.

oxygen rebound occurs, resulting in a rearranged alcohol framework, moving from the littorine to the hyoscyamine series. As depicted in Figure 5.10, the high valent oxo-iron abstracts one of the benzylic hydrogen atoms, leaving a benzyl radical (stabilized by delocalization into the phenyl ring) in the phenyl-lactyl moiety to attack the neighboring ester carbonyl group for one-electron transfer. This would generate a transient cyclopropyl alkoxy radical that can go backwards or, in the forward direction, undergo C–C bond migration to produce the branched chain alcohol with the unpaired electron density on the alcoholic carbon. If $OH^{\bullet}$ rebound occurs at this point the gem diol of the rearranged scaffold is produced. This is the hydrate of the hyoscyamine (scopolamine) aldehyde. Enzymatic reduction leads initially to the (−) alcohol but facile enolization can occur to give the racemic mixture, which is atropine, an infamous plant metabolite (Dewick 2009).

5.4.2 Tyrosine is the Entry Point for Several Complex Alkaloid Scaffolds

There is a short efficient route from the central amino acid metabolite tyrosine to the bicyclic core of the tetrahydroisoquinoline alkaloids (Dewick 2009). As demonstrated in Figure 5.11, *ortho*-hydroxylation and enzymatic decarboxylation of Tyr produce dopamine. One more oxygenase reaction and two SAM-dependent *O*-methylations yield the dimethylated aromatic ether. Condensation with pyruvate allows imine formation and then a variant of the Mannich reaction where the attacking carbanion equivalent forms a new ring system, termed a Pictet–Spengler reaction. In this instance, the carbanion is the *para* resonance contributor of the *para*-phenoxy methyl ether, as shown. Rearomatization yields the bicyclic tetrahydroisoquinoline 6,6 ring system. A series of oxidative decarboxylation, demethylation, and imine reduction yields the anhalonidine alkaloid.

The tyrosine-derived dopamine is an early common intermediate in the production of both opium alkaloid and a South American arrow poison. Condensation of dopamine with *para*-hydroxybenzaldehyde (itself a metabolite of tyrosine oxidative decarboxylation) occurs by just such a Pictet–Spengler reaction and, as shown in Figure 5.12, yields the tetrahydroisoquinoline *S*-norcoclaurine (Stadler, Kutchan *et al.* 1989, Ilari, Franceschini *et al.* 2009).

This is a branch point metabolite. As summarized in the figure, *S*-norcoclaurine can undergo a hydroxylation to the *S*-isomer of

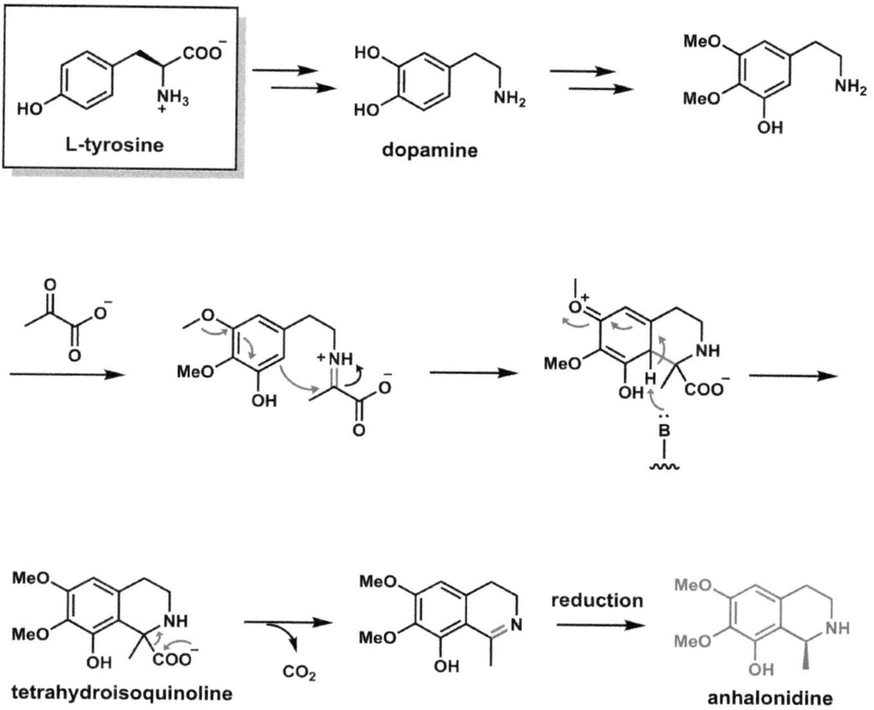

Figure 5.11 Tyrosine is progenitor of isoquinoline alkaloids. Dopamine is an early intermediate on the route to anhalonidine. A bicyclic framework is generated using a variant of the Pictet–Spengler reaction.

norreticuline and then two additional *O*-methylations to produce the opium metabolite papaverine. If a SAM-dependent *N*-methyltransferase intervenes and shunts *S*-norcoclaurine to the *N*-methyl metabolite, that is then a substrate for enzymatic generation of phenoxy radicals, presumably by the kind of iron-based oxygenase noted above in the hyoscyamine pathway. In this instance the delocalized phenoxy radicals, rather than being captured in oxygen rebound, undergo a regiocontrolled dimerization to the complex desmethyltubocurarine framework (Dewick 2009, Panjikar, Stoeckigt *et al.* 2012). *N*-Methylation to fix a positive charge on the quaternary amine creates the paralyzing neurotoxic tubocurarine, historically used as an arrow-coating poison by South American hunters. We will call out several additional examples in Chapter 9 of oxygen-dependent enzymes that generate carbon radical intermediates that are diverted to fates other than completion of the usual hydroxylation.

5.4.3 Tyrosine to *S*-Reticuline to Berberine

The *S*-norreticuline in the papaverine pathway of Figure 5.12 can be *N*-methylated to *S*-reticuline (Figure 5.13), where it serves as an intermediate to the pentacyclic berberine alkaloid family. Of special note is the enzymatic transformation of *S*-reticuline to *S*-scoulerine (Figure 5.14) in which a new C–C bond is fashioned between the *N*-methyl and the catechol ring. This enzyme, known as the berberine bridge enzyme, contains the redox active coenzyme FAD, and the mechanism of C–C bond formation is shown in Figure 5.14 (Winkler, Lyskowski *et al.* 2008). The availability of the carbon nucleophile *ortho*

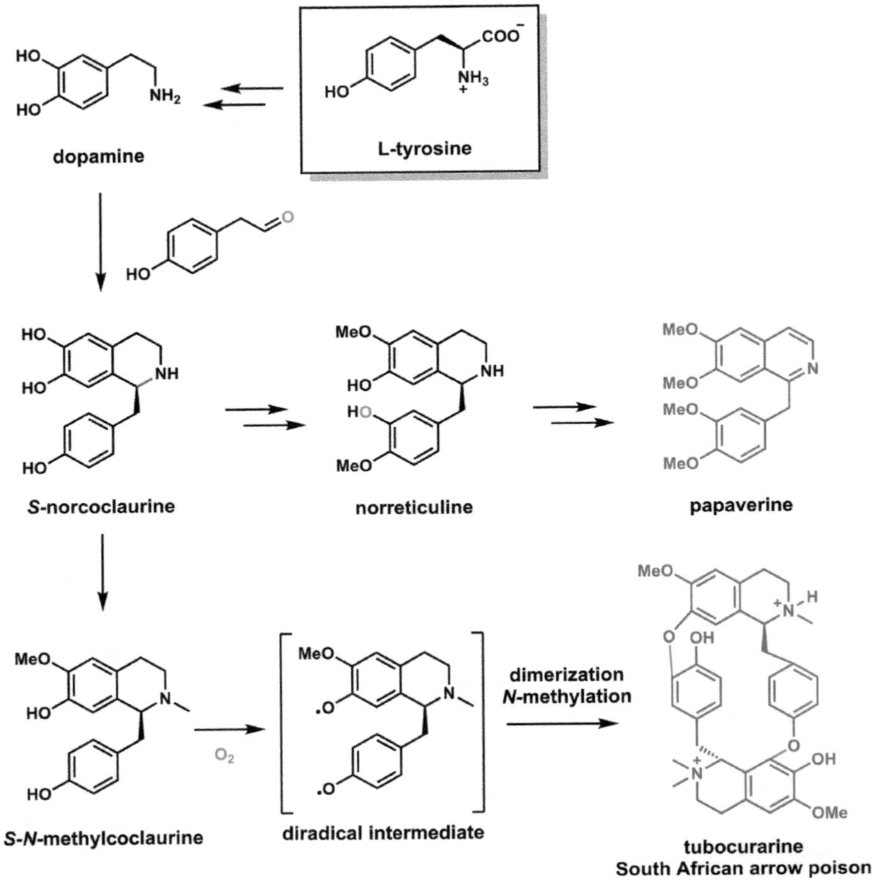

Figure 5.12 Tyrosine to the opium alkaloid papaverine or to the South American arrow poison tubocurarine; *S*-norcoclaurine is a metabolic branch point.

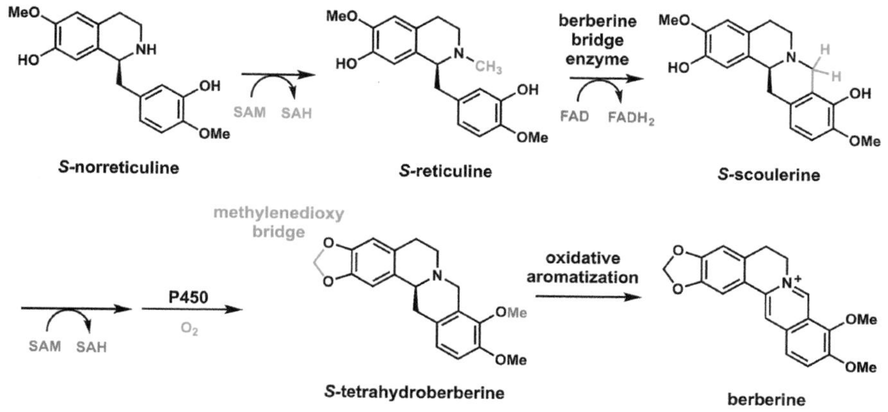

Figure 5.13 Tyrosine to *S*-reticuline to berberine. Methylation of nor-reticuline gives the *S*-reticuline isomer. The conversion of *S*-reticuline to the tetracyclic *S*-scoulerine is mediated by an FAD enzyme known as the berberine bridge enzyme.

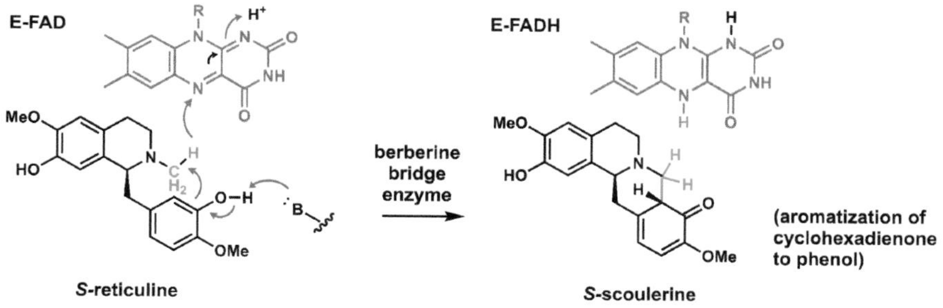

Figure 5.14 Mechanistic proposal for the berberine bridge enzyme in C–C bond formation.

to the phenol as nucleophile is not surprising, given the carbanion resonance structure on generation of the phenolate anion by an active site base. Much less obvious had been how the enzyme activates the methyl carbon as an electrophile. A concerted attack of the carbon nucleophile is proposed, with hydride ejection from the methyl group to the N_5 of oxidized FAD. Formation of the C–C bridge occurs as $FADH_2$ is generated. This is not a transformation that can be effected nonenzymatically. The initial adduct is the cyclohexadienone tautomer and it is facilely revertible to the aromatic phenol by proton abstraction, with consequent gain of resonance energy.

Returning to *S*-scoulerine (Figure 5.14) (Sato, Hashimoto *et al.* 2001) for a moment, it is processed next by a cytochrome P450 oxygenase to form the methylenedioxy bridge. This is a typical pattern for monomethyl catechols in aromatic natural products and the hydroxylation that occurs is cryptic. Oxygenation of the O–CH_3 to the O–CH_2OH proceeds by standard P450 enzyme chemistry. While that could decompose to formaldehyde and the free catechol unimolecularly, it appears that the neighboring OH– of the catechol ring is the kinetically competent nucleophile in those oxygenase active sites for intramolecular displacement of OH (otherwise an indifferent leaving group at best) to produce the methylene dioxy bridge (see Figure 9.20). The bridge is less polar and more compact than the starting *O*-methyl catechol. The final step on the way to berberine is oxidative aromatization of the central dihydropyridine to the fully heteroaromatic, positively charged pyridine ring. This is a facile process and could be a nonenzymatic autoxidation *in planta*.

5.4.4 Tyrosine to *R*-Reticuline to Morphine

In the Pictet–Spengler reaction that forms the tetrahydroquinoline bicyclic ring system of norcoclaurine (Figure 5.12), the ring-forming step creates a chiral center adjacent to the endocyclic amine. If the enzyme in question installs the *para*-OH-phenyl above the plane this is the *S*-isomer (as in Figure 5.12), if below the plane it is the *R*-isomer. The *R*-reticuline is in fact a natural product and is the isomer on pathway to thebaine, codeine, and morphine, as shown in Figure 5.15. An unusual P450-oxidoreductase (termed STORR for *S* to *R*-reticuline) interconverts the *S*- and *R*-isomers of reticuline by an oxidation/chiral re-reduction process (Winzer, Kern *et al.* 2015).

The enzyme taking *R*-reticuline forward is an iron-based cytochrome P450 oxygenase (Grobe, Zhang *et al.* 2009) (oxygenases are everywhere in alkaloid maturation pathways!) and there are functional mechanistic analogies to the formation of phenoxy radicals in the *S*-methylcoclaurine to desmethyltubocurarine shown in Figure 5.12. In that case two phenoxy radicals coupled intermolecularly to form a new set of C–C bonds. In the *R*-reticuline case there are two phenol rings in the molecule, and a phenoxy radical pair can be generated intramolecularly in tandem catalytic cycles (Figure 5.15B). Presumably they are energetically accessible and kinetically long lived enough to find each other to couple. The conformation of the *R*-reticuline diradical in the enzyme active site will determine the observed *ortho–para* regiospecificity of coupling.

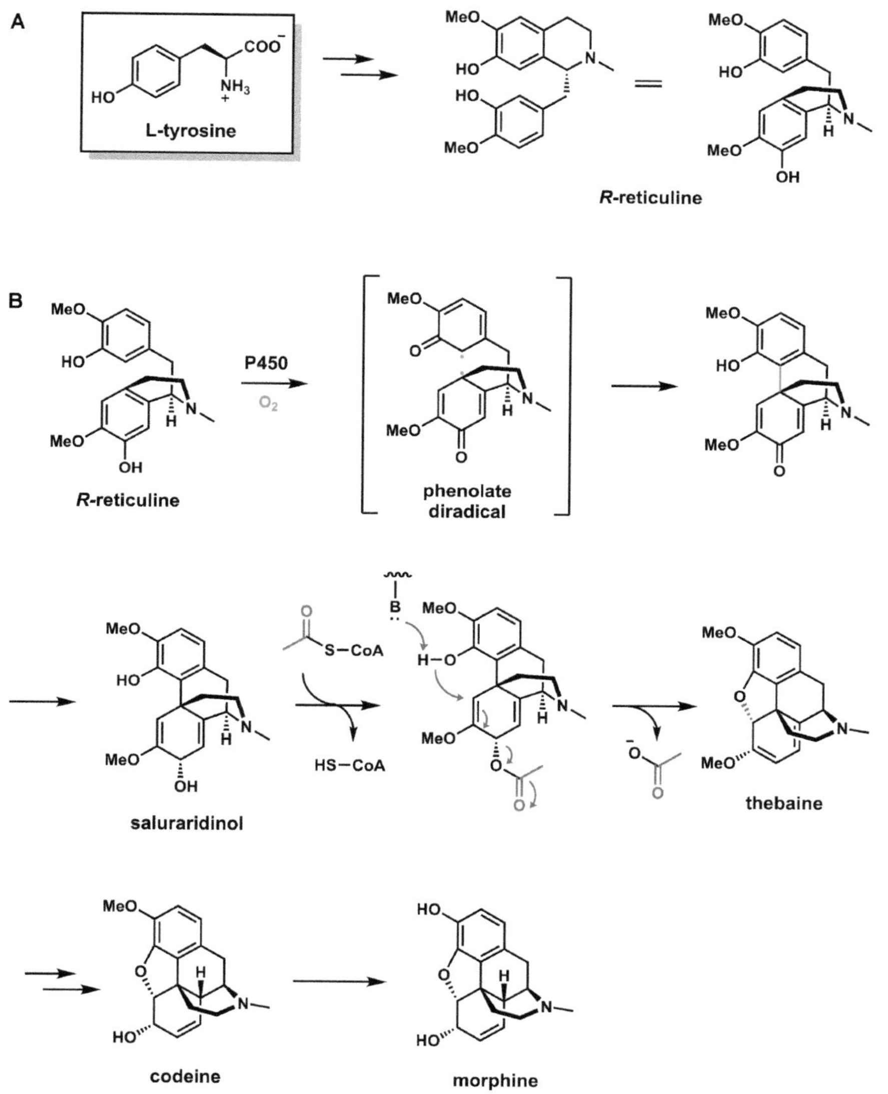

Figure 5.15 (A) Tyrosine can also give rise to the *R*-isomer of reticuline, which is on the biosynthetic route to morphine. (B) Proposed diradical coupling for C–C bond formation from *R*-reticuline to salutaridinol in the morphine biosynthetic pathway.

An alternative route would be generation of a single radical intermediate in one catalytic cycle and one electron transfer from the other phenol to create the C–C bond and a product radical that would then give an electron back to the active site iron. Enzymatic reduction of the resultant quinone to the alcohol generates salutaridinol. The homolytic formation of the new C–C bond has created the characteristic fused tetracyclic framework found in the morphine alkaloid series.

To finish the pathway to morphine, an acetyl transferase acts on the newly generated alcohol to convert it into a potential acetoxy leaving group (Figure 5.15B). This ensues when the phenolate anion of the top ring adds in a conjugated nucleophilic substitution with departure of the acetate moiety, carrying away that oxygen as thebaine results. Enzyme-mediated manipulation of the redox state of the bottom ring to the allyl alcohol produces codeine, before oxygenative demethylation arrives at the analgesic molecule morphine (Weld, Ziegler *et al.* 2004, Galanie, Thodey *et al.* 2015).

Vignette 5.2 Psychoactive Plant Metabolites: Cocaine (Cocoa), Mescaline (Peyote), Morphine (Poppy), THC (Cannabis), Scopolamine, Ephedrine.

Plants can make a variety of alkaloids where the basic nitrogens in heterocyclic rings can mimic the structure of mammalian neurotransmitters. As ligands they can sometimes function as agonists and sometimes as antagonists. The six plant-derived alkaloids have different mammalian central nervous system targets, reflecting the variety of alkaloid scaffolds (Dewick 2009).

As shown in Figure 5.V2, tetrahydrocannabinol (THC) can mimic the endogenous cannabinoid receptor ligand anandamide (Devane, Hanus *et al.* 1992). *Cannabis sativa* generates a range of cannabinoid ligands but THC is the molecule with the highest psychotropic activity. It has been used for decades to centuries in different cultures for the amelioration of various kinds of chronic pain as well as alleviating nausea and vomiting associated with cancer chemotherapy regimens. Figure 5.V1 shows trichomes (hairlike glandular outgrowths of plant epidermis) in cannabis flowers exuding THC-rich resin.

Ephedrine is a phenylethylamine (amphetamine scaffold) that can bind to epinephrine (adrenaline) receptors and intensify the effects of adrenaline. It can be used to treat asthma but pseudoephedrine has a better therapeutic index. Ephedrine was extracted historically from *Ephedra sinica* but its simple chemical structure has led to commercial production by total synthetic methods.

Mescaline, from peyote and other cacti, is a serotonin mimic that can function as an agonist at multiple subclasses of serotonin (5-hydroxytryptamine) receptors (5-HT$_A$, 5-HT$_c$). It has high hallucinogenic properties, comparable to lysergic acid diethylamide and the mushroom constituent

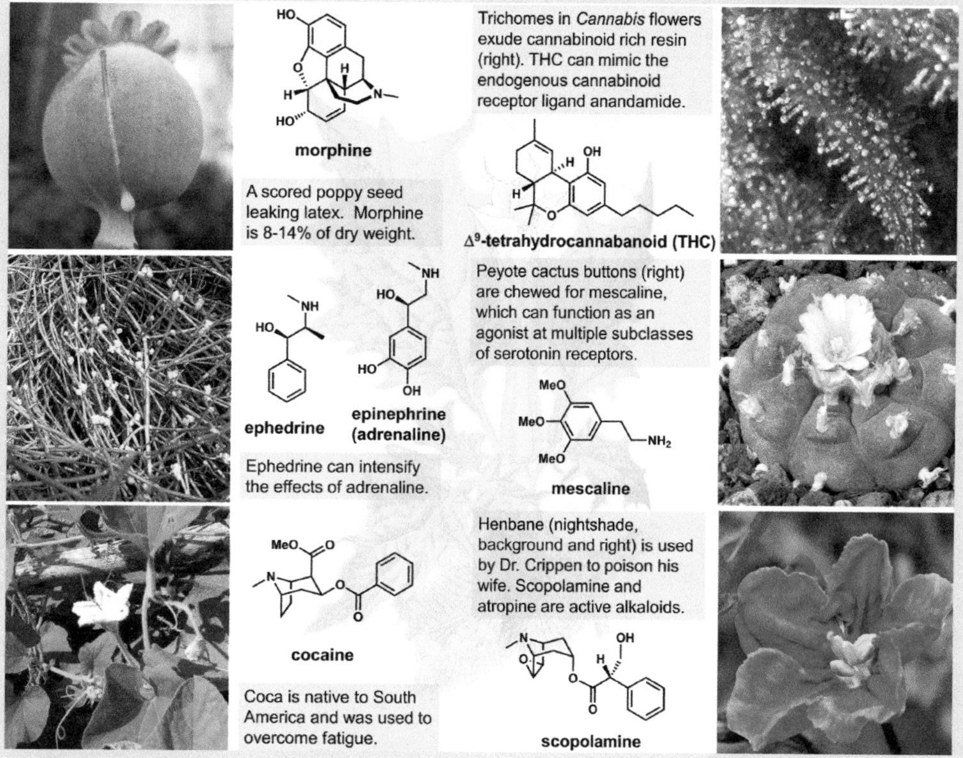

Figure 5.V2 Six psychoactive plant metabolites.
Photo credits: *Ephedra sinica*: Alex Lomas, https://www.flickr.com/photos/51464530@N00/2943621347, License: CC-BY-2.0 (https://creativecommons.org/licenses/by/2.0/); Coca plant and flower: Steven Damron, https://www.flickr.com/photos/sadsnaps/4604307658, License: CC-BY-2.0 (https://creativecommons.org/licenses/by/2.0/). *Lophophora williamsii*: Peter A. Mansfeld, https://commons.wikimedia.org/wiki/File:Lophophora_williamsii_pm.jpg, License: CC-BY-3.0 (https://creativecommons.org/licenses/by/3.0/deed.en).

psilocybin. The tops of peyote cacti are often sliced off, buttons reform and these are the parts of the cactus chewed for mescaline content (Figure 5.V2).

Morphine, an end product of poppy secondary metabolism, is a selective agonist for the μ-opioid receptors, with much lower affinity at the δ- and κ-receptors, and functions as an analgesic (pain-relieving) agent. It can also cause euphoria and is highly addictive. The poppy seed, when subjected to mechanical scoring or a gash, leaks latex which is alkaloid rich, with morphine comprising 8–14% of the dry weight.

The fifth molecule shown is scopolamine (also known as hyoscine), one of the belladonna alkaloids. It is an antagonist ligand of the class of muscarinic acetylcholine receptors in the parasympathetic nervous system. It has been used to treat motion sickness as well as postoperative nausea and vomiting. The henbane plant was reputedly the source of the atropine used by the notorious Dr Crippen in the poisoning of his wife in London in 1910 (Gaute and Odell 1996).

Cocaine, the acetyl and benzoyl diester of the bicyclic tropane alkaloid scaffold found at high levels in coca plant leaves, crosses the blood–brain barrier and acts as a competitive ligand for reuptake of neuroamine transmitters, most notably for dopamine in the mesolimbic system, but also including noradrenaline and serotonin. The acute accumulation of higher levels of synaptic dopamine and the other neuroactive amines accounts for the pleasure response, but can also be accompanied by altered regulation of neuroamine transporters at neuronal surfaces. In addition to acute psychoactive effects and subsequent craving and addiction, there are many physical symptoms including increased heart rate. Chronic adverse effects can include risks of stroke, autoimmune syndromes, kidney dysfunction including failure, and arrhythmia-induced cardiac deaths.

Vignette 5.3 Alkaloids Used as Arrow Poisons.

The six alkaloid molecules shown in Figure 5.V2 interact with central nervous system receptors to evoke their complex psychoactive effects. The six molecules in Figure 5.V3 also are neuroactive but with results that can be lethal to macrofauna, including humans (Pelletier 1983). Of these six alkaloids only one, batrachotoxin, is from an animal source, the skin glands of South American tree frogs. The other five are plant metabolites, from such infamous plants as *Strychnos nus vomica* (Hawaii tree), *Hyoscyamus niger* (henbane), *Atropa belladonna* (deadly nightshade), and *Mandragora officanarum* (mandrake).

Batrachotoxin, a steroidal alkaloid, targets voltage-gated sodium channels, opening them even at resting potentials. Neuromuscular paralysis ensues, at doses as low as 2 $\mu g\,kg^{-1}$ in mice.

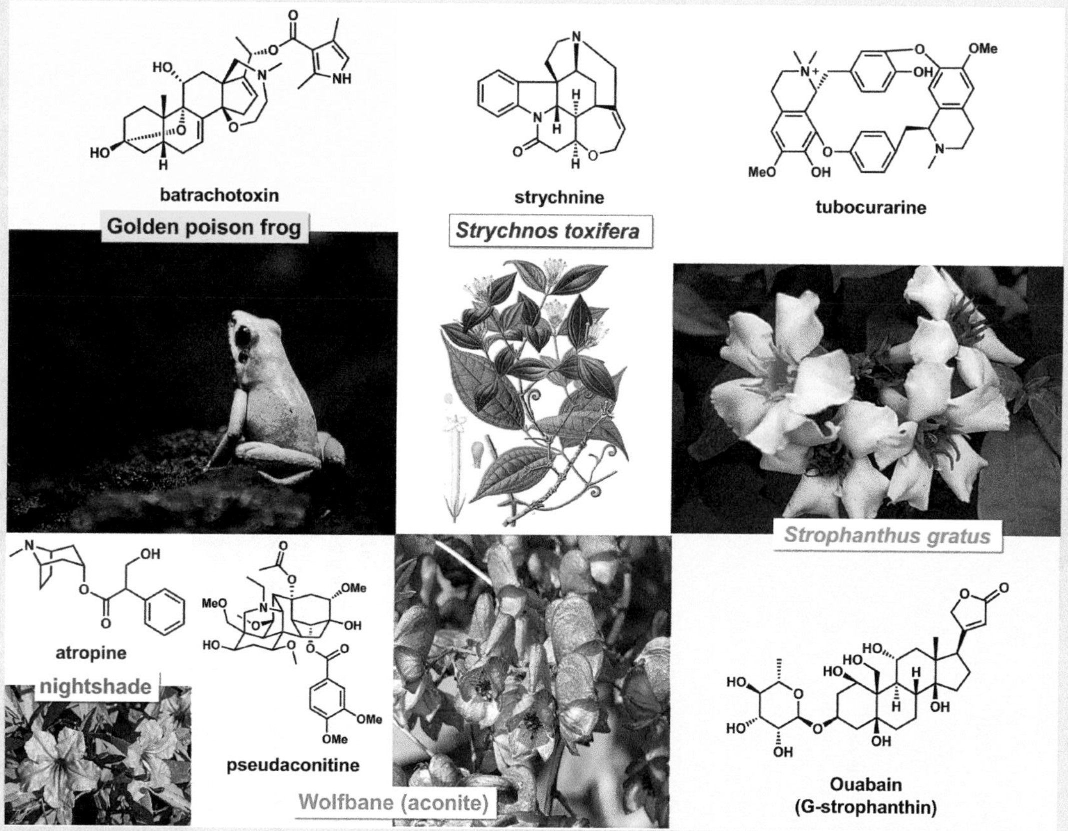

Figure 5.V3 Six alkaloids used as arrow poisons.
Photo credits: golden poison frog: kuhnmi, https://www.flickr.com/photos/31176607@N05/13152578983, License: CC-BY-2.0 (https://creativecommons.org/licenses/by/2.0/). *Strophanthus gratus*: Jayesh Patil, https://www.flickr.com/photos/54439360@N04/5542608003, License: CC-BY-2.0 (https://creativecommons.org/licenses/by/2.0/).

Strychnine is found in trees indigenous to the Malabar coast in India, and in Sri Lanka, and also Hawaii. Although it was known as early as 1818, the complex structure of strychnine was not determined until 1946 by Robert Robinson's group (Robinson 1952), with total synthesis by Woodward and colleagues following in 1954 (Woodward, Cava *et al.* 1963). Strychnine is an antagonistic ligand for glycine in inhibitory neuronal tracts. As a consequence, muscular convulsions ensue, become quite violent, and death results from respiratory paralysis.

Tubocurarine, a monocationic dimeric benzyisoquinolne alkaloid, is probably the most active toxin in curare preparations, extracted from the bark of *Chondrodendron tormentosum*, a South American plant. The amine cationic group allows tubocurarine to act as an acetylcholine agonist at nicotinic acetylcholine receptors. Binding leads to neuromuscular blockade and respiratory paralysis. The South American Indians who dipped their hunting arrows in curare pastes were able to eat the meat of dead prey because tubocurarine does not cross mucous membranes readily. The animals received the toxin intramuscularly or intraperitoneally.

Ouabain (strophanthin G), from the *Strophanthus gratus* and *Acokanthera schimperi* plants in east Africa, was used in Africa as an arrow poison. Ouabain is a member of the steroidal alkaloids that are cardio-toxic by virtue of binding to the Na, K-ATPase that serves as an ion channel. Sodium ions accumulate inside the cell as do calcium ions. Contractility increases, cardiac arrhythmias develop, and cardiac arrest follows. Less potent cardiac glycosides, such as digoxin, have a useful therapeutic ratio between effective contractility and toxic arrhythmias. Ouabain appears to be too potent to have such a therapeutic window.

The buttercup family includes the plant known as wolfsbane, presumably because it was thought to be a ward against werewolves (Lin, Chan *et al.* 2004). The prominent toxin in the plant is the diterpene alkaloid pseudaconitine. It can bind to acetylcholinesterase in synaptic clefts, leading to aprolonged rise in acetylcholine levels and consequent overstimulation of postsynaptic membrane receptors. In the Himalayas hunters used the root mixtures to hunt ibex, Japanese Ainu males reportedly hunted bear (Peissel 1984), and Aleutians speared whales with spear tips covered in pseudaconitine poisons ("A Pacific Eskimo invention in whale hunting in historic times eScholarship", escholarship.org. Retrieved 6 October 2014).

The sixth of the alkaloid poisons depicted in Figure 5.V3 is atropine, one of the alkaloids from the belladonna plant (named perhaps fancifully from the dilatation of pupils by the tropine alkaloid, causing a wide eyed gaze in beautiful young Italian women). Atropine is actually a mix of the D- and L-isomers of hyoscyamine. It shares with tubocurarine the property of being an antagonistic acetylcholine ligand but at a different subset, the muscarinic cholinergic receptors.

Nicotinic acetylcholine receptors (sensitive to nicotine) are ligand-gated ion channels whereas muscarinic receptors are not. Instead, the muscarinic family of acetylcholine receptors (sensitive to muscarine as a ligand) are seven G-protein-coupled transmembrane protein receptors. Because these are two very different protein superfamilies, the nicotinic and muscarinic acetylcholine receptors are inhibited differentially by small molecules, including natural products. Tubocurarine blocks the nicotinic receptors while atropine and the pure isomer scopolamine block muscarinic receptors.

5.5 Tryptophan as a Building Block for Alkaloids

5.5.1 Tryptophan to Harmane: β-Carbolines

Tryptophan is the most versatile and widely utilized building block in alkaloid biogenesis, based both on the number of plant end products and also the unique nucleophilicity of the indole ring as a carbanion equivalent at several sites around the bicyclic ring system. Chapter 8 is devoted to the highly abundant indole terpene class of alkaloids which uses the indole ring nucleophilicity as the dominant reaction paradigm.

Perhaps the simplest manifestation of the useful nucleophilicity of the indole side chain is in the generation of the tricyclic β-carboline ring system found in harmane alkaloids (Figure 5.16A) (Dewick 2009, Seigler 2012, Cseke, Kirakosyan *et al.* 2016). Enzymatic decarboxylation of tryptophan to tryptamine is followed by condensation with acetaldehyde. The imine is capturable in a Pictet–Spengler reaction. Proton abstraction from the initial adduct is driven by quenching the charge on the iminium form of the indole, and after further prototypic hydroxylation and *O*-methylation yields tetrahydroharmane. Subsequent enzymatic or nonenzymatic isomerization generates the heteroaromatic scaffold of harmane as the fully oxidized beta carboline.

5.5.2 Tryptophan to Strictosidine and Beyond: Ajmalacine, Camptothecin, Quinine

The more familiar and widely implemented Pictet–Spengler enzyme in plant alkaloid metabolism is the one operating on tryptamine and secologanin (Figure 5.16B) to create strictosidine (Maresh, Giddings *et al.* 2008, Brown, Clastre *et al.* 2015). In turn, as we will exemplify,

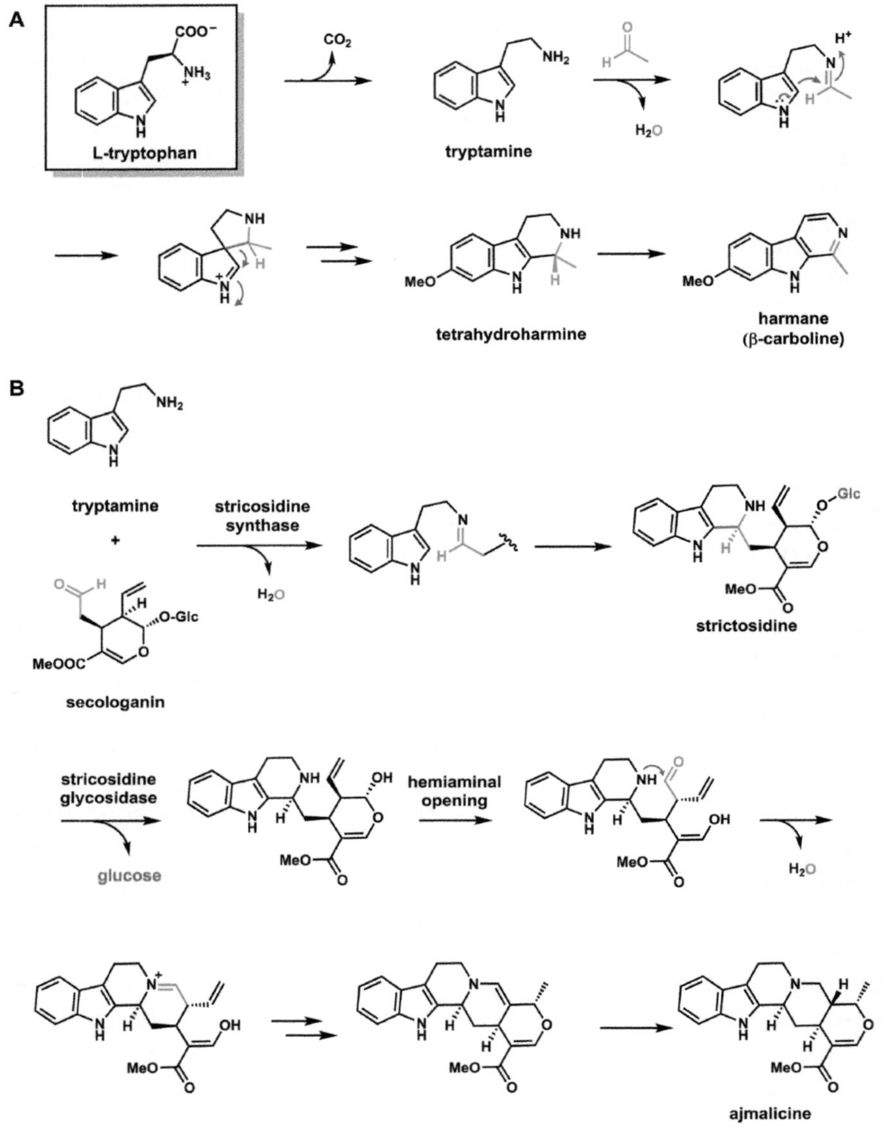

Figure 5.16 Tryptophan as alkaloid precursor. Decarboxylation to tryptamine is followed by condensation with aldehydes in formation of both the simple tricyclic β-carboline scaffold of harmane and the more complex strictosidine. The participation of the pyrrolic ring of the indole moiety as a carbanion equivalent puts these two Mannich reactions into the special category known as Pictet–Spengler condensations. Shown is the transformation of strictosidine to the antihypertensive ajmalacine.

strictosidine is the key intermediate to >1000 downstream alkaloid structures, generally of increased complexity.

We noted the formation of secologanin from geraniol at the end of the preceding chapter as an example of the heavy oxygenative morphing of the C_{10} isoprenyl alcohol geraniol (see Figure 4.38). That figure might be consulted again for this discussion. Additionally, we noted the use of glucose as a protecting group in the formation of an acetal that keeps a latent hemiacetal tied up and unreactive until a releasing glucosidase is subsequently employed. Secologanin does have a free aldehyde group and that is the required functionality for imine formation with tryptamine. As in the simple tetrahydroharmane example above, C_3 of the indole ring as carbanion to form the spirocycle that collapses with single bond migration to C_2 of the iminium moiety in the spirocycle to complete the Mannich/Pictet–Spengler C–C bond-forming reaction to set the tricyclic β-carboline core in the strictosidine product.

The strictosidine synthase enzyme from the plant *Rawolfia serpentina* has been overproduced, crystallized, and its molecular structure determined (Ma, Panjikar *et al.* 2006). The protein is organized as a six bladed β-propeller fold, each blade composed of four twisted beta sheets (Figure 5.17). The protein is synthesized with an N-terminal signal sequence that directs the nascent protein to the vacuolar membrane where it is proteolytically trimmed and moved into the vacuole. It is also *N*-glycosylated.

The structural studies have identified active site residues that on mutation have allowed conversion of methyl, methoxy, and halogenated tryptamines into strictosidine analogs. The structure of the product strictosidine bound to the synthase is also shown in Figure 5.17 (Loris, Panjikar *et al.* 2007). The following enzyme, the strictosidine glycosidase, has also been purified and characterized structurally, setting up structure–activity studies for downstream analog processing (Barleben, Panjikar *et al.* 2007).

As one example of the forward processing of strictosidine, glycosidase removal of the sugar protecting group gives the hemiacetal that can open spontaneously to the enol aldehyde. The aldehyde and the amine in the carboline ring system (originally the amino group of tryptamine) can form a cyclic imine, now in a tetracyclic framework (Dewick 2009). That imine can be captured in turn by the enol oxygen to form the pentacyclic ring system of cathenamine. Redox adjustment by enzymatic reduction of the endocyclic imine gives the end product ajmalacine (Figure 5.16).

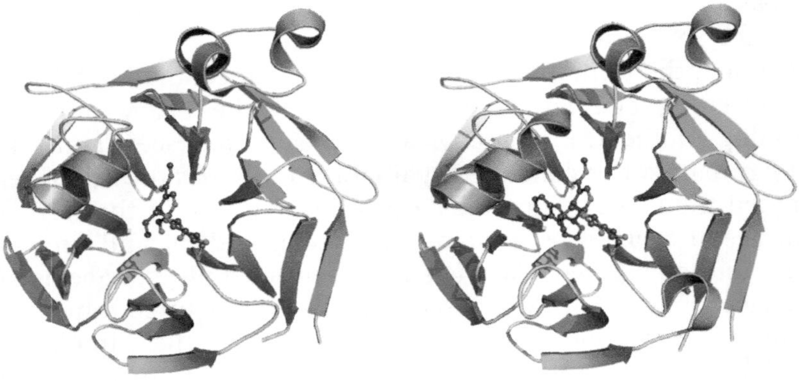

Figure 5.17 X-ray structure of *Rauvolfia* strictosidine synthase with bound secologanin (left, PDB code: 2FPC) and or strictosidine product (right, PDB code 2V91).
Left image data from Ma, X., S. Panjikar, J. Koepke, E. Lorris and J. Stöckigt (2006). "The Structure of *Rauvolfia serpentine* Strictosidine Synthase is a Novel Six-Bladed β-Propeller Fold in Plant Proteins". *Plant Cell* **18**(4): 907–920. Right image data from Loris, E. A., S. Panjikar, M. Ruppert, L. Barleben, M. Unger, H. Schübel and J. Stöckigt. "Structure-Based Engineering of Strictosidine Synthase: Auxiliary for Alkaloid Libraries". *Chem. Biol.* **12**(9): 979–985. Image credit: Yang Hai.

Figure 5.18 shows seven other alkaloid scaffolds formed by forward processing of strictosidine, but this is still the tip of the alkaloid biosynthetic iceberg of the one to two thousand or so molecules that are downstream of this β-carboline metabolite. Two additional pathway examples are noted in Figures 5.19 and 5.20.

The first outlines the strategy for converting strictosidine to camptothecin (Figure 5.19), a molecule with anticancer properties that has been the basis for the imitative useful clinical agent topotecan. This pathway exemplifies a different strategy for the initial processing of strictosidine. Rather than early glycosidase action, the first chemical step is thought to be a cyclization of the carboline amine on the *O*-methyl ester to produce a pentacyclic amide. Next, a dioxygenase-mediated cleavage across the C_2–C_3 centers of the indole ring is proposed. There are well characterized indole 2,3-dioxygenases in primary metabolism as precedent (Thackray, Mowat *et al.* 2008).

Such dioxygenase cleavage would create a nine-member dicarbonyl macrocyclic ring. Enolate formation yields a carbanion that can attack

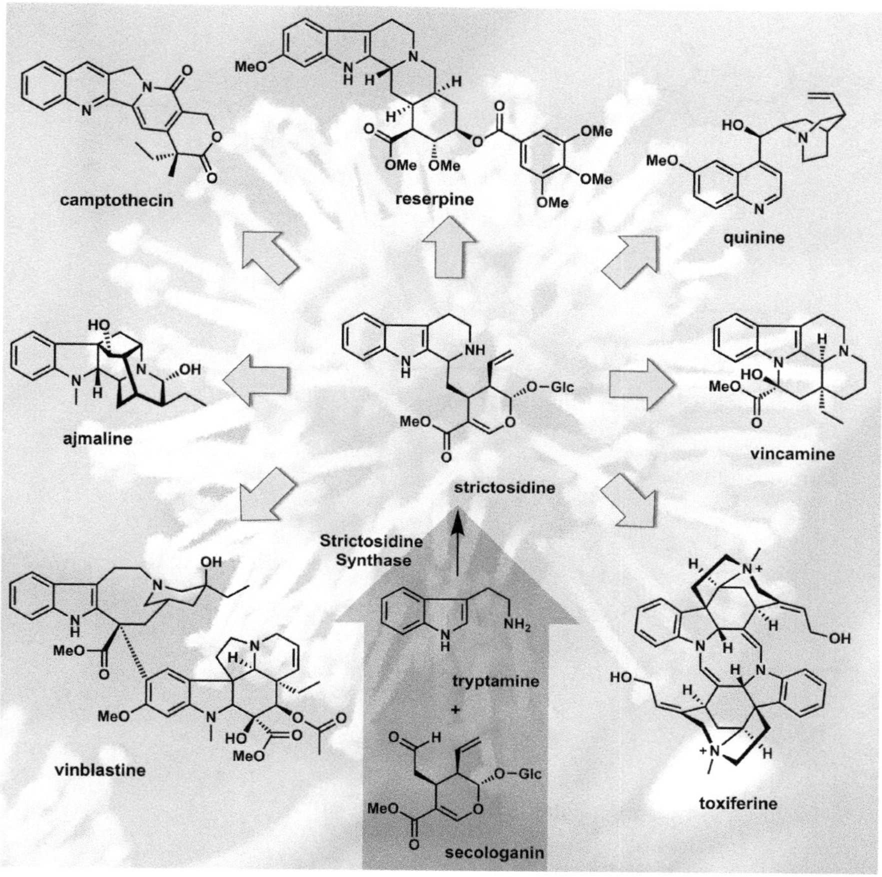

Figure 5.18 Eight (out of ~1000) alkaloid scaffolds downstream of strictosidine.
Data from Ma, X., S. Panjikar, J. Koepke, E. Loris and J. Stöckigt (2006). "The Structure of *Rauvolfia serpentina* Strictosidine Synthase is a Novel Six-Bladed β-Propeller Fold in Plant Protein". *Plant Cell* **18**(4): 907–920.

the opposite carbonyl in a transannular aldol condensation, followed by dehydration to produce a new pentacyclic scaffold. Note that the dioxygenase plus transannular aldol converts a 6-5-6-6-6 ring system to a 6-6-5-6-6, interconverting the sizes of the second and third rings as well as installation of a central enone group. This framework is the pumiloside molecule. Several redox adjustment enzymatic steps are required as well as late stage glucosidase removal of the hexose protecting group to arrive at the camptothecin molecule, but the

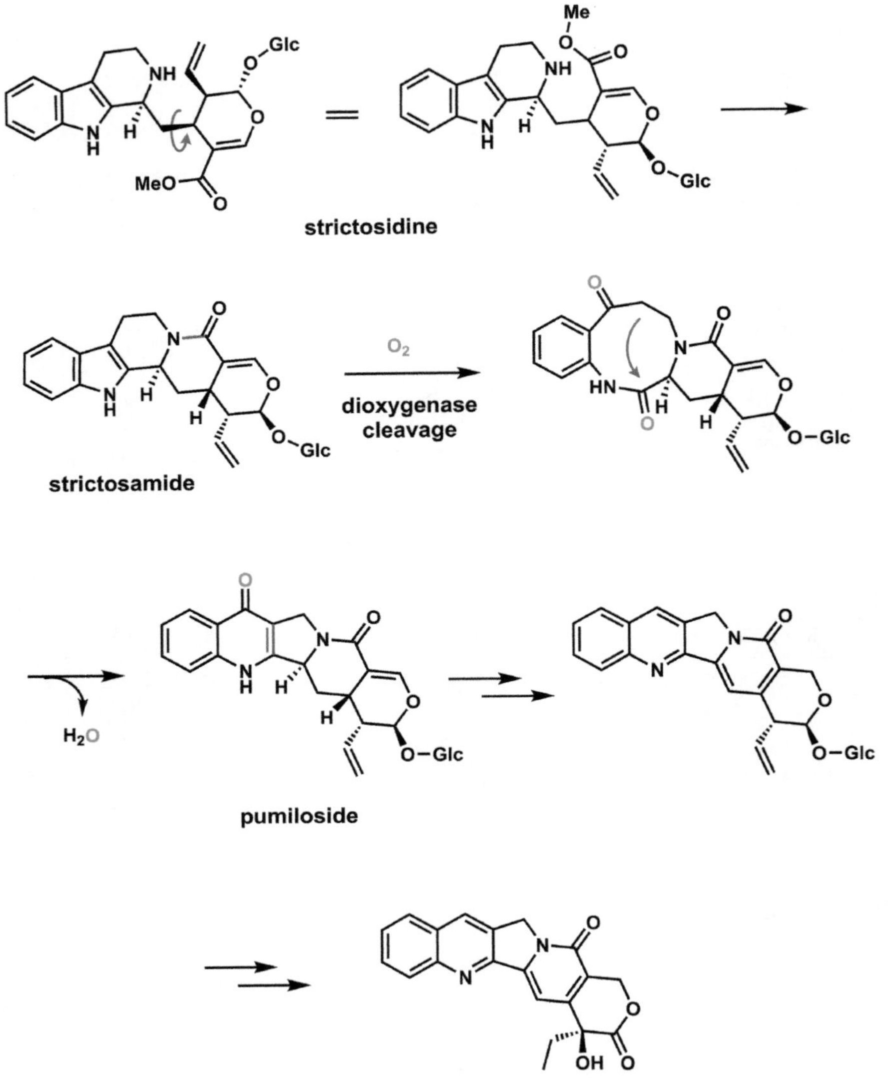

Figure 5.19 The central intermediacy of strictosidine: the route to camptothecin involves a proposed ring-cleaving dioxygenase.

essential framework changes have been accomplished by arrival at pumiloside.

Figure 5.20 indicates a third path for forward processing of strictosidine, on the way to cinchonidine and quinine (Dewick 2009).

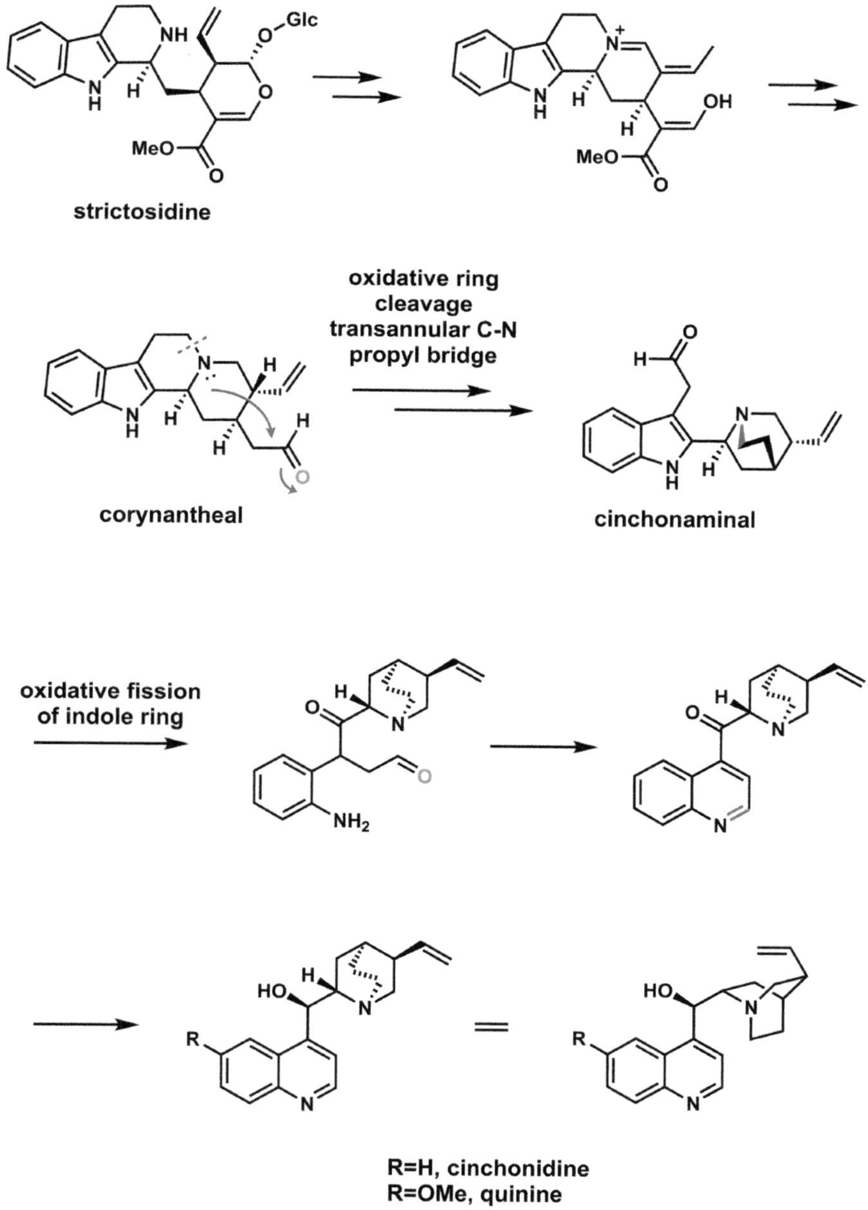

Figure 5.20 The central intermediacy of strictosidine: conversion of the aglycone to cinchonidine and quinine.

Again early formation of a tetracyclic scaffold is suggested. Glycosidase action must also happen at some relatively early point to generate the free hemiacetal (not shown). Then, with somewhat analogous logic to the camptothecin enzymology, oxygenative ring cleavage of the central tricyclic β-carboline is proposed, this time not across the pyrrole moiety of the indole but instead at the piperidine ring. This will generate an amino aldehyde. A second latent aldehyde exists in the form of the enol ether moiety in the hemiacetal. Unraveling that aldehyde and attack of the amine can, in transannular capture, yield the characteristic C–N propyl bridge of cinchona and quinine alkaloids. Methyl ester hydrolysis and decarboxylation of the β-aldehyde acid would be a low energy process if it happens before the transannular bridge forms.

Now a second oxygenative fission of an indole ring is proposed, but rather than the 2,3-regiochemistry in camptothecin, a 1,2-regiochemistry is shown (Dewick 2009), akin to intradiol *vs.* extradiol regioselectivities of well-studied aromatic ring-cleaving bacterial dioxygenases (Lipscomb 2008). The product will have a ketone from indole cleavage but still has the aldehyde from the first dioxygenative cleavage of the carboline. Imine formation between what had been the indole pyrrolic nitrogen and that aldehyde, followed by dehydration, generates the 6-6-bicyclic quinoline ring system that is a second hallmark of quinine. Reduction of the ketone to alcohol yields quinine and cinchonidine.

These three examples (out of a plethora that could have been chosen) reveal distinct and complementary strategies for morphing the core tricyclic β-carboline ring system into distinct heterocycle connectivities. These cases reveal four types of enzymatic reaction strategies for processing strictosidine. They also illustrate how the chemical features of the indole on one hand and the secologanin array of functional groups on the other are admirably and efficiently tuned for many reaction manifolds.

One such reaction is condensation of amines and carbonyl functional groups to imines and particular patterns of five- and six-membered ring fusions. A second is oxygenative carving up of the carboline core and refashioning it into alternative ring scaffolds. A third is the redox adjustment mediated by dehydrogenases and reductases to control nucleophilicity and electrophilicity of amines and imines. The fourth is the timing of glucosidase action to convert the stable enol acetal into the enol hemiacetal and its equilibrative opening to the enol which can react as nucleophile or electrophilic aldehyde. With such iterated enzymatic machinery the many chemical reactivity nuances of the strictosidine molecule are well suited to its protean role as precursor to many indole alkaloid structural variants.

5.5.3 Tryptophan to Lysergic Acid and Ergotamine

The ergot alkaloids from spores of the fungus *Claviceps purpurae*, a pathogen in grasses, among them cultivated rye, cause the set of symptoms known as ergotism that include gastrointestinal (GI), circulatory and neurologic problems, from vomiting to vasoconstriction to convulsions (Schiff 2006). The vasoconstriction from chronic consumption of ergot-contaminated rye can cause gangrene. Such patients were cared for by the Order of St. Anthony in the Middle Ages, and one name for ergotism was St. Anthony's fire (Dewick 2009). Over 50 alkaloids are produced in ergot, among them lysergic acid and the lysergyl tripeptide scaffold ergotamine. This modified tripeptide has oxytocin-mimetic effects for obstetric activity.

The production of ergotamine reflects a convergence of three types of natural product pathways: alkaloids, isoprenoids, and nonribosomal peptide assembly line enzyme machinery. The pathway begins with prenylation of tryptophan at C_4 of the indole ring, an indole terpene formation that is a reaction type examined in more detail in Chapter 8. The isoprenyl chain is subsequently hydroxylated transiently to set up an elimination of the elements of water that creates the diene indicated in Figure 5.21. Then a second oxygenation is effected, this time as epoxidation of the terminal olefin to generate the tricyclic ring system of the lysergic acid biosynthetic intermediate chanoclavine-I. A decarboxylative vinylogous epoxide opening is the proposed mechanism. This evokes memories of the epoxide opening cascades in the polyether subclass of polyketides noted in Chapter 2.

Further processing of chanoclavine is thought to involve oxidation of the alcohol to the aldehyde, formation of the cyclic imine and its reduction to the tetracyclic ring system (agroclavine). Two additional oxygenases then act to take the exocyclic methyl group up to the acid oxidation state found in lysergic acid (four oxygenases thus far in the pathway). The last step (from paspalic acid to lysergic acid) is thought to be nonenzymatic isomerization of the endocyclic olefin into conjugation with the indole ring.

Lysergic acid can function as an unusual starter unit for a four module NRPS assembly line in the producer fungi (Havemann, Vogel *et al.* 2014). LPS1 has one module, specific for activating and tethering the lysergic acid on peptidyl carrier protein domain-1. LSP2 is a nine domain, three module NRPS that in turn activates alanine, phenylalanine, and proline. The most downstream domain

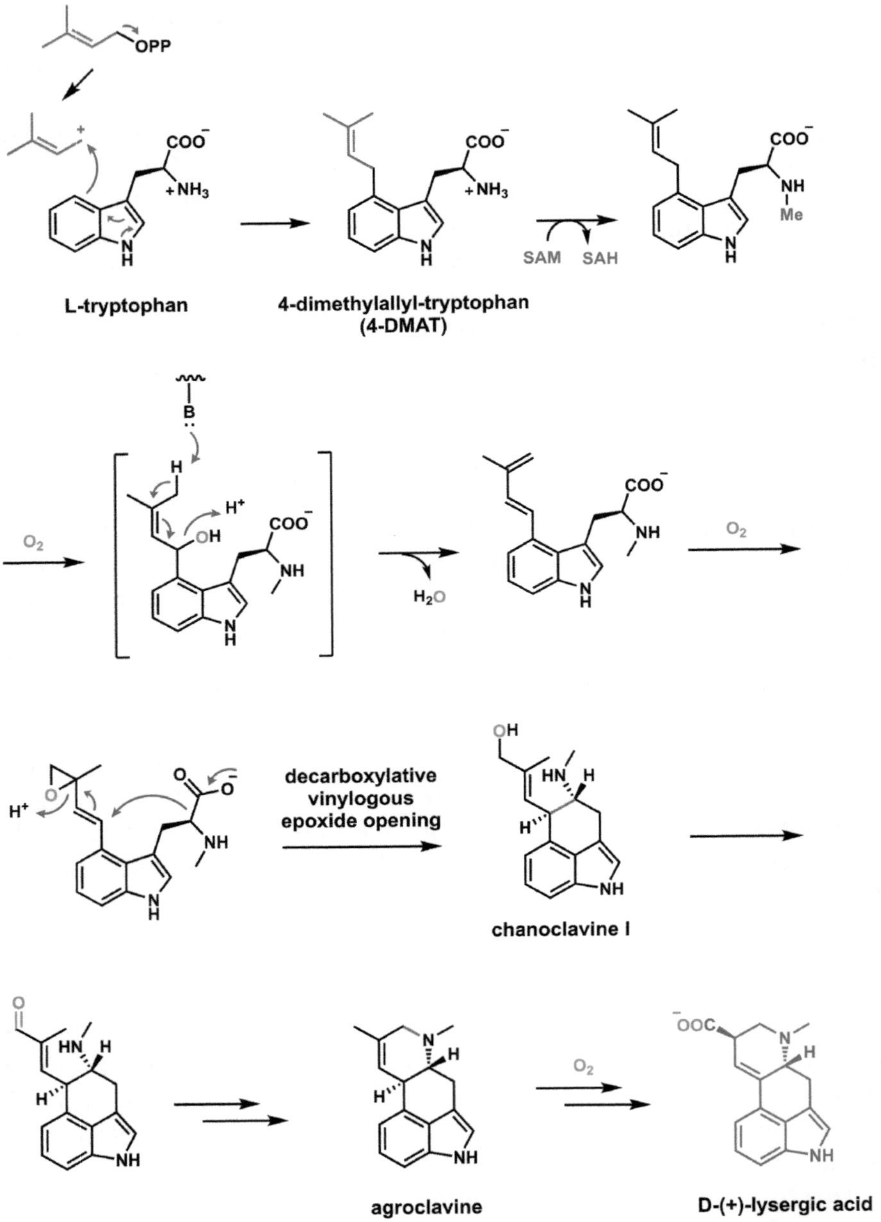

Figure 5.21 Isoprenylation of tryptophan at C_4 is the first step in the lysergic acid biosynthetic pathway. An epoxidation and then a decarboxylative vinylogous epoxide opening are proposed to form the third fused six-membered ring. Cyclic imine formation and isomerization to the conjugated imine complete the tetracyclic framework of D-lysergic acid.

in LPS2 is not a thioesterase enzymatic domain but a condensation domain, a typical arrangement in fungal NRPS assembly lines where macrocyclization is the release mode (Gao, Haynes *et al.* 2012). Indeed, in this four module ergotamine assembly line, chain release is proposed to occur by attack of the Phe_3-amide NH on the activated Pro_4 carbonyl to yield a version of an *N*-acylated Phe–Pro diketopiperazine (Figure 5.22).

This is not the final product. Post assembly line hydroxylation of the Ala_2 residue at its Cα sets up an intramolecular attack on the Pro_4 amide carbonyl (Havemann, Vogel *et al.* 2014). The accumulating ergotamine peptide exists as the strain-free five-membered orthoester/hemiaminal. The confluence of alkaloid, isoprenoid, and NRPS assembly line enzymatic machinery and chemical logic has created a natural product of high structural and functional group complexity.

5.6 Anthranilate as a Starter and Extender Unit for Fungal Peptidyl Alkaloids of Substantial Complexity

Anthranilate, formally *ortho*-aminobenzoate, is a primary metabolite in microbes and fungi, and in plants as a key building block for tryptophan. Figure 5.10 noted that anthranilate formation draws off some of the metabolic flux from chorismate (the precursor to Phe and Tyr) in an amination/aromatization enzymatic sequence. Anthranilate is then processed to indoleglycerol-3-P, in turn the immediate precursor of tryptophan. Thus, anthranilate is in the role of metabolic grandparent to Trp (Walsh, Haynes *et al.* 2012).

Although anthranilate is not directly utilized for protein synthesis (it is a beta-, not an alpha-amino acid) it does serve as progenitor to a wide range of natural products, including tricyclic phenazines, tomaymycin, and actinomycins (Figure 5.23). Here we only focus on its starter role in a nested set of alkaloids, such as asperlicin E (Figure 5.24), that use both anthranilate and its granddaughter primary metabolite tryptophan (Walsh, Haynes *et al.* 2013). The six peptidyl alkaloids shown in Figure 5.24 encompass bicyclic to heptacyclic alkaloid frameworks. All are made by short, efficient two to three enzyme pathways, using only three kinds of enzyme catalysts. First are two-module or three-module NRPS assembly lines

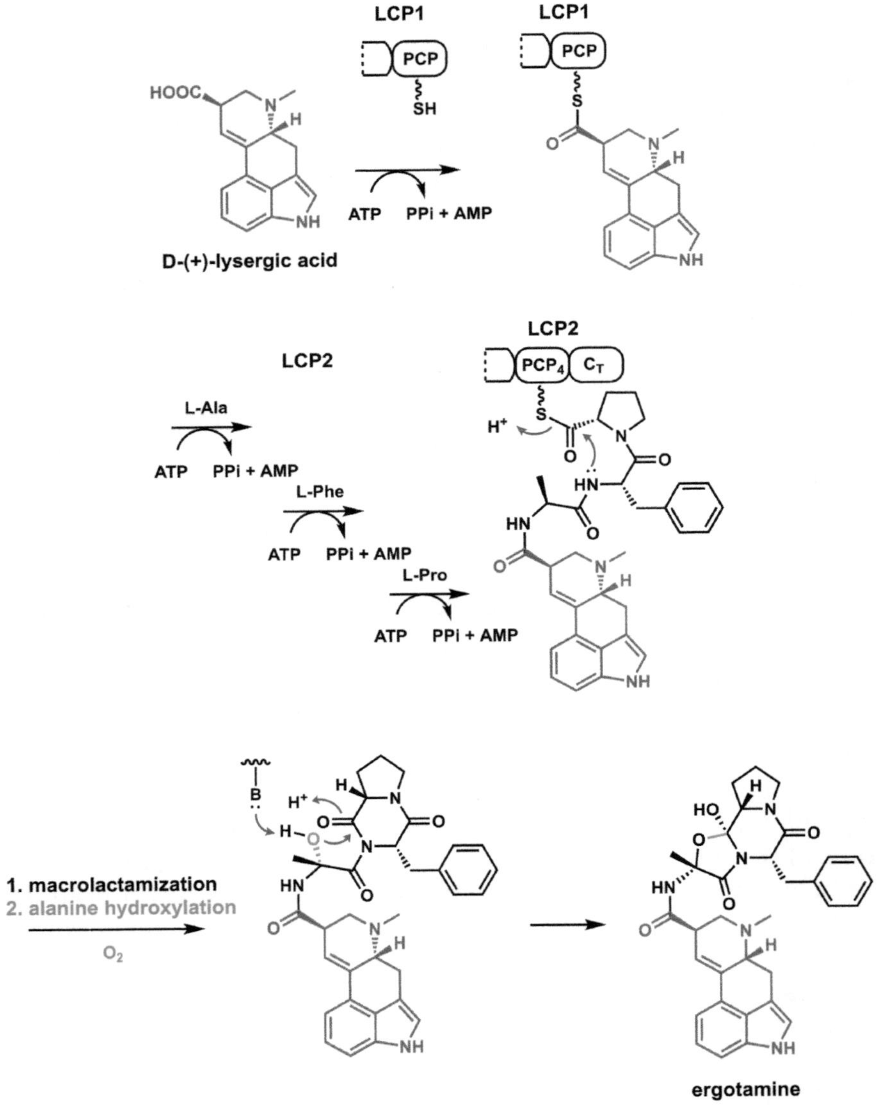

Figure 5.22 Lysergic acid is a unique starter unit for the four module ergotamine NRPS assembly line. Macrolactamizing chain release is followed by post assembly line Ala hydroxylation to the hydroxy-oxazolidinone ring of ergotamine.

where the first module is specific for anthranilate recognition, activation, and covalent tethering as the PCP thioester (Chapter 3). The second module, in turn, is specific for Ala or Trp.

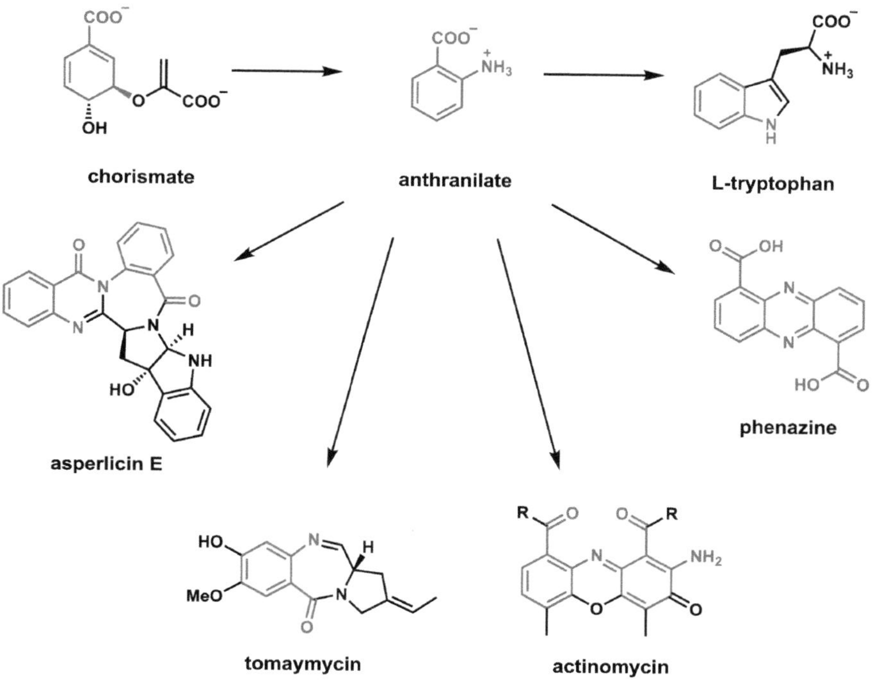

Figure 5.23 Anthranilate as a building block for an array of tricyclic to heptacyclic alkaloids.

5.6.1 Anthranilate to the Benzodiazepinedione Core

In the simplest example the two-module NRPS for aszonalenin makes an Ant-Trp-*S*-PCP_2 tethered thioester which is released by action of a terminal condensation domain by macrocyclization of the anthranilate (blue) free amine on the Trp_2 thioester carbonyl. Because anthranilate is a β- not an α-amino acid the released bicyclic amide, benzodiazepinedione, has an unusual seven-membered ring. Thus, anthranilate provides entry into the otherwise rare diazepine ring system (Figure 5.25). Only one additional enzyme is required to go from this 6-7-bicyclic ring system to the pentacyclic framework in aszonalenin, the fungal pathway end product. That is a prenyltransferase. It transfers the C_5 prenyl unit from dimethylallyl-PP to C_3 of the indole moiety of the benzodiazepinedione (Figure 5.25). The observed structure indicates that the allyl cation was captured at its tertiary $C_{3'}$ center rather than $C_{1'}$. Attack of the diazepine amide NH on that indoline adduct creates a new N–C bond and produces the 6-7-5-5-6 fused ring system of

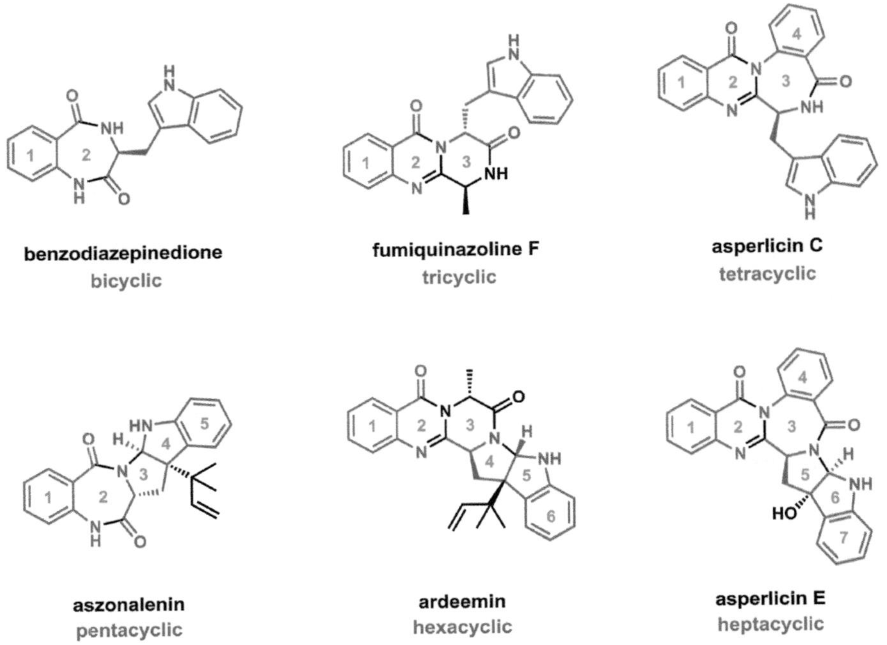

Figure 5.24 Anthranilate and tryptophan combine to form bicyclic to heptacyclic peptidyl alkaloid scaffolds in fungal metabolism.

aszonalenin: a remarkably efficient two enzyme system for product scaffold complexity.

5.6.2 Anthranilate to the Fumiquinazoline Core in Fungal Peptidyl Alkaloids

In addition to the benzodiazepinedione and its fully elaborated end product aszonalenin, Figure 5.24 also highlights in green the anthranilyl portion of the tricyclic benzoquinazolinone core of fumiquinazoline F. This is a constitutively produced secondary metabolite by the human pathogenic *Aspergillus fumigatus* (Ames, Haynes *et al.* 2011).

Fumiquinazoline F is fashioned on a trimodular *Aspergillus fumigatus* NRPS assembly line that activates anthranilate, tryptophan, and alanine in that order. The Trp residue that gets incorporated has the D-configuration, consistent with "on-assembly line" epimerization of the Ant-Trp-*S*-PCP intermediate (Chapter 3, see Figure 3.30). Figure 5.26 suggests that the trimodular NRPS system with macrocyclizing release should give out a 6,10-macrocycle as the nascent product. This is not detected: instead only the tricyclic 6-6-6 fumiquinazoline F

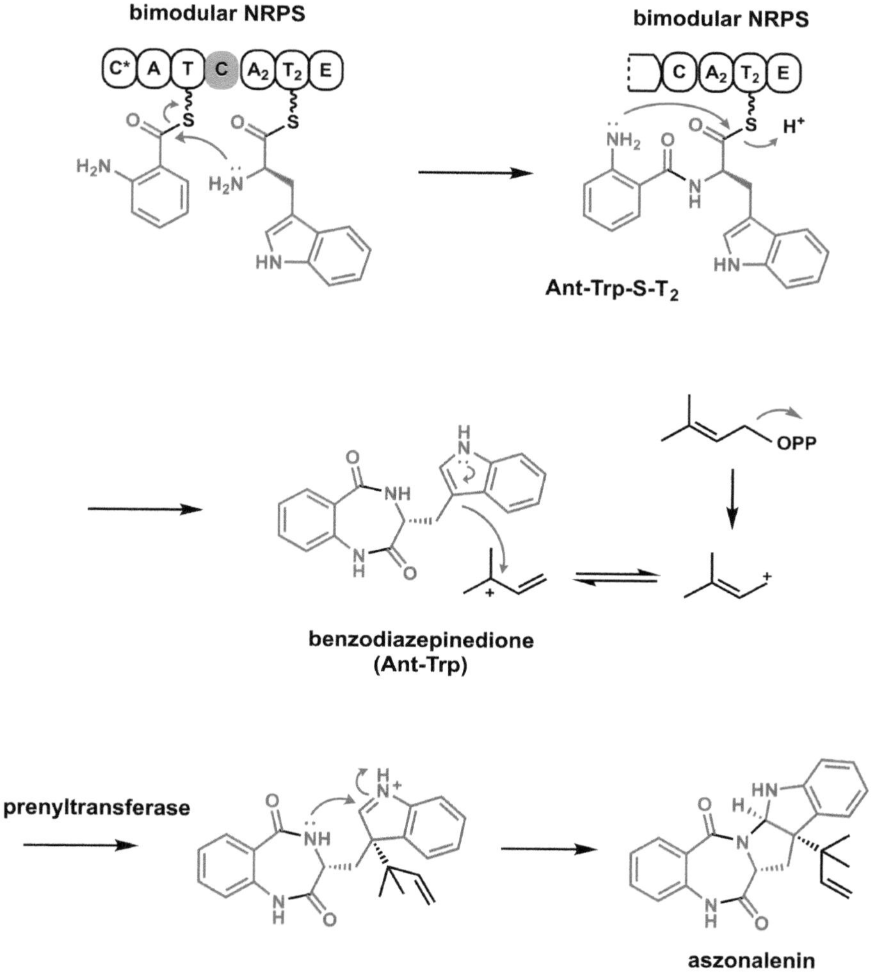

Figure 5.25 The benzodiazepinone 6,7-bicyclic framework is assembled on a two module fungal NRPS. Anthranilate is the chain-initiating unit. Chain release is by intramolecular macrolactamization. A prenyltransferase completes the biogenesis of aszonalenin by alkylation at C_3 of indole followed by cyclization of the diazepine NH on the iminium isomer of the indoline ring. The result is the fused 6-7-5-5-6 pentacyclic ring system of aszonalenin.

accumulates, even at the earliest time points. The proposal is therefore that the transannular attack of the amide NH on the Ala-derived carbonyl, followed by dehydration to give the FQF scaffold, is favored and occurs very rapidly. Close mechanistic analysis of this constitutive

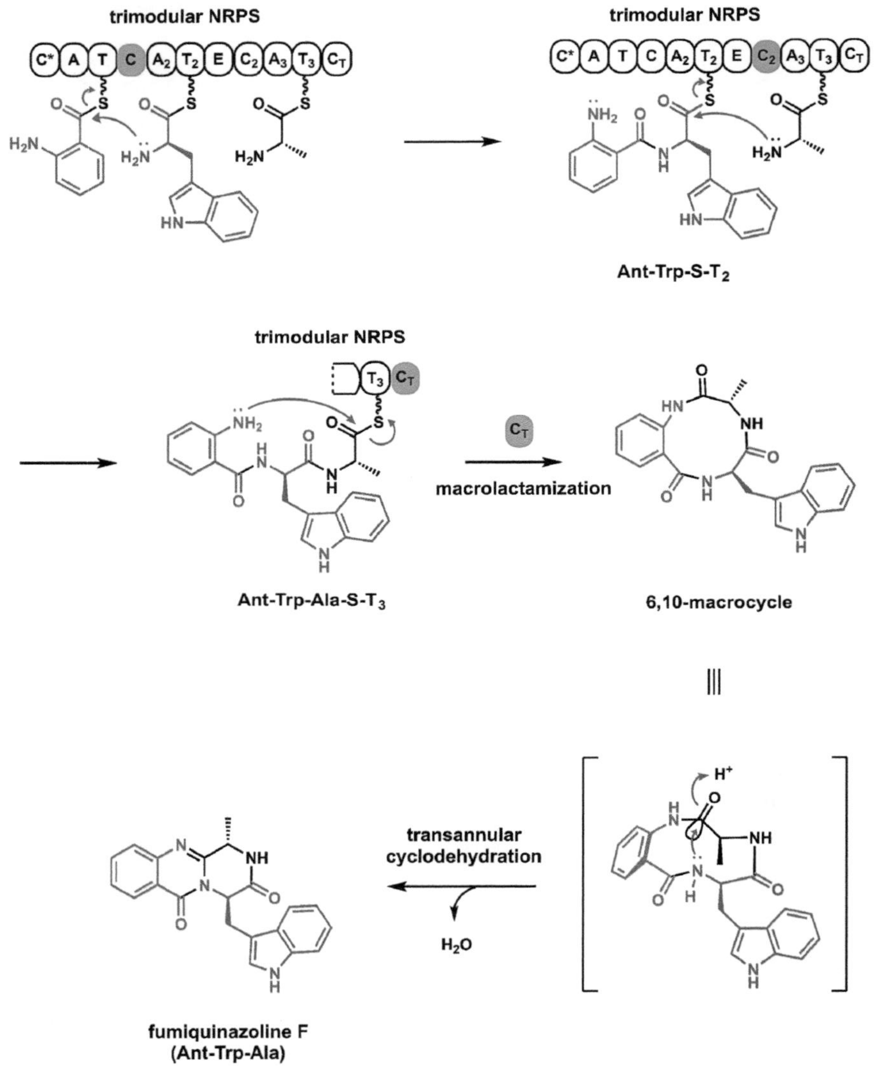

Figure 5.26 The trimodular NRPS activates anthranilate, Trp, and Ala, effecting head to tail macrolactamizing release. The nascent 6,10-macrocycle is proposed to undergo transannular cyclodehydration to the observed 6-6-6 tricyclic fumiquinazoline F.

fungal peptidyl alkaloid thus predicts cryptic medium ring formation and disappearance during product release.

Three further fumiquinazoline metabolites, fumiquinazoline A and fumiquinazolines C and D, are also elaborated by *A. fumigatus*

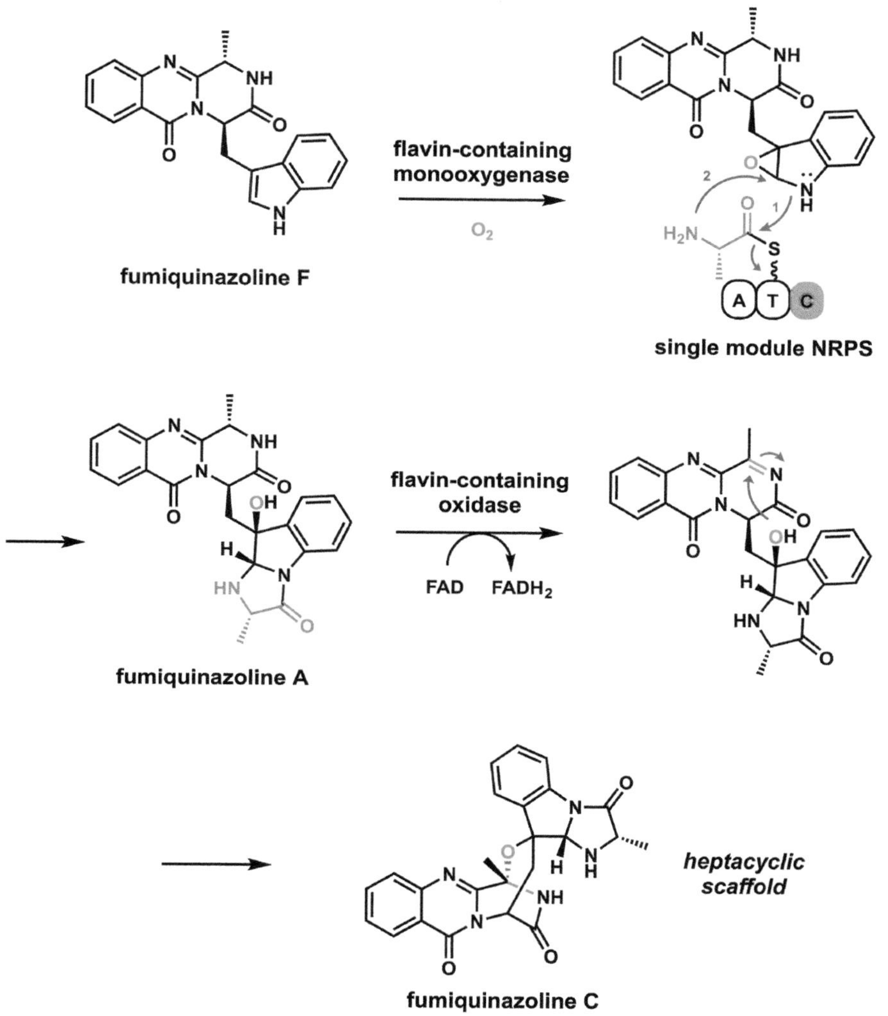

Figure 5.27 A three enzyme pathway takes fumiquinazoline F onto fumiquinazoline A by indole epoxygenation and annulation with another L-alanine to yield a tricyclic imidazolone-indole. The last enzyme in the pathway is a flavoprotein oxidase which generates an imine that is captured intramolecularly by the indicated –OH group to form the heptacyclic hemiaminal scaffold of fumiquinazoline C.

cultures and they have been found to arise from two-step enzymatic processing of fumiquinazoline F (Figure 5.27). The next enzyme is a monomodular NRPS that activates L-alanine as the Ala-*S*-PCP. It waits for a companion enzyme, an FAD-containing epoxygenase, to generate the 2,3-epoxide on the indole moiety of FQF. The amine of the activated Ala-*S*-PCP opens the epoxide and the indole nitrogen captures the activated Ala carbonyl. The result is a net annulation across N_1 and C_2 of the indole moiety with the epoxide oxygen ending up at C_3, converting the bicyclic indole to a 5-5-6 tricyclic ring system. This is fumiquinazoline A. The third enzyme then converts fumiquinazoline A to fumiquinazoline C. It is another FAD-containing enzyme. This one functions as an amine to imine oxidoreductase. The nascent imine is captured intramolecularly by the hydroxyl oxygen that in turn had been introduced as the transient epoxide to yield a stable six-membered hemiaminal. The resultant fumiquinazoline C is a heptacyclic peptidyl alkaloid. This molecular complexity is achieved in a short four enzyme fungal pathway (Walsh, Haynes *et al.* 2013).

Another fungal quinazoline in this class is the fused hexacyclic ardeemin, a blocker of the multidrug resistance efflux pump in tumor cells. A retrobiosynthetic analysis in Figure 5.28 suggests that a late

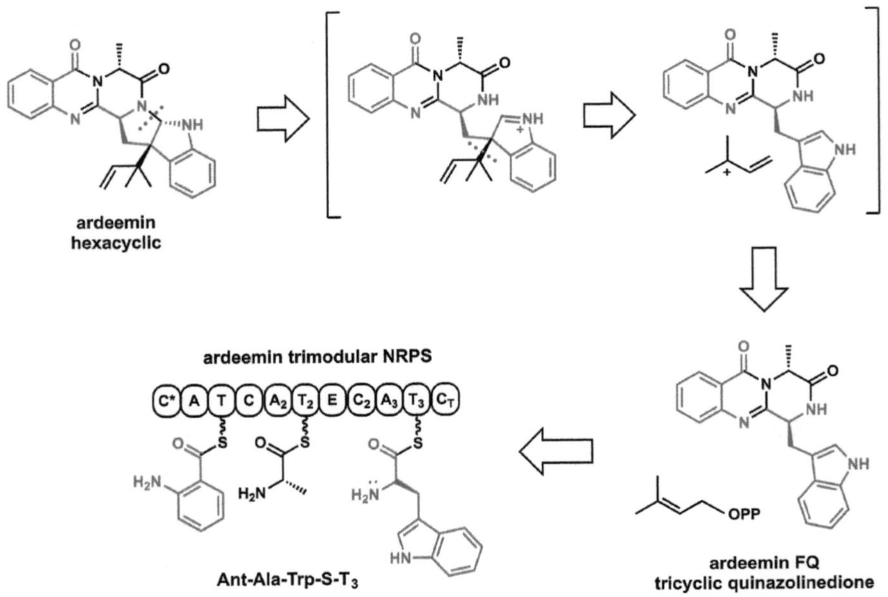

Figure 5.28 Retrobiosynthetic analysis of the hexacyclic ardeemin suggests a three module NRPS releasing an isomeric fumiquinazoline, followed by indole prenylation at C_3 and cyclization *via* attack of one of the quinazoline nitrogens on the indoline imine.

stage prenylation with "reverse" regiochemistry would generate the 5-5 connectivity as occurs in aszonalenin. The substrate for such prenylation would be a quinazoline isomeric to fumiquinazoline F. It would arise from a trimodular NRPS assembly line that reversed the order of selection of residues 2 and 3. The ardeemin NRPS in fact makes an Ant-D-Trp-L-Ala-tripeptidyl-*S*-enzyme. The release is presumed again to be two step, with a nascent 6,10 bicyclic product undergoing transannular capture and dehydration to the 6-6-6 tricyclic quinazoline.

Figure 5.29 summarizes a remarkable simplicity in the biosynthetic logic and catalytic machinery to this set of anthranilate- and tryptophan-containing fused polycyclic peptidyl alkaloids from fungi. Only three types of enzyme are used. The first stage involves two or three module NRPS assembly lines where the chain initiation module is selective for anthranilate. The post assembly line tailoring stage involves either a flavin-dependent indole epoxygenase or a prenyltransferase. In either case the nucleophilicity of C_3 of the indole moiety creates the attacking nucleophile and indole nitrogen alkylation follows, introducing scaffold complexity. One can generalize this as shown in Figure 5.30 as a ring-building strategy that utilizes indole chemical tendencies to react as a carbon nucleophile at C_3 and then as an electrophile at C_2 for net annulation.

5.7 Tryptophan to Indolocarbazole Alkaloids

Indolocarbazoles (Lounasmaa and Tolvanen 2000, Schmidt, Reddy *et al.* 2012) are bis-indole alkaloids connected by a benzene ring, as shown in Figure 5.31. Two molecules in this alkaloid class that have attracted much interest are staurosporine, isolated in 1977 (Nakano and Omura 2009), and rebeccamycin, a few years later (Sanchez, Mendez *et al.* 2006). These are typically bacterial and slime mold metabolites rather than higher plant alkaloids. Staurosporine is a potent inhibitor of many eukaryotic protein kinases, binding as an ATP competitor. It served as a first generation core scaffold for design of more selective kinase inhibitors, several of which have become first line agents in different therapeutic areas including cancer treatments. Rebeccamycin is not a protein kinase ligand but instead is an inhibitor of eukaryotic DNA topoisomerase I and is of interest as a scaffold for optimizing design of DNA-targeted anticancer agents.

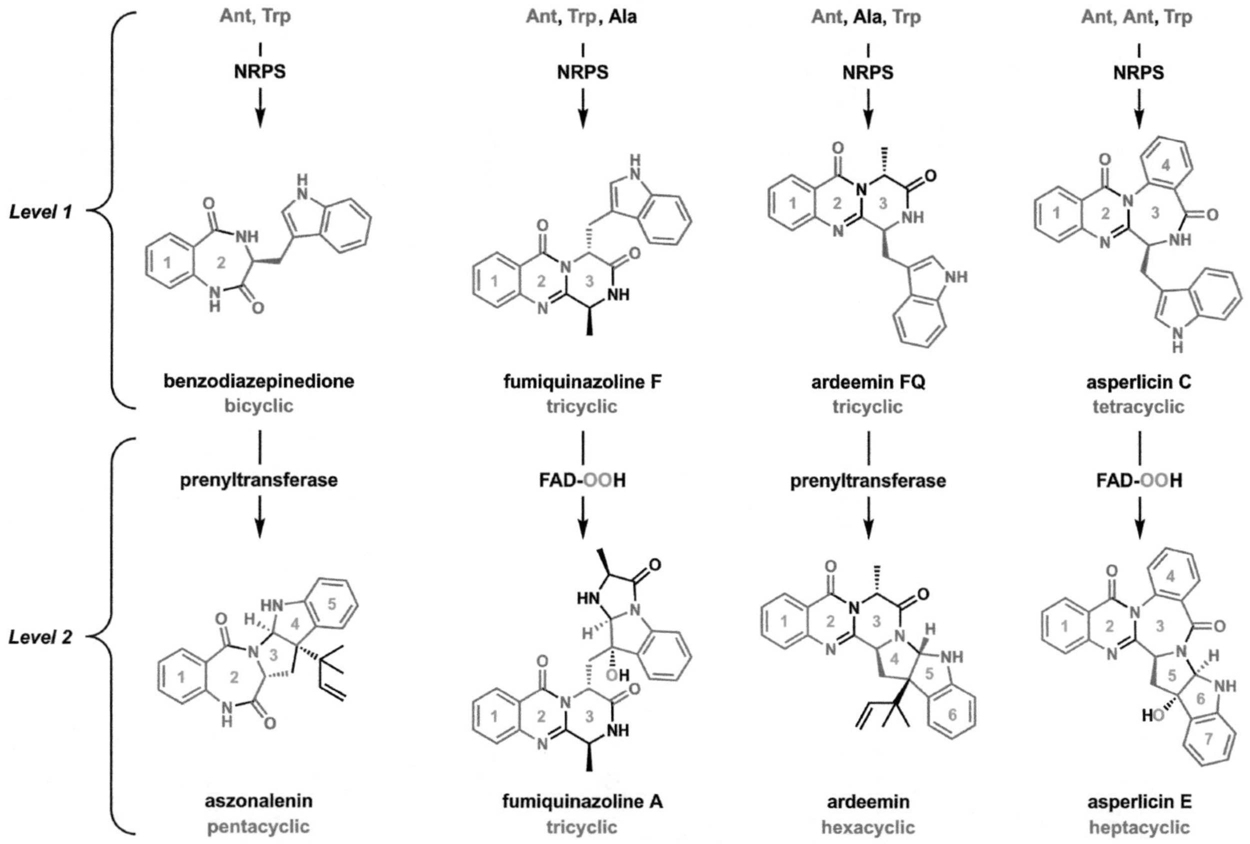

Figure 5.29 Simplicity in biosynthetic logic and catalytic machinery for building complex peptidyl alkaloids from anthranilate and tryptophan with a limited set of enzymes.

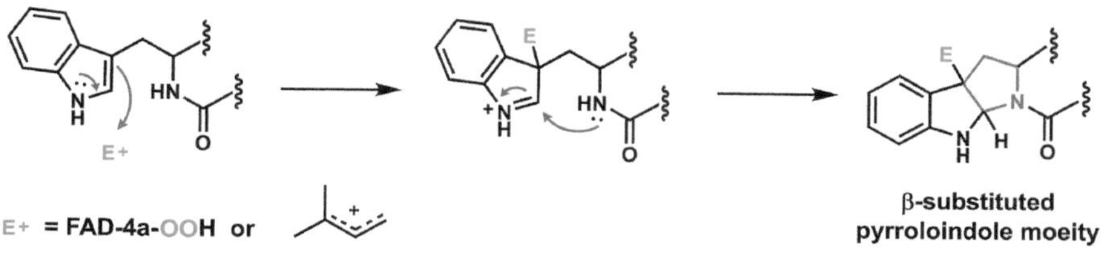

Figure 5.30 Biological ring building strategy by delivery of electrophiles at C_3 of the indole ring.

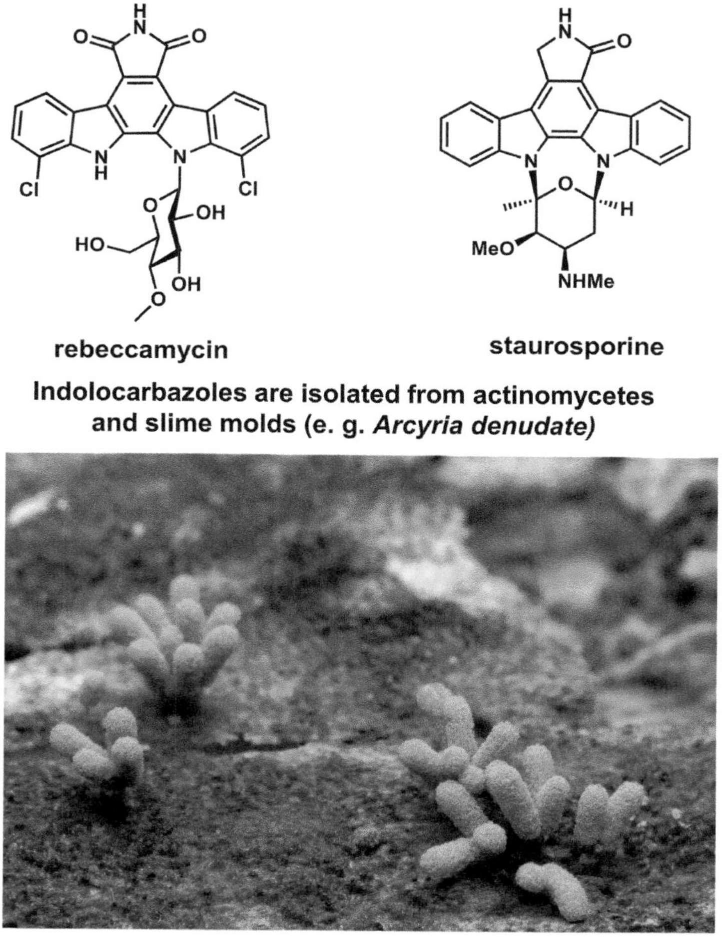

rebeccamycin
staurosporine
Indolocarbazoles are isolated from actinomycetes
and slime molds (e. g. *Arcyria denudate*)

Figure 5.31 Rebeccamycin and staurosporine: two representative indole-carbazoles isolated from bacteria and molds.
Photo credit: Pink slime mold, Katja Schulz, https://www.flickr.com/photos/treegrow/28904470335, License: CC-BY-2.0 (https://creativecommons.org/licenses/by/2.0/).

Rebeccamycin and staurosporine have the same pentacyclic indolocarbazole core but differ in three structural and functional aspects. First, the two indoles in rebeccamycin are chlorinated at the 7 position. Second, the five-member pyrrolidine ring sitting on top of the benzene bridge is at a different oxidation state in the two molecules, differing by four electrons. Third, rebeccamycin is *N*-glucosylated on one of the indole nitrogens while staurosporine has both indole nitrogens attached to a modified hexose (discussed in detail in Chapter 11).

Clearly, the logic of indolocarbazole biosynthesis must involve oxidative dimerization of tryptophan-derived indole rings. The chlorination step occurs at the level of free tryptophan by a two component $FADH_2$-dependent halogenase (Yeh, Garneau *et al.* 2005, Yeh, Cole *et al.* 2006). These are oxygen-dependent halogenation catalysts that are examined in detail in Chapter 9 (see Figure 9.42). The first committed enzyme in the indolocarbazole pathway is an FAD-dependent tryptophan amino acid oxidase, generating the indolepyruvate imine as initial product (Figure 5.32). Imino acids can undergo water attack and net hydrolysis to the ketone and ammonia. They can also isomerize to the enamine form which can act as a C_3 carbanion (and slows hydrolysis).

The second enzyme in the pathway (StaD or RebD) is a P450 type oxygenase (Howard-Jones and Walsh 2005). It condenses one molecule of indolepyruvate enamine as amine nucleophile with a second molecule of imine as electrophile before hydrolysis intervenes (Figure 5.32). The resultant dimeric adduct can expel NH_3. At this point in the catalytic cycle the one electron reaction manifold characteristic of P450 enzymes sets in. Abstraction of a hydrogen atom ($H^\bullet$) by the high valent oxo-iron center in the enzyme generates a carbon-centered radical that can close to a new C–C bond, on the way to the new pyrrole ring. Eventual transfer of a second $H^\bullet$ now to the Fe–OH heme completes formation of the heteroaromatic pyrrole, with subsequent double bond shifts into conjugation. This is chromopyrrolic acid. Two tryptophans have been oxidatively dimerized with one new C–C bond created by homolytic chemistry. Six electrons have been removed in proceeding from two tryptophan substrate molecules to the chromopyrrolic acid adduct.

The remaining two enzymes in the pathway, another FAD-enzyme and cytochrome P450 pair, RebC/StaC and RebP/StaP, convert the chromopyrrolic acids on to the final indolecarbazole structure (Figure 5.33). This involves one more C–C bond, again *via* radical chemistry mediated by the second P450 enzyme in the pathway, and a set of one or two oxidations of the newly generated pyrrole. The

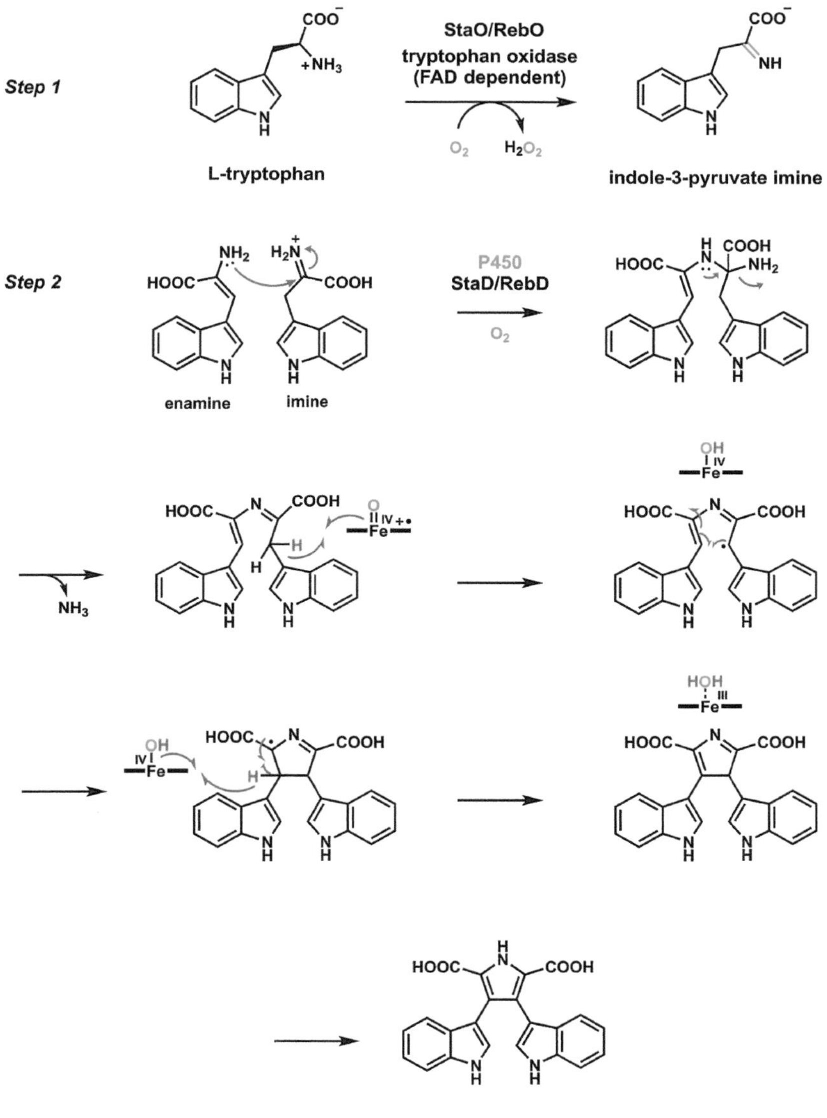

Figure 5.32 Tryptophan is enzymatically oxidized to indole-3-pyruvate. Dimerization occurs between the imine and enamine forms, involving homolytic C–C bond formation catalyzed by a P450 enzyme, yielding chromopyrrolic acid.

details of the conversion of dicarboxy-pyrrole to the maleimide in rebeccamycin or the half imide in staurosporine are not fully understood.

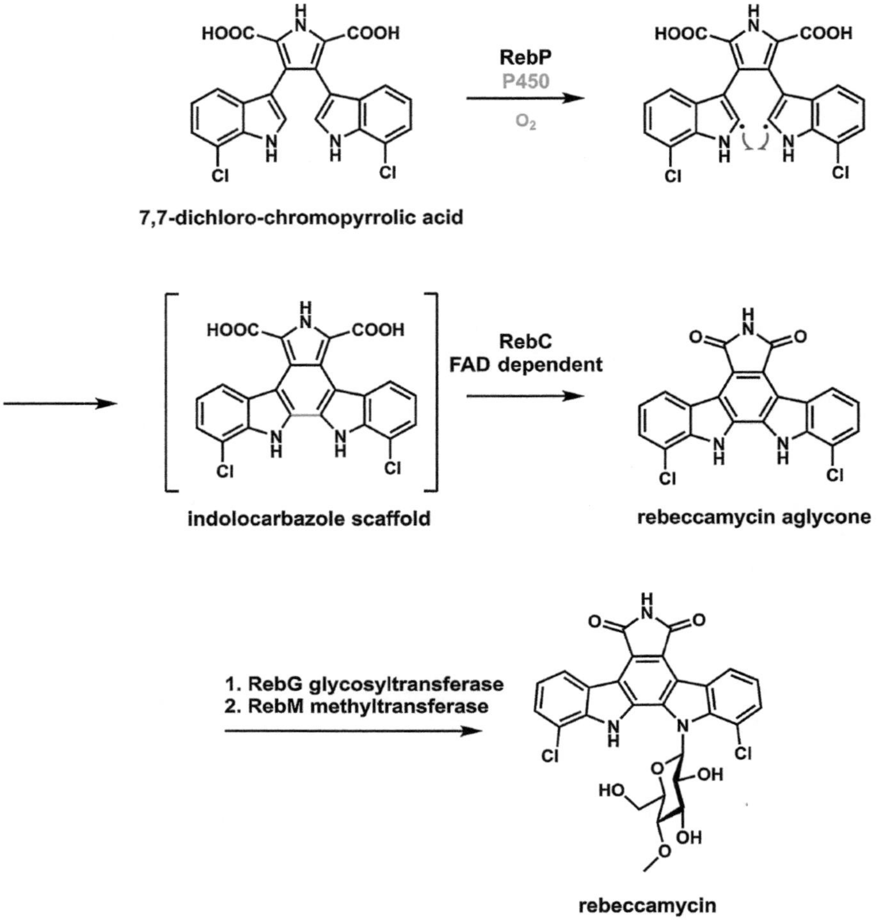

Figure 5.33 Conversion of the chromopyrrolic acid intermediate to the fused hexacyclic scaffold of indolocarbazoles requires another iron enzyme-catalyzed homolytic C–C bond formation. The aglycone is then converted to rebeccamycin by *N*-glycosylation.

The creation of the new C–C bond again is thought to proceed *via* one electron reaction manifolds, as diagrammed in Figure 5.33 in which a pair of C_2 indole radicals form the C–C bond. An alternate route would be creation of a single radical by H atom removal from C of one indole, formation of the C–C single bond bridge by participation of the 2,3-π electrons of the other indole and subsequent H atom transfer to iron. The two CYP P450s in the Reb and Sta pathway are responsible for the two new C–C bonds, *via* one electron reaction manifolds. In neither case does molecular oxygen, the required co-substrate, end up in product. These are two examples of the separation of substrate carbon radical formation from any OH$^{\bullet}$ capture (see Chapter 9 for fuller context).

5.8 Tryptophan Oxidative Dimerization to Terrequinone

A distinct mode of tryptophan oxidative dimerization logic and enzymatic machinery exists for the generation of the prenylated benzoquinone terrequinone A (Figure 5.34). Tryptophan is again oxidized at carbon 2 but this time *via* a PLP-dependent transaminase catalyst TdiD which directly generates indole pyruvate without the intermediacy of the imine (Balibar, Howard-Jones *et al.* 2007). This is one way of ruling out imine/enamine mediated dimerization chemistry. A second difference is that the *keto acid* is then selected, activated and tethered on a monomodular NRPS enzyme TdiA with three domains, adenylation-peptidyl carrier protein-thioesterase (A-PCP-TE). The selection and activation of a keto acid instead of an amino acid is observed in some other NRPS assembly line variants including cereulide and valinomycin synthases as well as bacillaene (Magarvey, Ehling-Schulz *et al.* 2006, Calderone, Bumpus *et al.* 2008).

Once an indolepyruvyl group has been moved from PCP to the TE domain as an oxoester, another indolepyruvate can be loaded onto the PCP domain. This pair can now undergo a directed intermolecular Claisen condensation, with enol as nucleophile, thioester as electrophile. This results in C–C bond formation. A second, intramolecular Claisen condensation generates a second C–C bond and leads to product release, shown as the bis-indolyl-tetraketocyclohexane, and then its favored isomerization to the quinone.

Thus, two features of this NRPS-mediated dimerization/release pathway are unusual. First, it is the keto acid not the amino acid that is the NRPS substrate and, second, this was the first example of a Claisen condensation/cyclization as release step. Two subsequent prenylations are catalyzed by TdiB and TdiE, one at C_2 of the benzoquinone and one at C_2 of one of the indole rings, to produce terrequinone A. While C_2 of the indole ring can act as carbon nucleophile toward the DMAPP-derived allyl cation, the quinone cannot. Indeed, there is an NADH-consuming reductase TdiC which generates the hydroquinone transiently. In that oxidation state C_2 is nucleophilic and can undergo prenylation.

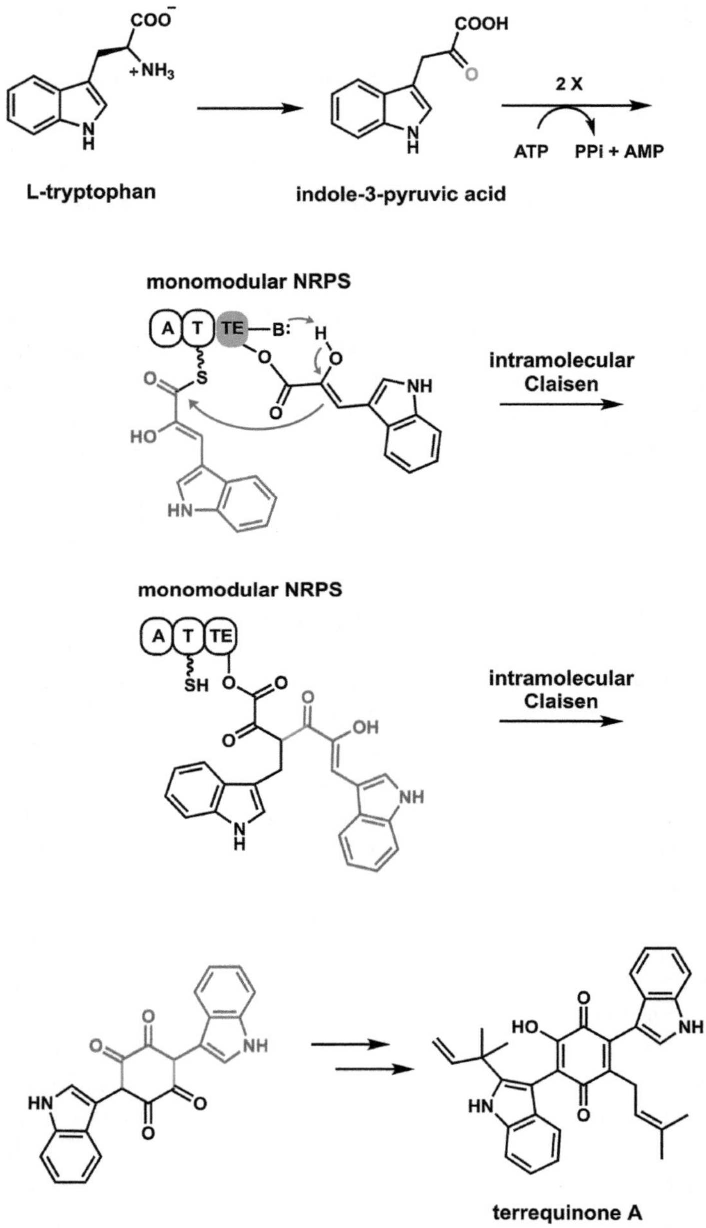

Figure 5.34 Tandem Claisen condensations on indole-3-pyruvates covalently tethered to TdiA yield the bis-indolyl benzoquinone as released product.

5.9 Additional Alkaloids: Steroidal Alkaloids

A substantial range of steroid alkaloids are known that derive metabolically from cholesterol. Solanidine (Figure 5.35), and its various glycosides, is the most predominant member of this subclass in potato flowers, leaves, and sprouts. The 6-6-6-5 tetracyclic framework of cholesterol has been converted to the hexacyclic 6-6-6-5-5-6 ring system in solanidine. Also, cholesterol is nitrogen free while the alkaloid product has incorporated one nitrogen, at the stage of

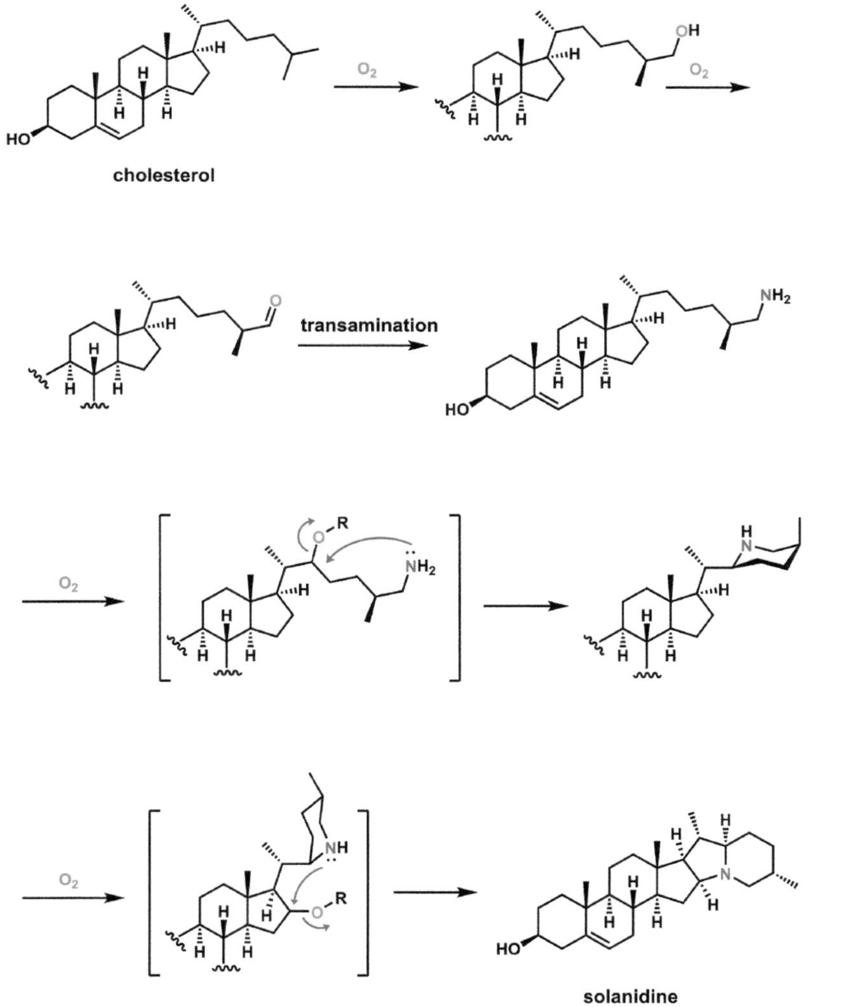

Figure 5.35 Steroidal alkaloids: solanidine formation from cholesterol.

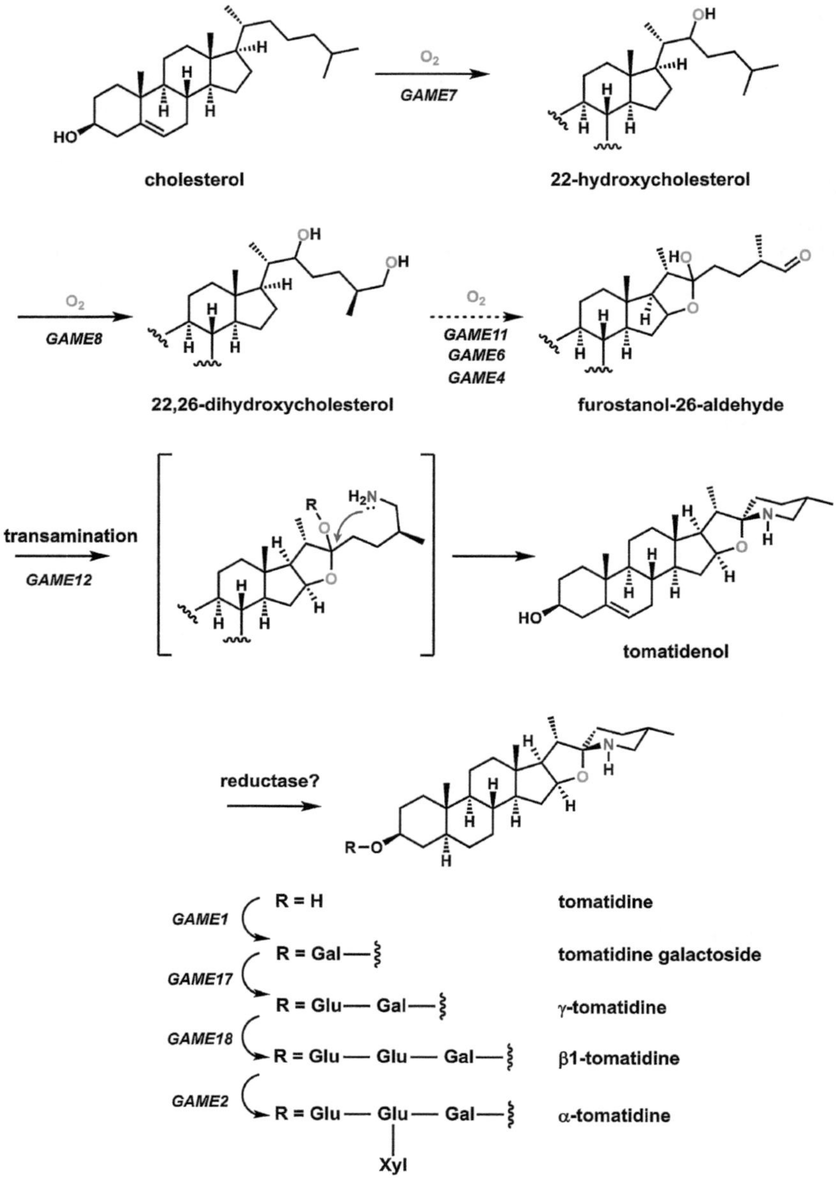

Figure 5.36 Steroidal alkaloids: formation of the tomatidines from cholesterol involves possibly clustered genes.

the aldehyde intermediate, by action of a transaminase with arginine as donor of the amine group. Two amine mediated displacements of –OH groups have been proposed (Dewick 2009) as the fifth and then the sixth ring-forming cyclization steps. It is not clear whether this formalism reflects direct –OH displacement or occurs only after conversion metabolically to a phosphate or acetyl ester to make a better set of leaving groups. Recent studies have identified a set of clustered genes encoding glycoalkaloid metabolizing enzymes (GAMEs) with predicted oxygenase, reductase, and glycosyltransferase activities that have led to the scheme shown in Figure 5.36 leading to the glycosylated α-,β-, and γ-tomatines.

A second infamous steroidal alkaloid is cyclopamine, from *Veratrum californicum* plants, which causes craniofacial defects in grazing sheep, most notably the monocular phenotype reminiscent of the Cyclops in the Odyssey. Figure 5.37 shows a suggested

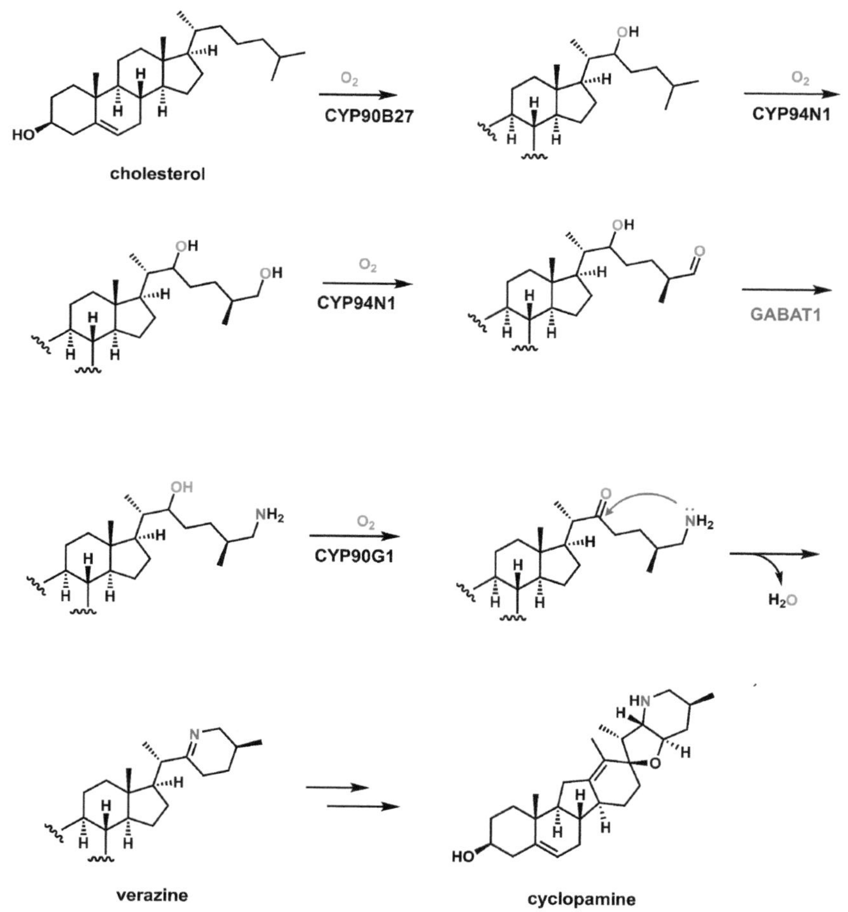

Figure 5.37 Steroidal alkaloids: cyclopamine formation from cholesterol.

biosynthetic route from cholesterol to cyclopamine (Augustin, Ruzicka *et al.* 2015). In some analogy to the solanidine case above, the cholesterol side chain is oxygenated up to the terminal aldehyde and then transaminated from γ-aminobutyric caid (GABA) to introduce the nitrogen atom. Further oxidation of C_{23} to the ketone sets up formation of the cyclic six-membered imine to give verazine on the way to the final hexacyclic scaffold in cyclopamine.

5.10 Summary

There are two main routes for incorporating amino acid building blocks into natural product scaffolds. The NRPS assembly lines of Chapter 3 use both proteinogenic and many nonproteinogenic amino acids as building blocks. Macrocycles with peptide backbones are common products, usually subject to post assembly line maturation by sets of dedicated tailoring enzymes.

The second route is alkaloid biogenesis, of a larger variety of product scaffolds than from the NRPS routes. In addition to shunting some of the cellular pools of proteinogenic Lys, Phe, Tyr, Trp, and His into nitrogen-containing heterocyclic alkaloids, the nonproteinogenic ornithine and anthranilate also serve as entry points.

Among the frequently used inventory of enzymatic transformations in alkaloid biosynthetic pathways are amino acid decarboxylases, Mannich condensation enzymes involving an amine, a carbonyl partner (typically an aldehyde) and a carbanion equivalent. Methyltransferases donate [CH_3^+] equivalents from SAM to nucleophilic N and O atoms. Acyl transferases (*e.g.* cocaine) add acetyl or benzoyl groups from the corresponding acyl-CoAs to nitrogen and oxygen atoms in cosubstrates.

Less common and mechanistically notable enzymatic transformations in alkaloid pathways include C–C bond forming enzymes. These include the berberine bridge enzyme, the Pictet–Spengler coupling in strictosidine synthase, and the proposed decarboxylative vinylogous epoxide opening in building the lysergic acid framework.

A variety of C–C bonds are formed by *homolytic mechanisms* under the catalytic action of cytochrome P450 enzymes. These include radical couplings in tubocurarine dimerization, in *R*-reticuline to salutaradinol, and in the steps to chromopyrrolic acid and then to the final indolecarbazoles in rebeccamycin and staurosporine construction. Integrated across the sweep of known alkaloid structures, this class of natural products encompasses essentially all the known nitrogen-containing heterocyclic scaffolds, from monocyclic to heptacyclic frameworks, as illustrated in this chapter and additionally in Chapter 8 for complex indole terpene scaffolds.

References

Ames, B. D., S. W. Haynes, X. Gao, B. S. Evans, N. L. Kelleher, Y. Tang and C. T. Walsh (2011). "Complexity generation in fungal peptidyl alkaloid biosynthesis: oxidation of fumiquinazoline A to the heptacyclic hemiaminal fumiquinazoline C by the flavoenzyme Af12070 from Aspergillus fumigatus". *Biochemistry* **50**(40): 8756–8769.

Augustin, M. M., D. R. Ruzicka, A. K. Shukla, J. M. Augustin, C. M. Starks, M. O'Neil-Johnson, M. R. McKain, B. S. Evans, M. D. Barrett, A. Smithson, G. K. Wong, M. K. Deyholos, P. P. Edger, J. C. Pires, J. H. Leebens-Mack, D. A. Mann and T. M. Kutchan (2015). "Elucidating steroid alkaloid biosynthesis in Veratrum californicum: production of verazine in Sf9 cells". *Plant J.* **82**(6): 991–1003.

Balibar, C. J., A. R. Howard-Jones and C. T. Walsh (2007). "Terrequinone A biosynthesis through L-tryptophan oxidation, dimerization and bisprenylation". *Nat. Chem. Biol.* **3**(9): 584–592.

Barleben, L., S. Panjikar, M. Ruppert, J. Koepke and J. Stockigt (2007). "Molecular architecture of strictosidine glucosidase: the gateway to the biosynthesis of the monoterpenoid indole alkaloid family". *Plant Cell* **19**(9): 2886–2897.

Brown, S., M. Clastre, V. Courdavault and S. E. O'Connor (2015). "De novo production of the plant-derived alkaloid strictosidine in yeast". *Proc. Natl. Acad. Sci. U. S. A.* **112**(11): 3205–3210.

Calderone, C. T., S. B. Bumpus, N. L. Kelleher, C. T. Walsh and N. A. Magarvey (2008). "A ketoreductase domain in the PksJ protein of the bacillaene assembly line carries out both alpha- and beta-ketone reduction during chain growth". *Proc. Natl. Acad. Sci. U. S. A.* **105**(35): 12809–12814.

Cseke, L. J., A. Kirakosyan, P. B. Kaufman, S. Warber, J. A. Duje and H. L. Brielmann (2016). *Natural Products from Plants*, 2nd edn, CRC Press.

Devane, W. A., L. Hanus, A. Breuer, R. G. Pertwee, L. A. Stevenson, G. Griffin, D. Gibson, A. Mandelbaum, A. Etinger and R. Mechoulam (1992). "Isolation and structure of a brain constituent that binds to the cannabinoid receptor". *Science* **258**(5090): 1946–1949.

Dewick, P. (2009). *Medicinal Natural Products, A Biosynthetic Approach*. UK, Wiley.
Galanie, S., K. Thodey, I. J. Trenchard, M. Filsinger Interrante and C. D. Smolke (2015). Complete biosynthesis of opioids in yeast''. *Science* **349**(6252): 1095–1100.
Gao, X., S. W. Haynes, B. D. Ames, P. Wang, L. P. Vien, C. T. Walsh and Y. Tang (2012). ''Cyclization of fungal nonribosomal peptides by a terminal condensation-like domain''. *Nat. Chem. Biol.* **8**(10): 823–830.
Gaute, J. and R. Odell (1996). *The New Murderer's Who's Who*. London, Harrap Books.
Grobe, N., B. Zhang, U. Fisinger, T. M. Kutchan, M. H. Zenk and F. P. Guengerich (2009). ''Mammalian cytochrome P450 enzymes catalyze the phenol-coupling step in endogenous morphine biosynthesis''. *J. Biol. Chem.* **284**(36): 24425–24431.
Havemann, J., D. Vogel, B. Loll and U. Keller (2014). ''Cyclolization of D-lysergic acid alkaloid peptides''. *Chem. Biol.* **21**(1): 146–155.
Herbert, R. (1989). *The Biosynthesisof Secondary Metabolites*, Springer.
Hesse, M. (2002). *Alkaloids: Nature's Curse or Blessing?*, John Wiley & Sons.
Hill, R. and A. Yudin (2006). ''Making Carbon-Nitrogen Bonds on Biological and Chemical Synthesis''. *Nat. Chem. Biol.* **2**: 284–287.
Howard-Jones, A. R. and C. T. Walsh (2005). ''Enzymatic generation of the chromopyrrolic acid scaffold of rebeccamycin by the tandem action of RebO and RebD''. *Biochemistry* **44**(48): 15652–15663.
Humphrey, A. J. and D. O'Hagan (2001). ''Tropane alkaloid biosynthesis. A century old problem unresolved''. *Nat. Prod. Rep.* **18**(5): 494–502.
Ilari, A., S. Franceschini, A. Bonamore, F. Arenghi, B. Botta, A. Macone, A. Pasquo, L. Bellucci and A. Boffi (2009). ''Structural basis of enzymatic (S)-norcoclaurine biosynthesis''. *J. Biol. Chem.* **284**(2): 897–904.
Jirschitzka, J., G. W. Schmidt, M. Reichelt, B. Schneider, J. Gershenzon and J. C. D'Auria (2012). ''Plant tropane alkaloid biosynthesis evolved independently in the Solanaceae and Erythroxylaceae''. *Proc. Natl. Acad. Sci. U. S. A.* **109**(26): 10304–10309.
Lin, C. C., T. Y. Chan and J. F. Deng (2004). ''Clinical features and management of herb-induced aconitine poisoning''. *Ann. Emerg. Med.* **43**(5): 574–579.
Lipscomb, J. D. (2008). ''Mechanism of extradiol aromatic ring-cleaving dioxygenases''. *Curr. Opin. Struct. Biol.* **18**: 644–649.
Loris, E. A., S. Panjikar, M. Ruppert, L. Barleben, M. Unger, H. Schubel and J. Stockigt (2007). ''Structure-based engineering of

strictosidine synthase: auxiliary for alkaloid libraries". *Chem. Biol.* **14**(9): 979–985.

Lounasmaa, M. and A. Tolvanen (2000). "Simple indole alkaloids and those with a nonrearranged monoterpenoid unit". *Nat. Prod. Rep.* **17**(2): 175–191.

Ma, X., S. Panjikar, J. Koepke, E. Loris and J. Stockigt (2006). "The structure of Rauvolfia serpentina strictosidine synthase is a novel six-bladed beta-propeller fold in plant proteins". *Plant Cell* **18**(4): 907–920.

Magarvey, N. A., M. Ehling-Schulz and C. T. Walsh (2006). "Characterization of the cereulide NRPS alpha-hydroxy acid specifying modules: activation of alpha-keto acids and chiral reduction on the assembly line". *J. Am. Chem. Soc.* **128**(33): 10698–10699.

Maresh, J. J., L. A. Giddings, A. Friedrich, E. A. Loris, S. Panjikar, B. L. Trout, J. Stockigt, B. Peters and S. E. O'Connor (2008). "Strictosidine synthase: mechanism of a Pictet-Spengler catalyzing enzyme". *J. Am. Chem. Soc.* **130**(2): 710–723.

Nakano, H. and S. Omura (2009). "Chemical biology of natural indolocarbazole products: 30 years since the discovery of staurosporine". *J. Antibiot.* **62**(1): 17–26.

Panjikar, S., J. Stoeckigt, S. O'Connor and H. Warzecha (2012). "The impact of structural biology on alkaloid biosynthesis research". *Nat. Prod. Rep.* **29**(10): 1176–1200.

Peissel, M. (1984). *The Ants' Gold. The Discovery of the Greek El Dorado in the Himalayas*. London, Harvill Press.

Pelletier, S. W. (1983). *Alkaloids: Chremical and Biological Perspectives*. New York, John Wiley & Sons.

Robinson, R. (1952). "Molecular structure of Strychnine, Brucine and Vomicine". *Prog. Org. Chem.* **1**: 2.

Sanchez, C., C. Mendez and J. A. Salas (2006). "Indolocarbazole natural products: occurrence, biosynthesis, and biological activity". *Nat. Prod. Rep.* **23**(6): 1007–1045.

Sato, F., T. Hashimoto, A. Hachiya, K. Tamura, K. B. Choi, T. Morishige, H. Fujimoto and Y. Yamada (2001). "Metabolic engineering of plant alkaloid biosynthesis". *Proc. Natl. Acad. Sci. U. S. A.* **98**(1): 367–372.

Schiff, P. L. (2006). "Ergot and its Alkaloids". *Am. J. Pharm. Educ.* **70**: 1–10.

Schmidt, A. W., K. R. Reddy and H. J. Knolker (2012). "Occurrence, biogenesis, and synthesis of biologically active carbazole alkaloids". *Chem. Rev.* **112**(6): 3193–3328.

Schmidt, G. W., J. Jirschitzka, T. Porta, M. Reichelt, K. Luck, J. C. Torre, F. Dolke, E. Varesio, G. Hopfgartner, J. Gershenzon and J. C. D'Auria (2015). "The last step in cocaine biosynthesis is catalyzed by a BAHD acyltransferase". *Plant Physiol.* **167**(1): 89–101.
Seigler, D. (2012). *Plant Secondary Metabolism*, Springer Science and Business Media: 759.
Stadler, R., T. Kutchan and M. H. Zenk (1989). "(S)-norcoclaurine is the central intermediate in benzylisoquinoline alkaloid biosynthesis". *Phytochemistry* **4**: 1083–1086.
Thackray, S. J., C. G. Mowat and S. K. Chapman (2008). "Exploring the mechanism of tryptophan 2,3-dioxygenase". *Biochem. Soc. Trans.* **36**(Pt 6): 1120–1123.
Walsh, C. T., S. W. Haynes and B. D. Ames (2012). "Aminobenzoates as building blocks for natural product assembly lines". *Nat. Prod. Rep.* **29**(1): 37–59.
Walsh, C. T., S. W. Haynes, B. D. Ames, X. Gao and Y. Tang (2013). "Short pathways to complexity generation: fungal peptidyl alkaloid multicyclic scaffolds from anthranilate building blocks". *ACS Chem. Biol.* **8**(7): 1366–1382.
Weld, M., J. Ziegler and T. Kutchan (2004). "Morphine biosynthesis in the opium poppy, Papaver somniferum". *Proc. Natl. Acad. Sci. U. S. A.* **101**: 13957–13962.
Winkler, A., A. Lyskowski, S. Riedl, M. Puhl, T. M. Kutchan, P. Macheroux and K. Gruber (2008). "A concerted mechanism for berberine bridge enzyme". *Nat. Chem. Biol.* **4**(12): 739–741.
Winzer, T., M. Kern, A. J. King, T. R. Larson, R. I. Teodor, S. L. Donninger, Y. Li, A. A. Dowle, J. Cartwright, R. Bates, D. Ashford, J. Thomas, C. Walker, T. A. Bowser and I. A. Graham (2015). "Plant science. Morphinan biosynthesis in opium poppy requires a P450-oxidoreductase fusion protein". *Science* **349**(6245): 309–312.
Woodward, R. B., M. P. Cava, W. D. Ollis, A. Hunger, H. U. Daeniker and K. Schenker (1963). "The total synthesis of strychnine". *Tetrahedron Lett.* **19**: 247–288.
Yeh, E., L. J. Cole, E. W. Barr, J. M. Bollinger, Jr., D. P. Ballou and C. T. Walsh (2006). "Flavin redox chemistry precedes substrate chlorination during the reaction of the flavin-dependent halogenase RebH". *Biochemistry* **45**(25): 7904–7912.
Yeh, E., S. Garneau and C. T. Walsh (2005). "Robust in vitro activity of RebF and RebH, a two-component reductase/halogenase, generating 7-chlorotryptophan during rebeccamycin biosynthesis". *Proc. Natl. Acad. Sci. U. S. A.* **102**(11): 3960–3965.

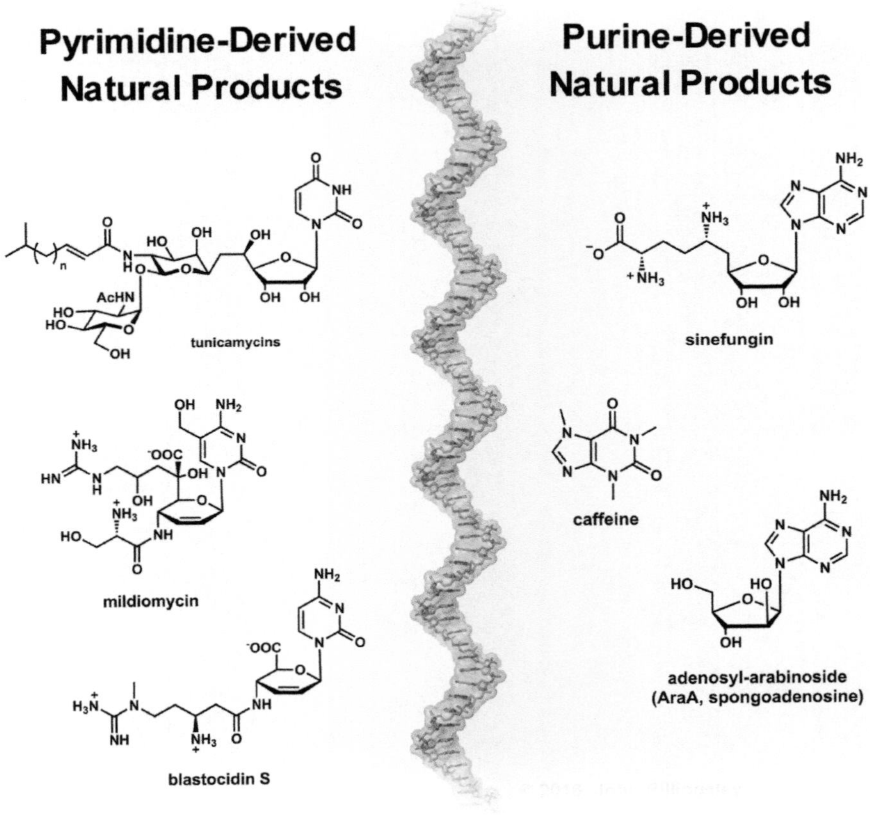

Purine and pyrimidine derived natural products.
Copyright (2016) John Billingsley.

6 Purine- and Pyrimidine-derived Natural Products

6.1 Introduction

From the perspective of the alkaloid natural product family just discussed in Chapter 5, the four molecules in Figure 6.1 also qualify as alkaloids in fitting the remarkably permissive definition of at least one nitrogen atom in a heterocyclic ring. Thus, replacement of one carbon atom in a benzene ring gives pyridine, while replacement of two in a 1, 3-relationship yields the pyrimidine ring. The fused bicyclic imidazolopyrimidine has the familiar name of purine while the 7-deazapurine would be a pyrrolopyrimidine ring system.

From a larger context of both primary and secondary metabolism, one might continue to focus on pyridine and pyrrolopyrimidine rings as quintessential alkaloid scaffolds, although pyridines are at the core of two of the most common coenzymes, nicotinamides and pyridoxal phosphate, of primary metabolism, so few biological chemists would name them first as alkaloid cores.

This is even more valid for the pyrimidine and purine ring systems which comprise the informational cores of both RNA and DNA macromolecules. So alkaloids they may be in a formal sense, but we treat them in this separate chapter, given their metabolic centrality. As noted below and throughout this chapter, the *de novo* biosynthetic assembly of pyrimidine and purine scaffolds goes through the ribonucleotides and deoxyribonucleotides. The free pyrimidines and purines do not tend to accumulate except as specific enzymatic

Natural Product Biosynthesis: Chemical Logic and Enzymatic Machinery
By Christopher T. Walsh and Yi Tang

Published by the Royal Society of Chemistry, www.rsc.org

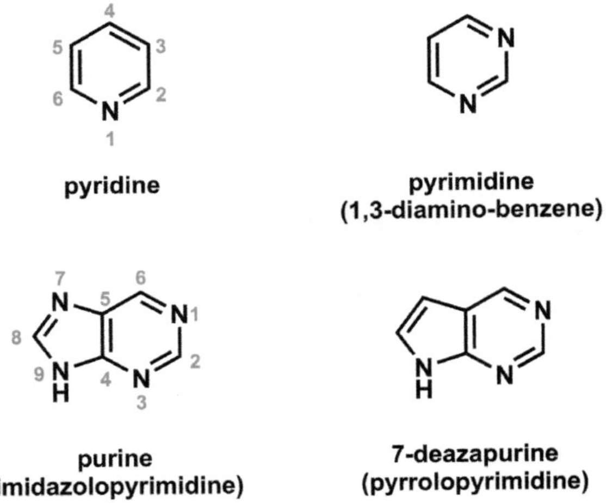

Figure 6.1 Structures of pyridine, pyrimidine, purine, and 7-deazapurine.

degradation products of enzymatic trimming: (deoxy) nucleotides to (deoxy) nucleosides and the nucleosides then onto the free bases and ribose.

6.2 Pairing of Specific Purines and Pyrimidines in RNA and DNA

The key biologic constraints on purine and pyrimidine abundance and composition are the sequences of specific DNA and RNA molecules. It was, of course, the set of deductions by Watson and Crick on the structure of DNA that explained the constant adenine : thymine and guanine : cytosine ratios as two sets of base pairs in the double helix. The information content in every gene is determined by the order and number of those two base pairs (Figure 6.2). In turn, the enzymatic transcription of any gene into RNA, coding, noncoding, regulatory, and structural functions, transmits the DNA sequence of one of the strands into an RNA copy.

Figure 6.3 shows the structure of the two purines found in DNA, and also in RNA: adenine (A) and guanine (G). The corresponding pyrimidines in RNA are uracil (U) and cytosine (C). In DNA, the 5-methyluracil ring system, thymine (T), replaces U while the 2′-OH of ribose is replaced by the 2′-deoxy hydrogen. Actually shown are the

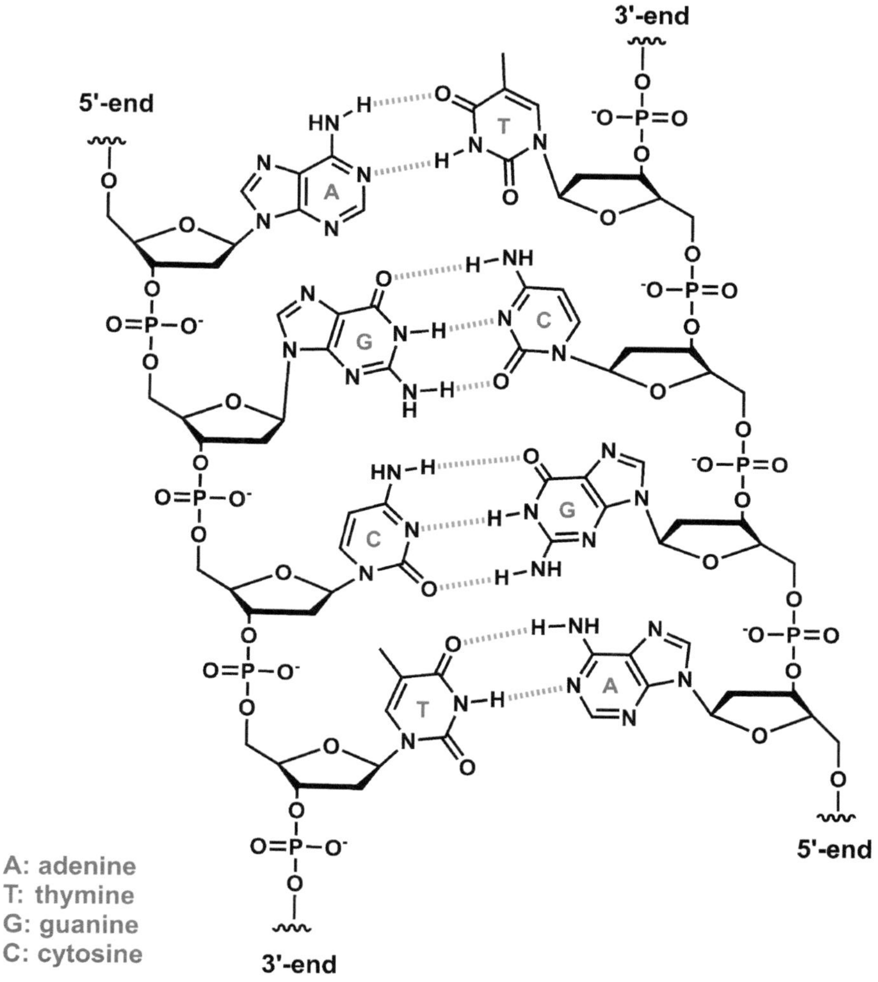

Figure 6.2 Purine and pyrimidine hydrogen-bonding lead to base-pairing in DNA and RNA.

nucleoside monophosphates for A (adenosine monophosphate, AMP), G (guanosine monophosphate, GMP) and the corresponding triphosphates for U (uridine triphosphate, UTP), C (cytidine triphosphate, CTP), and 2′-OH-deoxy T (2′-deoxythymidine triphosphate, dTTP). The nomenclature (Figure 6.4) for the free bases is adenine and uracil, the base–ribose combinations are known as nucleosides (adenosine and uridine), and the tripartite base–ribose–phosphate esters known

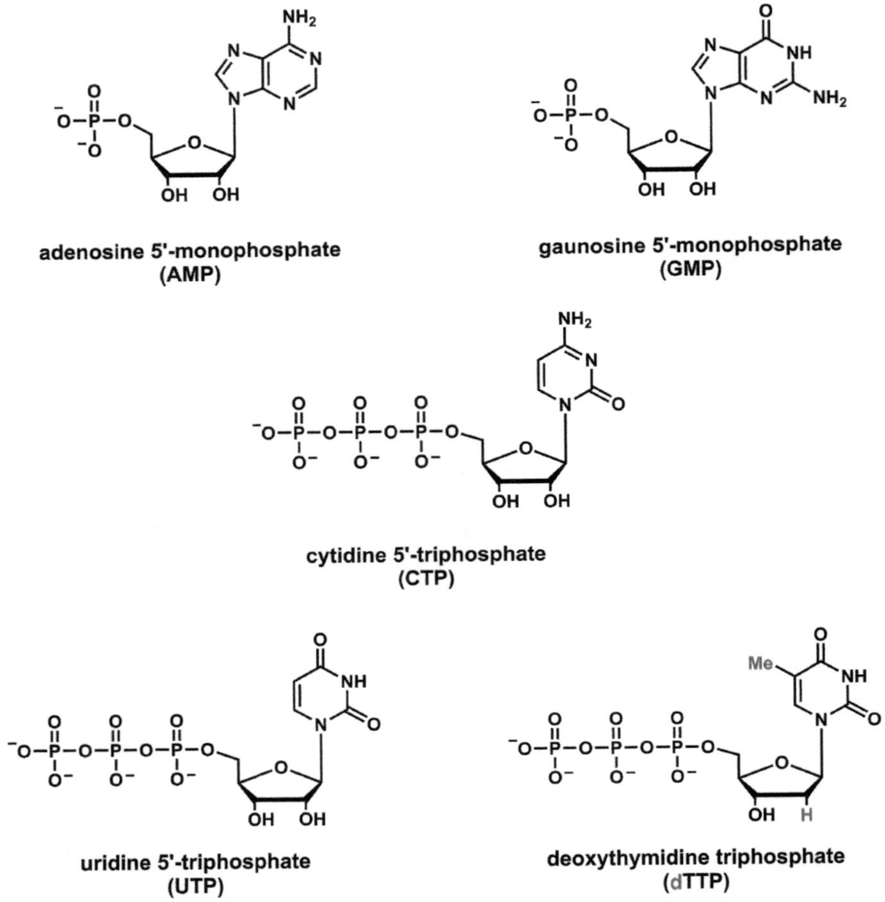

Figure 6.3 Purine and pyrimidines found in RNA and DNA.

as nucleotides (adenosine monophosphate and uridine monophosphate) complete the scaffolds for the nucleic acid building block components.

One final introductory comment concerns the stability of RNA *vs.* DNA. As shown in Figure 6.5, the internucleotide covalent 3′,5′-phosphodiester bonds in RNA are in principle labile to hydrolysis by intramolecular attack of the 2′-OH to give the 2′,3′ cyclic intermediate as the chain breaks and information is lost. By contrast, the use of 2′-deoxyribose moieties in the DNA internucleotide links removes that intramolecular lability. DNA is more stable, a better device for long term, stable storage of information.

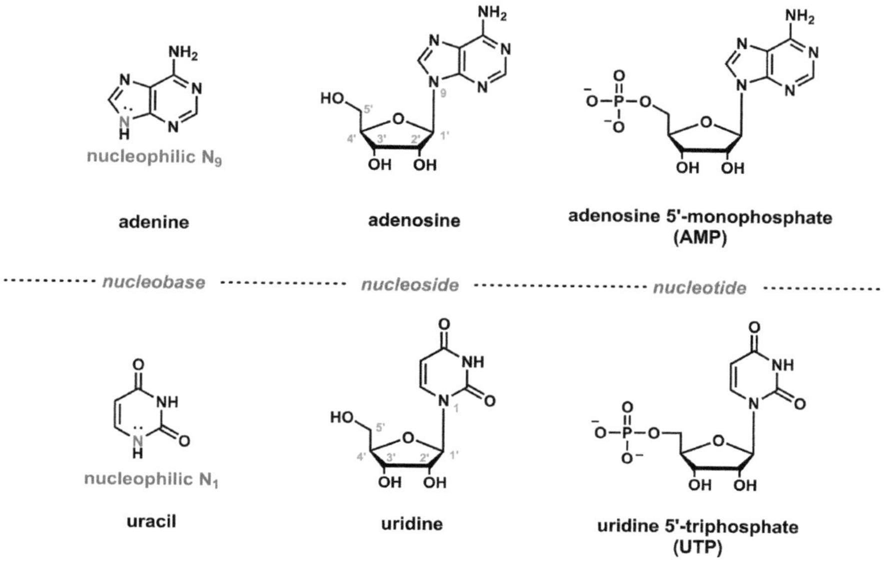

Figure 6.4 Purine and pyrimidine nucleobases, nucleosides, and nucleotides.

6.3 Remnants of an RNA World?

One of the prevailing hypotheses is that life started in organisms that stored genetic information in RNA rather than DNA (Cech 2012). In such a scenario the contemporary RNA viruses would be direct descendants of the primordial organisms. The rest of us will have converted to DNA-based, more reliable genetic information systems at some later date, perhaps driven by the above chemical logic (Figure 6.6).

To go from RNA to DNA requires not only reverse transcriptases, found in RNA viruses today, but also the enzyme ribonucleotide reductase. This key catalyst of the transition between RNA and DNA operates at the level of ribonucleoside diphosphates (Nordlund and Reichard 2006). It converts all four of ADP, GDP, UDP, and CDP to 2′-dADP, 2′-dGDP, 2′-CDP, and 2′-UDP. The second change from RNA to DNA requires the enzyme thymidylate synthase, acting at the level of dUMP to generate and insert the 5-methyl group of dTMP (= 5-methyl-dUMP) (Carreras and Santi 1995). As these enzymes are mainstays of primary metabolism but not secondary metabolism, we

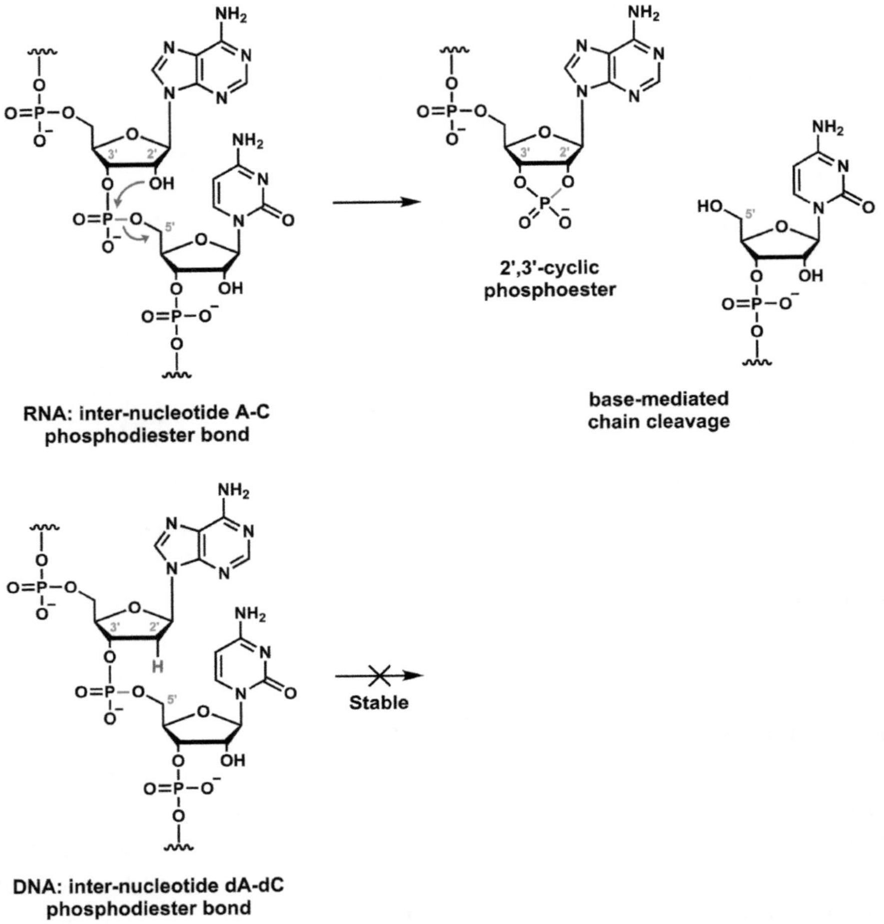

Figure 6.5 The internucleotide phosphodiester links in DNA are more stable than the ones in RNA: DNA is a superior choice for long term information storage.

do not delve into mechanisms here, while acknowledging the constraints this chemical logic places on the molecules that organisms make and accumulate.

Other indications of molecules that may be left over from an RNA world are the nitrogen heterocycles (Walsh 2015). These are not only the imidazoles, pyrimidines, and pyridines (NAD and pyridoxal-P) but also the adenosine moieties of ATP and *S*-adenosylmethionine, and coenzyme A (Figure 6.7). Figure 6.7 also reminds us that two of

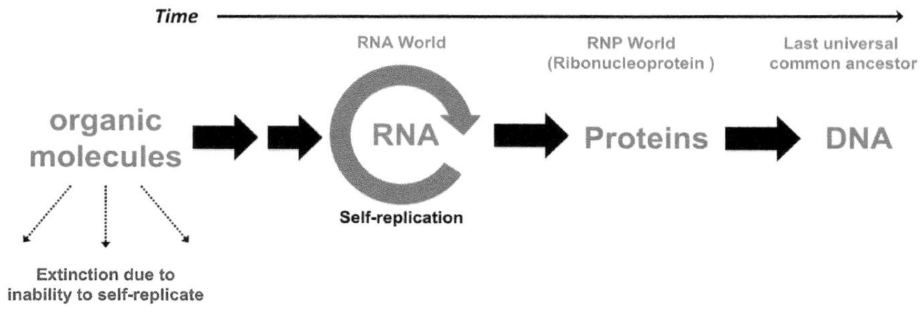

Figure 6.6 An evolution model in which RNA molecules emerged first as the central genetic and catalytic macromolecules, followed by proteins and DNA.
Data from Cech, T. R. (2012). The RNA Worlds in Context, *Cold Spring Harbor Perspect. Biol.* **4**.

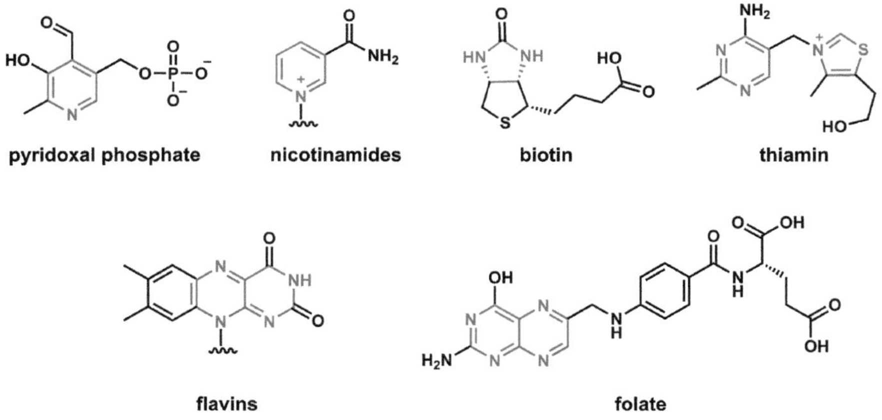

Figure 6.7 Coenzymes with nitrogen heterocycles as key functional elements.

the coenzymes of primary metabolism, thiamin-pyrophosphate and folate, have an explicit pyrimidine moiety as part of their heterocyclic cores. Although neither the pyrimidine moiety of TPP nor that of tetrahydrofolate engages in the specific chemistry mediated by these coenzymes, they are specificity determinants for recognition by partner proteins. As it happens, the pyrimidine moiety in TPP is assembled by a chemical logic distinct from the canonical route noted below.

6.4 Canonical Biosynthetic Routes to Purines and Pyrimidines

Both the single pyrimidine ring and the bicyclic purine ring building blocks for RNA and DNA are assembled on the framework of the primary sugar metabolite D-ribose-5-P. In the pyrimidine cases the ribose-5-P is added at the stage of orotate (Figure 6.8A), yielding OMP (orotidine monophosphate) *via* action of a phosphoribosyltransferase. Decarboxylation of OMP to UMP gives entry into one of the two pyrimidine building blocks. The second, cytosine, comes at the level of UTP by amination of the C_4-carbonyl *via* action of CTP synthase.

The key N_1–$C_{1'}$ bond generating the nucleotide in OMP is created by the basic N_1 of orotate acting as nucleophile on the activated 5-phosphoribosyl-1-pyrophosphate (PRPP). In Chapter 11, and later in this chapter in passing, we will note that the chemical logic for activation of hexoses as electrophiles at $C_{1'}$ is *via* nucleoside diphosphate derivatives, a complementary strategy to the –PP leaving group in PRPP.

For the substantially longer and more complex assembly of the bicyclic imidazolopyrimidine framework of purine building blocks (Figure 6.8B), the two heterocyclic rings are assembled on the ribose-5-P scaffold from the outset. The first committed enzyme converts PRPP into 5-phosphoribososyl-1′-amine. The pyrrole ring is built first (accounting for ribose attachment to this ring) and then the pyrimidine ring, to produce inosine monophosphate as the first metabolite with the intact purine scaffold. Enzymatic conversion to the 6-amino group in adenosine and the 2-amino group in guanosine ribonucleotides is achieved by tailoring enzymes.

One consequence of this biosynthetic chemical logic is that the ribonucleotides are the immediate biosynthetic end product. The corresponding nucleosides and free bases (adenine, guanine, cytosine, uracil) are NOT on the biosynthetic routes. Thus, any free heterocycle pyrimidine and purine metabolites are rare in both primary and secondary pathways and arise from hydrolytic enzymatic processing of nucleotides into nucleosides by phosphatases, and of nucleosides into ribose-5-P and the free purine and pyrimidine bases by nucleoside hydrolases (Figure 6.9). In turn, these are not at all common as the nucleosides and nucleotides are the precious building blocks for information transfer. Instead, salvage pathways operate in the reverse direction to capture any free bases and nucleosides to

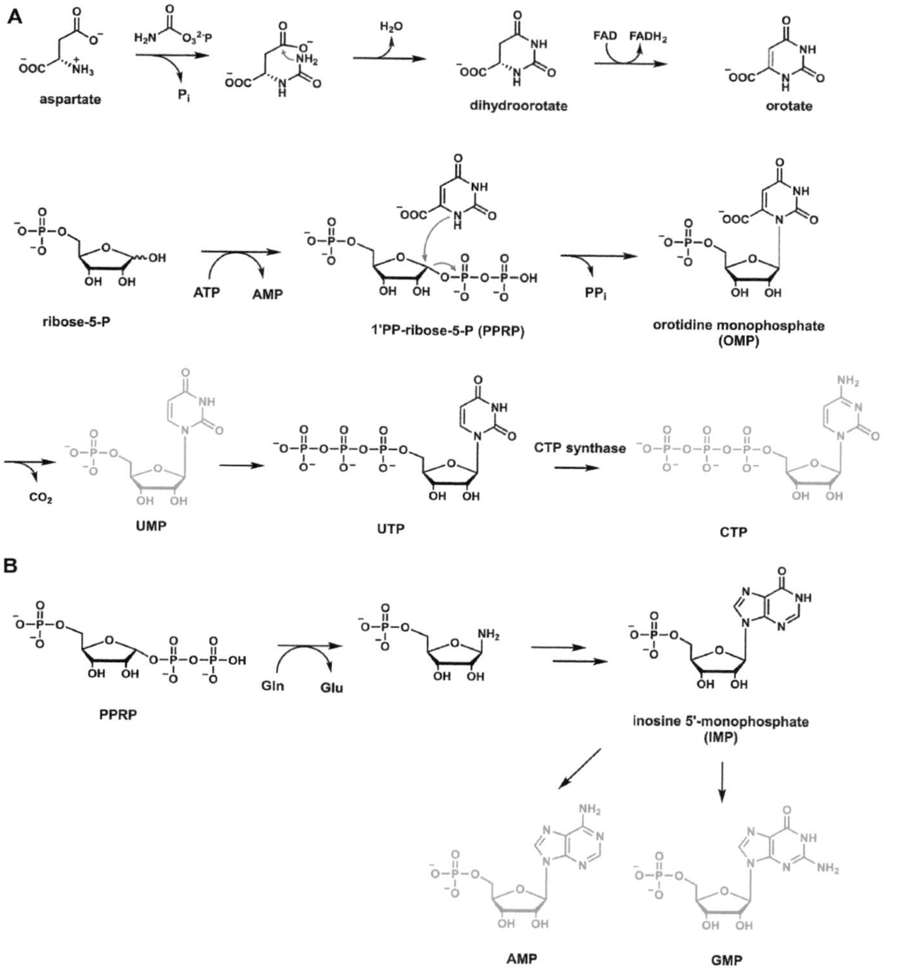

Figure 6.8 (A) The initial pyrimidine ring forming step occurs in dihydroorotate formation, followed by dehydrogenation to orotate, phosphoribosylation to OMP, and decarboxylation to UMP. The amination at C_4 to create the cytidine series only occurs at the level of UTP, not UMP or UDP. (B) AMP and GMP are downstream of the initial purine ribonucleotide inosine monophosphate (IMP).

avoid having to spend the energy for *de novo* construction of the millions to billions of RNA and DNA building blocks required in every cell division.

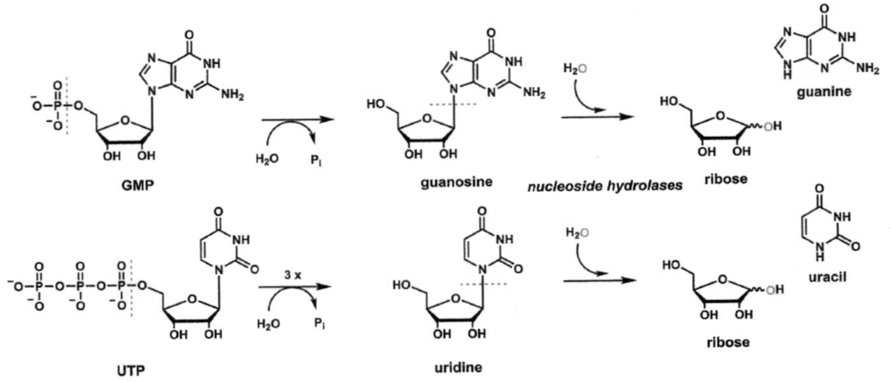

Figure 6.9 Purine and pyrimidine biosynthetic pathways give nucleotides as initial products. Nucleosides and free bases arise from degradative enzyme action.

6.5 Caffeine, Theobromine and Theophylline

One set of examples is provided by the purine alkaloids theobromine, theophylline, and caffeine, all pharmacologic stimulants (Dewick 2009). These three alkaloids are formed in a short pathway in coffee and coca plants from the purine pathway intermediate inosine monophosphate (IMP) (Figure 6.10). IMP is converted to the 2-oxo metabolite xanthosine monophosphate (XMP), and then the 5′O–PO_3 bond is cleaved by a hydrolytic phosphatase to yield the nucleoside xanthosine.

This nucleoside undergoes the first of three consecutive SAM-dependent methylations, at C_7, which labilizes the N_1–$C_{1'}$ bond to the ribose to hydrolytic displacement. Thus, the conversion of nucleoside to methylated free purine base is mediated by chemical methylation/bond labilization, not enzymatically by a hydrolase. The N_7-methylxanthine is methylated a second time, regiospecifically at N_1, to yield theobromine. A third methylation gives 1,3,7-trimethylxanthine, more familiarly known as caffeine. The other dimethyl xanthine regioisomer, 1,3-dimethylxanthine, is the metabolite theophylline. Caffeine is the most stimulatory of the three methylxanthines (Figure 6.11), while theophylline is pharmacologically useful as a smooth muscle relaxant for treatment of bronchial attacks. Caffeine is an antagonist at all subtypes of adenosine receptors, and at GABA-A receptors, and inhibits several molecular forms of phosphodiesterases, accounting for its widespread pharmacologic actions (Ribeiro and Sebastiao 2010).

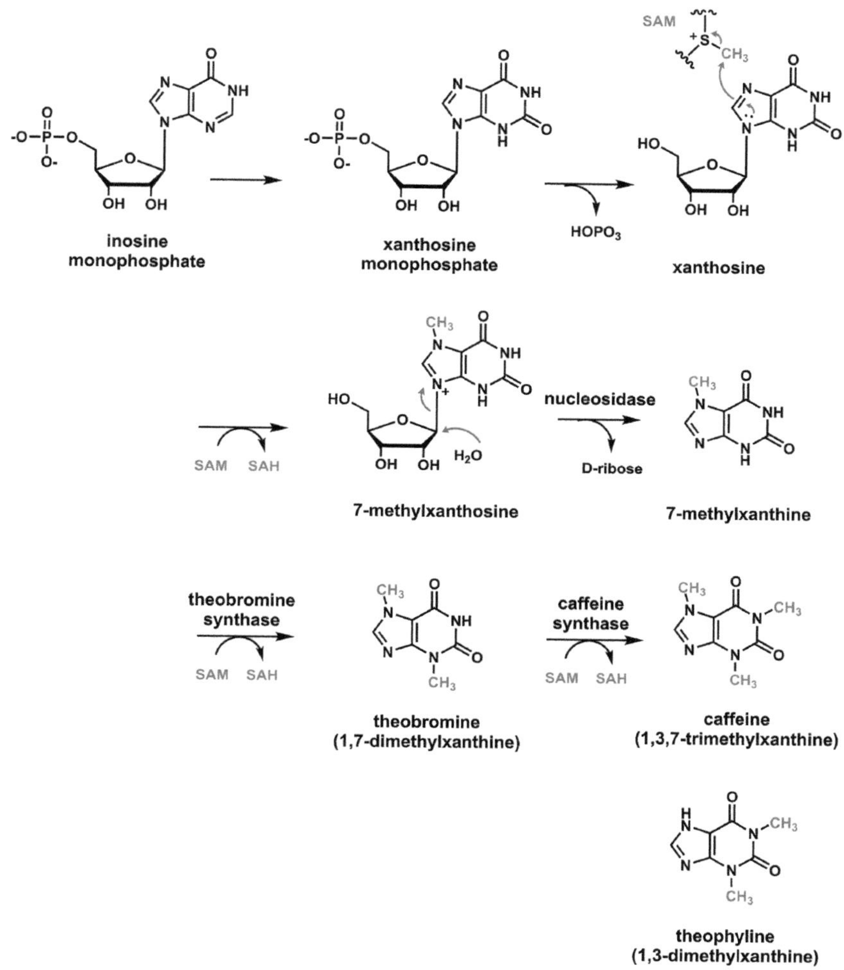

Figure 6.10 Caffeine, theobromine, and theophylline from enzymatic methylations of xanthosine monophosphate and then xanthine.

6.6 Plant Isopentenyl Adenine Cytokinins

A second example of metabolic processing of nucleotide primary metabolites to the purine base, devoid of sugar and phosphate moieties, occurs in the elaboration of isopentenyl adenines in the plant cytokinin class of hormones (Frebort, Kowalska *et al.* 2011). As summarized in Figure 6.12, ATP (or ADP or AMP) can serve as the

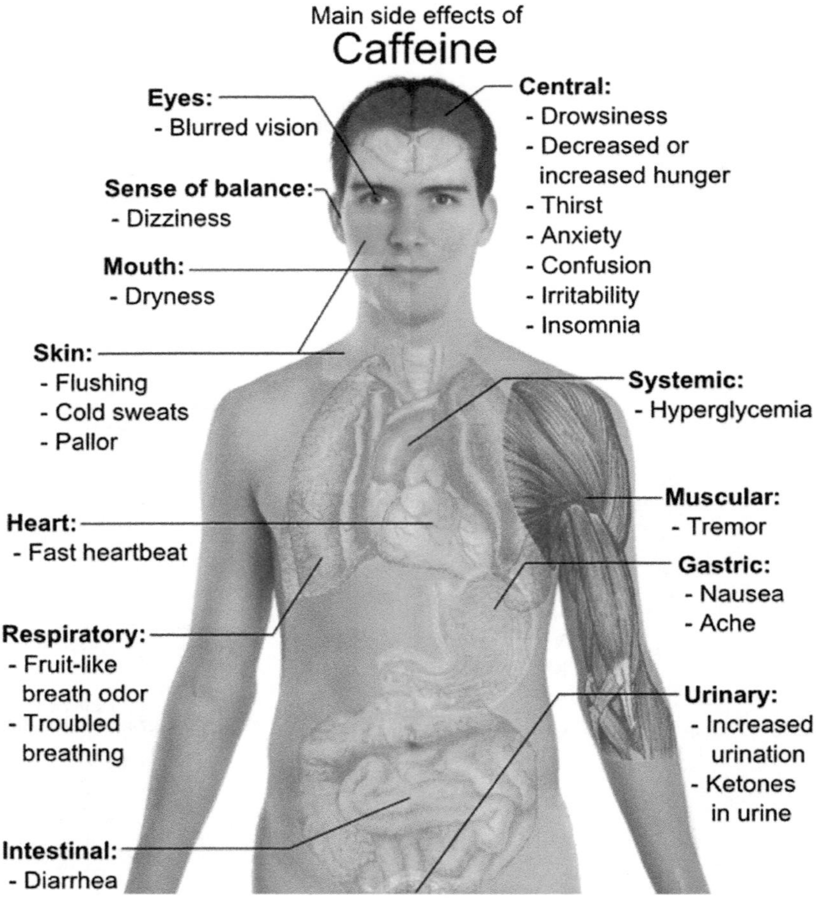

Figure 6.11 Caffeine is a pharmacologically active alkaloid. Photo credit: Mikael Häggström.

nucleophilic cosubstrate at the exocyclic amine group to attack the allyl carbocation arising from early dissociation of Δ^2-isopentenyl-PP. The initial product is the isopentenyl nucleotide. This undergoes phosphatase-mediated conversion to the nucleoside and then a second enzymatic cleavage, this time at the N_1–$C_{1'}$ ribosidic linkage, to release ribose and the isopentenyl adenine. A related cytokinin with plant growth and cell division activities is *trans*-zeatin, which arises from a cytochrome P450-catalyzed hydroxylation of one of the terminal methyl groups in the isopentenyl moiety (Einset 1986, Taiz, Zeiger *et al.* 2015).

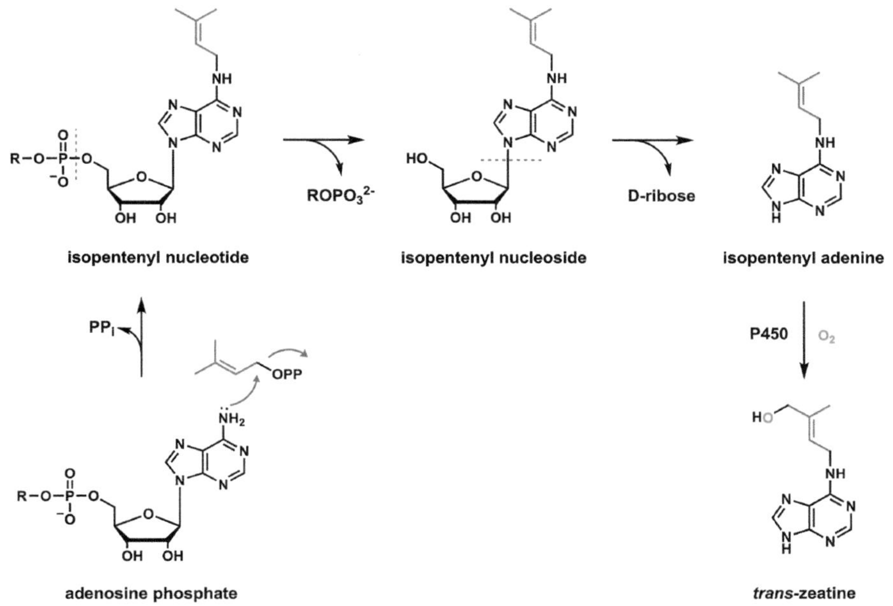

Figure 6.12 Formation of isopentenyl adenine plant hormones.

6.7 Maturation of Ribonucleotides to Modified Purine and Pyrimidine Natural Products

6.7.1 Heterocycle Modification

In a sense the above examples of isopentenyl adenine and *trans*-zeatin can be categorized as modifications to the purine or pyrimidine core followed by trimming away of the sugar and phosphates. The cytokinins reflect the nucleophilicity of the adenine exocyclic NH_2. Analogously, the exocyclic amine of guanine nucleotides is nucleophilic: for example it is the site of covalent modification of DNA by the fungal metabolite aflatoxin (D'Andrea and Haseltine 1978). Analogously, the N_3 of cytidine residues in DNA are the sites of methylation from SAM by a chemically straightforward, enzymatically specified transfer of $[CH_3^+]$ equivalents at CpG islands of DNA strands.

A less chemically obvious set of modifications to the purine ring system occurs in conversion of GTP to a set of 7-deazapurines, some of which are depicted in Figure 6.13. These are net conversions of the

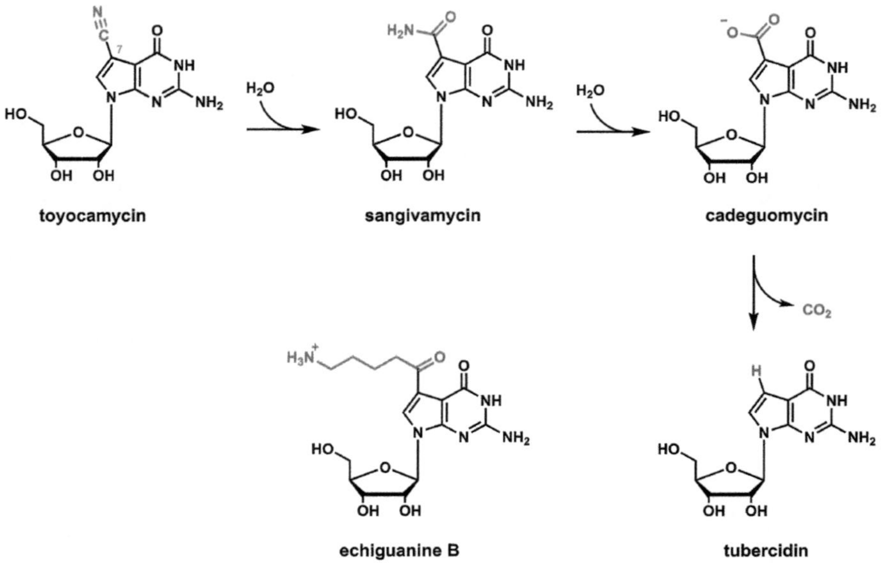

Figure 6.13 Naturally occurring 7-deazapurine nucleosides. Deazaguanine natural products have a C_7 in place of an N_7 atom.

imidazolopyrimidine to pyrrolopyrimidine, with extrusion of the N_7 nitrogen from the imidazole ring and its formal replacement by a CH group introduced as a one carbon fragment from *S*-adenosylmethionine. The progression from toyocamycin to sangivamycin to cadeguomycin to tubercidin is a straightforward hydration of the 7-deaza cyano group to the carboxamide, hydrolysis to the carboxylate, and then a decarboxylation to give tubercidin. The echiguanine series instead has a lysine derived ketone functionality that has been installed at C_7 of the pyrrolopyrimidine framework.

These conversions of GTP to these and other 7-deazaguanines are mechanistically complex transformations. First there is ring expansion to the 6,6-bicyclic pterin system with extrusion of C_8 as formate, the typical pattern for a GTP cyclohydrolase (also found in folate biosynthesis). Carbons 1′-3′, labeled in cyan in Figure 6.14, are retained. The $C_{3'}$ becomes the carboxylate carbon in 6-carboxy-H_4-pterin and then the cyano carbon in the 7-cyano-7-deazaguanine. The conversion of the 6,6-bicyclic tetrahydropterin ring system to the 6,5-deazagauanine ring system involves a distinct reactivity facet of SAM as a coenzyme/cosubstrate: the ability to act as a radical initiator (Bandarian and Drennan 2015). The 7-deazguanine forming pathway is taken up as one of several manifestations of this orthogonal

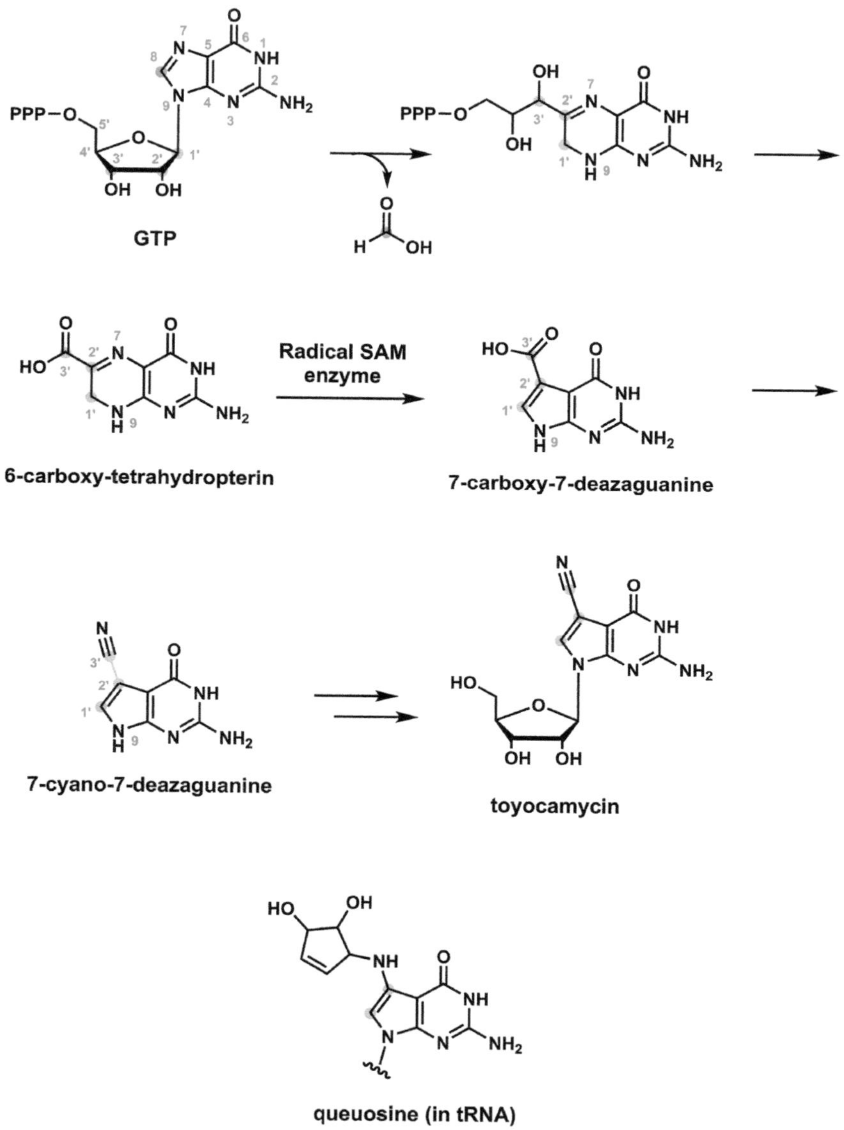

Figure 6.14 Path of carbon atoms in the conversion of GTP to 7-cyano-7-deazaguanine.

reactivity pattern of SAM, as source of 5′-deoxyadenosyl radical, in Chapter 10.

One further example of a modified base occurs in pyrazomycin, where only a five-membered heterocycle is attached to the ribose framework (Figure 6.15). The hydroxy-pyrazine carboxamide is an

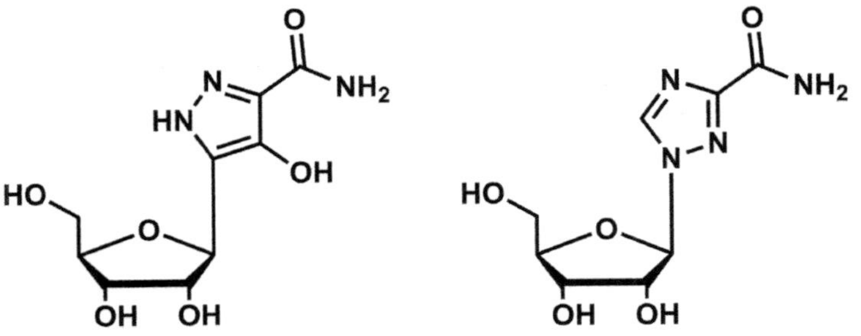

Figure 6.15 Pyrazinomycin and the synthetic antiviral drug ribavirin.

analog of the normal purine biosynthetic intermediate AICAR but has two nitrogens in a 1,2-relationship and it is a C–nucleoside (C–$C_{1'}$) rather than having a traditional N–$C_{1'}$ bond between base and ribose. The mechanism of biosynthesis of the N–N bond and the C–nucleoside link is as yet undetermined. It is precedent for a synthetic triazole analog, ribavirin (Thomas, Ghany *et al.* 2013), which was used for several years with interferon in combination therapy for hepatitis C before a more curative combination regimen was introduced.

6.7.2 Sugar Modifications

We noted in an earlier section of this chapter the combination of the chemical strategy to use 5-phosphoribosyl-α-1-pyrophosphate (PRPP) as the common metabolite for activating $C_{1'}$ of the ribosyl group as an electrophile. The stereochemical specificity to act with inversion of both orotate phosphoribosyltransferase in pyrimidine assembly and the 5-phosphoribosylamine-forming enzyme in the first step of the purine pathway means that all the nucleotides (and derived nucleosides) are purine-β- and pyrimidine-β-nucleosides. All the modified sugars in purine and pyrimidine natural products similarly have β-$C_{1'}$-stereochemistry.

Adenosyl arabinoside (AraA or spongoadenosine) is shown in Figure 6.15. It and its 3′-*O*-acetyl derivative were first synthesized 60 years ago and later isolated from the coral *Eunicella cavolini* (Cimino, de Rosa *et al.* 1984). Likewise, the marine spongouridine (Bergmann and Burke 1955) and spongothymidine (AraT) (Bergmann and Feeney 1950) have arabinose sugars attached to the

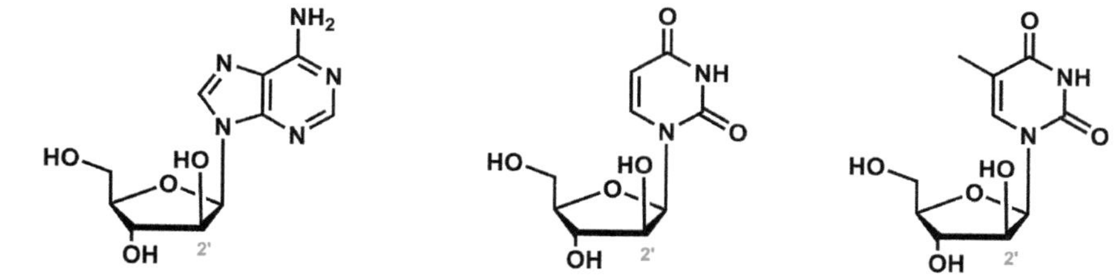

Figure 6.16 Pyrimidine arabinosides adenosyl arabinoside (AraA), spongouridine, and spongothymidine (AraT) isolated from marine sources.

pyrimidine rather than purine bases (Figure 6.16). D-Arabinose differs from D-ribose in the stereochemistry at the $C_{2'}$-OH (Huang, Chen *et al.* 2014).

Vignette 6.1 Nucleosides in Antiviral and Cancer Chemotherapy.

These natural arabinose nucleosides, AraA and AraT, prefigured the synthesis of AraC and its use in antiviral and cancer chemotherapy. AraC is still utilized as part of combination chemotherapy regimens for leukemias and lymphomas but AraA is rarely used because of its narrower therapeutic index. In a separate sense, the AraT framework was an inspiration for 3′-azido-T (AZT) which became an early clinical drug for HIV treatment, as noted in Figure 6.V1. Given the clinical success of AZT and related nucleoside analogs, a sustained burst of medicinal chemistry over the past two decades has created successive generations of deoxy nucleoside analogs that block different classes of viral replication from HIV, to human cytomegalovirus, to hepatitis B and C viruses, with differential effectiveness.

The general mechanism of action of the antiviral/antitumor (deoxy)-ribosides is uptake of the nucleosides by viruses or cancer cells and enzymatic phosphorylation up to the nucleoside triphosphates. At that stage they can compete as substrates for the physiologic dNTPs and NTPs for DNA and RNA synthesis, respectively. Once incorporated, however, they are chain terminators as they are incapable of reacting as the 3′-OH nucleophile for the next round of dNMP addition. Both the reverse transcriptase inhibitors as antiviral agents and AraC as antileukemic agent are typically given in combination with other agents that act by different mechanisms on the same or alternative targets, to reduce the frequency of resistance development.

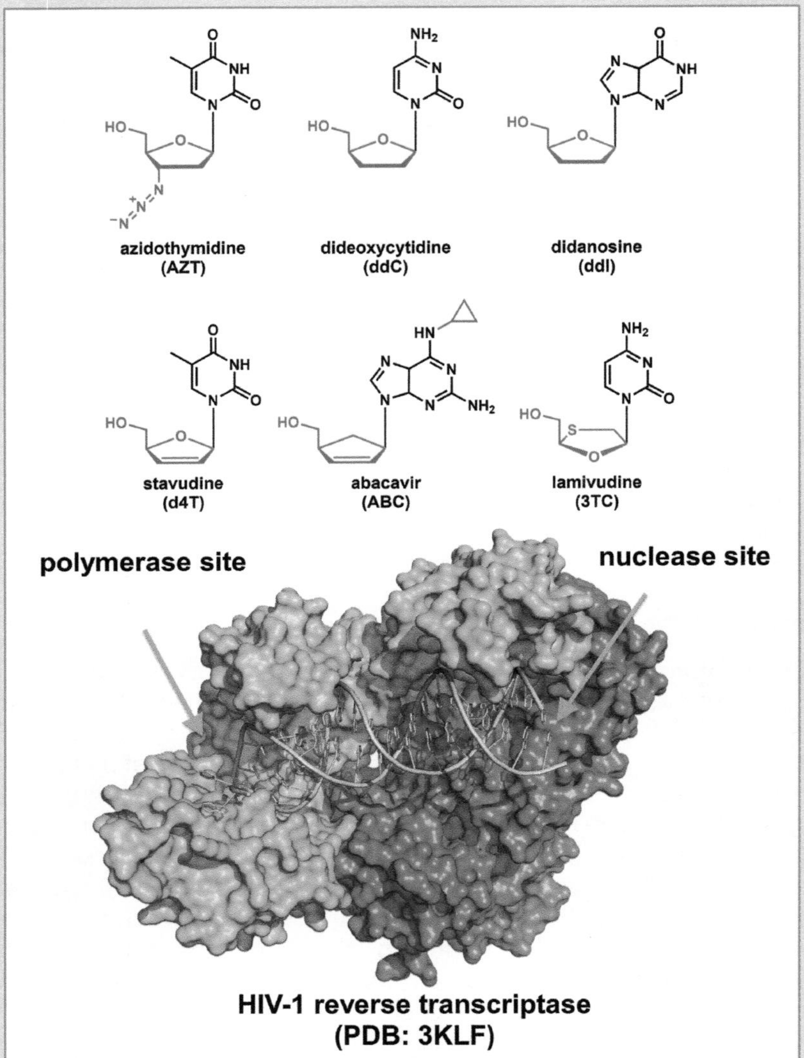

Figure 6.V1 Nucleoside analogs as reverse transcriptase inhibitors for antiviral and cancer chemotherapy.
Image credit: Yang Hai.

Vignette 6.2 A Three Enzyme Route to Didanosine (2′,3′-dideoxyinosine).

One of the early anti HIV drug candidates was the inosine nucleoside devoid of –OH groups at both the 2′- and 3′-positions of the D-ribose ring. This molecule, 2′, 3′-dideoxyinosine, known by the colloquial name didanosine (Figure 6.V2), once metabolized to the 5′-triphosphate and

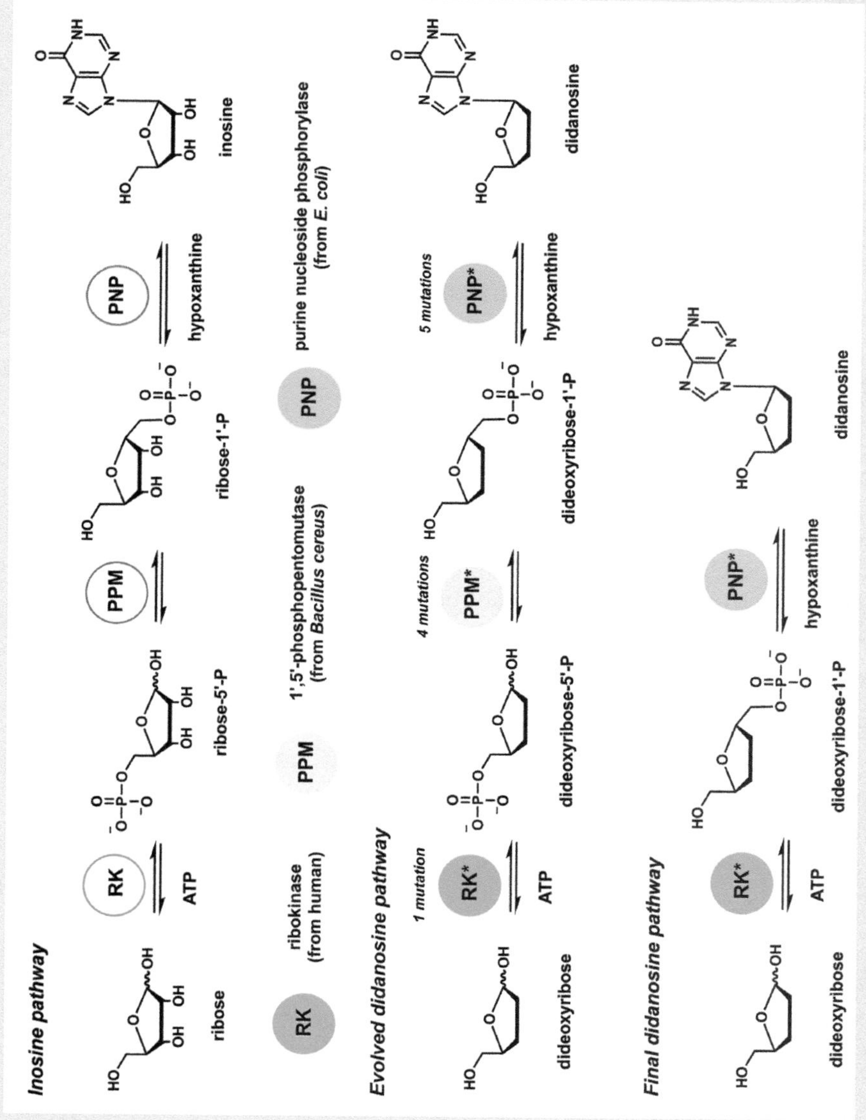

Figure 6.V2 Engineering the inosine biosynthetic pathway components towards enzymatic synthesis of the chain termination antimetabolite didanosine (2′,3′-dideoxyinosine).

incorporated into viral RNA, would clearly be incapable of further chain elongation and thus serve as a chain termination antimetabolite.

Bachmann and colleagues at Vanderbilt came up with a two enzyme scheme to convert synthetic 2′, 3′-deoxy-D-ribose into didanosine *in vitro* (Birmingham, Starbird *et al.* 2014). Working backwards, they needed to reengineer the specificity of purine nucleoside (PNP) phosphorylase, 1,5-phosphomutase (PPM), and finally ribokinase (RK) to decrease the fidelity of the three enzyme set. Concomitantly, the mutant forms of the enzymes needed to increase the flux of dideoxyribose to the dideoxyribose-5′-phosphate (by RK), the dideoxyribose-5′-P to the corresponding 1′-P (by PPM), and finally the displacement of the dideoxyribose-1′-P by N_9 of hypoxanthine (by PNP) to yield the antiviral nucleoside analog.

Through a series of structure-guided mutations and semi-random gene evolutions, a 9500-fold increase in nucleoside production selectivity ensued, in favor of didanosine over inosine. Fortuitously, the engineered RK was able to phosphorylate deoxyribose directly at the 1′ position to yield dideoxyribose-1′-P, thereby eliminating the need for a separate PPM. This successful directed evolution campaign of a two enzyme pathway (RK + PNP) *in vitro* is the first stage for deciding whether to move the gene variants into a suitable host and evaluate productivity and toxicity issues *in vivo* as prelude to further evaluation of flux and recovery issues of this water soluble 2′,3′-dideoxynucleoside. In principle, a variety of other antiviral nucleosides could be fashioned from short, engineered metabolic pathways by the "bioretrosynthetic" set of operations.

There are also natural nucleosides in the 2′-deoxyribose series that contain modified bases. One example, with a bulky modification, is in the molecule avinosol (Figure 6.17) isolated from a *Dysidea* sponge community (it is unknown whether the producer organism is the sponge or the associated microbial consortium) (Diaz-Marrero,

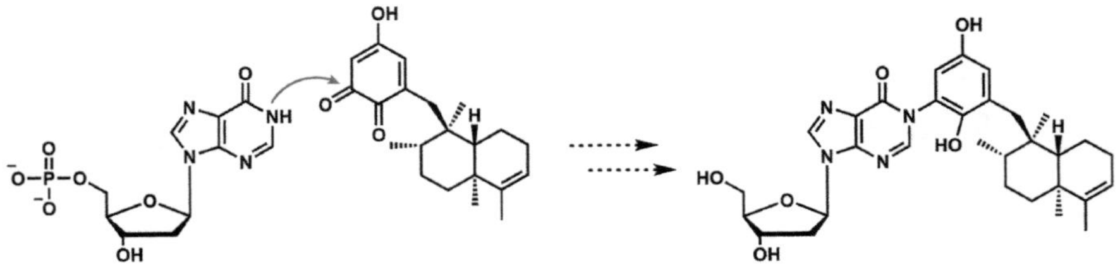

2′-deoxyinosine-5′-monophophate (dIMP)

avinosol

Figure 6.17 Avinosol, an N_3-alkykated 2′-deoxy inosine sponge natural product.

Austin *et al.* 2006). The core is an inosine deoxyriboside with a hydroquinone–decalin moiety attached to N_1. It may be that the quinone oxidation state was the electrophile for attack by dIMP. The decalin motif suggests a potential Diel–Alderase at some point in the biosynthesis of that part of the molecule but no details are known.

6.7.3 Cyclopentanetriol "Carba" Analogs of the D-Ribose Moiety in Neplanocin A and Aristeromycin

Streptomyces citricolor produces two carbacyclic analogs of adenosine. Neplanocin A has a 4′,5′-olefinic link while asterimycin has the fully saturated cyclopentanetriol ring analog of ribose (Jenkins and Turner 1995). The trihydroxycyclopentane framework arises metabolically from glucose by the mechanism proposed in Figure 6.18. Glucose is proposed to undergo isomerization to fructose and then formation of a five-membered cyclitol phosphate, which is followed by elimination of pyrophosphate and water to yield the indicated 5-keto, 2, 6-ene triol (Kudo, Tsunoda *et al.* 2016). Enzymatic reduction of the ketone would set up the cyclopentene for enzymatic conversion to the 1-pyrophosphate (perhaps as the 5-phosho-1-pyrophosphate) that could serve as the electrophilic partner in capture by free adenine, as shown. That would yield neplanocin A. It is known that an oxidoreductase can then take neplanocin A to the saturated carbacycle in aristeromycin *via* hydride addition from NADPH (Parry and Jiang 1994).

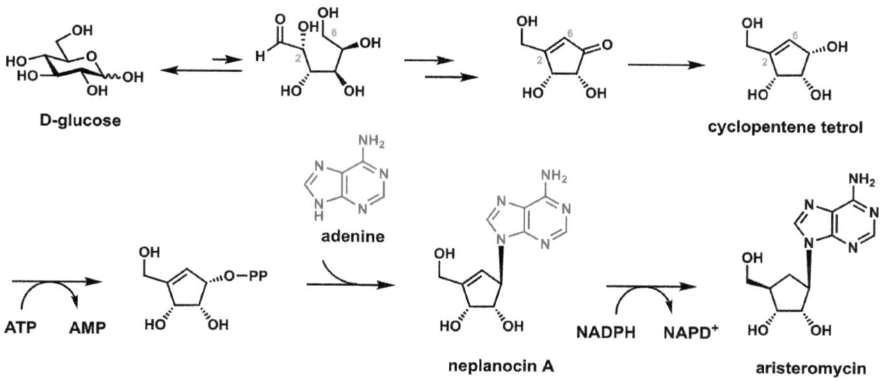

Figure 6.18 Proposed biosynthesis of neplanocin A and aristeromycin: a glucose-derived carbacycle rather than a ribose-derived core.

6.8 Peptidyl Nucleosides

6.8.1 Switch from Ribose to Hexose Sugars in Nucleotide Analogs

Culture filtrates of *Streptomyces gougeroti* were observed in 1962 to contain a molecule with weak antibiotic activity but an unusual structure. Gougerotin is a cytidine nucleoside with a hexose rather than a ribose sugar. It also has a C_6 carboxamide substituent and a substituted C_4 amine (Figure 6.19). The biosynthetic gene cluster

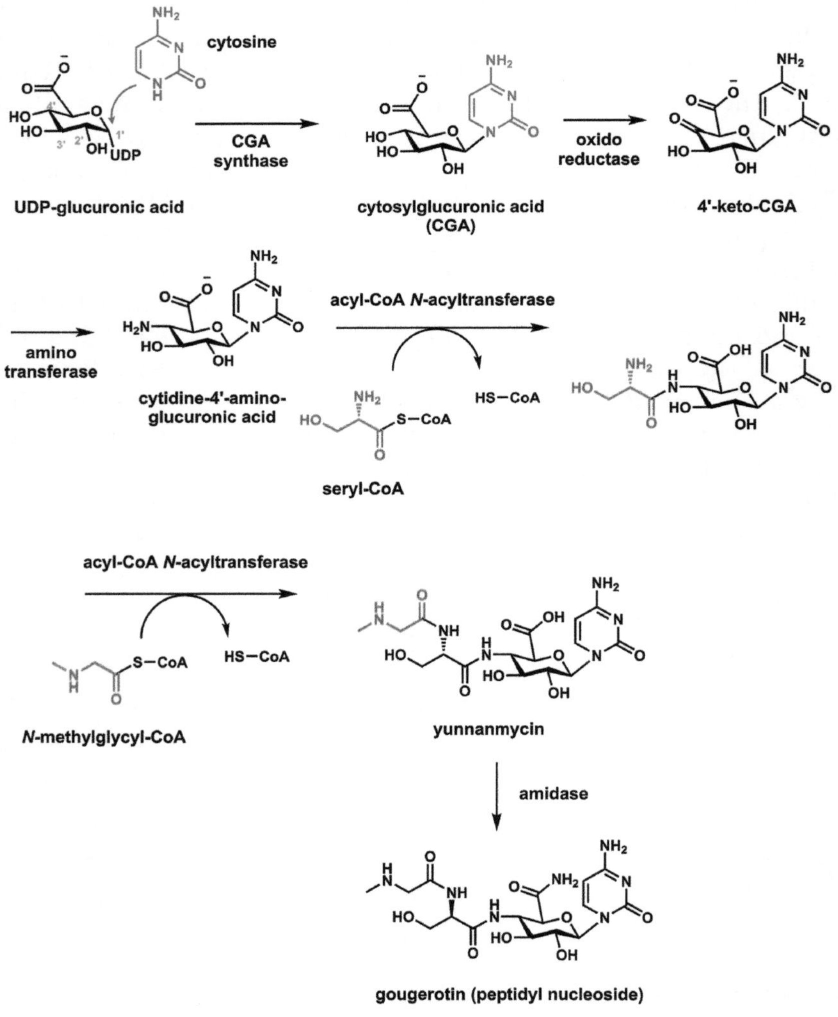

Figure 6.19 Gougerotin is a $C_{4'}$ amino cytidine peptidyl nucleoside from a UDP-glucuronate building block.

has been reported and a proposed route of formation advanced (Jiang, Wei *et al.* 2013). The proposed pathway starts with attack of N_1 of free base cytosine on UDP-glucuronic acid. We shall note in Chapter 10 that NDP-hexoses are the biological reagents for activation of hexosyl moieties as electrophiles at C_1. The use of an NDP leaving group with hexoses compares strategically to the use of the pyrophosphate group with ribose as electrophile. Oxidation of the $C_{4'}$–OH to the ketone and then transamination installs the 4′-amino group of cytidine-4′-aminoglucuronic acid. Peptide bond formation with serine and then *N*-methylglycine, followed by amination of the carboxylate, would complete the peptidyl nucleoside with the amino hexosamide sugar.

Two additional examples of nucleoside natural products with hexose rather than ribose cores are mildiomycin (Wu, Li *et al.* 2012) and blastocidin S (Cone, Yin *et al.* 2003, Thibodeaux, Melancon *et al.* 2007, Thibodeaux, Melancon *et al.* 2008) (Figure 6.20). The same starting logic applies as above for gougerotin, using either cytosine or hydroxymethylcytosine as nucleophilic free base to attack UDP-glucuronate at $C_{1'}$. For mildiomycin, the transformation of the glucuronate into the 2,3-dehydro sugar is proposed to involve radical intermediates involving SAM as initiator (see Chapter 10) to expel two equivalents of water from the 2′ and 3′ carbons. Attachment of the $C_{5'}$ chain is also not obvious on the way to mildiomycin. For blastocidin S, a similar route is proposed with transamination of a 4′-keto intermediate to set up peptide bond formation.

6.8.2 Modification of 4′-Substituents

The previous three examples, gougerotin, mildiomycin, and blastocidin S, are examples of peptidyl nucleosides, albeit unusual ones in containing a glucuronate rather than a ribose moiety as the central sugar. A substantial number of uridine-based peptidyl nucleosides and lipopeptidyl nucleosides bearing the common ribose core have been characterized for their biosynthetic logic (Walsh and Zhang 2011, Zhang, Ntai *et al.* 2011).

As depicted in Figure 6.21, these pathways begin with conversion of UMP to uridine-5′-aldehyde by a nonheme Fe(II)-dependent α-ketoglutarate dioxygenase (Van Lanen, Koichi *et al.* 2012). This can then be combined with glycine to form a 5′-C-glycyl adduct, catalyzed by a homolog of serine hydroxymethyltransferase (SHMT) in each pathway. The 5′-C-glycyluridine core is the common intermediate to caprezomycins, liposidomycins, and muraymycins (Barnard-Britson, Chi *et al.* 2012). All these (lipo)peptidyl nucleosides are inhibitory

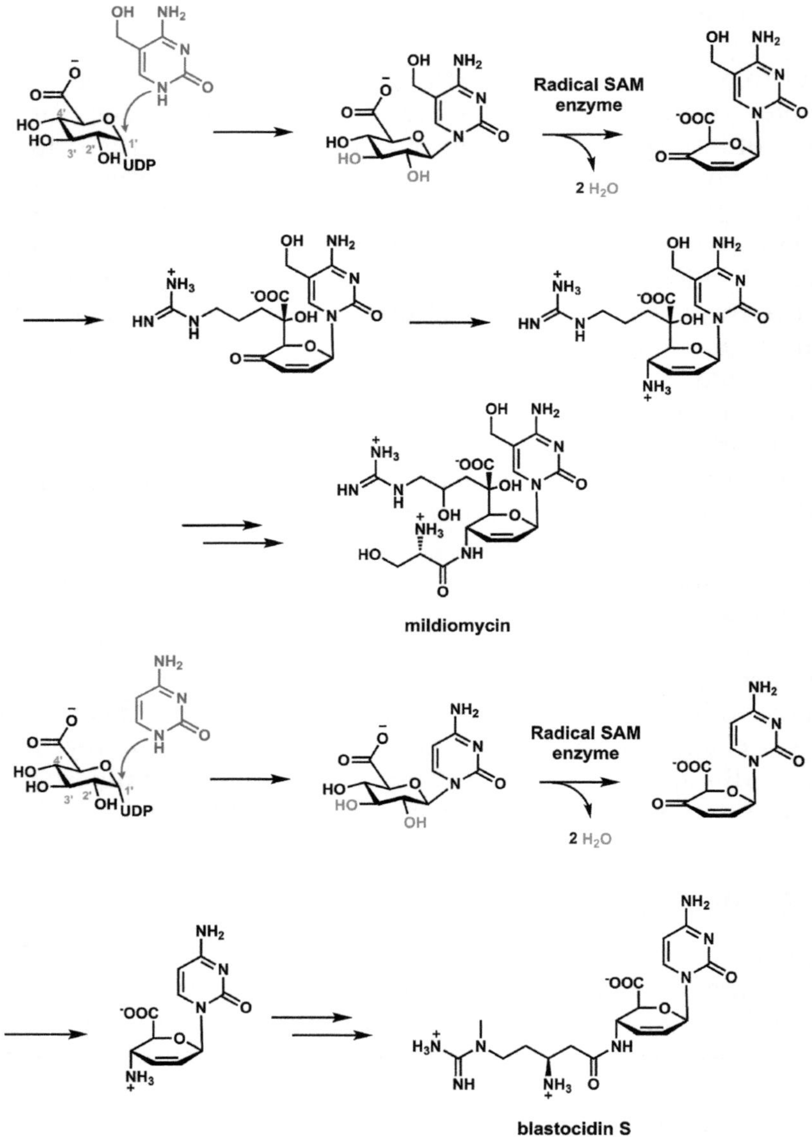

Figure 6.20 Mildiomycin and blastocidin S are peptidyl nucleosides with unsaturated hexose cores.

substrate analogs of the bacterial cell wall translocase MraY. MraY acts at the interface of the cytoplasmic phase and the membrane phase of bacterial peptidoglycan assembly, transferring the muramyl

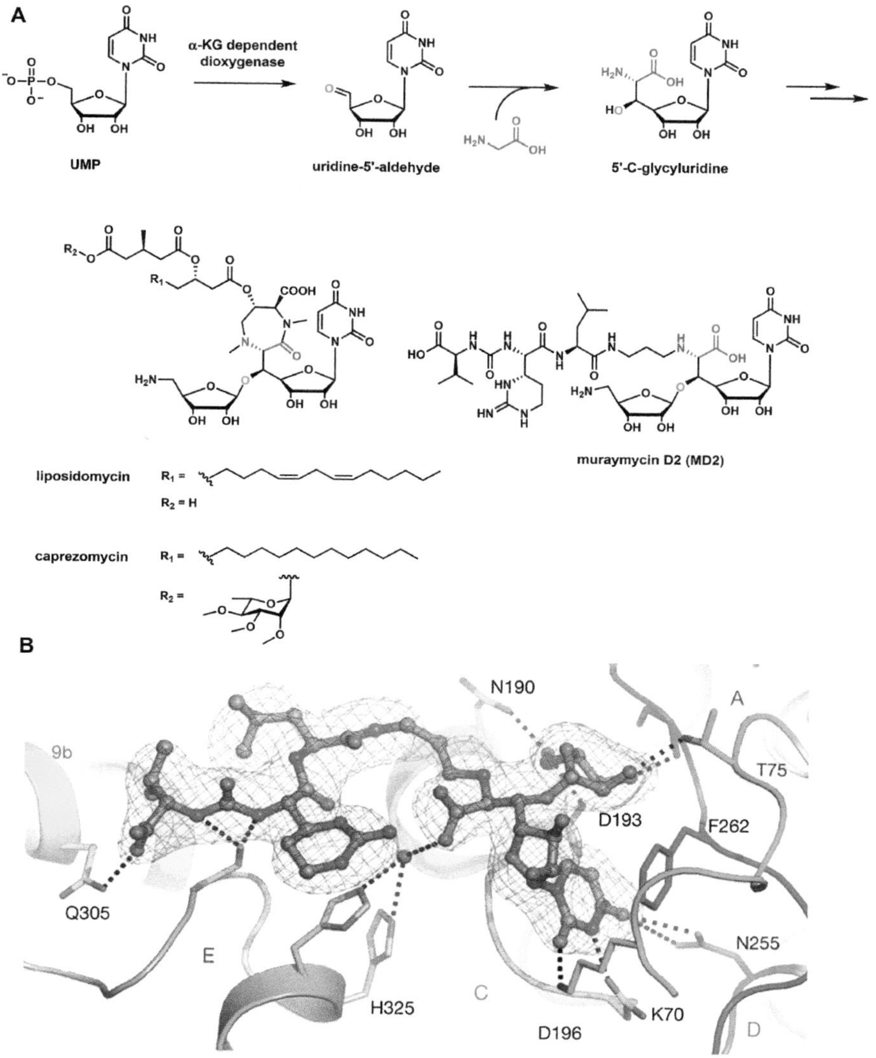

Figure 6.21 (A) Uridine to uridine 5′-aldehyde to uridine 5′-C-glycyl adduct: a common pathway to caprezomycins, liposidomycins, and muraymycins. (B) Cocrystal structure of the *Aquifex aeolicus* MraY with muraymycin D2.
Reprinted by permission from Macmillan Publishers Ltd: Nature (Chung, B. C., E. H. Mashalidis, T. Tanino, M. Kim, A. Matsuda, J. Hong, S. Ichikawa and S.-Y. Lee. "Structural insights into inhibition of lipid I production in bacterial cell wall synthesis". *Nature* **553**: 557–560), copyright (2016).

pentapeptide chain from UDP to the isoprenoid bactoprenol to bring the chain to the membrane.

A recent cocrystal structure of the *Aquifex aeolicus* MraY with muraymycin D2 (binding constant ~20 nM) (Chung, Mashalidis *et al.* 2016) gives new insight into the mode of inhibition of MraY's transferase activity by such peptidyl nucleosides (Figure 6.21B). The uracil and 5-aminoribose moieties, in their interaction with the enzyme's nucleotide binding pocket, are likened to the way "a two pronged electrical plug inserts into a socket" (Chung *et al.* 2016), with partial overlap with the substrate-binding region. The high affinity for a peptidyl nucleoside analog without a pyrophosphate substructure, which can cross into target bacterial cells, may suggest improvements in this natural product class by synthesis or pathway engineering to get to a useful antibiotic framework.

6.8.3 Sinefungin, Tunicamycin, Polyoxin and Jawsamycin

Sinefungin is a bacterial metabolite that is a general inhibitor of *S*-adenosylmethionine-utilizing methyltransferases (Zhang and Zheng 2016). Sinefungin is also named adenosyl ornithine, connected *via* C_5 of ornithine to $C_{5'}$ of the D-ribose moiety of the adenosine. The δ-amine is cationic at physiologic pH so sinefungin mimics the positive sulfonium center of SAM substrates and competes with it for binding at a large number of methyltransferases (Figure 6.22). The sinefungin sugar moiety is a conventional D-ribose but, unlike the peptidyl nucleosides in the previous section, $C_{5'}$ is connected to a carbon atom not a nitrogen substituent.

The proposed route to C_5–$C_{5'}$ bond formation is to form an imine between Nδ of ornithine and pyridoxal-P in an enzyme active site and make the stabilized carbanion, shown in Figure 6.22. That is the requisite nucleophile and is shown displacing the –OH group from $C_{5'}$ of adenosine (it is not clear that it is not AMP that is the substrate, with departure of $HOPO_3$ in this step). An additional dipeptide variant of sinefungin with valine and alanine in peptide linkage to the ornithine carboxylate has been found to have antitrypanosmal activity (Niitsuma, Hashida *et al.* 2010).

Tunicamyicns are fatty acyl uridine nucleosides, not amino acid constituents, with an 11-carbon tunicamine core (Price and Tsvetanova 2007). They are produced by strains of streptomycete bacteria and, like the peptidyl nucleosides in the caprezomycin and liposidomycin series, are also MraY inhibitors. Tunicamycins are also potent inhibitors of a comparable early step in eukaryotic

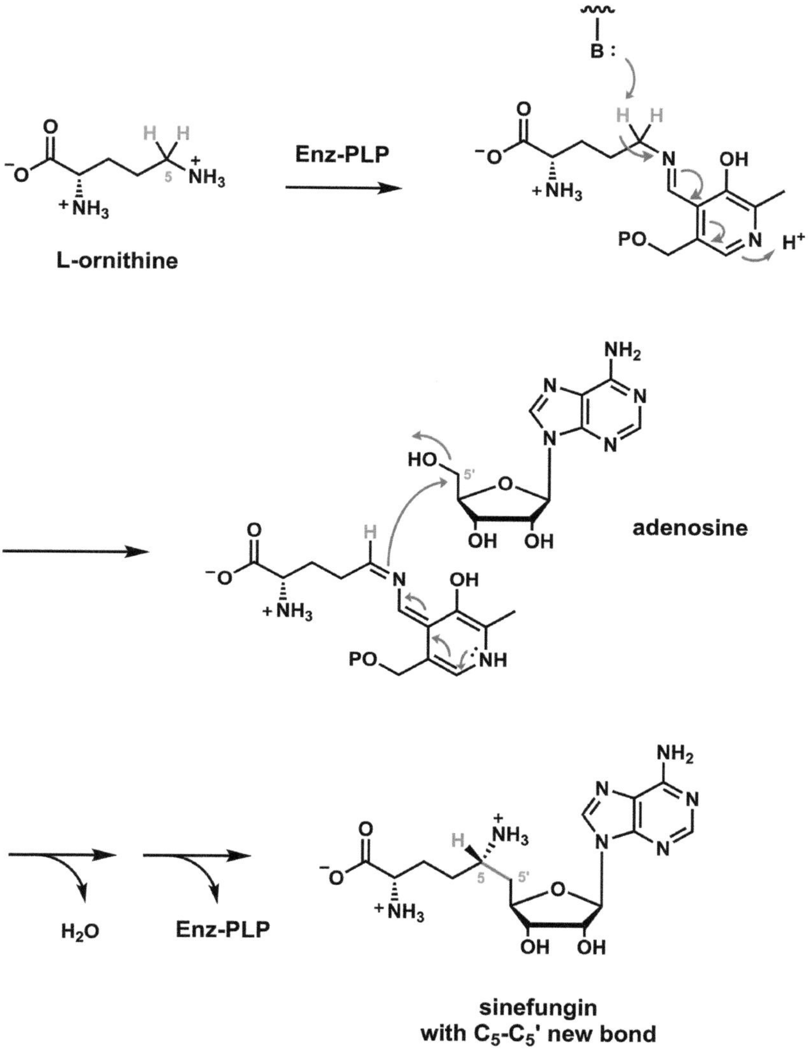

Figure 6.22 Ornithine and adenosine as building blocks for the C_5–$C_{5'}$ bond in sinefungin.

N-glycoprotein maturation and are too toxic for clinical use in humans. In some analogy to sinefungin, tunicamycin has a C–C bond connection from $C_{5'}$ of the uridine moiety to C_6 of a glucose-derived hexose unit. The route to this unusual linkage has been identified as proceeding through an *exo*-methylene glycal (Wyszynski, Lee *et al.* 2012), as shown in Figure 6.23.

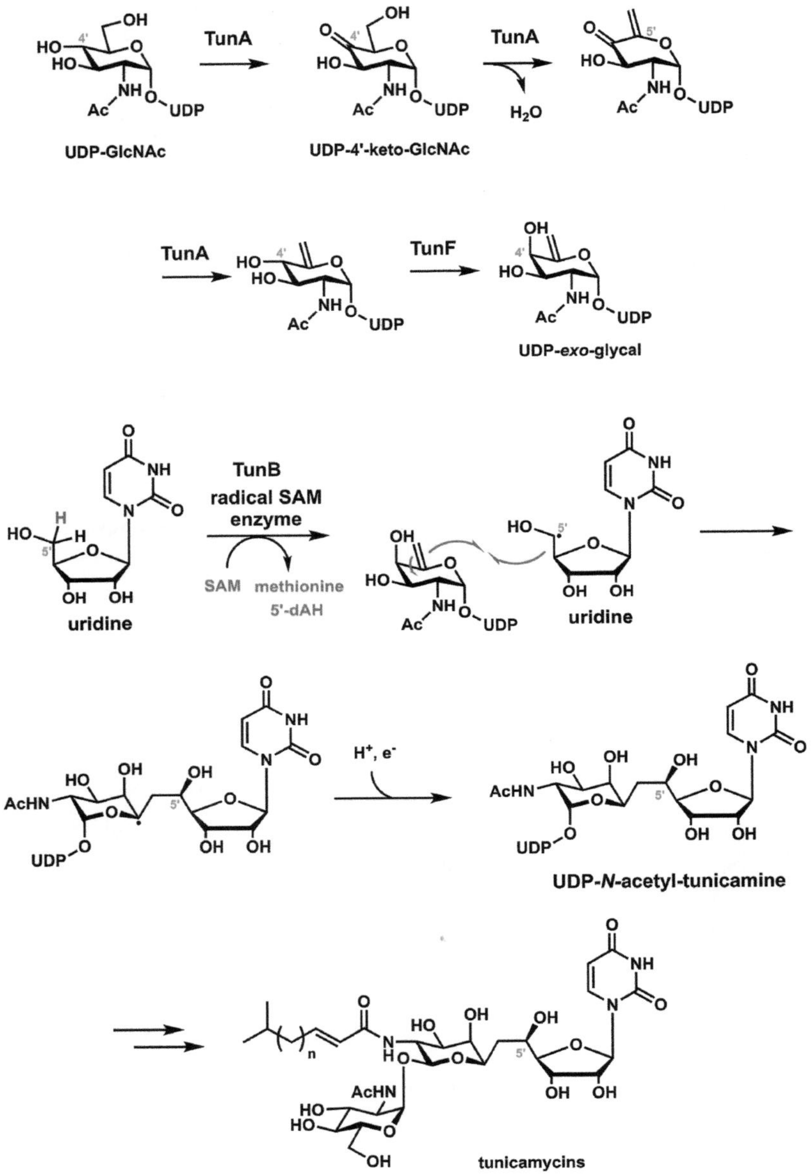

Figure 6.23 An *exo*-methylene glycal in the biosynthetic pathway to the tunicamine core of tunicamycins.

UDP-*N*-acetylglucosamine (UDP-GlcNAc) is oxidized to the 4′-keto intermediate from which loss of water generates the 4′,6′-ene-one. Re-reduction of the 4′-keto group followed by epimerization creates the UDP-*exo*-methylene glycal that is featured as the cosubstrate for coupling at C_5 of uridine. The enzyme TunB is predicted to use SAM as a radical initiator (see Chapter 10), presumably creating a transient radical at $C_{5'}$ of uridine for coupling to the glycal olefin. The resultant product is the featured UDP-tunicamine core set up for subsequent GlcNAc acylation and replacement of the acetyl moiety with a long chain fatty acyl group at later stages of tunicamycin formation.

A radical SAM-dependent route to C–C extension at $C_{5'}$ of uridine scaffolds also occurs in nikkomycins, polyoxins, and malayamycins. The malayamycins retain a bicyclic 5-6 octosyl acid ring system which appears to be an early common framework (as shown in Figure 6.24) (He, Wu *et al.* 2017). The first step is enzymatic conversion of the 3′-OH of UMP into the enolpyruvyl ether with phosphoenolpyruvate (PEP) as a cosubstrate (He, Wu *et al.* 2017, Lilla and Yokoyama 2016). Then SAM is utilized by PolH as a radical generator of the 5′-deoxyadenosyl radical in the active site of the enzyme. The resultant 5′-dA• is proposed to abstract one of the two prochiral hydrogens at $C_{5'}$ as a hydrogen atom, generating the observed product 5′-deoxyadenosine. The $C_{5'}$ radical is proposed to react with the 3′-ene to create the 5,6-bicyclic radical. Transfer of the unpaired electron to an active site Cys thiol produces the product octosyl acid phosphate in which the new C–C bond has been formed at $C_{5'}$ of the UMP scaffold.

The bicyclic ring can persist, as shown in malayamycin A. Alternatively, in polyoxin biosynthesis a pair of α-ketoglutarate-requiring oxygenases are proposed to oxidatively open and chain shorten the side chain to the uridine glyoxalate (He, Wu *et al.* 2017). A reductive transamination generates the aminohexuronate framework that is characteristic of polyoxins and nikkomycins.

The final molecule in this trio is jawsamycin (Hiratsuka, Suzuki *et al.* 2014) (Figure 6.25), named presumably because the five cyclopropane rings call to mind the shark teeth in the movie *Jaws*. It is a microbial product from *Streptoverticillium fervens* and has potent antifungal activity. The run of cyclopropane rings is a strikingly unusual feature. Jawsamycin represents a biosynthetic convergence of a uridine-5′-amine moiety with an acyl chain clearly derived from a polyunsaturated fatty acid-forming polyketide synthase assembly line. The uridine-5′-amine arises by formation of the uridine 5′-aldehyde as in above examples, either by an oxygenase or an oxidoreductase (both types of enzyme are precedented for this conversion) and subsequent transamination.

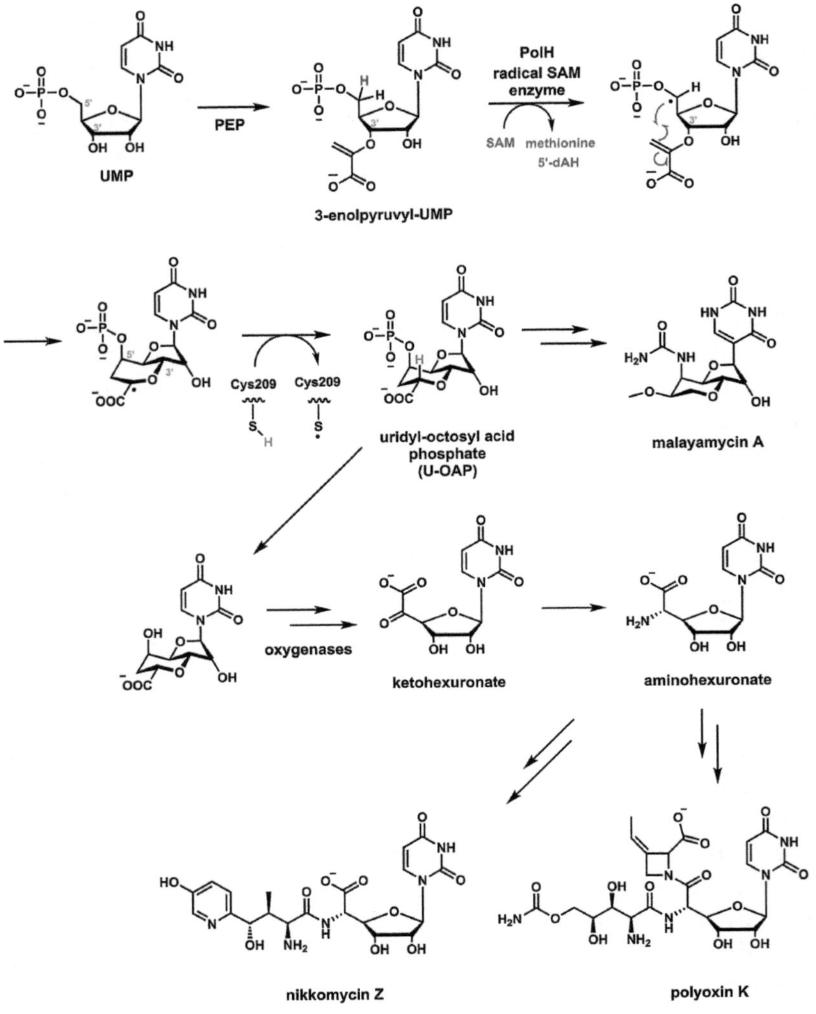

Figure 6.24 A radical SAM enzyme is involved in generating octosyl acid intermediate from UMP during the biosynthesis of nikkomycin and polyoxin peptidyl nucleosides.

The methyl group of SAM is the source for the one-carbon unit in each of the five cyclopropanes. Jaw5 is predicted to be another one of the radical SAM enzymes (three noted previously in this short chapter, pointing towards full mechanistic discussion in Chapter 10), and functions as the iterative cyclopropanation catalyst while the polyunsaturated fatty acyl chain is still on the polyketide synthase.

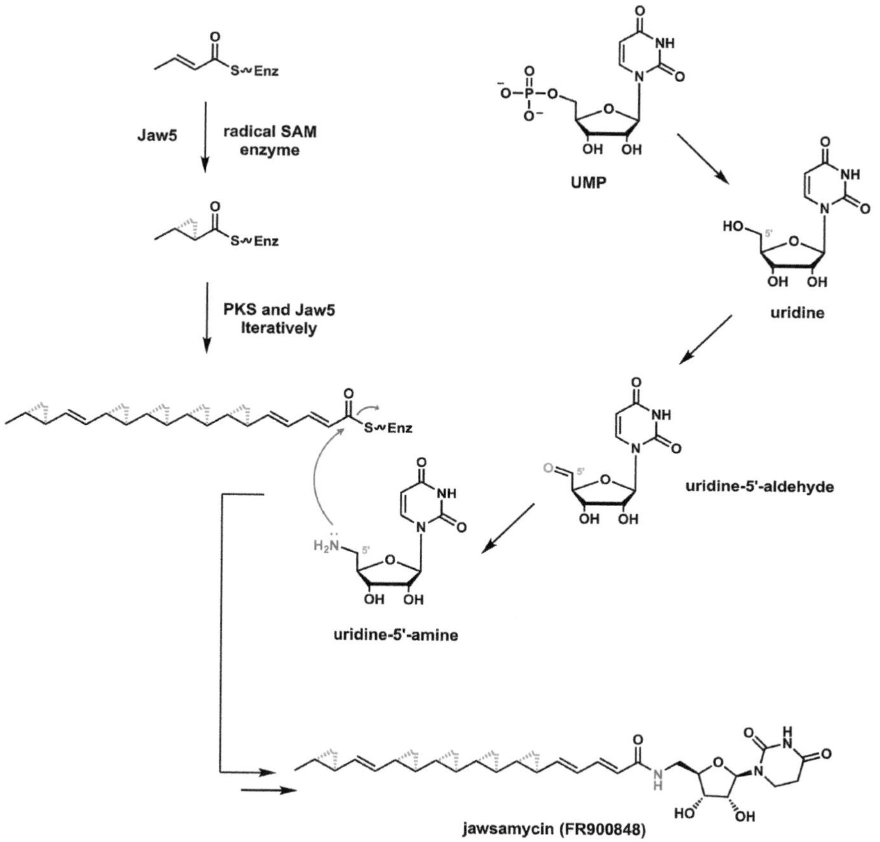

Figure 6.25 Proposed route to Jawsamycin.

6.9 Summary

The monocyclic pyrimidine and the bicyclic purine (imidazolopyrimidine) heterocycles are the key bits for both RNA and DNA informational macromolecules, providing the one-letter codes that specify each gene. They are presumed to be ancient, even prebiotic, nitrogen heterocycles. Contemporary organisms build these heterocycles into ribonucleotide and 2′-deoxyribonucleotide scaffolds as the biosynthetic products, immediately usable for RNA and DNA biosynthesis.

On the other hand, a relatively small number of nucleoside natural products are known that derive from controlled enzymatic hydrolysis of the nucleotide phosphoester linkages. An even smaller number of nucleosides are hydrolyzed to the free heterocycles. The purine alkaloids caffeine, theophylline, and theobromine are famous pharmacologically active examples.

Natural product variations can occur in any of the three parts of nucleotide frameworks: the heterocycle, the sugar, or the groups attached at the 4′ or 5′ carbons that create differentiated biological activities. Heterocyclic variations include the pyrrolopyrimidine ring systems in 7-deazaguanines as well as the *N*-alkylated plant cytokinin hormones. Sugar variations include carbacycles, and hexoses in place of the core ribose. They also include arabinose, the $C_{2'}$ epimer of ribose in nucleosides that were clues to synthetic antiviral nucleoside drugs. Substituents at the $C_{4'}$ and $C_{5'}$ carbons are found in peptidyl nucleosides of several classes and the tunicamycins.

References

Bandarian, V. and C. L. Drennan (2015). "Radical-mediated ring contraction in the biosynthesis of 7-deazapurines". *Curr. Opin. Struct. Biol.* **35**: 116–124.

Barnard-Britson, S., X. Chi, K. Nonaka, A. P. Spork, N. Tibrewal, A. Goswami, P. Pahari, C. Ducho, J. Rohr and S. G. Van Lanen (2012). "Amalgamation of nucleosides and amino acids in antibiotic biosynthesis: discovery of an L-threonine:uridine-5'-aldehyde transaldolase". *J. Am. Chem. Soc.* **134**(45): 18514–18517.

Bergmann, W. and D. Burke (1955). "Contributions to the study of marine products: The nucleosides of sponges .3. Spongothymidine and spongouridine". *J. Org. Chem.* **20**: 1501–1507.

Bergmann, W. and R. Feeney (1950). "The isolation of a new thymine pentoside from sponges". *J. Am. Chem. Soc.* **72**: 2809–2810.

Birmingham, W. R., C. A. Starbird, T. D. Panosian, D. P. Nannemann, T. M. Iverson and B. O. Bachmann (2014). "Bioretrosynthetic construction of a didanosine biosynthetic pathway". *Nat. Chem. Biol.* **10**(5): 392–399.

Carreras, C. W. and D. V. Santi (1995). "The catalytic mechanism and structure of thymidylate synthase". *Annu. Rev. Biochem.* **64**: 721–762.

Cech, T. R. (2012). "The RNA worlds in context". *Cold Spring Harbor Perspect. Biol.* **4**(7): a006742.

Chung, B. C., E. H. Mashalidis, T. Tanino, M. Kim, A. Matsuda, J. Hong, S. Ichikawa and S. Y. Lee (2016). "Structural insights into inhibition of lipid I production in bacterial cell wall synthesis". *Nature* **533**(7604): 557–560.

Cimino, G., S. de Rosa and S. de Stefano (1984). "Antiviral agents from a gorgonian, Eunicella cavolini". *Experientia* **40:** 339–340.

Cone, M. C., X. Yin, L. L. Grochowski, M. R. Parker and T. M. Zabriskie (2003). "The blasticidin S biosynthesis gene cluster from Streptomyces griseochromogenes: sequence analysis, organization, and initial characterization". *ChemBioChem* **4**(9): 821–828.

D'Andrea, A. D. and W. A. Haseltine (1978). "Modification of DNA by aflatoxin B1 creates alkali-labile lesions in DNA at positions of guanine and adenine". *Proc. Natl. Acad. Sci. U. S. A.* **75**(9): 4120–4124.

Dewick, P. (2009). *Medicinal Natural Products, A Biosynthetic Approach.* UK, Wiley.

Diaz-Marrero, A. R., P. Austin, R. Van Soest, T. Matainaho, C. D. Roskelley, M. Roberge and R. J. Andersen (2006). "Avinosol, a meroterpenoid-nucleoside conjugate with antiinvasion activity isolated from the marine sponge Dysidea sp". *Org. Lett.* **8**(17): 3749–3752.

Einset, J. W. (1986). "Zeatin biosynthesis from N6-(A2-isopentenyl)adenine in Actinidia and other woody plants". *Proc. Natl. Acad. Sci. U. S. A.* **83:** 972–975.

Frebort, I., M. Kowalska, T. Hluska, J. Frebortova and P. Galuszka (2011). "Evolution of cytokinin biosynthesis and degradation". *J. Exp. Bot.* **62**(8): 2431–2452.

He, N., P. Wu, Y. Lei, B. Xu, X. Zhu, G. Xu, Y. Gao, J. Qi, Z. Deng, G. Tang, W. Chen and Y. Xiao (2017). "Construction of an octosyl acid backbone catalyzed by a radical S-adenosylmethionine enzyme and a phosphatase in the biosynthesis of high-carbon sugar nucleoside antibiotics". *Chem. Sci.* **8:** 444–451.

Hiratsuka, T., H. Suzuki, R. Kariya, T. Seo, A. Minami and H. Oikawa (2014). "Biosynthesis of the structurally unique polycyclopropanated polyketide-nucleoside hybrid jawsamycin (FR-900848)". *Angew. Chem., Int. Ed.* **53**(21): 5423–5426.

Huang, R. M., Y. N. Chen, Z. Zeng, C. H. Gao, X. Su and Y. Peng (2014). "Marine nucleosides: structure, bioactivity, synthesis and biosynthesis". *Mar. Drugs* **12**(12): 5817–5838.

Jenkins, G. J. and N. J. Turner (1995). "The biosynthesis of carbocyclic nucleotides". *Chem. Soc. Rev.* 169–176.

Jiang, L., J. Wei, L. Li, G. Niu and H. Tan (2013). "Combined gene cluster engineering and precursor feeding to improve gougerotin production in Streptomyces graminearus". *Appl. Microbiol. Biotechnol.* **97**(24): 10469–10477.

Kudo, F., T. Tsunoda, M. Takashima and T. Eguchi (2016). "Five-Membered Cyclitol Phosphate Formation by a myo-Inositol

Phosphate Synthase Orthologue in the Biosynthesis of the Carbocyclic Nucleoside Antibiotic Aristeromycin". *ChemBioChem* **17**: 2143–2148.

Lilla, E. A. and K. Yokoyama (2016). "Carbon extension in peptidylnucleoside biosynthesis by radical SASAM enzymes". *Nat. Chem. Biol.* **12**: 905–907.

Niitsuma, M., J. Hashida, M. Iwatsuki, M. Mori, A. Ishiyama, M. Namatame, A. Nishihara-Tsukashima, A. Matsumoto, Y. Takahashi, H. Yamada, K. Otoguro, K. Shiomi and S. Omura (2010). "Sinefungin VA and dehydrosinefungin V, new anti-trypanosomal antibiotics produced by Streptomyces sp. K05-0178". *J. Antibiot.* **63**(11): 673–679.

Nordlund, P. and P. Reichard (2006). "Ribonucleotide reductases". *Annu. Rev. Biochem.* **75**: 681–706.

Parry, R. and Y. Jiang (1994). "The biosynthesis of aristeromycin. Conversion of neplanocin A to aristeromycin by a novel enzymatic reduction". *Tetrahedron Lett.* **35**: 9665–9668.

Price, N. P. and B. Tsvetanova (2007). "Biosynthesis of the tunicamycins: a review". *J. Antibiot.* **60**(8): 485–491.

Ribeiro, J. A. and A. M. Sebastiao (2010). "Caffeine and adenosine". *J. Alzheimer's Dis.* **20**(Suppl 1): S3–S15.

Taiz, L., E. Zeiger, I. Moller and A. Murphy, eds. (2015). *Plant Physiology and Development*, 6th edn, Sinauer Associates.

Thibodeaux, C. J., C. E. Melancon, 3rd and H. W. Liu (2008). "Natural-product sugar biosynthesis and enzymatic glycodiversification". *Angew. Chem., Int. Ed.* **47**(51): 9814–9859.

Thibodeaux, C. J., C. E. Melancon and H. W. Liu (2007). "Unusual sugar biosynthesis and natural product glycodiversification". *Nature* **446**(7139): 1008–1016.

Thomas, E., M. G. Ghany and T. J. Liang (2013). "The application and mechanism of action of ribavirin in therapy of hepatitis C". *Antiviral Chem. Chemother.* **23**(1): 1–12.

Van Lanen, S., N. Koichi, J. Unrine and Z. Yang (2012). "Fe(II)-Dependent, Uridine-5′-Monophosphate α-Ketoglutarate Dioxygenases in the Synthesis of 5′-Modified Nucleosides". *Methods Enzymol.* **516**: 153–168.

Walsh, C. T. (2015). "A chemocentric view of the natural product inventory". *Nat. Chem. Biol.* **11**(9): 620–624.

Walsh, C. T. and W. Zhang (2011). "Chemical logic and enzymatic machinery for biological assembly of peptidyl nucleoside antibiotics". *ACS Chem. Biol.* **6**(10): 1000–1007.

Wu, J., L. Li, Z. Deng, T. M. Zabriskie and X. He (2012). "Analysis of the mildiomycin biosynthesis gene cluster in Streptoverticillum remofaciens ZJU5119 and characterization of MilC, a hydroxymethyl cytosyl-glucuronic acid synthase". *ChemBioChem* **13**(11): 1613–1621.

Wyszynski, F. J., S. S. Lee, T. Yabe, H. Wang, J. P. Gomez-Escribano, M. J. Bibb, S. J. Lee, G. J. Davies and B. G. Davis (2012). "Biosynthesis of the tunicamycin antibiotics proceeds *via* unique exo-glycal intermediates". *Nat. Chem.* **4**(7): 539–546.

Zhang, J. and Y. G. Zheng (2016). "SAM/SAH Analogs as Versatile Tools for SAM-Dependent Methyltransferases". *ACS Chem. Biol.* **11**(3): 583–597.

Zhang, W., I. Ntai, M. L. Bolla, S. J. Malcolmson, D. Kahne, N. L. Kelleher and C. T. Walsh (2011). "Nine enzymes are required for assembly of the pacidamycin group of peptidyl nucleoside antibiotics". *J. Am. Chem. Soc.* **133**(14): 5240–5243.

Lignin derived from the phenylpropanoid pathways is a principle component of wood.

7 Phenylpropanoid Natural Product Biosynthesis

7.1 Introduction

The class of molecules known as phenylpropanoids derive from the single amino acid phenylalanine and typically retain its nine carbon atoms, the six in the phenyl ring and the three in the propane side chain. As we shall note below, the first reaction to all the phenylpropanoid scaffolds involves enzyme-mediated loss of the elements of ammonia across the C_2C_3 bond of L-Phe, so the side chain is immediately at the propenoic rather than the propanoic oxidation state.

Figure 7.1 indicates that two early central intermediates in phenylpropanoid biogenesis are cinnamic acid and its 4-hydroxy derivative *para*-coumaric acid (Vogt 2010, Fraser and Chapple 2011). The figure also notes that phenylalanine, cinnamate, and 4-OH-coumarate can yield suites of volatile metabolites as well as the more complex scaffolds that are nonvolatile. The volatiles such as the simple alcohols, aldehydes, esters such as methyl benzoate, and the terminal olefins eugenol and methyl eugenol are components of plant essential oils and retain fewer than the canonical nine carbons by virtue of oxidative metabolism.

Figure 7.1 also reveals that phenylalanine and *para*-coumarate are progenitors to most of the subclasses of phenylpropanoid secondary metabolites, including olefinic stilbenes (resveratrol), chalcones, and flavonoids (*e.g.* naringenin) (Pled-Zehavi, Oliva *et al.* 2015). In turn, the flavonoids are precursors to isoflavonoids (*e.g.* genistein) that serve in plant defense, and to the anthocyanin pigments in flowers as

Natural Product Biosynthesis: Chemical Logic and Enzymatic Machinery
By Christopher T. Walsh and Yi Tang

Published by the Royal Society of Chemistry, www.rsc.org

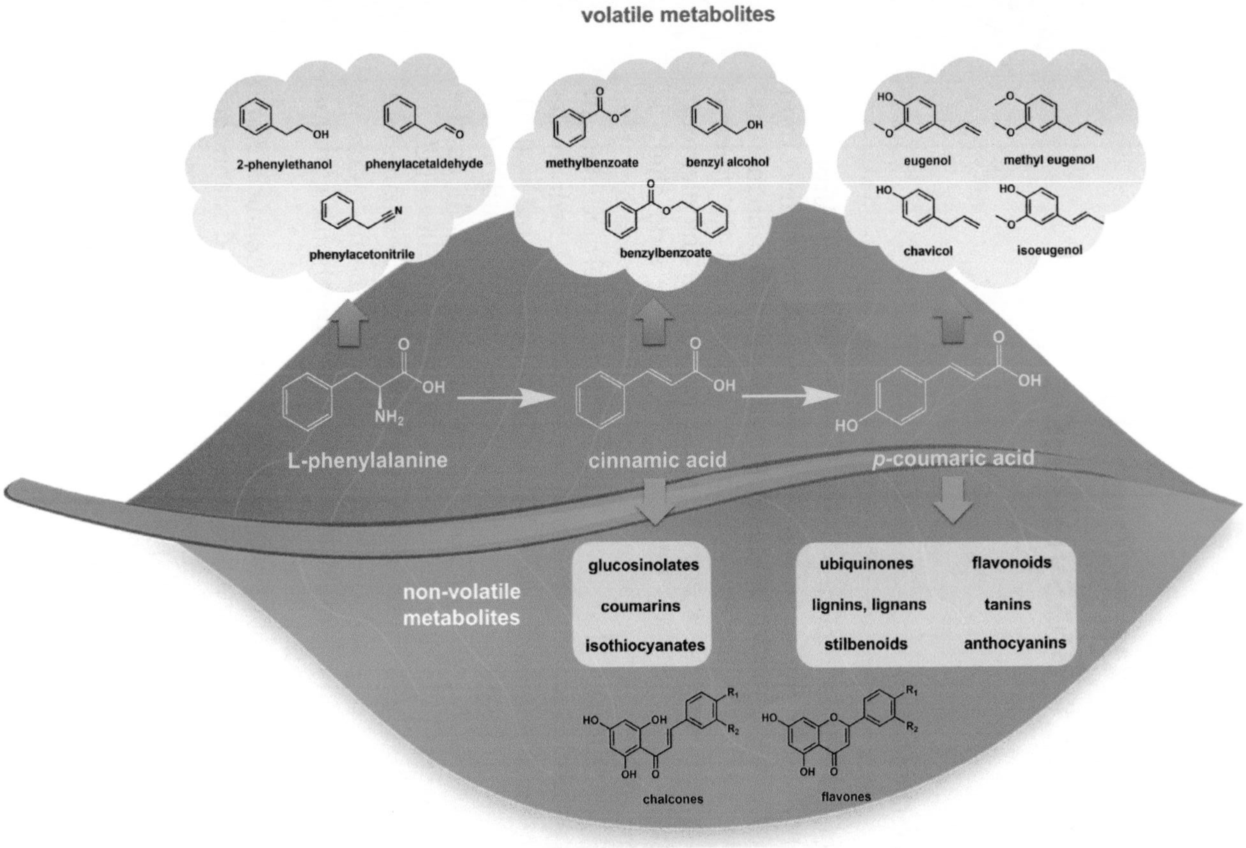

Figure 7.1 Phenylalanine is the entry point to all phenylpropanoid metabolites. Cinnamic aid and then 4-OH-coumaric acid represent branch points for subclasses of phenylpropanoids.
Adapted from Peled-Zehavi, H., M. Oliva, Q. Xie, V. Tzin, M. Oren-Shamir, A. Aharoni and G. Galili (2015). "Metabolic Engineering of the Phenylpropanoid and Its Primary, Precursor Pathway to Enhance the Flavor of Fruits and the Aroma of Flowers". *Bioengineering* **2**(4): 204–212. Published under the terms and conditions of the Creative Commons Attribution License (http://creativecommons.org/licenses/by/4.0/).

well as condensed tannins. A distinct, ramifying pathway from *para*-coumarate leads to a set of three alcohols, *p*-coumaryl alcohol, coniferyl alcohol, and sinapyl alcohol. These are often termed monolignols to reflect their ability to undergo two oligomerization fates. They can be dimerized to 8-8′ connected lignans such as pinoresinol. They can also undergo more extensive, peroxidative oligomerization to insoluble lignins which play key structural roles in plants, ensuring water transport through the different parts of plants (Dewick 2009).

Figure 7.2 schematizes different distributions of phenylpropanoid metabolites in plant tissues and distinct functions. These include

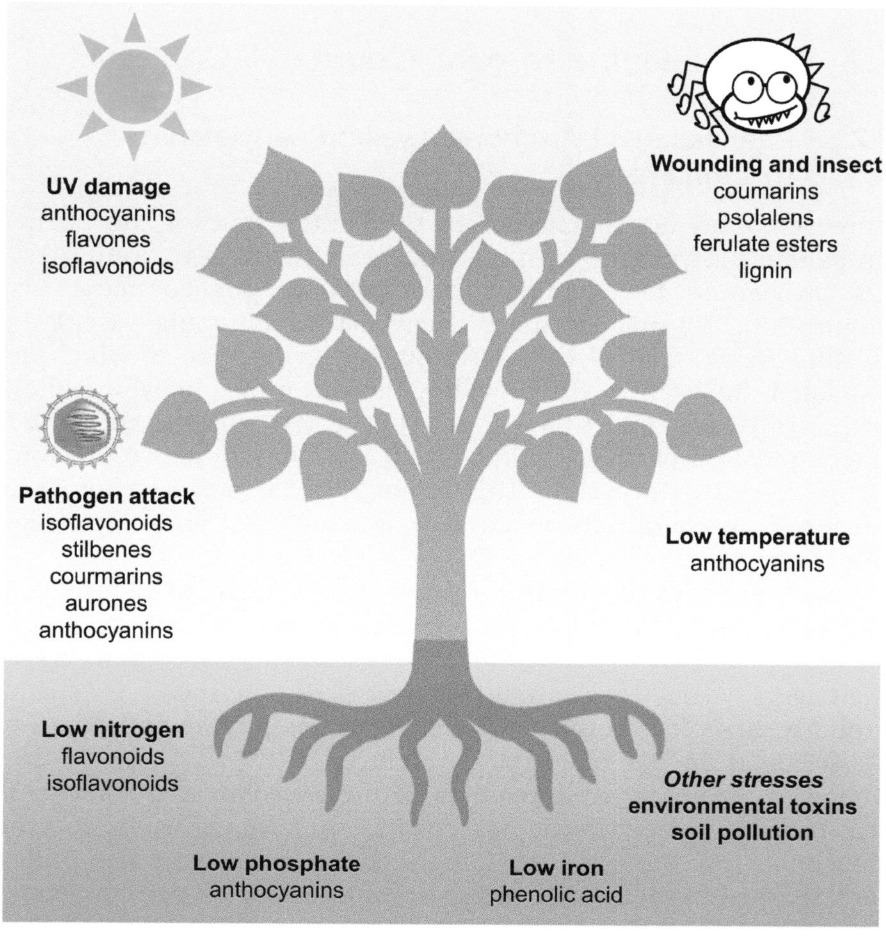

Figure 7.2 Multiple roles for a wide range of phenylpropanoid metabolites in plant physiology.
Adapted from http://www.phytostem.com. Copyright 2011 PHYTO-STEM and/or its suppliers.

protection against UV radiation damage in leaves, defense against insect attack, for example *via* antifeedant properties, and tissue invasion, and signaling functions to sense limiting nitrogen, phosphate, or iron in soils. In the flavonoid subclass alone, there are 10 000 characterized molecules (Dictionary of Natural Products, dnp.chemnetbase.com) (Harborne 1999) that span a range of properties from flower color (anthocycanins and aurones), UV protection (flavanols), root nodule formation (isoflavones), and plant insect herbivory protection (flavanols). Also documented are pollination functions *via* flower colors, and pollen fertility (mix of flavonoids).

7.2 Phenylalanine to *para*-Coumaryl-CoA

7.2.1 Phenylalanine Ammonia Lyase: Phenylalanine to Cinnamate

Three enzymes act in tandem in the central lane of the phenylpropanoid biosynthetic highway to convert the primary amino acid L-phenylalanine to cinnamate and then to *para*-coumaryl CoA (Figure 7.3). The first enzyme is phenylalanine ammonia lyase (PAL). In *Arabidopsis thaliana* there are four *pal* genes, two of which are regulated consistently with providing flux down the phenylpropanoid pathways (Fraser and Chapple 2011). The PAL enzyme carries out the elimination of the C_2 amine group and one of the two prochiral C_3 hydrogens as a proton. This is not a trivial process chemically because C_3 is chemically unactivated, making C_3–H proton loss high in energy.

Consistent with this chemical reactivity challenge, the active form of phenylalanine ammonia lyase has an unusual prosthetic group, 5-methylidine-imidazolone (MIO), in its active site which lowers the energy barrier for catalysis (Figure 7.4). It is termed a prosthetic group because it is covalently tethered to the enzyme and does not dissociate at the end of a catalytic cycle.

One of the remarkable features of the covalent MIO cofactor is that it is not present in the nascent PAL protein as it comes off the ribosome. Instead, once the (single chain) proenzyme has folded into its three-dimensional native conformation, or perhaps attendant to reaching that final folded conformer, three adjacent residues, Gly202, Ser203, and Ala204, in a loop undergo autocataytic maturation to produce MIO as shown (Walsh 2005). The amide NH of Ala204 is featured to attack the neighboring amide carbonyl of

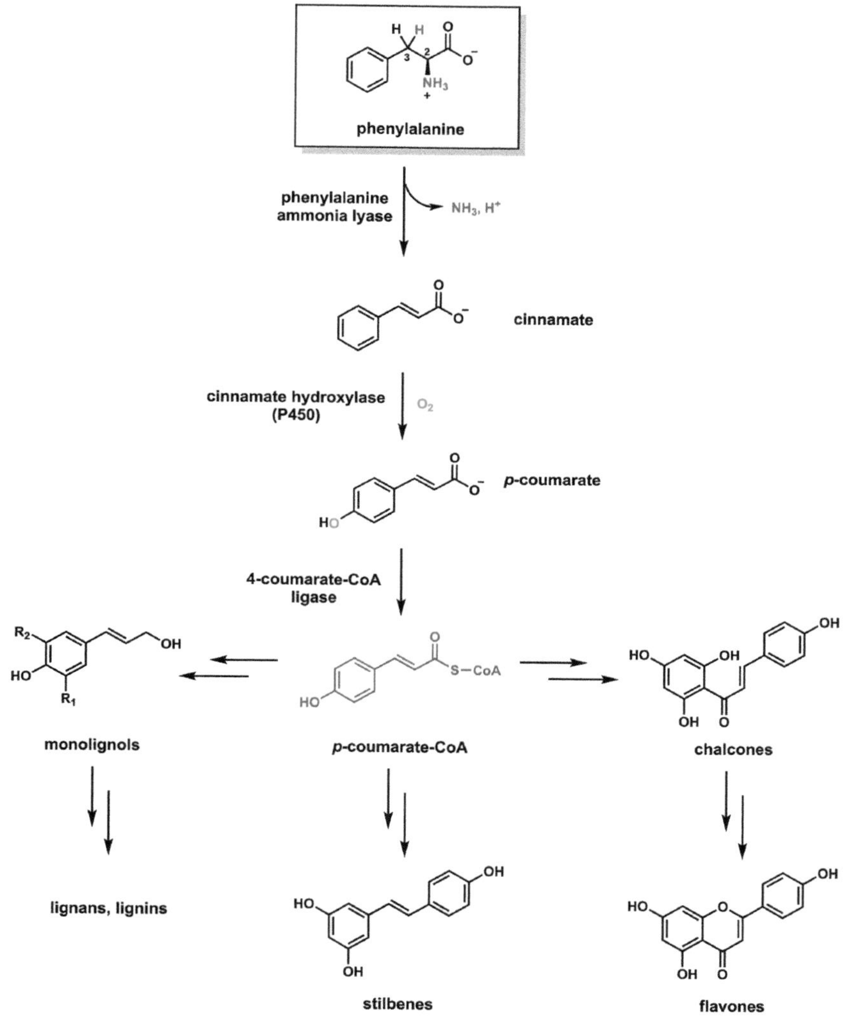

Figure 7.3 Phenylalanine to *para*-coumaryl-CoA is carried out by three enzymes that constitute a central early pathway to phenylpropanoids.

Ser202, followed by consecutive loss of two water molecules, the second across the C_2, C_3 bond of the Ser203 residue to yield the exocyclic methylidene.

That olefin is conjugated to the imidazolone carbonyl and can function as an electrophile. Both mechanistic and X-ray structural analyses support the mechanism shown in Figure 7.5. Attack of the bound Phe-amino group on the olefinic terminus of MIO produces

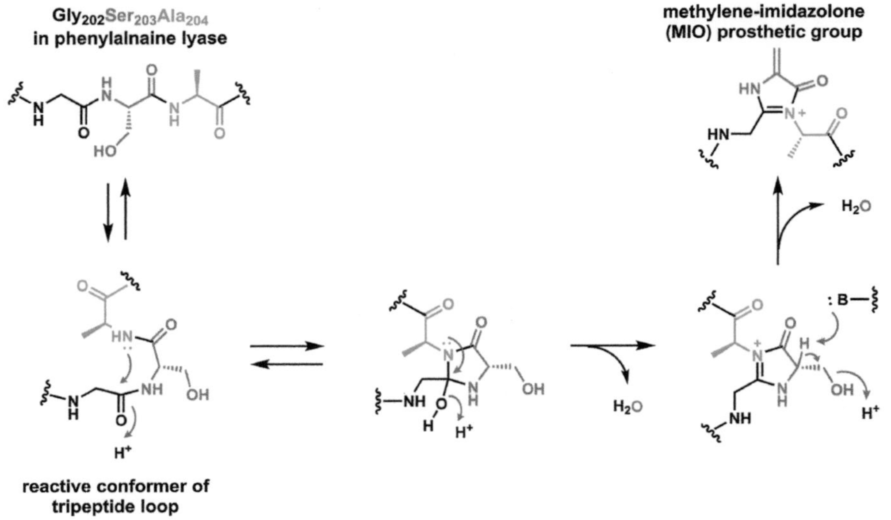

Figure 7.4 Autoconversion of Gly_{202}–Ser_{203}–Ala_{204} to the methylene-imidazolone (MIO) prosthetic group (an unusual invention in enzyme catalysis) as active site electrophile after autoactivation of phenylalanine lyase.

a covalent intermediate. In that structure the barrier to C_3–H cleavage is lowered, as is that for cleavage of the C_2–N bond. This completes the fragmentation of Phe into cinnamate, which diffuses out of the active site. Meanwhile the Phe-derived NH_2 is still covalently connected to the MIO prosthetic group. It can be extruded by isomerization of the enol form of MIO back to the heteroaromatic imidazolone ring system, setting PAL back in its original MIO adduct state, ready for the next catalytic cycle.

The MIO group is an unusual invention in protein-mediated catalysis. It forms autocatalytically within the folded protein's active site and therein provides an unusual electron sink to enable deamination of Phe (a comparable MIO group also forms in the folded state of histidine deaminase (Walsh 2005)). Thus, a unique chemical strategy and enzyme machinery sit at the entryway to all the phenylpropanoid metabolites of plants. It is also worth noting that, while amino acid basic nitrogens are key functional groups in all the alkaloids noted in Chapters 5 and 8, the amine of phenylalanine is jettisoned in the very first step. Phenylpropanoid natural products are not about basic nitrogens but rather about the olefins conjugated to aromatic rings that have been hydroxylated to allow facile phenoxy radical chemistry.

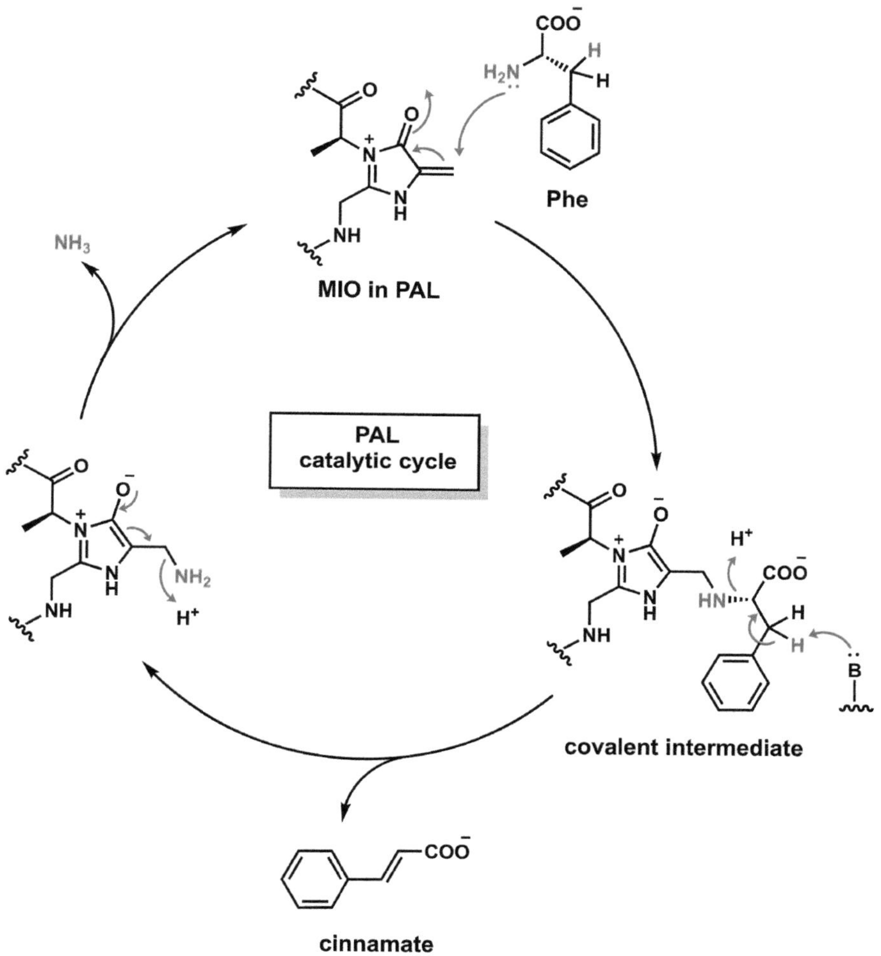

Figure 7.5 Mechanistic proposal for phenylalanine lyase utilizing the MIO prosthetic group as electrophile for removal of ammonia.

One additional comment about metabolic flux demands inclusion. A substantial amount of the L-phenylalanine pool is shunted away from protein synthesis and down the phenylpropanoid pathway, including to the insoluble structural polymer lignin. It has been estimated that lignin is the second most abundant biopolymer on the planet after cellulose. In turn, this requires a significant flux to phenylalanine from the central aromatic biosynthetic pathway (Dixon, Achnine *et al.* 2002). As Figure 7.6 reminds us, this is the shikimate pathway that leads to chorismate, prephenate, and

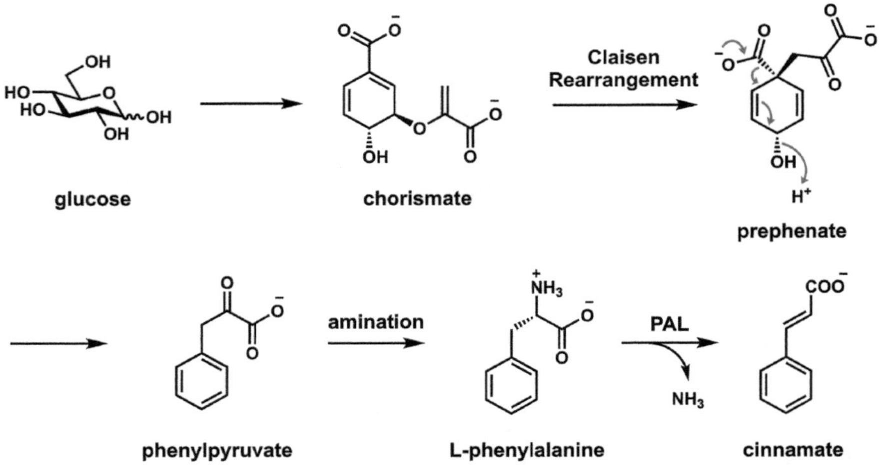

Figure 7.6 A significant amount of flux through the shikimate pathway to phenylalanine is required to provide the material needed to create the massive amounts of lignan in woody plants.

phenylpyruvate before transamination yields phenylalanine. In the two-step process from phenylpyruvate to phenylalanine to cinnamate, the amino group is installed in one step and discarded in the next. On the other hand, there is no other obvious way to get from phenylpyruvate to cinnamate. We will return to shikimate in the guise of caffeoyl-shikimate esters as a metabolic signal for forward conversion of carbon into lignan and lignin biosynthesis in a subsequent section.

7.2.2 Cinnamate Hydroxylase: Cinnamate to *para*-Coumarate

The second of the three enzymes in the central arm of the phenylpropanoid pathway is the first of three oxygenases of the cytochrome P450 family that act in tandem with methyltransferases and reductases in the lignan and lignin arms downstream (Vogt 2010, Fraser and Chapple 2011). The cinnamate hydroxylase appears to be a traditional monooxygenase with *para*-regiospecificity on the phenyl ring, most likely proceeding *via* an arene oxide intermediate that opens with a 1,2-hydride shift to the 4-OH-cyclohexadiene tautomer that then rearomatizes to the phenol (Rupasinghe, Baudry *et al.* 2003).

7.2.3 *para*-Coumarate CoA Ligase: *para*-Coumarate to *para*-Coumarate CoA

The third enzyme in the early common stage is an ATP-dependent acyl-CoA ligase. A molecule of ATP is spent to activate the substrate carboxylate, typically as the acyl-AMP mixed anhydride. This is thermodynamically activated but kinetically unstable to hydrolysis as a freely diffusing intermediate. Capture by the thiolate of coenzyme A converts the acyl-AMP to the acyl thioester, retaining the thermodynamic activity and now gaining sufficient kinetic stability to survive as a freely mobilizable metabolite in cellular milieus (see Figure 7.3).

7.3 Monolignol, Lignan and Lignin Biosynthesis

7.3.1 *para*-Coumaryl-CoA to Three Monolignols

The strategy for the next phase of the pathway, from *para*-coumaryl-CoA to three building blocks, monolignols (*p*-coumaryl alcohol, coniferyl alcohol, and sinapyl alcohol), for both dimeric lignan and oligomeric lignin formation, utilizes three kinds of enzyme. One is a two enzyme set comprising the second and third cytochromes P450 (the first being the cinnamate to 4-OH-coumarate) in the pathway. They are regioselective hydroxylases, acting at C_3 and C_5 of the aromatic ring on their respective substrates. Although it had been initially proposed that all three P450s would work on the free carboxylic acid forms of cinnamate, coumarate, and ferulate, this turns out not to be the case (Vanholme *et al.* 2010).

Surprisingly, the three P450s, although in the same P450 subfamily, discriminate on the basis of the chemical form of C_1 of the substrate. P450 #2 does not work on either coumarate or the coumaryl-CoA thioester. Instead the *para*-coumaryl group is transferred enzymatically to one of the hydroxyl groups of shikimate. The coumaryl-shikimate ester (Figure 7.7) is the substrate for this C_3-specific P450 oxygenase. Then the 3,4-dihydroxy acyl group is transferred enzymatically back to CoA-SH. The caffeoyl-CoA is substrate for a SAM-utilizing 3-*O*-methyl transferase to give feruloyl-CoA. This molecule is now a substrate for the thioester to aldehyde oxidoreductase. This coniferyl aldehyde (coniferaldehyde) is the substrate for cytochrome P450 #3, regiospecifically hydroxylating carbon 5. In turn this is substrate for the next SAM-dependent 5-*O*-methyltransferase, yielding sinapyl aldehyde (sinapaldehyde).

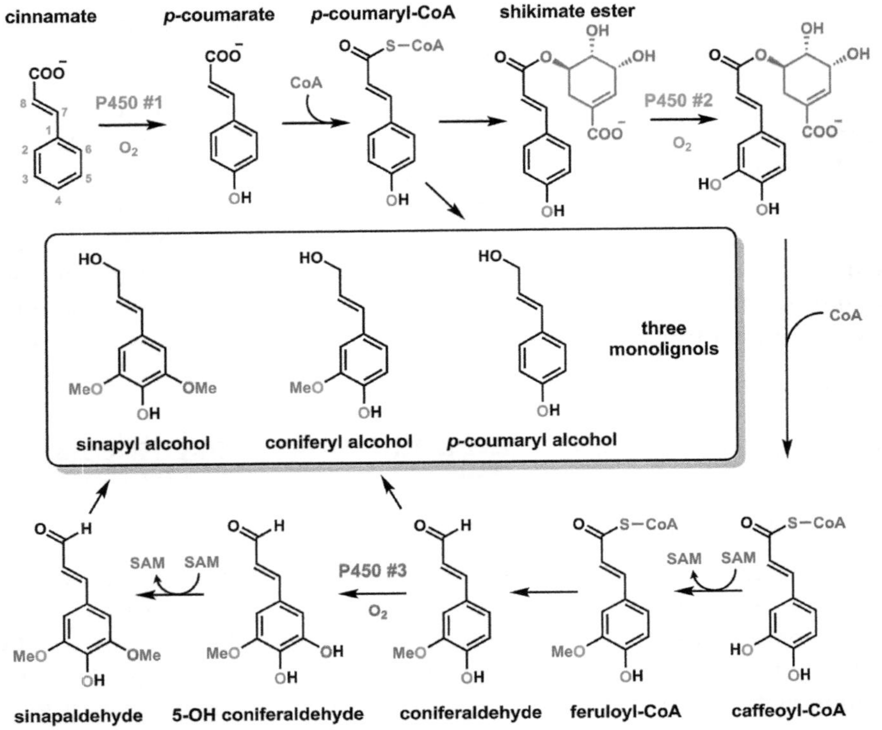

Figure 7.7 Cinnamate to the three monolignols. Three P450 oxygenases work on different forms of the substrates.

The rationale for alternating oxygenase and *O*-methyltransferase is not obvious but may be the metric for building in selective hydroxylation and methylation patterns. The interposition of a coumaryl-shikimate ester may be the means by which plant cells and tissues know whether there is enough carbon available (Figure 7.6) to send it down this highly ramifying secondary metabolic pathway.

7.3.2 Dimerization of Monolignols to Lignans

All three of the monolignols noted above have one free phenol hydroxyl group. The *O*-methylations in the coniferyl and sinapyl scaffolds have left only the *para*-OH free to react in one-electron oxidative dimerization modes. In plants where all three monolignols are present, homotypic and heterotypic coupling of long lived

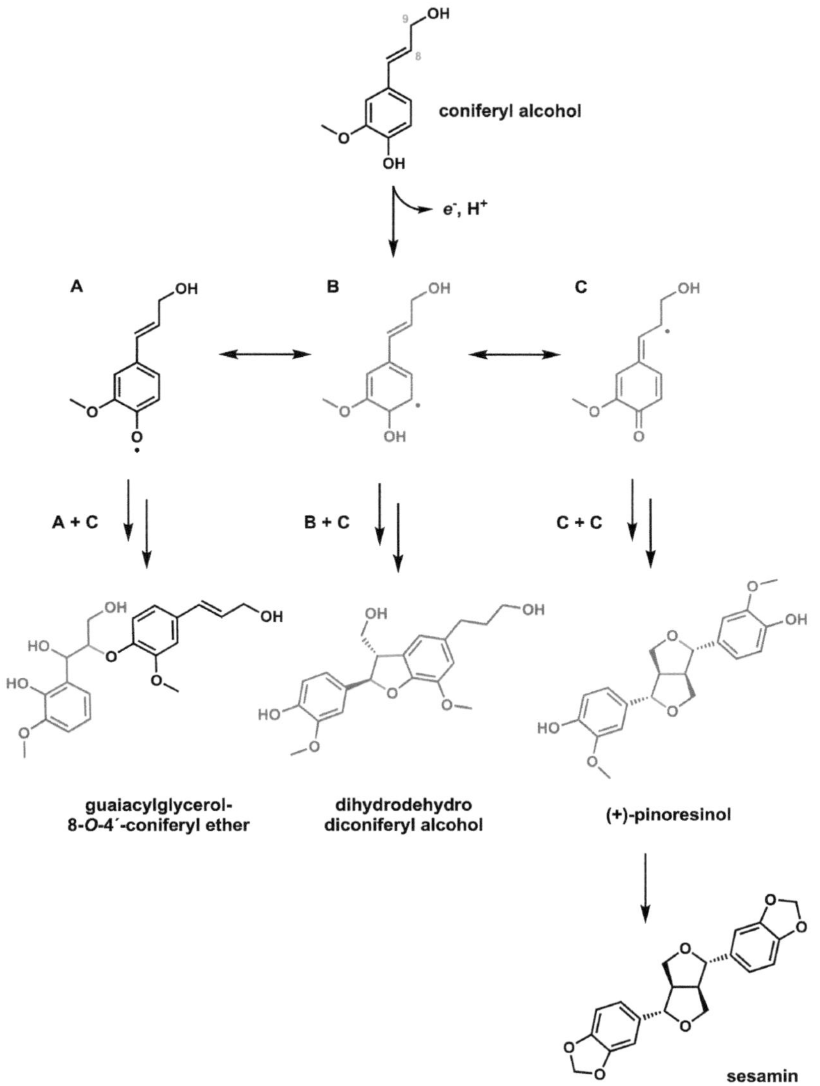

Figure 7.8 Oxidative dimerization between different pairs of the three monolignols generates the dimeric lignans *via* phenoxy radicals that yield different coupling regiochemistry outcomes.

phenoxy radicals are observed to yield an array of dimers (Davin and Lewis 2000, Davin and Lewis 2003). The dimers are all coupled through carbons 8 and 8′ (Figure 7.8), but can have substantial structural variation around the new 8-8′ C–C bond. For example,

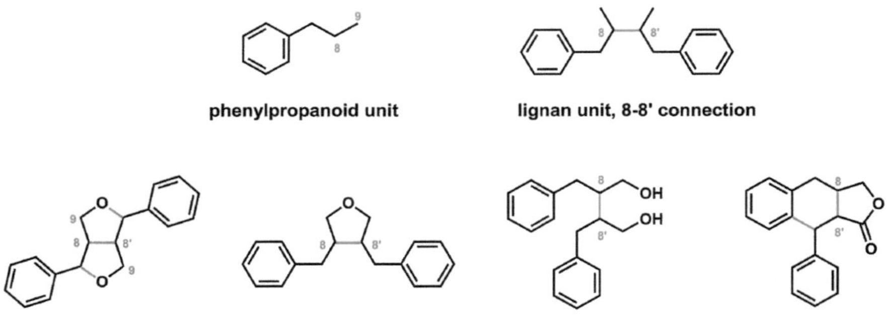

Figure 7.9 Some lignan subclass frameworks.

Figure 7.9 shows four lignan structural subclasses, furanofuran, furan, dibenzylbutane, and aryltetralin. We will illustrate how each of these forms in subsequent analysis of podophyllotoxin biogenesis. As shown in Figure 7.8, the homotypic radical coupling of three resonance contributors of a coniferyl alcohol radical yields three structural product forms, including the furanofuran pinoresinol. Noncatalytic dirigent proteins bind the monomers and direct regiochemistry of coupling in at least some cases (Davin and Lewis 2000). Further metabolism of each of these subclasses can give appreciation of how >300 lignan dimers and derivatives have been found to date. They have a variety of functional activities, including insect antifeeding and arrested juvenile development functions for insect pathogens (Harmatha and Dinan 2003, Satake, Ono *et al.* 2013). The conversion of pinoresinol to the bis methylenedioxy-bridged lignan sesamin in sesame seeds creates the antifeedant lignan sesamin (see Figure 7.11) (Mizutani and Sato 2011).

7.3.3 Pinoresinol to Podophyllotoxin

Perhaps the most famous of the subsequently elaborated lignan metabolites is podophyllotoxin (Figure 7.10) because of its antitumor activity and the inspiration for the closely related and clinically useful topoisomerase inhibitor etoposide (Janick and Whipkey 2002). The enzymology and heterologous expression of the podophyllotoxin enzymes have recently been achieved *via* expression in tobacco plants (Lau and Sattely 2015) (discussed in Chapter 13).

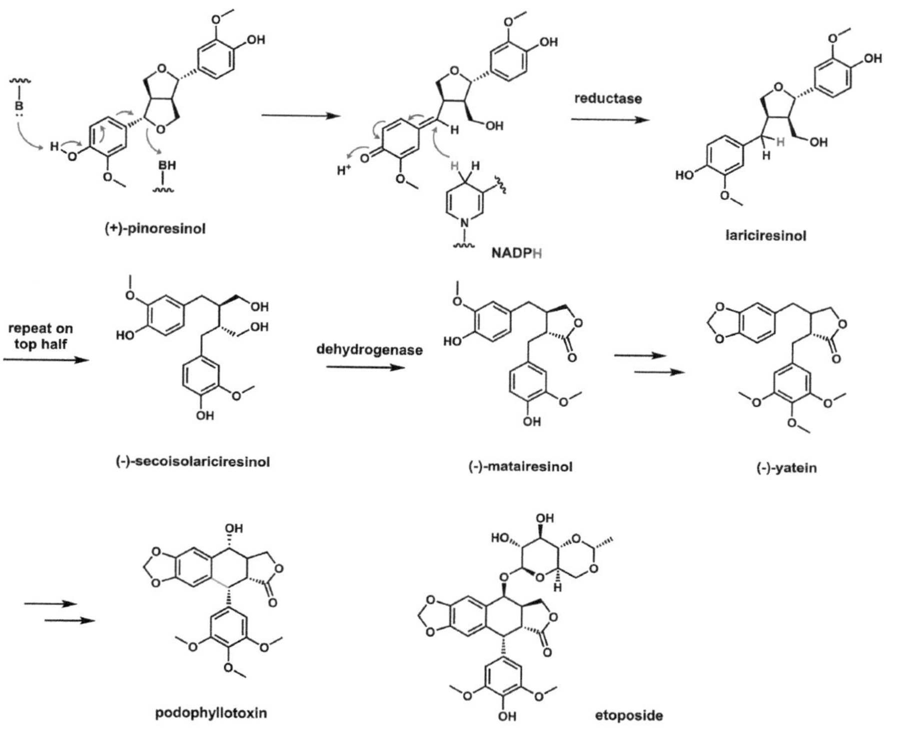

Figure 7.10 Enzymatic processing of pinoresinol to podophyllotoxin, which is an inspiration for the clinically used topoisomerase inhibitor etoposide.

Tandem reductions of the furan rings of pinoresinol by hydride ions from NADPH are proposed to go *via* quinone methide intermediates as shown and thereby to generate a pair of alcohols (Fujita, Gang *et al.* 1999) (see Figure 6.13). This is an unusual reaction sequence, enabled by the *para*-OH group in the phenol moiety. Indeed, much of the pathway to this point and beyond reveals the strategy that has evolved to carry out controlled phenolate two electron and one electron reaction manifolds on the monolignols and the dimeric lignans.

In terms of lignan subclasses of scaffolds (Figure 7.10), this oxidoreductase pair converts the furanofuran to a furan and then the dibenzylbutane scaffold. Double oxidation of one of the alcohols up to the acid sets up intramolecular lactonization to create matairesinol. The final product podophyllotoxin has an aryl tetralin core.

To proceed to podophyllotoxin, a P450-forming methylenedioxy bridge oxygenase, two aryl ring P450 hydroxylases, and two SAM-utilizing *O*-methyltransferases must act. Most notably, a new C–C bond must form to create the nonaromatic six-membered central ring of the 5-6-6-5 tetracyclic framework of podophyllotoxin. This could occur through phenol radical coupling (not shown). From matairesinol to podophyllotoxin four P450s would be at work. Add in the three P450s that generate the monolignols and the lignan dimer-forming enzyme and that totals *eight oxygenases*, going from cinnamate to podophyllotoxin (Lau and Sattely 2015). Figure 7.10 also shows how fundamental the podophyllotoxin framework is to the antitumor agent etoposide.

Vignette 7.1 Podophyllotoxin and Etoposide.

Etoposide is a semisynthetic molecule built on the core of the epimer of podophyllotoxin, a phenylpropanoid metabolite from the Mayapple, *Podophyllum peltatum*. Etoposide is a potent inhibitor of human topoisomerase II, and blocks ligation of cleaved DNA, leading to accumulating single and double strand breaks and apoptotic cell death. It kills rapidly growing cells and is used in chemotherapy regimens to treat small cell lung cancers, recurrent testicular cancers, and ovarian cancers.

Podophyllotoxin is found in the *P. peltatum* roots, presumably as antimicrobial defense molecules (Figure 7.V1). Conversion to etoposide involves synthetic chemical steps of epimerization and demethylation. About 300 000 lb of mayapple roots have been estimated to be collected annually to provide sufficient podophyllotoxin as advanced chemical intermediate for the final etoposide drug.

Podophyllotoxin is a member of the lignan class of phenylpropanoids, arising in an enzymatically induced dimerization of coniferyl alcohol through phenoxy radical coupling. The regioselective coupling to the (+)-pinoresinol dimeric scaffold is mediated by a product-determining protein cofactor termed a dirigent protein, presumed to orient the reacting coniferyl alcohol phenoxy radicals in one favorable orientation for radical coupling. Six remaining steps involving hydroxylations, methylations, and a ring-forming C–C bond to set the epipodophylltoxin scaffold have recently been achieved by Lau and Sattely (Lau and Sattely 2015) by expression of the full pathway in tobacco plants as a heterologous host, with product detection by mass spectrometry. Reconstitution in tobacco leaves allows for structure–activity variation work on future etoposide analogs and secures a biosynthetic pathway approach to production that will not require mass collections of Mayapple roots.

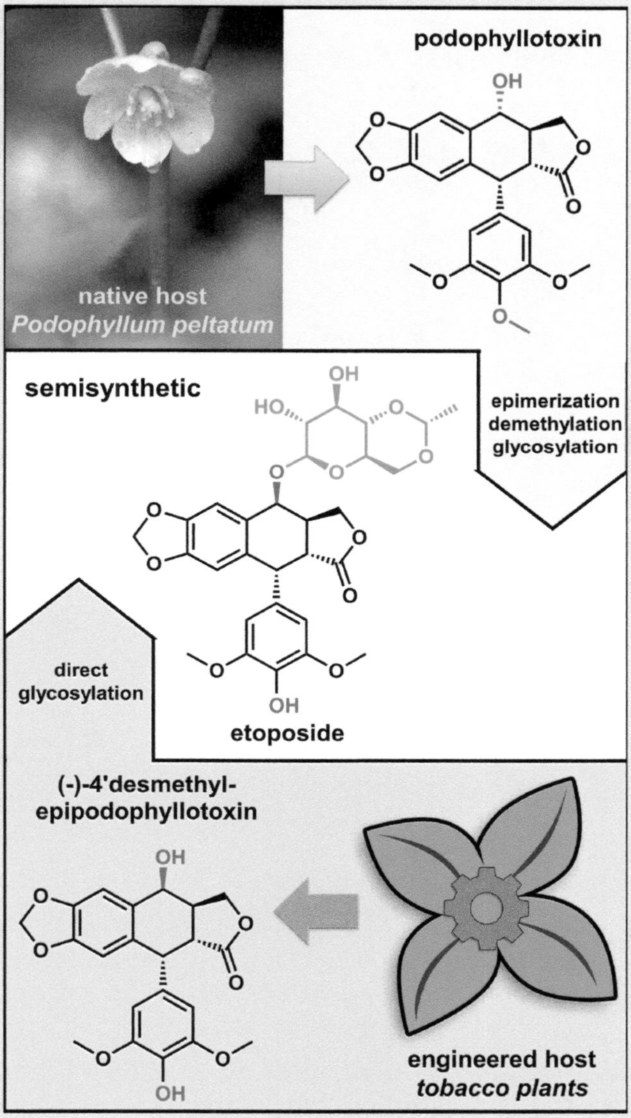

Figure 7.V1 Established and emerging semisynthetic routes to etoposide.

7.3.4 Lignin Formation and Function: Peroxidases and Laccases

The monolignol building blocks that are oxidatively dimerized to the many dimeric lignan variants noted above can also be oxidatively

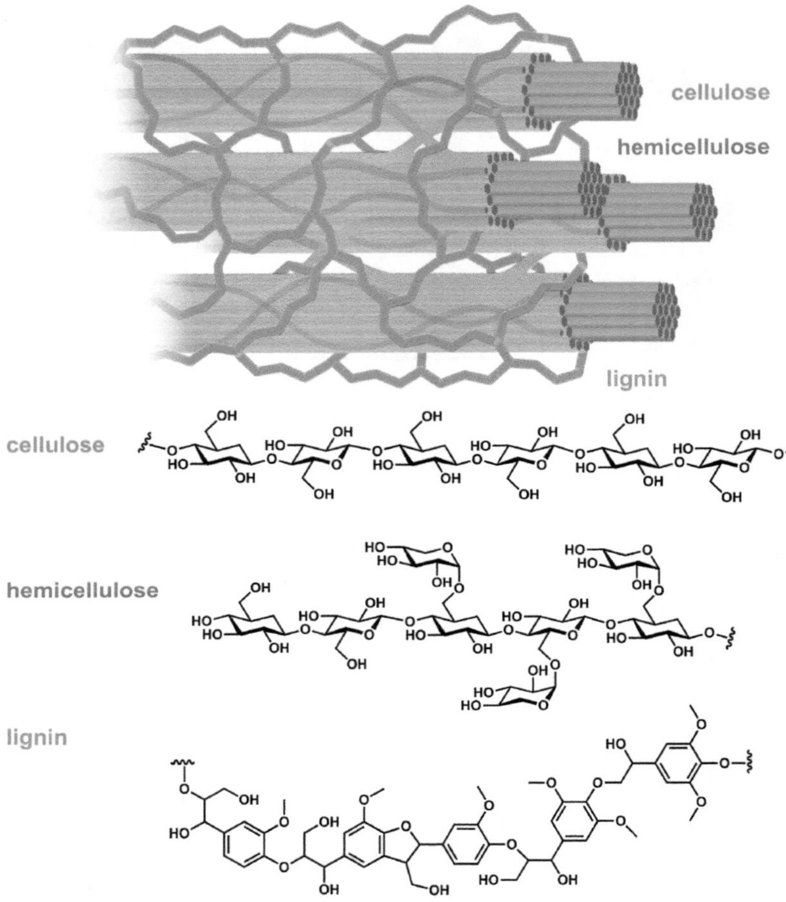

Figure 7.11 Cellulose, hemicellulose, and lignin as distinct extracellular mechanically strengthening polymers in plants. Image credit: John Billingsley.

oligomerized outside the plant cells to insoluble *lignin* polymers that serve as wall strengthening structural and protective elements (Figures 7.11 and 7.12) (Boerjan, Ralph *et al.* 2003, Vanholme, Demedts *et al.* 2010).

The three monolignols, 4-hydroxy-coumaryl, sinapyl, and coniferyl, produce lignin subtypes known as H-, S-, and G-, respectively, where H is for hydroxy-coumaryl, S for sinapyl, and G for guaiacyl, respectively (Figure 7.13). Lignins in gymnosperms are essentially all from G-units, those from angiosperm dicots a mix of G- and S-units. H-units are minor elements in most plants, reflecting either

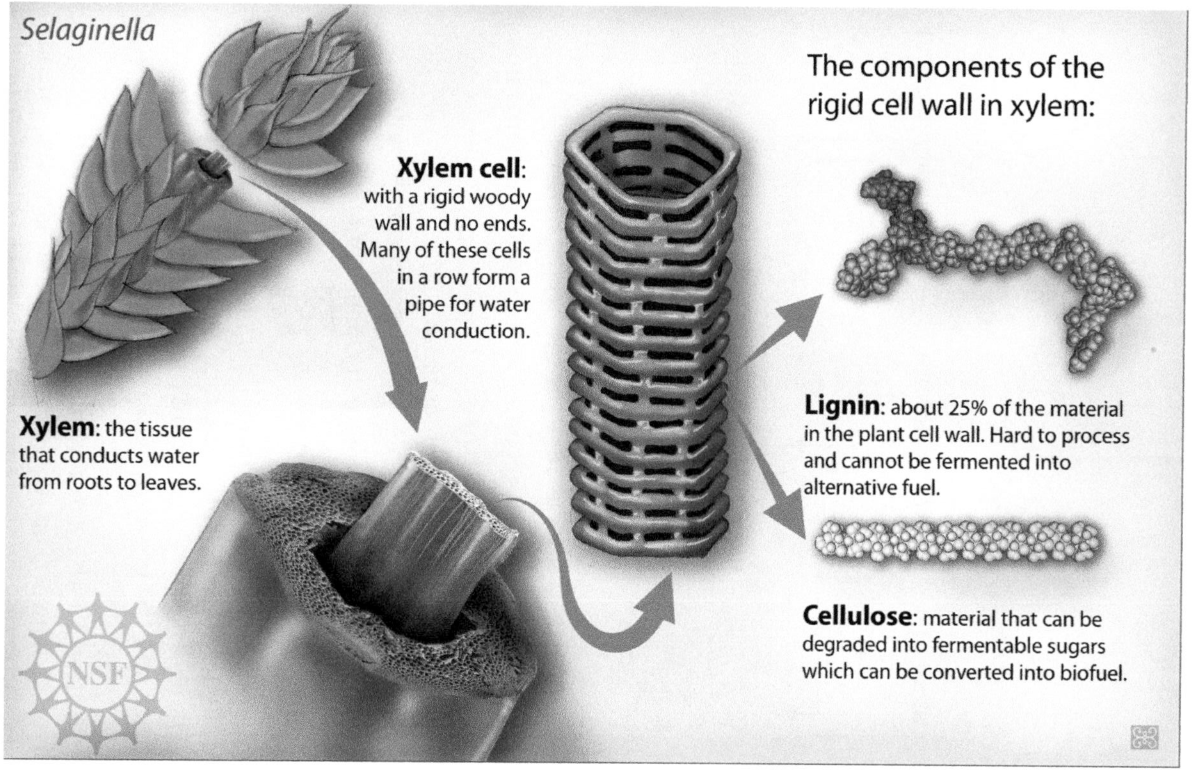

Figure 7.12 Schematic of lignin in plant leaves.
Courtesy: National Science Foundation (nsf.gov).

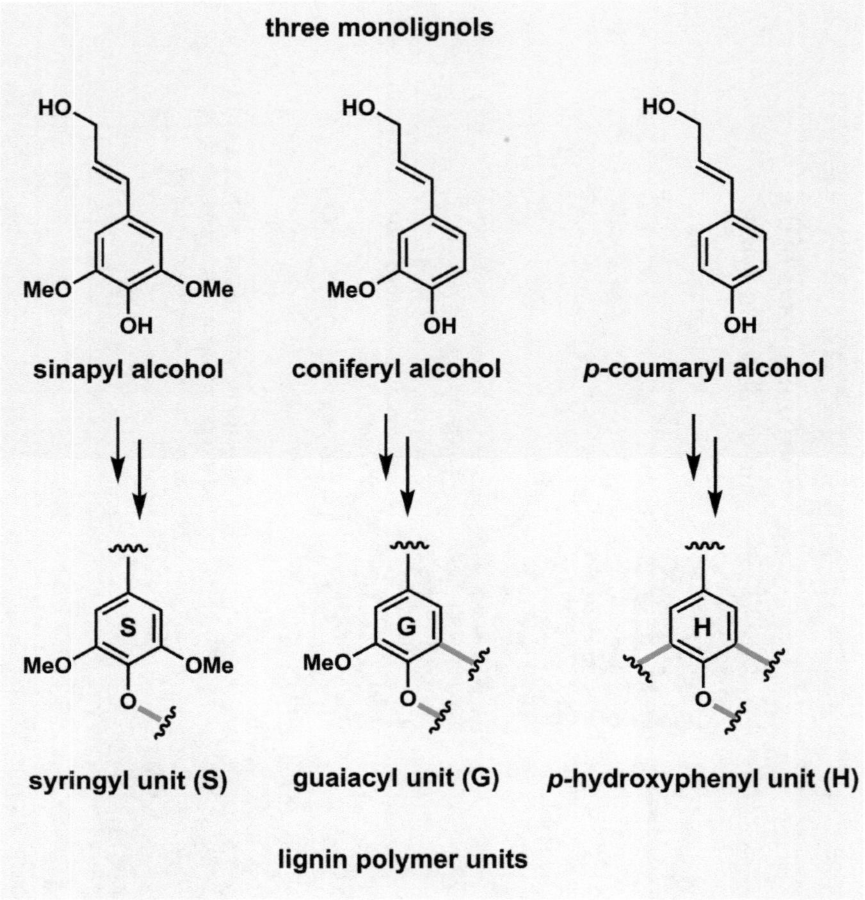

Figure 7.13 Different lignin subtypes that can be generated from the three monolignols.

differential abundance of monolignols, lessened transport across the cell walls to polymerization extracellular sites, or discrimination by the oxidative enzymes that create the phenoxy radicals that couple to each other and the growing lignin chains (Vanholme, Storme *et al.* 2012).

Polymerization in the extracellular space is initiated by secreted peroxidative enzymes. The *Arabidopsis thaliana* genome encodes 73 laccase-like proteins, consistent with multiple oxidative metabolic functions (Fraser and Chapple 2011). Laccases are copper-containing oxygen-activating enzymes that can initiate one electron transfer from substrates to O_2. The phenoxy radicals in the

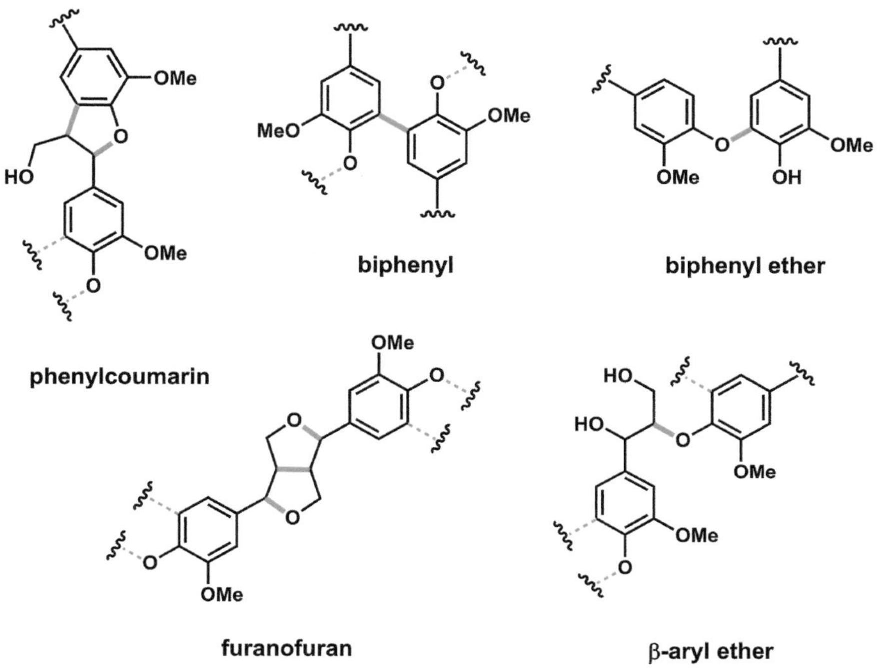

Figure 7.14 Major linkages in lignins derived from the three subtypes.

monolignols and lignin chains must be long-lived enough to pair up with another monolignol radical species or a corresponding one in a neighboring region of an existing lignin chain (see Figure 7.8). Figure 7.14 depicts some of the common linkages found in sections of lignin polymers and compares them to major subclasses of lignan scaffolds. Because lignins are polymerized free in solution, not in enzyme active sites (in contrast to lignan dimerizations), they are achiral and have a mixture of building blocks from copolymerization of different monomers and addition of monomeric monolignol radicals to growing lignin chains. Figure 7.15 shows a representation of a section of a lignin molecule from poplar trees and shows four of the linkage types schematized in Figure 7.13 (Vanholme, Demedts *et al.* 2010, Lochab, Shukla *et al.* 2014).

The genes for converting cinnamic acid and coumaryl-CoA into lignans and lignins are under several levels of regulation, both in normal development in different cell types and in response to stress and other external challenges such as insect attacks and tissue wounding (Vanholme, Storme *et al.* 2012). Given the high global abundance of lignin, the most abundant aromatic biopolymer on the

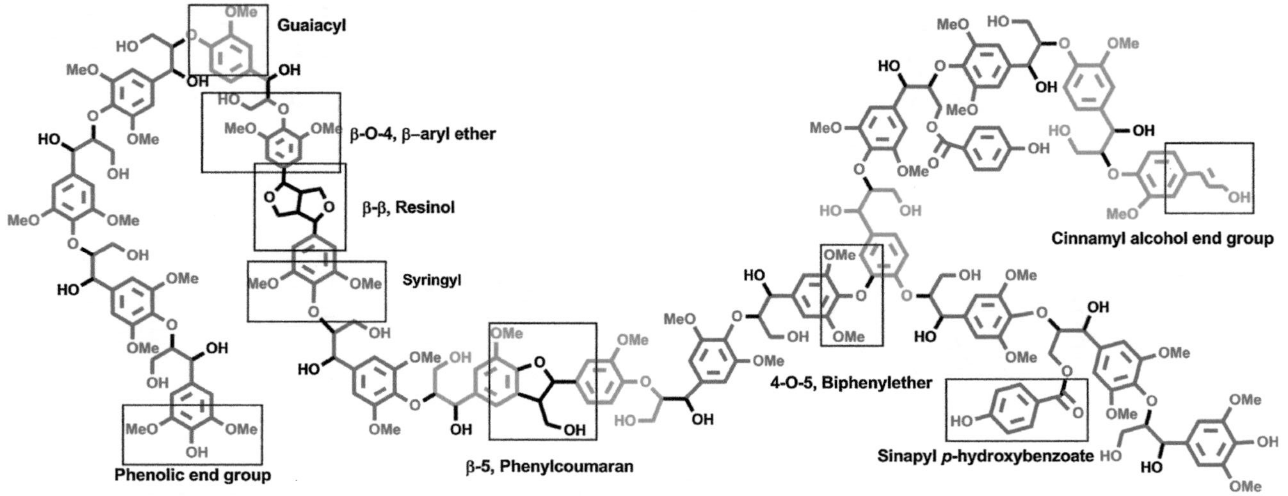

Figure 7.15 Representation of a region of lignin from poplar trees.
Reproduced from Lochab, B., S. Shukla and I. K. Varma (2014). "Naturally occurring phenolic sources: monomers and polymers". *RSC Adv.* **4**: 21712, with permission from The Royal Society of Chemistry.

planet, there is much interest in engineering biosynthesis to give more homogeneous lignin polymers to facilitate degradation and also in discovery of enzymes that allow efficient degradation of lignin into usable products.

7.4 *para*-Coumaryl-CoA to All the Other Classes of Phenylpropanoids

Although the lignin pathway consumes an enormous amount of the flux and mass from phenylalanine to cinnamate to *para*-coumaryl-CoA (4-coumaryl CoA) in woody plants, there are a plethora of other important phenylpropanoid metabolites and molecular frameworks that branch off that common intermediate *para*-coumaryl CoA.

Prominent among them are the metabolites arising from intersection of phenylpropanoid and polyketide synthase chemical logic. The polyketide synthases in plants are distinct from those in bacteria and fungi in *not utilizing specialized acyl carrier proteins.* For this difference they are termed type III polyketide synthases (Austin and Noel 2002). (Recall from Chapter 2 that type I PKS are modular, as in deoxyerythronolide synthase, while type II PKS are iterative, as exemplified by tetracycline synthases.) The chain elongation chemistry is the same for all three PKS types – decarboxylative Claisen condensation *via* malonyl thioesters – but the type III PKS chain elongation thioesters are all CoA derivatives (malonyl-CoA). Similarly, the starter unit is also an acyl-CoA, most notably 4-coumaryl-CoA or feruloyl-CoA for the examples detailed below. However other aryl-CoA molecules are also starter units for related scaffolds.

7.4.1 Type III PKS with One Malonyl-CoA Chain Extension

Among the simplest versions of plant type III polyketide synthases using 4-coumaryl CoA as a starter unit are those carrying out only a single chain extension with malonyl-CoA (Dewick 2009). As noted in Figure 7.16, addition of a two-carbon unit to feruloyl-CoA and its subsequent coupling to a second molecule of feruloyl-CoA give rise to curcumin, the diketo-dimeric spice characteristic of turmeric. Analogously, transfer of the PKS product to hexanoyl-CoA yields 6-gingerol and its dehydration product 6-shogaol, volatile flavor elements in ginger.

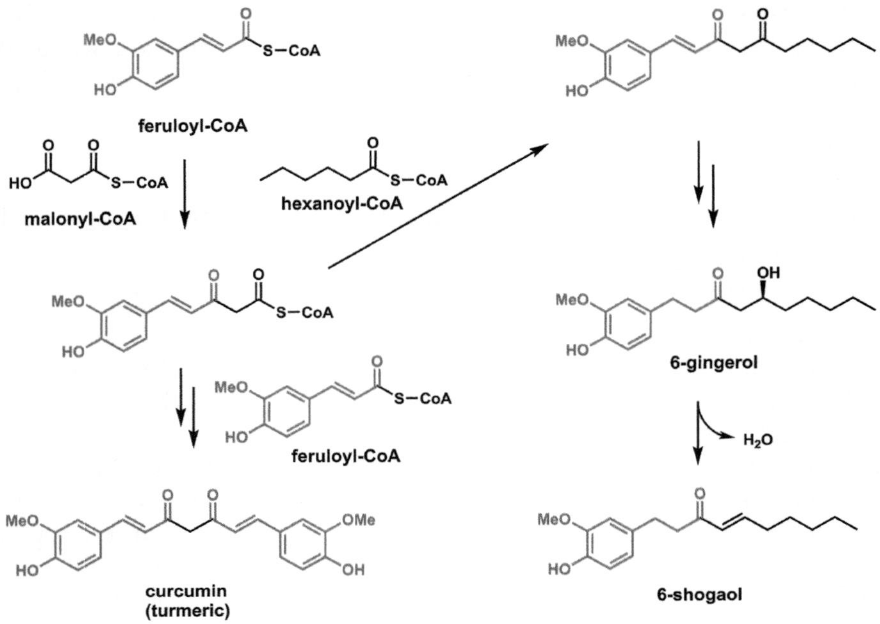

Figure 7.16 Type III PKS with feruloyl-CoA as starter unit: one chain extension with malonyl-CoA leads to curcumin, 6-gingerol, and 6-shogaol.

7.4.2 Stilbene Synthases *vs.* Chalcone Synthases: Rerouting Substrates to Different Products after Three Chain Extensions

Related type III PKS enzymes can do up to 10 malonyl-CoA chain extensions on aryl CoA starter units to create more complex polyketonic nascent products. The best studied are a related set of enzymes that carry out three malonyl chain extensions on 4-coumaryl CoA but yield distinct products (Figure 7.17). One set processes the nascent triketonic thioester products by a C_6–C_1 Claisen condensation to yield scaffolds known as chalcones (Jez, Bowman *et al.* 2000, Jez and Noel 2000, Abe and Morita 2010). Also shown in the figure is how a chalcone synthase generates naringenin chalcone. Some 900 of these enzyme members are known in plant proteomes (Fraser and Chapple 2011). The tetrahydrocannabinol scaffold noted in Chapter 4 is assembled by a variant of a chalcone synthase (see Figures 2.45 and 4.37).

The second enzyme subset folds the triketone nascent product differently in the active sites and instead carries out a C_2–C_7 cyclizing

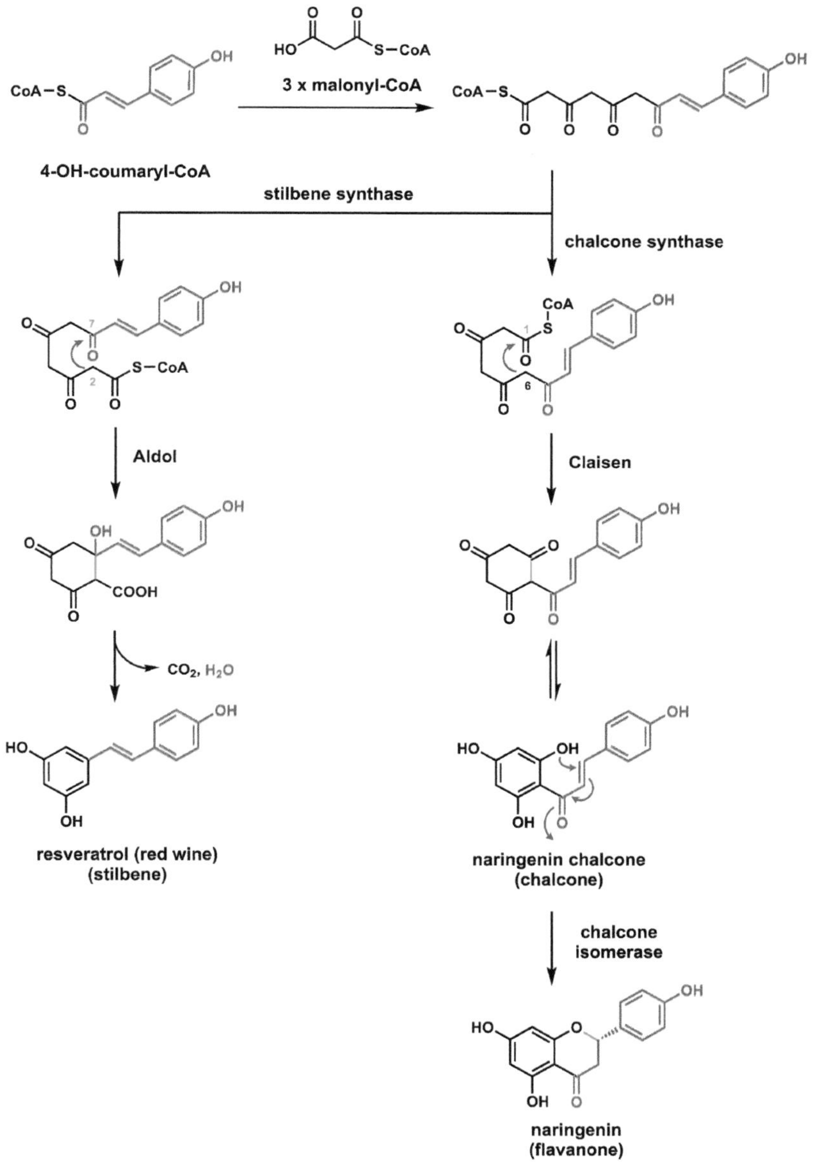

Figure 7.17 Type II PKS with coumaryl-CoA as starter units. Three chain extensions can lead to stilbenes or chalcones.

aldol condensation to produce stilbenes, *e.g.* resveratrol (Schroder and Schroder 1990). Stilbene synthases can leak up to 3% chalcone product while, correspondingly, chalcone synthases can leak 2–4% of the product as stilbenes. These presumably reflect dynamics in the linear triketonic foldamers in the active sites and the relatively low energy barriers for the alternate cyclization modes. It is likely that stilbene synthases evolved from chalcone synthases (Parage, Tavares *et al.* 2012). Figure 7.18 notes two additional starter acyl-CoAs for

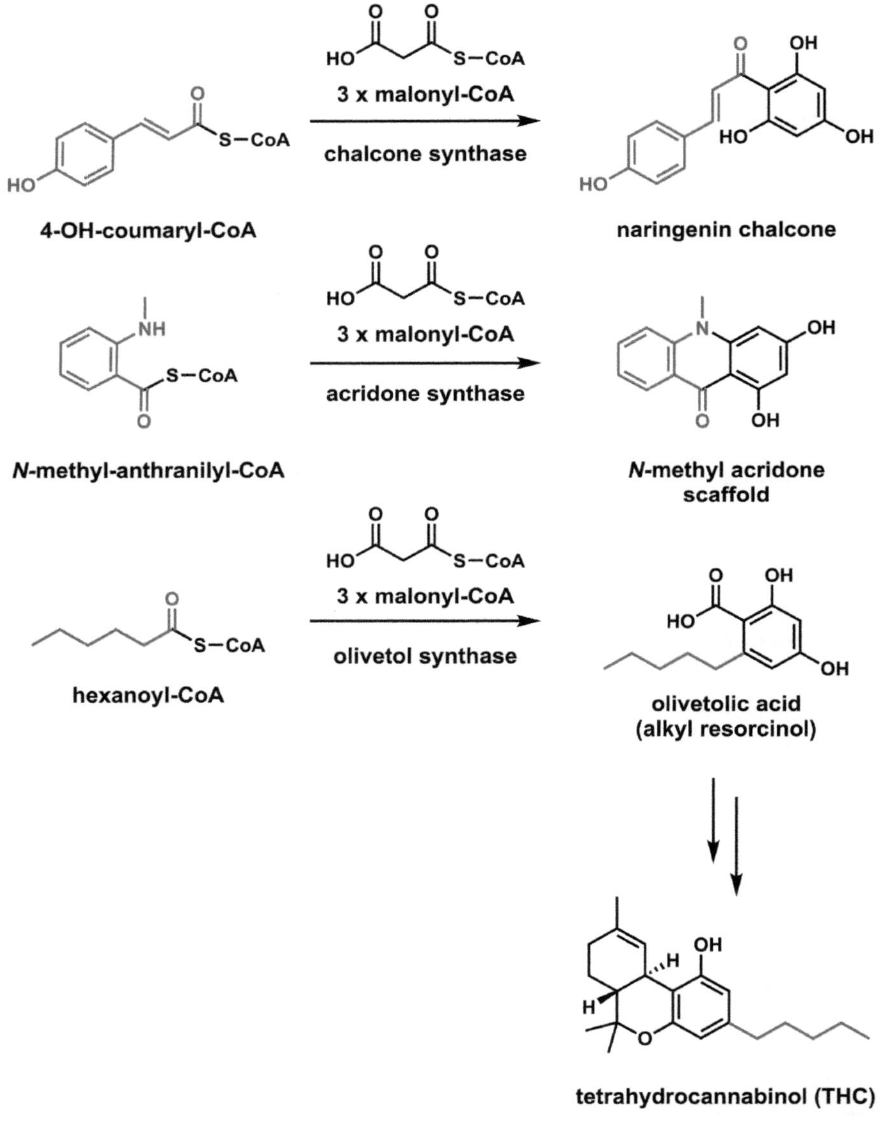

Figure 7.18 Type III PKS that carry out three chain elongations with malonyl-CoA but use distinct starter aryl-CoAs.

particular chalcone synthase-like catalysts. *N*-Methylanthranilyl-CoA, extended by three malonyl units, generates the tricyclic *N*-methylacridone scaffold. Olivetol synthase, the first committed enzyme in the tetrahydrocannabinol pathway, uses hexanoyl-CoA as starter unit. Three decarboxylative condensations with malonyl-CoA yield olivetolic acid (Taura, Tanaka *et al.* 2009).

Vignette 7.2 Alkyl Catechols, Alkyl Resorcinols, Poison Ivy, and Japanese Lacquer.

Alkyl resorcinols are a subgroup of polyketides (see Figure 2.1) typically made from long or medium chain acyl-CoAs as starter units and three malonyl-CoAs as chain extender units, catalyzed by type III polyketide synthases (Baerson, Rimando *et al.* 2008). These enzymes are close variants to the stilbene synthases and also olivetolic acid (5-pentylresorcinol) synthase (Figures 7.17 and 7.18) involved in tetrahydrocannabinol assembly. The chain extension reactions lead to alkyl tetraketides which undergo cyclization, decarboxylation, and dehydrative aromatizations to generate the alkyl substituted aromatic meta-phenol (resorcinol) ring systems.

Presumably related biosynthetically are alkyl catechols where the two –OH substituents are in an *ortho* rather than *meta* relationship on the aromatic ring, notorious among them urushiol (Figure 7.V2), named for the Japanese word for "sap" to signify its high content in Japanese lacquer trees. Urushiol's importance as the starting material for polymerized Japanese lacquerware is matched by its notoriety as the irritating allergen in poison ivy and poison oak.

The biosynthetic route for urushiol is not yet determined but a type III PKS with the tri-olefinic C_{16} acyl-CoA as starter and the prototypic three elongation cycles with malonyl-CoA would yield the indicated tetraketidyl-CoA. The NAD(P)H-dependent reduction of one of the keto groups would set up intramolecular aldol condensation, dehydrative aromatization, and decarboxylation, in analogy to alkyl resorcinol formation (see Figures 2.1 and 7.17) to yield the alkenyl monophenol. Phenol to catechol oxygenation would then be the likely province of an FAD-dependent flavin-linked monooxygenase.

Urushiol has a C_{15} tri-olefinic alkyl chain (from the corresponding trienyl C_{16} acyl-CoA starter unit), can penetrate skin, and then targets proteins on Langerhans cells (whether covalent adducts form on the cell surface is not yet determined). In turn this cell surface-presented antigen is recognized by T cells. The T cells initiate an immune cascade by releasing cytokines and chemokines, recruiting macrophages, and setting off a hyper-robust inflammatory response with the characteristic pruritus and redness that marks poison ivy/oak exposure in responders.

The physiologic role of urushiol in producer plants has been hypothesized to function as a phytoanticipin or phytoalexin type of defense

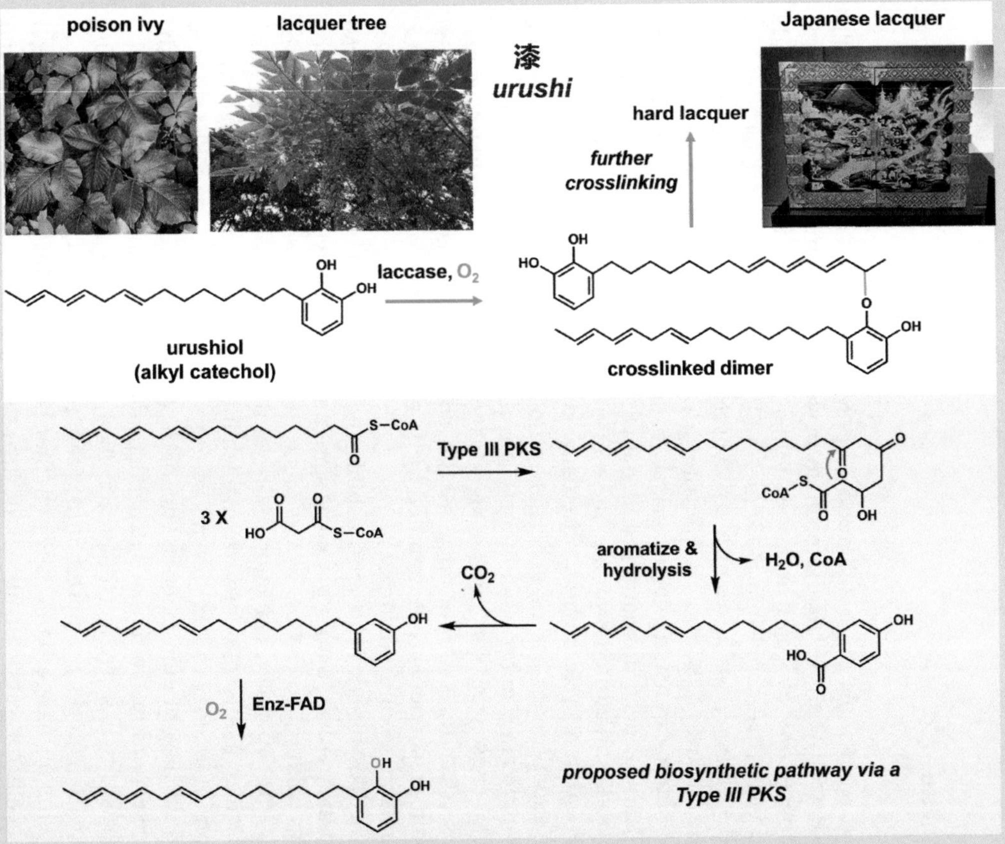

Figure 7.V2 Structure of urushinol and a proposed biosynthetic route.

molecule. In the Japanese lacquer sap, urushiol is about 60–65% of the mass, with another 25% water and minor components, including the copper oxidase laccase. We have noted earlier in this chapter the role of laccases in polymerization of lignans into lignin. The lacquer laccase also catalyzes oxidative polymerization of the urushiol-rich sap, possibly *via* orthoquinone species, first to the indicated dimer, and then onto a cross-linked polymer. Up to 30 layers of the urushiol sap, artfully applied, build up the hard, clear lacquer prized in Japanese and Chinese furniture and artworks.

7.5 Chalcone to Flavanones and Beyond

7.5.1 Chalcone Isomerases Convert Chalcones to Flavanones

The entry point to the downstream flavonoids is the enzyme chalcone isomerase (Ngaki, Louie *et al.* 2012). It speeds up an otherwise nonenzymatic Michael addition of one of the trihydroxyphenol hydroxyl groups into the olefinic terminus of the conjugated enone in the chalcone framework (Figure 7.19). The enzymatic process occurs with stereochemical control to yield 2*S*-naringenin in the example shown. This is now a flavanone scaffold and it can be diversified by several enzymatic routes to major subcategories of flavonoids (Falcone Ferreyra, Rius *et al.* 2012), shown in Figure 7.20. The flavone, flavonol, and flavandiol subclasses arise straightforwardly. The isoflavones, anthocyanidins, and aurones bear some discussion. The condensed proanthocyanidins are trimers found in

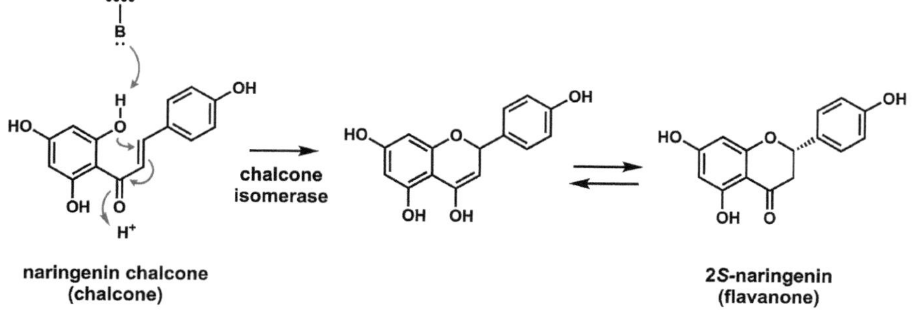

Figure 7.19 Chalcone isomerase catalyzes the ring-forming Michael addition to create chalcone scaffolds, in this example naringenin chalcone.

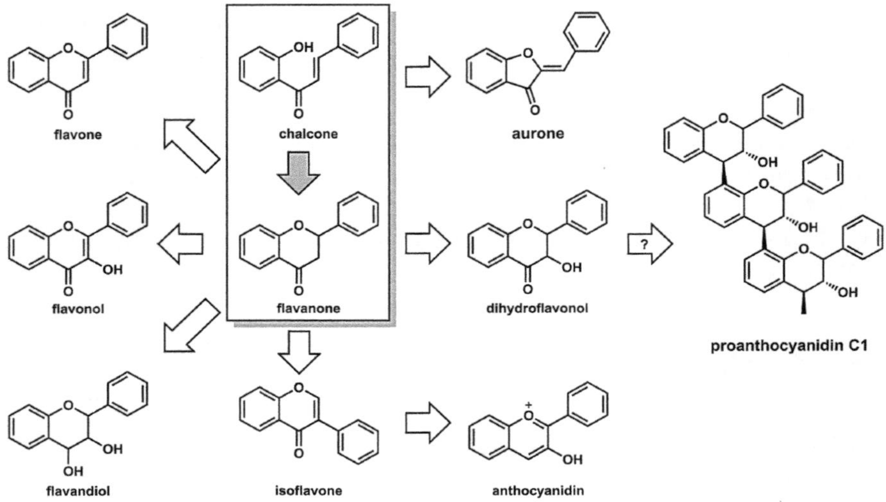

Figure 7.20 Major categories of plant flavonoids.

high amounts in apples and in red wine (91 mg per 150 ml) and are astringent flavor polyphenols in those red wines. They are thought to serve defense roles against plant predators (Cos, De Bruyne *et al.* 2004). Little is known about their mechanism of oligomerization *in planta*.

Figure 7.21A schematizes the conversion of a hydroxy flavanone (a dihydroflavonol) to the flavone quercetin, a molecule highlighted in Mediterranean diets as an antioxidant (Winkel-Shirley 2001). Its highest concentration is in capers but it is widely distributed in many edible plants. To increase flavonoid solubility and enable transport to different cells and tissues *in planta* a significant fraction of these molecules undergo enzymatic glycosylation. The structure of the prevalent quercetin disaccharide rutin (the sugars are rhamnose and rutinose, see Chapter 11) is shown.

The pulse chase [$^{13}CO_2$] feeding approach noted in Chapter 4 (Figure 4.8) to distinguish the mevalonate *vs.* methylerythritol routes to isoprene scaffolds has also been used in the analysis of flavonoid trisaccharides such as kaempferol-3-*O*-β-sophoroside-7-*O*-β-glucoside in *Papaver nudicaule* (Tatsis, Eylert *et al.* 2014). The labeling patterns accord with flux through the shikimate pathway (*via* erythrose-4-phosphate, E4P) to phenylalanine, then to the coumaryl-CoA starter unit and chain extensions with malonyl-CoA building blocks.

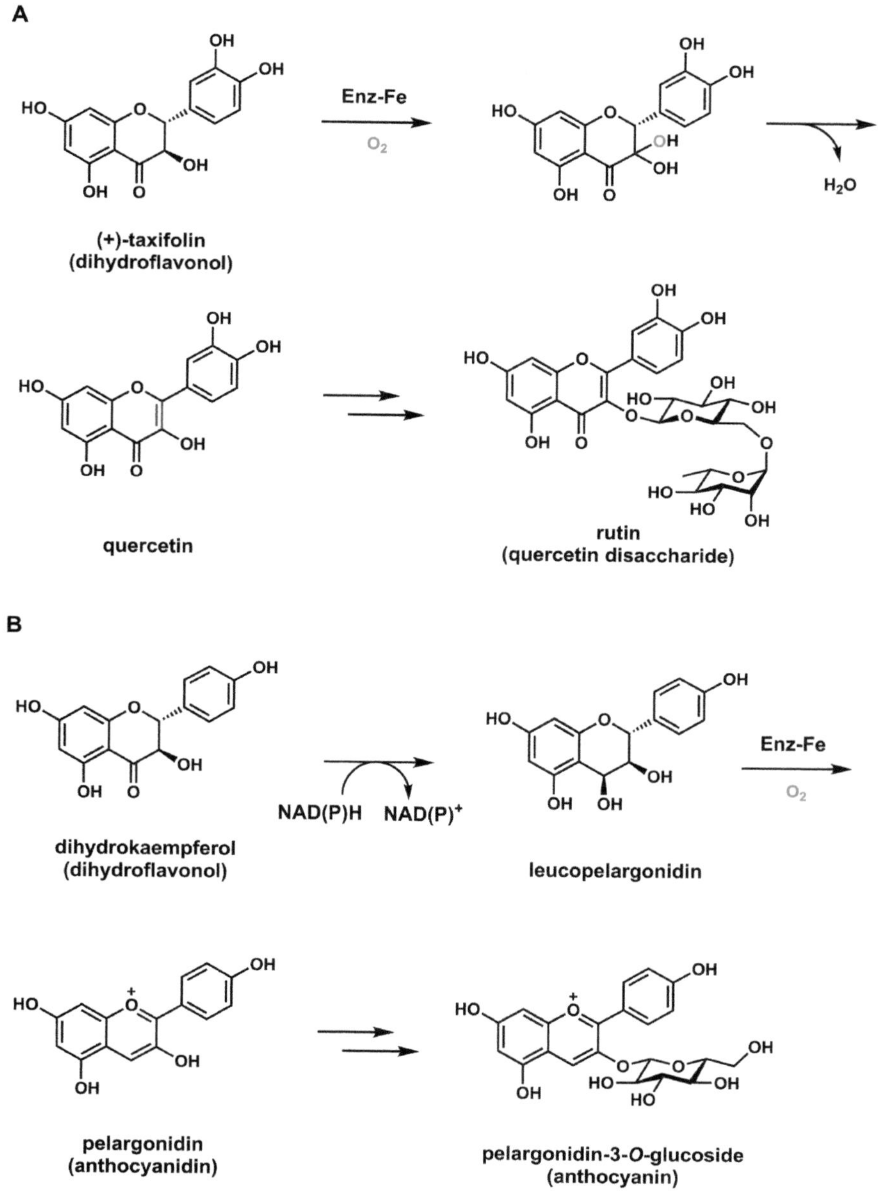

Figure 7.21 Conversion of dihydroflavonols to the (A) flavone quercetin or to the (B) anthocyanidin flower pigment pelargonidin-3-*O*-glucoside. Enzymatic glycosylation is a common modification of flavonoids for water solubility and for transport. Rutin is a quercetin disaccharide.

7.5.2 Flavones to Anthocyanidins

Figure 7.21B shows the reductive conversion of a dihydroflavonol (dihydrokaempferol) to leucopelargonidin and then the oxidative desaturation of the pyrone ring to the anthocyanidin pelargonidin responsible for orange flower pigmentation in geraniums. The color arises from the conjugated oxonium ion (flavylium = 2-phenylchromneylium ion). Pelargonidin is also present in berries (raspberries, blueberries, blackberries, and strawberries), as well as at high levels in kidney beans. It can also accumulate as the more soluble glucoside. Such glycosides are termed anthocyanins (Winkel-Shirley 2001, Veitch and Grayer 2008).

A range of substituent patterns (–H, –OH, $–OCH_3$) around the six positions of the two phenyl rings alter the coloration of the anthocyanidins. Figure 7.22 depicts two other anthocyanidins. Dihydroquercetin generates the pink pigment cyanidin, while dihydromyricetin produces the purple color of delphinidin. A set of flavanone 3′ and 3′,5′-hydroxylases control the pigment color two stages downstream in the final anthocyanidin oxonium ions. Figure 7.23 lists six anthocyanidins that are central to food colors, including cyanidin (blackcurrants, raspberries), pelargonidin (strawberries), and delphinidin (blueberries). The other three are malvidin (grapes), peonidin (cranberries), and petunidin (chokeberries).

7.5.3 Chalcones to Aurones

One more type of related colored pigment arises by a distinct route. Aurones (Figure 7.24) are yellow colored molecules that arise not from the flavone scaffolds but by a rerouting of chalcones through action of aureosidin synthase. The enzyme oxygenates the righthand phenol of a chalcone to a catechol and can oxidize it to an ortho-quinone (Nakayama, Yonekura-Sakakibara *et al.* 2000, Nakayama, Sato *et al.* 2001, Ono, Hatayama *et al.* 2006, Vogt 2010). The quinone makes the exocyclic olefin now electrophilic at the opposite terminus of the olefin from the pattern seen in chalcone isomerase action. Michael addition occurs not to yield a fused 6,6-bicyclic ring system but rather a fused 6,5-system. Rearomatization of the quinone back to the catechol yields the conjugated aurone with its yellow color. Figure 7.24 also shows the yellow color in some snapdragon flowers that arises from the aurone molecules (Sato, Nakayama *et al.* 2001).

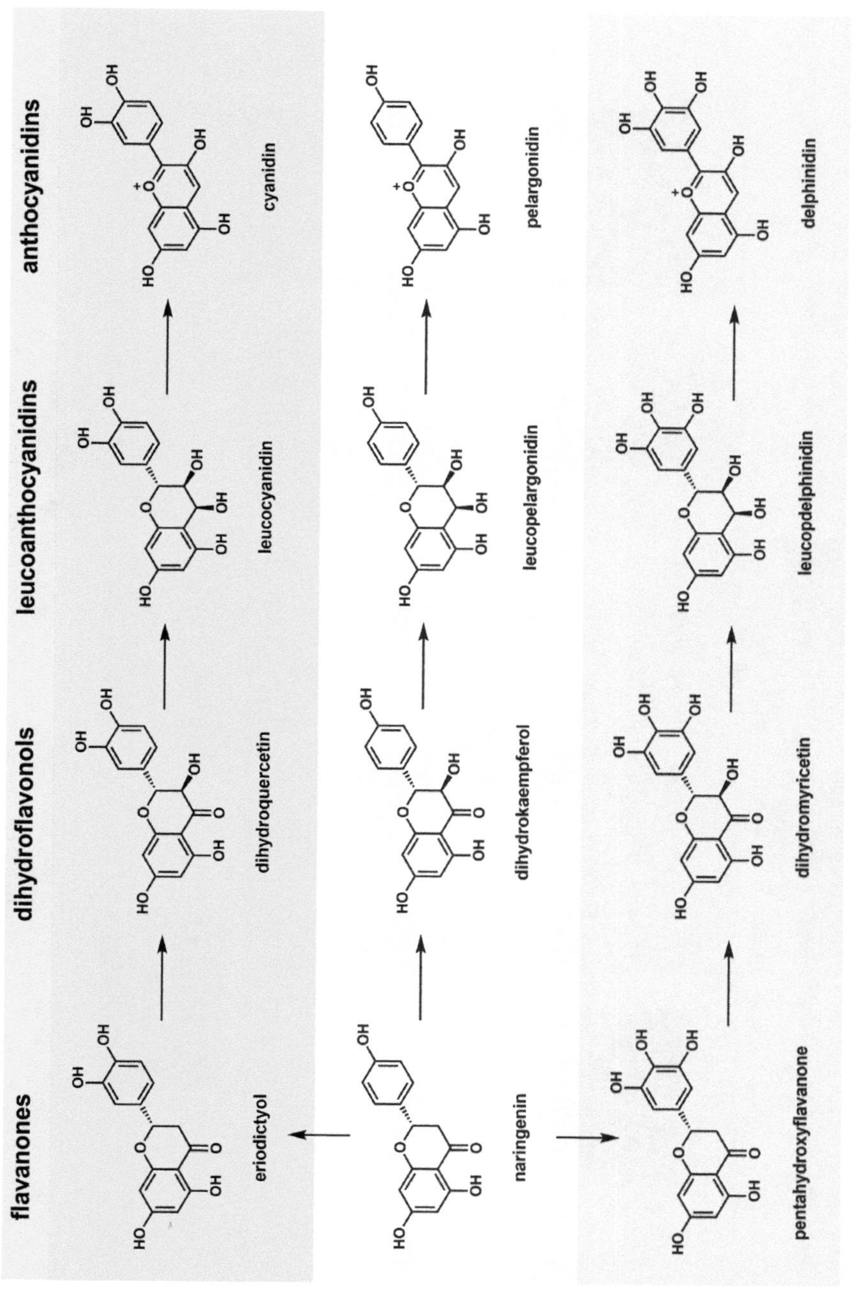

Figure 7.22 Anthocyanidins as flower pigments: chalcones to flavanones to dihydroflavonols to anthocyanidins.

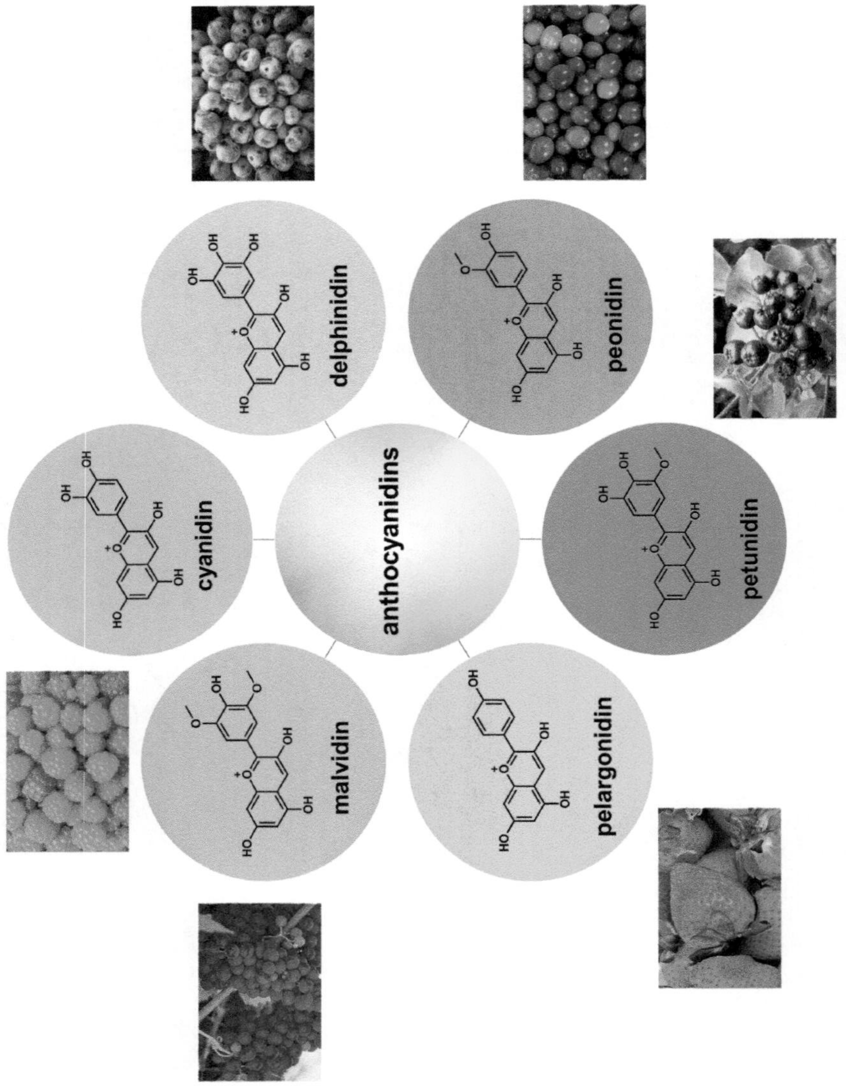

Figure 7.23 Six anthocyanidins central to food colors.

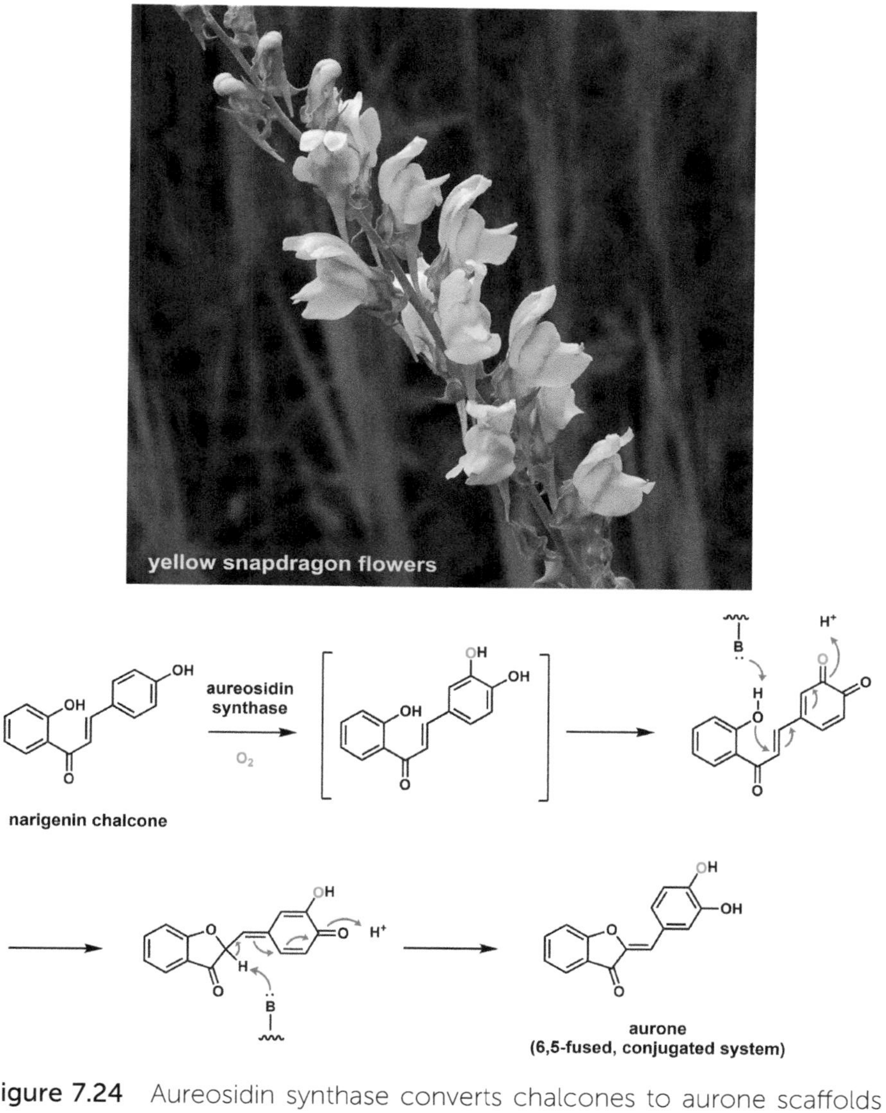

Figure 7.24 Aureosidin synthase converts chalcones to aurone scaffolds to generate yellow pigments.

7.5.4 Flavanones to Isoflavones to Phytoalexins

Another unusual enzymatic transformation enables entry of flavanones into isoflavone chemical space. Figure 7.25 diagrams a cytochrome P450-mediated conversion of naringenin to genistein. This is formulated as a 1,2-aryl shift. This P450 is thus in the class of

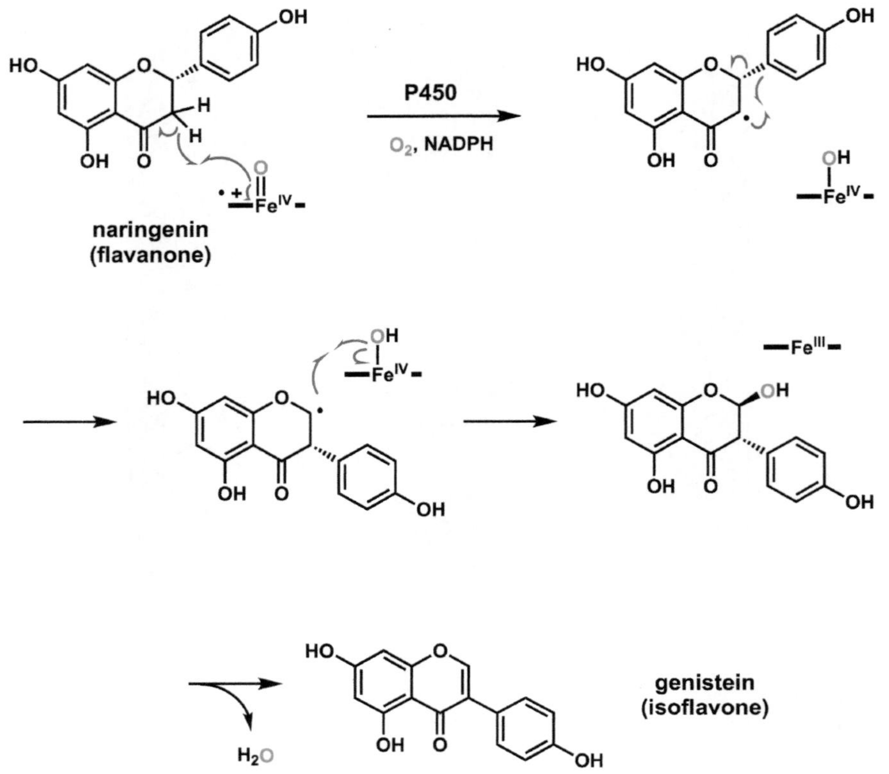

Figure 7.25 Flavonoids to isoflavonoids. Cytochrome P450-mediated 1,2-aryl shift *via* radical intermediates: naringenin to genistein.

oxygenases which carry out the first half reaction, generation of one or more carbon-centered radicals, but do not complete the second half reaction, OH• transfer (see Chapter 9 for more context) before some rearrangement in the substrate occurs (Vogt 2010, Mizutani and Sato 2011). Instead, the carbon radical reacts intramolecularly, in this instance by migration of the phenol moiety as a radical. This forms a new C–C bond and leaves an unpaired electron at the site of the C–C bond just broken. Now the enzyme transfers the OH• equivalent to that site to form a transient hydroxy isoflavone. Elimination of water, the just introduced –OH group, yields the enone moiety in genistein. Other isoflavones are processed the same way for 1,2-aryl group migration *via* one electron reaction manifolds.

Isoflavones on their own have some antifeedant activity but are also progenitors to phytoalexins (plant "warding off" scaffolds) (VanEtten,

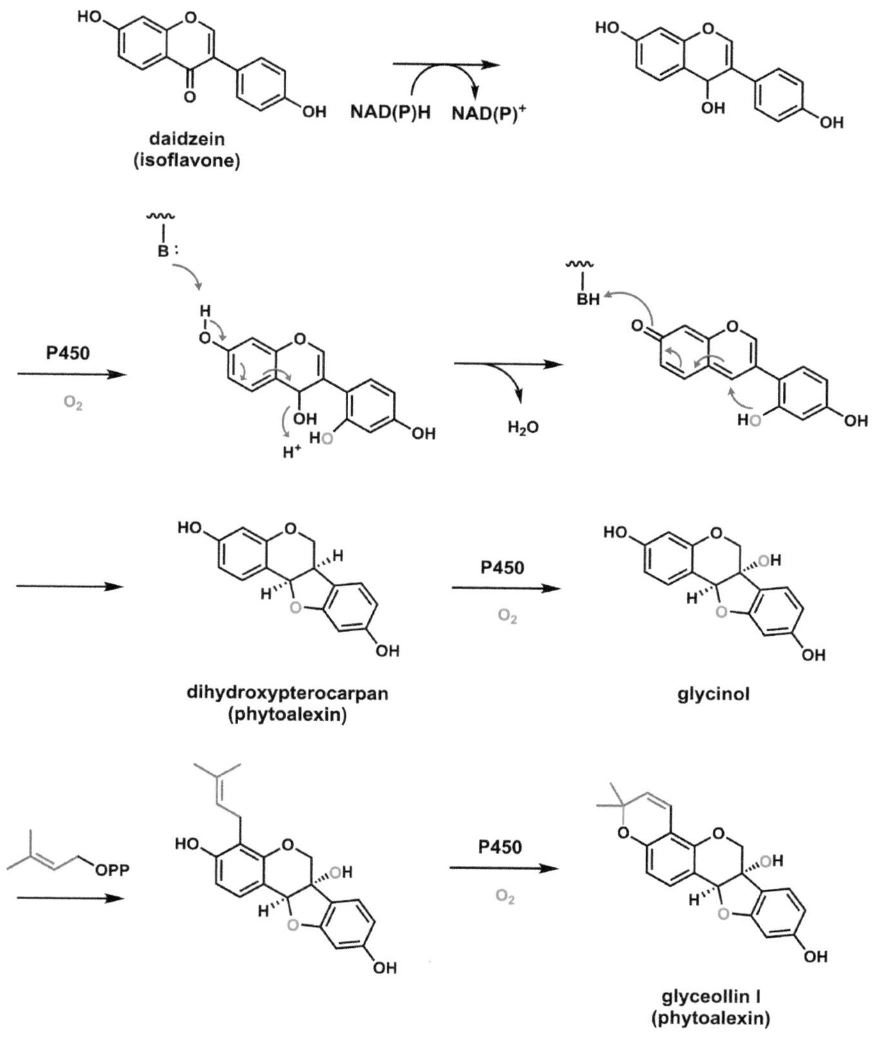

Figure 7.26 Enzymatic conversion of an isoflavone to the tetracyclic framework of the phytoalexin dihydroxypterocarpan and glyceolin in alfalfa.

Mansfield *et al.* 1994), molecules which are synthesized and accumulate at sites of pathogen (typically fungi) infection, show diverse structures, and have various defense mechanisms. One such pathway is displayed in Figure 7.26 where the isoflavone daidzein is processed enzymatically to tetracyclic dihydroxypterocarpan and pentacyclic glyceolin (Guo, Dixon *et al.* 1994). The generation of the bridging furan ring in dihydroxypterocarpan, a pterocarpin framework example, is proposed to involve quinone methide formation to aid in elimination of the indicated –OH group. This is followed by capture of the quinone methide by one of the OH groups in the 2,4-diphenol moiety. Dihydroxypterocarpan can subsequently undergo prenylation at a carbon *ortho* to the phenolic-OH and then undergo oxygenase-mediated (one electron pathway) cyclization to the pentacyclic ring system of glyceollin. Hundreds of isoflavones are available by different patterns of hydroxylation and methylation on the isoflavone core (Gonzalez-Lamothe, Mitchell *et al.* 2009).

The genes for isoflavone and phytoalexin biosynthesis are up-regulated in response to pathogens, consistent with the assignment of phytoalexin function (Kurusu, Hamada *et al.* 2010, Lin, Shih *et al.* 2014). Many of the isoflavones are present *in planta* as glycosides for storage and solubility (Richelle, Pridmore-Merten *et al.* 2002) and the aglycones are releasable by glycosidase action (Figure 7.27) for rapid response that does not need to await *de novo* metabolite biogenesis.

7.5.5 Isoflavones to Rotenone

Cyclization of a methoxyisoflavone is proposed to involve transient oxygenation of the methoxy CH_3 to CH_2OH (Figure 7.28). Water loss would yield a methylene oxonium ion that could undergo an oxy-Cope rearrangement to create the tricyclic system of rotenone with an oxonium ring. Hydride transfer from NAD(P)H would quench the charge on oxygen and yield the tetracyclic desmethyl-munduserone (Dewick 2009, Crombie and Whiting 1998). Prenylation at the carbon *ortho* to the phenolic OH and cyclization to the five-membered ring produces rotenone. Rotenone can function both as an insecticide in plants and also for fishermen as a piscicide by blocking mitochondrial oxidative phosphorylation at the level of electron transfer from complex I to ubiquinone carriers.

7.5.6 Phytoestrogens

Isoflavones have been termed phytoestrogens because they are weak ligands for vertebrate estrogen receptors (Figure 7.29) (Patisaul and Jefferson 2010). For example, soybeans have significant levels of daidzein and genistein (Figures 7.25 and 7.26) both free and as the *O*-7-glycosides (Figure 7.27).

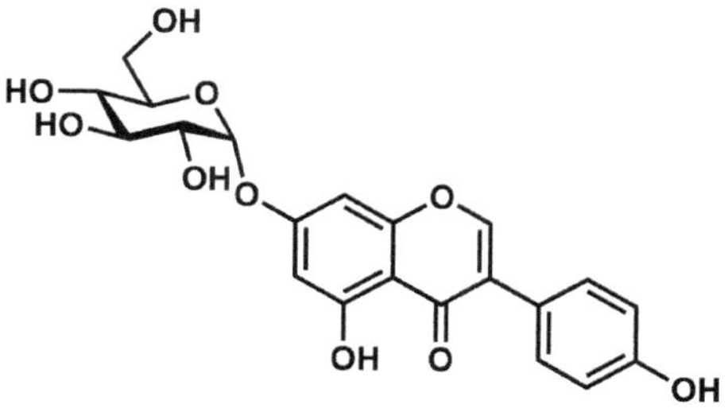

genistin = genistein glucoside

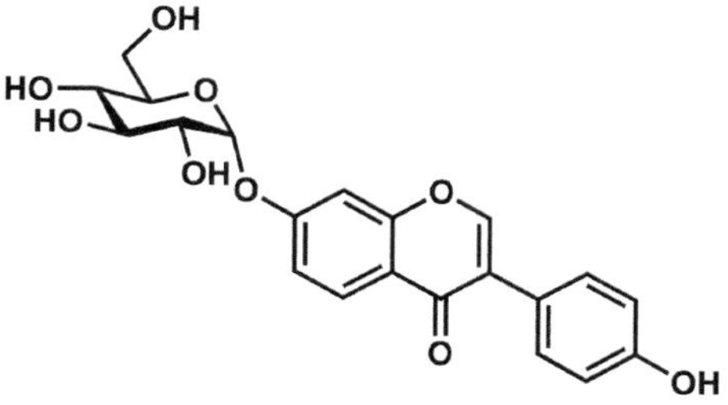

daidzein glucoside

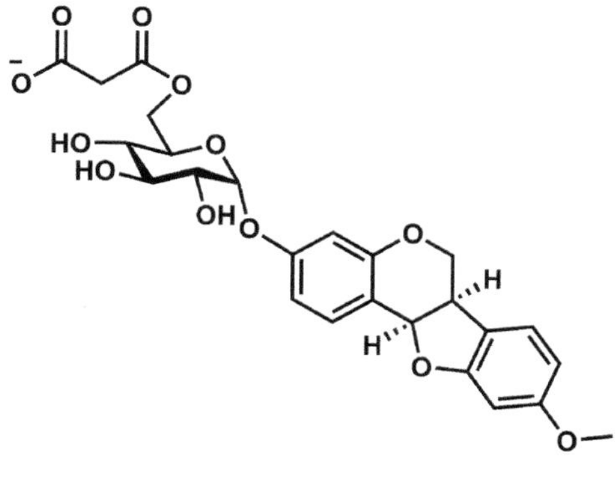

medicarpin 3-*O*-glucoside-6'malonate

Figure 7.27 Isoflavone and phytoalexin glycosides as rapid response storage forms of phytoalexins on glycosidase action.

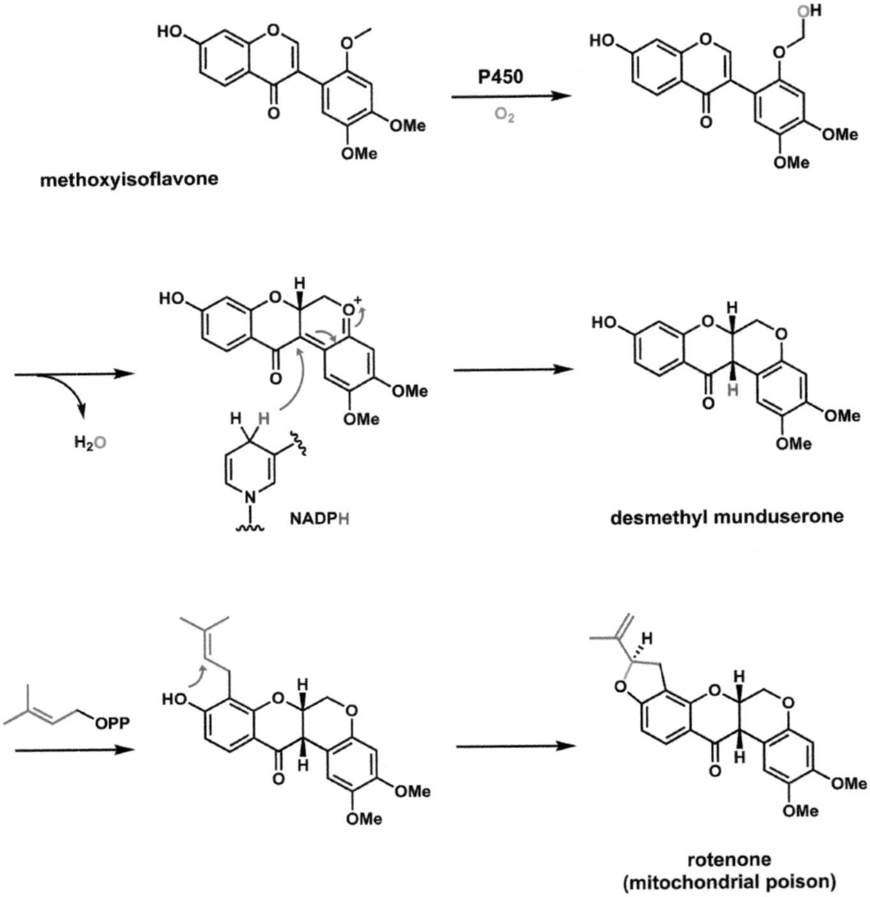

Figure 7.28 Isoflavone to rotenone: intersection of phenylpropanoid and isoprenoid routes.

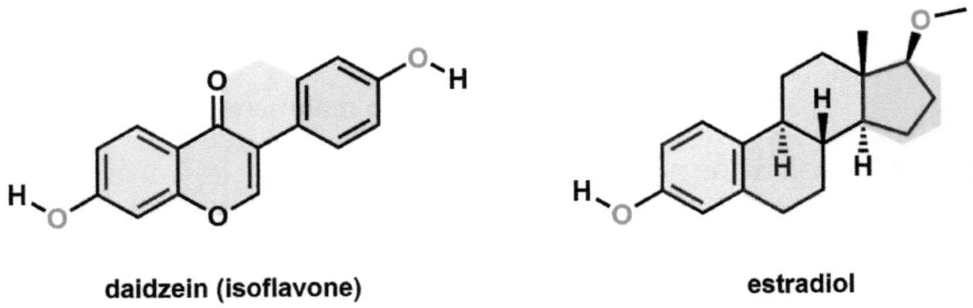

Figure 7.29 Phytoestrogens include some isoflavone scaffolds.

7.6 Cinnamate Derived Phenylpropanoids

7.6.1 Cinnamate to Phenylpropenes and Phenylpropenals

Cinnamon bark contains cinnamyl alcohol esterified to acetate, and this can be the source of cinnamaldehyde both as a volatile flavoring agent by enzymatic hydrolysis and enzymatic oxidation of the alcohol to the aldehyde. Higher levels of the corresponding coniferyl acetate than cinnamyl acetate in cinnamon bark (Dewick 2009) lead to elimination *via* quinone methide, followed, as in many other cases noted in these chapters, by hydride transfer to the methide double bond (Figure 7.30). This yields the phenylpropene eugenol, a flavorful component not only in cinnamon oil but also in oil of cloves, used as a sometime dental anesthetic agent. Cytochrome P450 conversion of

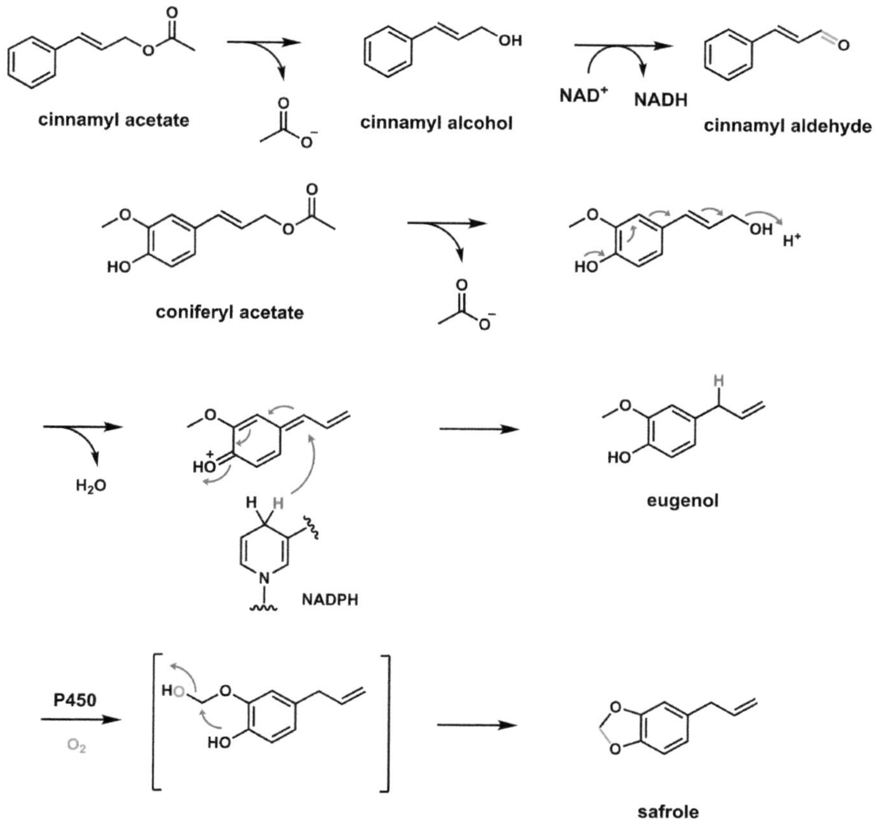

Figure 7.30 Phenylpropene and phenylpropenal essential oils.

the adjacent methoxyphenol grouping to the methylenedioxy bridge generates safrole, the major component of oil of sassafras.

7.6.2 Cinnamate to Coumarins

The bicyclic 6-6 ring system of coumarins is also available by a short pathway from cinnamate (Dewick 2009). Hydroxylation *ortho* or *para* followed by *E*- to *Z*-olefin isomerization sets up 2-hydroxy cinnamate to close to the parent coumarin system (1,2-benzopyrone) (Figure 7.31) and the 4-hydroxy-cinnamate (coumarate) to umbelliferone (7-hydroxy-coumarin). In turn umbelliferone is a precursor to the phototoxic psoralen and subsequent psoralen metabolites. Prenylation of umbelliferone is followed by two P450 oxygenases (Bourgaud, Hehn *et al.* 2006). The first generates the furanocoumarin tricyclic ring system and the second mediates a one-electron pathway for the elimination of the elements of acetone as the conjugated psoralen system is completed.

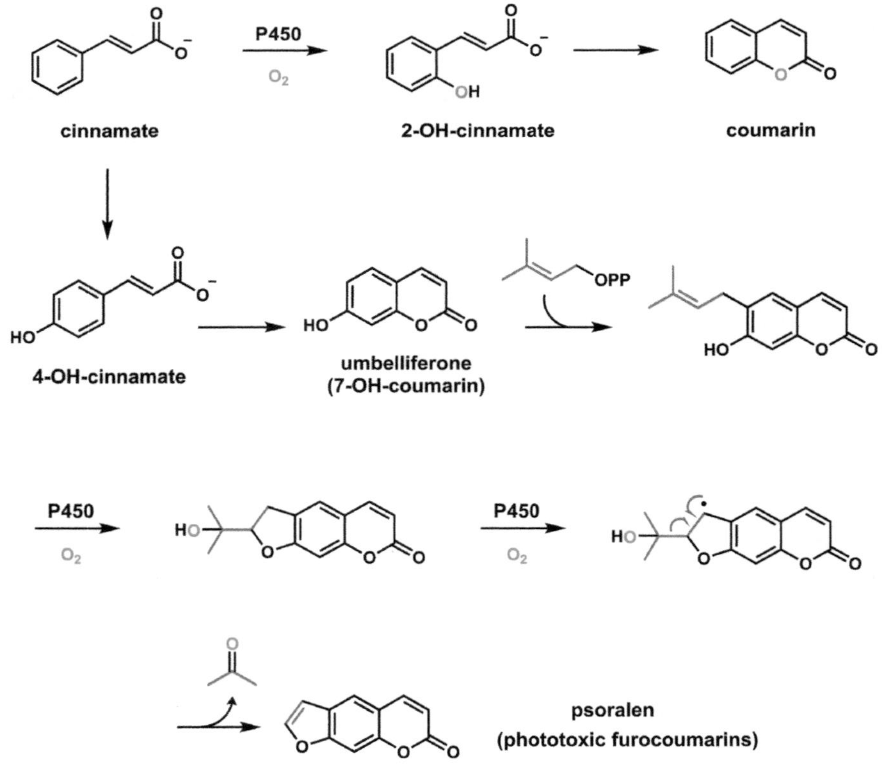

Figure 7.31 Metabolism of cinnamate to coumarins, such as umbelliferone, which can be prenylated and oxidized to psoralen.

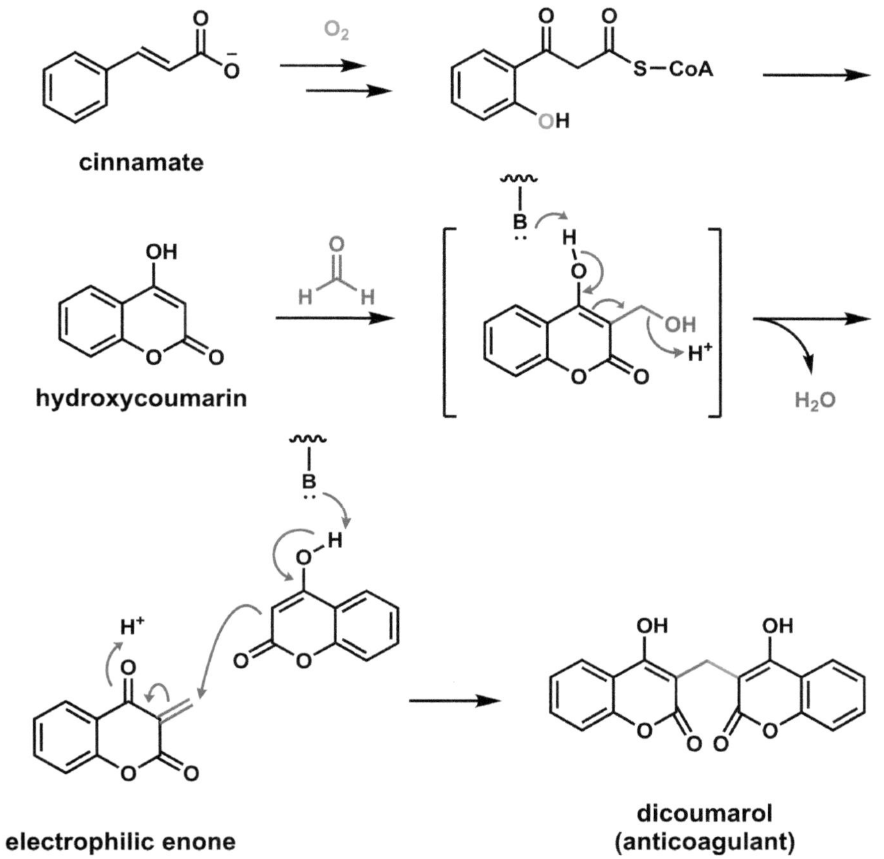

Figure 7.32 Metabolism of cinnamate to the anticoagulant dicoumarol.

Hydration of cinnamate or cinnamyl-CoA and oxidation to the ketone set up the hydroxycoumarin ring formation (Figure 7.32). Addition of a formaldehyde equivalent (Bourgaud, Hehn *et al.* 2006, Dewick 2009) to create an electrophilic enone that can be attacked by the enol of the hydroxycoumarin partner would generate the bridged dicoumarol, an anticoagulant compound in clover, by virtue of blocking the vitamin K dependent reductase in the blood coagulation cascade.

7.6.3 Ferulate to Piperine and Capsaicin

The 3-methoxy-4-hydroxy cinnamate, ferulate, featured as one of the key precursors to monolignols earlier in this chapter, can also undergo enzymatic conversion to two amides that are the main spices in black peppers and chili peppers, respectively (Dewick 2009). As depicted in Figure 7.33A, feruloyl-CoA can be operated

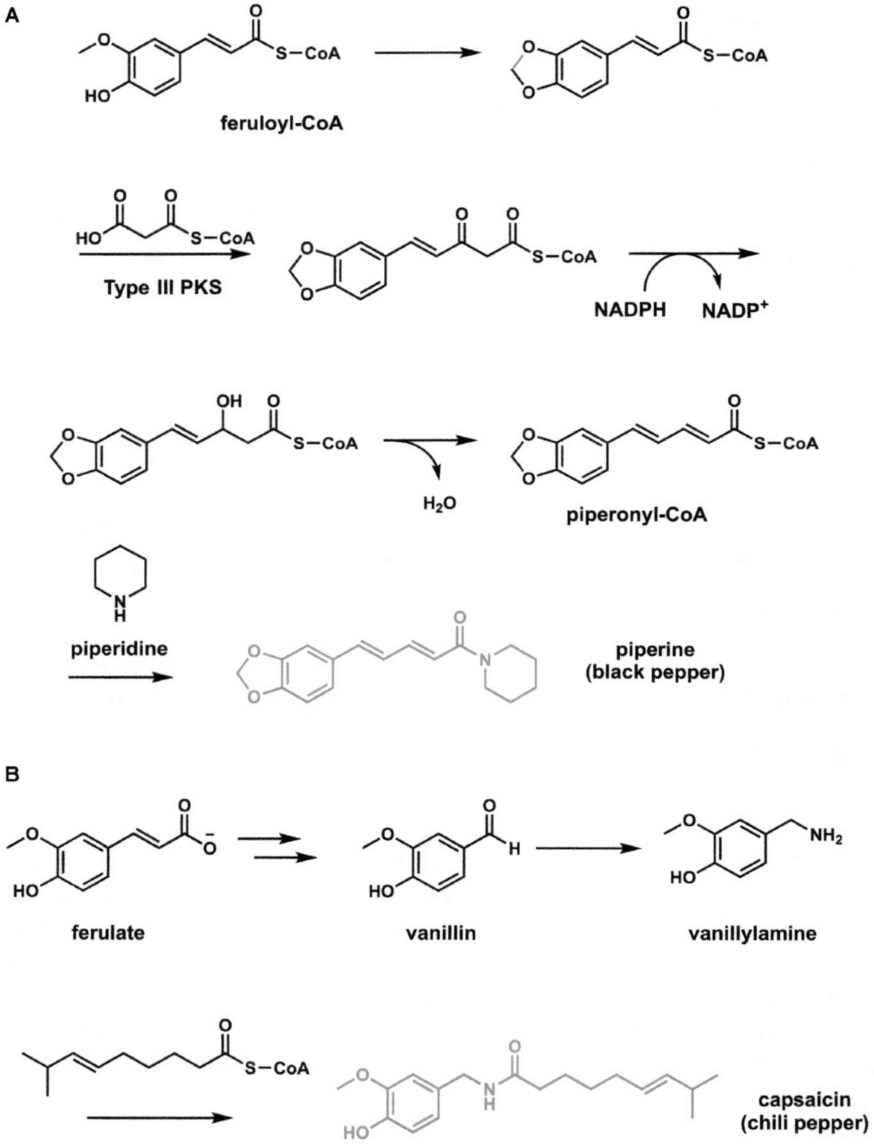

Figure 7.33 Conversion of ferulyl-CoA and ferulate to piperine and capsaicin, spicy molecules in black peppers and chili peppers, respectively.

on by a classic methylenedioxy bridge-forming cytochrome P450. Then a type III PKS catalyzing a single chain elongation *via* malonyl-CoA (a reminder of gingerol formation noted early in this chapter) yields the β-ketoacyl-CoA product. This can undergo fatty acid synthase/modular polyketide synthase processing logic to the bis-olefinic thioester (piperonyl-CoA). Now, a third biosynthetic logic converges, transfer of the acyl group to the structurally simple alkaloid piperidine. Piperidine is available from lysine decarboxylation, oxidative cyclization and Δ^1-piperidine reduction (Chapter 5). Action of a piperine-forming amide synthase completes the biosynthesis of this spice, one of the earliest natural products purified (in 1816), as noted in Chapter 1.

Ferulate rather than the CoA thioester is the entry point for capsaicin formation, the spicy flavoring in chili peppers (Figure 7.33B) (Kang, Jung *et al.* 2005). Capsaicin exerts its heat-causing sensations by acting as a ligand for TRP (transient receptor potential) ion channels (Caterina, Schumacher *et al.* 1997). Nonoxidative chain shortening to the substituted benzaldehyde yields the flavoring molecule vanillin and, after enzymatic transamination, vanilylamine (Dewick 2009). This will be the attacking nucleophile in a corresponding amide synthase reaction, utilizing a branched chain C_{10}-CoA (from a polyketide synthase assembly line) to create the capsaicin end product.

Vignette 7.3 Capsaicin and Caspaicinoids.

Capsaicin was originally isolated in impure form from *Capsicum* (hence the name) in 1816 and purified through the years to finally yield the pure compound 82 years later in 1898 (Figure 7.V3). As early as the 1870s, capsaicin was known to cause burning sensations on contact with epithelial cells and mucous membranes. As noted in this chapter, capsaicin arises from the convergence of two pathways. The phenylpropanoid pathway yields vanillylamine after transamination of the aldehyde precursor. The other half of the molecule comes from a branched 10-carbon acyl-CoA, a constituent of fatty acid metabolism in capsicum plants. The two halves are stitched together by an amide ligase.

Capsaicin and related capsaicinoids are thought to serve as antifeeding, deterrent molecules to herbivores and also as antifungal agents. It is an active ingredient in commercial pepper sprays. In humans capsaicin serves as an activating ligand for two related neuronal ion channels, TRPA1 and TRPV1, which signal through spinal dorsal root ganglia to the brain and give the burning sensation end readout (Caterina, Schumacher *et al.* 1997, Bessac and Jordt 2008, Escalera, von Hehn *et al.* 2008, Clapham 2015).

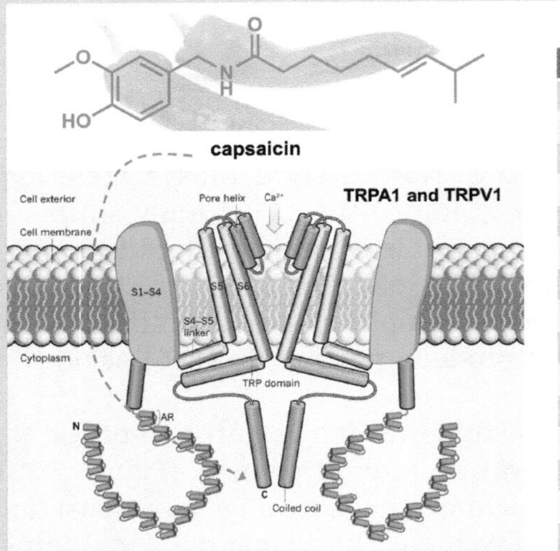

Scoville Scale of Pungency

Heat Units	Types of Peppers
15,000,000	Pure capsaicin
4,000,000	Pepper spray
1,000,000	Bhut Jolokia
577,000	Red Savina
200,000-350,000	Habanero
100,000-250,000	Chiltepin
30,000-50,000	Cayenne
15,000-30,000	Arbol
8,000-23,000	Serrano
5,000-8,000	Yellow Hot
2,500-4,000	Guajillo
1,500-2,500	Chilaca
500-2,000	Anaheim
500-750	Red Chile
0	Sweet Bells

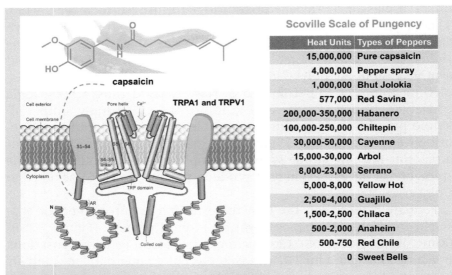

Figure 7.V3 Capsaicin and caspaicinoids.
Illustration of TRPA1 reprinted by permission from Macmillan Publishers Ltd: Nature (Clapham, D. E. "Structural biology: Pain-Sensing TRPA1 Channel Resolved". *Nature* **520**: 439–441), copyright (2015).

Capsaicinoids are present in different levels in distinct species of chili peppers, which can be ranked in terms of spiciness or "heat". Wilbur Scoville in 1912 invented the Scoville scale, an organoleptic test in which three tasters evaluate spiciness of a given amount of dissolved pepper solids *via* a series of aqueous dilution tests (Scoville 1912). The "hottest" peppers are in the range of 1–2 million Scoville units, compared to 15 000 000 for pure capsaicin (Figure 7.V3).

7.7 A Closing Look at a Different Phenylpropanoid Route: Tyrosine as Precursor to Plastoquinines and Tocopherols

One final biosynthetic look involves biogenesis of plant plastoquinones and the tocopherols of the vitamin E series. The isoprenoid side chains are assembled by the kinds of logic and enzymatic machinery discussed in Chapter 4. Our attention here is on the generation of the benzoquinone and tetrahydrochromane ring systems, respectively (Dewick 2009).

In this rare instance it is not phenylalanine that is the amino acid portal but instead the nitrogen-free prephenate, routed to *para*-OH-phenylpyruvate rather than phenylpyruvate. As shown in Figure 7.34, *para*-hydroxyphenylpyruvate is substrate for a nonheme, mononuclear iron-based oxygenase which carries out a *para*-hydroxylation with attendant side chain decarboxylation and 1,2-migration. The ketoacid functionality in the substrate plays the equivalent

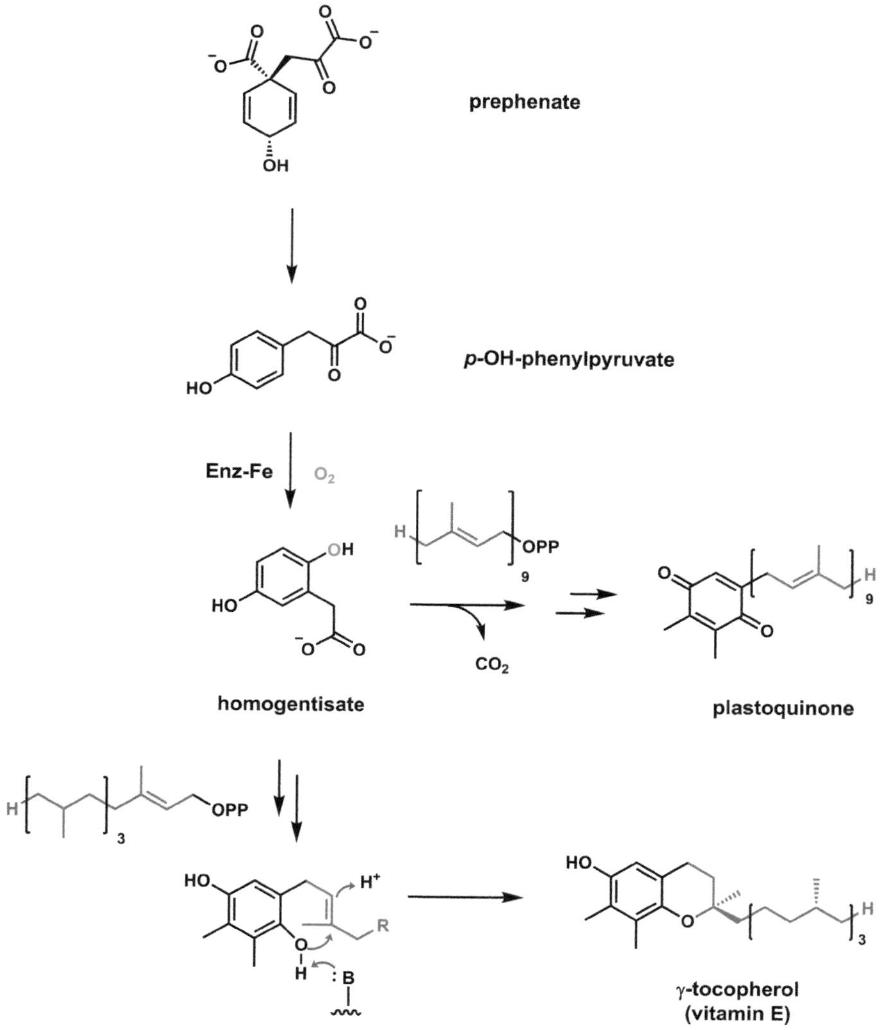

Figure 7.34 Phenylpropanoids are precursors to electron transfer quinones and vitamin E.

intramolecular role in oxygen activation that 2-oxoglutarate does in intermolecular cases (see Chapter 9). The product molecule is homogentisate and it is a common precursor both for plastoquinones and the chromane ring systems after methylations and decarboxylation.

Vignette 7.4 An Enzymatic 1,3-Dipolar Cycloaddition Mechanism in Aromatic Decarboxylase Action?

In recent years, a series of enzyme-catalyzed [4 + 2] cycloadditions have become better defined, as noted in Chapter 2, including most recently the AbyU enzyme which catalyzes formation of the cyclohexene ring in the spirotetronate moiety during abyssomycin biosynthesis (Byrne, Lees *et al.* 2016). Evidence has begun to accrete that at least some of these enzymes are bona fide Diels–Alderases. The 3,3-Claisen type rearrangement for chorismate mutase has been known and studied for decades (Marti, Andres *et al.* 2004). (That enzyme will also catalyze a Cope rearrangement with carba-chorismate.)

Adding to the list of biological cycloaddition mechanisms is the recent proposal for a [3 + 2] cycloaddition pathway in the action of a pair of microbial decarboxylases in secondary metabolic pathways. One of the enzymes decarboxylates 3-prenyl-4-hydroxybenzoate on the way to the benzoquinone electron transfer coenzyme ubiquinone-*n* ($n = 1$–12), also known as coenzyme Q (Dewick 2009). The second enzyme is ferulate decarboxylase (ferulate is 3-methoxy-4-hydroxycinnamate, one of the central metabolites in lignan and lignin formation, noted in this chapter) (Figure 7.V4).

Each of these decarboxylases needs a partner enzyme that converts the dihydro form of free flavin coenzyme FMN into a prenylated tetracyclic form that has gained new modes of reactivity, in particular as a precursor to a participant in 1,3-dipolar cycloadduct formation and reversal (White, Payne *et al.* 2015). As shown in Figure 7.V5, the bacterial UbiX or the homologous fungal enzyme Pad1 can cyclize the five carbons of Δ^2-isoprenyl monophosphate (rather than the ubiquitous prenyl diphosphate), *via* C_1 and C_3 onto C_6 and N_5 of $FMNH_2$.

This is presumed to be released from UbiX/Pad1 and to diffuse to the active site of UbiD/Fdc1 and undergo O_2-mediated oxidation to the N_5 iminium adduct, in which there is azomethine ylid character, appropriate for function as a 1,3-dipole. The cinnamic/ferulic acid substrates have the conjugated α/β-unsaturated carbonyl functionalities to act as dipolarophiles. Although, one could write a stepwise Michael addition mechanism (Payne, White *et al.* 2015), the investigators argued strongly for the 1,3-dipolar cycloaddition route shown. The resultant adduct

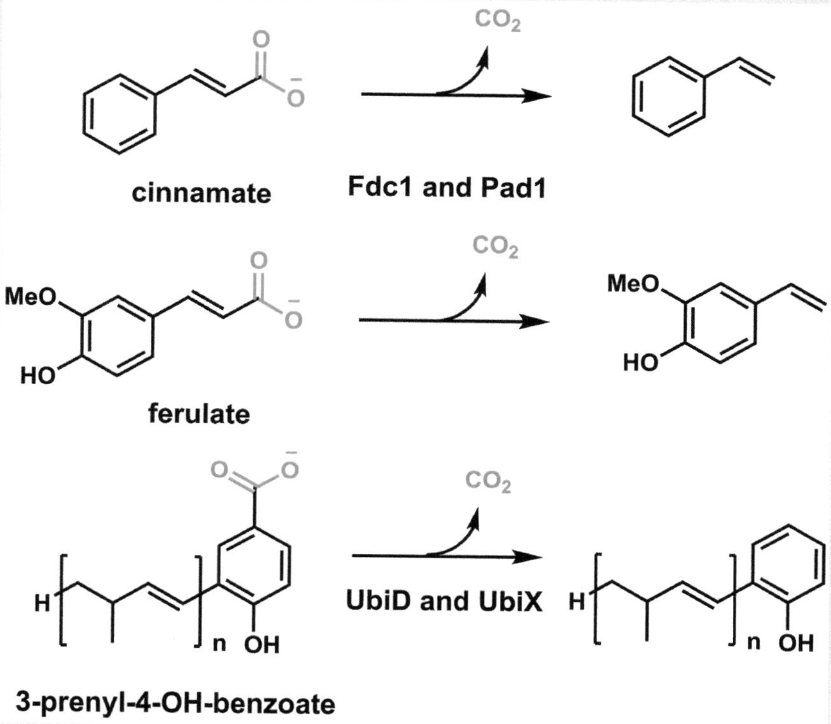

Figure 7.V4 Reactions catalyzed by microbial decarboxylases.

could lose CO_2 as shown and then undergo a retro 1,3-dipolar addition to yield the observed terminal alkenes in the styrene framework. An analogous mechanism would be in play for the ubiquinone decarboxylation step.

This study reveals new facets of biosynthetic chemistry in both the cinnamoyl class of phenylpropanoid metabolites and the polyprenyl ubiquinone electron transfer coenzymes. It uncovers a novel prenylated tetracyclic coenzyme form of vitamin B2 (riboflavin) with an expanded reach of enzymatic chemistry that may now turn up in other enzymes. Third, this study raises the prospect of the first example of a [3 + 2] electrocyclic reaction mode in natural product assembly and has stirred continued mechanistic investigation (Ferguson, Arunrattanamook *et al.* 2016) of this decarboxylase class. Fourth, the ability to purify and investigate specific biosynthetic enzymes for chemical mechanism is likely to continue to turn up reactions known in organic chemistry but not yet mapped onto biology.

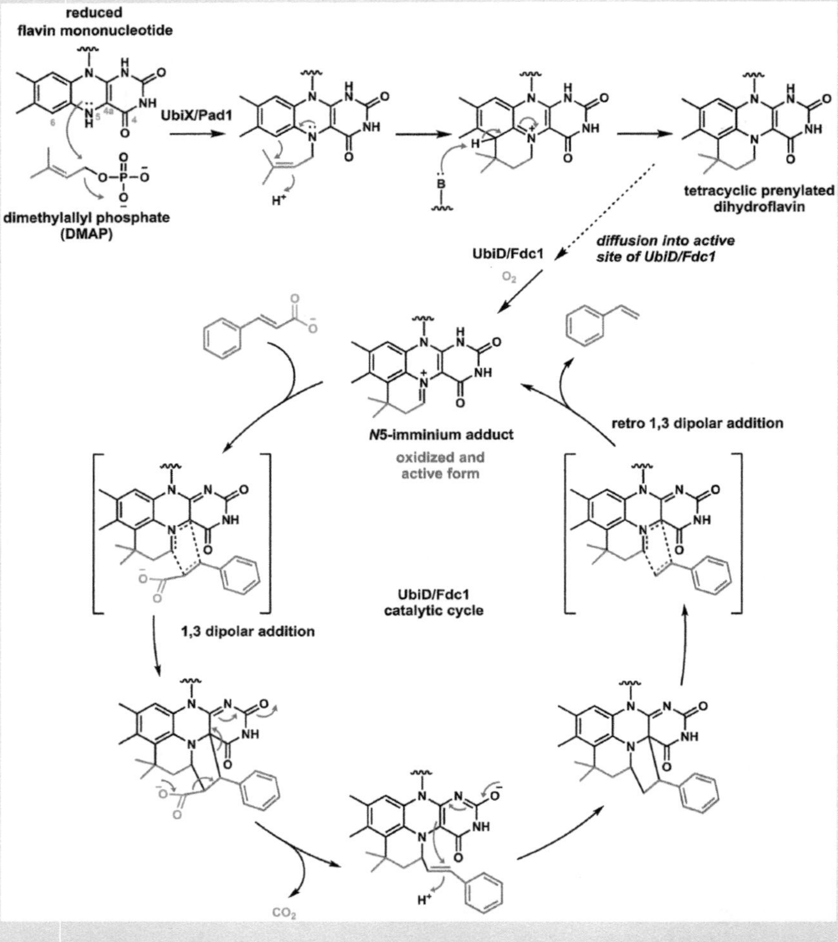

Figure 7.V5 The proposed mechanism of the flavin-dependent decarboxylases involves a 1,3-dipolar addition step.

7.8 Summary

The C_6–C_3 carbon scaffold of phenylalanine gives rise to an enormous variety of structures and functions in the phenylpropanoid family of natural products. This is essentially a nitrogen-free chemical universe with the amino group of phenylalanine jettisoned in the first committed step of the pathways.

In the lignan and lignin branch, aromatic ring hydroxylations and covering methylations leave a phenolic-OH group that is the initiator of a pervasive set of phenoxy radical dimerizations and oligomerizations. The strategically placed *O*-methylations have the net effect of controlling sites of radical density and thereby regioselectivity in dimerizations and polymerizations.

Oxygenases also sculpt the scaffolds in many of the non lignan/lignan arms of phenylpropanoid pathways. The many methylenedioxy bridge formations are one measure, while the 1,2-aryl shifts to send flavonoid flux to isoflavonoids and phytoalexins demonstrate another footprint of oxygenase one electron reaction manifolds.

Colored anthocyanidin/anthocyanin flower pigments result from flavanone oxidations to conjugated systems while regioselective methylations control wavelength maxima. The chalcone/stilbene partition also creates scaffold diversity, as do the 900 type II polyketide synthases in plants that produce spices and essential oils.

References

Abe, I. and H. Morita (2010). "Structure and function of the chalcone synthase superfamily of plant type III polyketide synthases". *Nat. Prod. Rep.* **27**(6): 809–838.

Austin, M. and J. P. Noel (2002). "The chalcone synthase superfamily of type III polyketide synthases". *Nat. Prod. Rep.* **20**: 79–110.

Baerson, S. R., A. M. Rimando and Z. Pan (2008). "Probing allelochemical biosynthesis in sorghum root hairs". *Plant Signaling Behav.* **3**(9): 667–670.

Bessac, B. F. and S. E. Jordt (2008). "Breathtaking TRP channels: TRPA1 and TRPV1 in airway chemosensation and reflex control". *Physiology* **23**: 360–370.

Boerjan, W., J. Ralph and M. Baucher (2003). "Lignin biosynthesis". *Annu. Rev. Plant Biol.* **54**: 519–546.

Bourgaud, F., A. Hehn, R. Larbat, S. Doerper, E. Gontier, S. Kellner and U. Matern (2006). "Biosynthesis of coumarins in plants: a major pathway still to be unravelled for cytochrome P450 enzymes". *Phytochem. Rev.* **5**: 293–308.

Byrne, M. J., N. R. Lees, L. C. Han, M. W. van der Kamp, A. J. Mulholland, J. E. Stach, C. L. Willis and P. R. Race (2016). "The Catalytic Mechanism of a Natural Diels-Alderase Revealed in Molecular Detail". *J. Am. Chem. Soc.* **138**(19): 6095–6098.

Caterina, M. J., M. A. Schumacher, M. Tominaga, T. A. Rosen, J. D. Levine and D. Julius (1997). "The capsaicin receptor: a heat-activated ion channel in the pain pathway". *Nature* **389**(6653): 816–824.

Clapham, D. E. (2015). "Structural biology: Pain-sensing TRPA1 channel resolved". *Nature* **520**(7548): 439–441.

Cos, P., T. De Bruyne, N. Hermans, S. Apers, D. V. Berghe and A. J. Vlietinck (2004). "Proanthocyanidins in health care: current and new trends". *Curr. Med. Chem.* **11**(10): 1345–1359.

Crombie, L. and D. A. Whiting (1998). "Review article number 135 biosynthesis in the rotenoid group of natural products: applications of isotope methodology". *Phytochemistry* **49**(6): 1479–1507.

Davin, L. B. and N. G. Lewis (2000). "Dirigent proteins and dirigent sites explain the mystery of specificity of radical precursor coupling in lignan and lignin biosynthesis". *Plant Physiol.* **123**(2): 453–462.

Davin, L. B. and N. G. Lewis (2003). "An historical perspective on lignan biosynthesis: Monolignol, allylphenol and hydroxycinnamic acid coupling and downstream metabolism". *Phytochem. Rev.* **2**: 257–283.

Dewick, P. (2009). *Medicinal Natural Products, a Biosynthetic Approach.* UK, Wiley.

Dixon, R. A., L. Achnine, P. Kota, C. J. Liu, M. S. Reddy and L. Wang (2002). "The phenylpropanoid pathway and plant defence-a genomics perspective". *Mol. Plant Pathol.* **3**(5): 371–390.

Escalera, J., C. A. von Hehn, B. F. Bessac, M. Sivula and S. E. Jordt (2008). "TRPA1 mediates the noxious effects of natural sesquiterpene deterrents". *J. Biol. Chem.* **283**(35): 24136–24144.

Falcone Ferreyra, M. L., S. P. Rius and P. Casati (2012). "Flavonoids: biosynthesis, biological functions, and biotechnological applications". *Front. Plant Sci.* **3**: 222.

Ferguson, K. L., N. Arunrattanamook and E. N. Marsh (2016). "Mechanism of the Novel Prenylated Flavin-Containing Enzyme Ferulic Acid Decarboxylase Probed by Isotope Effects and Linear Free-Energy Relationships". *Biochemistry* **55**(20): 2857–2863.

Fraser, C. M. and C. Chapple (2011). "The phenylpropanoid pathway in Arabidopsis". *Arabidopsis Book* **9**: e0152.

Fujita, M., D. R. Gang, L. B. Davin and N. G. Lewis (1999). "Recombinant pinoresinol-lariciresinol reductases from western red cedar (Thuja plicata) catalyze opposite enantiospecific conversions". *J. Biol. Chem.* **274**(2): 618–627.

Gonzalez-Lamothe, R., G. Mitchell, M. Gattuso, M. S. Diarra, F. Malouin and K. Bouarab (2009). "Plant antimicrobial agents and their effects on plant and human pathogens". *Int. J. Mol. Sci.* **10**(8): 3400–3419.

Guo, L., R. A. Dixon and N. L. Paiva (1994). "The 'pterocarpan synthase' of alfalfa: association and co-induction of vestitone reductase and 7,2'-dihydroxy-4'-methoxy-isoflavanol (DMI) dehydratase, the two final enzymes in medicarpin biosynthesis". *FEBS Lett.* **356**(2-3): 221–225.

Harborne, J. (1999). *The Handbook of Natural Flavonoids*, Wiley/ Blackwell.

Harmatha, J. and L. Dinan (2003). "Biological activities of lignans and stilbenoids associated with plant-insect chemical interactions". *Phytochem. Rev.* **2**: 321–330.

Janick, J. and A. Whipkey, eds. (2002). The American Mayapple and its Potential for Podophyllotoxin Production. *Trends in New Crops and New Uses*. Alexandria, Virginia, ASHS Press.

Jez, J. M., M. E. Bowman, R. A. Dixon and J. P. Noel (2000). "Structure and mechanism of the evolutionarily unique plant enzyme chalcone isomerase". *Nat. Struct. Biol.* 7(9): 786–791.

Jez, J. M. and J. P. Noel (2000). "Mechanism of chalcone synthase. pKa of the catalytic cysteine and the role of the conserved histidine in a plant polyketide synthase". *J. Biol. Chem.* **275**(50): 39640–39646.

Kang, S. M., H. Y. Jung, Y. M. Kang, J. Y. Min, C. S. Karigar, J. K. Yang, S. W. Kim, Y. R. Ha, S. H. Lee and M. S. Choi (2005). "Biotransformation and impact of ferulic acid on phenylpropanoid and capsaicin levels in Capsicum annuum L. cv. P1482 cell suspension cultures". *J. Agric. Food Chem.* **53**(9): 3449–3453.

Kurusu, T., J. Hamada, H. Nokajima, Y. Kitagawa, M. Kiyoduka, A. Takahashi, S. Hanamata, R. Ohno, T. Hayashi, K. Okada, J. Koga, H. Hirochika, H. Yamane and K. Kuchitsu (2010). "Regulation of microbe-associated molecular pattern-induced hypersensitive cell death, phytoalexin production, and defense gene expression by calcineurin B-like protein-interacting protein kinases, OsCIPK14/ 15, in rice cultured cells". *Plant Physiol.* **153**(2): 678–692.

Lau, W. and E. Sattely (2015). "Six enzymes from mayapple that complete the biosynthetic pathway to the etoposide aglycone". *Science* **349**: 1224–1228.

Lin, Y. M., S. L. Shih, W. C. Lin, J. W. Wu, Y. T. Chen, C. Y. Hsieh, L. C. Guan, L. Lin and C. P. Cheng (2014). "Phytoalexin biosynthesis genes are regulated and involved in plant response to Ralstonia solanacearum infection". *Plant Sci.* **224**: 86–94.

Lochab, B., S. Shukla and I. K. Varma (2014). "Naturally occurring phenolic sources: monomers and polymers". *RSC Adv.* **4**: 21712–21752.

Marti, S., J. Andres, V. Moliner, E. Silla, I. Tunon and J. Bertran (2004). "A comparative study of claisen and cope rearrangements catalyzed by chorismate mutase. An insight into enzymatic efficiency: transition state stabilization or substrate preorganization?". *J. Am. Chem. Soc.* **126**(1): 311–319.

Mizutani, M. and F. Sato (2011). "Unusual P450 reactions in plant secondary metabolism". *Arch. Biochem. Biophys.* **507**(1): 194–203.

Nakayama, T., T. Sato, Y. Fukui, K. Yonekura-Sakakibara, H. Hayashi, Y. Tanaka, T. Kusumi and T. Nishino (2001). "Specificity analysis and mechanism of aurone synthesis catalyzed by aureusidin synthase, a polyphenol oxidase homolog responsible for flower coloration". *FEBS Lett.* **499**(1–2): 107–111.

Nakayama, T., K. Yonekura-Sakakibara, T. Sato, S. Kikuchi, Y. Fukui, M. Fukuchi-Mizutani, T. Ueda, M. Nakao, Y. Tanaka, T. Kusumi and T. Nishino (2000). "Aureusidin synthase: a polyphenol oxidase homolog responsible for flower coloration". *Science* **290**(5494): 1163–1166.

Ngaki, M. N., G. V. Louie, R. N. Philippe, G. Manning, F. Pojer, M. E. Bowman, L. Li, E. Larsen, E. S. Wurtele and J. P. Noel (2012). "Evolution of the chalcone-isomerase fold from fatty-acid binding to stereospecific catalysis". *Nature* **485**(7399): 530–533.

Ono, E., M. Hatayama, Y. Isono, T. Sato, R. Watanabe, K. Yonekura-Sakakibara, M. Fukuchi-Mizutani, Y. Tanaka, T. Kusumi, T. Nishino and T. Nakayama (2006). "Localization of a flavonoid biosynthetic polyphenol oxidase in vacuoles". *Plant J.* **45**(2): 133–143.

Parage, C., R. Tavares, S. Rety, R. Baltenweck-Guyot, A. Poutaraud, L. Renault, D. Heintz, R. Lugan, G. A. Marais, S. Aubourg and P. Hugueney (2012). "Structural, functional, and evolutionary analysis of the unusually large stilbene synthase gene family in grapevine". *Plant Physiol.* **160**(3): 1407–1419.

Patisaul, H. B. and W. Jefferson (2010). "The pros and cons of phyto estrogens". *Front. Neuroendocrinol.* **31**(4): 400–419.

Payne, K. A., M. D. White, K. Fisher, B. Khara, S. S. Bailey, D. Parker, N. J. Rattray, D. K. Trivedi, R. Goodacre, R. Beveridge, P. Barran, S. E. Rigby, N. S. Scrutton, S. Hay and D. Leys (2015). "New cofactor supports alpha,beta-unsaturated acid decarboxylation *via* 1,3-dipolar cycloaddition". *Nature* **522**(7557): 497–501.

Pled-Zehavi, H., M. Oliva, Q. Xie, V. Tzin, M. Oren-Shamir, A. Aharoni and G. Galli (2015). "Metabolic Engineering of the Phenylpropanoid and Its Primary, Precursor Pathway to Enhance the Flavor of Fruits and the Aroma of Flowers". *Bioengineering* **2**: 204–212.

Richelle, M., S. Pridmore-Merten, S. Bodenstab, M. Enslen and E. A. Offord (2002). "Hydrolysis of isoflavone glycosides to aglycones by beta-glycosidase does not alter plasma and urine isoflavone pharmacokinetics in postmenopausal women". *J. Nutr.* **132**(9): 2587–2592.

Rupasinghe, S., J. Baudry and M. A. Schuler (2003). "Common active site architecture and binding strategy of four phenylpropanoid P450s from Arabidopsis thaliana as revealed by molecular modeling". *Protein Eng.* **16**(10): 721–731.

Satake, H., E. Ono and J. Murata (2013). "Recent advances in the metabolic engineering of lignan biosynthesis pathways for the production of transgenic plant-based foods and supplements". *J. Agric. Food Chem.* **61**(48): 11721–11729.

Sato, T., T. Nakayama, S. Kikuchi, Y. Fukui, K. Yonekura-Sakakibara, T. Ueda, T. Nishino, Y. Tanaka and T. Kusumi (2001). "Enzymatic formation of aurones in the extracts of yellow snapdragon flowers". *Plant Sci.* **160**(2): 229–236.

Schroder, J. and G. Schroder (1990). "Stilbene and chalcone synthases: related enzymes with key functions in plant-specific pathways". *Z. Naturforsch., C: J. Biosci.* **45**(1–2): 1–8.

Scoville, W. (1912). "Nite on Capsicums". *J. Am. Pharm. Assoc.* **1**: 453–454.

Tatsis, E. C., E. Eylert, R. K. Maddula, E. Ostrozhenkova, A. Svatos, W. Eisenreich and B. Schneider (2014). "Biosynthesis of Nudicaulins: A (13) CO2 -pulse/chase labeling study with Papaver nudicaule". *ChemBioChem* **15**(11): 1645–1650.

Taura, F., S. Tanaka, C. Taguchi, T. Fukamizu, H. Tanaka, Y. Shoyama and S. Morimoto (2009). "Characterization of olivetol synthase, a polyketide synthase putatively involved in cannabinoid biosynthetic pathway". *FEBS Lett.* **583**(12): 2061–2066.

VanEtten, H. D., J. W. Mansfield, J. A. Bailey and E. E. Farmer (1994). "Two Classes of Plant Antibiotics: Phytoalexins versus "Phytoanticipins"". *Plant Cell* **6**(9): 1191–1192.

Vanholme, R., B. Demedts, K. Morreel, J. Ralph and W. Boerjan (2010). "Lignin biosynthesis and structure". *Plant Physiol.* **153**(3): 895–905.

Vanholme, R., V. Storme, B. Vanholme, L. Sundin, J. H. Christensen, G. Goeminne, C. Halpin, A. Rohde, K. Morreel and W. Boerjan (2012). "A systems biology view of responses to lignin biosynthesis perturbations in Arabidopsis". *Plant Cell* **24**(9): 3506–3529.

Veitch, N. C. and R. J. Grayer (2008). "Flavonoids and their glycosides, including anthocyanins". *Nat. Prod. Rep.* **25**(3): 555–611.

Vogt, T. (2010). "Phenylpropanoid biosynthesis". *Mol. Plant* **3**(1): 2–20.

Walsh, C. T. (2005). *Posttranslational Modification of Proteins: Expanding Nature's Inventory*. Englewood, Colorado, Roberts and Company.

White, M. D., K. A. Payne, K. Fisher, S. A. Marshall, D. Parker, N. J. Rattray, D. K. Trivedi, R. Goodacre, S. E. Rigby, N. S. Scrutton, S. Hay and D. Leys (2015). "UbiX is a flavin prenyltransferase required for bacterial ubiquinone biosynthesis". *Nature* **522**(7557): 502–506.

Winkel-Shirley, B. (2001). "Flavonoid biosynthesis. A colorful model for genetics, biochemistry, cell biology, and biotechnology". *Plant Physiol.* **126**(2): 485–493.

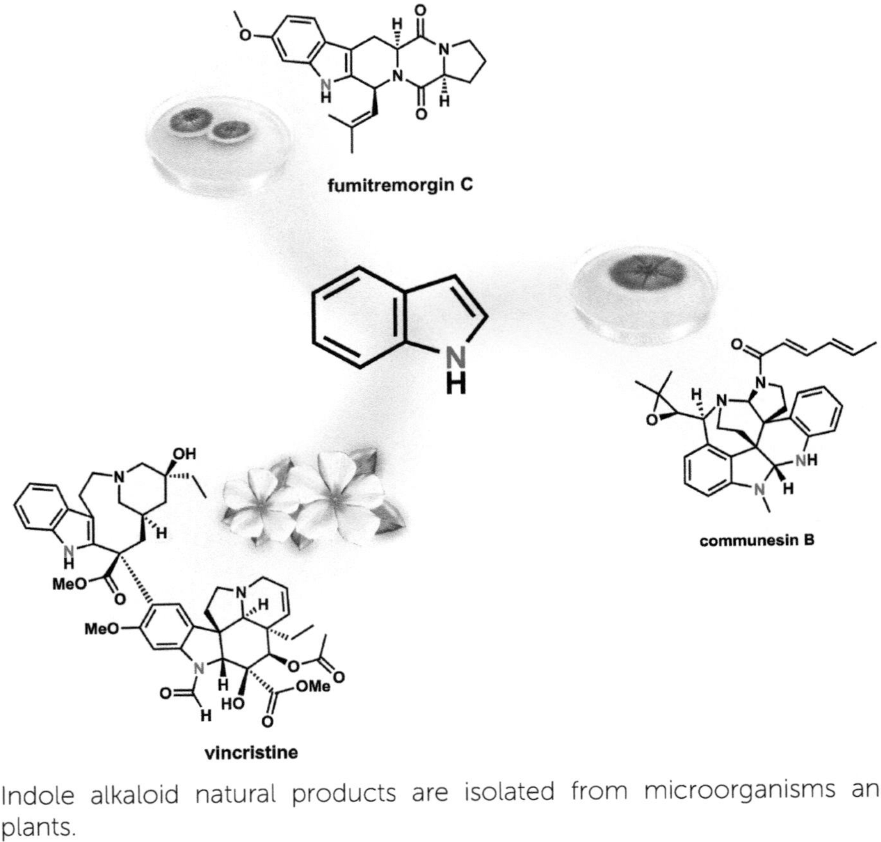

Indole alkaloid natural products are isolated from microorganisms and plants.
Copyright (2016) John Billingsley.

8 Indole Terpenes: Alkaloids II

8.1 Introduction

Chapter 5 introduced the alkaloid class of natural products and gave examples of different amino acid building blocks as entry points for particular classes of alkaloid heterocyclic products (Roberts and Wink 1998). Among the half dozen alkaloid progenitor amino acids, tryptophan was featured in three examples. The first is the conversion to the 6-5-6 tricyclic ring system of β-carbolines in formation of the simple scaffold of harmane which introduced a variant of Pictet–Spengler ring-forming chemistry.

The second example was the more familiar Pictet–Spengler catalyst strictosidine synthase, joining tryptamine and secologanin to yield the glycosyl-protected strictosidine. Since secologanin derives ultimately from geraniol oxidation, strictosidine is an indole monoterpene (O'Connor and Maresh 2006). In turn the thousand or more alkaloids to which strictosidine is enzymatic progenitor fall under a broad indole terpene rubric.

The third example was conversion of tryptophan to lysergic acid and ergotamine. The alkylation and rearrangement of the isoprenyl chain is a key step in creating chanoclavine and the framework of lysergic acid.

In this second pass through the reaction of tryptophan with prenyl donors we focus more on the reactivity of the indole ring as carbanion equivalent for enzyme-directed alkylation at any of six sites: two-electron reaction manifolds. We also focus on microbial rather than plant indole terpene metabolites and examine the role of O_2 as cosubstrate in enabling one-electron reaction manifolds and a set of

Natural Product Biosynthesis: Chemical Logic and Enzymatic Machinery
By Christopher T. Walsh and Yi Tang

Published by the Royal Society of Chemistry, www.rsc.org

rearrangements and insertion reactions that are unique and mechanistically instructive.

8.2 Two Routes to Tricyclic Scaffolds from Trp: β-Carbolines and Pyrroloindoles

Enzymatic machinery exists to convert the bicyclic indole ring of tryptophan and its decarboxylation metabolite tryptamine into two different sets of tricyclic ring systems (Seigler 2012), the 6-5-6 carboline and the 6-5-5 pyrroloindole systems. As diagrammed in Figure 8.1, β-carboline formation results when an aldehyde reacts with the amine of tryptophan/tryptamine to yield the carbinolamine and then the iminium ion. An intramolecular Mannich reaction can occur with C_3 of the indole as carbon nucleophile attacking the side chain imine to yield the 5,5-spirocycle. Migration of the indicated C–C single bond to C_2 of the indoline imine yields the β-carboline tricyclic scaffold. This need not involve prenyl/terpene cosubstrates, unless the prenyl cosubstrate contains the aldehyde group: that is the case for secologanin, and strictosidine synthase does create a β-carboline framework (see Figure 5.16).

The generation of the 6-5-5 pyrroloindole framework (Figure 8.2) can occur *via* capture of an electrophile, not at C_2 of the indole as above, but at C_3, also a nucleophilic site as we will explicitly note below. The two most common electrophiles captured at C_3 are the allyl cations from Δ^2-prenyl substrates or –OH from FAD-4a-OOH (see Figure 5.29 and Chapter 9)(Walsh, Haynes *et al.* 2013). (However, we note in the physostigmine biosynthetic route in Figure 8.3 that a methyl cation equivalent from *S*-adenosylmethionine can serve as the ring-closing electrophile.) The resultant indoline iminium ion can be captured by neighboring nitrogen atoms as schematized in Figures 8.2 and 8.3.

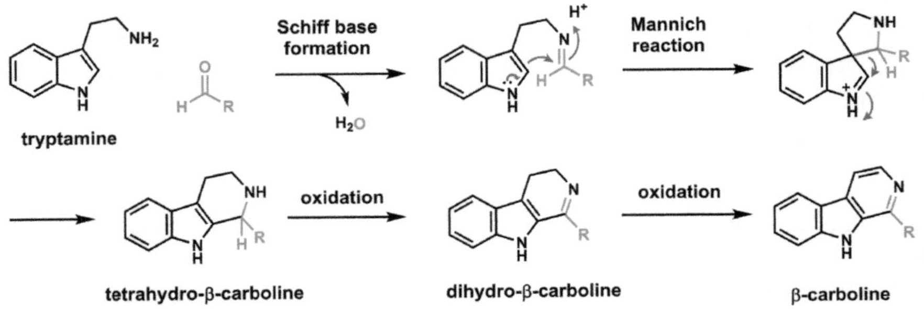

Figure 8.1 Schematic for conversion of bicyclic indole to tricyclic β-carboline rings.

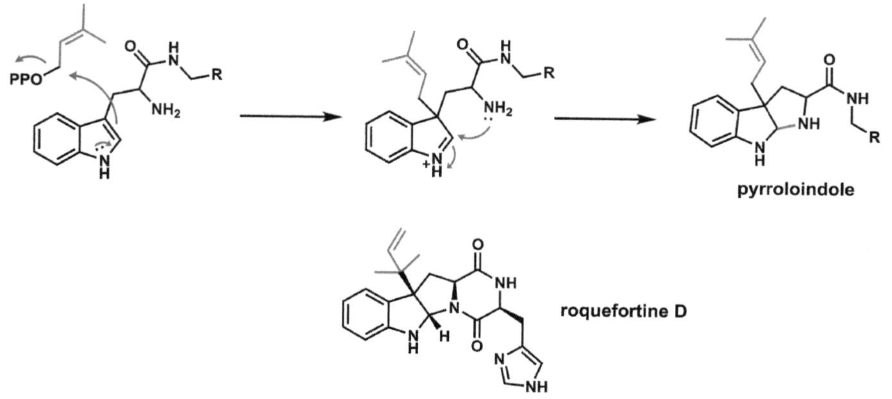

Figure 8.2 Schematic for conversion of indoles into tricyclic pyrroloindole rings.

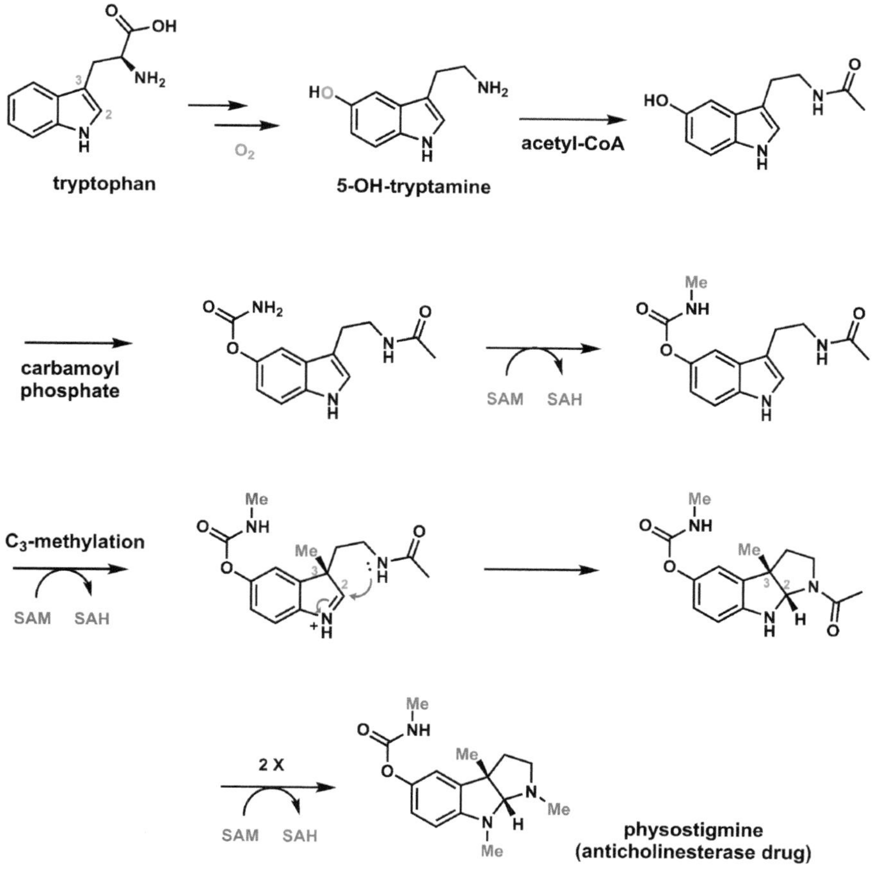

Figure 8.3 Four SAM-dependent methylations in conversion of tryptamine to the pyrroloindole physostigmine.

The structure of roquefortine D indicates that the pyrroloindoline substructure has arisen by capture of the DMAPP-derived prenyl cation at $C_{3'}$ rather than $C_{1'}$ of the delocalized cation. We will note below that even amide nitrogens, with lessened intrinsic basicity, are sufficiently good nucleophiles to close to pyrroloindole substructures.

Figure 8.3 shows the recently characterized pathway (Liu, Ng *et al.* 2014) to physostigmine, an anticholinergic metabolite used in eye drops. Tryptophan is hydroxylated to 5-OH-Trp and then reacted with acetyl-CoA to yield the *N*-acetyl-5-OH-Trp. Carbamylation and *N*-methylation of the 5-OH group then occur before the pyrroloindole-forming step. Of particular note, *S*-adenosylmethionine, acting in its canonical role as a donor of a $[CH_3^+]$ equivalent to a cosubstrate nucleophile, has been attacked by C_3 of the indole moiety. The adjacent amide nitrogen, weak nucleophile that it may be, can close on the iminium ion and create the pyrroloindole framework. Physostigmine itself is two more *N*-methylations away. All told, four equivalents of SAM are utilized to transfer four methyl groups, three to nitrogen and one to a nucleophilic indole C_3 carbon site, in this short biosynthetic pathway.

Figure 8.4 shows the structures of six prenylated indole natural products. These indole terpenes reflect different regiochemistries for prenylation on the indole bicyclic ring. Echinulin has been prenylated three times, at C_2, C_5, and C_7, roquefortine C at C_3 in pyrroloindole formation, and cyclomarin C on the indole nitrogen of the macrocyclic peptide. Ergotamine has its lysergic acid moiety prenylated at C_4, while fumitremorgin C was prenylated at C_2 and then further modified. The brevianamide B framework represents a highly morphed C_2-prenylated indole system. Clearly the producing fungi have a suite of indole prenyltransferases that are specific for distinct sites on the indole ring of particular substrates.

8.3 Trp-Xaa Diketopiperazine NRPS Assembly Line Products as Substrates for Regioselective Prenylations

All six structures in Figure 8.4 are built on nonribosomal peptide synthetase assembly lines (see Chapter 3). Cyclomarin C is a hexapeptide bristling with four nonproteinogenic amino acid residues, implicating a hexamodule NRPS assembly line. Analogously, we have noted in Figure 5.22 that ergotamine uses lysergic acid as a starter unit in a four module NRPS assembly line. The other four molecules

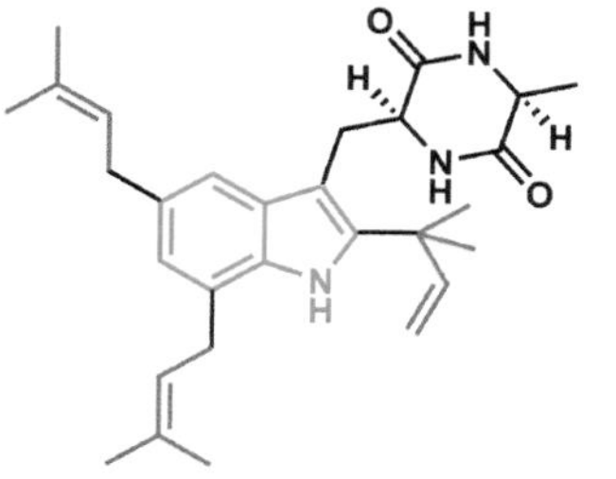

echinulin

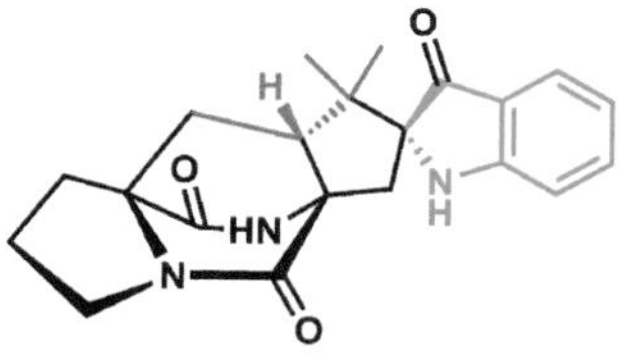

brevianamide B

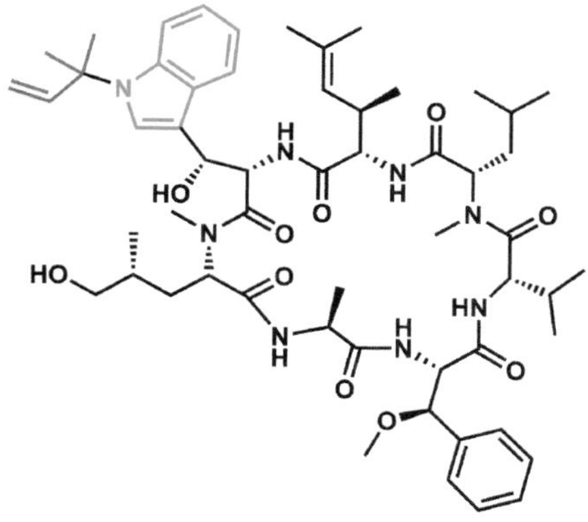

cyclomarin C

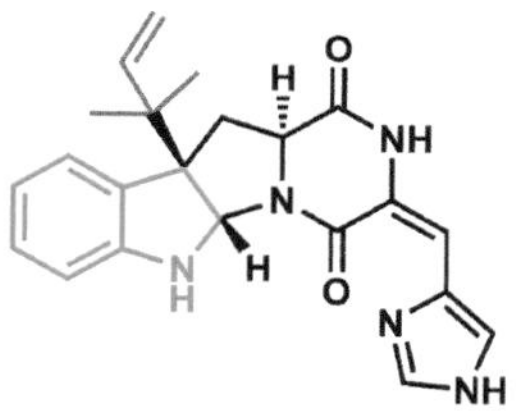

roquefortine C

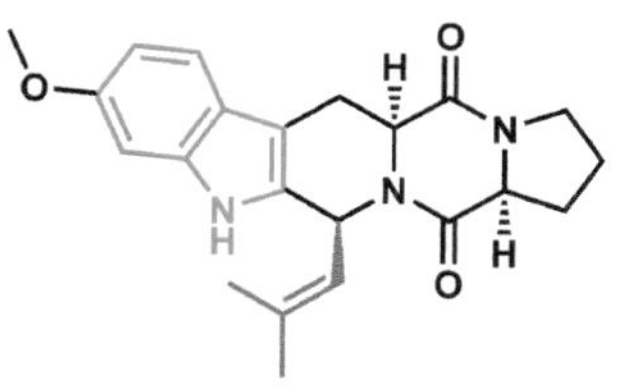

fumitremorgin C

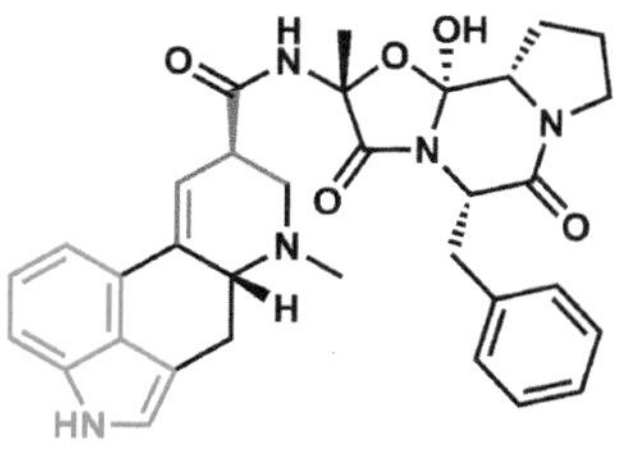

ergotamine

Figure 8.4 Prenylated indole scaffolds, assembled by distinct logic and enzymatic machinery.

are clearly diketopiperazines, echinulin from Trp-Ala, roquefortine from Trp-His, and fumitremorgin and brevianamide from Trp-Pro diketopiperazines (Gu, He *et al.* 2013).

Figure 8.5 shows how those three diketopiperazines (DKPs) are formed from dipeptidyl-*S*-carrier proteins on bimodular NRPS assembly lines. The free amine on residue 1 of the dipeptidyl-*S*-PCP

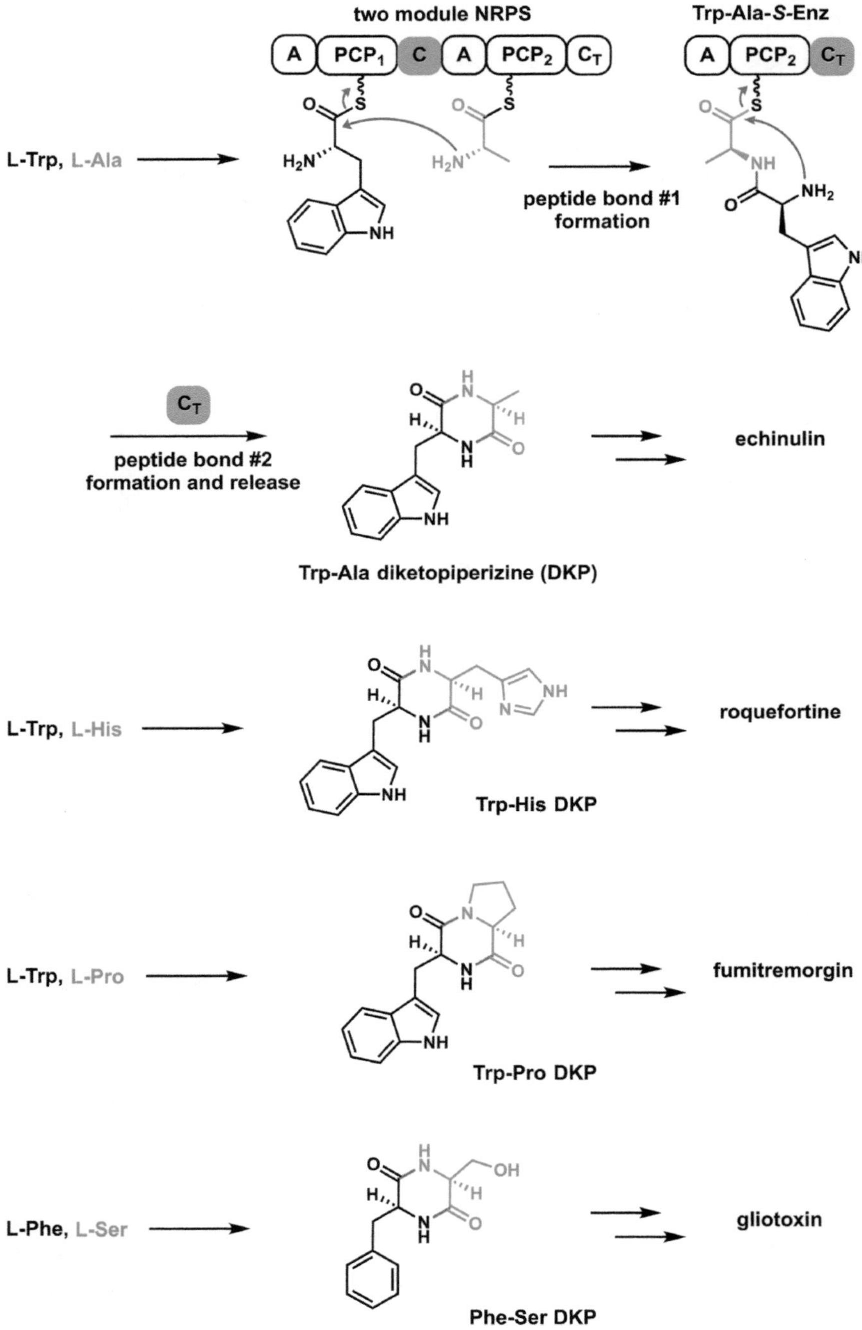

Figure 8.5 Fungal bimodular NRPS assembly lines release diketopiperazines as the free amine of residue 1 attacks the activated thioester carbonyl of residue 2 in intramolecular lactam formation.

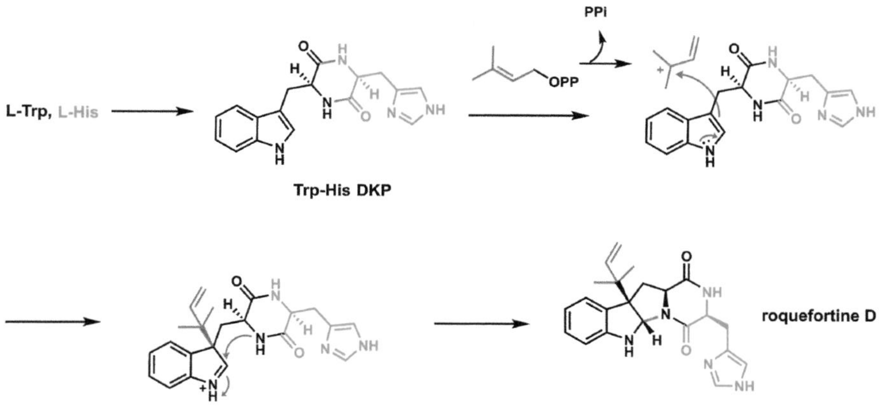

Figure 8.6 Conversion of Trp-His DKP to the fused tetracyclic framework of roquefortine D by C_3 "reverse" prenylation and pyrroloindole ring closure.

attacks the thioester carbonyl intramolecularly to form the six-membered diketopiperazine and release that cyclic product from its pantetheinyl tethering arm. Figure 8.5 also shows the analogous Phe-Ser DKP that is platform to gliotoxin, one of the most notorious fungal peptidyl alkaloid toxins (Balibar and Walsh 2006). Figure 8.6 explicitly shows how the Trp-His DKP can yield roquefortine D in the presence of just one additional enzyme, a C_3-selective indole prenyltransferase.

Vignette 8.1 Roquefortine.

The intersection of indole alkaloid and prenylation biosynthetic pathways occurs frequently in fungal secondary metabolism. Roquefortine C, named for its presence in Roquefort cheese, is the result of one such convergence (Figure 8.V1). The biosynthetic pathway is quite short. First, a two module nonribosomal peptide synthetase couples tryptophan and histidine, and internal capture of the His thioester carbonyl by the amino group of Trp releases the cyclic diketopiperazine. A second enzyme is a prenyltransferase selective for adding the allyl cation derived from Δ2-IPP to C_3 of the indole ring of the Trp-His-DKP. Capture of the resultant indole iminium adduct by the neighboring NH of the DKP creates the 6-5-5-6 tetracyclic framework of roquefortine (Figure 8.6). A third enzyme is presumed to introduce the exocyclic olefin and create the indicated isomer of the dehydro-His side chain.

Roquefortine is generated by *Penicillium roquefortii* growing on the nascent cheese, which is made from the milk of the Lacaune sheep in a

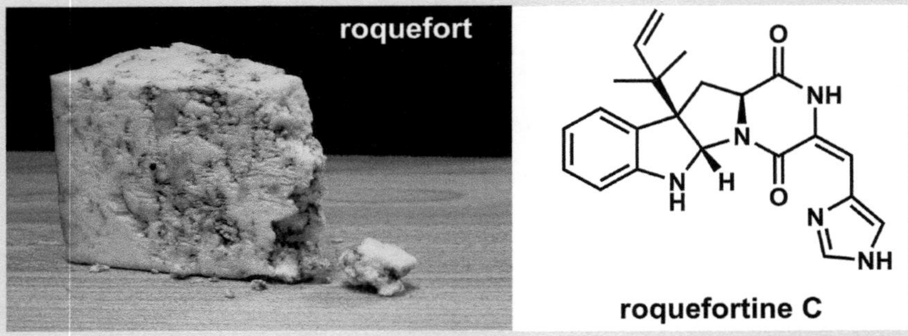

Figure 8.V1 Roquefort and roquefortine alkaloids.

specific area of the Auverne region of France. Roquefortine is typically present at levels of 0.05–1.4 mg kg^{-1} in Roquefort (sometimes termed the "king of cheeses"), Stilton, and Gogonozola forms of blue cheeses. At high concentrations roquefortine is actually a neurotoxin but the endemic levels in the cheeses have been ruled safe for human consumption. One of the main flavoring agents in Roquefort is actually the natural C4 acid butyric acid.

8.4 Seven Nucleophilic Sites on the Indole Ring: A Cornucopia of Possibilities

Indole can serve as a functional carbanion equivalent at all non-bridgehead carbons as well as the indole nitrogen (Walsh 2014, Tanner 2015). These are two-electron reaction manifolds but in a subsequent section on lyngbyatoxin biogenesis we will note one-electron manifolds as well. (We note that all the P450 mediated indole hydroxylations are most probably one-electron reaction manifolds.) Figure 8.7 shows resonance contributors that emphasize that the carbanion electron density can be placed at C_2, C_3, C_4, C_5, C_6, or C_7 by using the lone pair on nitrogen, although outside of enzyme active sites C_3 is chemically observed to be the most reactive as nucleophile.

The almost irresistible, carbocation-generating substrate partners in cell metabolism are the Δ^2-prenyl diphosphate donors. Fungi have exploited this reactivity potential by evolving a suite of prenyltransferases that can couple the prenyl cations with each of the seven reactive sites on indole rings, either in free tryptophan or more often on Trp-X-diketopiperazines, as noted in the examples above (Steffan, Grundmann *et al.* 2009). Figure 8.8 reminds us that the

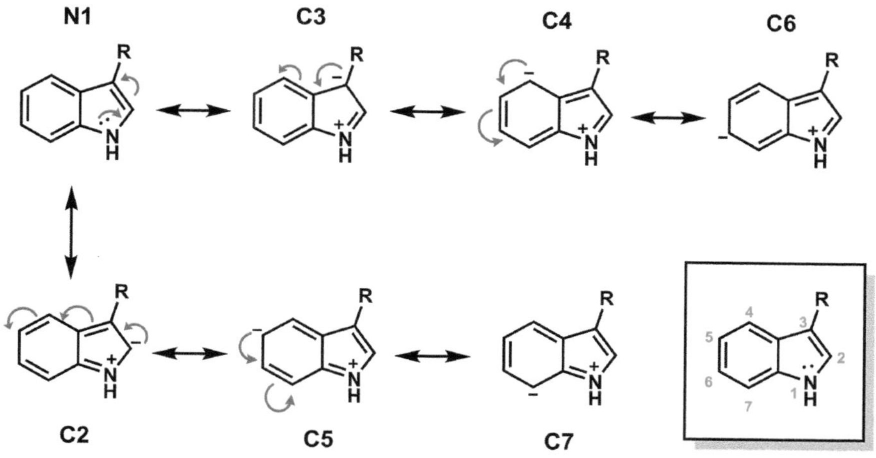

Figure 8.7 The indole ring as a source of multiple carbon nucleophile equivalents. The indole nitrogen also undergoes enzymatic prenylation.

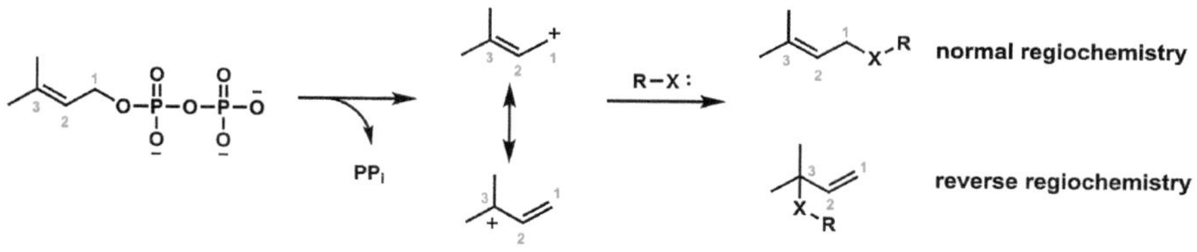

Figure 8.8 The Δ^2-prenyl-PP substrates can undergo early dissociation of the C_1–OPP bond to generate the allyl cations that are capturable by cosubstrate nucleophiles either at C_1 ("normal" regiochemistry of alkylation) or C_3 ("reverse" regiochemistry).

delocalized allyl cations can be captured either at C_1 ("normal" prenylation regiochemistry) or at C_3 (often termed "reverse" prenylation regiochemistry in the primary literature).

We have noted specific enzymatic prenylations at C_2 of indole for brevianamide F (=Trp-Pro DKP; Figure 8.5) and at C_3 of indole in roquefortine D formation (Figure 8.6). Figure 8.9 gives examples of specific prenylations at C_4, C_5, C_6, and C_7 and the product metabolites, chanoclavine, flustramine D, 6-dimethylallylindole-3-carbaldehyde, and mellamide, respectively (Walsh 2014). These nascent prenylated molecules can be converted to more complex architectures by further processing in their respective pathways.

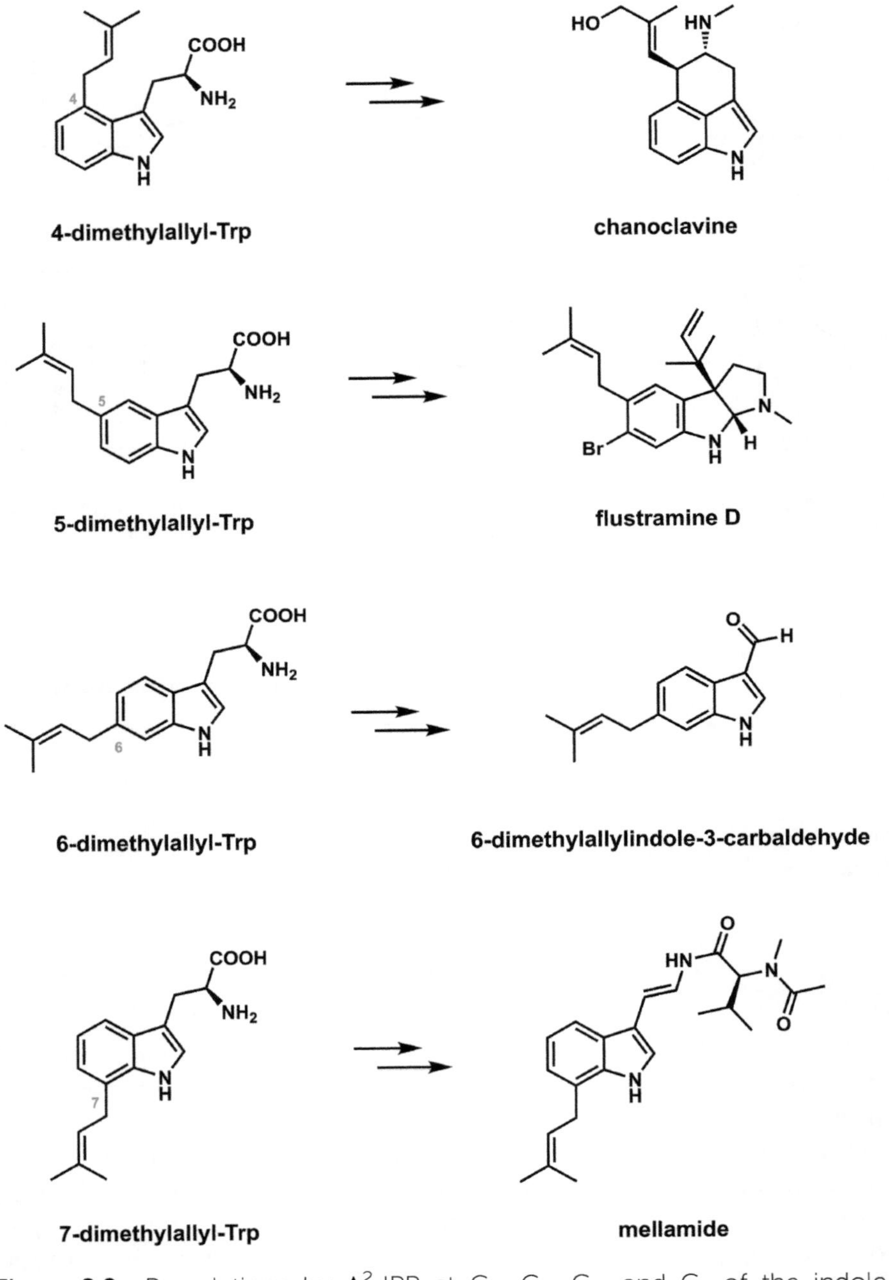

Figure 8.9 Prenylations by Δ^2-IPP at C_4, C_5, C_6, and C_7 of the indole nucleus in specific secondary metabolite pathways.

From time to time the question of the mechanism of enzymatic indole prenylations has been examined (Walsh 2014, Tanner 2015). Two alternatives are the direct prenylation route (as suggested in Figure 8.6) *vs.* indirect routes that feature rearrangements of initial adducts. Figure 8.10 shows three such possible indirect routes from an initial C_3 adduct. Part A raises the possibility that the observed 4-prenyl-Trp-Xaa-diketopyrazines could arise from prior reverse prenylation at C_3 (as in roquefortine) and then a Cope rearrangement to give normal prenylation regiochemistry at C_4. Analogously, a

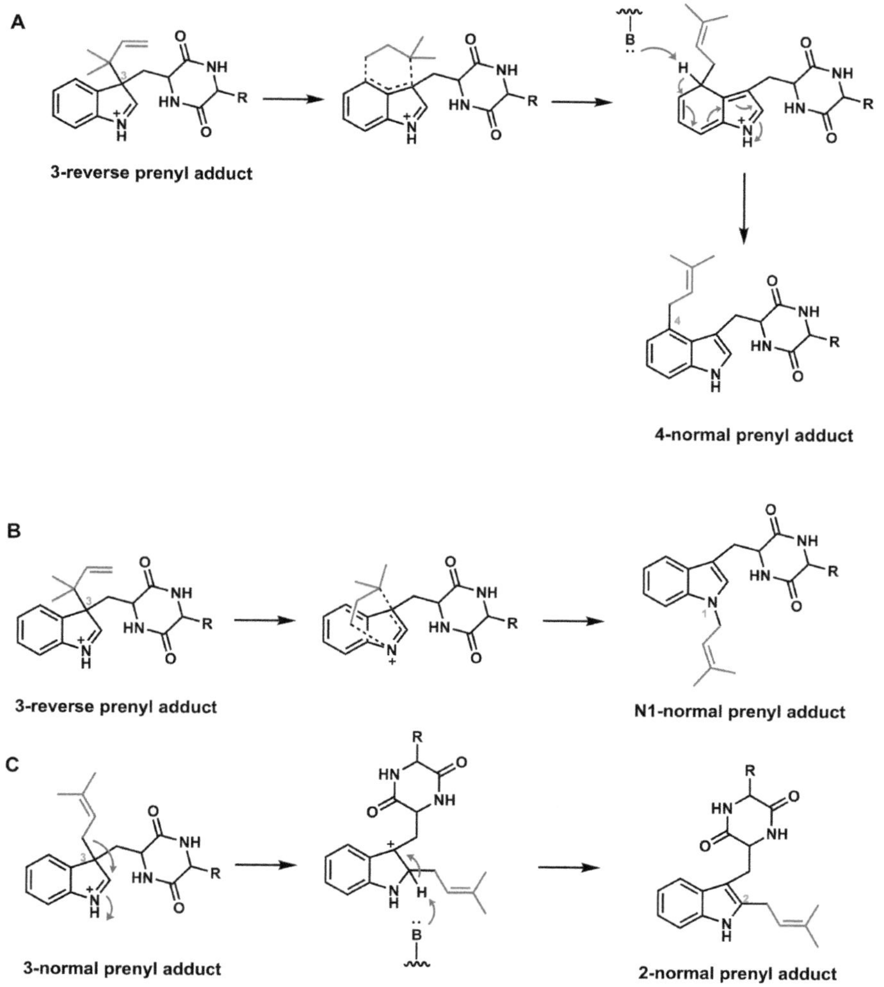

Figure 8.10 Possible indirect routes to regiospecifically prenylated Trp-Xaa-diketopiperazines.

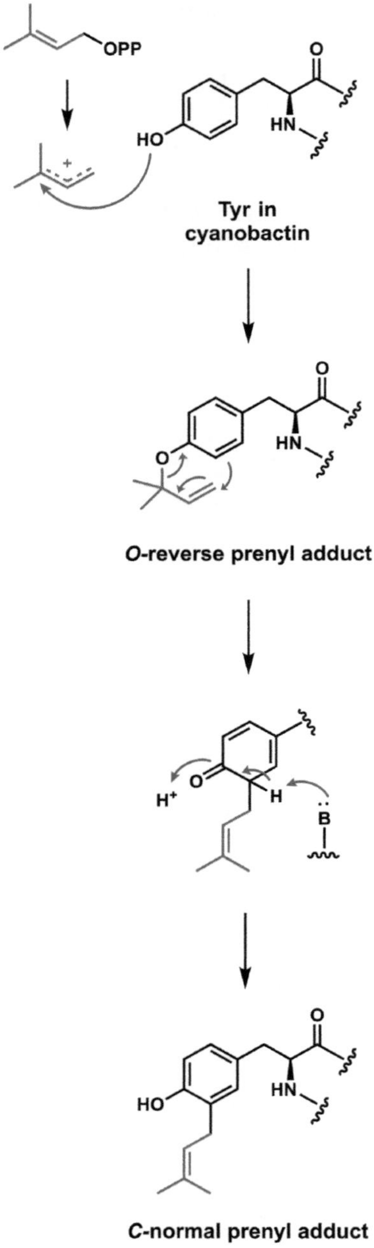

Figure 8.11 The C-prenyl-tyrosine residue in cyanobactins arises *via* rearrangement of an initial *O*-prenyl-tyrosine product.

transannular rearrangement/migration of the reverse prenyl group from C_3 to N_1 could occur (part B), and even migration of the same initial C_3 reverse prenyl adduct to C_2 (brevianamide) (part C). There is no conclusive evidence to argue for indirect transfer as a general mechanism but in the cyanobacterial natural product cyanobactins an initial *O*- "reverse" prenylation of a tyrosine residue is observed to rearrange to the adjacent C_3 C-prenylated tyrosine derivative (Figure 8.11) (Donia and Schmidt 2011, Martins and Vasconcelos 2015).

8.5 Fungal Generation of Tryptophan Derived Alkaloids from DKP

8.5.1 Tryptophan to Fumitremorgin C to Fumitremorgin B to Verruculogen to Fumitremorgin A

A more complex morphing of the Trp-Pro DKP platform occurs in the fungal peptidyl alkaloids, in the tremorgenic metabolites from *Aspergillus fumigatus* molds (Li 2011, Tsunematsu, Ishikawa *et al.* 2013, Walsh 2015). Fumitremorgin A is one of the fumitremorgin pathway end products. On its way from Trp-Pro DKP, also known as brevianamide F, it has undergone two hydroxylations, an endoperoxide insertion (this peroxide is reminiscent of artemisinin), and three prenylations, at least one of which looks not to be straightforward. As schematized in Figure 8.12 brevianamide F undergoes a first prenylation by FtmPT1 to yield the 2-prenylated metabolite known as tryprostatin B. (We noted above how prenylation at C_2 of indole could be direct or indirect *via* rearrangement for a C_3 reverse adduct.) Tryprostatin B is processed on to 12,13-dihydroxyfumitremorgin C. Although two iron-based oxygenases are clearly required for the C_{12} and C_{13} hydroxylations, it also appears that there is a requirement for a third one that could be functioning cryptically as an oxygenase. The dihydro fumitremorgin C has a new N–C bond between the DKP amide nitrogen and C_2 of the prenyl group originally installed on the indole ring (Kato, Suzuki *et al.* 2009).

It is not obvious how a two-electron reaction manifold generates an electrophilic center at that prenyl group (green) C_2 position. However, one electron pathways *via* oxygenase action could perform C_2–H abstraction as $H^\bullet$ and yield a delocalized allyl radical intermediate. Recombination with one electron from the DKP amide nitrogen before oxygen rebound (Figure 8.13) would build the N–C bond, leaving

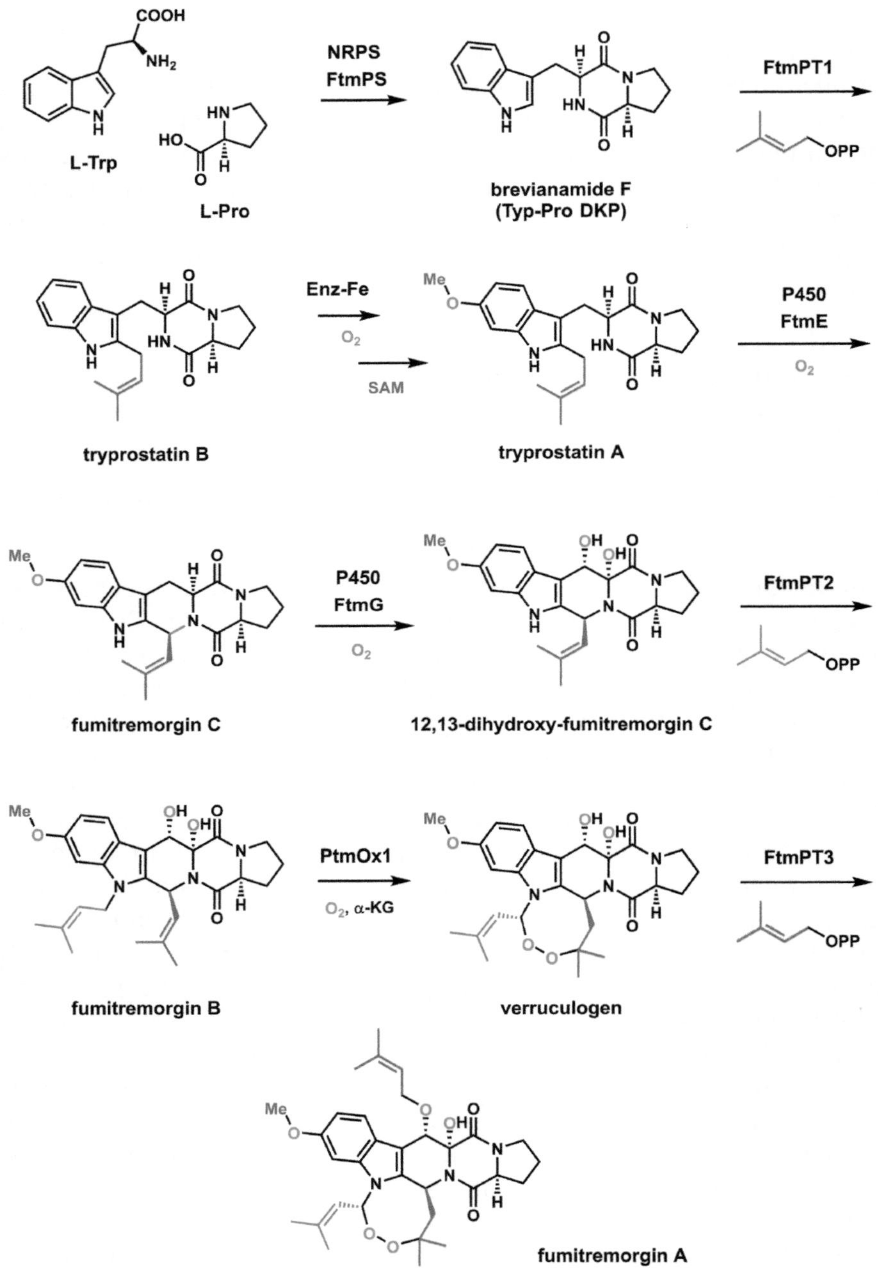

Figure 8.12 Progression to complex scaffolds: a two module DKP-forming NRPS, prenyltransferases, and oxygenases in the verruculogen pathway.

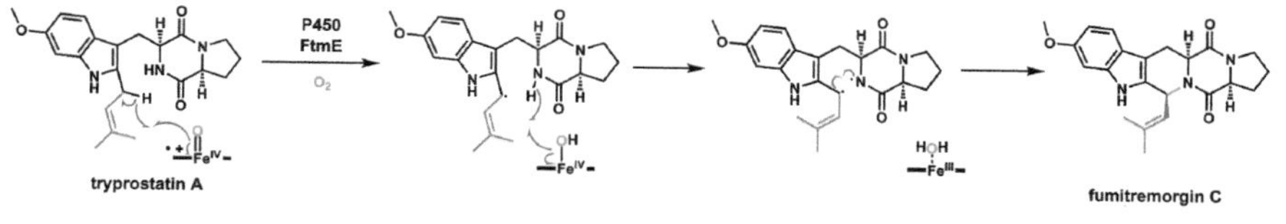

Figure 8.13 Tryprostatin A to fumitremorgin C: cytochrome P450-mediated radical closure onto C_2 of a prenyl substituent.

an unpaired electron on nitrogen which could be back transferred to an Fe(III)-OH equivalent, to complete formation of the dihydro fumitremorgin C pentacyclic scaffold. This would be a one-electron variant of the two-electron manifolds for DKP amide nitrogen noted above for roquefortine D.

The second prenyltransferase then prenylates N_1 of the indole with normal regiochemistry (pink in Figure 8.12) to produce fumitremorgin B. This is substrate for a remarkable endoperoxide-forming reaction sequence that leads to verruculogen (Kato, Suzuki *et al.* 2011). The endoperoxide has been formally inserted between C_2 of the just introduced prenyl group and what had been C_3 of the original prenyl moiety. The crystal structure of the responsible enzyme FtmOX1, which is a nonheme iron α-KG-dependent oxygenase, was recently solved and this led to a mechanistic proposal as shown in Figure 8.14. One important finding is that a tyrosine residue (Y224) was confirmed to participate in the radical-initiated catalytic cycle during endoperoxide formation (Yan, Song *et al.* 2015).

Also noted in Figure 8.12, verruculogen is one prenylation away from the end product fumitremorgin A (Li, 2011). This utilizes a third prenyltransferase, FtmPT3, and the product is the 12-*O*-prenyl metabolite, distinct from the two prior prenylations that have occurred at C_2 and N_1 of the indole moiety (Li, 2011). While the fumitremorgins have been named on the basis of their pharmacology, as *A. fumigatus* tremor-inducing metabolites in mice (molecular mechanism unclear), their physiologic functions in the producer aspergilli are not identified.

8.5.2 Fumitremorgins to Spirotryprostatin Framework

It is known that the fumitremorgin scaffold is also progenitor to the rearranged spirocyclic framework of the spirotrypostatins (Tsunematsu, Ishikawa *et al.* 2013). Figure 8.15 shows how a desmethoxy version of fumitremorgin C can be processed to the 11,12-dihydroxy intermediate by sequential hydroxylation at C_{13} and

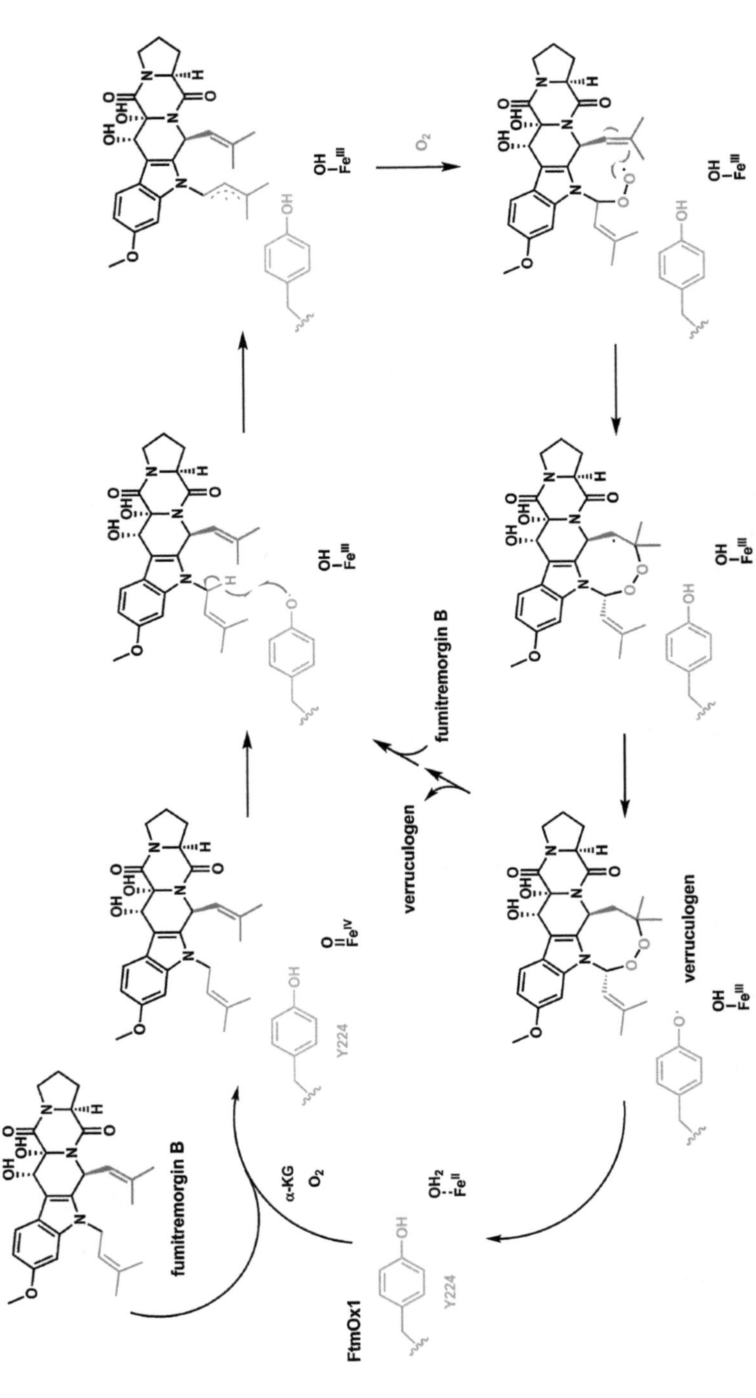

Figure 8.14 Fumitremorgin B to verruculogen: insertion of a peroxide bridge into two pendant prenyl chains.

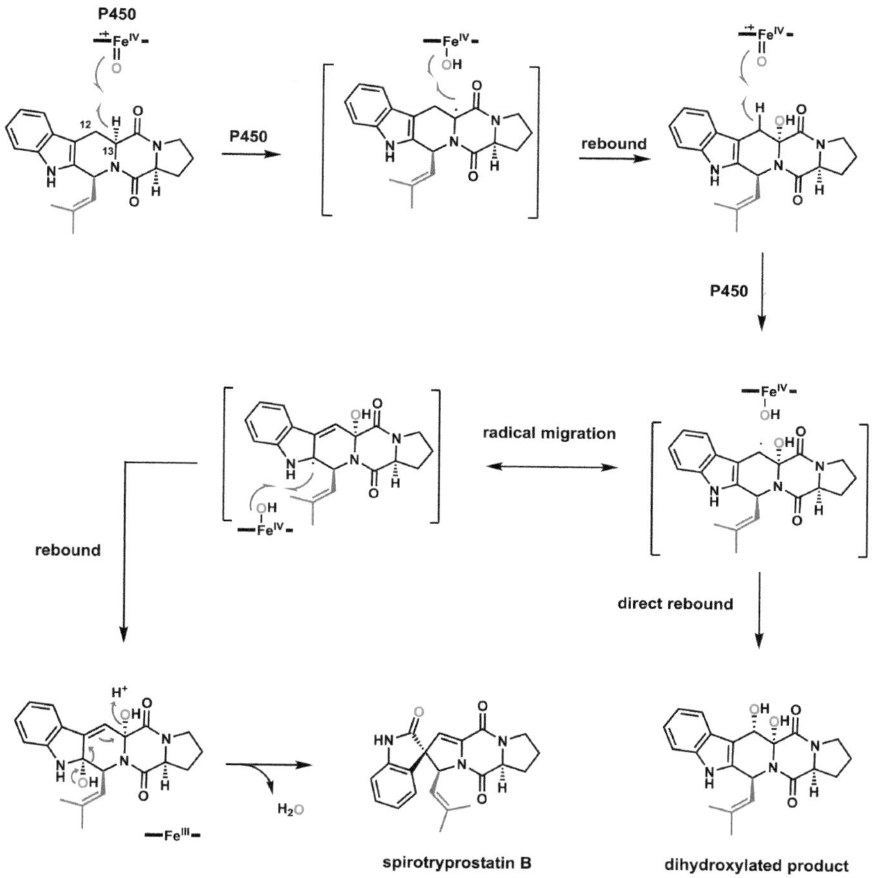

Figure 8.15 Fumitremorgin framework to spirotryprostatin scaffold.

then C_{12}. The radical intermediate featured for the second hydroxylation could eliminate water and rearrange to an alternative, allyl radical at the ring B,C-bridgehead and undergo hydroxylation there. This bridgehead hydroxyl could convert to the ketone with cleavage of one of its carbon substituents, including loss of the C_{13}-OH group in a semi-pinacol type rearrangement. This involves C–C bond migration as shown and would convert the 5-6 central rings into the spiro 5-5 pattern of spirotryprostatin B. Verruculogen and spirotryprostatin B assembly reveal allyl radical intermediates in the indicated P450 mediated rearrangements.

A mechanistic alternative to an allyl radical intermediate would be an actual oxygenation across the 2,3-double bond of the indole moiety, creating a transient epoxide (Figure 8.16) (Walsh, 2015). This

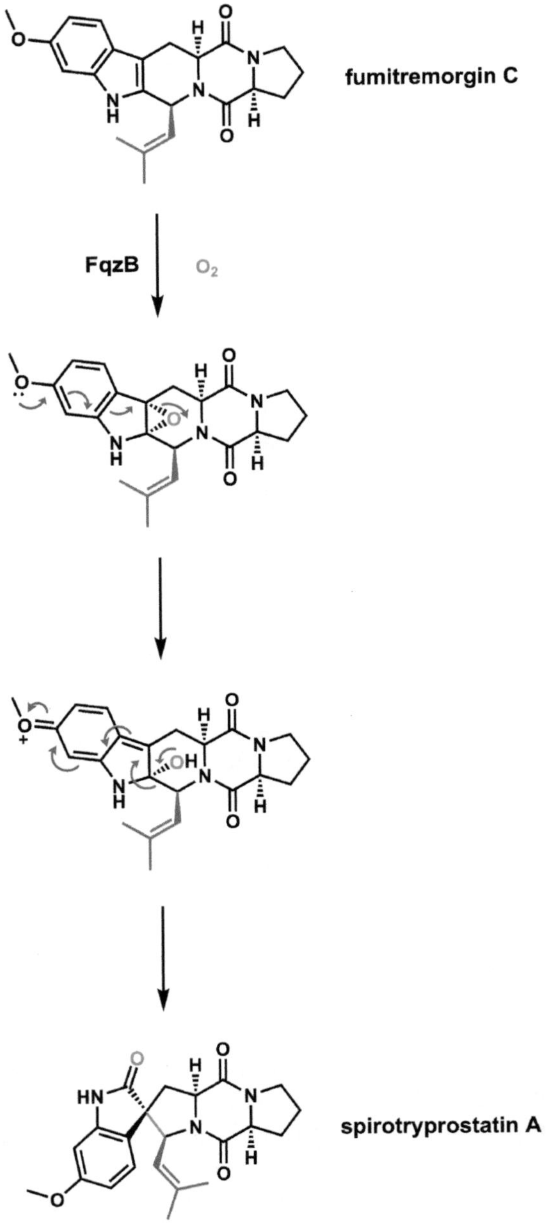

Figure 8.16 Fumitremorgin C to spirotryprostatin A: rearrangement of an indole epoxide intermediate.

appears to be the case in formation of spirotryprostatin A rather than spirotryprostatin B, where an FAD-linked oxygenase carries out the rearrangement of fumitremorgin C. Recall that FAD-dependent oxygenases deliver less activated oxidizing equivalents and an [OH^+] rather than an [$OH^•$] is more likely, arguing for epoxidation. Subsequent neighboring group participation by the 4-oxygen of the *O*-methyl substituent could provide directionality to epoxide opening. The resultant C_2 oxy anion could ketonize and drive the rearrangement of the fused 5-6 ring to the spiro 5-5 product: yet one more example of the power of oxygenases to morph natural product scaffolds.

8.5.3 Brevianamide F to Notoamide D

It is instructive to look at one more set of enzymatic maturations on the brevianamide F (Trp-Pro) DKP scaffold (Li, Anand *et al.* 2012) (Figure 8.17). Two prenyltransferases acting *ad seriatim* act first to carry out a "reverse" prenylation at C_2 and a normal prenylation at C_7, with a C5-hydroxylation step by a hydroxylase in between. The next step is an intramolecular attack of the C_5-OH on C_3 of the adjacent prenyl substituent to yield notoamide E, and this could go *via* one-electron reaction manifolds to form the cyclic ether linkage. Now an FAD-dependent oxygenase, similar to that just featured in spirotryprostatin A formation, would generate the indole 2,3-epoxide, or its hydroxy iminium tautomer, that is captured by the DKP amide NH to produce notoamide D (Li, Finefield *et al.* 2012). This five-enzyme sequence creates the fused hexacyclic platform of notoamide D, building complexity by an economically small set of enzymatic transformations. These are reminiscent of the efficient complexity-generating peptidyl alkaloid scaffolds from anthranilate noted in Chapter 5 (*e.g.* Figures 5.27–5.29).

8.6 Bacterial Generation of Pentacyclic Indolecarbazoles

Endophytic bacteria that inhabit mangrove trees make some novel prenylated indole scaffolds, including xiamycin B and sespenine (Figure 8.18), perhaps as part of the antibiotic reservoir of the mangrove community (Xu, Baunach *et al.* 2012). Whether this represents some horizontal gene transfer of plant biosynthetic genes to the *Streptomyces* spp. is not yet known.

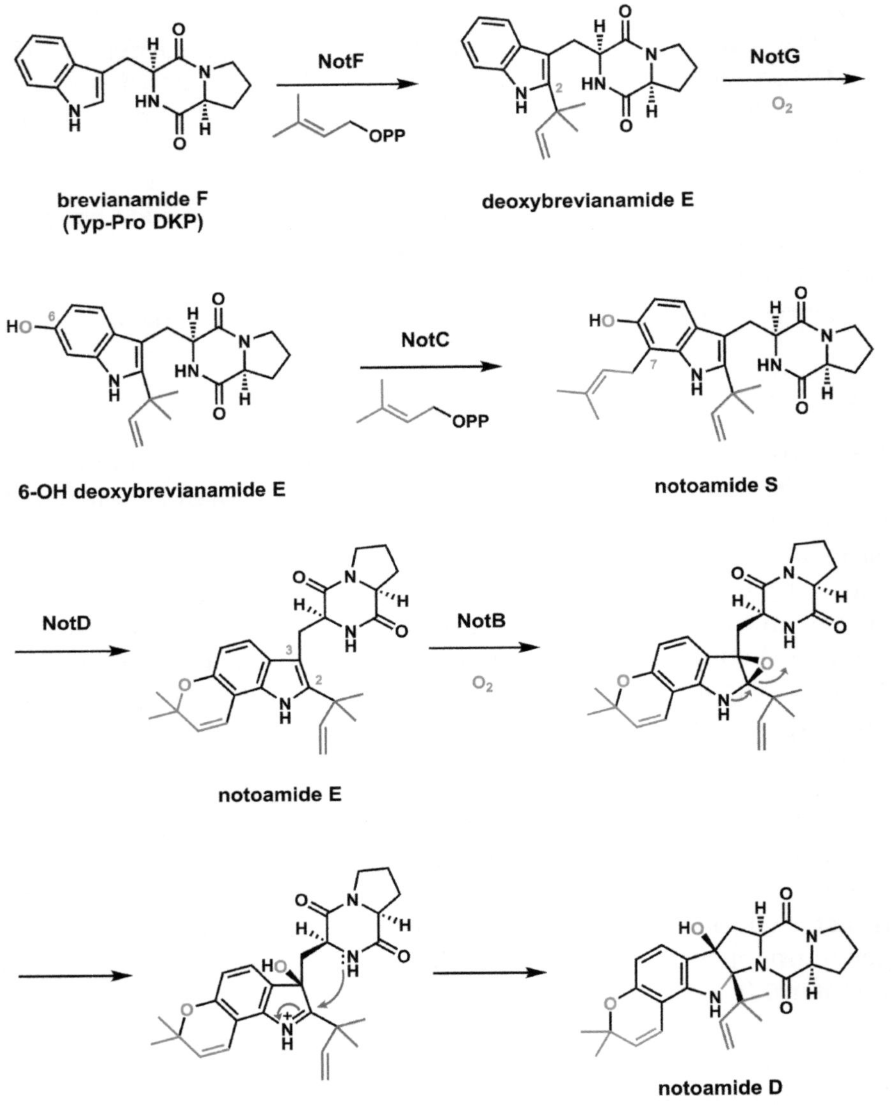

Figure 8.17 Brevianamide F to notoamide D. Two prenylations, indole epoxygenation, and pyrroloindole formation.

The proposed pathway depicted in Figure 8.18 indicates an indole sesquiterpene as an early intermediate, perhaps from attack of free indole on the C_1 allyl cation from farnesyl-PP. Free indole could arise either by way of indole-glycerol-3-P aldolase or the bacterial

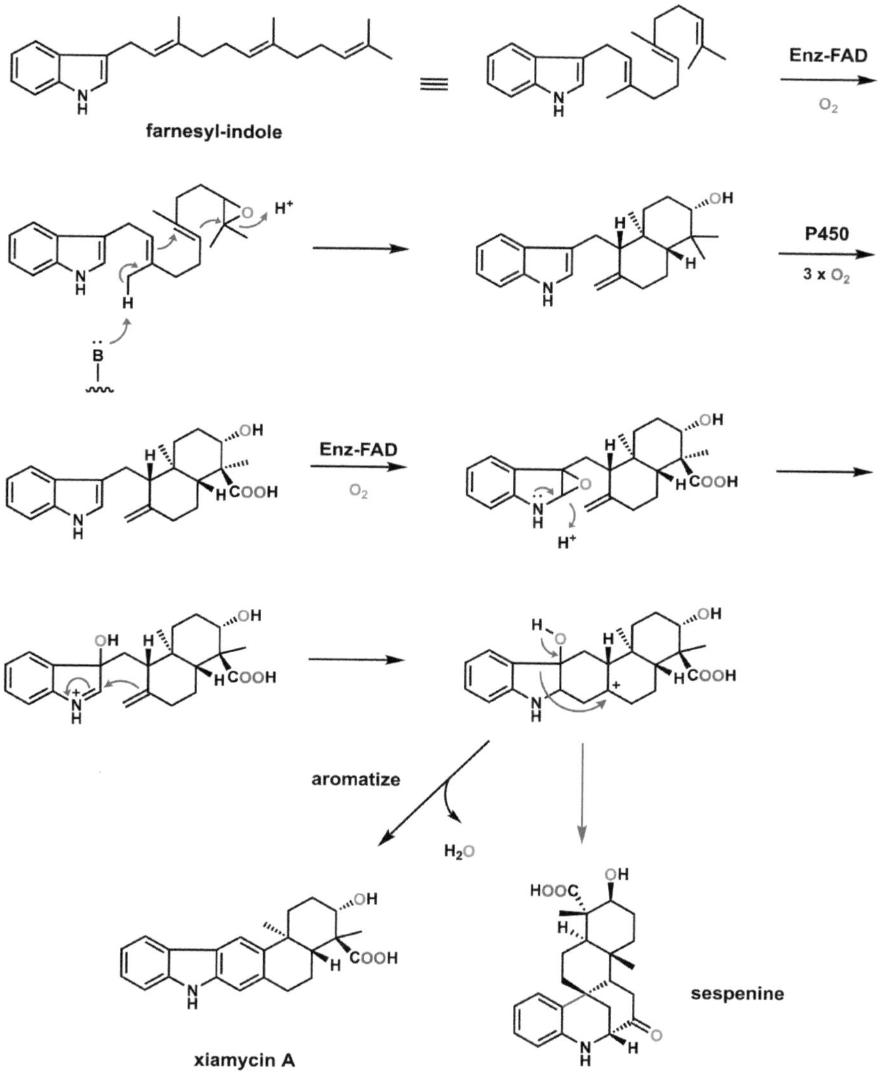

Figure 8.18 Bacterial indole sesquiterpenes: tandem epoxidations on the farnesyl and indole moieties drive rearrangements to xiamycin and sespenine.

tryptophan lyase enzymes. This is analogous to the initial step of indole diterpene biosynthesis as shown for paxilline in Figure 4.37.

Most notably, subsequent processing involves two flavoprotein epoxidases, one for the farnesyl moiety, reminiscent of squalene epoxidase (Chapter 4), and then the second one for the 2,3-double bond of the indole ring, akin to the ones just described in the prior sections.

The attendant cyclizations following those epoxidations yield the unusual fused pentacyclic indole carbazole system that is a dehydrative aromatization away from xiamycin A. A second route forward involves rearrangement as the 3-OH on the indoline ring ketonizes and the indicated single bond migrates. The bicyclic 6-5 indole goes to a 6,6-ring system as the connectivity in the pentacyclic scaffold is altered in sespenine.

8.7 Vinca Alkaloids: Strictosidine to Tabersonine to Vindoline

The most famous of the indole terpene producing plants is probably the Madagascar periwinkle, originally classified as *Vinca rosea* and then reclassified as *Catharanthus rosea*. More than 70 alkaloids have been identified from the plant extracts (van Der Heijden, Jacobs *et al.* 2004). The most attention has been paid to vincristine and the related metabolic precursor vinblastine (see Figure 8.21) (Dewick, 2009). These two vinca secondary metabolic end products are among the most complex alkaloid scaffolds known, and dramatically more complex than the simple alkaloids from ornithine and lysine described in the early parts of Chapter 5. Vincristine is a potent inhibitor of human tubulin and has been approved as an anticancer drug, trade name oncovorin, since 1960. It is a front line component of combination therapy for acute leukemia, and Hodgkin's and non-Hodgkin's lymphomas (Weber 2015).

The pathway to vinblastine and vincristine goes through strictosidine, as depicted in Figure 8.19. It is known that strictosidine undergoes glycosidase action to convert the acetal to hemiacetal and then an open chain aldehyde which can cyclize to the indicated imine. The enzymes in this part of the pathway, from strictosidine to tabersonine, are not yet known and characterized, so the proposed schematic in Figure 8.19 (after Dewick, 2009) represents only one possible set of transformations. The dashed lines indicate they bonds that must be broken as strictosidine proceeds to preakummacin, stemmadenine, and dehydrosecodine. A proposed Diels–Alder [4 + 2] cyclization would get to the tabersonine scaffold.

By contrast, the next six enzymes that convert tabersonine to vindoline (Figure 8.20) are well characterized at this point (Qu, Easson

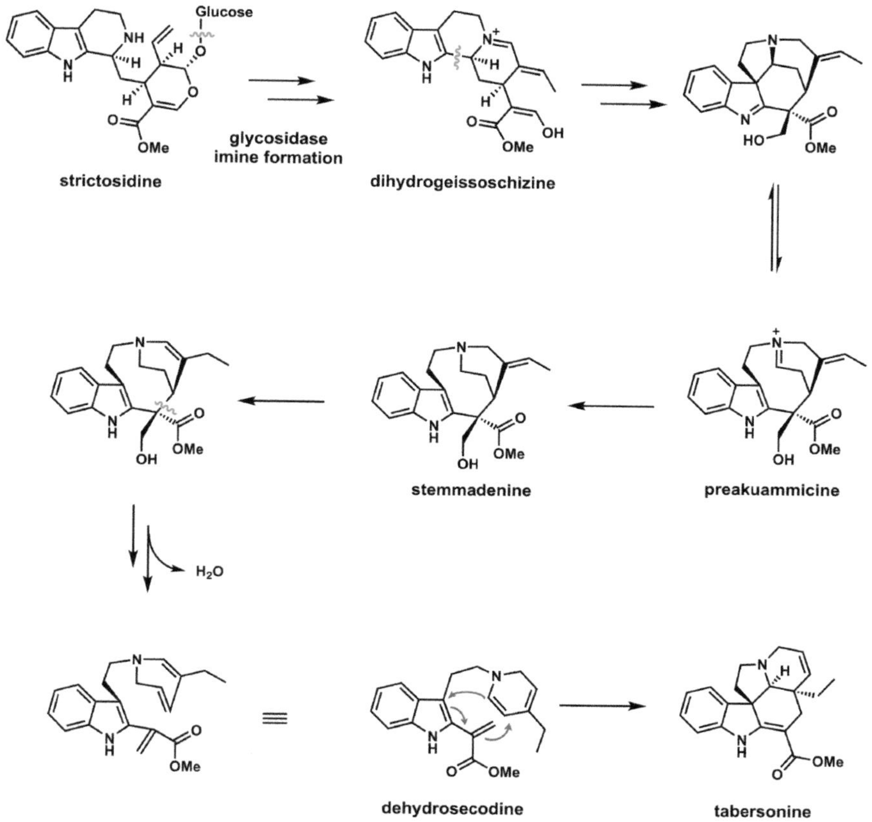

Figure 8.19 Strictosidine to tabersonine (enzymes unknown). Dashed lines represent proposed formal bond disconnections to go through preakummacine, stemmadenine, and dehydrosecodine to the fused pentacyclic ring system of tabersonine. Data from Dewick, P. M. (2009). Front Matter, in *Medicinal Natural Products: A Biosynthetic Approach*. 3rd edn, Chichester, UK, John Wiley & Sons, Ltd.

et al. 2015) (Kries and O'Connor 2016). They reflect by now standard hydroxylations, *O*- and *N*-methylations, and *O*-acetylation, as shown (also see Figure 13.19).

Vindoline is one of the two partners in the asymmetric dimerization reaction that yields the vinblastine scaffold (Figure 8.21). The other is catharanthine, also derived from strictosidine (pathway not shown) (Zhu, Wang *et al.* 2015). The coupling enzyme is not yet identified but catharanthine must undergo a transformation to become electrophilic at the bridgehead carbon attached to C_2 of the indole. One

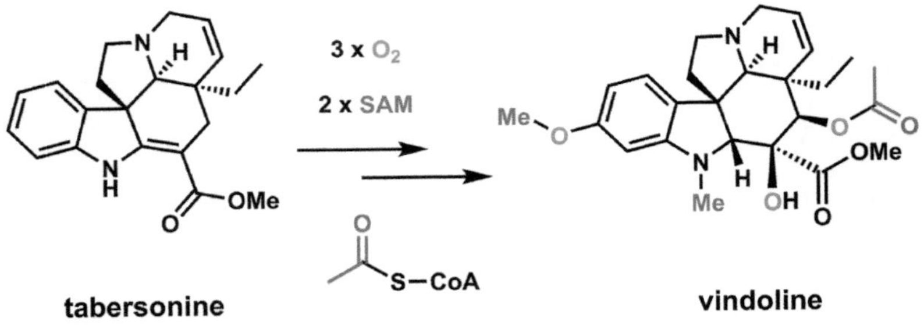

Figure 8.20 Tabersonine to vindoline involves familiar iron-based enzymatic hydroxylations, *O*- and *N*-methylations, and *O*-acetylation.

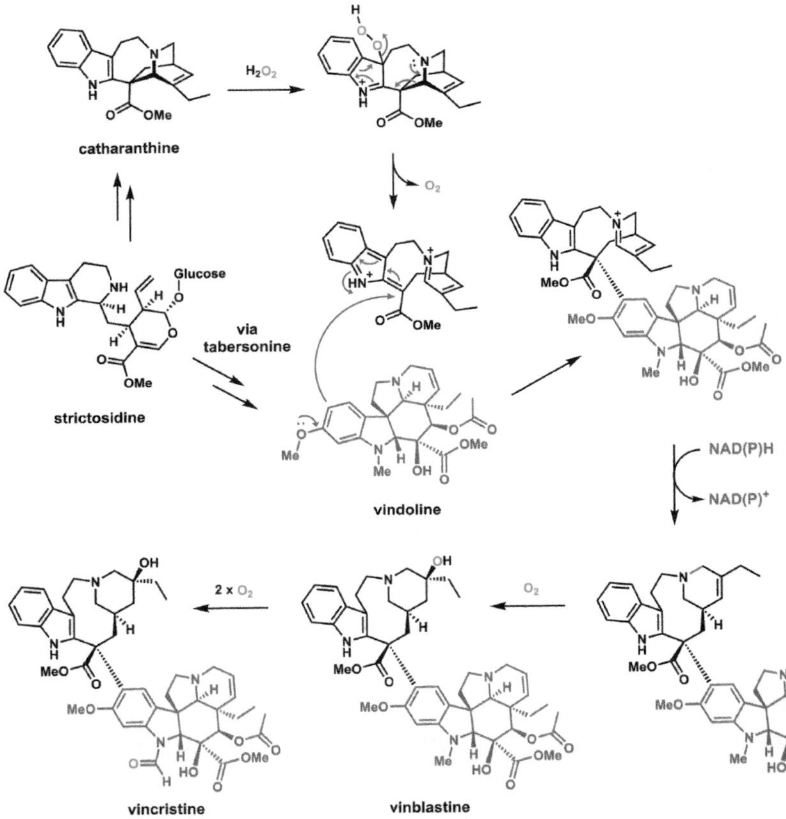

Figure 8.21 Vinblastine and vincristine arise from coupling of vindoline and an oxidized intermediate from catharanthine.
Data from Dewick, P. M. (2009). Front Matter, in *Medicinal Natural Products: A Biosynthetic Approach*. 3rd edn, Chichester, UK, John Wiley & Sons, Ltd.

possible route (Dewick, 2009) would be a hydroperoxylation of the indole to generate a 3-hydroperoxyindoline (or equivalent) that could fragment as indicated to the iminium species. This could serve as an electrophilic partner as the methoxy group on the benzene ring of the vindoline activates the *ortho* carbon as nucleophile. Minor adjustment of the redox state of the dimeric product gives vinblastine. Vincristine differs only in the oxidation state of the one-carbon unit attached to the bottom indole nitrogen. It is *N*-methyl in vinblastine but has been oxidized by four electrons to the aldehyde in vincristine, likely *via* oxygenase action as depicted.

8.8 Lyngbyatoxin: One- and Two-electron Reaction Manifolds in Indole in a Single Biosynthetic Pathway

We have spent much of this chapter emphasizing the ability of the indole ring to serve as a carbon nucleophile in two-electron pathways. It can also function by one-electron redox routes as well. The two examples of this subsection illustrate that biological utility.

Lyngbyatoxin is a cyanobacterial metabolite from *Lyngbya majuscula* that appears to serve a defense function against predators. It is also the active agent in skin irritation and the condition known as "swimmer's itch" (Cardellina, Marner *et al.* 1979). The biosynthetic pathway is short, consisting of only three enzymes (Figure 8.22) (Edwards and Gerwick 2004, Ongley, Bian *et al.* 2013). The first is a bimodular NRPS activating Trp and Val with an unusual C-terminal domain in the assembly line. It is an NAD-dependent reductase and carries out a two step four-electron reductive release (Read and Walsh 2007). First the *N*-methyl-Val-Trp-S-PCP_2 thioester is reduced to the thiohemiacetal, which unravels to the aldehyde. Before release this is reduced by a second equivalent of NADPH to the *N*-methyl-valyltryptaphanol. This dipeptide alcohol is the released product and becomes the substrate for the next enzyme.

The second enzyme is a cytochrome P450 oxygenase. Like many others noted in these pages, it carries out the first half reaction but not the second (no net delivery of an $OH^\bullet$ equivalent to the substrate carbon centered radical). Hydrogen atom removal from C_4 of the indole ring and also from the valyl amide nitrogen would allow a properly folded diradical to couple and form the new N–C single bond in the observed product indolactam-V. This would constitute a

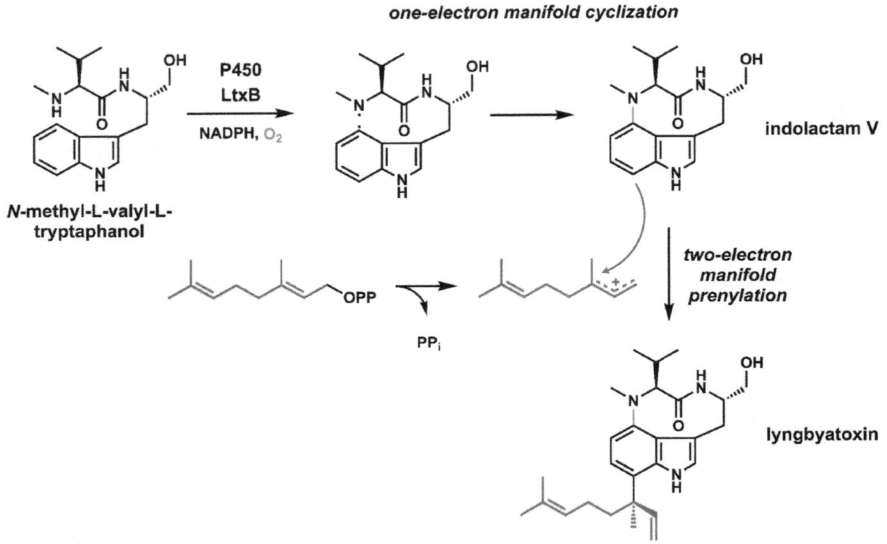

Figure 8.22 Lyngbyatoxin: one-electron and two-electron reaction manifolds during a short biosynthetic pathway from Val-Trp.

one-electron reaction manifold. It is also possible that hydroxylation occurs transiently and then the –OH is displaced by the indole ring acting as a C_4 carbanion equivalent (Tang, Zou *et al.* 2017). The third enzyme carries out a two-electron reaction manifold, acting as a regioselective prenyltransferase, coupling C_7 of the indole ring in indolactam V to the allyl cation from the C_{10} isoprenoid geranyl-PP to give the "reverse" prenylated lyngbyatoxin.

Other indications of facile one-electron enzymatic chemistry of indoles are provided by the dimeric dibrevianamide F and di-tryptophenaline molecules, most readily ascribed to regioselective dimerization of indolyl radical species (Figure 8.23) (Saruwatari, Yagishita *et al.* 2014) (Walsh, 2014).

Vignette 8.2 Communesins: Short Pathways to Scaffold Complexity.

The communesin class of 11 dimeric indole alkaloids exemplify complex scaffolding effected in remarkably efficient biosynthetic pathways. The communesins contain four nitrogens, seven rings, two aminals, four contiguous stereocenters, and a pair of vicinal quaternary carbons (Siengalewicz, Gaich *et al.* 2008, Lin, Chiou *et al.* 2015, Lin, McMahon *et al.* 2016).

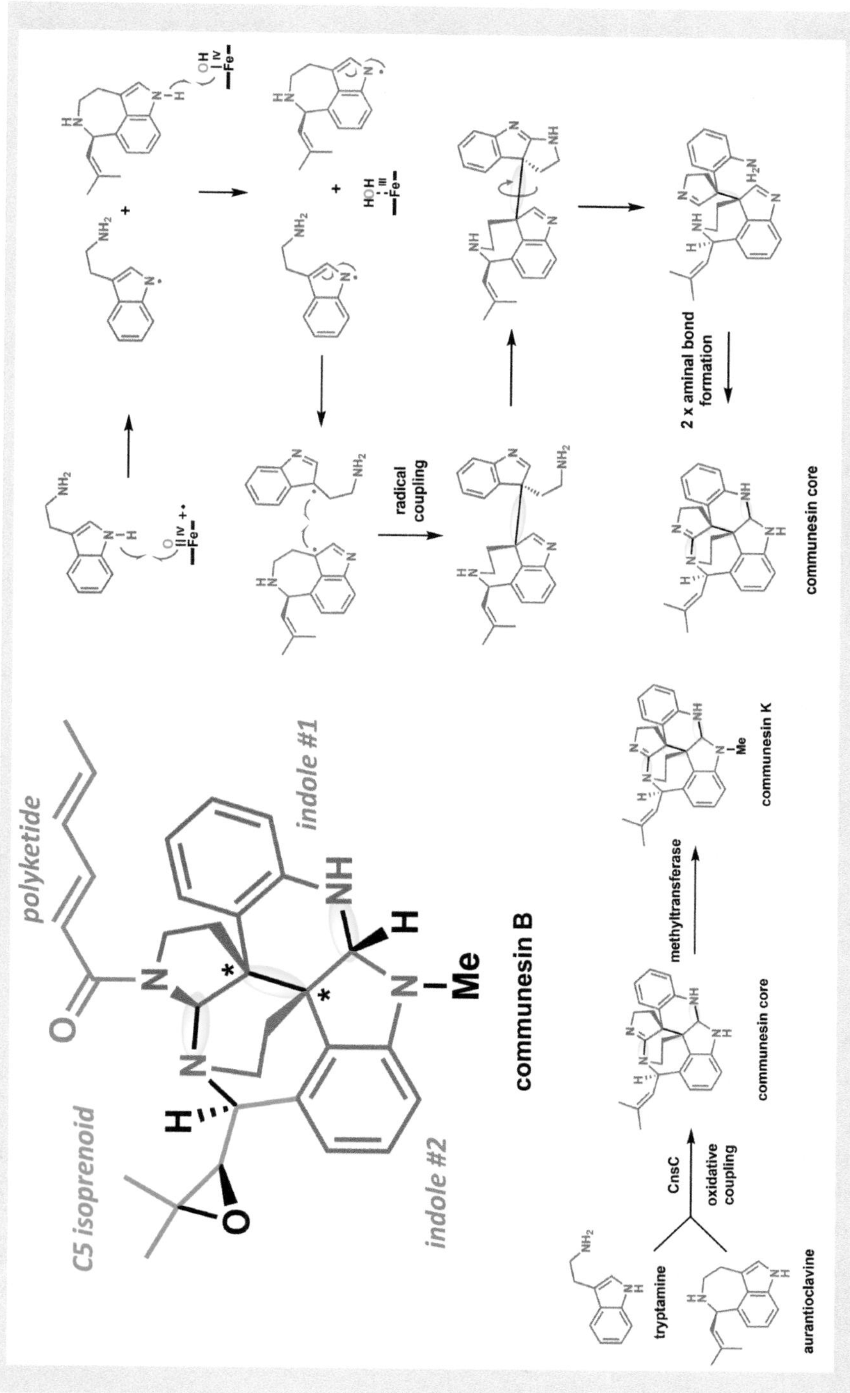

Figure 8.V2 Biosynthesis of the heterodimeric indole alkaloid communesin involves a P450-catalyzed coupling step.

These frameworks have been of considerable interest to the synthetic community, especially as the total synthesis led to revision of an initially misassigned structure of nomofungin. The Stoltz group total synthesis (May, Zeidan *et al.* 2003) of tryptamine and aurantioclavine fragments *via* a 3-3′ coupling turns out to prefigure the biosynthetic route.

The key coupling enzyme in the *Penicillium expansum* mold producer is a cytochrome P450 CnsC (Lin, McMahon *et al.* 2016). Coexpression of CnsC and its NADPH dependent reductase to provide electrons, along with the SAM-dependent methyltransferase to create the *N*-methylindole moiety known to stabilize the communesin scaffold, yields communesin K (Figure 8.V2), with three new bonds and control of the four stereogenic centers.

P450 enzymes use O_2 to generate high valent oxo-iron species that generate carbon radicals in bound substrates. We will note in Chapter 9 many examples where the carbon radicals escape subsequent hydroxylation and instead react with other carbon centers (Lin, McMahon *et al.* 2016). That is the communesin mechanistic proposal for this C3–C3′ coupling on the C3 aza-allyl radicals derived from one-electron oxidation of each of the two substrates in the enzyme's active site, as shown in Figure 8.V2. The formation of the two aminals, seven-membered and six-membered respectively, are then proposed to be directed regiospecifically in the active site microenvironment and conditioned by the substituent patterns on the initial 3-3′ adduct. The product that emerges from the enzyme is the heptacyclic dimeric mature communesin framework.

The ability of one enzyme catalyst, CnsC, to generate two vicinal quaternary centers and then set a microenvironment for regio- and stereospecific maturation to the heptacyclic scaffold of communesin K on the way to additional family members is a remarkably efficient route to complexity generation. These findings augur for deeper examination of this fungal catalyst class for complex synthetic transformations made easy.

8.9 Tryptophan to Cyclopiazonic Acid

One final example of a prenylated tryptophan metabolite with a biosynthetic twist is the fungal natural product cyclopiazonic acid (Holzapfel 1968) from *Penicillium cyclopium* and *Aspergillus* spp. It is a potent neuromuscular toxin due to nanomolar potency as an inhibitory ligand for the sarcoplasmic reticulum ATPase in muscle cells (Figure 8.24) (Moncoq, Trieber *et al.* 2007). The biosynthetic pathway to this complex fused pentacyclic toxin is again short and efficient, with only three enzymes required.

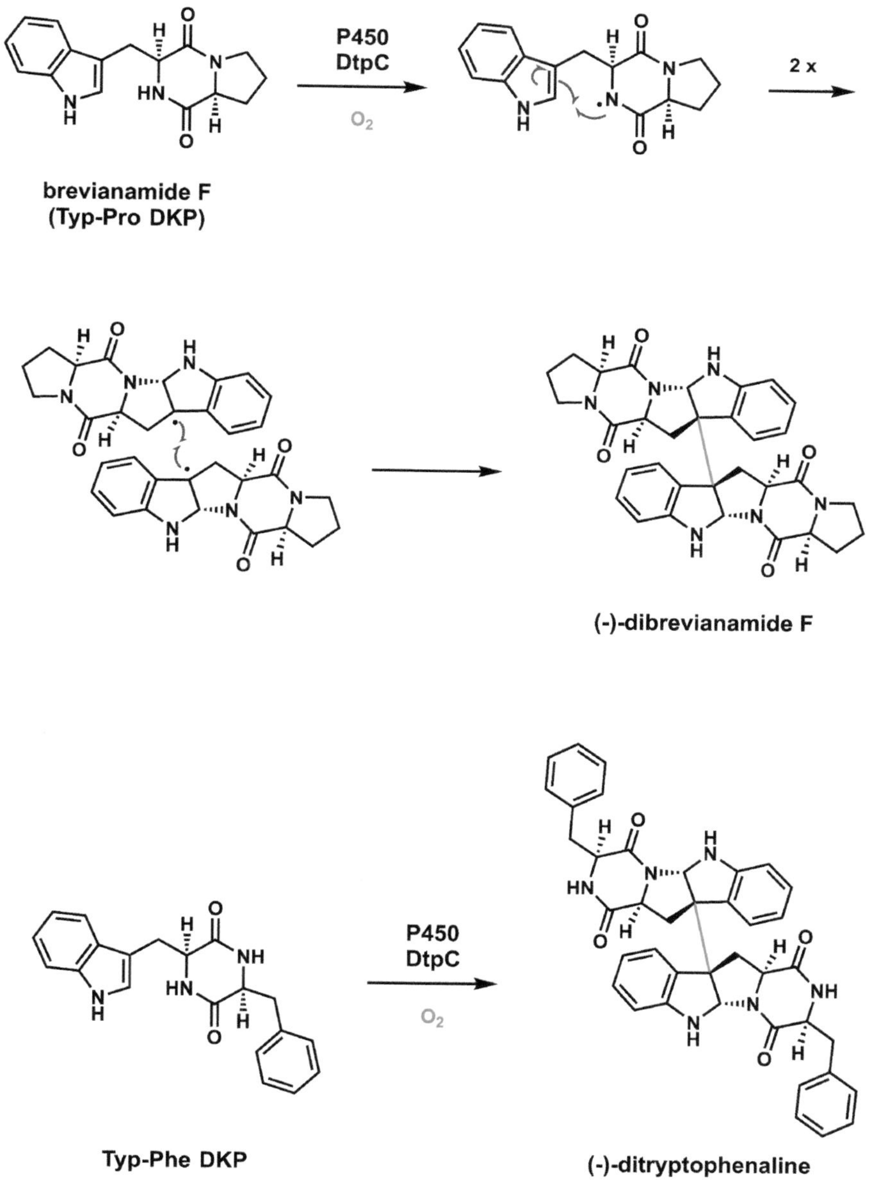

Figure 8.23 Dibrevianamide F and ditryptophenaline are proposed to undergo regioselective dimerization *via* long-lived delocalized radical species.

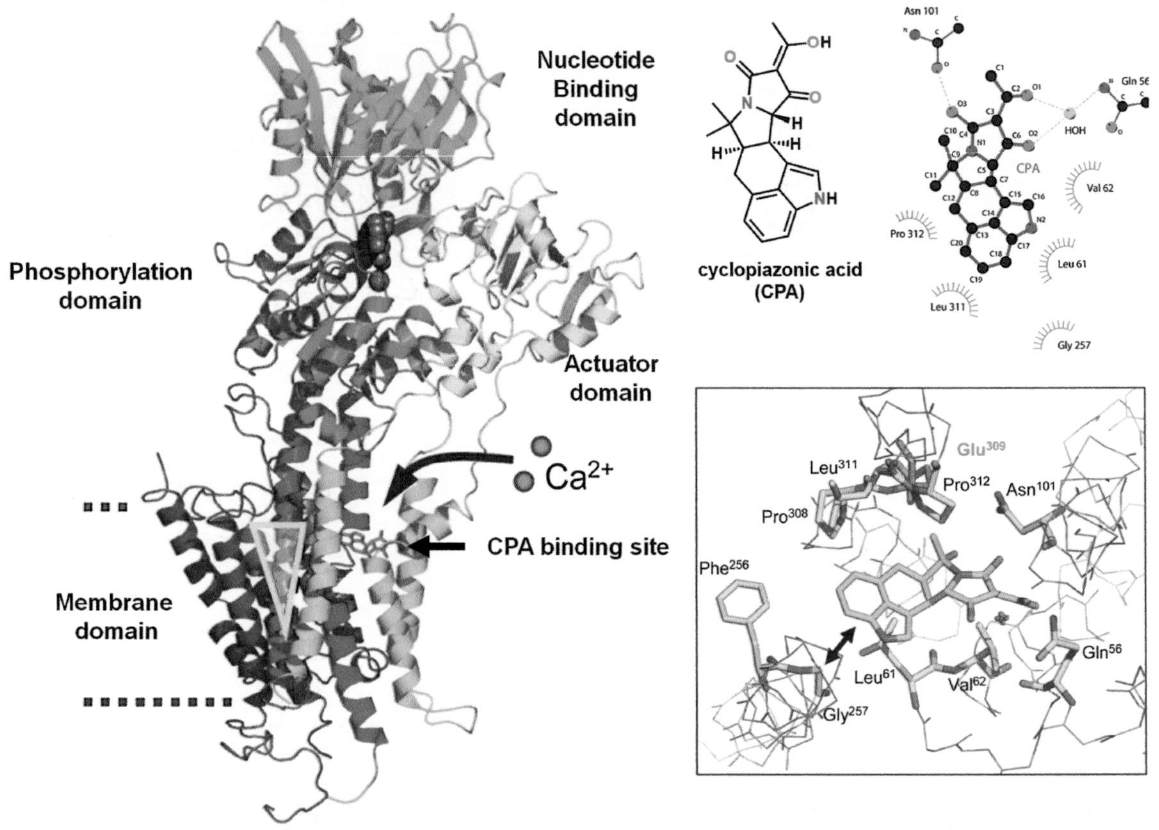

Figure 8.24 The fungal mycotoxin cyclopiazonic acid (CPA) is a nanomolar inhibitor of muscle cell sarcoplasmic reticulum Ca^{++}-ATPase.
Reprinted from K. Moncoq, C. A. Trieber and H. S. Young (2007). The molecular basis for cyclopiazonic acid inhibition of the sarcoplasmic reticulum calcium pump. *J. Biol. Chem.* 282: Copyright (2007) The American Society for Biochemistry and Molecular Biology.

The first is a hybrid PKS–NRPS bimodular assembly line enzyme. An acetoacetyl-*S*-acyl carrier protein becomes tethered on module 1 while Trp is tethered as thioester to PCP_2 (Figure 8.25) (Liu and Walsh 2009a). Attack of the free NH_2 of Trp constitutes the chain elongation

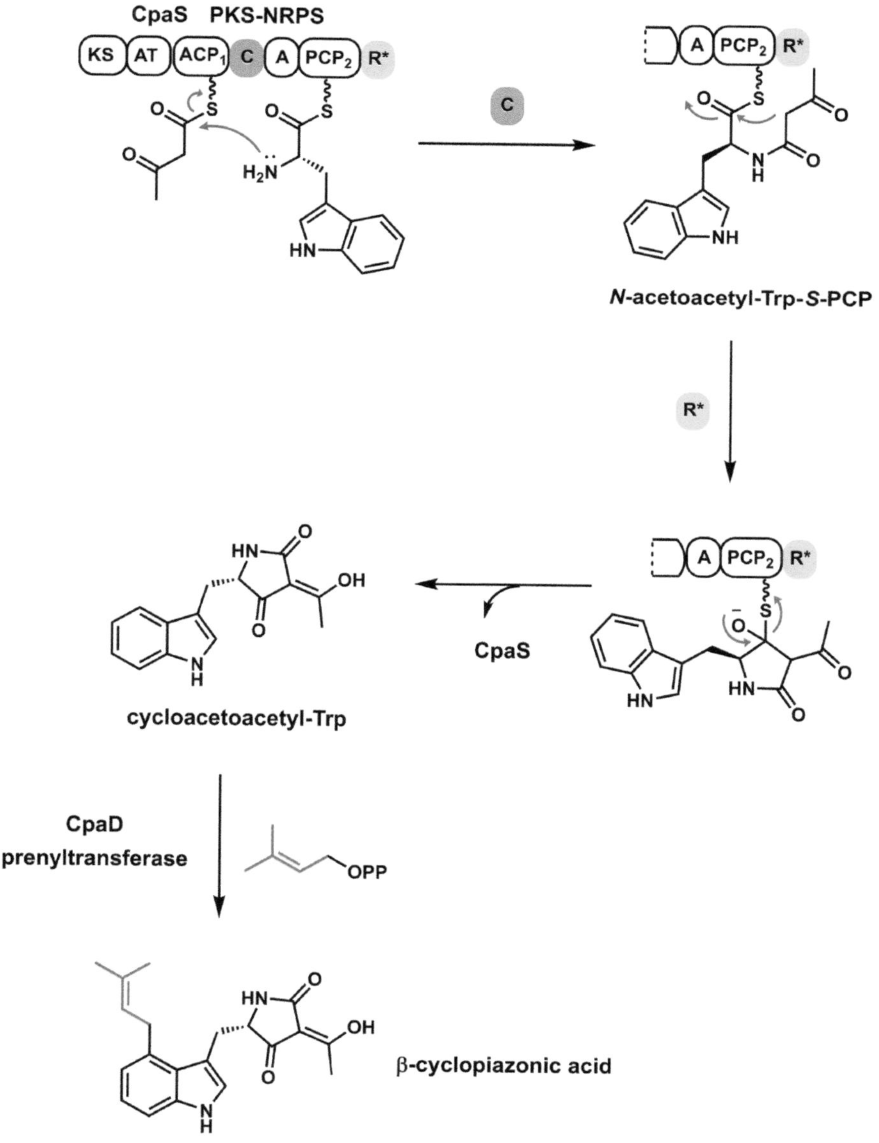

Figure 8.25 Cyclopiazonate biosynthesis: a hybrid PKS–NRPS releases an *N*-acetoacetyl-Trp chain by Dieckmann cyclization.

step, catalyzed by the condensation (C) domain. The *N*-acetoacetyl-Trp-*S*-PCP is then released by an unusual terminal domain in the NRPS module. It has the hallmarks of a reductase domain, as in the lyngbyatoxin assembly line noted above, but lacks key catalytic residues. Instead this R* domain acts as a Dieckmann cyclase (see Chapter 3, Figure 3.35). The C_3 carbanion of the acetoacetyl moiety attacks the thioester carbonyl, releasing cycloacetoacetyl-Trp with a newly formed tetramic acid ring. The second enzyme is a conventional prenyltransferase, installing the C_5 prenyl unit at C_4 of the indole ring to give β-cyclopiazonate (Liu and Walsh 2009b).

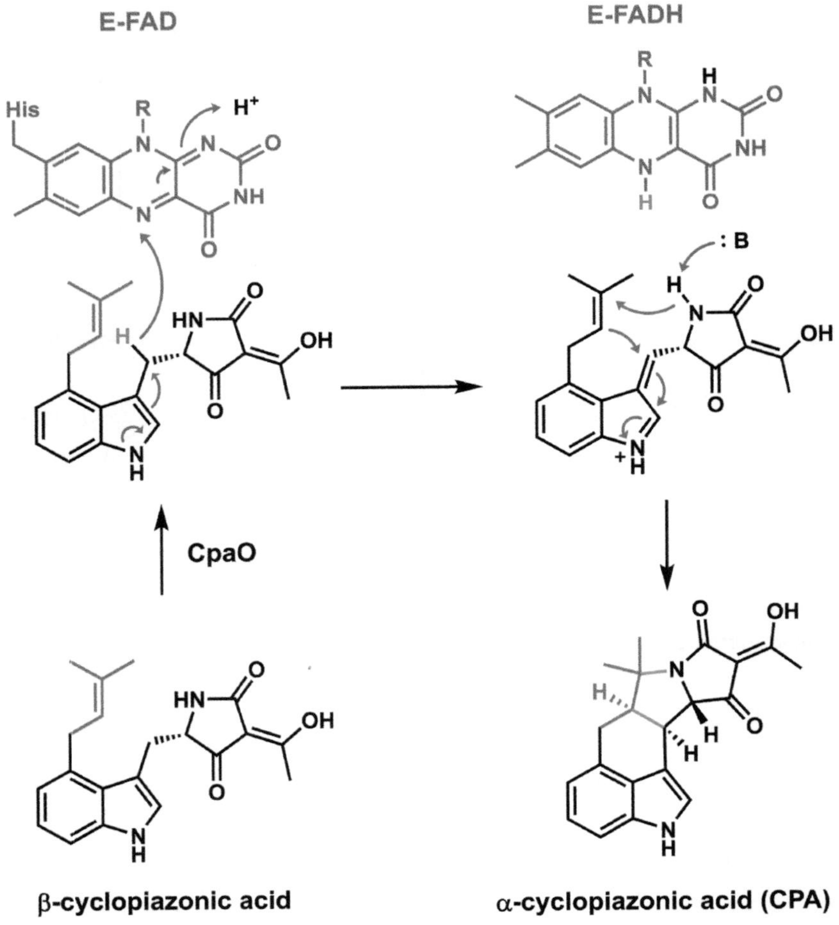

Figure 8.26 Formation of the final two 6-5 rings in cyclopiazonate by a flavoenzyme desaturase.

Conversion to the active α-toxin is the task of the third enzyme, an FAD-containing oxidoreductase. As shown in Figure 8.26, catalysis is proposed to initiate by removal of one of the hydrogens on the CH_2 group adjacent to C_3 of the indole ring in β-CPA as a hydride ion. This is achieved by neighboring group participation of the C_{2-3} double bond of the indole ring to assist in expulsion of that hydride to N_5 of the FAD coenzyme in the active site. (This has analogies to the berberine bridge enzyme mechanism noted in Figure 5.14.) Now the methylene–indoline is an electron sink and could assist the pictured amide nitrogen vinylogous attack that would form two C–C bonds in one transition state as the iminium ion in the indole was quenched. The product is the potent toxin α-cyclopiazonate. The FAD-containing oxidoreductase has created the pentacyclic scaffold by creating two rings at once, a substantial chemical achievement. This short efficient pathway represents three natural product biosynthetic strands coming together: polyketides, nonribosomal peptides, and isoprenoids around a tryptophan core.

8.10 Summary

The indole terpene family of alkaloids encompass some of the most complex scaffold architectures. They pair the underlying chemistry of the indole ring as regio-promiscuous carbon nucleophile with the facile carbocation formation of the prenyl-diphosphates. Cyclizations and oxidations are common tailoring reactions that add complexity to the scaffolds. As elsewhere in alkaloid biosynthetic pathways (Chapters 5 and 9), oxygenases are prominent catalysts in both functional group insertion and adjustment, and in driving re-arrangements *via* epoxide and carbon radical intermediates.

References

Balibar, C. J. and C. T. Walsh (2006). "GliP, a multimodular nonribosomal peptide synthetase in Aspergillus fumigatus, makes the diketopiperazine scaffold of gliotoxin". *Biochemistry* **45**(50): 15029–15038.

Cardellina, 2nd, J. H., F. J. Marner and R. E. Moore (1979). "Seaweed dermatitis: structure of lyngbyatoxin A". *Science* **204**(4389): 193–195.

Dewick, P. (2009). *Medicinal Natural Products, a Biosynthetic Approach.* UK, Wiley.
Donia, M. S. and E. W. Schmidt (2011). "Linking chemistry and genetics in the growing cyanobactin natural products family". *Chem. Biol.* **18**(4): 508–519.
Edwards, D. J. and W. H. Gerwick (2004). "Lyngbyatoxin biosynthesis: sequence of biosynthetic gene cluster and identification of a novel aromatic prenyltransferase". *J. Am. Chem. Soc.* **126**(37): 11432–11433.
Gu, B., S. He, X. Yan and L. Zhang (2013). "Tentative biosynthetic pathways of some microbial diketopiperazines". *Appl. Microbiol. Biotechnol.* **97**(19): 8439–8453.
Holzapfel, W. (1968). "The isolation and structure of cyclopiazonic acid, a toxic metabolite of Penicillium cyclopium Westling". *Tetrahedron Lett.* **24**: 2101–2119.
Kato, N., H. Suzuki, H. Takagi, Y. Asami, H. Kakeya, M. Uramoto, T. Usui, S. Takahashi, Y. Sugimoto and H. Osada (2009). "Identification of cytochrome P450s required for fumitremorgin biosynthesis in Aspergillus fumigatus". *ChemBioChem* **10**(5): 920–928.
Kato, N., H. Suzuki, H. Takagi, M. Uramoto, S. Takahashi and H. Osada (2011). "Gene disruption and biochemical characterization of verruculogen synthase of Aspergillus fumigatus". *ChemBioChem* **12**(5): 711–714.
Kries, H. and S. E. O'Connor (2016). "Biocatalysts from alkaloid producing plants". *Curr. Opin. Chem. Biol.* **31**: 22–30.
Li, S., K. Anand, H. Tran, F. Yu, J. M. Finefield, J. D. Sunderhaus, T. J. McAfoos, S. Tsukamoto, R. M. Williams and D. H. Sherman (2012). "Comparative analysis of the biosynthetic systems for fungal bicyclo[2.2.2]diazaoctane indole alkaloids: the (+)/(-)-notoamide, paraherquamide and malbrancheamide pathways". *MedChemComm* **3**(8): 987–996.
Li, S., J. M. Finefield, J. D. Sunderhaus, T. J. McAfoos, R. M. Williams and D. H. Sherman (2012). "Biochemical characterization of NotB as an FAD-dependent oxidase in the biosynthesis of notoamide indole alkaloids". *J. Am. Chem. Soc.* **134**(2): 788–791.
Li, S. M. (2011). "Genome mining and biosynthesis of fumitremorgin-type alkaloids in ascomycetes". *J. Antibiot.* **64**(1): 45–49.
Lin, H. C., G. Chiou, Y. H. Chooi, T. C. McMahon, W. Xu, N. K. Garg and Y. Tang (2015). "Elucidation of the concise biosynthetic pathway of the communesin indole alkaloids". *Angew. Chem., Int. Ed.* **54**(10): 3004–3007.

Lin, H.-C., T. McMahon, A. Patel, M. Corsello, A. Simon, W. Xu, M. Zhao, K. N. Houk, N. K. Garg and Y. Tang (2016). "P450-Mediated Coupling of Indole Fragments To Forge Communesin and Unnatural Isomers". *J. Am. Chem. Soc.* **138**: 4002–4005.

Liu, J., T. Ng, Z. Rui, O. Ad and W. Zhang (2014). "Unusual acetylation-dependent reaction cascade in the biosynthesis of the pyrroloindole drug physostigmine". *Angew. Chem., Int. Ed.* **53**(1): 136–139.

Liu, X. and C. T. Walsh (2009a). "Characterization of cyclo-acetoacetyl-L-tryptophan dimethylallyltransferase in cyclopiazonic acid biosynthesis: substrate promiscuity and site directed mutagenesis studies". *Biochemistry* **48**(46): 11032–11044.

Liu, X. and C. T. Walsh (2009b). "Cyclopiazonic acid biosynthesis in Aspergillus sp.: characterization of a reductase-like R* domain in cyclopiazonate synthetase that forms and releases cyclo-acetoacetyl-L-tryptophan". *Biochemistry* **48**(36): 8746–8757.

Martins, J. and V. Vasconcelos (2015). "Cyanobactins from Cyanobacteria: Current Genetic and Chemical State of Knowledge". *Mar. Drugs* **13**(11): 6910–6946.

May, J. A., R. K. Zeidan and B. M. Stoltz (2003). "Biomimetic approach to communesin B (a.k.a. nomofungin)". *Tetrahedron Lett.* **44**(6): 1203–1205.

Moncoq, K., C. A. Trieber and H. S. Young (2007). "The molecular basis for cyclopiazonic acid inhibition of the sarcoplasmic reticulum calcium pump". *J. Biol. Chem.* **282**(13): 9748–9757.

O'Connor, S. E. and J. J. Maresh (2006). "Chemistry and biology of monoterpene indole alkaloid biosynthesis". *Nat. Prod. Rep.* **23**(4): 532–547.

Ongley, S. E., X. Bian, Y. Zhang, R. Chau, W. H. Gerwick, R. Muller and B. A. Neilan (2013). "High-titer heterologous production in E. coli of lyngbyatoxin, a protein kinase C activator from an uncultured marine cyanobacterium". *ACS Chem. Biol.* **8**(9): 1888–1893.

Qu, Y., M. L. Easson, J. Froese, R. Simionescu, T. Hudlicky and V. De Luca (2015). "Completion of the seven-step pathway from tabersonine to the anticancer drug precursor vindoline and its assembly in yeast". *Proc. Natl. Acad. Sci. U. S. A.* **112**(19): 6224–6229.

Read, J. A. and C. T. Walsh (2007). "The lyngbyatoxin biosynthetic assembly line: chain release by four-electron reduction of a dipeptidyl thioester to the corresponding alcohol". *J. Am. Chem. Soc.* **129**(51): 15762–15763.

Roberts, M. and M. Wink, eds. (1998). *Alkaloids: Biochemistry, Biology, and Medical Applications*, Springer.
Saruwatari, T., F. Yagishita, T. Mino, H. Noguchi, K. Hotta and K. Watanabe (2014). "Cytochrome P450 as dimerization catalyst in diketopiperazine alkaloid biosynthesis". *ChemBioChem* **15**(5): 656–659.
Seigler, D. (2012). *Plant Secondary Metabolism*, Springer Science and Business Media: 759.
Siengalewicz, P., T. Gaich and J. Mulzer (2008). "It all began with an error: the nomofungin/communesin story". *Angew. Chem., Int. Ed.* **47**(43): 8170–8176.
Steffan, N., A. Grundmann, W. B. Yin, A. Kremer and S. M. Li (2009). "Indole prenyltransferases from fungi: a new enzyme group with high potential for the production of prenylated indole derivatives". *Curr. Med. Chem.* **16**(2): 218–231.
Tang, M.-C., Y. Zou, C. T. Walsh and Y. Tang (2017). "Oxidative Cyclization in Natural Product Biosynthesis". *Chem. Rev.*, DOI: 10.1021/acs.chemrev.6b00478.
Tanner, M. E. (2015). "Mechanistic studies on the indole prenyltransferases". *Nat. Prod. Rep.* **32**(1): 88–101.
Tsunematsu, Y., N. Ishikawa, D. Wakana, Y. Goda, H. Noguchi, H. Moriya, K. Hotta and K. Watanabe (2013). "Distinct mechanisms for spiro-carbon formation reveal biosynthetic pathway crosstalk". *Nat. Chem. Biol.* **9**(12): 818–825.
van Der Heijden, R., D. I. Jacobs, W. Snoeijer, D. Hallard and R. Verpoorte (2004). "The Catharanthus alkaloids: pharmacognosy and biotechnology". *Curr. Med. Chem.* **11**(5): 607–628.
Walsh, C. T. (2014). "Biological matching of chemical reactivity: pairing indole nucleophilicity with electrophilic isoprenoids". *ACS Chem. Biol.* **9**(12): 2718–2728.
Walsh, C. T. (2015). "A chemocentric view of the natural product inventory". *Nat. Chem. Biol.* **11**(9): 620–624.
Walsh, C. T., S. W. Haynes, B. D. Ames, X. Gao and Y. Tang (2013). "Short pathways to complexity generation: fungal peptidyl alkaloid multicyclic scaffolds from anthranilate building blocks". *ACS Chem. Biol.* **8**(7): 1366–1382.
Weber, G. (2015). *Molecular Therapies of Cancer*, Springer.
Xu, Z., M. Baunach, L. Ding and C. Hertweck (2012). "Bacterial synthesis of diverse indole terpene alkaloids by an unparalleled cyclization sequence". *Angew. Chem., Int. Ed.* **51**(41): 10293–10297.

Yan, W., H. Song, F. Song, Y. Guo, C. H. Wu, A. Sae Her, Y. Pu, S. Wang, N. Naowarojna, A. Weitz, M. P. Hendrich, C. E. Costello, L. Zhang, P. Liu and Y. J. Zhang (2015). "Endoperoxide formation by an alpha-ketoglutarate-dependent mononuclear non-haem iron enzyme". *Nature* **527**(7579): 539–543.

Zhu, J., M. Wang, W. Wen and R. Yu (2015). "Biosynthesis and regulation of terpenoid indole alkaloids in *Catharanthus roseus*". *Pharmacogn. Rev.* **9**(17): 24–28.

Section III

Key Enzymes in Natural Product Biosynthetic Pathways

This section takes up three topics that cut across the biosynthesis of multiple classes of the natural products discussed in the previous section. These are oxygenases, glycosylations, and *S*-adenosylmethionine as a radical initiator.

Oxygenases are pervasive in all of the natural product class biosynthetic routes. In bacterial and fungal biosynthetic gene clusters, dedicated oxygenases with tailoring functions are often encoded along with the key sets of chain elongation enzymes. A particularly notable set is found in the maturation of taxadiene to taxol. The C_{20} hydrocarbon taxadiene undergoes oxygenative processing to add eight oxygen atoms to the periphery on the way to taxol. Another remarkable run of tandem oxygenase action occurs in the conversion of geraniol to secologanin, a key substrate for strictosidine synthase. Strictosidine in turn is precursor to more than a thousand downstream indole terpene metabolites.

Iron-based oxygenases have learned how to reductively activate O_2 by one-electron pathways, generating high valent oxoiron intermediates which break C–H bonds in cosubstrates homolytically. The resultant carbon radicals can often rearrange *intramolecularly* in competition with *intermolecular* hydroxylation by an iron-bound [$OH^{\bullet}$] equivalent.

Thus, a substantial number of oxygen-consuming enzymes in this category do not transfer an OH to a coproduct. Instead, intramolecular radical chemistry for C–C bond formation intervenes. This is the case in several key metabolic pathway steps, including reticuline processing to salutaridinol in the morphine pathway, and

Natural Product Biosynthesis: Chemical Logic and Enzymatic Machinery
By Christopher T. Walsh and Yi Tang

Published by the Royal Society of Chemistry, www.rsc.org

in the formation of isopenicillin N from an acyclic tripeptide precursor. It is also the outcome for the O_2-consuming "expandase" enzyme, converting the 4,5 ring system of penicillin N to the 4,6-ring system in cephalosporins.

A second route to carbon radicals that can be used for C–C bond formation, among other outcomes, occurs at the other end of the aerobic spectrum, in the anaerobic microenvironments of cells. *S*-Adenosylmethionine is famous for its function in aerobic milieus for transfer of $[CH_3^+]$ equivalents to nucleophilic cosubstrates. However, it can also serve as a radical initiator when bound to enzymes that contain air-sensitive Fe_4/S_4 clusters. One-electron transfer from iron to a coordinated SAM in the active site leads to fragmentation of the C5′–S bond and creates the 5′-deoxyadenosyl radical. This can abstract an $H^\bullet$ from bound cosubstrates, and set off carbon-centered radical reactions. In some circumstances SAM functions coezymatically and is regenerated at the end of each catalytic cycle. In other cases the 5′-deoxyadenosine escapes as a coproduct while intermediate carbon radicals have coupled in scaffold rearrangements. In yet other cases, a second molecule of SAM is utilized for $[CH_3^\bullet]$ transfer to unactivated carbon sites on the cosubstrate.

The third topic in this section is enzymatic glycosylation of natural products of every class. These are typically late stage modifications. In some contexts the glycosylations are reversible and the ratio of glycoside to free aglycone metabolites reflects a balance between net glycosyltransferase synthetic activity and net glycosidase hydrolytic activity. In many other cases the glycosylations are irreversible and represent key binding elements for biological activity of the natural products, exemplified in the antibiotic erythromycin and the antifungal polyether nystatin. In this latter set, the hexoses are often not readily available glucose but rather a diverse set of deoxy-, aminodeoxy-, and methylated deoxyhexoses that provide an altered hydrophobic/hydrophilic balance to the hexose moieties. These modified hexoses arise from enzymatic processing of TDP-glucose and the enzymes are often encoded, along with dedicated Gtfs, in the biosynthetic gene clusters.

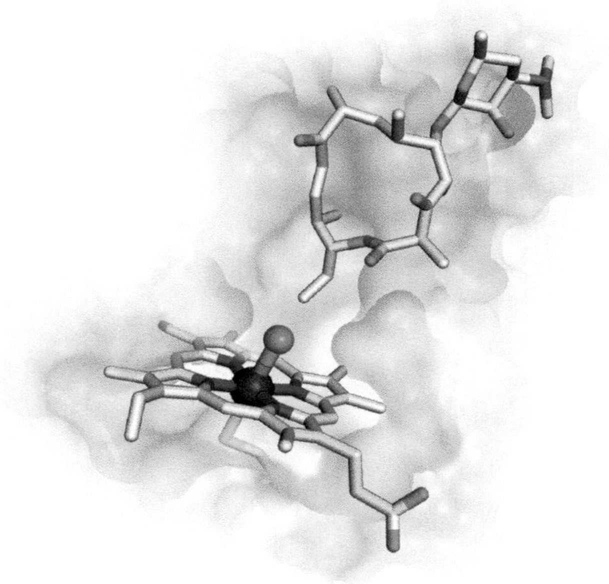

Crystal structure of the active site of the P450 PikC involved in pikromycin biosynthesis (PDB ID: 2VZ7).

9 Carbon-based Radicals in C–C Bond Formations in Natural Products

A. Oxygenases

B. Oxygen-dependent Halogenases

9.1 Introduction

In the various classes of natural products discussed in the preceding chapters, scaffold complexity tends to increase as secondary metabolite pathways branch off from primary metabolism. The building blocks, such as amino acids, acetyl-CoA and malonyl-CoA, cinnamate in the phenylpropanoids, the Δ^2 and Δ^3-isoprenyl-PP isomers, all get elaborated into higher molecular weight intermediates and natural product end molecules. These occur with various new C–C, C–N, and C–O bond formations as scaffold complexity increases.

Most bonds formed and broken in both primary and secondary metabolisms go *via* two-electron pathways: *heterolytic* cleavage and formation of the particular C–H, C–C, O–C, N–C bonds that break and reform (Figure 9.1). All such reactions need pathways to both accessible carbanions and electropositive carbon equivalents in bond-forming (and breaking) steps, as we noted in the introductory chapter.

Natural Product Biosynthesis: Chemical Logic and Enzymatic Machinery
By Christopher T. Walsh and Yi Tang

Published by the Royal Society of Chemistry, www.rsc.org

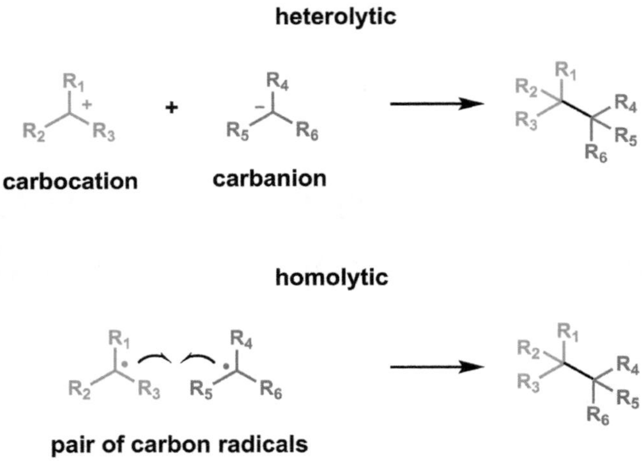

Figure 9.1 Two limiting mechanistic options for C–C bond formation in assembly of natural product scaffolds: a heterolytic mechanism involving joining of carbon nucleophiles and carbon electrophiles *vs.* A homolytic mechanism joining two carbon radical centers.

In primary metabolism, including glycolysis and the citrate cycle, the pentose phosphate pathway, nucleic acid biosynthesis and degradation, and protein biosynthesis at the ribosome, a first pass scan of the structures of the large arrays of primary metabolites and the types of enzymes and cofactors in play suggests that two-electron reaction manifolds dominate. The exceptions are in those cases where flavin coenzymes (FAD and FMN) are involved in enzyme-catalyzed redox transformations. For example, succinate dehydrogenase likely removes the two hydrogens and two electrons in one transition state as succinate is oxidized to fumarate by a proton/hydride mechanism as far as succinate is concerned. The $FADH_2$ thereby generated passes electrons to iron/sulfur clusters that are obligate one-electron transfer cofactors. The FAD coenzyme thus passes through the one-electron oxidized semiquinone state on the way back up to the two electrons oxidized, air stable FAD starting state (Figure 9.2). Among the two common redox coenzymes of primary metabolism, FAD and NAD(P)H, the nicotinamides (NAD/NADP) are restricted to two electrons only (hydride) transfer mechanisms because of the high energy of pyridinyl radicals.

In contrast, the flavin semiquinone is kinetically accessible and thermodynamically stable enough to be functional in two subclasses of flavoenzymes: those that pass electrons to or receive them from iron atoms or pass them to molecular oxygen (Walsh and Wencewicz 2013). The two-electron/one-electron stepdown redox transforming

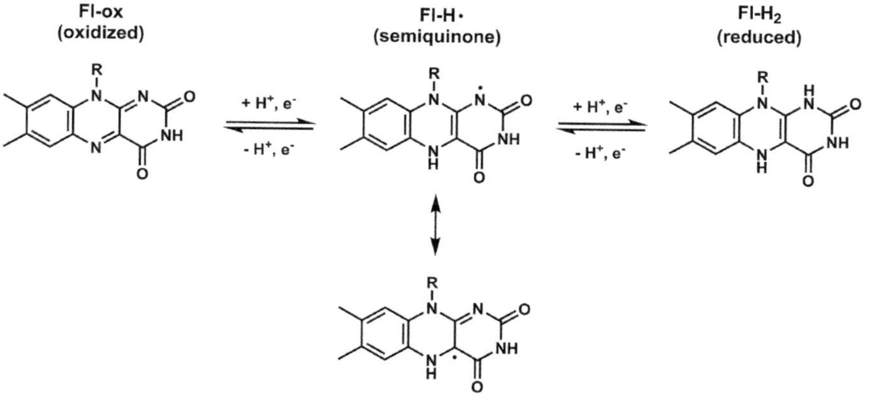

Figure 9.2 Flavin coenzymes as two-electron/one-electron stepdown redox transfer agents.

property arises from the tricyclic isoalloxazine business end of the flavin coenzymes (Walsh 1980). As we have seen in the prior chapters and will codify in this one, the oxygen reactivity of the dihydroflavin ($FMNH_2$, $FADH_2$) forms of the tightly bound flavin coenzyme places it square in the oxygenase frame.

In this chapter we delve into the less common *homolytic* bond-breaking and -making routes (Figure 9.1). While these are indeed rare in primary metabolism, they are substantially more common in secondary metabolic pathways, as we have commented in preceding chapters and will bring to the fore in this one.

Molecular oxygen, a ground state triplet molecule (see below), is by far the best known of the biologically central molecules that react by one-electron only pathways. It can abstract hydrogen atoms from C–H bonds, yielding carbon-centered radicals. Thus, one set of carbon-based radicals, able to engage in biosynthetic bond rearrangements and net bond formations, occur *aerobically*.

About 5000 natural products contain halogen atoms (Gribble 2004). Over the past two decades two kinds of enzymes, responsible for a large fraction of biosynthetic halogenations, have been discovered and characterized as oxygen-requiring flavoenzymes or nonheme iron enzymes (Vaillancourt, Yeh *et al.* 2006). They appear to be halogenases that have evolved from oxygenases in these two classes. As in the two large oxygenase families, whether FAD or iron is utilized for concomitant oxygen and halide activation depends on the electronic demands of the cosubstrates to be halogenated. We will take up these mechanistic variants of oxygen enzymology in the second part of this chapter.

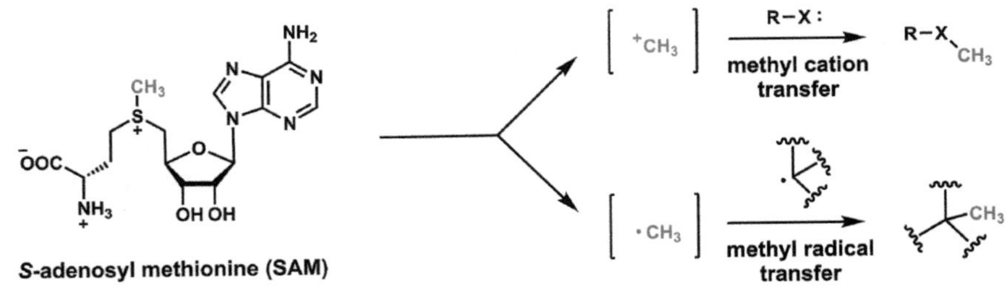

Figure 9.3 Enzyme-mediated biosynthetic radical chemistry without oxygens: SAM as a generator of the 5′-deoxadenosyl radical. Methyl radical transfers require iron/sulfur clusters as one electron initiator for SAM homolytic cleavage.

Some 15–20 years ago, *S*-adenosylmethionine (SAM), one of the most widely used cofactors in both primary and secondary metabolism, was regarded as the preeminent, perhaps the sole donor of methyl groups with the polarity of methyl cations [CH_3^+] to cosubstrate nucleophiles (Loenen 2006). Indeed, *O*-methylation and *N*-methylation reactions populate many of the alkaloid and phenylpropanoid biosynthetic routes, as we have seen in Chapters 5–8.

In an unanticipated turn of events, it also transpires that some SAM-utilizing enzymes can transfer the one carbon methyl fragment as a methyl radical equivalent [$CH_3^{\bullet}$]. These radical SAM enzymes all depend on a 4Fe/4S inorganic cluster as radical initiator to cleave SAM homolytically. The resulting 5′-deoxyadenosyl radical (Figure 9.3) can go on to abstract a hydrogen atom for substrate C–H bonds and thereby generate carbon radical species. Because of the Fe/S clusters that are sensitive to oxygen-mediated oxidative decomposition, radical SAM enzymes tend to be fastidiously anaerobic. So those radical reactions proceed under anaerobic conditions. This will be the subject of Chapter 10 (Broderick, Duffus *et al.* 2014, Mehta, Abdelwahed *et al.* 2015).

Thus, the radical reactions, homolytic scissions, and bond-forming processes can occur *either* aerobically or anaerobically, depending on whether O_2 or SAM is the one electron donating/receiving cosubstrate.

9.2 Oxygenases in Primary *vs.* Secondary Metabolism

A scan of the major metabolic pathways of primary metabolism, particularly in eukaryotes, indicates that oxygenases, enzymes that reductively activate O_2 and incorporate one oxygen (or both) atom(s) into a coproduct, *are relatively rare.* We noted in Chapter 4 that the main exceptions are found in enzymes that process cholesterol to the various steroid-based hormones in adrenal cells and those that generate the androgens and estrogens. Vitamin D, a metabolite from 7-dehydrocholesterol (see Figure 4.30), also undergoes three cytochrome P450 hydroxylations to reach the active form of the calcium-mobilizing hormone.

This paucity of reductive oxygen metabolism may reflect early evolution of primary metabolic pathways in organisms under anaerobic or microaerophilic conditions. In turn, that would suggest that aerobic metabolism of steroids is a later development. As depicted in Chapter 4, the cyclization of squalene to hopenes and hopanols occurs by an oxygen-independent route, in contrast to the parallel one initiated from squalene-2,3-epoxide, reminding us that the cyclization cascades do not need O_2 as cosubstrate.

As we have noted in previous chapters and will focus on in this one, the iron-containing set of oxygenases generate carbon radicals. Carbon radical chemistry did/does happen in anaerobic organisms, but with much of the flux *via* radical SAM reactions, not with O_2.

9.2.1 Some Features of Molecular Oxygen

The molecular orbital diagram for O_2 (Figure 9.4) shows that in the outer shell the last two electrons fill π^*_x and $^*\pi^*_y$ orbitals singly in the lowest energy state. Thus, molecular oxygen is in a triplet ground state with two unpaired electrons; O_2 reacts sluggishly with two-electron-only transfer reagents (most organic metabolites in cells) and rapidly with one-electron-only cosubstrates. There is a spin-paired single state that is capable of facile reaction with spin-paired organic molecules but it is 22 kcal mol^{-1} above the triplet ground state. This barrier is larger than the ones enzymes can typically surmount or lower, and there is no indication that spin-paired singlet molecular oxygen has any physiologic role in primary or secondary metabolic pathways (Emsley 2001).

Aerobic organisms have learned how to reduce O_2 efficiently, in the terminal stage of the mitochondrial respiratory chain where four electrons and four protons are delivered from the iron- and copper-containing cytochrome oxidase to reduce molecular oxygen all the way to two molecules of water (Figure 9.5). The respiratory chain ultimately gets much of its electron flux from reduced nicotinamide coenzymes (reduction potential for two electrons, $E^{0\prime} = -320$ mV). The

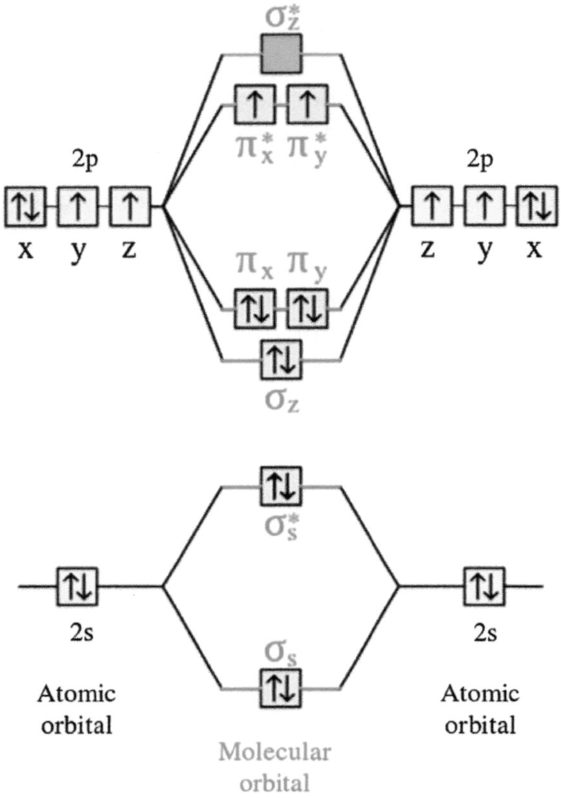

Figure 9.4 Molecular orbital diagram of O_2: single occupancy of *pi* orbitals means O_2 is a ground state triplet.

corresponding reduction potential for O_2 to H_2O is $E^{0\prime} = +820$ mV. Therefore, electrons pass through a potential drop of 1.14 V from NADH to O_2. This corresponds to a K_{eq} of $\sim 10^{38}$ in favor of reduction. Reduction of O_2 by cytochrome oxidase is accompanied by anisotropic expulsion of protons across the inner mitochondrial membrane. They can flow back down their electrochemical gradient through a channel in the transmembrane ATP synthase, allowing harvesting of the thermodynamic potential into ATP as the major cellular chemical energy store.

The situation then is that reduction of O_2 to water, as organic metabolites are oxidized ultimately to CO_2 as all hydrogens are stripped away, is enormously favorable thermodynamically. The kinetic incompatibility for reaction with spin-paired, two-electron-only, organic molecules under physiologic conditions allows the atmosphere to contain 20% O_2 while reduced organic molecules, including people, can coexist with O_2.

9.2.2 The Nature of the Four-electron Reduction: Superoxide, Peroxide, Hydroxyl Radicals. The Surveillance Role of SOD and Catalase

One other feature of the single electron at a time, four step process that reduces O_2 to H_2O is that all three of the intermediates are potentially (differentially) toxic if they were to get free (Figure 9.5). The discrete formation of superoxide anion, hydrogen peroxide, and hydroxyl radical ($OH^•$) is avoided by the cytochrome oxidase system. Organisms produce scavenger enzymes for the first intermediate, the one-electron reduced superoxide, anionic at physiologic pH ($pK_a = 4.8$), and for two-electron reduced hydrogen peroxide. These are superoxide dismutases (McCord and Fridovich 1988) and catalase (Chelikani, Fita *et al.* 2004), respectively (Figure 9.6). The first enzyme

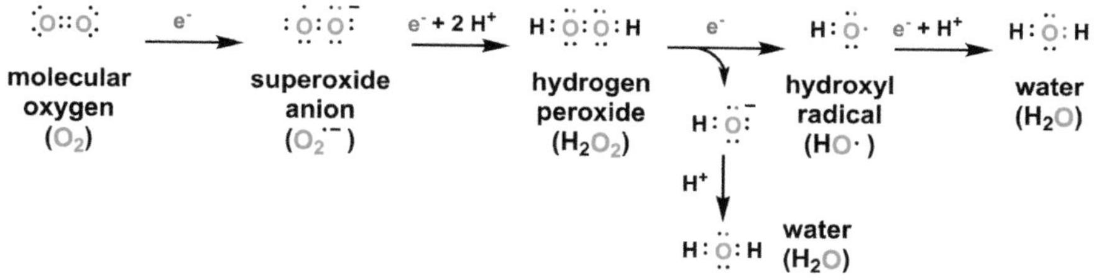

Figure 9.5 Reductive oxygen metabolism: thermodynamically favored, one electron pathways proceed *via* superoxide anion, hydrogen peroxide, and hydroxyl radical [$OH^•$] on the way to H_2O.

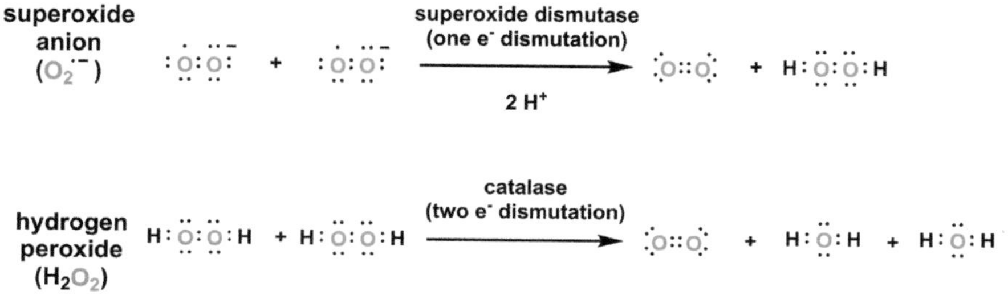

Figure 9.6 Superoxide dismutase and catalase are surveillance enzymes to deal with one-electron-reduced and two-electron-reduced forms of O_2 by one-electron and two-electron dismutation reactions, respectively.

mediates a one-electron dismutation, producing O_2 and H_2O_2, while the second enzyme does a two-electron dismutation, yielding O_2 and water. These successfully detoxify superoxide and peroxide. There is no comparable enzyme surveillance and interception strategy for the three-electron reduced hydroxyl radical which would be the most damaging of the partially reduced O_2 intermediates.

9.3 Oxidases *vs.* Oxygenases

9.3.1 Oxidases

There are a variety of enzymes in primary metabolism that function as oxidases, carrying out a net two-electron reduction of O_2 to H_2O_2 while a cosubstrate is oxidized by a comparable two electrons. The stoichiometry is depicted in Figure 9.7. Such enzymes include L-amino acid and D-amino acid oxidases (to the respective imino acids, see Figure 5.32 at the start of the rebeccamycin and staurosporine pathways), glucose oxidase (glucose to gluconolactone), monoamine oxidase, and sulfhydryl oxidases.

These are all flavin-containing enzymes (Walsh and Wencewicz 2013). The specific substrate is oxidized in the first half reaction, yielding product which can diffuse away and $FADH_2$ which remains enzyme bound. This form of the enzyme can react rapidly with O_2 by one electron transfer to give superoxide anion and the FAD semiquinone. Radical recombination yields the FAD-4a-OOH from which hydrogen peroxide is eliminated as oxidized FAD reforms.

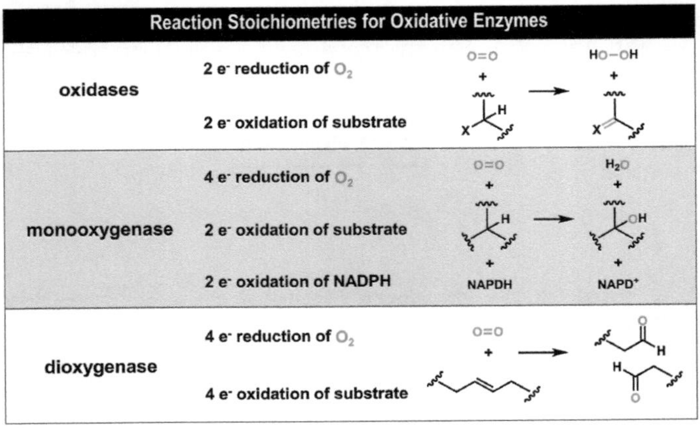

Figure 9.7 Reaction stoichiometries for oxidases, monooxygenases, and dioxygenases.

There are *no* NAD-containing oxidases where NAD(P)H is generated and directly reoxidized by O_2 because of the incompatibility of the hydride transfer coenzyme (two-electron manifold only) reacting with O_2 (one-electron manifold only). For this reason, NAD(P)H can serve as a diffusible coenzyme in cells in the presence of oxygen. On the other hand, $FADH_2$ is autoxidized in less than a second by O_2 and so is kept sequestered in flavoenzyme active sites. Flavin cofactors have $K_D < 10^{-8}$ M for their apoproteins and none of the FAD, FADH•, or $FADH_2$ oxidation states is typically found as a freely diffusing redox coenzyme form. Iron is also typically not utilized in simple oxidase two-electron transfer enzymes although it is the predominant cofactor in oxygenases.

9.3.2 Oxygenases

Figure 9.7 also shows stoichiometries for generalized monooxygenase and dioxygenase reactions. Both oxygenase modes reflect not a two-electron reduction of O_2 as in the oxidase mode but rather a four-electron reduction of O_2. A substrate undergoing monooxygenation, such as a hydroxylation event, undergoes only a net two electron oxidation. Therefore another two electron donor is required for monooxygenase catalysis to ensue and the equation to balance. Typically NADPH is the most abundant and readily available (diffusible to oxygenase active sites) packet of two electron cellular redox currency.

A key issue arises with the use of NAD(P)H as the common electron donor: the incompatibility of this hydride transfer coenzyme to deliver one electron at a time to O_2. The strategy nature has evolved is to use flavin coenzymes as 2/1 stepdown redox transformers (Figure 9.2), placed between NAD(P)H and O_2 (Walsh 1980, Walsh and Wencewicz 2013). In turn we will note that some monooxygenases are simple flavoproteins. The bound FAD is reduced in a two-electron hydride transfer manifold. The $FADH_2$ thereby generated can react *via* one electron transfer to form superoxide and FADH• and then recombine to the FAD4a-OOH. We will note below that this is the active oxygen transfer agent in this oxygenase subclass.

Other monooxygenases have iron as the oxygen-reactive agent, either in a heme or a nonheme-based array, as will be noted below. The iron atoms are particularly effective one electron conduits to bound O_2. The same objection for NAD(P)H arises for one electron transfer to Fe atoms in enzyme active sites as to O_2 itself. Fe(III) and Fe(II) are not rapid two electron transfer agents. The dozens of cytochrome P450 hydroxylases we have catalogued in passing in Chapters 3–8 all fall into this category.

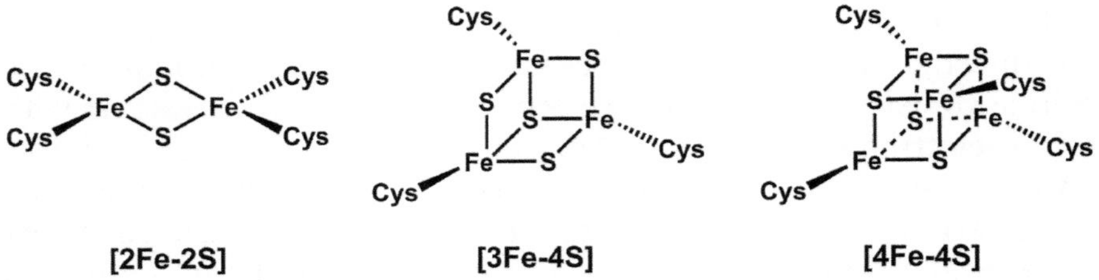

Figure 9.8 Simple architectures for Fe/S clusters that serve as one electron conduits to iron at the active site of monooxygenases.

The general solution is that there are feeder enzymes that interface between the readily available form of two electron redox currency, NAD(P)H, and the iron-based monooxygenases. Those feeder enzymes contain FAD and FMN. The FAD directly accepts the hydride from NAD(P)H and then the electrons can disproportionate between $FADH_2$ and FMN to give the semiquinone forms, $FADH^{\bullet}$ and $FMNH^{\bullet}$, which can feed electrons one at a time to the heme center in the P450 oxygenase. They also contain iron atoms, typically in either 2Fe/2S or 4Fe/4S clusters (Figure 9.8). The iron/sulfur clusters may be in a distinct partner protein. In any case the direction of electron transfer is as shown in Figure 9.9: NADPH to FAD to Fe/S to cytochrome P450 to O_2. The FAD serves as the splitting coenzyme between two electron input and one electron output (Jensen and Moller 2010, Pandey and Fluck 2013, Riddick, Ding *et al.* 2013). We will analyze the catalytic scope of FAD *vs.* iron-containing monooxygenases in the next sections.

9.3.3 Oxygenase/Oxidase/Oxygenase in Tirandamycin Biosynthesis

In the biosynthesis of the tirandamycin class of antibiotics a heme P450 enzyme TamI and a flavoenzyme TamL (Figure 9.10) act tandemly with O_2 as cosubstrate (Carlson, Li *et al.* 2011). First TamI acts on the TirC molecule to create an allylic hydroxyl moiety, the first oxygenation event. Then TamL utilizes O_2 while the alcohol group in TirE is oxidized to the ketone of TirD. This is NOT an oxygenase reaction but merely an oxidase. The O_2 is reduced to H_2O_2. Now TamI goes back to work as an oxygenase for two more cycles. It epoxidizes the double bond with one molecule of O_2 cosubstrate and then converts the methyl group to a hydroxymethyl with yet another O_2, to generate TirB as end product. Four molecules of O_2 have been consumed. One has been reduced by two electrons to H_2O_2, the other

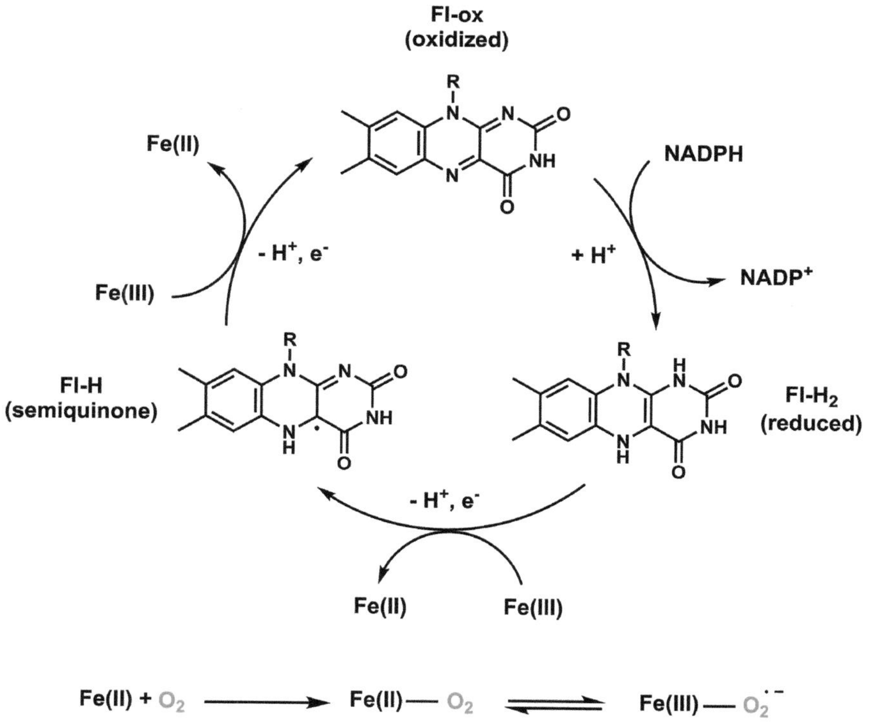

Figure 9.9 FAD as two electron/one electron stepdown redox transformer between NADPH and Fe(III) centers, including Fe/S clusters.

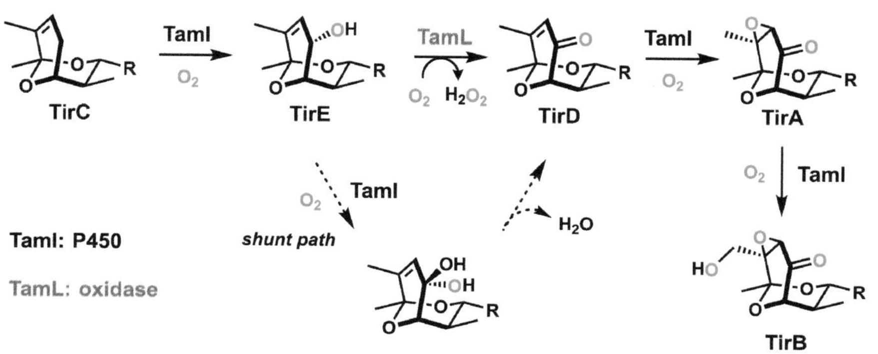

Figure 9.10 Tandem action of cytochrome P450 TamI and flavoenzyme TamL in tirandamycin biosynthetic oxygenations.

three to water and oxygenated substrate: these are four electron reductions of those three O_2 molecules.

9.3.4 Dioxygenases

The stoichiometry of dioxygenation of natural product substrates is distinct from the monooxygenations (Figure 9.7). In dioxygenase catalysis, both atoms from the same O_2 end up in the coproduct(s) (Bugg 2003). A classic case that we noted in Chapter 4 is in the late stages of isoprenoid metabolism where β-carotene is cleaved symmetrically by a dioxygenase to give two molecules of vitamin A aldehyde (see Figure 4.33) (Harrison and Bugg 2014).

9.4 Organic *vs.* Inorganic Cofactors for Oxygenase Catalysis

As noted above, Nature has evolved two mechanisms for reductive activation of O_2 by one-electron paths. The first mechanism uses redox active metal cations, occasionally copper, but most predominantly iron, in the active site of O_2-reducing enzymes. Iron and copper can cycle between redox states spaced by one electron (*e.g.* Fe(II) and Fe(III); Cu(I) and Cu(II)) and readily transfer one electron to O_2. The most prominent of the copper-based monooxygenases is dopamine β-hydroxylase, oxygenating the benzylic position of dopamine on the way to noradrenaline and adrenaline neurotransmitters (Kapoor, Shandilya *et al.* 2011).

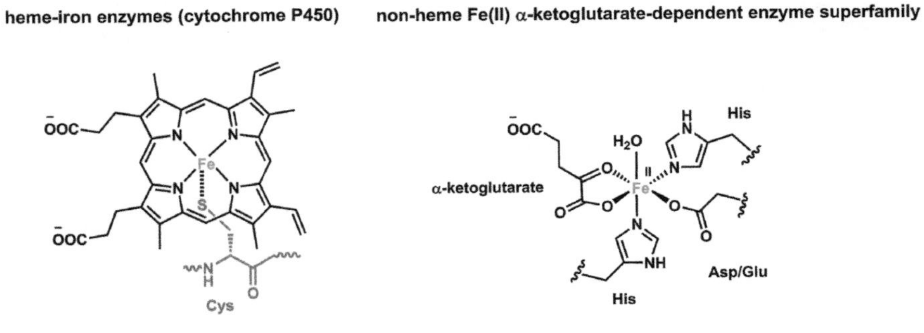

Figure 9.11 Two active site environments for iron-based monooxygenase enzymes: heme-based cytochrome P450 with bottom axial thiolate ligand; nonheme two His/one carboxylate ligand set.

More relevant to the oxygenative tailoring of natural product scaffolds are the iron-based enzymes. Two subfamilies have evolved (Figure 9.11). The most abundant contain iron embedded in the equatorial plane of the heme cofactor. Of the various subtypes of heme proteins, it is the ones with the bottom axial ligand as a thiolate provided from a cysteine side chain in the proteins and an open top axial position able to accommodate binding of O_2 and then its reductive activation (Ortiz de Monteallano 2015) that act as oxygenases. These are the famous cytochrome P450 family members, named on the basis of the λ_{max} of the soret band absorbance of the heme in the Fe(II)–CO complex. The strong red band at 450 nm is diagnostic for the thiolate coordinated Fe(II)–CO complex and a reliable identifier of P450 oxygenase activity. As one measure of abundance in plant secondary metabolism, the *Arabidopsis thaliana* genome encodes ~250 predicted P450 enzymes.

The second category of iron-based monooxygenases have evolved independently from the P450 type cytochromes. This second group contains a mononuclear iron atom, usually in the Fe(II) resting state, with two histidines and an Asp/Glu side chain carboxylate providing three of the anticipated six ligands to the bound iron (Que and Ho 1996, Solomon, Decker *et al.* 2003, Kovaleva and Lipscomb 2008, Martinez and Hausinger 2015). The fourth and fifth ligands to iron are usually provided by α-ketoglutarate, a reaction cosubstrate that coordinates through the keto group and one of the carboxylate oxygens. As will be noted below, the α-KG undergoes an oxidative decarboxylation process that leads to generation of the high valent oxoiron intermediate during each catalytic cycle.

The sixth coordination site in the nonheme mononuclear iron enzyme's first shell is occupied by a water molecule in the resting enzyme. On binding of the cosubstrate to be hydroxylated, the water is displaced by O_2 to generate the catalytically competent form of these oxygenases.

The second general solution to oxygen activation and catalytic reductive transfer of one oxygen atom to cosubstrates that has evolved does not utilize an inorganic transition metal as an electron conduit. Instead, the solution is to conscript a specific vitamin, by way of vitamin B2-based flavin coenzymes (Walsh and Wencewicz 2013). FAD is the predominant form (FMN is less common) and the business end of this redox active coenzyme is the tricyclic isoalloaxine ring system. We note in Figure 9.12 that the one-electron reduced semiquinone and the two-electron fully reduced oxidation state flavin coenzymes react with O_2 by one electron transfer.

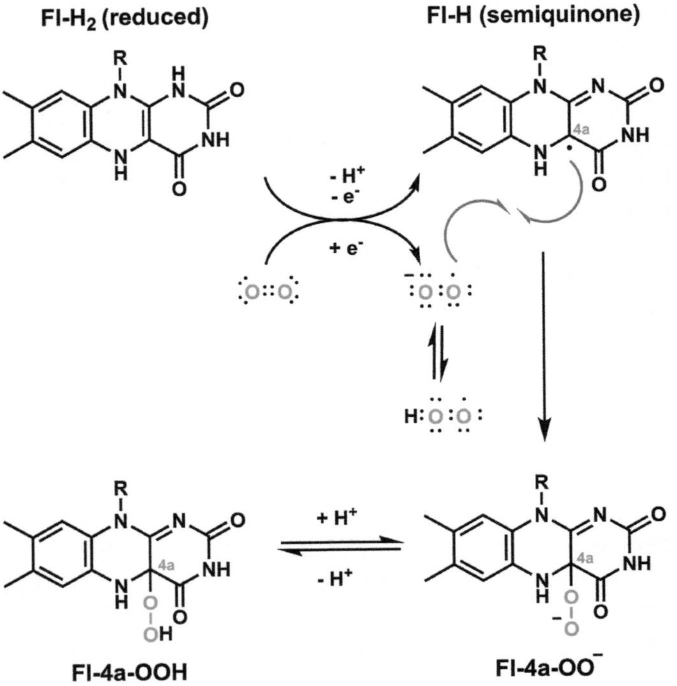

Figure 9.12 Reaction of one-electron-reduced and two-electron-reduced flavin coenzymes with O_2.

9.5 Scope and Mechanism of Oxygenations Catalyzed by Iron-based *vs.* Flavin-based Oxygenases

9.5.1 Flavin-based Oxygenases

The two types of monooxygenases, FAD-containing *vs.* iron-based, show clear differences in the types of cosubstrates they will oxygenate. These differences arise in large part due to the difference in oxidizing power of the proximal oxygen transfer agents within the active sites of these two enzyme classes. Flavoproteins use the FAD-4a-OOH as oxygenating agent. This organic hydroperoxide functions in most examples to *deliver an [OH^+] equivalent to a nucleophilic substrate*. As shown in Figure 9.13 this involves attack by a substrate carbanion equivalent on the distal oxygen of the Fl–OOH. The resultant Fl–OH

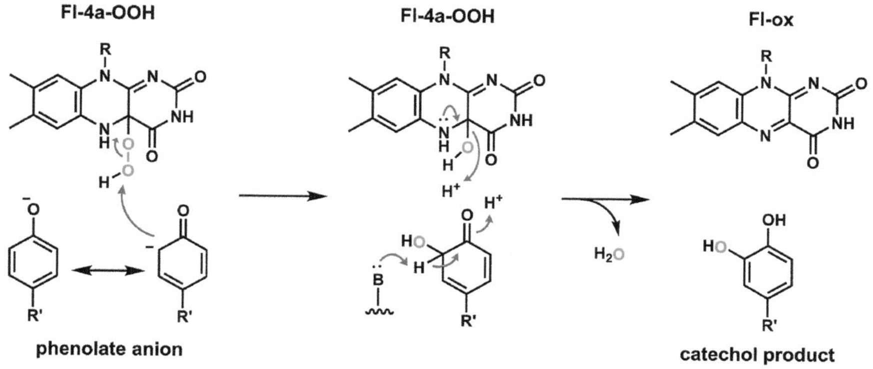

Figure 9.13 Transfer of [OH$^+$] equivalent from flavin-4a-OOH to substrate carbanion equivalent.

can decompose intramolecularly to give water and regenerate the oxidized FAD in the enzyme active site. The substrates that give rise to kinetically accessible carbanions in natural product assembly are commonly found in three electron rich aromatic rings: phenols, pyrroles, and indoles. *Ortho*- and *para*-hydroxylation of phenolic rings are the common patterns for FAD-linked oxygenases consistent with sites of carbanion formation.

An alternative oxygen transfer outcome for flavoenzymes is epoxidation of double bonds. In those cases the nucleophiles are the π-electrons of the olefin undergoing epoxidation. The most famous example encountered to date is squalene-2,3-epoxidase, which is noted in Figure 4.20 and is preamble to all the sterol metabolites from the oxidosqualene cyclase reaction manifold (Laden, Tang *et al.* 2000).

The second most famous case of flavoprotein epoxidases occurs in the late stages of polyether biosynthesis from polyketide frameworks, for example Lsd18 in lasalocid maturation (Minami, Shimaya *et al.* 2012, Suzuki, Minami *et al.* 2014). As noted in Figures 2.43 and 2.44, for lasalocid and ciguatoxin the epoxidation of sets of polyketide-introduced olefins are preconditions for the subsequent cascade of epoxide openings that form the furan and pyran rings that are the hallmarks of those metabolite classes. We have also noted that FAD enzymes also epoxidize the 2,3-double bond of indole in construction of the complex peptidyl alkaloid scaffolds in both the fumiquinazoline A families and the spirotrypostatins (see Figures 8.16 and 8.17).

An orthogonal mode of mono-oxygen transfer from O_2 to cosubstrate occurs in a subgroup of FAD enzymes that carry out Baeyer–Villiger type insertion of oxygen into ketones (Walsh and Chen 1988, Faber 2011). When the ketone is part of a ring system, this is a net conversion of a ketone to a lactone. The mechanism was worked out with a simple cyclohexanone substrate (Figure 9.14A), confirming a stereospecific migration of the C–C bond during ring expansion and consistent with the FAD–OOH acting, not as an electrophilic agent giving up [OH^+], but as the nucleophilic peroxide anion, Fl–OO^-, as shown.

Figure 9.14B shows the additional example of ketone to lactone conversion in a much more complex substrate scaffold, premithramycin B (Beam, Bosserman *et al.* 2009). The nascent seven-membered lactone product undergoes ring-opening hydrolysis that allows isomerization of what had been a trapped ene-diol to the hydroxy diketone in mithramycin DK.

Figure 9.15A also depicts an intriguing and novel outcome by flavoenzyme Baeyer–Villiger catalysis during maturation of the cytochalasin scaffold. A late stage ketone intermediate is ring expanded first to the lactone (ester). Then in an iterated process the lactone is again oxygenated on the other side to produce cytochalasin Z16, a vinyl carbonate ester (Hu, Dietrich *et al.* 2014), before further oxygenative processing by a P450 enzyme to cytochalasin E. It also appears that some cytochromes P450 can also function as Baeyer–Villiger catalysts, for example Cyp85A2 in oxidation of castasterone to brassinolide, as noted below (Kim, Hwang *et al.* 2005). In that reaction mode it is likely that an Fe(III)–OO anion, in analogy to the FAD–OO^- anion, is the initiating nucleophile. Figure 9.15B depicts proposed actions of SdnN in biosynthesis of the antifungal agent sordarin as a flavoprotein Baeyer–Villigerase that creates a dihydroxy lactone that is hydrolytically opened to the conjugated aldehydo acid. This metabolite can assume a conformation where the diene and dienophile are oriented productively for a net [4 + 2] cyclization (Kudo, Matsuura *et al.* 2016). This would generate the core cyclohexene ring in sordaricin. Action of a GDP-6-deoxy-D-altrose glycosyltransferase completes assembly of sordarin.

Enzymatic oxidation of intermediates to set up the requisite diene and dienophile is a common strategy in the several cases where net [4 + 2] cycloadditions occur, for spinosyn, abyssomycin, thiopeptides such as thiocillins, among others (Tang, Zou *et al.* 2017).

In contrast to the oxo-iron active site reagents discussed below, the flavin-OOH is a considerably weaker oxidant. There are no known instances where it can hydroxylate an unactivated carbon center.

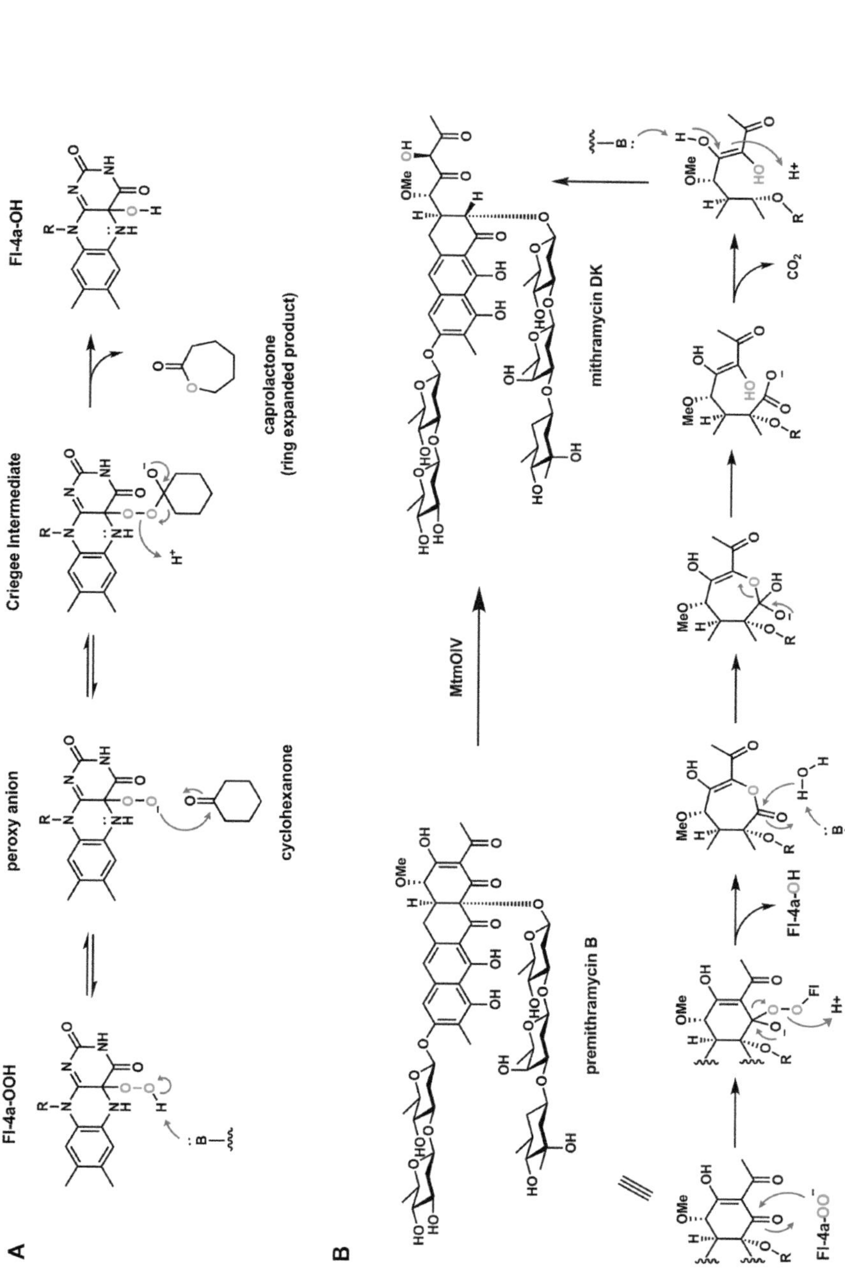

Figure 9.14 Flavoenzymes oxygenases as Baeyer–Villiger catalysts: (A) cyclohexanone monooxygenase and (B) ketone to lactone ring expansion during mithramycin biosynthesis.

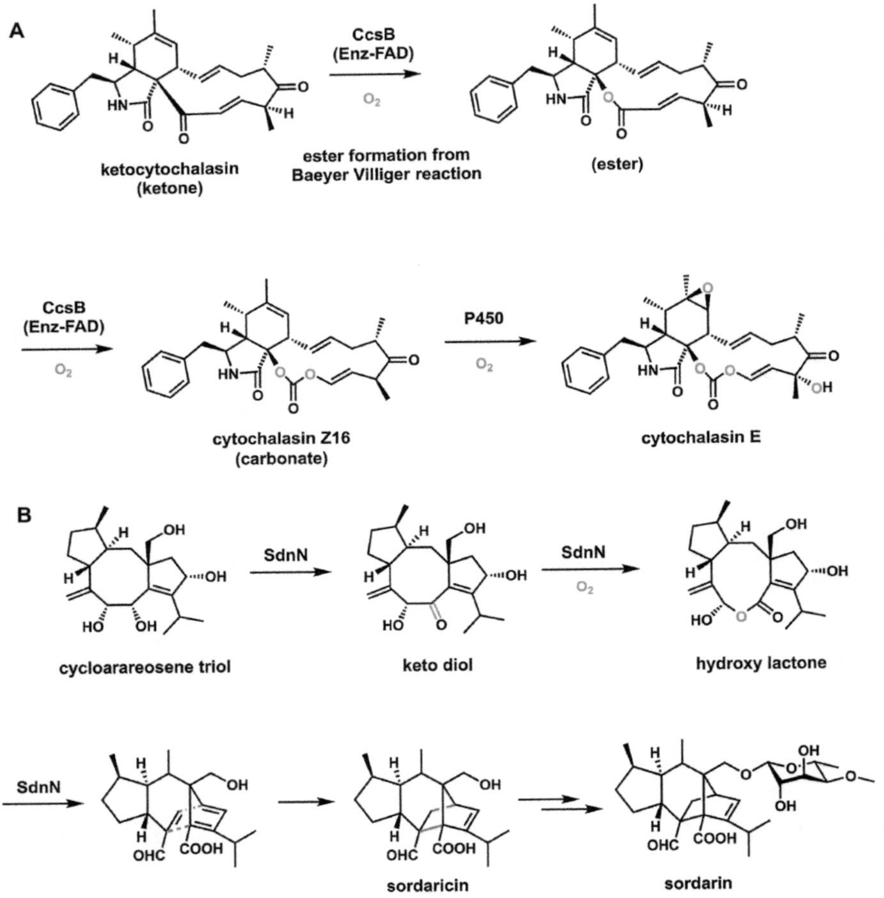

Figure 9.15 (A) A tandem FMO catalyzed oxygen insertion creates a vinyl carbonate ester in cytochalasin Z16 *en route* to cytochalasin E; (B) a multifunctional flavoenzyme SdnN is proposed to catalyze the conversion of cycloaraeosene triol to sordaricin.

In turn this reflects the inability of the FAD-4a-OOH to abstract a hydrogen atom from an unactivated C–H bond. This is a significant limitation to the biosynthetic niches in which flavoprotein oxygenases can function.

9.5.2 Iron-based Oxygenases

The heavy artillery for biological oxygenation is provided by the iron-based monooxygenases, in both heme and nonheme varieties. Both

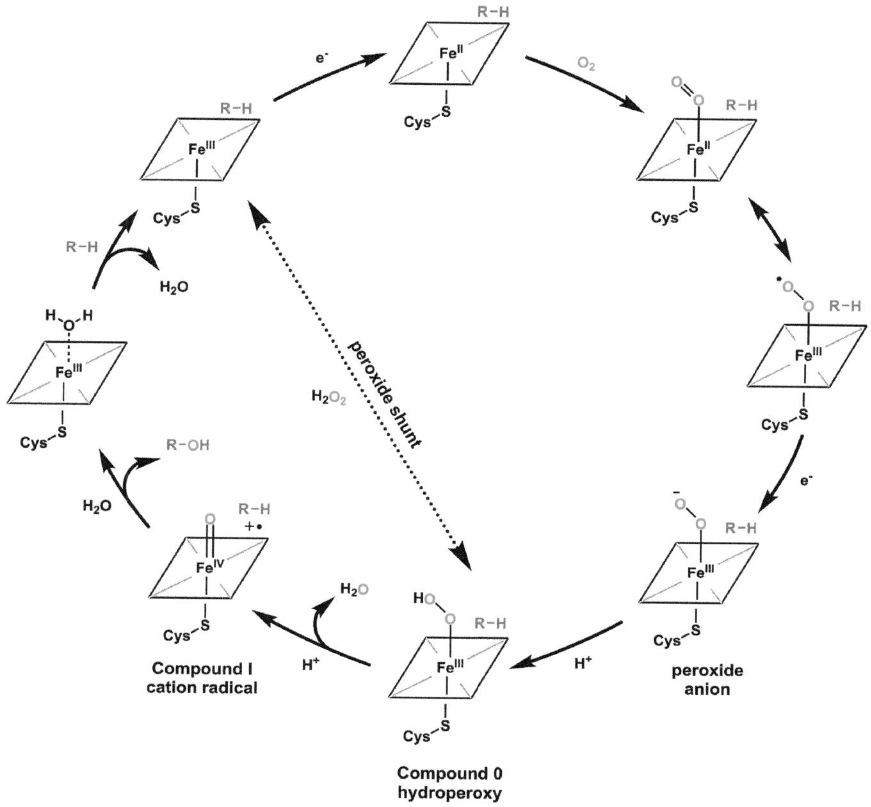

Figure 9.16 A prototypic catalytic cycle for a cytochrome P450 monooxygenase. The high valent oxo-iron oxygen transfer catalyst is pictured as the Fe(IV) porphyrin cation radical. It is a resonance contributor to an Fe(V)=O structure.

subclasses have been studied extensively for how electrons are passed from iron to liganded O_2 and for the reaction with bound substrates. Figure 9.16 shows a schematic for a canonical cytochrome P450 reaction cycle (Ortiz de Monteallano 2015). One electron from the feeder Fe/S cluster protein reduces the resting ferric heme to the ferrous Fe(II) oxidation state and O_2 becomes a top axial ligand. The Fe(II)–O_2 complex can be in equilibrium with an Fe(III)–superoxide form where one electron has passed from iron to oxygen. Transfer of a second electron in from the feeder system gives Fe(III)-peroxide anion. This is the species which may carry out P450-mediated Baeyer–Villiger transformations as noted above. An example is given in Figure 9.17

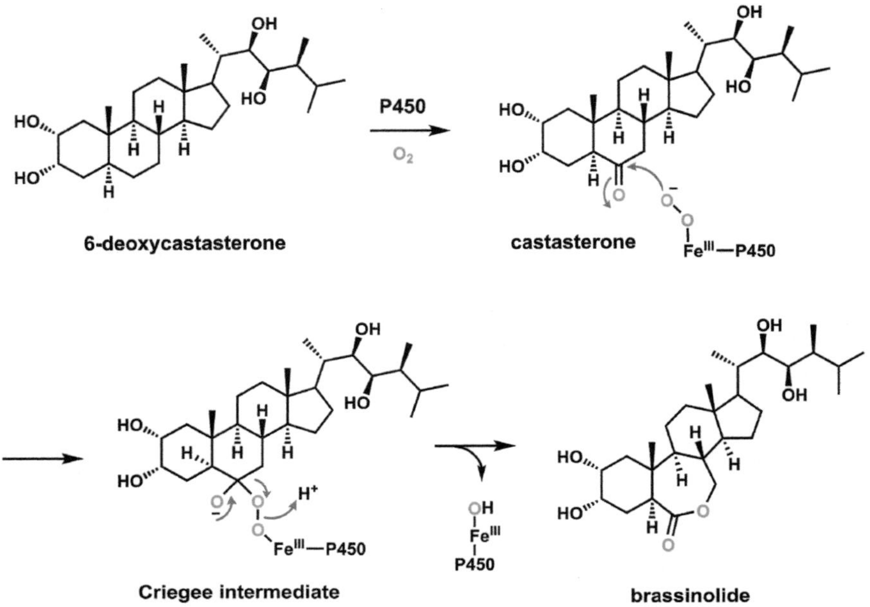

Figure 9.17 Cytochrome P450s in *Arabidopsis* and tomato secondary metabolism: tandem conversion of the B ring of the triterpene 6-deoxycastasterone to the ketone and then Baeyer–Villigerase expansion to the seven-member lactone.

from castasterone metabolism. After the 6-deoxyprecursor is hydroxylated twice to yield the 6-keto group in castasterone, this undergoes P450-mediated Baeyer–Villiger expansion of the ketone to the lactone in the pathway end product brassinolide.

Loss of water from the Fe(III)–O–OH, with cleavage of the weak O–O single bond, would generate a high valent oxo-iron species (compound I), retaining one of the two original oxygens from O_2. That can be written as an Fe(V)=O or as the Fe(IV)=O porphyrin cation radical shown, with stabilization from electron delocalization throughout the conjugated heme tetrapyrrole macrocycle. This high valent oxo-iron and the related one from the nonheme iron enzymes discussed below are strongly oxidizing agents. They can readily abstract a hydrogen atom (H$^•$) from unactivated sp^3 carbon atoms of bound substrate molecules. This would produce Fe(III)–OH and the corresponding substrate carbon radical (Figure 9.18). For a simple hydroxylation outcome, the carbon radical could be quenched by [OH$^•$] transfer. This can be sufficiently rapid to be described as an "oxygen rebound"

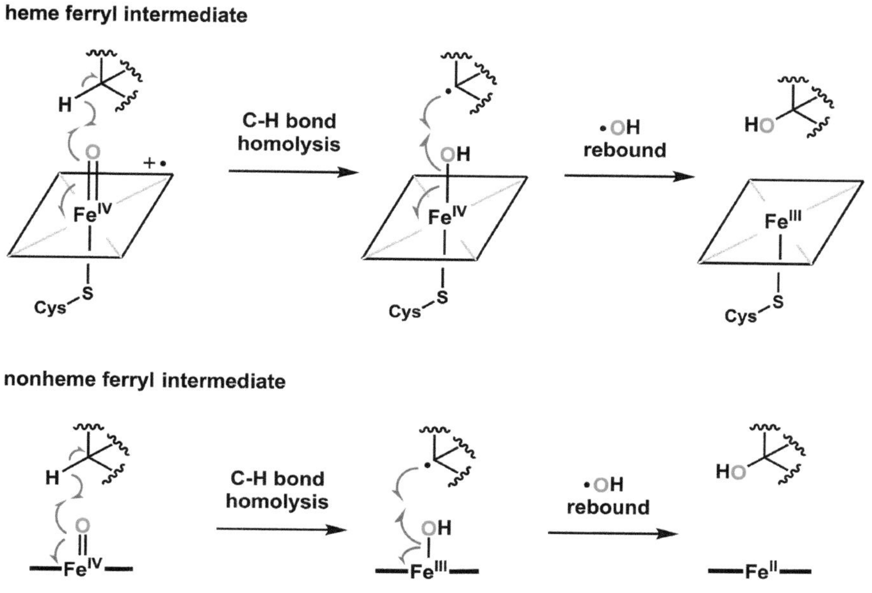

Figure 9.18 A typical iron-based oxygenase cycle involves substrate C–H bond homolysis and then $OH^•$ radical rebound.

event. However, we highlight several examples from previous chapters in which the carbon radical generation can be uncoupled from subsequent capture by [$OH^•$] transfer and intramolecular events can outcompete (intermolecular) hydroxylation.

The P450s have great catalytic scope and range of cosubstrate functional groups oxygenated. These include aromatic ring hydroxylation, *via* arene oxide intermediates, olefin epoxidations, and *N*- and *S*-oxygenations (Ortiz de Monteallano 2015) (Figures 9.19 and 9.20). P450s are the most abundant form of iron oxygenases in plant secondary metabolic pathways. A total of about 250 P450s are proposed to be encoded in the *A. thaliana* genome and some 400 in the wild rice genome (Bak, Beisson *et al.* 2011).

The chemical logic for generation of a high valent iron oxide in the mononuclear nonheme iron oxygenase subclass is related to the P450 reaction sequence. The iron-ligated α-ketoglutarate gets oxidatively decarboxylated to CO_2 and succinate while serving as the source of electrons for fragmenting O_2 and producing an Fe(IV)=O. In a formal sense this is one electron less oxidized than the P450 Fe(V)=O but is still potent enough to carry out homolysis of unactivated C–H bonds

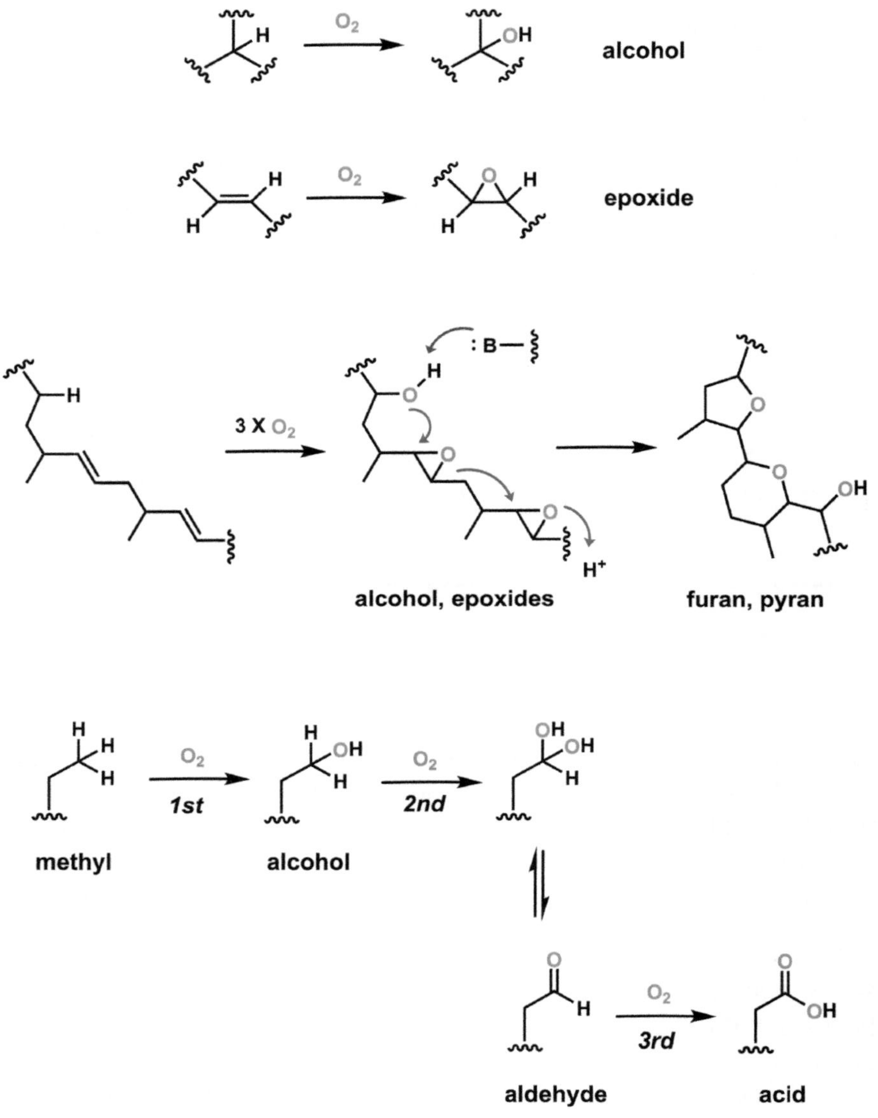

Figure 9.19 Oxygenases create a diversity of functional groups in natural product scaffolds (part 1).

in bound cosubstrates (Solomon, Decker *et al.* 2003). This starts an analogous mechanistic path to carbon centered radicals that could be captured by [OH•] rebound to yield hydroxylated products (Figure 9.18).

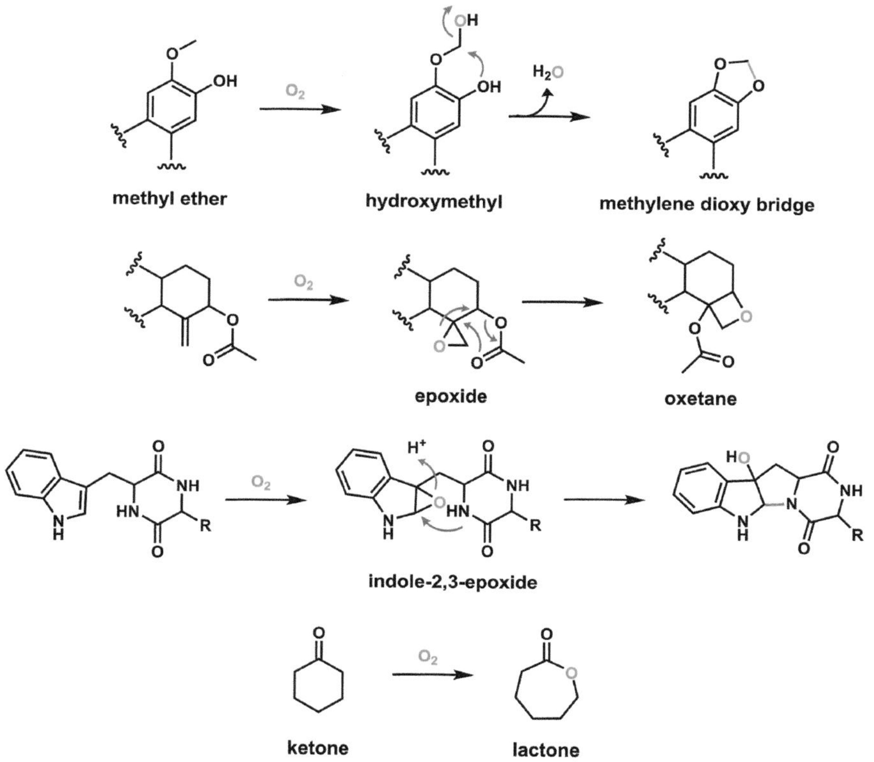

Figure 9.20 Oxygenases create a diversity of functional groups in natural product scaffolds (part 2).

9.6 Oxygenases in Specific Natural Product Pathways

In contrast to the very limited use of oxygenases in primary metabolic pathways, oxygenases are critical tailoring and maturation catalysts in all the natural product classes examined in Chapters 2–8 of this book. Each of the half dozen natural product classes relies on key oxygenation events to attain the final framework and array of functional groups essential to biologic activity. We first note examples where the oxygen incorporation into product structures has been completed. Then we separately turn to the consequential subgroup of P450s in which substrate radicals are generated but they are *rerouted to intermediates and products that do not complete the oxygenation process.* That set is all about the carbon radical generation strategies.

9.6.1 Polyketides

Polyketide-derived natural products, depending on the particular "on assembly line" tailoring enzyme inventory, can in principle retain one oxygen atom in each elongation cycle. The oxygens in the cores of both macrolide and polycyclic aromatic scaffolds arise from the acyl thioester carbonyls during assembly line action. A scan of erythromycin and the DEBS assembly line, for example, reveals that six of the eight oxygen atoms in this macrolide antibiotic have that origin. The other two are incorporated by post assembly line P450 oxygenases, selective for introducing the 6-OH (Andersen, Tatsuta *et al.* 1993) and the 12-OH groups (Lambalot, Cane *et al.* 1995), by homolytic, one electron pathways (see Figure 2.23).

Figure 9.21 indicates that two hemeprotein cytochrome P450 enzymes deliver an epoxide oxygen to the substrate double bond in pimaricin epoxidation (Kells, Ouellet *et al.* 2010) and epothilone epoxidation (Tang, Shah *et al.* 2000, Kern, Dier *et al.* 2015). These dedicated tailoring P450s use substrate olefin π-electrons as the attacking species. In this figure the Fe(III)–O–O$^-$ rather than Fe(V)=O was proposed by the authors as the proximal oxygen transfer agent.

As a third type of example we have noted above the Lsd18 flavoenzyme-mediated epoxidation of the bis-olefin released from the lasalocid assembly line and the tri-olefin from the monensin assembly line. Opening of these epoxides in a cascade by the epoxide hydrolase Lsd19 sets the furan/pyran framework characteristic of this polyether class (see Figure 2.36).

One more example of mechanistic note is in late stages of aurovertin E assembly (Figure 9.22). Eight chain elongations of starter unit propionyl-CoA yield the hexa-olefinic diketo linear acyl chain tethered to the ACP domain. Cyclizing chain release yields the hexa-olefinic pyrone. It is then proposed that the last of the three *E*-double bonds in the terminal triene portion is isomerized to the *Z*-configuration olefins by the kind of mechanism we discussed in Chapter 4 (see Figure 4.33) for phytoene. At this juncture, bis-epoxidation by P450s of the two terminal olefins and water-initiated ring openings would yield the dihydroxyfuran (echoes of the lasalocid, monensin cascades) (Mao, Zhan *et al.* 2015). A third iteration of olefin epoxidation by AurC would set up intramolecular opening of that epoxide by one of the –OH groups in a proposed 6-*endo*-tet cyclization regiochemistry. The triple epoxidation reaction manifold creates the

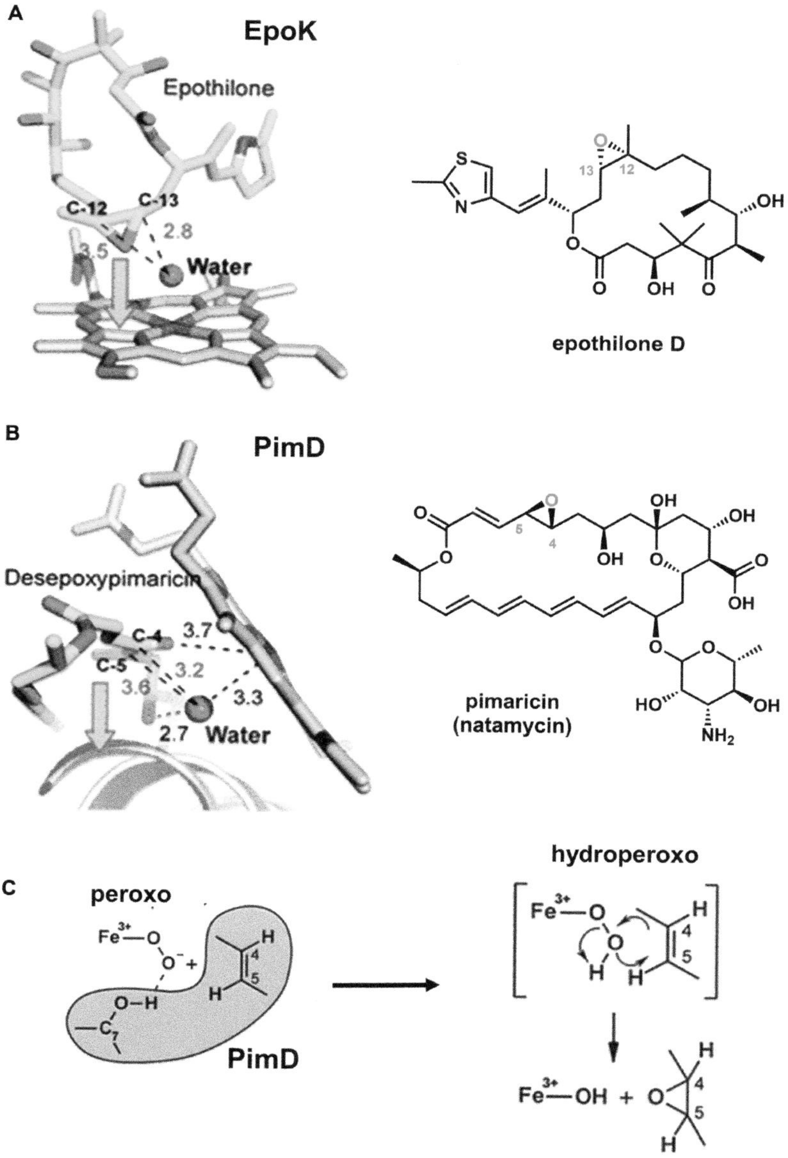

Figure 9.21 Epoxide formation by P450s in the (A) epothilone and (B) pimaricin pathways. (C) Mechanism of epoxide formation in PimD. Enzyme structures reproduced from Podust, L. M. and D. H. Sherman (2012). "Diversity of P450 Enzymes in the Biosynthesis of Natural Products". *Nat. Prod. Rep.* **29**: 1251 with permission from The Royal Society of Chemistry.

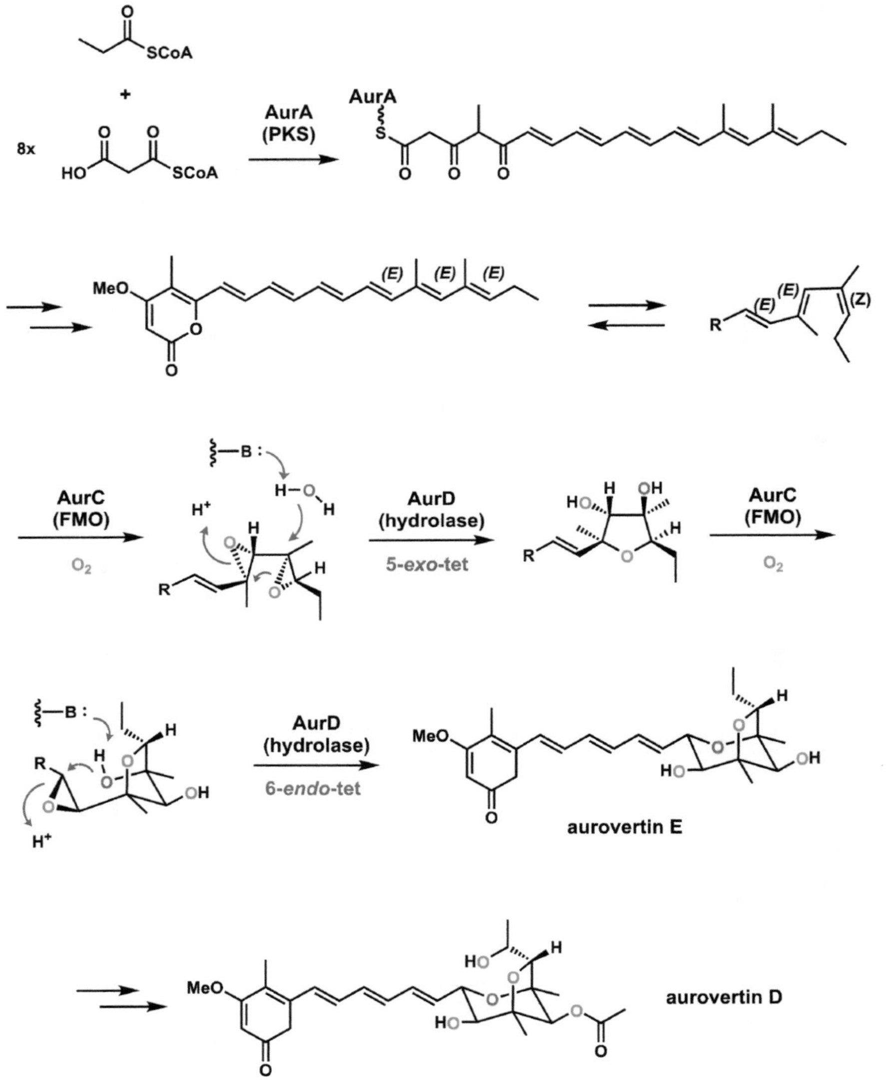

Figure 9.22 Oxygenative tailoring of the terminal triene to a 2,6-dioxabicyclo[3.2.1]octane in aurovertin formation: three successive "disappearing" epoxidations. FMO: flavin monooxygenase.

2,6-dioxabicyclo[3.2.1]octane ring in aurovertin E: efficient complexity generation by tandem oxygenations. These examples are elegant illustrations of "disappearing" epoxide groups in morphing polyketide scaffolds.

9.6.2 Nonribosomal Peptides

A number of β-hydroxy amino acid residues are found in nonribosomal peptides, including β-OH-His in nikkomycin (Chen, Thomas *et al.* 2001), β-OH-Tyr in novobiocin and vancomycin (Cryle, Meinhart *et al.* 2010), β-OH-Trp in echinomycins, β-OH-Pro in echinocandins, and β-OH-Phe, β-OH-O-methyl-Tyr, and β-OH-Leu in skyllamycin. These are generated, after loading as aminoacyl-*S*-peptidyl carrier proteins, by action of classical P450 hydroxylation catalysts. For the *Streptomyces* cyclic undecapeptidolactone skyllamycin (Pohle, Appelt *et al.* 2011), an inhibitor of platelet derived growth factor signaling in eukaryotic cells (Figure 9.23), one P450 services three amino acid residues, Phe_5, *O*-methyl-Tyr_7 and Leu_{11}, to hydroxylate each of these on the growing undecapeptidyl assembly line (Uhlmann, Sussmuth *et al.* 2013) at the respective β-carbons, an unusual case of both timing and promiscuity from a single oxygenase (Pohle, Appelt *et al.* 2011).

9.6.3 Isoprenoid Scaffolds

As we noted in Chapter 4, the prenyl chain elongation reactions of five-carbon units of nucleophilic Δ^3-prenyl-PP undergoing addition to elongating chains of electrophilic Δ^2-prenyl cosubstrates yield products that are highly hydrophobic. If the –OPP chain has been jettisoned at the end, the products are the apolar hydrocarbons. Plants induce a number of committed oxygenases to add polar, water-solubilizing oxygen functionalities. Several cases are instructive.

From a therapeutic perspective the biosynthesis of taxol is of special interest. The C20 linear geranylgeranyl-PP is converted, *via* a series of cation rearrangements, to a tricyclic scaffold. Final quenching by proton removal yields the C20 bis olefin taxadiene. A set of incompletely characterized P450 oxygenases (Croteau, Ketchum *et al.* 2006, Kaspera and Croteau 2006) then go to work to render this scaffold more polar and water soluble. In total, eight of the oxygen atoms (shown in red in Figure 9.24) are introduced by oxygenase action. Four of those are hydroxyl groups that are then acylated by acetyl, benzoyl, and an α-hydroxy-β-amino homo-Phe acyl group. Of the remaining four oxygens introduced from O_2, one is at the level of a ketone and one as an unusual four-membered oxetane. This latter ring is proposed to form from olefin epoxidation and acetyl migration as shown (Dewick 2009).

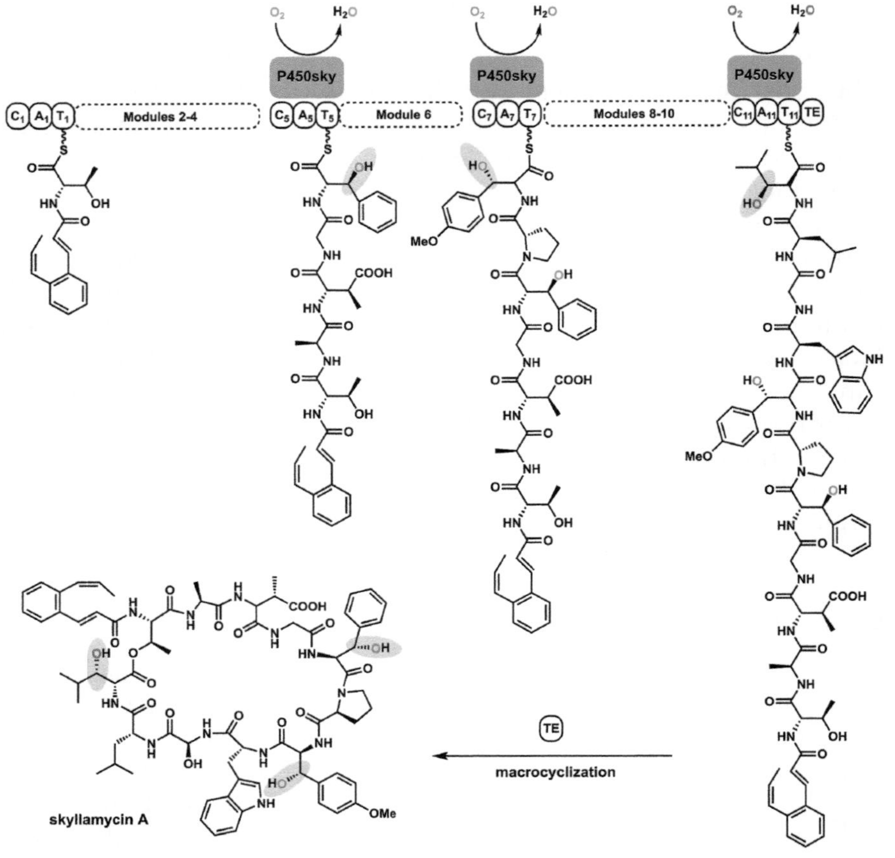

Figure 9.23 The cyclic undecapeptidolactone skyllamycin undergoes "on assembly line" β-hydroxylation at three different residues by one iron-based oxygenase.

Another metabolic niche where a burst of oxygenase action occurs on an isoprene substrate is in the transformation of geraniol to secologanin. Four oxygenases go to work sequentially. The highly oxidized secologanin is then condensed with tryptamine by the Pictet–Spenglerase strictosidine synthase and the progenitor of thousands of indole terpenes is set loose for further metabolism. The mechanistically notable transformation is the last one, cleavage of loganin to the aldehyde and olefin functionality in the secologanin product, which can be written as a radical induced fragmentation of the hydroxypentane carbacycle (Mizutani and Sato 2011) (see Figure 4.39).

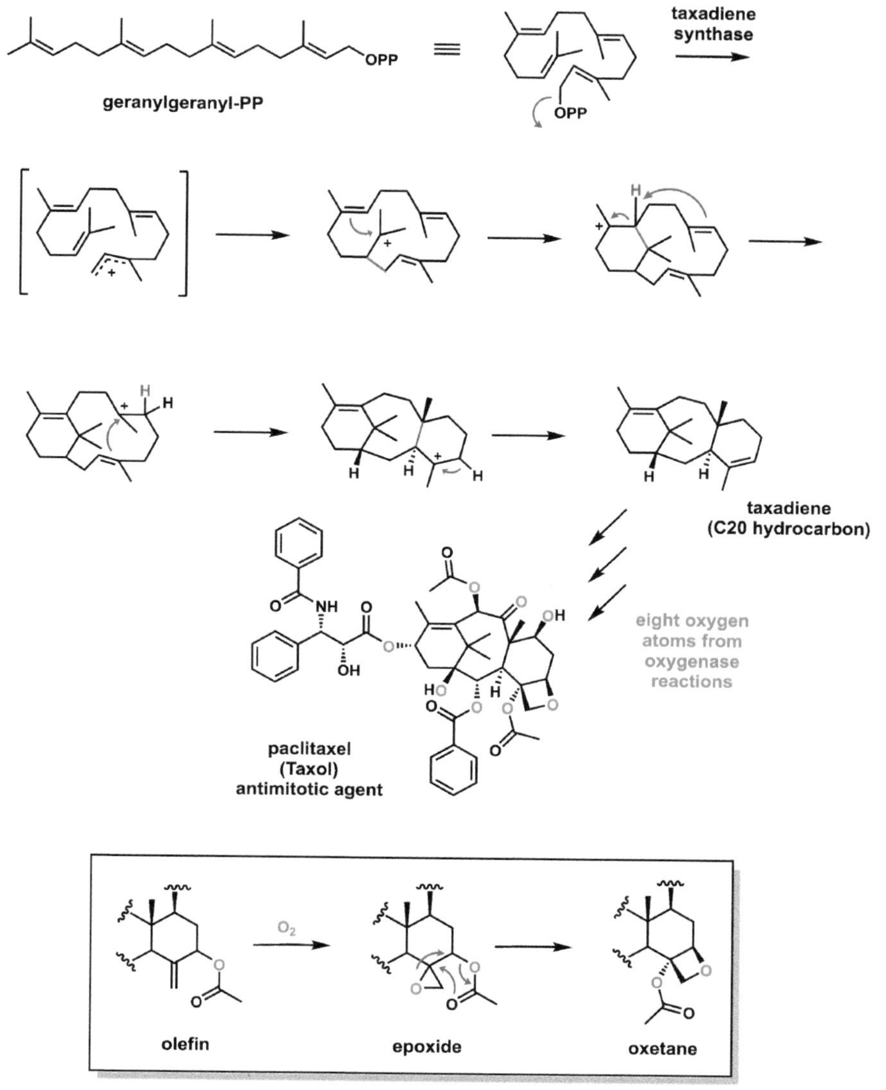

Figure 9.24 Introduction of eight oxygen atoms to the taxadiene scaffold on the way to taxol. Note the proposed conversion of olefin to epoxide to oxetane.

In the plant metabolism of campesterol to the root hormones of the brassinosteroid class, there are six oxygenations of the tetracyclic triterpene scaffold (Figure 9.25). Specific plant P450s have been identified for the oxygenation steps, including the Baeyer–Villiger cyclic ketone to ring-expanded lactone. This is a deep commitment to

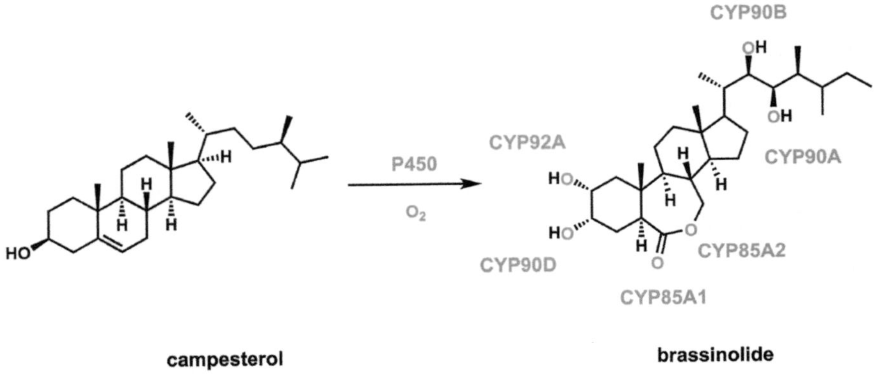

Figure 9.25 A set of six P450-type oxygenases convert campesterol to the framework of brassinosteroid root hormones. There are five hydroxylations and one ketone to lactone ring expansion step.

oxygenative terpene maturation to increase the polarity of the hydrocarbon framework and optimize interaction with the gene regulatory machinery (Mizutani and Ohta 2010).

An additional set of multiple epoxidations of terpenoid/isoprenoid frameworks shows how chemical rearrangements can be driven by this prior oxygenation strategy. There are four oxygenation steps *ad seriatim* in fumagillin biosynthesis (Figure 9.26) by *Aspergillus fumigatus* cells. The sesquiterpene precursor farnesyl-PP can be converted into the indicated 5,6-bicyclic hydrocarbon which then undergoes processing by one P450 enzyme Af510. The first O_2-consuming cycle creates the bridgehead alcohol, unexceptionally. The second cycle fragments the bicyclic skeleton and yields the indicated epoxyketone monocyclic product, *via* a suggested radical route. Then Af510 acts a third time to introduce a second epoxide on the newly generated exocyclic methylene group. This scaffold is one additional hydroxylation away from the natural product ovalicin. One more hydroxylase action is required, followed by *O*-methylation, ketone reduction, and diacyl side chain addition, to yield fumagillin (Lin, Chooi *et al.* 2013, Lin, Tsunematsu *et al.* 2014).

One final example in the terpene metabolite world of mechanistic and structural interest occurs in the biosynthetic pathway to the gibberellic acid family of plant hormones (Zi, Mafu *et al.* 2014). We have noted the formation of *ent*-kaurene from farnesyl-PP in Chapter 4 (Figure 4.16). In further processing, one of the A ring methyl groups gets oxidized up to the carboxylate (Figure 9.27). This is characteristic of the reactions that also happen as lanosterol goes to cholesterol at

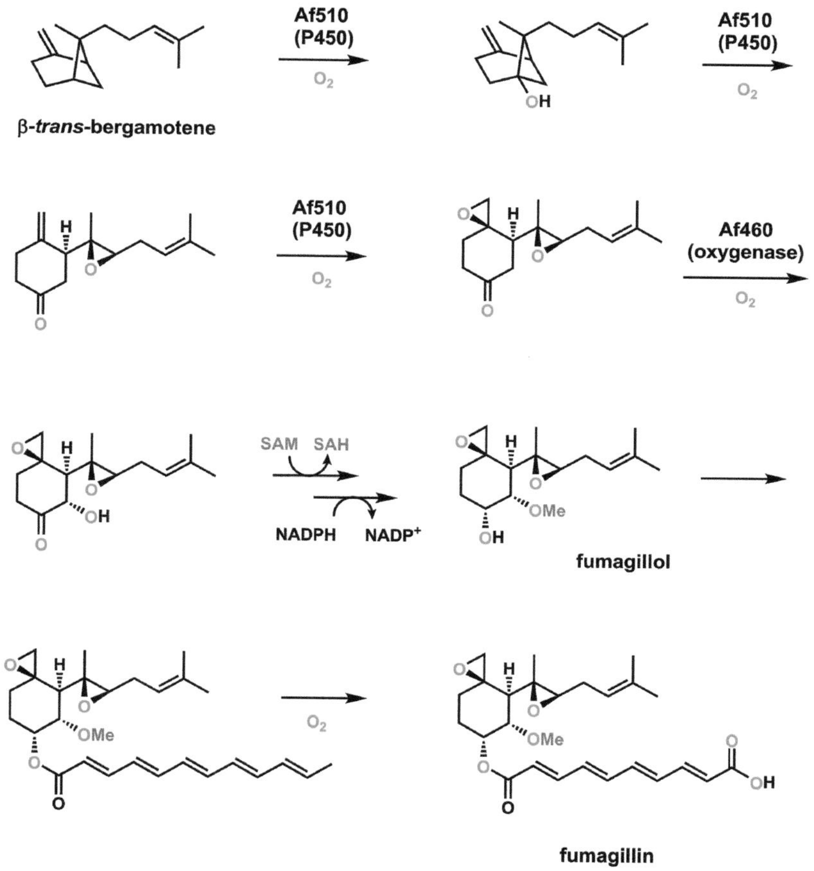

Figure 9.26 Two oxygenases suffice to convert a bicyclo [3.2.1] sesquiterpene to the bis-epoxy cyclohexanol core (fumagillol) of fumagillin.

the same carbon center and is mediated by three consecutive P450 cycles (–CH_3 to CH_2OH, to $CH(OH)_2$ = CHO, to COO)(Morrone, Chen *et al.* 2010). But the particular point to note here is the subsequent reaction catalyzed by CYP88A in which the hydroxylated, six-membered B ring is oxygenatively converted to the five-membered aldehyde (Helliwell, Chandler *et al.* 2001). Radical chemistry is proposed to account for the ring-contracting, C–C single bond migration to a product-like radical which then, and only then, undergoes the [OH•] rebound. The chance for intramolecular radical rearrangement to occur before OH• transfer captures the initial radical indicates that carbon-centered radicals can have a finite existence in P450 enzyme

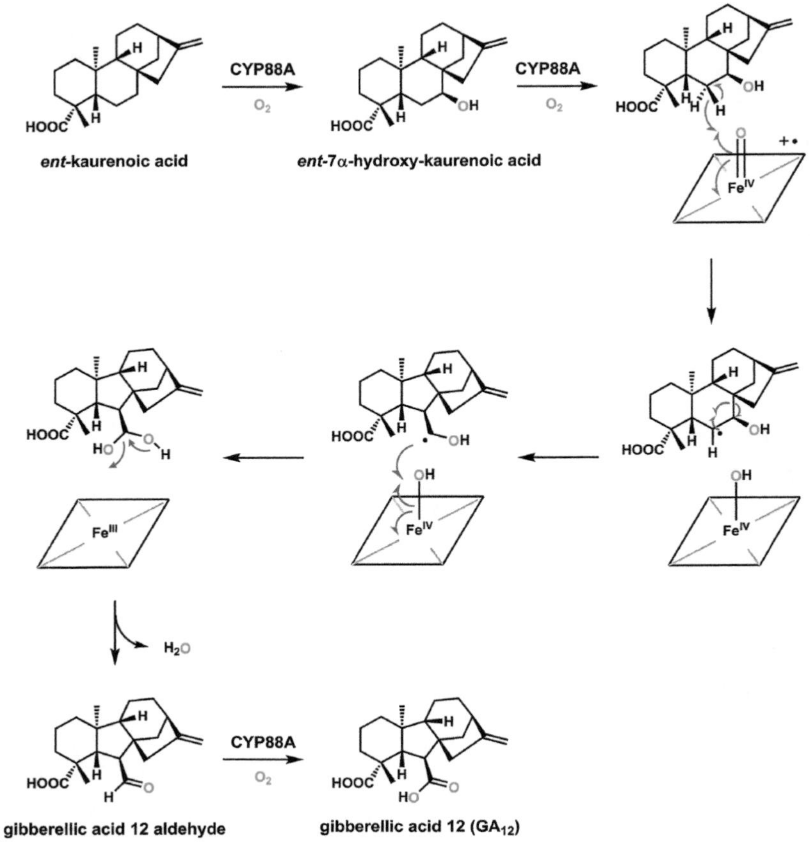

Figure 9.27 *ent*-Kaurenoic acid can undergo P450-mediated hydroxylation to the alcohol and then again to the aldehyde on the way to the dicarboxylate gibberellic acid 12. Concomitant with oxidation of alcohol to aldehyde is a radical-based rearrangement that leads to ring contraction and extrusion of one ring carbon as the exocyclic aldehyde.

active sites and is prelude to a whole set of P450 and nonheme iron enzymes in the next section where the carbon radicals never get captured by [OH•].

9.6.4 Phenylpropanoids

Oxygenases, acting explicitly to introduce alcohol functionalities, or cryptically to create substrate radicals which do intramolecular chemistry without completion of an oxygen transfer [OH•], are prevalent in multiple arms of phenylpropanoid biosynthetic pathways (Mizutani and Sato 2011). Phenylalanine-derived carbon flux both to lignans and most abundantly to the lignin insoluble structural polymers is directed down that pathway arm by three successive cytochrome P450-directed hydroxylations in what started as the phenyl ring of phenylalanine.

The first, the 4-hydroxylase, occurs on cinnamate as the free acid, the second, the 3-hydroxylase, on the coumaryl-shikimate oxoester, and the third (the 5-hydroxylase) at the level of coniferaldehyde (Figure 9.28). The trihydroxyphenol is never actually generated because regiospecific methyl transferases intervene such that only the 4-hydroxyl group is ever free. This regiochemistry is no doubt strategic to control/influence subsequent radical dimerizations and oligomerizations as the three monolignols (coumaryl alcohol, coniferyl alcohol, sinapyl alcohol) are copolymerized.

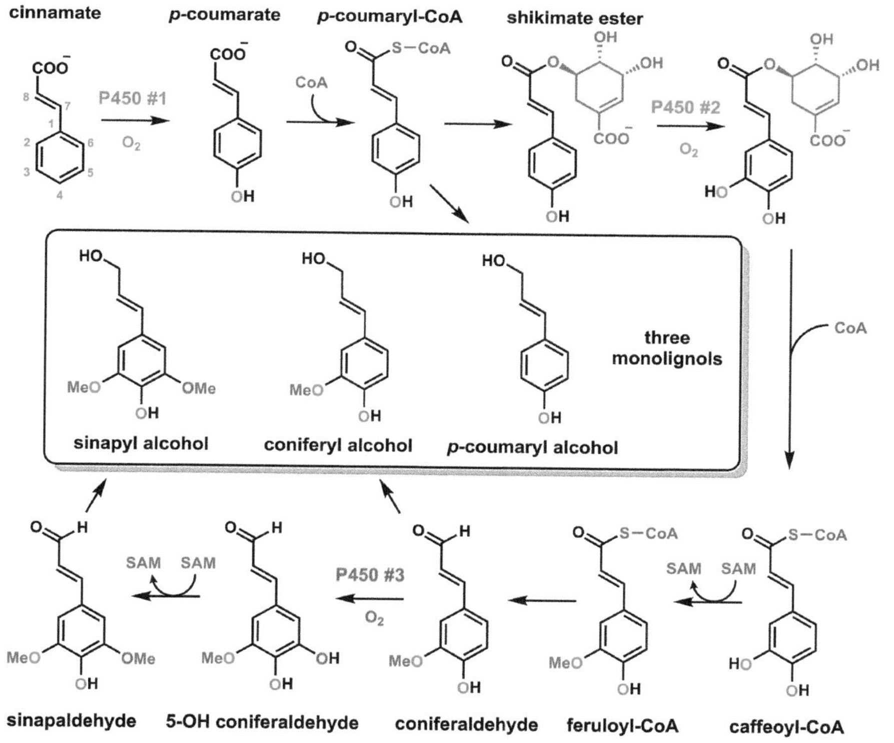

Figure 9.28 Cinnamate to the three monolignols: Three P450s work on three distinct forms of the aryl substrates to oxygenate at C_3, C_4, and C_5 of the phenyl ring.

One of the best studied lignans is pinoresinol. In sesame seeds it is precursor to sesamin (Figure 9.29) which can function as a defensive phytoalexin molecule (Kim, Ono *et al.* 2009). Sesamin differs from pinoresinol in having two methylene dioxy bridges replacing the 3-*O*-methyl-4-hydroxy substituents on the original phenyl ring. Although no oxygen atoms from O_2 show up in sesamin, the transformations

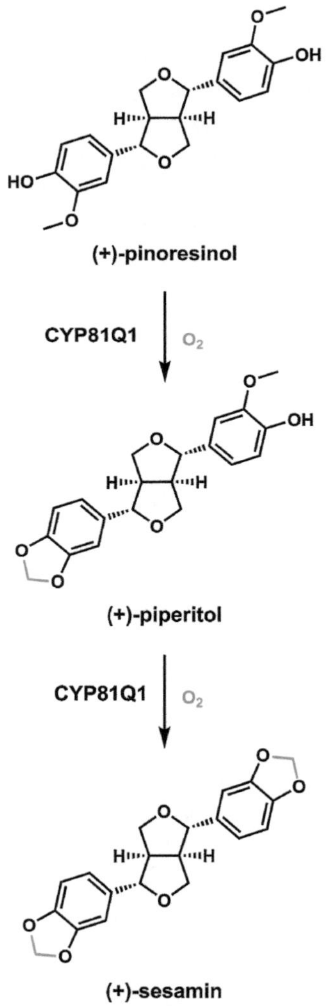

Figure 9.29 Pinoresinol processing to sesamin in sesame seeds: generation of two methylenedioxy bridges *via* P450 oxygenase action.

are mediated by a classic P450 enzyme, CYP81Q1 (Mizutani and Sato 2011). Oxygenation may well have occurred, but is cryptic in the observed product and the intermediate piperitol. Hydroxylation of the aryl–O–CH_3 methyl carbons (*via* radical intermediates) would yield the –OCH_2OH as nascent product. This could decompose unimolecularly in a net dealkylation reaction to formaldehyde and the catechol. But this is not observed. Instead it appears that the kinetically favored outcome is neighboring group attack by the 4-OH with expulsion of the just added –OH group to produce the methylenedioxy five-membered ring. This must happen on both ends of pinoresinol in sesamin formation. (It is possible that oxygen transfer to the 3-O–CH_3 does not actually proceed: that the reaction manifold gets to the 3-O–$CH_2^{\bullet}$ and the 4-$O^{\bullet}$ phenyl radical by two single electron removal events and those couple directly.)

An intriguing example of the consequence of epoxidation, to add to the accumulating list, is the proposal for formation of the key intermediate chanoclavine-I in the lysergic acid biosynthetic pathway (see Figure 5.21) (Herbert 1989). The mechanistic proposal is that epoxidation of the terminal olefin of a diene species leads to decarboxylation and use of the resultant carbanion in a vinylogous opening of the epoxide (Kozikowski, Chen *et al.* 1993). This reaction sequence creates the tricyclic core in chanoclavine.

9.6.5 Alkaloids: Ring Cleaving Oxygenases in Camptothecin, Quinine, and Cinchonidine Biosynthesis

Although the requisite genes and encoded enzymes for camptothecin, quinine, and cinchonidine formation from strictosidine are not yet characterized, chemically reasonable biosynthetic schemes suggest the participation of oxygenases that cleave rings in the core (Dewick 2009).

In the camptothecin pathway, the proposal would be a dioxygenative cleavage across the 2,3 double bond of the pyrrolic moiety of the indole (to give *N*-formyl kynurenine). Such indole 2,3 dioxygenases are known in other metabolic contexts and would give the pair of carbonyls disposed across the new macrocyclic ring from each other (see Figure 5.19). Reclosure is postulated to convert the 6-5-6 tricyclic core of the carboline moiety of strictosidine to the 6-6-5-distinct tricyclic core of pumiloside (and subsequently camptothecin). Such dioxygenases, as well as the one in carotene fragmentation, are thought to proceed *via* four-membered dioxetane.

The proposed oxygenative fragmentation in the core of deglucosylated strictosidine in the quinine/cinchonidine pathway would not be at the pyrrole ring of the indole but rather the adjacent six-membered ring, as noted in Figure 5.20. That could be a monooxygenase reaction. The resulting carbinolamine could unravel to the indicated acyclic aldehyde and the amine, which is then the nucleophile in transannular capture of the other aldehyde, on the way to imine and its reduction to the characteristic bicyclic amine moiety.

9.6.6 Endoperoxide Linkages in Artemisinin and Verruculogen

Several natural products contain hydroperoxide or endoperoxide links. Probably the best known endoperoxide in mammalian metabolism is prostaglandin H_2 (PGH2), formed by the enzyme cyclooxygenase (COX)-2 (Figure 9.30). The antimalarial artemisinin is the most celebrated of the plant endoperoxides. We have noted verruculogen and fumitremorgin A in the previous chapter (see Figure 8.14). The insertion mode for the endoperoxide is not known in artemisinin and verruculogen but in PGH2 formation COX-2 most likely generates allyl radicals that combine with superoxide anion (van der Donk, Tsai *et al.* 2002). An acyclic hydroperoxide may be on the way to the PGH2 endoperoxide linkage.

Conversion of a hydroperoxide to an endoperoxide may also be on path in artemisinin assembly. Figure 9.31 shows a possible biosynthetic route from the sesquiterpene amorpha-4,11-diene and its oxygenative conversion, presumably occurring in three oxygenase-mediated steps, of the angular methyl to the carboxylate of artemisinic acid. Yet another oxygenation could yield the indicated allylic hydroperoxide. Ring expansion with weak O–O bond cleavage could lead to formation of a seven-membered hemiacetal which would be in equilibrium with the indicated keto aldehyde. Another oxygenase-mediated formation of a hydroperoxide could then engender intramolecular attack on the ketone, producing the hydroxy endoperoxide. The hydroxyl group generated in the resultant tetrahedral adduct could set up an acetal/hemiacetal cascade to yield the heavily oxygenated artemisinin tricyclic framework.

We have speculated on endoperoxide formation in verruculogen/fumitremorgin A assembly in Figure 8.14. If O_2 rather than H_2O_2 is the cosubstrate then radical based intermediates from the two olefins could couple to both ends of a reduced oxygen molecule.

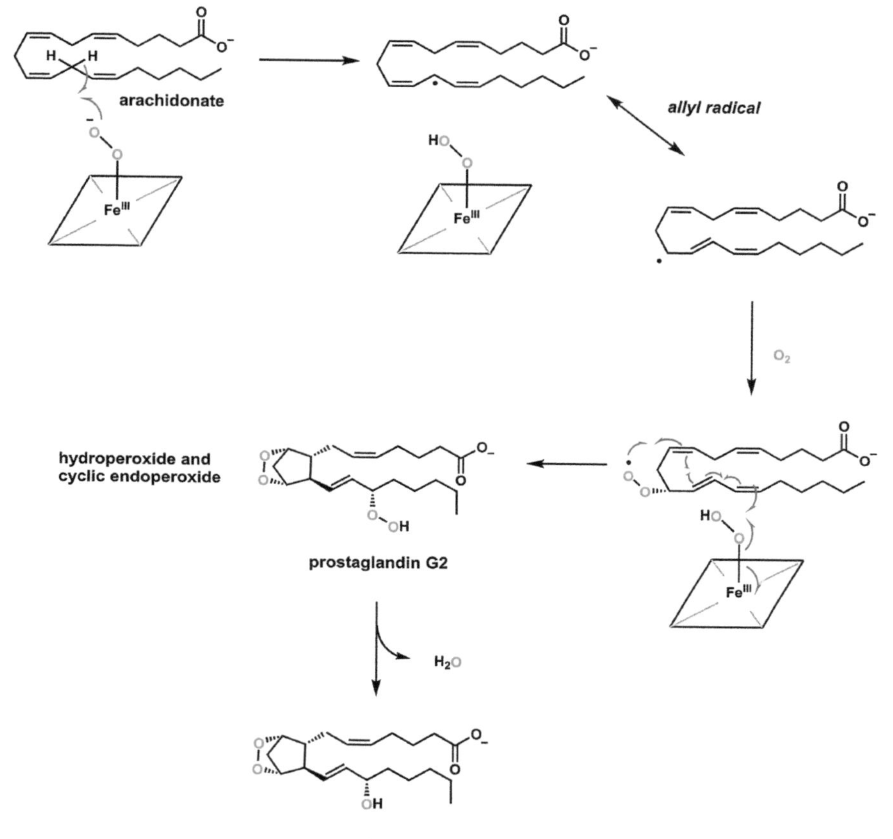

Figure 9.30 Prostaglandin cyclooxygenase provides a metabolic and mechanistic precedent for an iron-based hydroperoxidase and endoperoxidase.

9.7 Uncoupling of Carbon Radicals from OH Capture: Sidelight or Central Purpose of Natural Product Biosynthetic Iron Enzymes?

A large subset of iron-based oxygenases that play crucial roles in natural product biosynthetic pathways never complete the oxygenation half-reaction (Mizutani and Sato 2011). As we shall note in the examples below, some collected from previous chapters, these enzymes, both heme and nonheme iron catalysts, generate the *high*

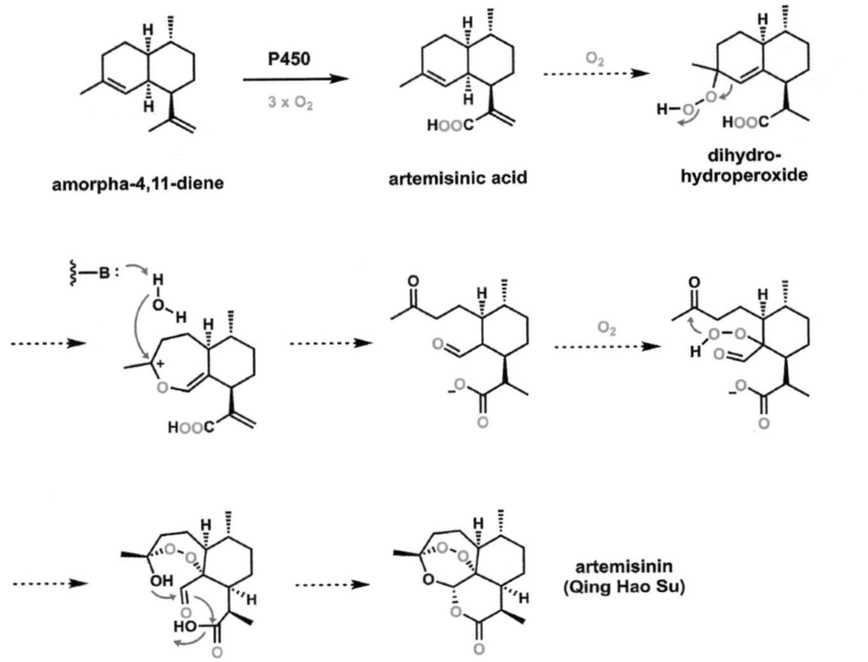

Figure 9.31 Proposal for insertions of oxygen into the amorphadiene scaffold to yield the antimalarial agent artemisinin.

valent oxo-iron species that are the key oxidants for homolytic cleavage of C–H bonds of bound substrates. However, the subsequent carbon radicals undergo internal reactions rather than capture iron-bound hydroxyl radical equivalents to complete the oxygenation half reaction. Consequently, although O_2 is consumed, producing two molecules of water, *no oxygens get incorporated into products.*

9.7.1 Polyketides

One such P450 example from the polyketide family occurs in generation of the key spirocycle during griseofulvin maturation. Figure 9.32 shows that a pentaketonyl-*S*-ACP intermediate undergoes aldol and Claisen intramolecular cyclizations to yield the bicyclic pentahydroxy product. Subsequent chlorination (the second half of this chapter delves into such halogenations) and bis *O*-methylations set up the substrate form that undergoes P450-mediated spirocyclization.

The generation of the Fe(v)=O oxidant is the key to the first half reaction, which generates a phenoxy radical in the substrate. While one could formulate a second catalytic cycle, without [OH•] transfer occurring in either of the two catalytic cycles to quench the carbon radicals which could then recombine, a more straightforward reaction sequence is shown in Figure 9.32. The phenoxy radical generated by GsfF can couple as indicated to yield the spirocycle with a *para*-quinoid radical. One electron transfer yields the desmethyl-dehydro species, a reduction and methylation away from griseofulvin. This direct radical attack route is supported by density functional theory computation as also lower in energy than an alternative epoxide route (Grandner, Cacho *et al.* 2016).

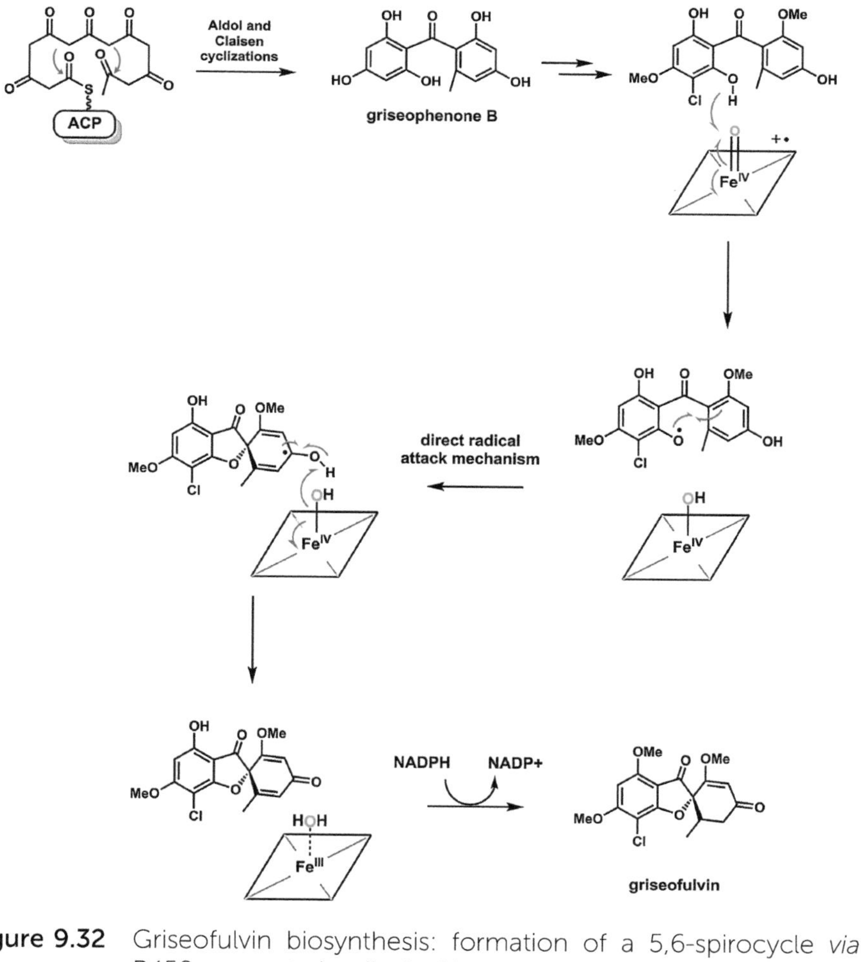

Figure 9.32 Griseofulvin biosynthesis: formation of a 5,6-spirocycle *via* P450-generated radical without intervening oxygen transfer.

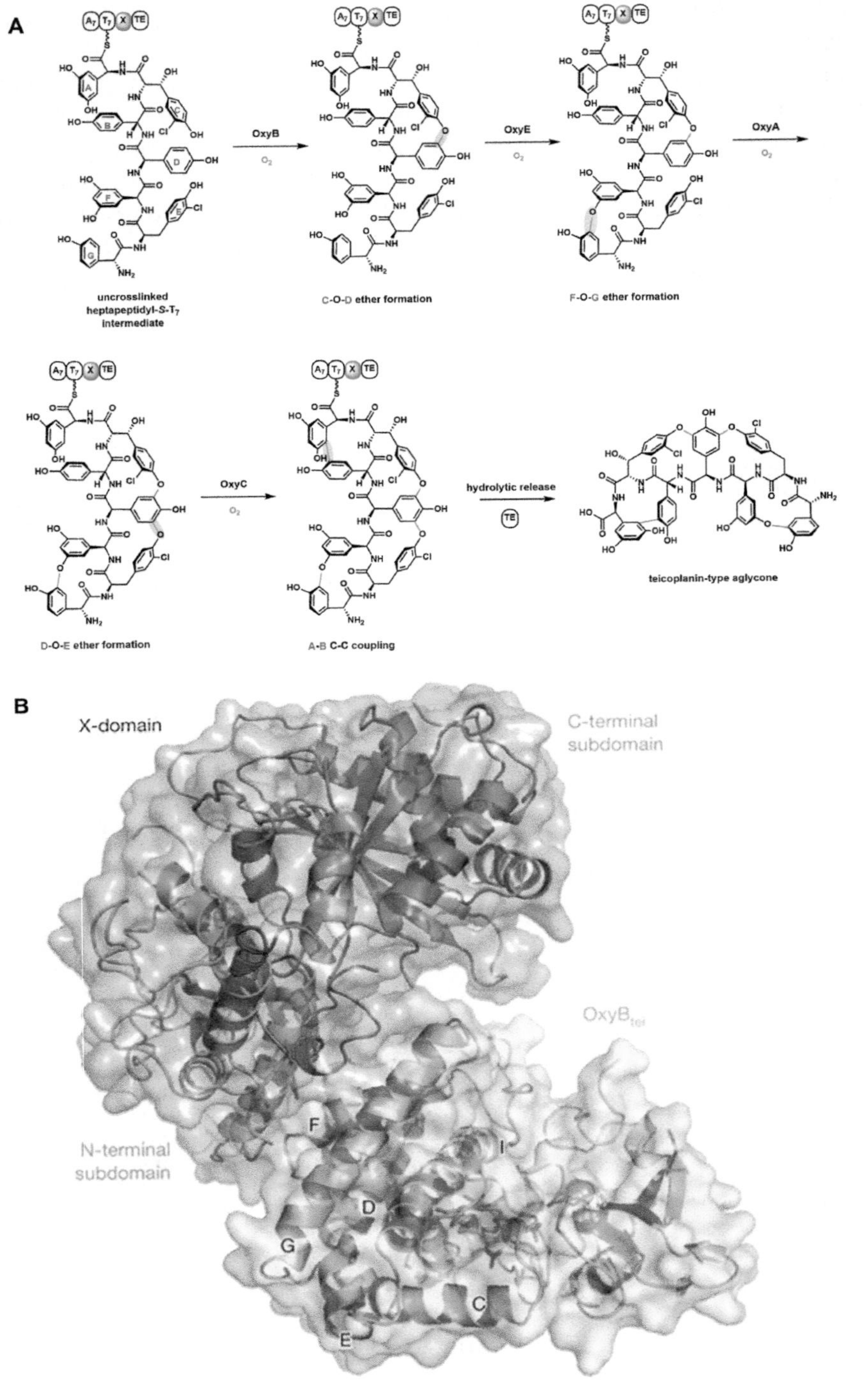
A
OxyB
O_2
OxyE
O_2
OxyA
O_2
uncrosslinked heptapeptidyl-*S*-T_7 intermediate
C-O-D ether formation
F-O-G ether formation
OxyC
O_2
hydrolytic release
TE
D-O-E ether formation
A-B C-C coupling
teicoplanin-type aglycone
B
X-domain
C-terminal subdomain
$OxyB_{tei}$
N-terminal subdomain
F
I
D
G
C
E

9.7.2 Nonribosomal Peptides

Three cases in the nonribosomal peptide chemical space illustrate comparable iron-based oxygenase strategies where the second half reaction, $OH^{\bullet}$ transfer, does not occur. The first involves action of three such cytochromes P450 (OxyA, B, C) that act consecutively to create comparable phenoxy radicals on the heptapeptidyl chain tethered to the last peptidyl carrier protein domain of the vancomycin (and congeners) synthetase assembly line (Zerbe, Pylypenko *et al.* 2002, Pylypenko, Vitali *et al.* 2003). Analogously, there are four such P450s (OxyB, E, A, C) involved in crosslinking all seven side chains of the teicoplanin heptapeptide scaffold.

Figure 9.33 diagrams the order of action of the four P450s, with construction of the 4,6 (OxyB) and then 2,4-aryl ether bonds (OxyE) between 4-OH-PheGly$_4$ to Cl-Tyr$_6$ and then 4-hydroxyPheGly$_4$ and Cl-Tyr$_2$, respectively. The third linkage introduced by OxyA connects the side chains of residues 1 and 3 in yet another aryl ether linkage. Then, the last P450 (OxyC) makes a direct C–C link between 4-OH-PheGly$_5$ and 3,5-dihydroxyPheGly$_7$. All seven residues, the Phe-Glys at positions 1, 3, 4, 5, and 7, and the chloro-Tyrs at 2, and 6, have phenol groups that are readily oxidized by one electron by the Fe(v)=O active site oxidants generated in each of the OxyA, B and C active sites. These must last long enough for phenoxy radical pairs to accumulate and then be coupled through the appropriate resonance contributors, particularly evident in the C–C bond formation between residues 5 and 7. The crosslinkages of the glycopeptide backbones are the crucial architectural constraints that enable these highly morphed peptides to block bacterial cell wall assembly and act as clinically central

Figure 9.33 Four cytochromes P450 act in tandem in crosslinking the side chains of the teicoplanin heptapeptidyl-*S*-PCP$_7$ species to rigidify the peptide scaffold *via* phenoxy radical intermediates. (A) Four crosslinking cytochromes P450 act on the teicoplanin heptapeptidyl chain while tethered on PCP$_7$. (B) The X domain in the final module of the teicoplanin synthetase assembly line is necessary and sufficient to recruit the four oxy P450s to control the timing of crosslinks. Shown is the T$_7$-X---OxyB cocomplex.

Reprinted by permission from Macmillan Publishers Ltd: Nature (Haslinger, K., M. Peschke, C. Brieke, E. Maximowitsch and M. J. Cryle (2015). "X-domain of peptide synthetases recruits oxygenases crucial for glycopeptide biosynthesis". *Nature* **521**: 105–109), copyright (2015).

antibiotics. The X-ray structure of OxyB shows an additional domain, the X domain, which creates interaction between the P450 and the terminal module of the teicoplanin assembly line to enable oxygenative cross-linking while the heptapeptidyl chain is still attached to PCP_7 (T_7).

As a second set of nonribosomal peptide examples, the most famous pair of iron-based oxygenases in which the oxygenase function is cryptic are isopenicillin N synthase (IPNS), (Burzlaff, Rutledge *et al.* 1999) and the desacetoxy cephalosporin synthase (DAOCS) (Valegard, van Scheltinga *et al.* 1998). This pair of enzymes act in tandem to convert the acyclic tripeptide aminoadipoyl-cysteinyl-D-valine (ACV) to isopenicillin N and desacetoxycephalosporin, respectively (Figure 9.34).

They are both mononuclear nonheme iron enzyme catalysts and utilize O_2 to generate Fe(IV)=O oxidants (Cox 2014, Walsh and Wencewicz 2016). A mechanism for IPNS is shown in Figure 9.35 and for DAOCS in Figure 9.36. IPNS catalysis is featured to begin with H• transfer from one of the Cys-β-prochiral hydrogens of bound ACV to the oxoiron intermediate. N–C bond formation generates the β-lactam. Next the hydrogen at the β-carbon of the Val residue is abstracted as H•. The resultant carbon radical combines with a transient sulfur radical to form a C–S bond, and thereby the thiane ring, and complete the 4,5-fused bicyclic scaffold of the penicillins. Note that one O_2 has been reduced by four electrons as two water molecules are produced. IPNS thus escapes the need for an exogenous

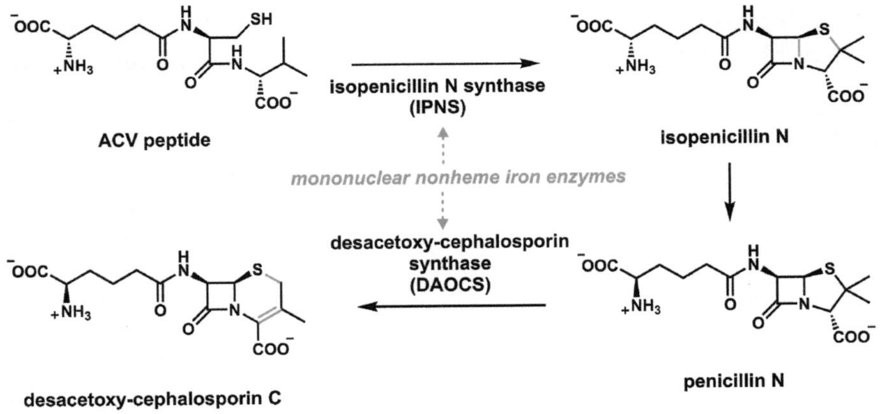

Figure 9.34 Schematic for conversion of ACV tripeptide to the 4,5-fused bicyclic rings in isopenicillin N, and for conversion of penicillin N to 4,6-fused desacetoxy-cephalosporin C, both by mononuclear nonheme iron enzymes.

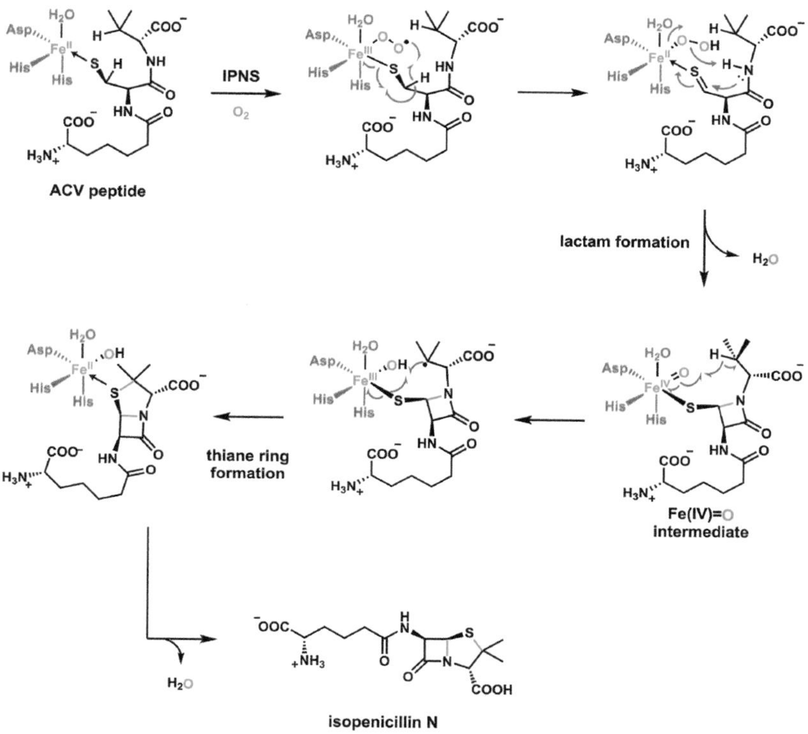

Figure 9.35 Oxidative cyclization of ACV tripeptide to isopenicillin N: formation of the two rings of the lactam scaffold by radical intermediates derived during O_2 reductive activation. The four-membered lactam forms followed by five-membered thiane ring formation *via* Fe(IV)=O one electron transfer chemistry.

α-ketoglutarate as cosubstrate by four electron oxidation of the starting ACV tripeptide.

The desacetoxycephalosporin synthase (DAOCS) enzyme that converts the 4,5-bicyclic ring system of penicillins to the fused 4,6 rings in the cephalosporin antibiotic series has been known colloquially as "expandase", an apt descriptor. In this nonheme iron O_2-consuming enzyme, radical chemistry is again the catalytic reaction manifold. Catalysis starts with homolytic cleavage of one of the C–H bonds of the exocyclic methyl groups of penicillin N (isopenicillin N to penicillin N involves prior epimerase action on the aminoadipoyl moiety). The carbon radical is again proposed to combine with a sulfur odd electron to give a new C–S bond, in this case as an episulfide. Homolytic

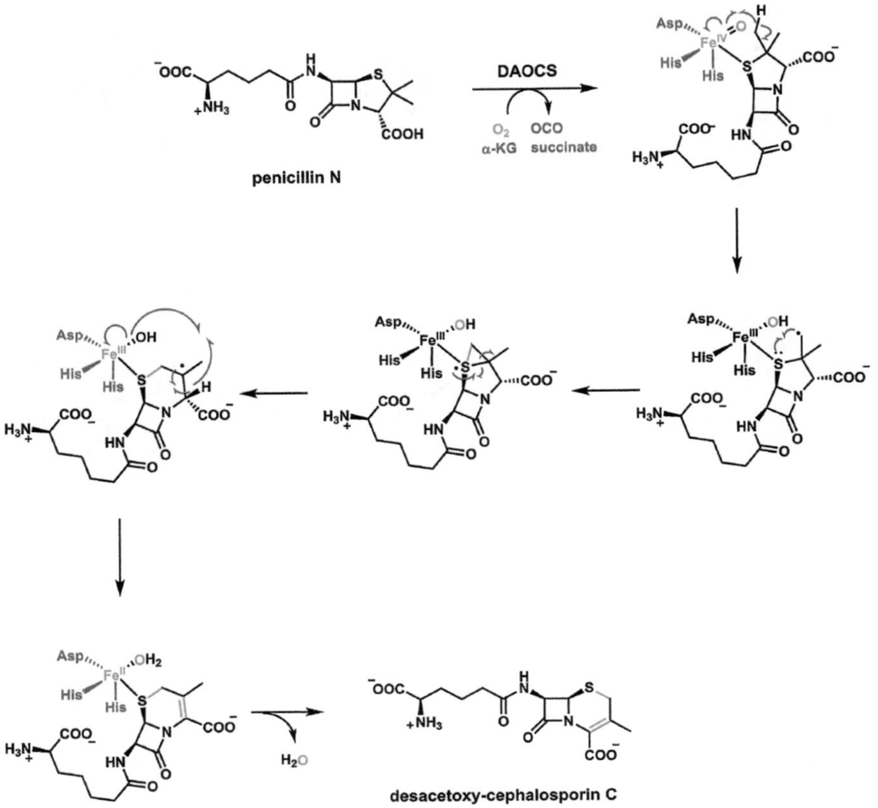

Figure 9.36 Oxidative expansion of the 4,5-penicillin core to the 4,6-cephalosporin core: an oxygenase catalyst that fails to insert oxygen.

fragmentation of the C–C bond in the three-membered episulfide constitutes the ring expansion step. The resultant carbon radical can be quenched by adjacent H• transfer to the iron center and formation of the characteristic double bond in the cephem nucleus.

Together the penicillins and cephalosporins dominate the $40 billion annual global market for antibiotics. The two cryptic oxygenases IPNS and DAOCS, the key post assembly line tailoring enzymes, are the *sine qua non* catalysts for this central antibiotic class.

9.7.3 Oxidized Diketopiperazines

A third case harks back to the diketopiperazine scaffolds noted in Chapter 8 that are part of indole terpene biosynthetic frameworks.

Those DKPs were released from bimodular NRPS assembly lines. A separate parallel strategy exists for generating DKPs from aminoacyl-tRNAs by action of cyclodepsipeptide synthases (Gondry, Sauguet *et al.* 2009, Sauguet, Moutiez *et al.* 2011). Two examples are depicted in Figure 9.37A. The first occurs in *Mycobacterium tuberculosis* metabolism where two molecules of tyrosyl-tRNA are rerouted from the normal destination of ribosomal peptide biosynthesis and instead coupled to make cyclo-Tyr-Tyr. The first Tyr-tRNA transfers its activated Tyr moiety to an active site seryl side chain of the synthase. Then a second Tyr-tRNA binds and peptide bond formation followed by intramolecular attack of the free amino group on the Tyr-Tyr-O-enzyme ester linkage releases the Tyr-Tyr-DKP. The DKP is then substrate for a mycobacterial cytochrome P450 (CYP121) which carries out one electron phenoxy radical chemistry on both phenolate side chains and then couples them in a direct C–C bond (Belin, Le Du *et al.* 2009).

The generation of a pair of delocalized, and presumably long-lived, phenoxy radicals is a common strategy for this P450 class in which oxygen atoms from O_2 do not enter the product, failing to compete with internal fates of the carbon radicals. This is a fate that occurs in other natural product scaffolds (Mizutani and Sato 2011). The result is the mycobacterial end product mycocyclosin, with no molecular oxygen atoms incorporated into the product.

Streptomyces noursei instead couples Leu and Phe to the corresponding DKP and then utilizes a flavin-linked oxidase to do a double desaturation of the cyclo-Phe-Leu (cFL) to the dehydro metabolite albonoursin (Sauguet, Moutiez *et al.* 2011) (Figure 9.37B). Both tailoring enzymes oxidize the initial DKP products, neither incorporates oxygen atoms, and they give different product architectures.

9.7.4 Alkaloids

We have commented on at least five examples in alkaloid biosynthetic pathways in Chapters 5 and 8 where, interspersed among the many oxygenases that complete oxygen transfer to cosubstrates, there are P450s that make the necessary high valent oxoiron oxidant to generate cosubstrate carbon-centered radicals *but do not complete the oxygen transfer*. These cryptic O_2-consuming transformations include the radical dimerization of the *S*-isomer of *N*-methyl norcoclaurine to tubocurarine (see Figure 5.12) and also the conversion of *R*-reticuline to salutaridinol (Figure 5.15), *via* C–C bond formation, on the way to morphine. Likewise, the last step in the creation of the

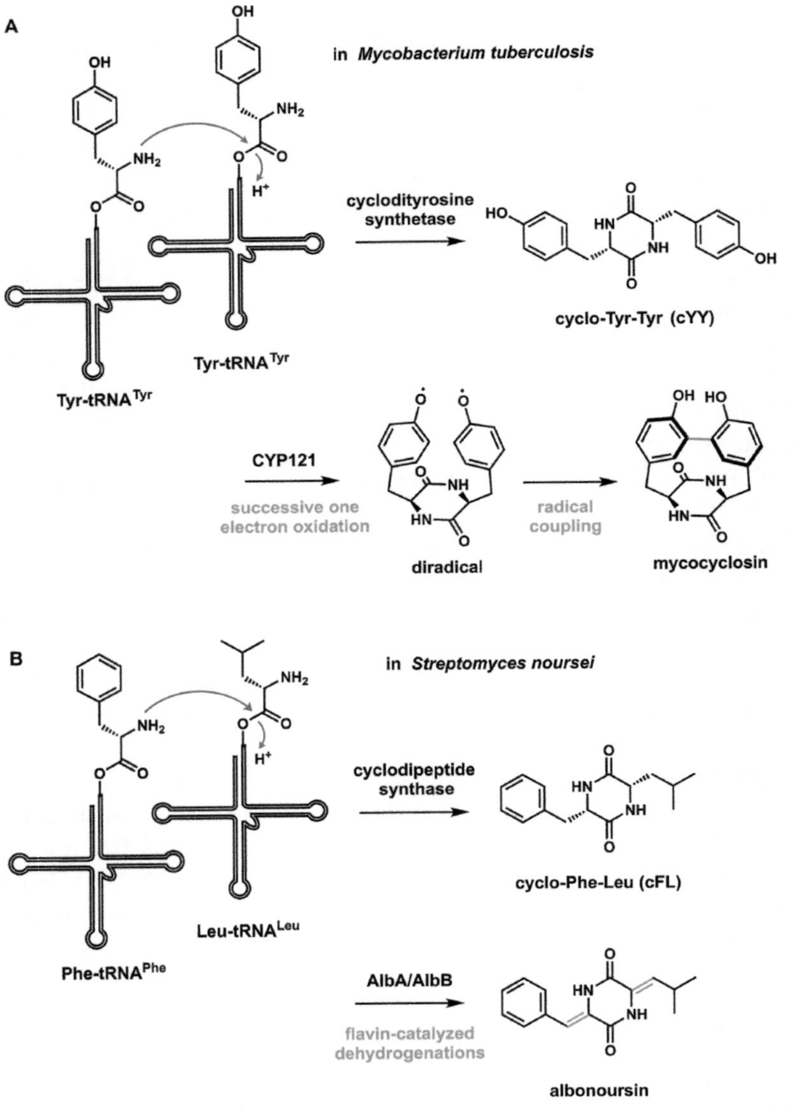

Figure 9.37 tRNA-dependent synthesis of diketopiperazine natural products: (A) mycocyclosin and (B) albonoursin. In the mycocyclosin pathway, a P450 (CYP121) is involved in successive one electron oxidation of the cyclo-Tyr-Tyr (cYY) to generate the diradical, which can undergo C–C coupling to yield the rigidified structure. In the albonoursin pathway, the product cyclo-Phe-Leu (cFL) is oxidized by the flavoenzymes AlbA and AlbB to yield the desaturated final product. In neither pathway is oxygen introduced into the final product, but it is required for the oxidation reactions.

indolecarbazole core by the StaP and RebP cytochrome P450 enzymes involves C–C bond formation by indole radical coupling (see Figures 5.32 and 5.33). We have also noted in Chapter 8 the conversion of tryprostatin A to fumitremorgin A with homolytic C–N bond formation between a DKP nitrogen and a prenyl group double bond (Figure 8.13). The fifth example in the indole terpenes structural class involved the formation of indolactam-V during lyngbyatoxin formation, where a new C–N bond is created *via* a radical pair in the Trp-valinol framework (Figure 8.22).

9.7.5 Phenylpropanoids

Two prominent examples in the phenylpropanoid sweep of secondary metabolites in which cryptic P450s consume O_2, generate substrate radicals, but fail to insert an oxygen atom into product are in the dimerization of monolignols to lignans and in the 1,2-aryl migration that occurs during enzymatic conversion of flavones to isoflavones (see Figures 7.8 and 7.25).

9.8 Oxygen-dependent Halogenases

More than 4000 natural products are known to contain carbon-halogen bonds (Gribble 2004, Vaillancourt, Yeh *et al.* 2006). The large majority are chlorinated, but some marine metabolites are brominated, reflecting higher amounts of dissolved bromide salts in sea water. There are few iodo metabolites and even fewer fluorinated ones (Figure 9.38). Figure 9.39 shows five halogenated metabolites, including two polyketides, chlortetracycline and calicheamycin, the nonribosomal heptapeptide vancomycin, the indole alkaloid dimeric rebeccamycin, and a recently characterized marine streptomycete salinosporamide.

Figure 9.40 indicates that there are four major classes of halogenating enzymes (Vaillancourt, Yeh *et al.* 2006). The vanadyl(v) and the heme enzymes use hydrogen peroxide as cosubstrate, so they are termed haloperoxidases. The other two, $FADH_2$-dependent and mononuclear iron enzyme classes, use O_2 as a cosubstrate. Tetraiodothyronine is a mammalian hormone known as T4 that is acted on by an iodoperoxidase. The iodination and aryl ether formation from two Tyr residues happens in the context of a large protein, thyroglobulin, that is then digested to release T4. Hypoiodite

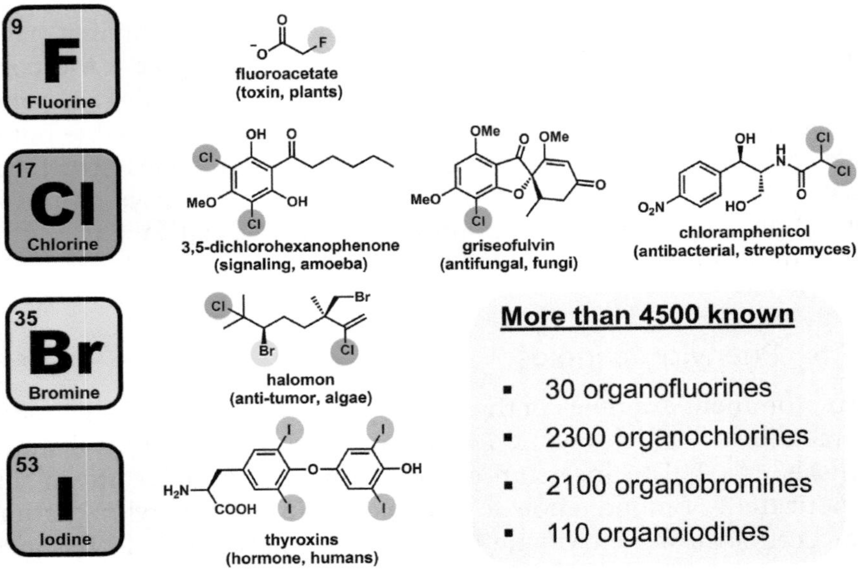

Figure 9.38 Halogenated metabolites: chlorinated molecules predominate; brominated molecules are found in marine organisms; iodo and fluoro metabolites are rare.
Data from Gribble, G. W. (2004). "Natural Organohalogens: A New Frontier for Medicinal Agents?" *J. Chem. Educ.* **81**(10): 1441.

coordinated as the top axial heme ligand is the donor of an $[I^+]$ equivalent to the carbons *ortho* to the phenol ring of the tyrosine residues.

The vanadyl peroxidases have a vanadate in the +5 oxidation state. It can bind HOOH, replacing two water ligands, as noted in Figure 9.41. Attack of bromide ion on the coordinated peroxo moiety yields hypobromite (HOBr). This is the active halogenating agent. As with the heme iodo peroxidase, the HOX hypohalite is a donor of X^+, in this instance Br^+. Figure 9.41 shows a mechanism for $[Br^+]$ addition to yield the terpene alcohol isomers α- and β-snyderol (Butler and Carter-Franklin 2004, Carter-Franklin and Butler 2004).

The other two halogenase classes, the flavin- and iron-containing enzymes, borrow the related oxygenase logic and machinery almost exactly. Thus, the flavin-dependent halogenases can only chlorinate activated substrates that can generate carbon nucleophiles, the

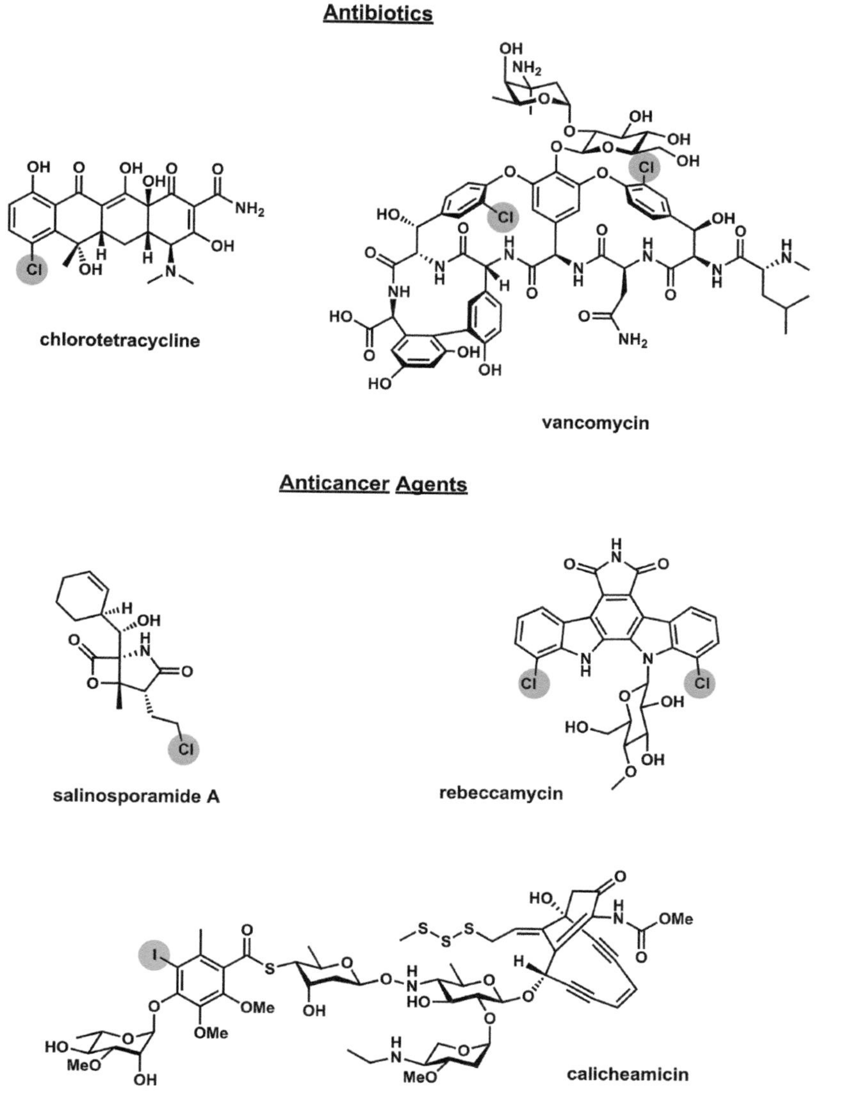

Figure 9.39 Halogenated natural products include polyketides, nonribosomal peptides, and indolecarbazoles.

same scope as the flavin-dependent oxygenases (Figure 9.42). As shown in Figure 9.42, tryptophan 7-halogenase generates the typical FAD-4a-OOH intermediate, but rather than being directly attacked by substrate, chloride ion attacks and yields nascent HOCl. This step has reversed the polarity from the anionic chloride ion to a Cl^+ equivalent

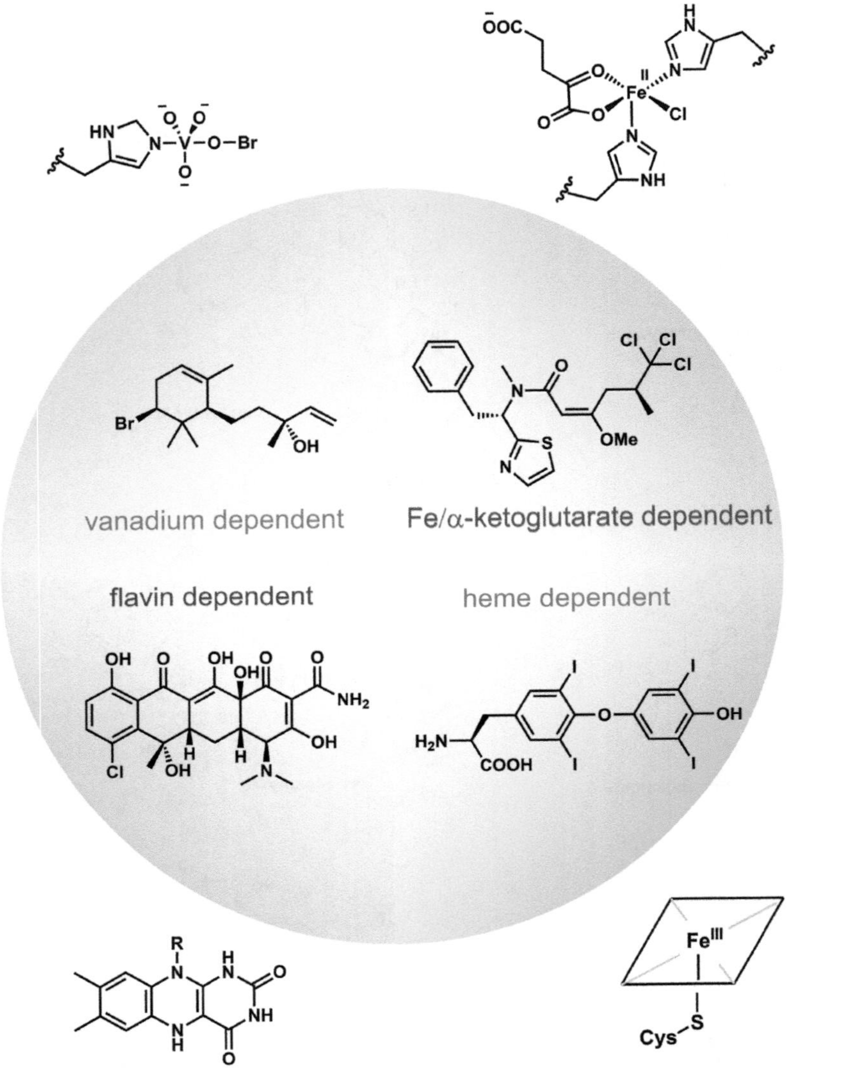

Figure 9.40 Four different types of halogenases with distinct cofactors, all of which convert halide ion to hypohalite species with change in polarity of the halogen from anion to cation equivalent.

in HOCl. There is indirect evidence that the HOCl is intercepted by the side chain of an essential Lys residue in the active site (Yeh, Garneau *et al.* 2005, Yeh, Cole *et al.* 2006) (Figure 9.43). The N_6-chloro-Lys is thought to be the proximal donor of Cl^+.

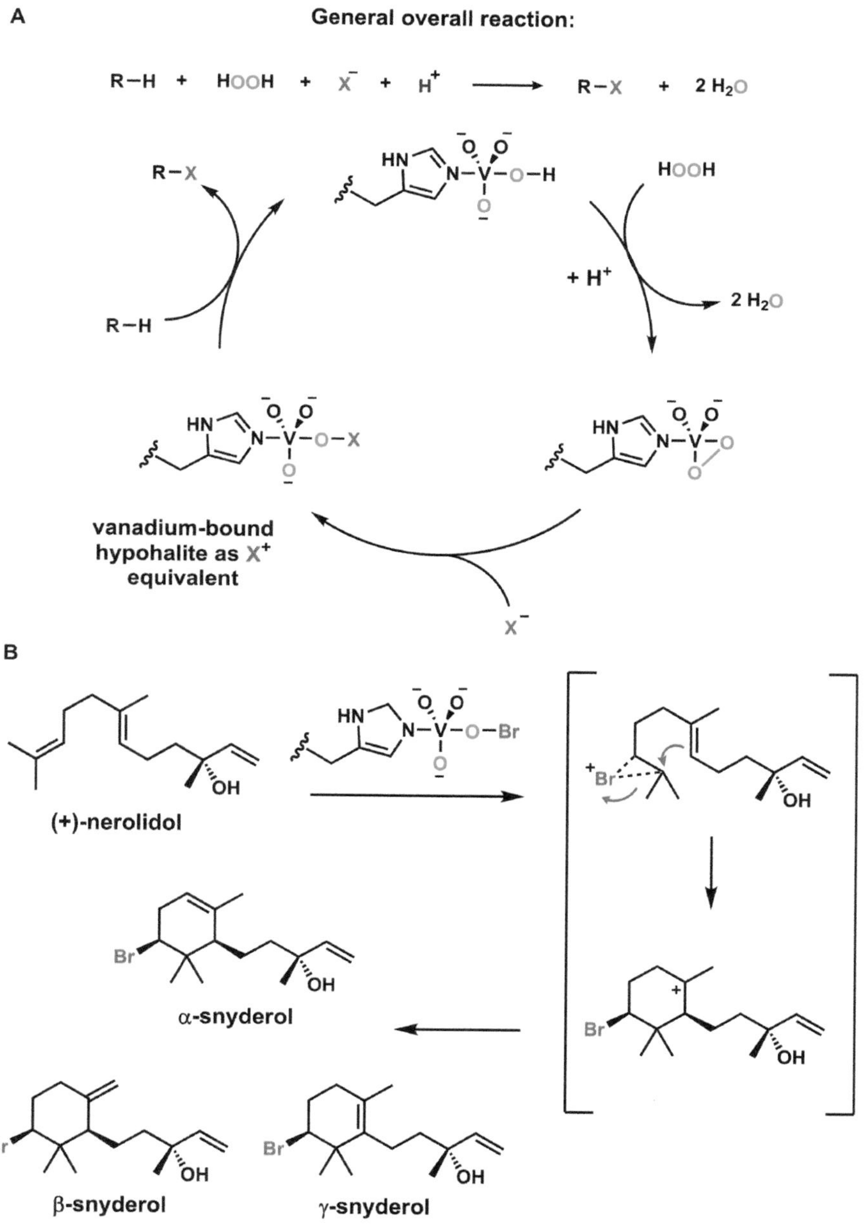

Figure 9.41 (A) Vanadium haloperoxidases bind H_2O_2 and halide ions, then form bound HOX coordinated to vanadium(V) and deliver Br^+ in the snyderol example (B).

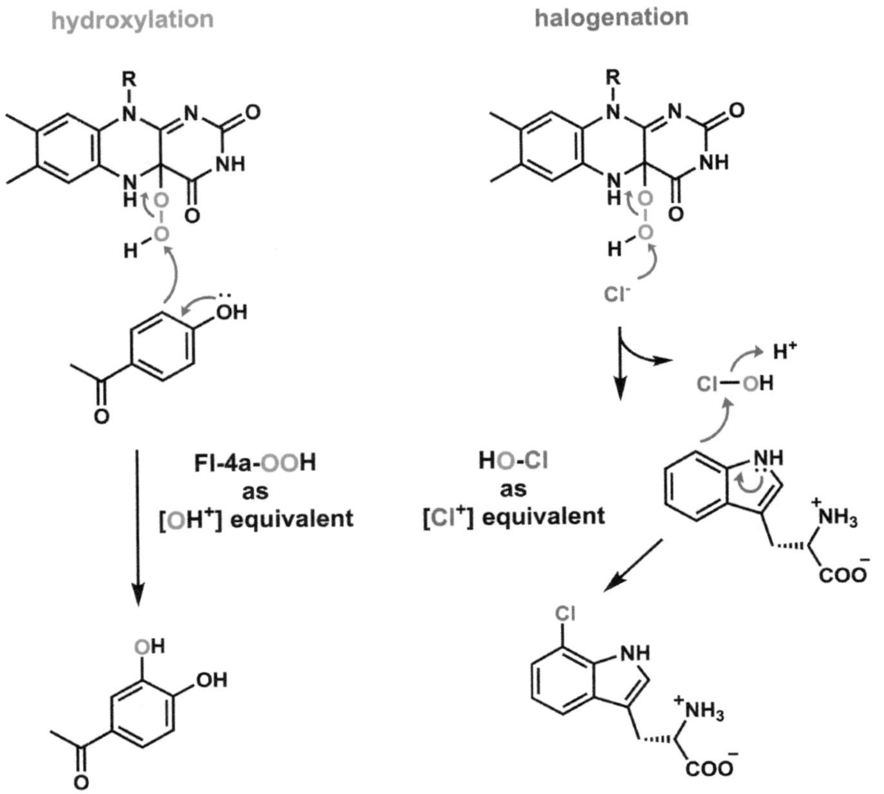

Figure 9.42 Flavin-dependent monooxoygenases: partitioning of the flavin-4a-OOH intermediate for $[OH^+]$ or $[Cl^+]$ transfer to activated cosubstrates.

Halogenases are known that catalyze chlorination at residues 5, 6, or 7 of free tryptophan. Also, pyrrole can be chlorinated at C_2 or C_3. Analogously, the 3-chlorotyrosines in vancomycin, the chlorophenol in griseofulvin (see Figure 9.32), the chlorinated aromatics in rebeccamycin, and calicheamycin all arise *via* the flavin oxygenase/ halogenase machinery. Figures 9.42 and 9.43 show why the $FADH_2$-dependent halogenases must consume O_2. Cosubstrate chloride ion attacks the flavin hydroperoxide to be converted to the required Cl^+ equivalent.

The nonheme mononuclear enzymes that act as halogenases are also variants of the oxygenases. A major distinction is that the Fe coordination sphere in the halogenases is missing the Asp ligand (Blasiak, Vaillancourt *et al.* 2006) (Figure 9.44). It is typically replaced

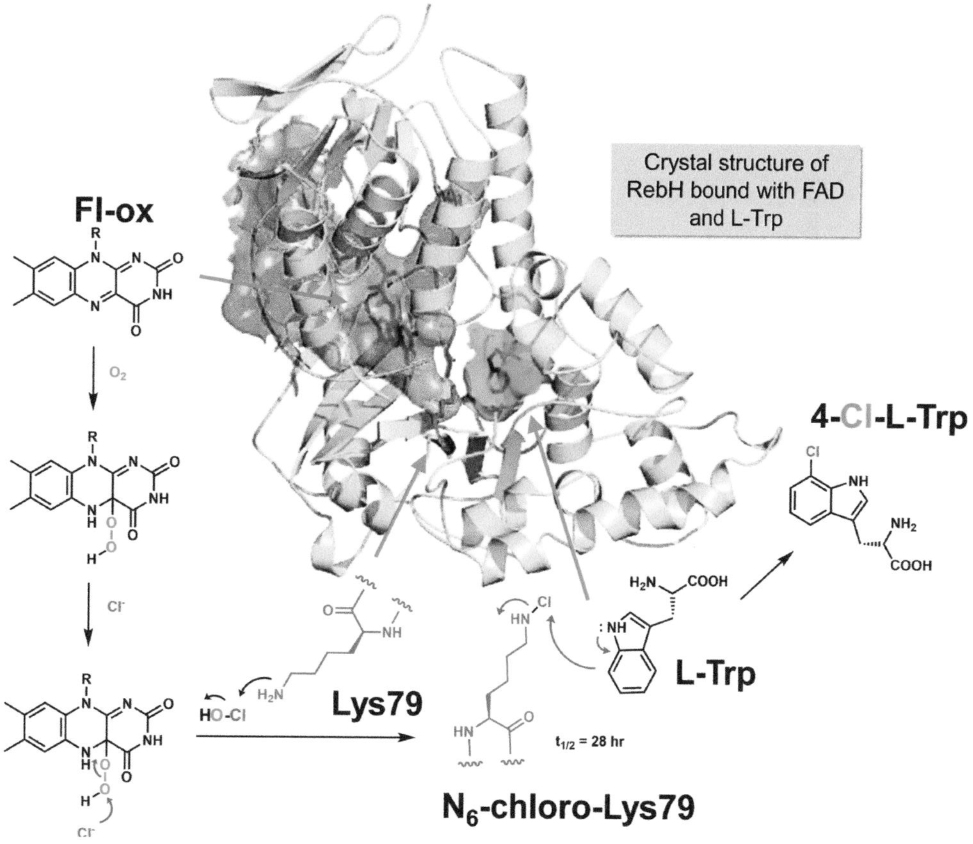

Figure 9.43 The proximal halogenating agent in tryptophan 7-halogenase is proposed to have an active site N_6-Cl-lysine residue. Crystal structure reprinted with permission from Yeh, E., L. C. Blasiak, A. Koglin, C. L. Drennan and C. T. Walsh (2007). "Chlorination by a long-lived intermediate in the mechanism of flavin-dependent halogenases". *Biochemistry* **46**(5): 1284–1292. Copyright (2007) American Chemical Society.

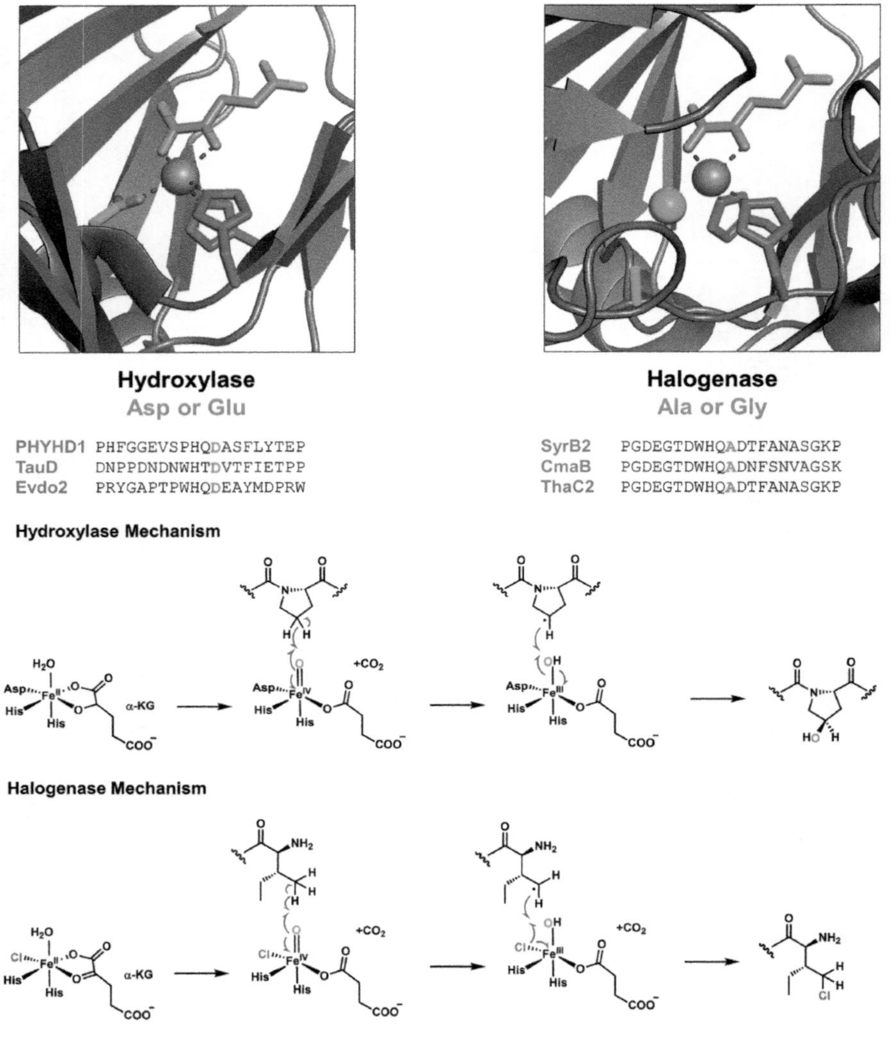

Figure 9.44 Nonheme iron halogenases have an open ligand position for binding chloride or bromide ion in the Fe(II) coordination sphere because of mutation of an Asp/Glu to an Ala/Ser in the active site. Schematic comparison of active site ligands to iron in oxygenase *vs.* halogenase catalysts is shown. Once O_2 has been utilized to generate the high valent oxo-iron intermediate, a halogenase *vs.* hydroxylase outcome depends on whether a halide is bound in the first coordination shell of the oxo-iron intermediate and positioned for $Cl^{\bullet}$ transfer to outcompete $OH^{\bullet}$ transfer. The hydrolase structure shown is for PHYHDA (PDB ID: 3OBZ). The chlorinase structure shown is for SyrB2 (PDB ID: 2FCT). Image credit: Allena Goren.

by Ala or Ser, a shorter chain that allows space for Cl to become a replacement ligand in the first coordination shell of iron. When the high valent Fe(IV)=O acts to abstract the substrate C–H, instead of [OH•] rebound, [Cl•] is transferred selectively to form the C–Cl bond homolytically. Thus, as shown in Figure 9.44, unactivated carbon sites can be either hydroxylated (Pro to 3-OH-Pro) or halogenated (L-*allo*-Ile to 4-chloro-*allo*-Ile), depending on whether the enzyme active site can bind chloride ion as ligand in its first coordination sphere. Unlike the flavoenzyme halogenases which transfer a Cl^+ or Br^+ equivalent to a substrate carbanion equivalent, the iron enzymes can transfer Cl• or Br• to unactivated carbon sites in cosubstrates, including the methyl side chain of a threonyl moiety on the NRPS assembly line to syringomycin (Vaillancourt, Yin *et al.* 2005) or at a secondary carbon in welwitindolinone maturation (Hillwig and Liu 2014) (Figure 9.45).

A net cryptic chlorination occurs in the formation of the cyclopropyl ring of the coronamic acid unit in coronatine, a mimic of the phytohormone jasmonic acid that is produced by phytopathogeneic pseudomonads (Figure 9.46). Chlorination of an *allo*-isoleucyl-*S*-peptidyl carrier protein at the unactivated δ-carbon is followed by intramolecular expulsion of chloride ion by the α-carbon thioester enolate to form the amino cyclopropyl thioester (Vaillancourt, Yeh *et al.* 2005). This unit is incorporated intact in the final coronatine scaffold.

9.9 Fluorination of Substrates by a Nonoxidative Route: Fluorinase

As noted in Figure 9.38, carbon–fluorine bonds are known in natural products but they are rare. Presumably this is because fluoride ion, in contrast to chloride, bromide, and iodide ion, is so electronegative it resists oxidation by enzymatic oxidants. Neither F• nor F^+ is accessible in biological reaction manifolds. Therefore any C–F bonds must arise from fluoride ion as nucleophile on an electrophilic carbon source. One such enzyme, termed a fluorinase (it will also act as a chlorinase), has been found in microbes (O'Hagan and Deng 2015). It uses *S*-adenosylmethionine as cosubstrate. Fluoride ion in the enzyme active site must be desolvated sufficiently to act as a free nucleophile at $C_{5'}$ of the ribose, displacing methionine and generating the 5′-fluoro (5′-deoxy) adenosine as coproduct

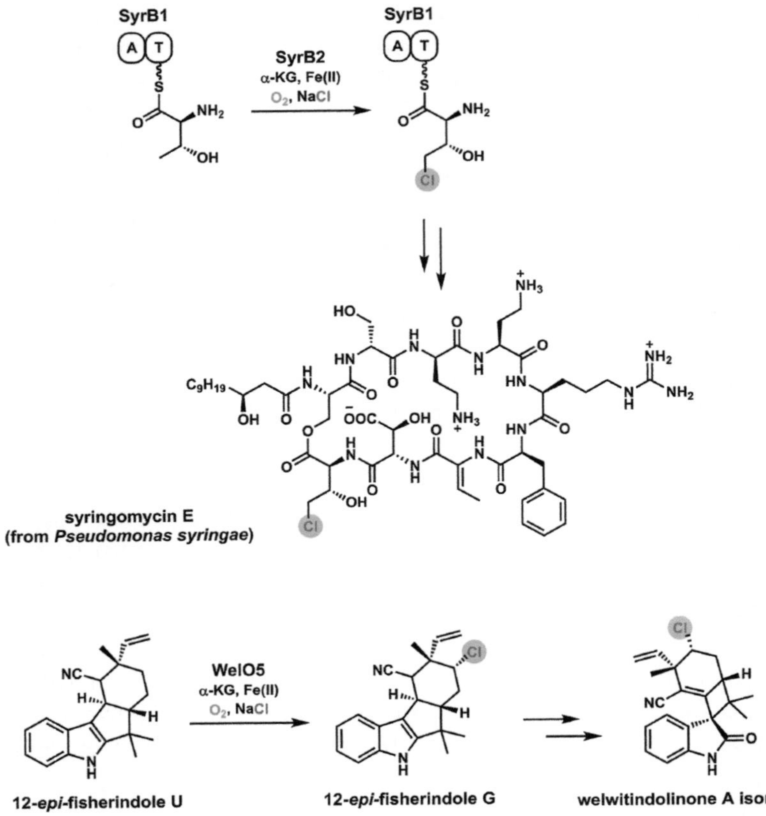

Figure 9.45 Unlike the flavoenzyme halogenases which transfer a Cl^+ or Br^+ equivalent to a substrate carbanion equivalent, the iron enzymes can transfer Cl• or Br• to unactivated carbon sites in cosubstrates, including the methyl side chain of a threonyl moiety on the NRPS assembly line to syringomycin (Vaillancourt, Yin *et al.* 2005) or at a secondary carbon in welwitindolinone maturation (Hillwig and Liu 2014).

(Figure 9.47A). Correspondingly, chloride ion can be utilized to generate 5′-chloroadenosine.

Both the fluoro- and chloroadenosines can be further metabolized eventually to haloethylmalonyl-CoA metabolites. These are on pathway to the natural products salinosporamide (chloro substituent) and fluorosalinosporamide (Eustaquio, McGlinchey *et al.* 2009) (Figure 9.47). In turn these are of interest as potent, irreversible inhibitors of

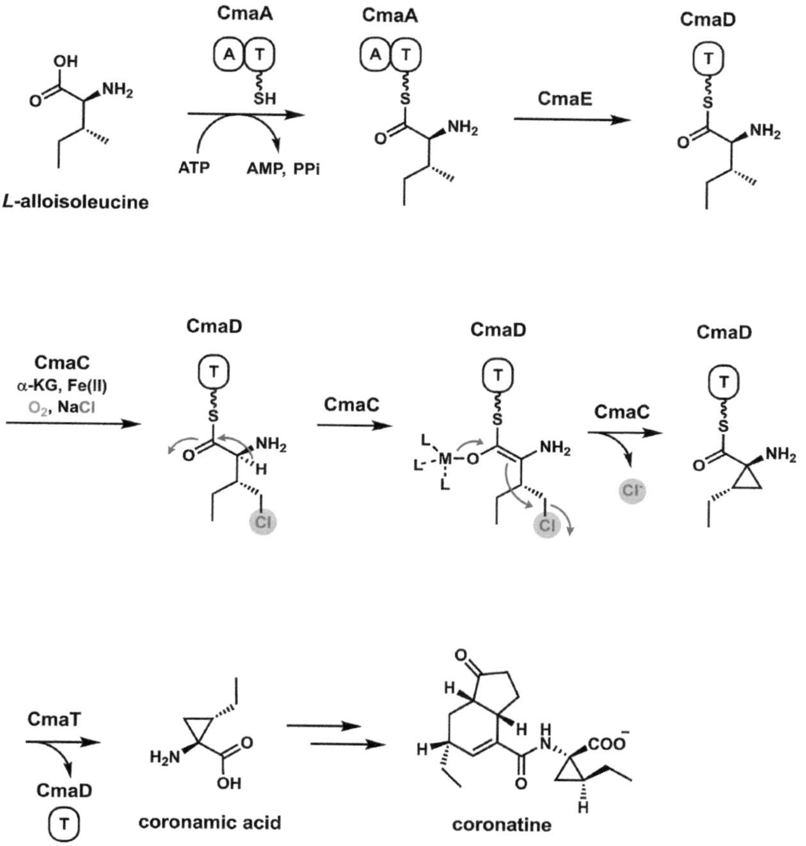

Figure 9.46 Cryptic chlorination of C4 of *allo*-L-Ile-*S*-PCP as prelude to cyclopropyl ring formation. The coronamic acid thus generated is enzymatically coupled to the polyketide fragment to yield coronatine, a plant hormone mimic elaborated by phytopathic strains of *Pseudomonas syringae*.

proteasomes. The active site threonine –OH of the proteasome attacks and opens the reactive beta lactone. The initial acyl enzyme now has a free –OH group which can act intramolecularly to displace the fluorine or chlorine substituent and form a new ring system that makes the acyl enzyme intermediate particularly long-lived (Figure 9.47B). This is the second example of enzymatic use of halide as a facile leaving group in internal substitution reactions (the coronamic acid formation above is the first).

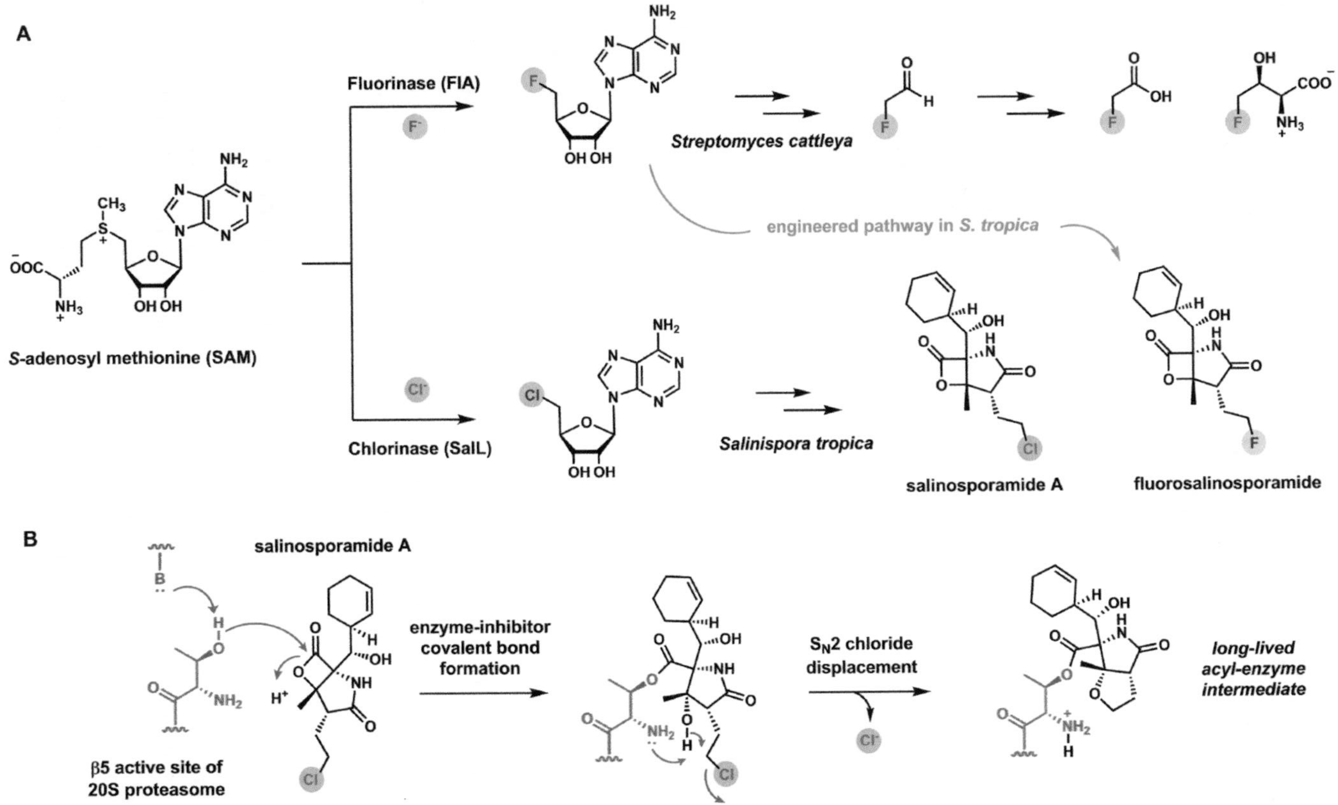

Figure 9.47 (A) The C_5 ribose carbon in SAM can serve as the electrophilic carbon for nucleophilic halide addition, including by the *Streptomyces cattley* fluorinase (FIA). The product can be converted into fluoroacetate. The same reaction with the chlorinase SalL is found in the salinosporamide A biosynthetic pathway in *Salinospora tropica*. Introducing the fluorinase into *S. tropica* led to the biosynthesis of fluorosalinosporamide. (B) Salinosporamide is a potent inhibitor of 20S proteasomes through formation of the long-lived acyl-enzyme intermediate. This is an example of enzymatic use of halide as a facile leaving group in internal substitution reactions.

9.10 Summary: The Chemical Versatility of Ferryl (High Valent Oxo-iron) Reaction Intermediates

The invention/evolution of iron enzymes that can reductively activate O_2 and split it to a water molecule and a high valent oxo-iron species, termed a ferryl intermediate, gave an enormous range of reactivity to the apoproteins as catalysts (Figure 9.48). The first half reaction that follows from ferryl formation is typically abstraction of a cosubstrate hydrogen atom to yield a carbon radical.

The second half reactions that can follow are remarkably varied. We have just compared hydroxylation and halogenation pathway parallels. The sp^2 center epoxidations are akin to the sp^3 carbon

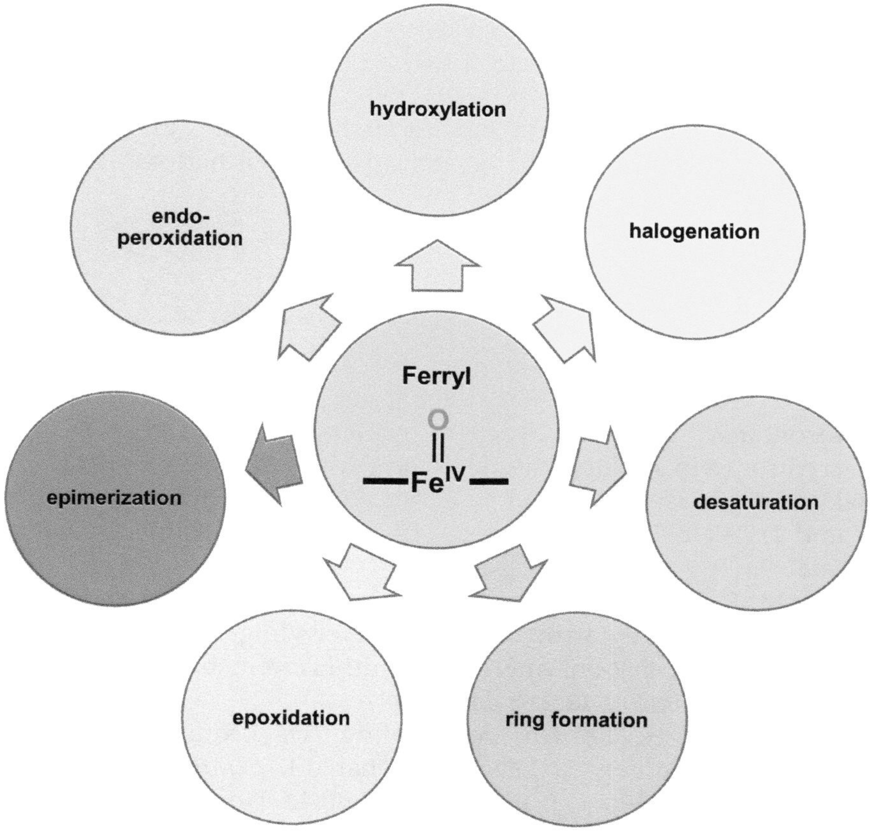

Figure 9.48 The chemical versatility of ferryl (high valent oxo-iron) reaction intermediates.

hydroxylation mechanisms, while endoperoxidation is well studied in the prostaglandin systems, albeit less so in artemisinin and fumitremorgin biosyntheses. Three of the seven outcomes in Figure 9.48 do not incorporate oxygen atoms into coproducts. In penicillin biosynthesis and expandase action to the cephem antibiotics, ring formations are the outcome. Desaturations and epimerizations are outcomes in carbapenem and clavulanate biosyntheses. We have additionally noted several phenoxy radical dimerization outcomes in this chapter as well (Mizutani and Sato 2011).

From the perspective of reductive activation and cleavage of O_2, plants are clearly committed to using oxygen to increase polarity and water solubility of products. From the point of view of carbon radicals in biosynthetic pathways, the dozens and dozens of plant and microbial P450s conduct these hundreds of biosynthetic steps by one-electron reaction manifolds. While the P450s and nonheme iron oxygenases may have evolved to deliver an oxygen atom efficiently and with high yield to substrate carbon radicals, the leakage into reaction manifolds where $OH^\bullet$ rebound is not kinetically competent has had a powerful benefit of mechanistic and structural diversion to create novel structures and functions in those molecules.

References

Andersen, J. F., K. Tatsuta, H. Gunji, T. Ishiyama and C. R. Hutchinson (1993). "Substrate specificity of 6-deoxyerythronolide B hydroxylase, a bacterial cytochrome P450 of erythromycin A biosynthesis". *Biochemistry* **32**(8): 1905–1913.

Bak, S., F. Beisson, G. Bishop, B. Hamberger, R. Hofer, S. Paquette and D. Werck-Reichhart (2011). "Cytochromes p450". *Arabidopsis Book* **9**: e0144.

Beam, M. P., M. A. Bosserman, N. Noinaj, M. Wehenkel and J. Rohr (2009). "Crystal structure of Baeyer-Villiger monooxygenase MtmOIV, the key enzyme of the mithramycin biosynthetic pathway". *Biochemistry* **48**(21): 4476–4487.

Belin, P., M. H. Le Du, A. Fielding, O. Lequin, M. Jacquet, J. B. Charbonnier, A. Lecoq, R. Thai, M. Courcon, C. Masson, C. Dugave, R. Genet, J. L. Pernodet and M. Gondry (2009). "Identification and structural basis of the reaction catalyzed by CYP121, an essential cytochrome P450 in Mycobacterium tuberculosis". *Proc. Natl. Acad. Sci. U. S. A.* **106**(18): 7426–7431.

Blasiak, L. C., F. H. Vaillancourt, C. T. Walsh and C. L. Drennan (2006). "Crystal structure of the non-haem iron halogenase SyrB2 in syringomycin biosynthesis". *Nature* **440**(7082): 368–371.
Broderick, J. B., B. R. Duffus, K. S. Duschene and E. M. Shepard (2014). "Radical S-adenosylmethionine enzymes". *Chem. Rev.* **114**(8): 4229–4317.
Bugg, T. D. (2003). "Dioxygenase enzymes: catalytic mechanisms and chemical models". *Tetrahedron Lett.* **59**: 7075–7101.
Burzlaff, N. I., P. J. Rutledge, I. J. Clifton, C. M. Hensgens, M. Pickford, R. M. Adlington, P. L. Roach and J. E. Baldwin (1999). "The reaction cycle of isopenicillin N synthase observed by X-ray diffraction". *Nature* **401**(6754): 721–724.
Butler, A. and J. N. Carter-Franklin (2004). "The role of vanadium bromoperoxidase in the biosynthesis of halogenated marine natural products". *Nat. Prod. Rep.* **21**(1): 180–188.
Carlson, J. C., S. Li, S. S. Gunatilleke, Y. Anzai, D. A. Burr, L. M. Podust and D. H. Sherman (2011). "Tirandamycin biosynthesis is mediated by co-dependent oxidative enzymes". *Nat. Chem.* **3**(8): 628–633.
Carter-Franklin, J. N. and A. Butler (2004). "Vanadium bromoperoxidase-catalyzed biosynthesis of halogenated marine natural products". *J. Am. Chem. Soc.* **126**(46): 15060–15066.
Chelikani, P., I. Fita and P. C. Loewen (2004). "Diversity of structures and properties among catalases". *Cell. Mol. Life Sci.* **61**(2): 192–208.
Chen, H., M. G. Thomas, S. E. O'Connor, B. K. Hubbard, M. D. Burkart and C. T. Walsh (2001). "Aminoacyl-S-enzyme intermediates in beta-hydroxylations and alpha,beta-desaturations of amino acids in peptide antibiotics". *Biochemistry* **40**(39): 11651–11659.
Cox, R. J. (2014). "Oxidative rearrangements during fungal biosynthesis". *Nat. Prod. Rep.* **31**(10): 1405–1424.
Croteau, R., R. E. Ketchum, R. M. Long, R. Kaspera and M. R. Wildung (2006). "Taxol biosynthesis and molecular genetics". *Phytochem. Rev.* **5**(1): 75–97.
Cryle, M. J., A. Meinhart and I. Schlichting (2010). "Structural characterization of OxyD, a cytochrome P450 involved in beta-hydroxytyrosine formation in vancomycin biosynthesis". *J. Biol. Chem.* **285**(32): 24562–24574.
Dewick, P. (2009). *Medicinal Natural Products, a Biosynthetic Approach*. UK, Wiley.
Emsley, J. (2001). *Nature's Building Blocks: An A-Z Guide to the Elements*. Oxford, England, Oxford University Press.
Eustaquio, A. S., R. P. McGlinchey, Y. Liu, C. Hazzard, L. L. Beer, G. Florova, M. M. Alhamadsheh, A. Lechner, A. J. Kale, Y. Kobayashi,

K. A. Reynolds and B. S. Moore (2009). "Biosynthesis of the salinosporamide A polyketide synthase substrate chloroethylmalonyl-coenzyme A from S-adenosyl-L-methionine". *Proc. Natl. Acad. Sci. U. S. A.* **106**(30): 12295–12300.

Faber, K. (2011). *Biotransformations in Organic Chemistry*. Springer Science and Business Media.

Gondry, M., L. Sauguet, P. Belin, R. Thai, R. Amouroux, C. Tellier, K. Tuphile, M. Jacquet, S. Braud, M. Courcon, C. Masson, S. Dubois, S. Lautru, A. Lecoq, S. Hashimoto, R. Genet and J. L. Pernodet (2009). "Cyclodipeptide synthases are a family of tRNA-dependent peptide bond-forming enzymes". *Nat. Chem. Biol.* **5**(6): 414–420.

Grandner, J. M., R. A. Cacho, Y. Tang and K. N. Houk (2016). "Mechanism of the P450-Catalyzed Oxidative Cyclization in the Biosynthesis of Griseofulvin". *ACS Catal.* **6**(7): 4506–4511.

Gribble, G. (2004). "Natural Organohalogens: A new frontier for medicinal agents?". *J. Chem. Educ.* **81**: 1441–1449.

Harrison, P. J. and T. D. Bugg (2014). "Enzymology of the carotenoid cleavage dioxygenases: reaction mechanisms, inhibition and biochemical roles". *Arch. Biochem. Biophys.* **544**: 105–111.

Helliwell, C. A., P. M. Chandler, A. Poole, E. S. Dennis and W. J. Peacock (2001). "The CYP88A cytochrome P450, ent-kaurenoic acid oxidase, catalyzes three steps of the gibberellin biosynthesis pathway". *Proc. Natl. Acad. Sci. U. S. A.* **98**(4): 2065–2070.

Herbert, R. (1989). *The Biosynthesisof Secondary Metabolites*. Springer.

Hillwig, M. L. and X. Liu (2014). "A new family of iron-dependent halogenases acts on freestanding substrates". *Nat. Chem. Biol.* **10**(11): 921–923.

Hu, Y., D. Dietrich, W. Xu, A. Patel, J. A. Thuss, J. Wang, W. B. Yin, K. Qiao, K. N. Houk, J. C. Vederas and Y. Tang (2014). "A carbonate-forming Baeyer-Villiger monooxygenase". *Nat. Chem. Biol.* **10**(7): 552–554.

Jensen, K. and B. L. Moller (2010). "Plant NADPH-cytochrome P450 oxidoreductases". *Phytochemistry* **71**(2-3): 132–141.

Kapoor, A., M. Shandilya and S. Kundu (2011). "Structural insight of dopamine beta-hydroxylase, a drug target for complex traits, and functional significance of exonic single nucleotide polymorphisms". *PLoS One* **6**(10) e26509.

Kaspera, R. and R. Croteau (2006). "Cytochrome P450 oxygenases of Taxol biosynthesis". *Phytochem. Rev.* **5**(2–3): 433–444.

Kells, P. M., H. Ouellet, J. Santos-Aberturas, J. F. Aparicio and L. M. Podust (2010). "Structure of cytochrome P450 PimD suggests epoxidation of the polyene macrolide pimaricin occurs *via* a hydroperoxoferric intermediate". *Chem. Biol.* **17**(8): 841–851.

Kern, F., T. K. Dier, Y. Khatri, K. M. Ewen, J. P. Jacquot, D. A. Volmer and R. Bernhardt (2015). "Highly Efficient CYP167A1 (EpoK) dependent Epothilone B Formation and Production of 7-Ketone Epothilone D as a New Epothilone Derivative". *Sci. Rep.* **5**: 14881.

Kim, H. J., E. Ono, K. Morimoto, T. Yamagaki, A. Okazawa, A. Kobayashi and H. Satake (2009). "Metabolic engineering of lignan biosynthesis in Forsythia cell culture". *Plant Cell Physiol.* **50**(12): 2200–2209.

Kim, T. W., J. Y. Hwang, Y. S. Kim, S. H. Joo, S. C. Chang, J. S. Lee, S. Takatsuto and S. K. Kim (2005). "Arabidopsis CYP85A2, a cytochrome P450, mediates the Baeyer-Villiger oxidation of castasterone to brassinolide in brassinosteroid biosynthesis". *Plant Cell* **17**(8): 2397–2412.

Kovaleva, E. G. and J. D. Lipscomb (2008). "Versatility of biological non-heme Fe(II) centers in oxygen activation reactions". *Nat. Chem. Biol.* **4**(3): 186–193.

Kozikowski, A., C. Chen, J.-P. Wu, M. Shibuya, C. G. Kim and H. G. Floss (1993). "Probng Alkaloid Biosynthesis: Intermediates in the Formation of Ring C". *J. Am. Chem. Soc.* **115**: 2482–2488.

Kudo, F., Y. Matsuura, T. Hayashi, M. Fukushima and T. Eguchi (2016). "Genome mining of the sordarin biosynthetic gene cluster from Sordaria araneosa Cain ATCC 36386: characterization of cycloaraneosene synthase and GDP-6-deoxyaltrose transferase". *J. Antiobiot* . DOI:10.1038/ja.2016.40 10.1038/ja.2016.40.

Laden, B. P., Y. Tang and T. D. Porter (2000). "Cloning, heterologous expression, and enzymological characterization of human squalene monooxygenase". *Arch. Biochem. Biophys.* **374**(2): 381–388.

Lambalot, R. H., D. E. Cane, J. J. Aparicio and L. Katz (1995). "Overproduction and characterization of the erythromycin C-12 hydroxylase, EryK". . *Biochemistry* **34**(6): 1858–1866.

Lin, H. C., Y. H. Chooi, S. Dhingra, W. Xu, A. M. Calvo and Y. Tang (2013). "The fumagillin biosynthetic gene cluster in Aspergillus fumigatus encodes a cryptic terpene cyclase involved in the formation of beta-trans-bergamotene". *J. Am. Chem. Soc.* **135**(12): 4616–4619.

Lin, H. C., Y. Tsunematsu, S. Dhingra, W. Xu, M. Fukutomi, Y. H. Chooi, D. E. Cane, A. M. Calvo, K. Watanabe and Y. Tang (2014). "Generation of complexity in fungal terpene biosynthesis: discovery of a multifunctional cytochrome P450 in the fumagillin pathway". *J. Am. Chem. Soc.* **136**(11): 4426–4436.

Loenen, W. A. (2006). "S-adenosylmethionine: jack of all trades and master of everything?". *Biochem. Soc. Trans.* **34**(Pt 2): 330–333.
Mao, X. M., Z. J. Zhan, M. N. Grayson, M. C. Tang, W. Xu, Y. Q. Li, W. B. Yin, H. C. Lin, Y. H. Chooi, K. N. Houk and Y. Tang (2015). "Efficient Biosynthesis of Fungal Polyketides Containing the Dioxabicyclo-octane Ring System". *J. Am. Chem. Soc.* **137**(37): 11904–11907.
Martinez, S. and R. P. Hausinger (2015). "Catalytic Mechanisms of Fe(II)- and 2-Oxoglutarate-dependent Oxygenases". *J. Biol. Chem.* **290**(34): 20702–20711.
McCord, J. M. and I. Fridovich (1988). "Superoxide dismutase: the first twenty years (1968-1988)". *Free Radical Biol. Med.* **5**(5-6): 363–369.
Mehta, A. P., S. H. Abdelwahed, N. Mahanta, D. Fedoseyenko, B. Philmus, L. E. Cooper, Y. Liu, I. Jhulki, S. E. Ealick and T. P. Begley (2015). "Radical S-adenosylmethionine (SAM) enzymes in cofactor biosynthesis: a treasure trove of complex organic radical rearrangement reactions". *J. Biol. Chem.* **290**(7): 3980–3986.
Minami, A., M. Shimaya, G. Suzuki, A. Migita, S. S. Shinde, K. Sato, K. Watanabe, T. Tamura, H. Oguri and H. Oikawa (2012). "Sequential enzymatic epoxidation involved in polyether lasalocid biosynthesis". *J. Am. Chem. Soc.* **134**(17): 7246–7249.
Mizutani, M. and D. Ohta (2010). "Diversification of P450 genes during land plant evolution". *Annu. Rev. Plant Biol.* **61**: 291–315.
Mizutani, M. and F. Sato (2011). "Unusual P450 reactions in plant secondary metabolism". *Arch. Biochem. Biophys.* **507**(1): 194–203.
Morrone, D., X. Chen, R. M. Coates and R. J. Peters (2010). "Characterization of the kaurene oxidase CYP701A3, a multifunctional cytochrome P450 from gibberellin biosynthesis". *Biochem. J.* **431**(3): 337–344.
O'Hagan, D. and H. Deng (2015). "Enzymatic fluorination and biotechnological developments of the fluorinase". *Chem. Rev.* **115**(2): 634–649.
Ortiz de Monteallano P. ed. (2015). *Cytochrome P450*, 4th edn. Germany, Springer.
Pandey, A. V. and C. E. Fluck (2013). "NADPH P450 oxidoreductase: structure, function, and pathology of diseases". *Pharmacol. Ther.* **138**(2): 229–254.
Pohle, S., C. Appelt, M. Roux, H. P. Fiedler and R. D. Sussmuth (2011). "Biosynthetic gene cluster of the non-ribosomally synthesized cyclodepsipeptide skyllamycin: deciphering unprecedented ways of unusual hydroxylation reactions". *J. Am. Chem. Soc.* **133**(16): 6194–6205.

Pylypenko, O., F. Vitali, K. Zerbe, J. A. Robinson and I. Schlichting (2003). "Crystal structure of OxyC, a cytochrome P450 implicated in an oxidative C-C coupling reaction during vancomycin biosynthesis". *J. Biol. Chem.* **278**(47): 46727–46733.
Que Jr., L. and R. Y. Ho (1996). "Dioxygen Activation by Enzymes with Mononuclear Non-Heme Iron Active Sites". *Chem. Rev.* **96**(7): 2607–2624.
Riddick, D. S., X. Ding, C. R. Wolf, T. D. Porter, A. V. Pandey, Q. Y. Zhang, J. Gu, R. D. Finn, S. Ronseaux, L. A. McLaughlin, C. J. Henderson, L. Zou and C. E. Fluck (2013). "NADPH-cytochrome P450 oxidoreductase: roles in physiology, pharmacology, and toxicology". *Drug Metab. Dispos.* **41**(1): 12–23.
Sauguet, L., M. Moutiez, Y. Li, P. Belin, J. Seguin, M. H. Le Du, R. Thai, C. Masson, M. Fonvielle, J. L. Pernodet, J. B. Charbonnier and M. Gondry (2011). "Cyclodipeptide synthases, a family of class-I aminoacyl-tRNA synthetase-like enzymes involved in non-ribosomal peptide synthesis". *Nucleic Acids Res.* **39**(10): 4475–4489.
Solomon, E. I., A. Decker and N. Lehnert (2003). "Non-heme iron enzymes: contrasts to heme catalysis". *Proc. Natl. Acad. Sci. U. S. A.* **100**(7): 3589–3594.
Suzuki, G., A. Minami, M. Shimaya and H. Oikawa (2014). "Analysis of Enantiofacial Selective Epoxidation Catalyzed by Flavin-containing Monooxygenase Lsd18 Involved in Ionophore Polyether Lasalocid Biosynthesis". *Chem. Lett.* **43**: 1779–1781.
Tang, L., S. Shah, L. Chung, J. Carney, L. Katz, C. Khosla and B. Julien (2000). "Cloning and heterologous expression of the epothilone gene cluster". *Science* **287**(5453): 640–642.
Tang, M.-C., Y. Zou, K. Watanabe, C. T. Walsh and Y. Tang (2017). "Oxidative Cyclization in Natural Product Biosynthesis". *Chem. Rev.*, DOI: 101021/acs.chemrev.6b00478.
Uhlmann, S., R. D. Sussmuth and M. J. Cryle (2013). "Cytochrome p450sky interacts directly with the nonribosomal peptide synthetase to generate three amino acid precursors in skyllamycin biosynthesis". *ACS Chem. Biol.* **8**(11): 2586–2596.
Vaillancourt, F. H., E. Yeh, D. A. Vosburg, S. Garneau-Tsodikova and C. T. Walsh (2006). "Nature's inventory of halogenation catalysts: oxidative strategies predominate". *Chem. Rev.* **106**(8): 3364–3378.
Vaillancourt, F. H., E. Yeh, D. A. Vosburg, S. E. O'Connor and C. T. Walsh (2005). "Cryptic chlorination by a non-haem iron enzyme during cyclopropyl amino acid biosynthesis". *Nature* **436**(7054): 1191–1194.

Vaillancourt, F. H., J. Yin and C. T. Walsh (2005). "SyrB2 in syringomycin E biosynthesis is a nonheme FeII alpha-ketoglutarate- and O2-dependent halogenase". *Proc. Natl. Acad. Sci. U. S. A.* **102**(29): 10111–10116.

Valegard, K., A. C. van Scheltinga, M. D. Lloyd, T. Hara, S. Ramaswamy, A. Perrakis, A. Thompson, H. J. Lee, J. E. Baldwin, C. J. Schofield, J. Hajdu and I. Andersson (1998). "Structure of a cephalosporin synthase". *Nature* **394**(6695): 805–809.

van der Donk, W. A., A. L. Tsai and R. J. Kulmacz (2002). "The cyclooxygenase reaction mechanism". *Biochemistry* **41**(52): 15451–15458.

Walsh, C. T. (1980). "Flavin-Coenzymes: At the Crossroads of Biological Redox Chemistry". *Acc. Chem. Res.* **13**: 148–155.

Walsh, C. T. and Y. J. C. Chen (1988). "Baeyer-Villiger Oxidations by Flavin-Dependent Baeyer-Villiger Monooxygenases". *Angew. Chem., Int. Ed.* **27**: 333–343.

Walsh, C. T. and T. Wencewicz (2016). *Antibiotics Challeneges, Mechanisms, Opportunities*. Washington DC, ASM Press.

Walsh, C. T. and T. A. Wencewicz (2013). "Flavoenzymes: versatile catalysts in biosynthetic pathways". *Nat. Prod. Rep.* **30**(1): 175–200.

Yeh, E., L. J. Cole, E. W. Barr, J. M. Bollinger Jr., D. P. Ballou and C. T. Walsh (2006). "Flavin redox chemistry precedes substrate chlorination during the reaction of the flavin-dependent halogenase RebH". *Biochemistry* **45**(25): 7904–7912.

Yeh, E., S. Garneau and C. T. Walsh (2005). "Robust in vitro activity of RebF and RebH, a two-component reductase/halogenase, generating 7-chlorotryptophan during rebeccamycin biosynthesis". *Proc. Natl. Acad. Sci. U. S. A.* **102**(11): 3960–3965.

Zerbe, K., O. Pylypenko, F. Vitali, W. Zhang, S. Rouset, M. Heck, J. W. Vrijbloed, D. Bischoff, B. Bister, R. D. Sussmuth, S. Pelzer, W. Wohlleben, J. A. Robinson and I. Schlichting (2002). "Crystal structure of OxyB, a cytochrome P450 implicated in an oxidative phenol coupling reaction during vancomycin biosynthesis". *J. Biol. Chem.* **277**(49): 47476–47485.

Zi, J., S. Mafu and R. J. Peters (2014). "To gibberellins and beyond! Surveying the evolution of (di)terpenoid metabolism". *Annu. Rev. Plant Biol.* **65**: 259–286.

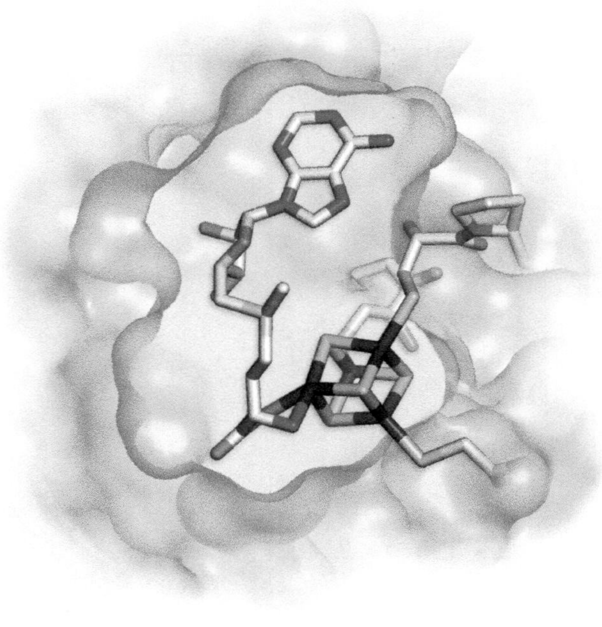

Crystal structure of the active site of the radical SAM enzyme biotin synthase (PDB ID: 1R30).

10 *S*-Adenosyl Methionine: One Electron and Two Electron Reaction Manifolds in Biosyntheses

10.1 Introduction

S-Adenosyl methionine (Figure 10.1) is one of the most widely used coenzymes in both primary and secondary metabolism (Chiang, Gordon *et al.* 1996, Fontecave, Atta *et al.* 2004). It is most fully appreciated as the biological methyl group donor, with hundreds of methyl group transfers to nucleophilic cosubstrate nitrogen, oxygen, and nucleophilic carbon atoms, mediated by a host of specific methyltransferases that act on both macromolecules (DNA > RNA, proteins, membrane phospholipids) and a variety of small molecule metabolites (Struck, Thompson *et al.* 2012). Among the most visible in recent times have been the methyltransferases that target the lysine-rich tails of histone proteins in chromatin, thereby affecting recruitment of partner proteins that mediate transcription of genes (Trievel 2004, Zhang, Wen *et al.* 2012).

The concentrations of intracellular SAM have been estimated from as low as 1 μM to as high as 228 μM (when the SAM synthetase is overexpressed in logarithmically growing *E. coli* cells), and there may be hundreds of thousands (to millions) of SAM-dependent alkylations per cell division cycle. The biosynthesis, utilization, and regeneration of SAM in cellular metabolism are therefore quite extensive (Fontecave, Atta *et al.* 2004) and are outlined in Figure 10.1.

Natural Product Biosynthesis: Chemical Logic and Enzymatic Machinery
By Christopher T. Walsh and Yi Tang

Published by the Royal Society of Chemistry, www.rsc.org

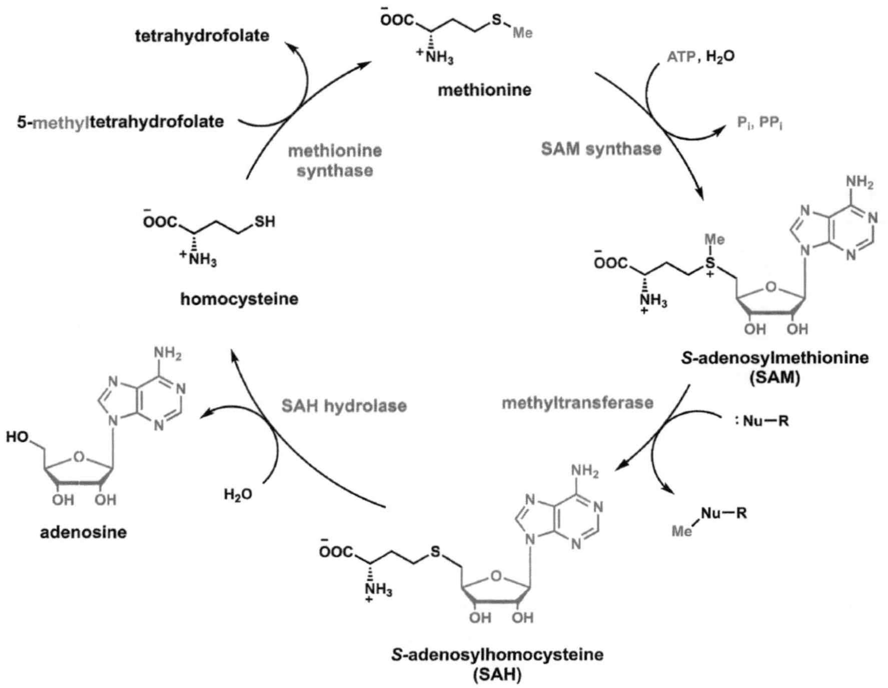

Figure 10.1 Biosynthesis, utilization, and regeneration of *S*-adenosylmethionine (SAM).

SAM is built enzymatically from ATP and methionine by SAM synthetase (Komoto, Yamada *et al.* 2004) (Figure 10.2). The trivalent sulfonium cation, formed when the methionine sulfur is alkylated by $C_{5'}$ of the ribose of ATP, is thermodynamically activated for attack by nucleophiles. SAM is sufficiently uphill thermodynamically that one of the "high energy" phosphoric anhydride bonds in an ATP side chain is broken along with the $C_{5'}$–O bond in each catalytic cycle to drive accumulation of SAM (Figure 10.2). The mechanism of C–S bond formation is not fully clear but can be written as attack of an electron pair of the methionine thioether sulfur atom on $C_{5'}$ of the ribose ring of ATP. The nascent inorganic triphosphate is hydrolyzed to PP_i and P_i before release, pulling on the forward equilibrium. The conformation of the SAM mimic *S*-adenosylethionine bound in the active site of SAM synthetase is shown in Figure 10.3 (Murray, Antonyuk *et al.* 2016).

In principle, SAM is activated for cleavage of any of the three C–S bonds by nucleophilic attack. In the previous chapters, and

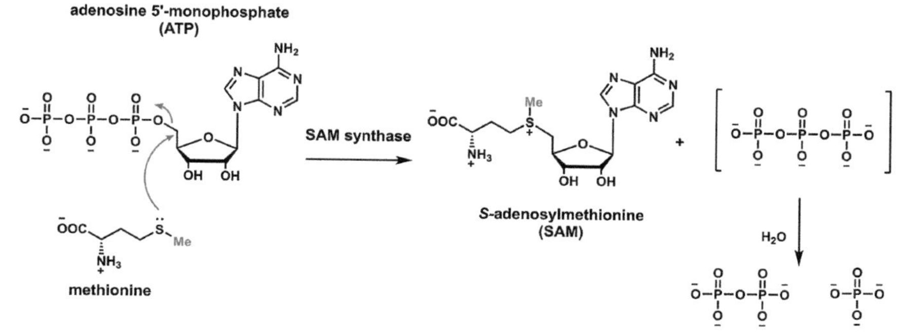

Figure 10.2 Biosynthesis of SAM from ATP and methionine. Cleavage of one of the side chain phosphoric anhydride bonds in ATP is required to drive the unfavorable biosynthetic equilibrium, emphasizing the thermodynamic activation of the sulfonium scaffold.

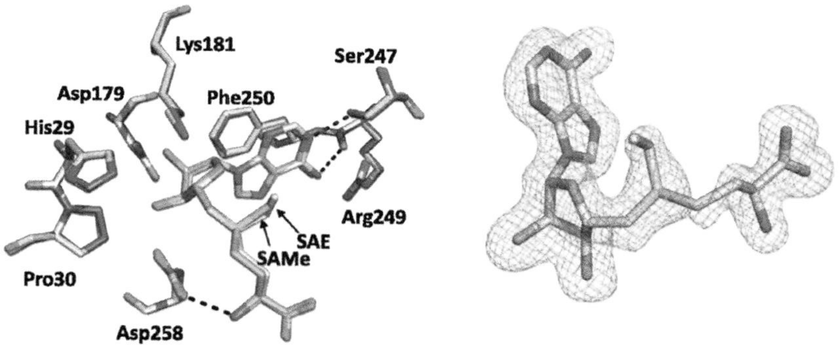

Figure 10.3 X-ray of SAM synthetase and conformation of the SAM analog *S*-adenosylethionine bound in the active site.
Reprinted with permission from Murray, B., S. V. Antonyuk, A. Marina, S. C. Lu, J. M. Mato, S. S. Hasnain and A. L. Rojas (2016). "Crystallography captures catalytic steps in human methionine adenosyltransferase enzymes". *Proc. Natl. Acad. Sci. U. S. A.* **113**(8): 2104.

particularly among the alkaloids, and phenylpropanoids, we have noted dozens of examples of methyl transfers, largely *O*- and *N*-methylations. This is the predominant cleavage mode for the great bulk of SAM that gets metabolized in eukaryotic and prokaryotic cells.

However, Figure 10.4 notes that nothing goes to waste in the SAM scaffold (Fontecave, Atta *et al.* 2004). In addition to cleavage of

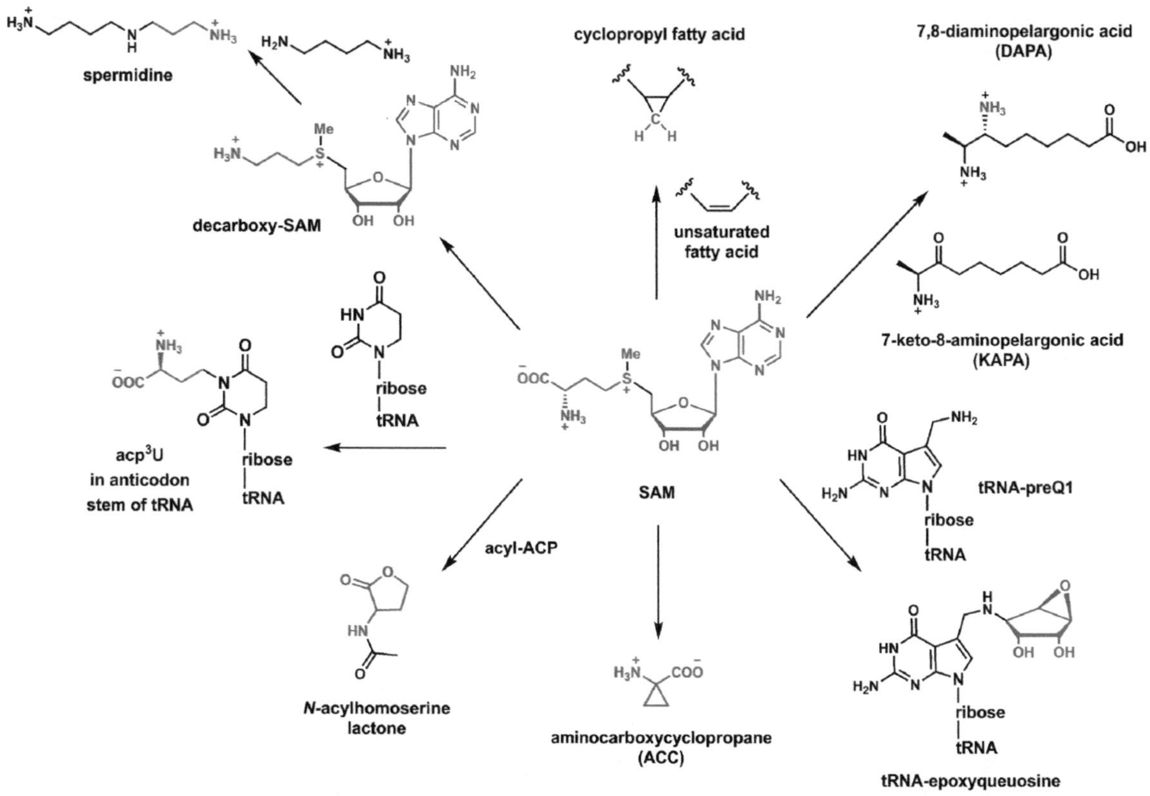

Figure 10.4 Multiple moieties of SAM are transferrable to cellular metabolites and tRNA.
Data from Fontecave, M., M. Atta and E. Mulliez (2014) "S-adenosylmethionine: nothing goes to waste". *Trends Biochem. Sci.* **29**: 243–249.

the S–CH_3 bond of SAM, the S–Cγ bond is cleaved as the methionyl arm is transferrable to N_3 of uridine in tRNA molecules. In Gram-negative bacteria the *N*-acylhomoserine lactones are quorum signaling molecules. They arise from the same kind of methionyl unit transfer to an acyl-*S*-ACP.

Also, after action of SAM decarboxylase, the resultant aminopropyl group is transferrable to putrescine (available from enzymatic decarboxylation of ornithine as noted in Chapter 5), forming spermidine. This flux to polyamine metabolism is significant as polyamines can neutralize negative charges on free DNA, especially in prokaryotes. The amino group of SAM is utilized in a transaminase reaction, specifically in the biosynthesis of the diaminopelargonic scaffold during assembly of the vitamin/coenzyme biotin. Nothing goes to waste (Roje 2006).

Figure 10.4 also depicts two routes to cyclopropane rings from fragments of SAM. In the cyclopropanation of Δ^9-olefinic fatty acids in *E. coli* (*e.g.* oleic acid), the CH_2 moiety of SAM's CH_3 group is transferred to the π electrons of the 9,10-double bond of the unsaturated fatty acid substrate. The formation of aminocarboxycyclopropane (ACC) is a more complex process and will be discussed explicitly below in the context of an aerobic two enzyme pathway to the fruit ripening hormone ethylene.

The last route in Figure 10.4 is fragmentative transfer of the ribose moiety of SAM to the aminomethyl side chain of prequeuosine tRNA (McCarty and Bandarian 2012, McCarty, Krebs *et al.* 2013). A speculative mechanistic route is offered in Figure 10.5, starting with abstraction of one of the two prochiral $C_{5'}$ hydrogens as a proton to enact cleavage of the $C_{1'}$–N connection to the adenine moiety of SAM. The epoxide moiety comes from what had been the acetal $C_{1'}$-oxygen in the step that eliminates the methionyl chain.

10.2 Aerobic Radical Chemistry for SAM

Although the remainder of this chapter will deal with one electron reaction manifolds, radical chemistry of SAM *under anaerobic conditions*, there is one important metabolic transformation in plants, conversion of SAM to the fruit-ripening hormone ethylene, that occurs aerobically. This is an example of nonheme iron oxygenase one electron chemical logic of the type we have described in Chapter 9.

Plants make ethylene in response to a number of signals, including wounding, and stress. Ethylene is also involved in signaling leaf

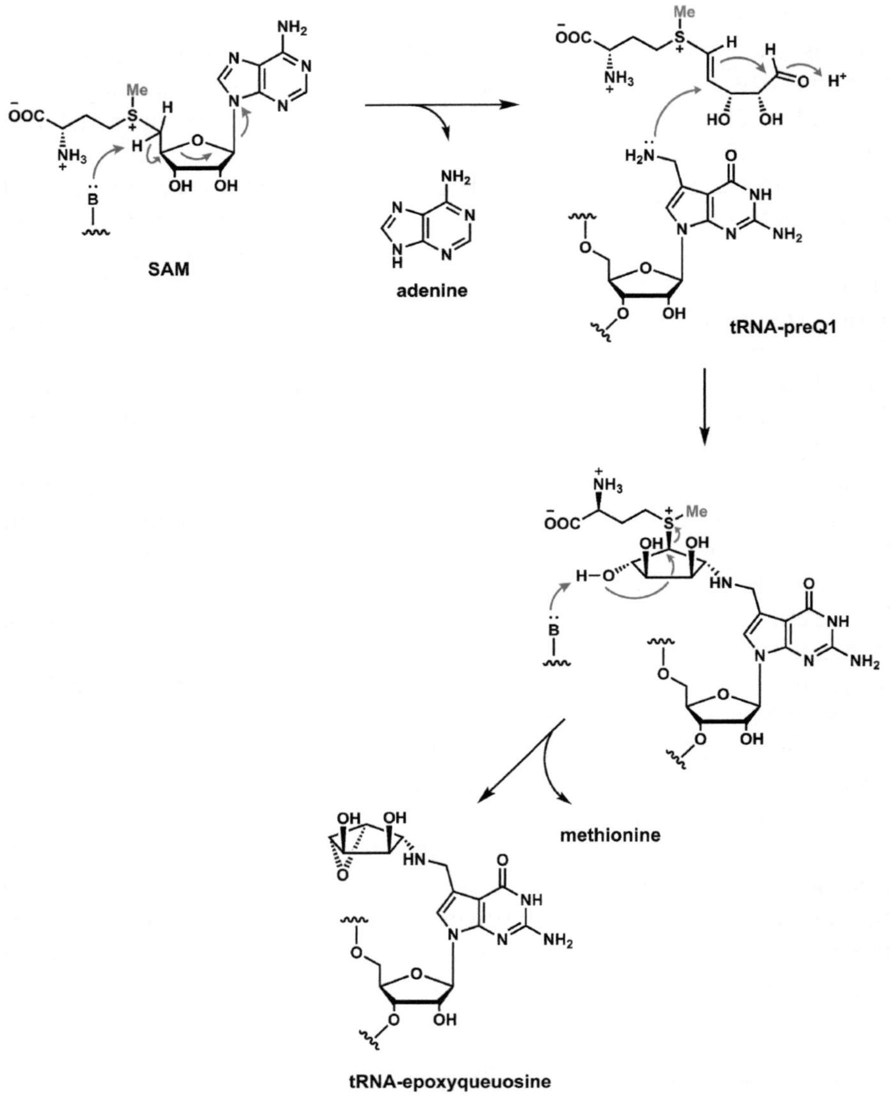

Figure 10.5 SAM as ribosyl group donor in epoxyqueuosine biosynthesis in tRNA molecules.

abscission and fruit ripening. Ethylene biosynthesis in plants was established in 1984 and the chemical logic and enzymatic machinery worked out in large part by the laboratory of S. F. Yang (Peiser, Wang *et al.* 1984, Yang and Hoffman 1984, Byers, Carbaugh *et al.* 2000, Alexander and Grierson 2002).

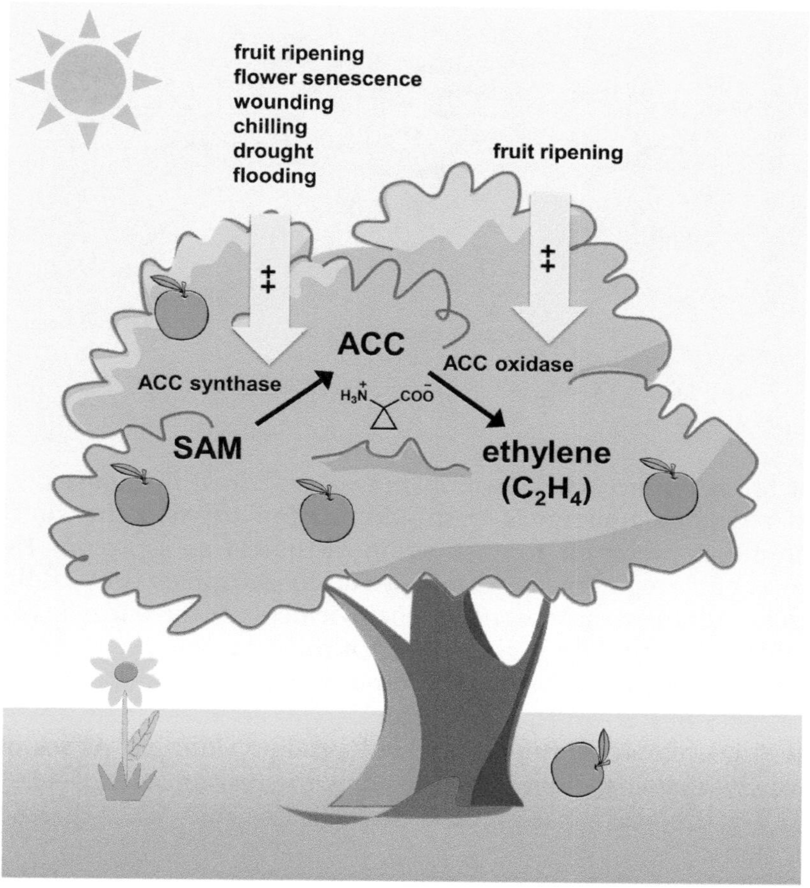

Figure 10.6 Multiple input signals regulate production of the fruit ripening hormone ethylene from SAM.

Ethylene production is regulated by a number of signals in plants (Figure 10.6) and is read out in the amounts and activity levels of two enzymes, aminocyclopropane carboxylate (ACC) synthase (Wang, Li *et al.* 2002) and aminocyclopropane carboxylate oxidase (Charng, Chou *et al.* 2001), respectively (Figure 10.7). Ethylene derives ultimately from C_β and C_γ (C_3 and C_4) of the methionyl arm of SAM by way of the 1-amino-1-carboxy cyclopropane (which represents the full methionyl arm of SAM) (Yang and Hoffman 1984).

Each of these two enzymes offers some mechanistically intriguing chemistry. ACC synthase contains pyridoxal-P in imine linkage to an active site lysine N_6-amine (Figure 10.8). The catalytic logic of this

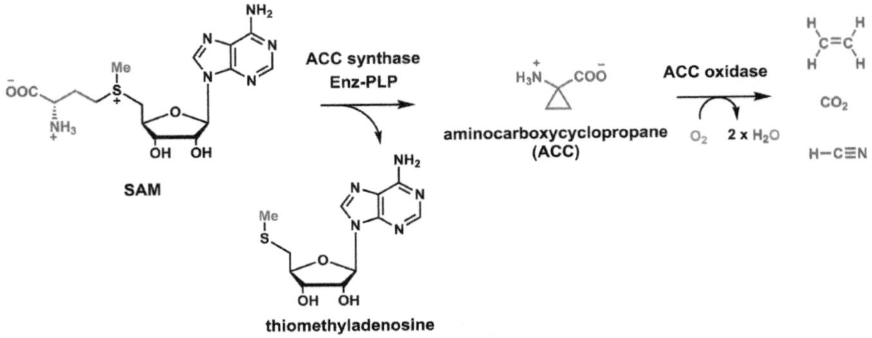

Figure 10.7 SAM to ethylene: ACC synthase and ACC oxidase.

cofactor in amino acid metabolism was described in Chapter 5. The SAM substrate undergoes transaldimination to yield the substrate aldimine as preamble to abstraction of the Hα as a proton. This is a low energy step in the substrate-pyridoxal phosphate (PLP) aldimine as the resultant α-carbanion can be delocalized at the benzylic carbon or all the way into the pyridine ring of the coenzyme. The carbanion density at Cα can be used as internal nucleophile to attack Cγ and break the Cγ–sulfur bond. This expels thiomethyladenosine and generates the cyclopropyl ring in ACC-PLP product aldimine. Attack of the active site Lys side chain amine restores the resting state of ACC synthase for the next catalytic cycle and releases free ACC (Zhang, Ren *et al.* 2004).

The ACC oxidase is a member of the nonheme mononuclear iron oxygenase family described in detail in Chapter 9. ACC and O_2 are cosubstrates. The products are ethylene, cyanide (from C_α carbon and amino group), CO_2 (from the carboxylate group), and two molecules of water. A prototypic Fe(IV)=O is featured as a one electron oxidant of coordinated ACC. This is a remarkable fragmentation. The exact details of the mechanism are not fully clear but radical intermediates are proposed to cause homolytic fragmentation of one of the C–C bonds of the cyclopropyl ring while the other becomes the ethylene product (Figure 10.9). The single bond connecting C_2–N in the ACC amino acid substrate is converted to the CN triple bond of coproduct cyanide ion, for which the mechanism is still obscure (see Vignette 10.V1 for a second route to cyanide in iron-hydrogenases) (Murphy, Robertson *et al.* 2014). Although O_2 is presumed to generate high valent oxo-iron species as the initiating oxidant of one electron chemistry, no oxygen atoms end up in the products.

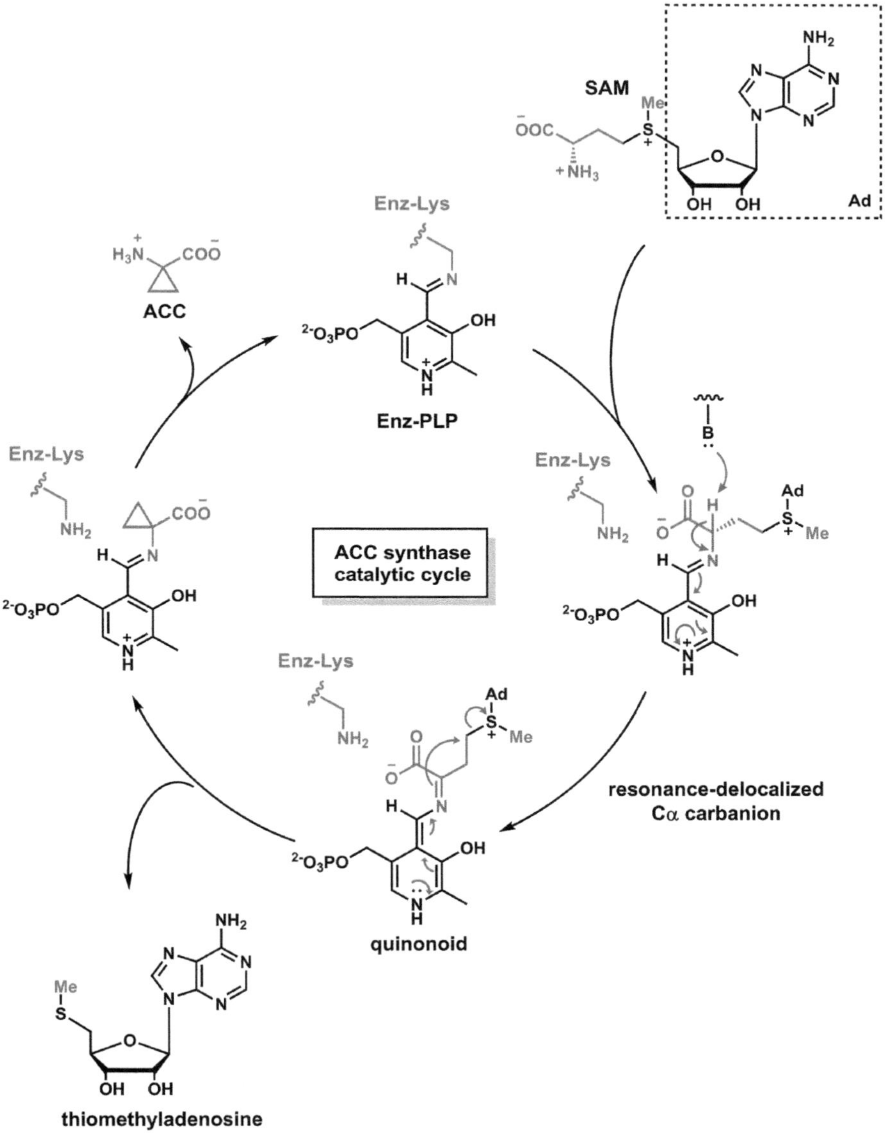

Figure 10.8 SAM to ACC: internal attack of the substrate α-carbanion equivalent on Cγ forms the cyclopropane ring as thiomethyladenosine is expelled.

There are five ethylene receptors in the model plant *Arabidopsis thaliana*, all of them thought to be transmembrane proteins in the endoplasmic reticulum of plant cells (Lacey and Binder 2014). The

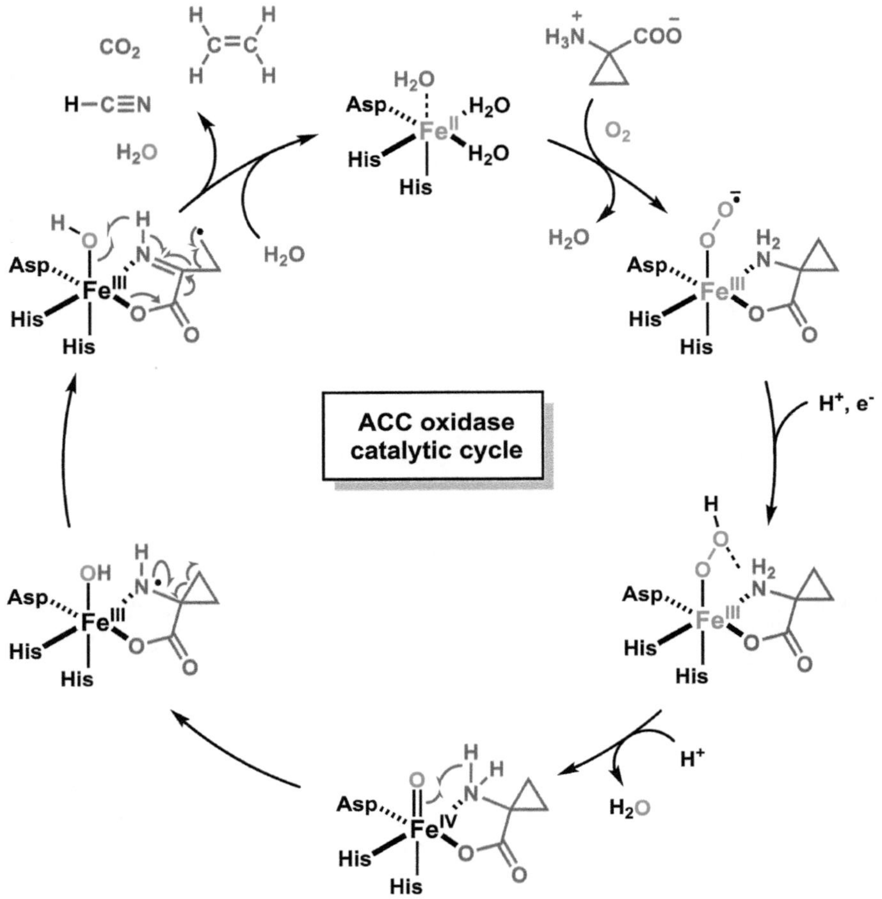

Figure 10.9 A proposed radical path under aerobic conditions *via* a high valent oxo-iron intermediate to fragment ACC to ethylene, cyanide, and CO_2.

receptors contain cuprous ion [Cu(I)] as a metal for coordinating ethylene with nanomolar affinity. Although the receptors have features of transmembrane histidine or serine/threonine kinases, it is not clear that phosphoryl transfer relays convey the downstream signals, which are thought to mediate relief of transcriptional repression of hundreds of genes.

The control of ethylene release from plants, including ripening fruits, has substantial practical market consequences. Two kinds of molecules have been commercialized as retardants of ethylene-induced hormonal effects. One, under the trade name ReTain, is the amino acid aminoethoxyvinylglycine which inhibits the first enzyme, the one generating ACC, by covalent modification of the enzyme active site. The second molecule, EthylBloc, is 1-methylcyclopropene and it competes with ethylene for its receptors. The vinylglycine molecule is used preharvest, Ethylbloc is used postharvest (Alexander and Grierson 2002).

10.3 Anaerobic Radical Chemistry for SAM

10.3.1 Two Routes for Generating Carbon Radicals

Nature has evolved two routes to carbon radical intermediates: molecular oxygen aerobically, as just noted for ACC oxidase, or radical SAM enzymes anaerobically (Figures 10.10 and 10.11). In contrast to the single known aerobic oxygenase that works on the aminobutyryl arm derived from SAM, just noted, there are tens of thousands of entries in genetic data bases for genes with radical SAM imputed functions (Sofia, Chen *et al.* 2001). A small number of these have been purified and characterized for functional activity and one electron reaction pathways with a diverse array of substrates over the past two decades.

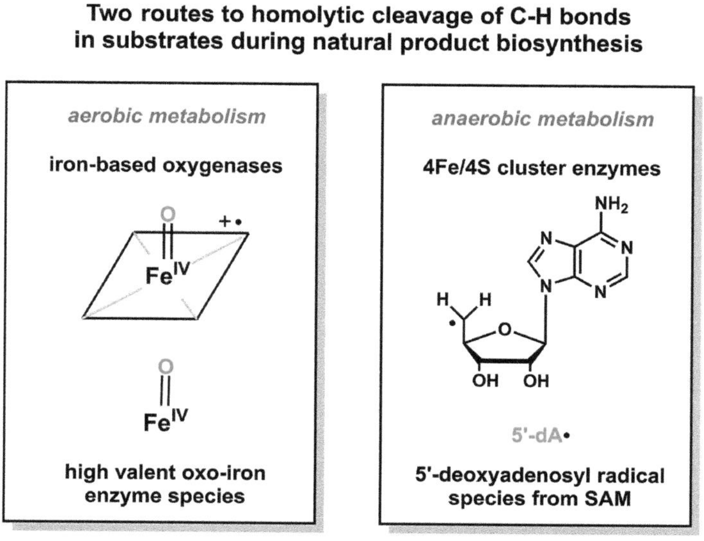

Figure 10.10 Radicals in secondary metabolism: Two routes to homolytic cleavage of C–H bonds of substrates.

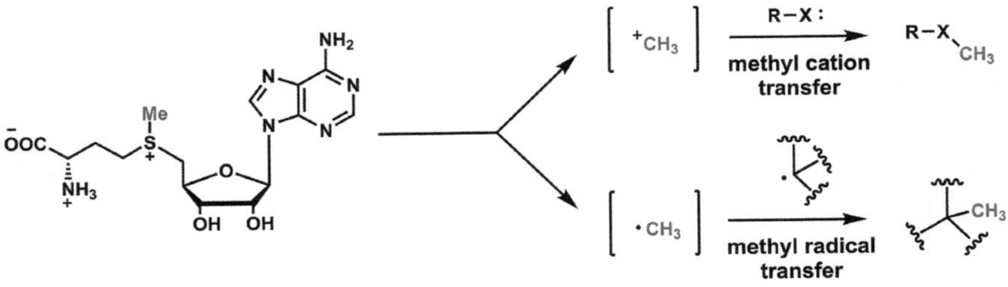

Figure 10.11 SAM can transfer [CH_3^+] or [$CH_3^\bullet$] equivalents, depending on the enzyme and cosubstrate context.

10.3.2 4Fe/4S Cluster as Radical Initiator: 5′-Deoxyadenosyl Radical as Proximal Reagent

For this putatively large class of enzymes that use SAM as a radical generator, and in some presumably large subset of those cases an agent for transfer of $CH_3^\bullet$ rather than CH_3^+, the telltale predictor is a sequence of three cysteines CX_3CX_2C that comprise the three cysteine ligands of a redox active 4Fe/4S cluster. The absence of a typical fourth cysteine ligand allows SAM to bind as a bidentate ligand to one of the iron atoms in the cluster (Figure 10.12) (Broderick, Duffus *et al.* 2014).

Iron–sulfur clusters function in single electron transfers in essentially every protein context where they are found, and the donation of one electron to ligated SAM is the first chemical step in the radical SAM enzyme class. The electron transfer is from the specific iron liganded to SAM to the antibonding orbital of the sulfonium cation–$C_{5'}$ bond (Figure 10.13). The result is conversion of the iron–sulfur 4Fe/4S cluster net oxidation state from +1 to +2 and fragmentation of the S–$C_{5'}$ bond. Methionine remains coordinated but the 5′-deoxyadenosine radical is now free to interact with cosubstrate.

One could wonder whether any of the other two bonds to sulfur in SAM could similarly be cleaved homolytically rather than heterolytically. Figure 10.14 shows that the enzyme Dph2 can in fact cleave the Cγ–S bond homolytically to yield thiomethyladenosine and the corresponding C_4-aminobutyryl carbon radical that is ultimately transferred to a histidine residue of the protein synthesis elongation factor 2 in posttranslational modification that forms diphthamide (Zhang, Zhu *et al.* 2010).

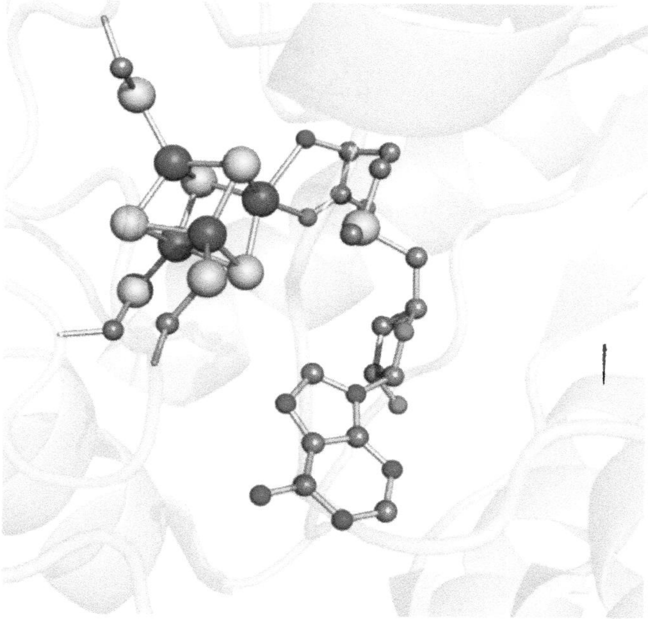

Figure 10.12 SAM as radical reagent requires coordination to the open ligand site in a 4Fe/4S cluster to set up the one electron transfer from 4Fe/4S (PDB ID: 3IIZ).
Data from Broderick, J. B., B. R. Duffus, K. S. Duschene and E. M. Shepard (2014). "Radical S-Adenosylmethionine Enzymes". *Chem. Rev.* **114**: 4229–4317, image created by Yang Hai.

10.3.3 Anaerobiosis Requirement

Essentially all of the radical SAM enzymes characterized to date are extremely sensitive to inactivation and iron–sulfur cluster decomposition by oxygen (Frey, Hegeman *et al.* 2008, Broderick, Duffus *et al.* 2014). The 5′-deoxyadenosyl radical (5-dA•) is likewise susceptible to auto-oxidative diversion. The inbuilt mechanism for radical generation in this enzyme class, iron–sulfur cluster to SAM fragmentation to 5-dA• to substrate radical to product radical, suggests that these mechanisms *evolved in anaerobic organisms.*

It may well be that this class of >50 000 ORFs in the protein data bases represent novel organic chemistry, that is/was difficult to accomplish by two electron reaction manifolds. The early evolution of the heterocyclic frameworks of vitamins noted in the next paragraph

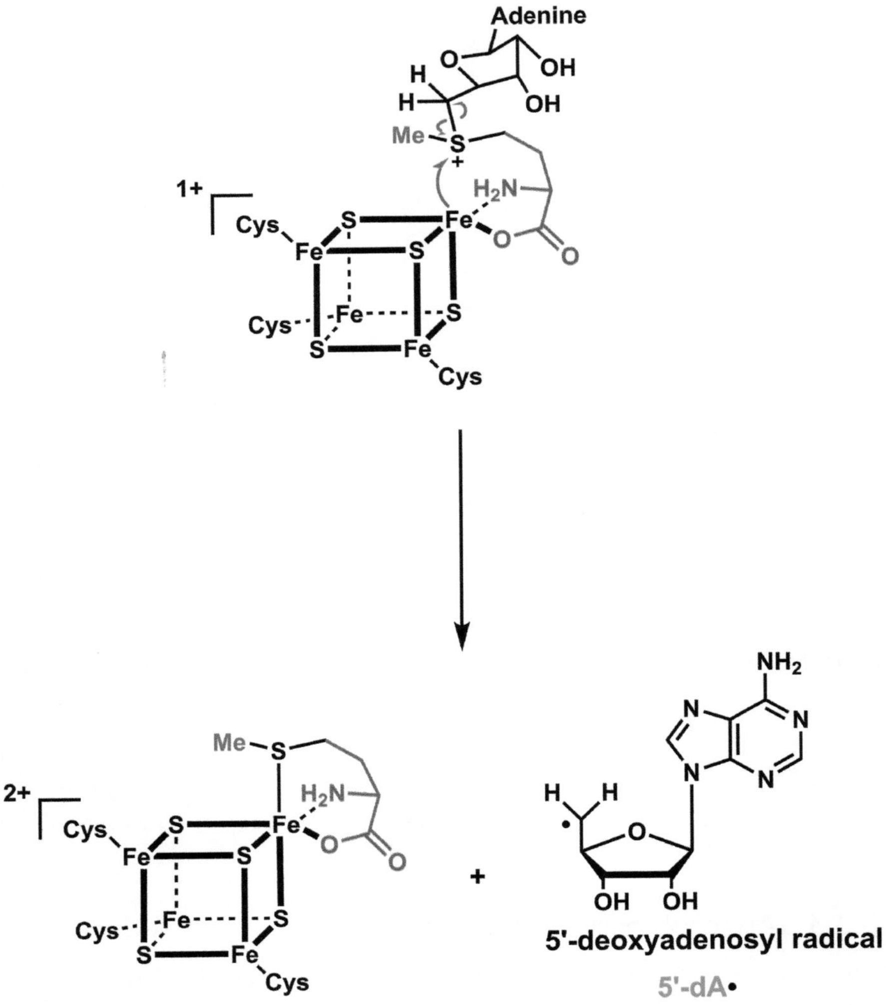

Figure 10.13 One electron transfer by an inner sphere mechanism, from the 4Fe4/4S cluster to the coordinated SAM sulfur atom, cleaves the $C_{5'}$–S bond to release methionine and generate the 5′-deoxyadenosyl radical.

to serve as coenzymes for reactions that apoproteins cannot catalyze on their own suggests that one electron chemistry *via* SAM and its constituent deoxyadenosyl radical may have been centrally important in anaerobic microenvironments (Figure 10.15). They may have allowed establishment of the present inventory of vitamins and coenzymes made and used by contemporary micro- and

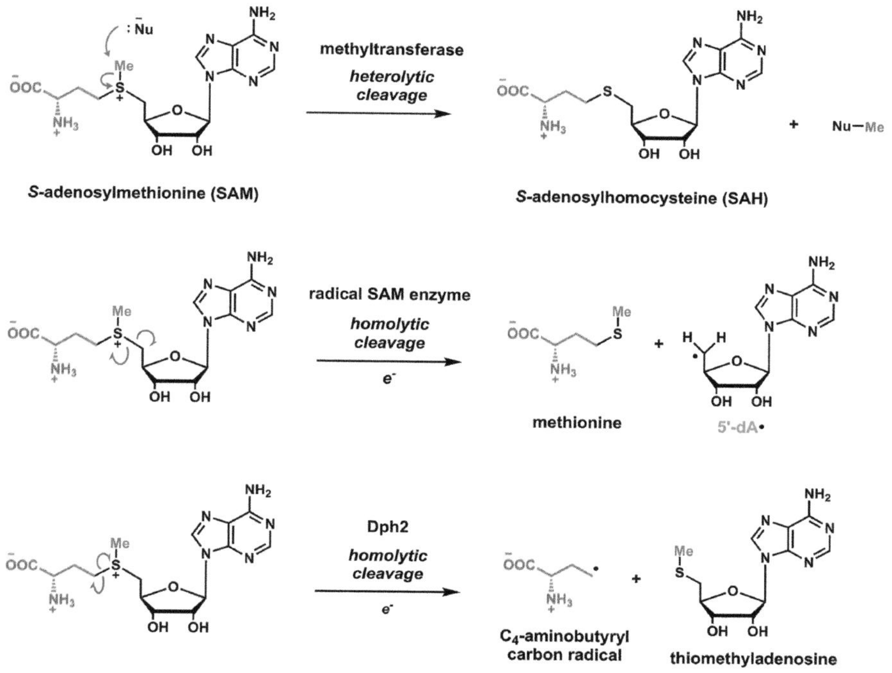

Figure 10.14 Two homolytic cleavage modes *vs.* one heterolytic cleavage mode for SAM.

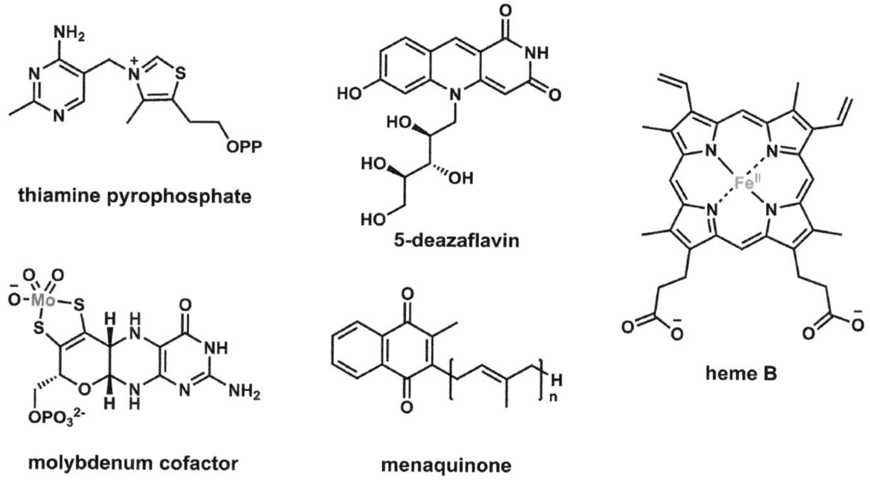

Figure 10.15 Five vitamins and coenzymes that are built utilizing radical SAM enzyme chemistry. This may suggest evolution of these metabolic pathways during anaerobic conditions.

macroorganisms. Clearly, many contemporary microbes can create oxygen-free interiors in enzyme cavities to conduct one electron deoxyadenosyl radical-initiated chemistry.

10.4 Scope of Reactions of Radical SAM Enzymes

Once the 5′-deoxyadenosyl radical has been set loose (Jarrett 2003) it can enable a wide range of homolytic reactions in bound cosubstrates. A subset of these are noted in Figure 10.16. Many of the radical SAM enzymes characterized early on have been involved in the

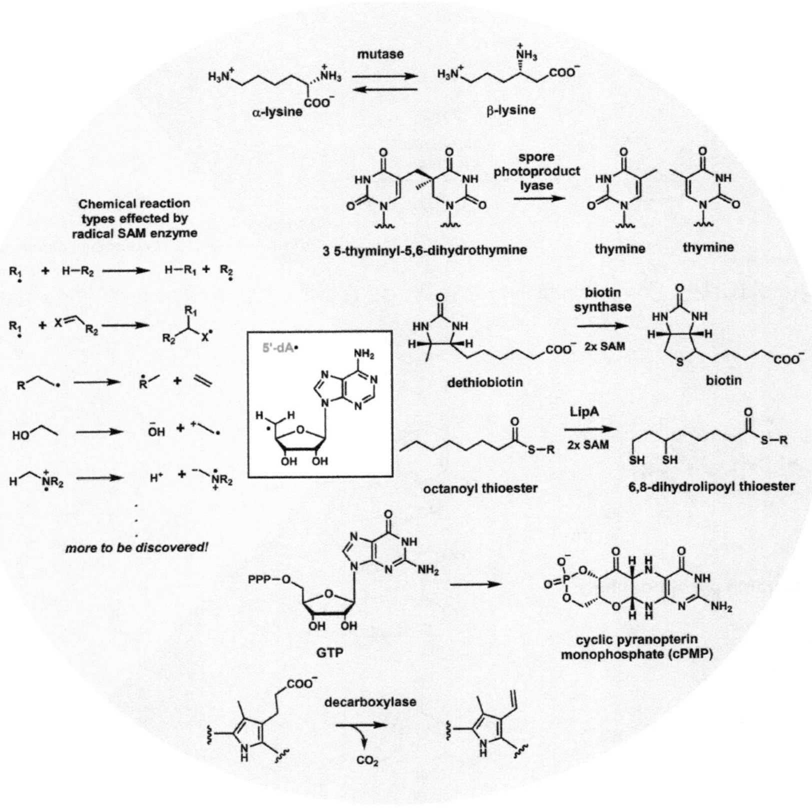

Figure 10.16 Chemical reaction types effected and some reactions catalyzed by radical SAM enzymes. In lysine mutase SAM functions as radical source catalytically. In the other reactions it serves as cosubstrate, cleaved to deoxyadenosine and methionine.

biosynthesis of several types of vitamins, including biotin, lipoic acid, menaquinone, and thiamine (Mehta, Abdelwahed *et al.* 2015). This is also the case for maturation steps of the heme moiety in chlorophyll and bacteriochlorophyll formation, the molybdopterin cofactor, 5-deazaflavin coenzymes, as well as protein bound pyrroloquinoline quinone cofactors (Broderick, Duffus *et al.* 2014). Figure 10.17 shows a more generalized set of radical transformations of this enzyme class to date, with only the surface scratched of the enormous number of as yet uncharacterized ORFs identified bioinformatically.

One useful way to categorize the radical SAM enzyme family is by the fate of SAM and the attendant reaction stoichiometry (Figure 10.17). In the first two reactions of Figure 10.16, lysine mutase and spore product lyase, SAM functions as a coenzyme. It is not consumed. It does not show up in the reaction stoichiometry. It is known that SAM does undergo cleavage to the 5′-deoxyradical in those enzyme active sites (Jarrett 2003) so it must be reformed by the end of the one electron reaction manifolds that convert substrates to products. We will examine how this outcome is achieved in lysine aminomutase and spore product lyase catalysis (Frey, Hegeman *et al.* 2008).

The second major class is where SAM does function as a substrate and undergoes irreversible cleavage during the course of the reaction (Figure 10.17). The products from SAM fragmentations are methionine and 5′-deoxyadenosine. Investigation of the source of the hydrogens at the 5′-methyl of deoxyadenosine for such transformations shows that one of the methyl hydrogens was transferred from substrate to the coproduct, 5′-deoxyadenosine (Hutcheson and Broderick 2012, Broderick, Duffus *et al.* 2014). We shall illustrate with examples below.

There is an additional variant to the radical SAM class where SAM is *consumed* as substrate. This is the subset of enzymes which carry out net methylation of a cosubstrate. In these cases two SAM are consumed, to give two distinct sets of products (Figure 10.17). The first SAM has the same role as above, undergoing fragmentation to methionine and 5′-deoxyadenosyl radical. That generates a substrate radical with $H^•$ transfer to give the 5′-deoxyadenosine product. The substrate radical can then react with the second molecule of SAM for transfer of a $[CH_3^•]$ equivalent. The *S*-adenosylhomocysteine radical thereby generated can give an electron back to the +2 oxidation state of the 4Fe/4S cluster, returning it to its resting +1 oxidation state and yielding the SAH product. We will take up specific examples of this two SAM reaction stoichiometry. The methylations that transfer $[CH_3^•]$ instead of the canonical $[CH_3^+]$ unit from SAM are typically to non-nucleophilic, unactivated carbon sites incompetent to instigate a methyl cation transfer.

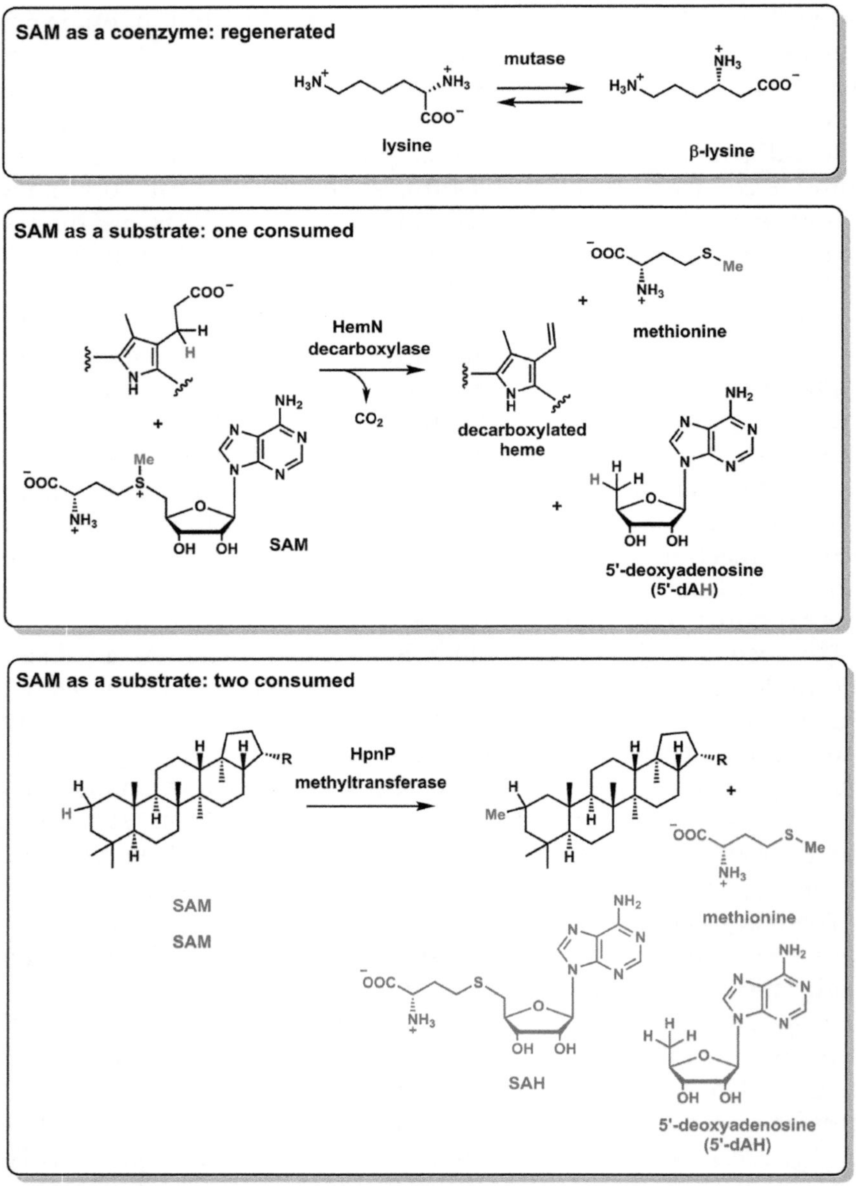

Figure 10.17 SAM can generate 5′-deoxyadenosyl radicals during a catalytic cycle and suffer two fates. It can function coenzymatically and be regenerated at the end of each catalytic cycle. Or it can be consumed as a substrate, undergoing irreversible cleavage to methionine and 5′-deoxyadenosine. In the subset where *C*-methylation occurs, two SAM molecules function as substrates, undergoing cleavage to two sets of mechanistically diagnostic products.

10.5 SAM as Coenzyme

10.5.1 Lysine-2,3-aminomutase

Certain strains of bacteria incorporate the nonproteinogenic amino acid β-lysine into peptide antibiotic scaffolds. The enzyme 2,3-lysine aminomutase supplies the β-lysine from the cellular pool of the primary metabolite α-lysine. As noted in Figure 10.18,

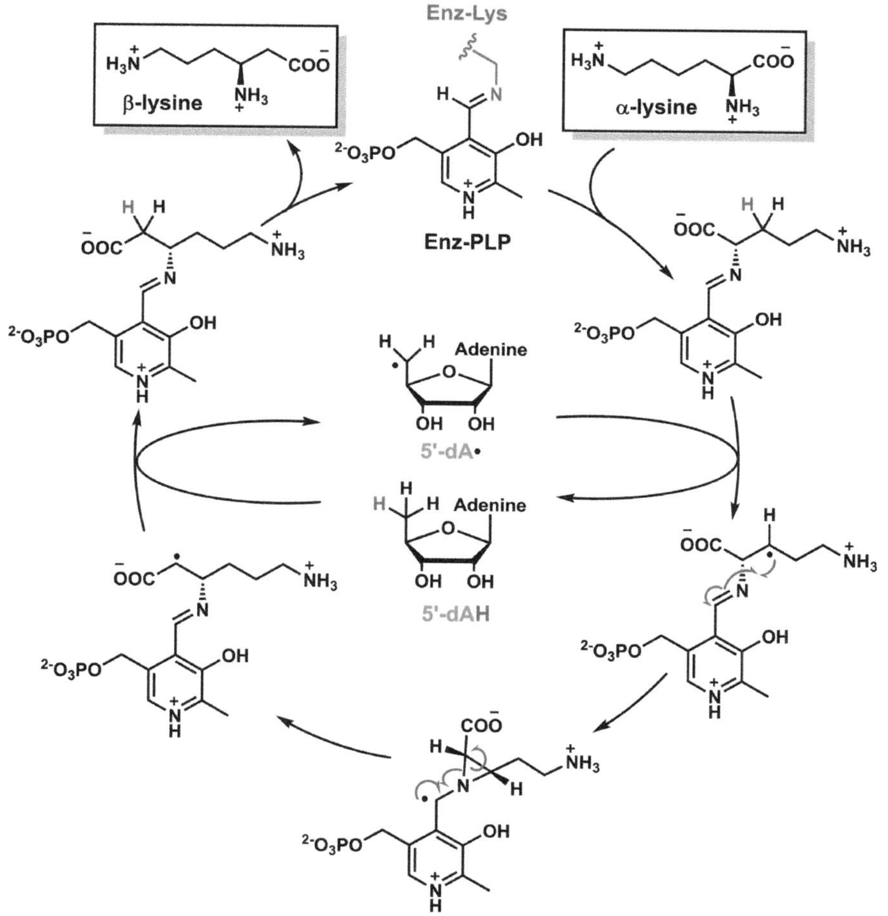

Figure 10.18 Lysine-2,3-aminomutase is a radical SAM enzyme that also contains a pyridoxal-phosphate coenzyme docked at its active site in an imine linkage to a Lys-side chain. 5′-Deoxyadenosyl radical is the initiator, removing a substrate beta hydrogen as H•. An aziridinyl benzyl radical on the coenzyme precedes the β-Lys α radical that abstracts an H• back from 5′-deoxyadenosine.

Lys-2,3-aminomutase is a radical SAM catalyst and purifies, in an active form under anaerobic conditions, with an intact 4Fe/4S cluster. It also contains one equivalent of the aldehyde form of vitamin B6, pyridoxal-phosphate, in an aldimine to the side chain of an active site lysine residue in the resting state. The substrate in the front direction (the reaction is reversible) is α-Lys. Finally, the enzyme shows an absolute requirement for a catalytic amount of SAM, which is not consumed during the mutase reaction (Frey, Hegeman *et al.* 2008).

The reaction begins with the standard transaldimination in which substrate Lys displaces the enzyme side chain to produce the Lys-PLP aldimine that is the default starting point for all PLP-dependent enzymes. Now the one electron transfer manifold kicks in. Electron transfer from the 4Fe/4S cluster cleaves coordinated SAM into methionine and the 5′-deoxyadenosyl radical. One of the prochiral hydrogens at the C_3 methylene group of the Lys-PLP adduct is abstracted as $H^•$ and forms 5′-deoxyadenosine, which remains bound in the active site. Radical density can be detected by EPR measurements at C_3 and C_2 of Lys during catalytic turnover with appropriate substrates and analogs and also at the benzylic carbon of PLP. Thus, the aziridinyl radical is featured as an intermediate that can open in the direction from which it came or open also to the C_2 radical form of Lys-PLP.

This has accomplished the migration of the amino group from C_2 to C_3, from α-Lys to β-Lys framework. The C_2 radical can be quenched by back transfer of an $H^•$ from 5′-deoxyadenosine. At this stage β-Lys formation is complete and transaldimination would give the resting enzyme PLP imine and released amino acid. Before that happens the SAM gets regenerated by reversal of the initial one electron transfer step. The methionine is still coordinated to the iron–sulfur cluster. The 5′-deoxyadenosyl radical can give back one electron to the 4Fe/4S cluster in the +2 oxidation state as the C–S bond to C_γ of methionine reforms. The catalytic role of SAM is thus not evident from inspection of the reaction stoichiometry.

10.5.2 Spore Photoproduct Lyase

A second example of *fully reversible* cleavage of SAM to Met and 5′-deoxyA radical occurs during photo-repair of the DNA lesion generated by UV crosslinking of two thymine residues in DNA. Figure 10.19 sketches a mechanistic path for hydrogen atom transfer to a transient 5′-deoxyadenosyl radical (Cheek and Broderick 2001). The thymine 4,4-dimer radical can now fragment as indicated, restoring one thymine and yielding the exocyclic methylene radical of the other. Back transfer of

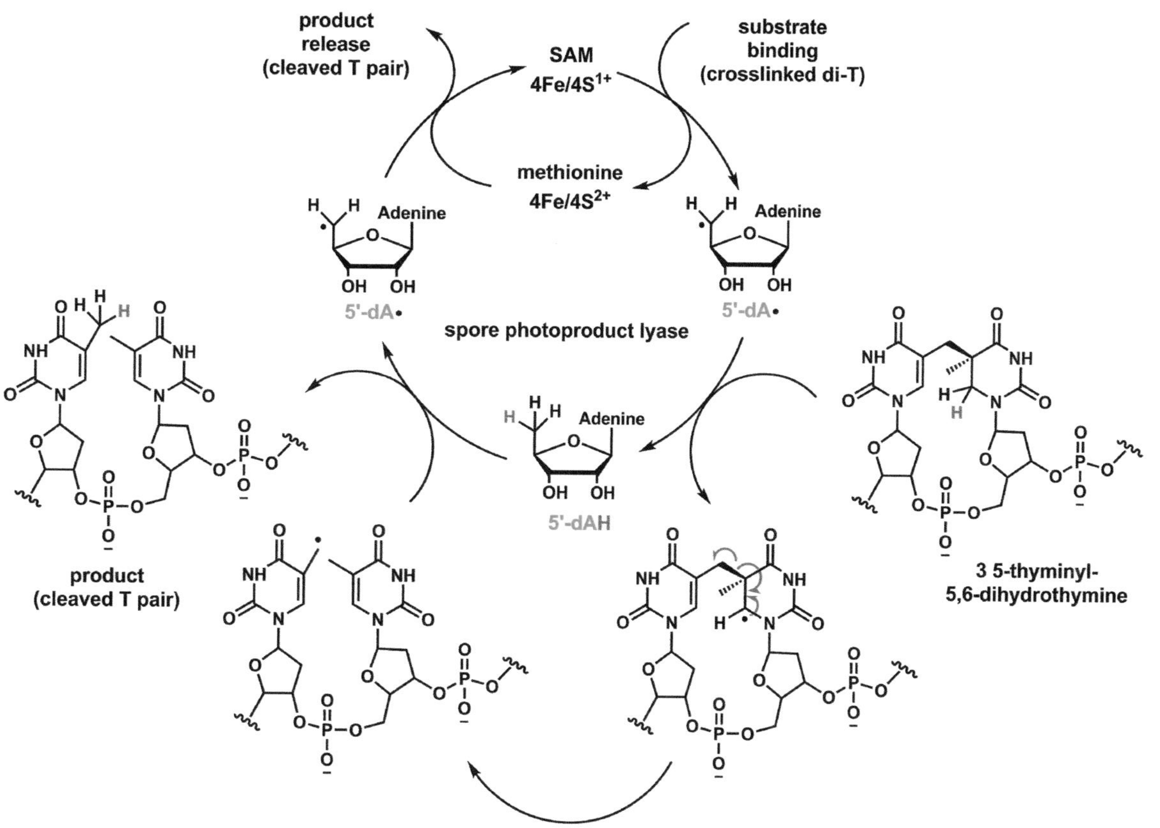

Figure 10.19 Repair of interstrand 4,4′-crosslink to release two restored thymine *via* a one electron pathway, initiated by the 5′-deoxyadenosyl radical.

one of the hydrogens from 5′-deoxyA as an H• atom forms the second, repaired thymine residue in the DNA strand. It also yields the 5′-deoxyA radical for transfer of an electron back to the $4Fe/4S^{+2}$ oxidation state to regenerate SAM to function in the next catalytic cycle.

10.5.3 Pyrrolysine and Queuosine Biosynthesis

Two more examples of SAM undergoing reversible one electron pathway cleavage and the resultant 5′-deoxyadenosyl radical acting as the proximal catalyst are exemplified in two scaffold rearrangements in biosynthesis of pyrrolysine and the 7-deazaguanine framework of queueosine, respectively. Figure 10.20 shows a schematic for rearrangement of L-lysine to 3-methyl-D-ornithine *en route* to pyrrolysine, the 22nd proteinogenic amino acid in some microbes (Gaston, Zhang *et al.* 2011).

As in the lysine amino mutase reaction, a Lys-β radical is formed but this time as the free amino acid. A homolytic fragmentation yields a glycyl radical and an olefin fragment. Recombination at the other terminus of the olefin produces the branched chain ornithinyl $CH_2^{\bullet}$. Back transfer of H• from 5′-deoxyadenosine yields the rearranged 3-methyl-D-ornithine product and the 5′-dA• is set up to recombine with methionine and regenerate SAM.

Figure 10.21 depicts a pathway for conversion of 6-carboxypterin to the 7-deazaguanine scaffold in queuosine, a modified tRNA base (also see Figure 6.14). A ring contraction of a 6-6 to a 6-5 bicyclic system is the key step, featured as initial cleavage of the C_6–H bond of the pterin substrate homolytically to set up ring opening and then ring contraction to the exocyclic amine radical. H• back transfer from 5′-dA gives the neutral amine product and 5′-dA• for regeneration of SAM. Loss of the amine group as ammonia is postulated to occur by neighboring group assistance by the pyrimidine exocyclic imine. Tautomerization yields the rearranged carboxy deazaguanine product (McCarty, Somogyi *et al.* 2009, Dowling, Bruender *et al.* 2014).

10.6 SAM as Consumable Substrate: No Methyl Transfers

10.6.1 Biotin Synthase and Dihydrolipoate Synthase

Biotin synthase and lipoic acid synthase are examples in this radical SAM class that have been studied for the longest intervals (Booker 2009) (Broderick, Duffus *et al.* 2014).

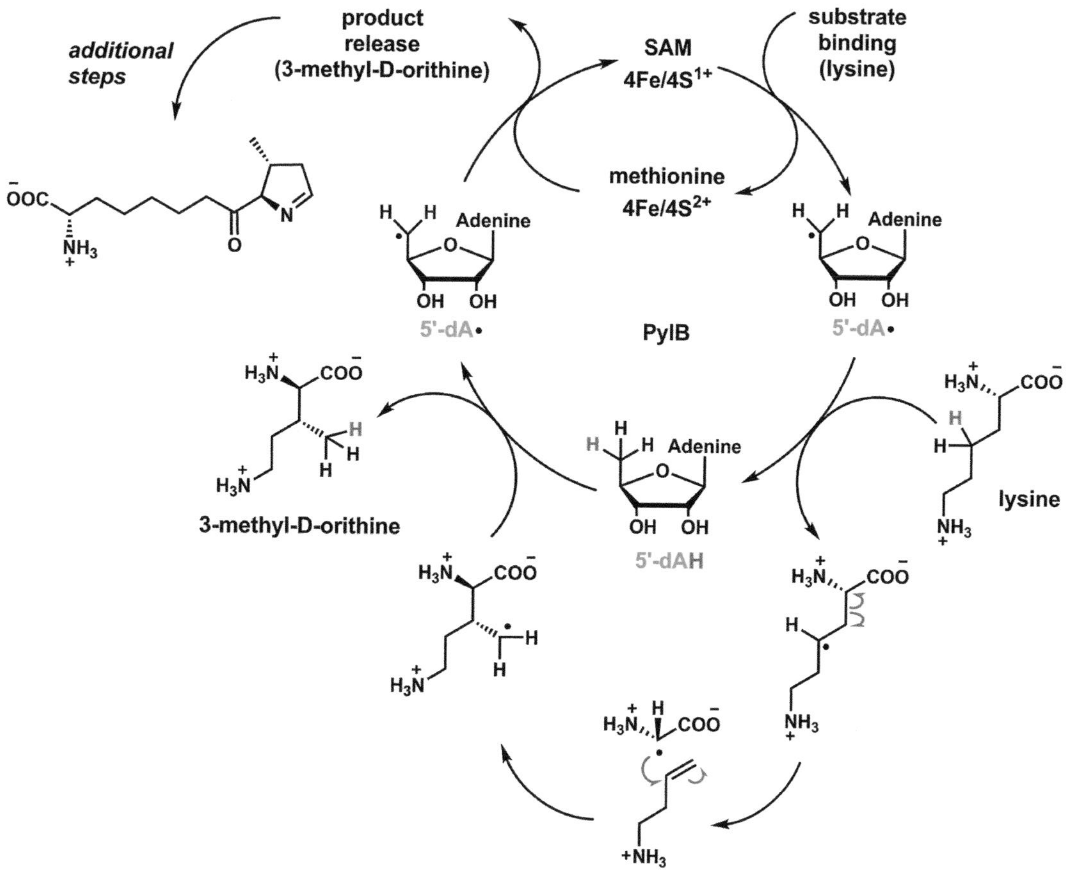

Figure 10.20 Formation of pyrrolysine from lysine by a radical fragmentation and recombination process initiated by an enzyme-bound 5′-deoxyadenosyl radical.

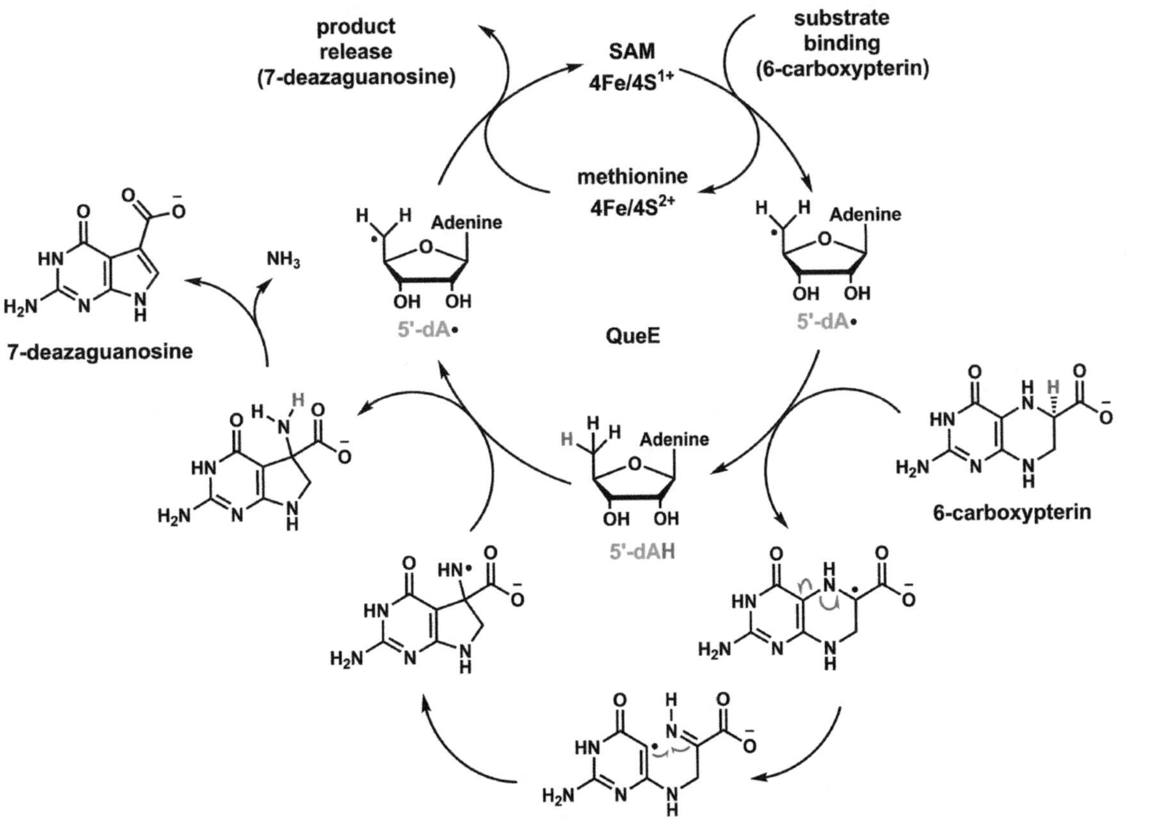

Figure 10.21 6-Carboxypterin to 7-deazguanine in the queuosine pathway; SAM as radical coenzyme, regenerated in each catalytic cycle.

In completion of the thiane ring of the biotin coenzyme the sulfur source is one of the sulfurs in a 2Fe/2S cluster. Two tandem radical reactions occur, each consuming a SAM cosubstrate and fragmenting it to 5′-deoxyadenosine and methionine, the product profile indicative of 5′-dA$^{\bullet}$ intermediacy for substrate hydrogen atom abstraction. The first such abstraction is from the pendant CH_3 on the desthiobiotin substrate, allowing C–S bond formation as the $-CH_2^{\bullet}$ combines with a sulfur atom being extracted from the 2Fe/2S cluster (Figure 10.22). To close the thiane ring requires another carbon radical *via* a second catalytic cycle and one electron transfer from the –SH substituent as the second C–S bond forms (Booker, Cicchillo *et al.* 2007).

The generation of two C–SH bonds in dihydrolipoate follows equivalent one electron logic (see Figure 10.16) (Cicchillo, Iwig *et al.* 2004). To build each of the C–S(H) bonds consumes one SAM cosubstrate molecule as the methylene carbon at C_6 and the methyl carbon at C_8 of the octanoyl thioester cosubstrate undergo $H^{\bullet}$ atom transfer to the transient 5′-dA$^{\bullet}$ species. The C_6 $CH^{\bullet}$ and the C_8 $CH_2^{\bullet}$ form C–SH bonds by homolytic coupling with sulfur from the Fe/S cluster.

10.6.2 Decarboxylation of Coproporphinogen III to Protporyphyrin IX in Heme and Chlorophyll Biosynthesis

In organisms that generate chlorophyll for photosynthesis there are 15 enzymes in the pathway, the first nine in common with heme assembly and the next six specific to chlorophyll maturation from the heme core. Two separate types of coproporphyrinogen oxidative decarboxylation enzymes in the common early part of the pathway for heme and chlorophyll have evolved.

HemF is an oxygen-dependent enzyme that catalyzes the conversion of two of the propionyl side chains (in rings A and B) to the vinyl side chains in protoporphyrinogen IX (Breckau, Mahlitz *et al.* 2003). The mechanism of HemF is not clear (one could suppose hydroxylation and water elimination among multiple possibilities). In parallel, HemN, *e.g.* from *E. coli*, carries out the same reaction under *anaerobic* conditions. Rather than being O_2-dependent, HemN is inactivated by O_2 by oxidative destruction of its 4Fe/4S cluster. Electrons are fed into the 4Fe/4S cluster (to allow subsequent single electron passage one at a time to coordinated SAM) of HemN from an NADH-reducing flavoprotein and a flavodoxin as two/one electron stepdown machinery. Figure 10.23 indicates stereospecific abstraction of one of the

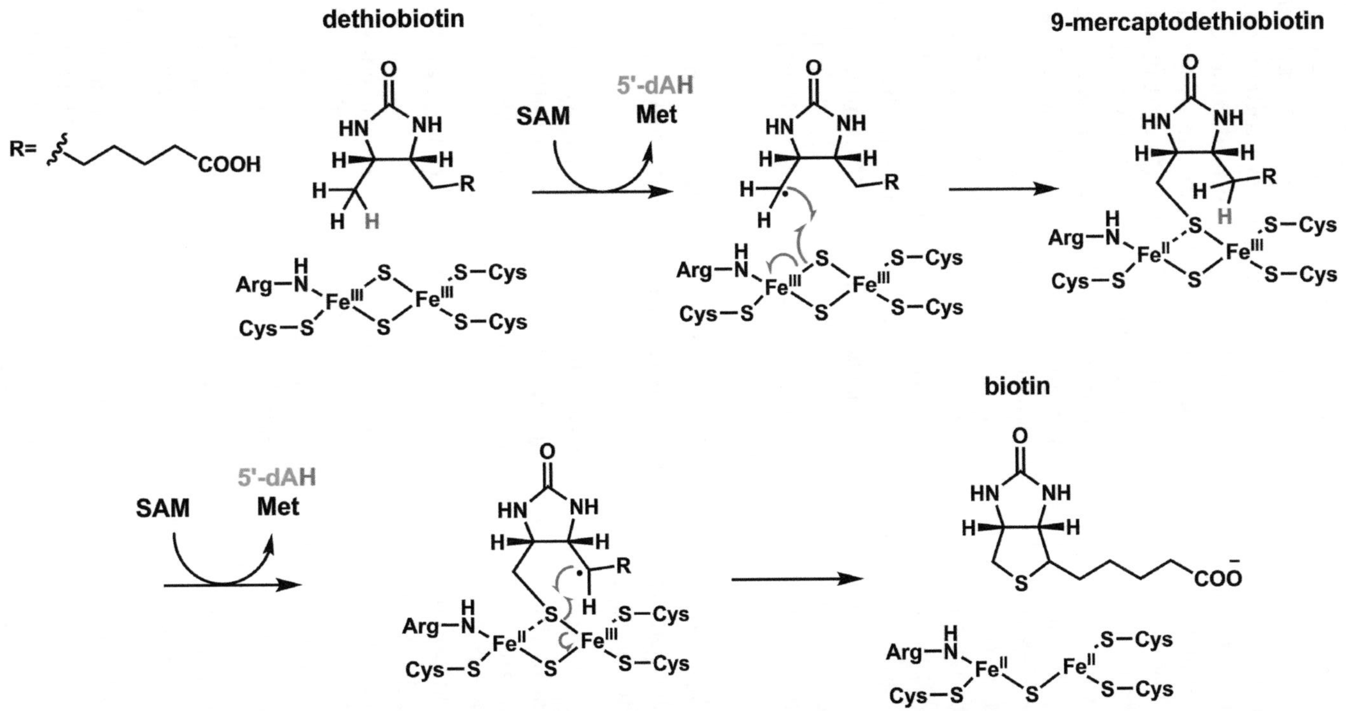

Figure 10.22 Sulfur insertion in biotin formation. Each of the two C–S bonds formed consumes one SAM *via* 5′-deoxyadenosyl radicals that are quenched by transfer of H• from substrate.

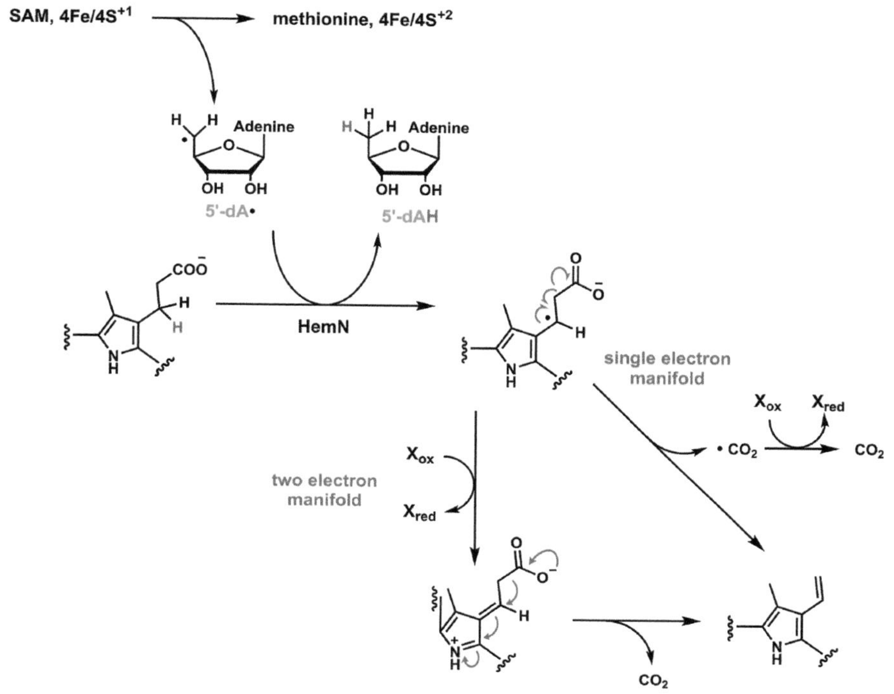

Figure 10.23 A radical pathway for decarboxylation of propionyl to vinyl side chains in protoporphyrinogen IX formation.

two prochiral C_2 hydrogens on a propionyl side chain (*e.g.* on ring A of substrate) as a hydrogen atom by 5′-dA•. Possible one electron and two electron reaction manifolds are shown but there is evidence for radical density favoring the one electron route. The unpaired electron is handed back to the 4Fe/4S cluster to reset the pyrrole aromaticity. This catalytic cycle is repeated for B ring decarboxylation as well, with consumption of one SAM to methionine and 5′-deoxyadenosine in each cycle (Layer, Moser *et al.* 2003, Layer, Kervio *et al.* 2005).

Vignette 10.1 Tyrosine Fragmentation to CO and CN.

The lowest molecular weight natural product is hydrogen gas, H_2. Microbial hydrogenases use nickel–iron, iron–iron, and iron clusters to convert protons and electrons to H_2 and *vice versa*. The Fe–Fe hydrogenases can display catalytic rate constants for H_2 production of 10 000/second. The active site has a conventional 4Fe/4S cluster linked to an unusual 2Fe cluster that has three carbon monoxide ligands (CO) and two cyanide ion ligands (CN) and an incompletely characterized bridging thiolate ligand (Figure 10.V1) (Mulder, Boyd *et al.* 2010).

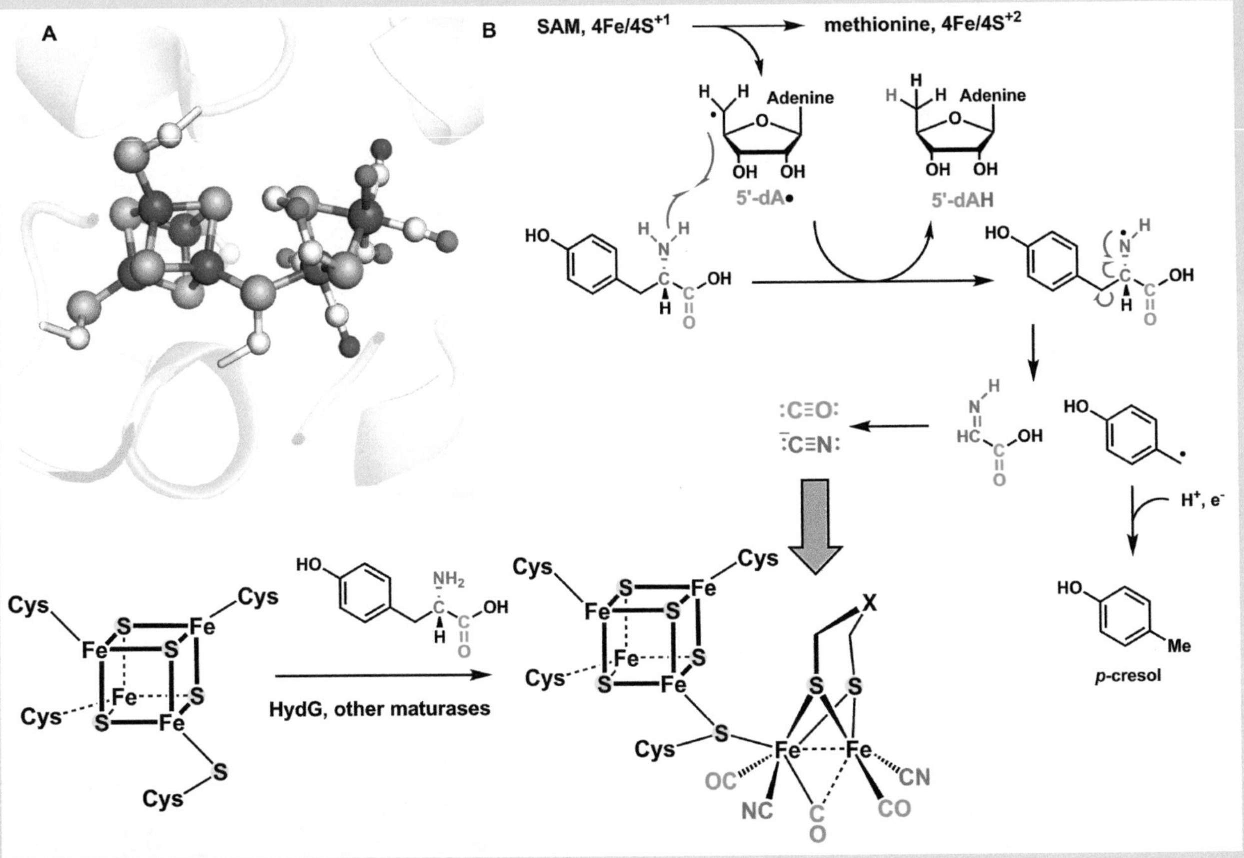

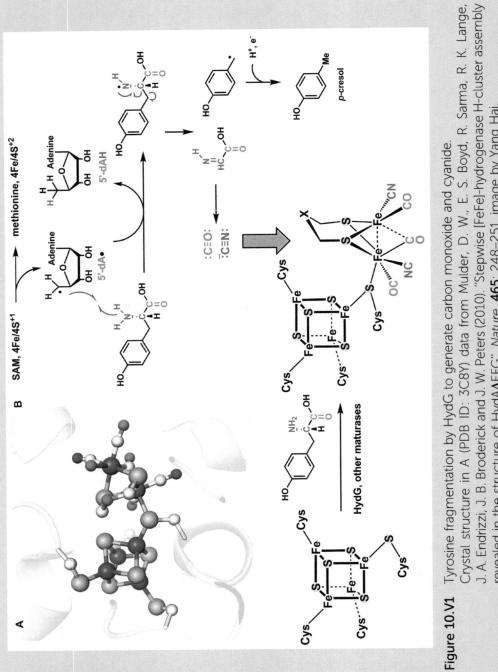

Figure 10.V1 Tyrosine fragmentation by HydG to generate carbon monoxide and cyanide. Crystal structure in A (PDB ID: 3C8Y) data from Mulder, D. W., E. S. Boyd, R. Sarma, R. K. Lange, J. A. Endrizzi, J. B. Broderick and J. W. Peters (2010). "Stepwise [FeFe]-hydrogenase H-cluster assembly revealed in the structure of HydAΔEFG". *Nature*. **465**: 248–251, image by Yang Hai.

The cyanide and CO molecules are generated by action of a maturase, HydG, *e.g.* from *Shewanella oneinodensis*, for building the 2Fe cluster in the Fe–Fe hydrogenases (Kuchenreuther, Myers *et al.* 2013). This is a radical SAM enzyme that cleaves a molecule of *S*-adenosylmethionine (SAM) ligated to one of two 4Fe/4S clusters in HydG. The N-terminal 4Fe/4S cluster delivers an electron to bound SAM that cleaves into methionine and a 5′-deoxyadenosyl radical. Fragmentation of the C_2–C_3 bond of that Tyr• with intramolecular assistance from the amine nitrogen would yield the para-quinonoid radical form of para-cresol and an iminoglyxoylate bound to the second 4Fe/4S cluster (not shown). One electron transfer and protonation give cresol. Meanwhile, fragmentation of the imino glyoxalate across the C_1–C_2 bond is proposed to give both CO and CN as ligands coordinated to the 4Fe/4S cluster and presumably subsequently transferable from HydG to the Fe–Fe hydrogenase.

10.6.3 Futalosine on the Way to Naphthoquinones

The naphthoquinone menaquinone, a key constituent of electron transport chains in bacteria, can be synthesized in prokaryotes by two independent routes from the central prearomatic metabolite chorismate. In addition to the long known classical *Men* pathway (MenFDHCEB), a novel four enzyme pathway (MqnABCD) to 1,4-dihydroxy-2-naphthoate was recently discovered (Hiratsuka, Furihata *et al.* 2008) (Figure 10.24). The MqnABCD route involves a nucleoside futalosine, appears to be widespread in anaerobes, may have preceded evolution of the aerobic pathway (Zhi, Yao *et al.* 2014), and has three predicted radical SAM enzymes, MqnA, MqnC, and MqnE (Mahanta, Fedoseyenko *et al.* 2013).

MqnA generates and reacts the 5′-deoxyadenosyl radical with the double bond of the enolpyruvyl ether moiety of 3-hydroxybenzoate to join the deoxyadenosine framework to the chorismate framework. A set of one electron rearrangements proposed in Figure 10.25 yield futalosine. This is the first example where SAM is not only split to generate the 5′-deoxyadenosyl radical as initiator of substrate-mediated one electron reaction manifolds but the 5′-deoxyadeosine moiety is also covalently incorporated into a reaction intermediate. Later in the pathway, after loss of the adenine base hydrolytically, MqnC uses the nascent 5′-dA• for $C_{4'}$-hydrogen atom abstraction on the ribose ring (Figure 10.25). Radical propagation leads to the 6,6-bicyclic ring system required in pathway product naphthoquinone.

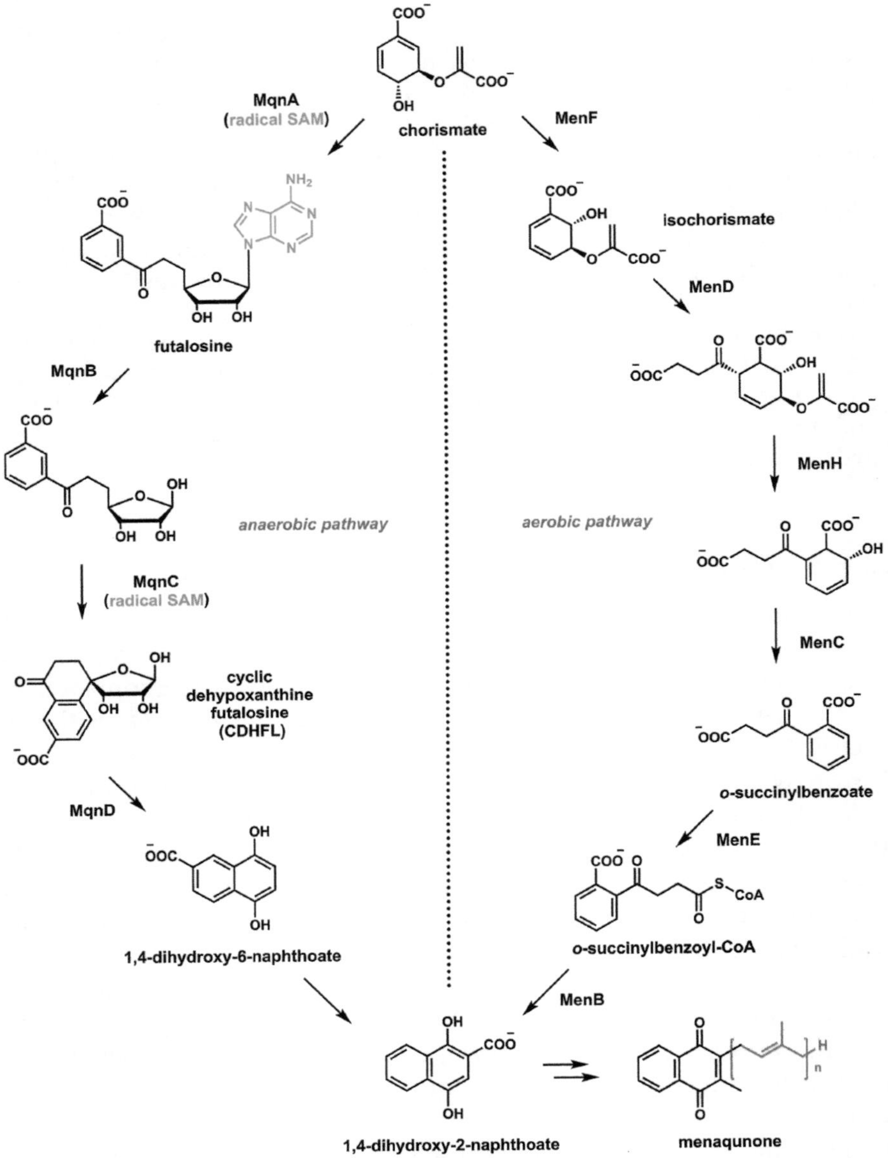

Figure 10.24 Two biosynthetic pathways to menaquinone: the aerobic *Men* pathway and the anaerobic *Mqn* pathway.

10.6.4 Additional Examples of SAM as Radical Initiator

Before leaving this survey of reactions in which SAM serves as progenitor to the 5′-deoxyadenosyl radical, we note that two of

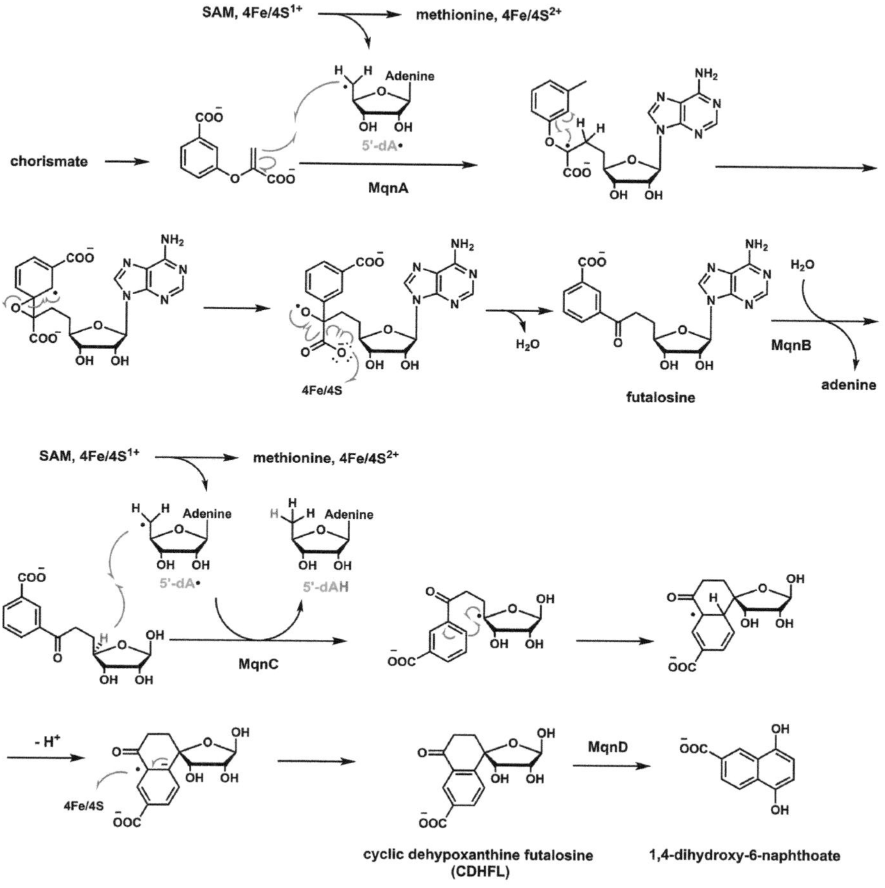

Figure 10.25 MqnA generates futalosine by combining the 5′-deoxyadenosyl radical with 3-enolpyruvyl-benzoate, with subsequent rearrangements. MqnC generates the 5′-deoxyadenosyl radical as initiator to create the bicyclic framework of menaquinone.

the transformations in Chapter 6 fall into this category. SAM is fragmented irreversibly and ultimately yields methionine and 5′-deoxyadenosine as reaction coproducts. One example involves the radical based C–C bond formations at $C_{5'}$ of uridine ribose to yield tunicamycins (see Figure 6.23) or octosyl acid phosphate en route to the malayamycin, polyoxin, and nikkomycin families of peptidyl nucleoside antibiotics (Figure 6.24). The second example is the iterative formation of the cyclopropyl rings in Jawsamycin (Figure 6.25) where each cyclopropane arrives in a radical SAM-mediated step.

10.7 Methylations at Unactivated Carbon Centers: Consumption of Two SAMs to Two Distinct Sets of Products

A subset of radical SAM enzymes actually carry out cosubstrate methylations (Zhang, van der Donk *et al.* 2012, Broderick, Duffus *et al.* 2014). Almost invariably these are at carbon atoms that are not chemically activated, cannot function as nucleophiles toward a transferring $[CH_3^+]$, and so are methylated by $[CH_3^\bullet]$ equivalents from SAM. These include the six molecules shown in Figure 10.26 as well as C_2-methyl and C_8-methyl adenines found in 23S rRNA of bacterial ribosomes that are antibiotic resistant.

The distinguishing feature of this subset of radical SAM enzymes from the one just covered above is that this latter set consumes not one but *two SAM molecules per catalytic cycle*. As noted in Figure 10.27, each of the methionines gives rise to different products. The first one (green) fulfills the role discussed all through this chapter: as a source of 5′-deoxyadenosyl radical to abstract a particular C–H bond from a cosubstrate as an $H^\bullet$ to generate the corresponding substrate carbon radical. The second SAM (purple) is attacked by that carbon radical: homolysis of the CH_3–S bond of this second SAM completes a radical-based C-methylation and transiently leaves an SAH radical on sulfur. That can be transferred back to the 4Fe/4S cluster to complete the catalytic cycle.

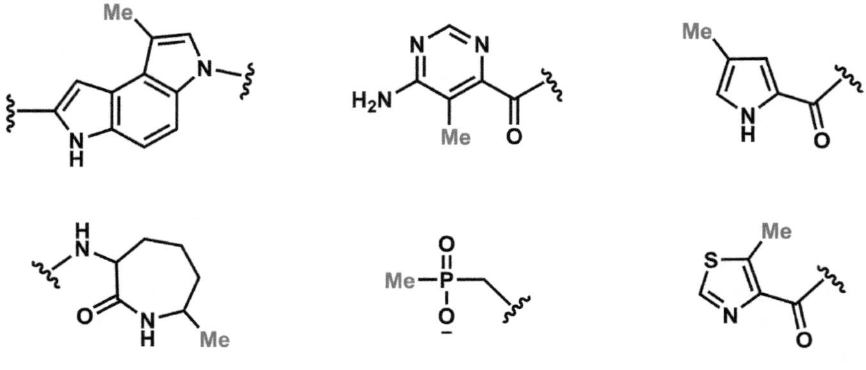

Figure 10.26 Natural products that are *C*-methylated by transfer of $[CH_3^\bullet]$ equivalents from the second SAM consumed in the reaction.

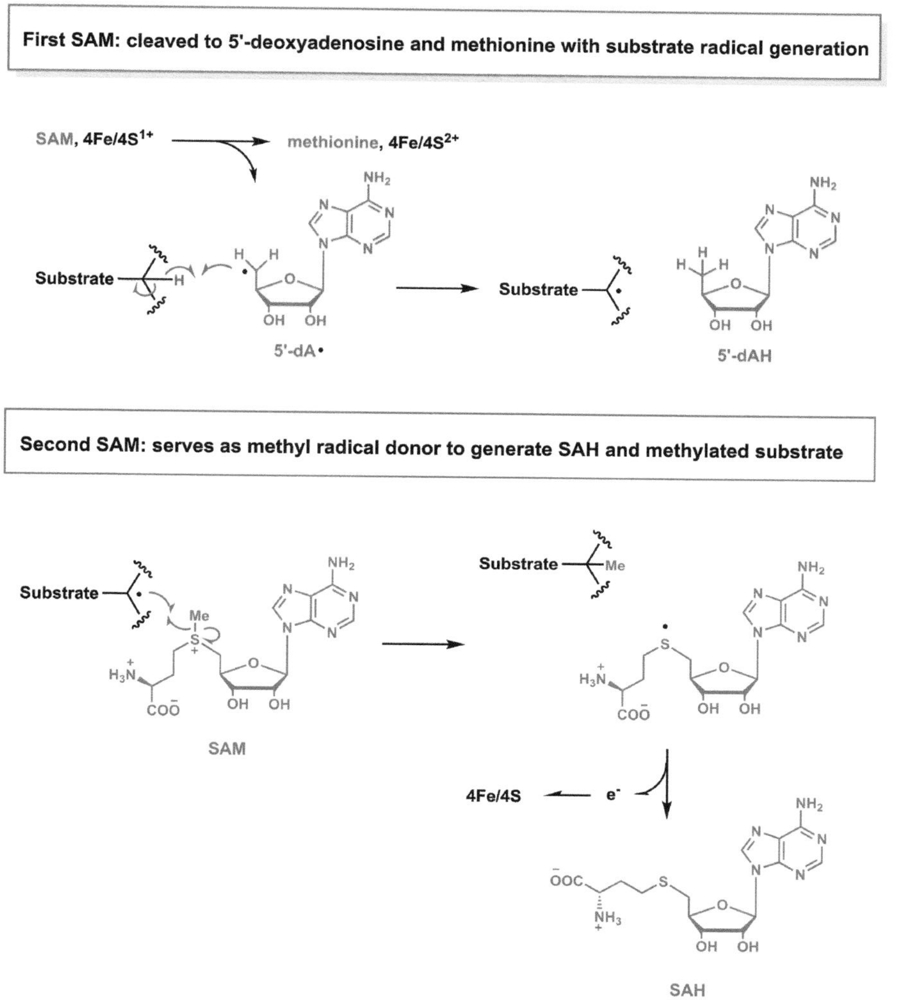

Figure 10.27 The first SAM is cleaved to methionine and 5′-deoxyadenosine. The second SAM is the [$CH_3^{\bullet}$] donor and gets cleaved to SAH.

Four examples give some sense of the scope of systems that can be methylated by this two SAM radical process (Broderick, Duffus *et al.* 2014). The first is in the methanogenic bacterial cofactor methanopterin, the functional analog of folate. Figure 10.28 shows the structure of the normal coenzyme folate in most prokaryotes and eukaryotes with the bicyclic pterin ring system. Also shown is tetrahydromethanopterin. The two methyl groups drawn in blue at

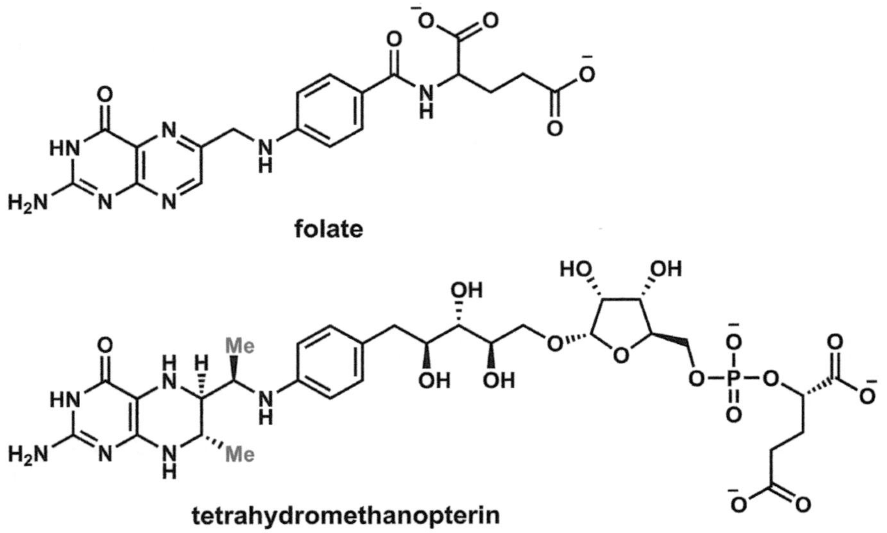

Figure 10.28 Methyl groups at C_7 and C_9 of tetrahydromethanopterin are introduced by radical SAM [$CH_3^\bullet$] transfers.

carbons 7 and 9 arise from the radical SAM enzymes that consume two SAM groups for each of the methyl groups installed at C_7 and C_9.

The next example is in aminoglycoside antibiotic biosynthetic maturation. As noted in Figure 10.29, the enzyme GenK converts gentamicin X_2 to G418 on the way to gentamicin C_1. The G418 molecule has acquired a C–CH_3 linkage (shown in purple) at the C_6 alcohol carbon of the GlcNAc ring. This proceeds by radical intermediates, consumes two SAMs, yielding SAH, 5′-deoxyadenosine, and methionine in addition to G418 (Kim, McCarty *et al.* 2013). GenK utilizes a coenzyme B12 molecule as the ultimate methyl donor to substrate, perhaps in analogy to the route shown in Figure 10.31 below.

A third example occurs in bacterial hopene metabolism. Figure 10.30 shows C-methylation at C_2 of the A ring of the pentacyclic hopene scaffold. This is clearly an unactivated carbon center. One would predict that it would take a radical SAM enzyme to effect this transformation, with two SAM molecules consumed in the process (Welander, Coleman *et al.* 2010).

The fourth example is from phosphinothricin biosynthesis. This antibiotic is notable, *inter alia*, for having two C–P bonds, a naturally occurring phosphinate. The first C–P bond was installed by the gateway enzyme PEP mutase (see Chapter 13 for C–P bond

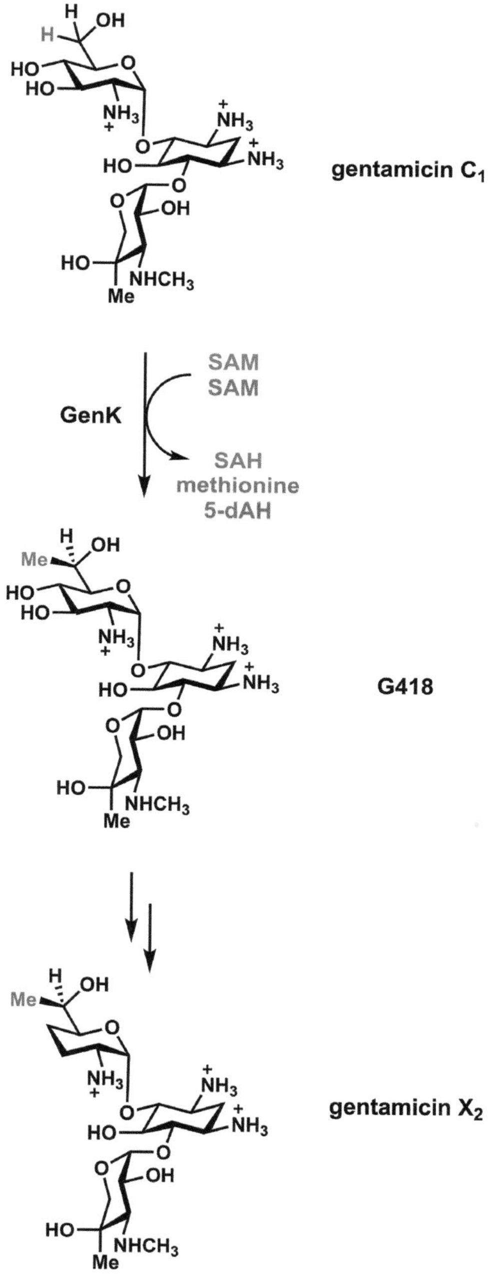

Figure 10.29 Radical *C*-methylation occurs during gentamicin maturation at a non-nucleophilic carbon.

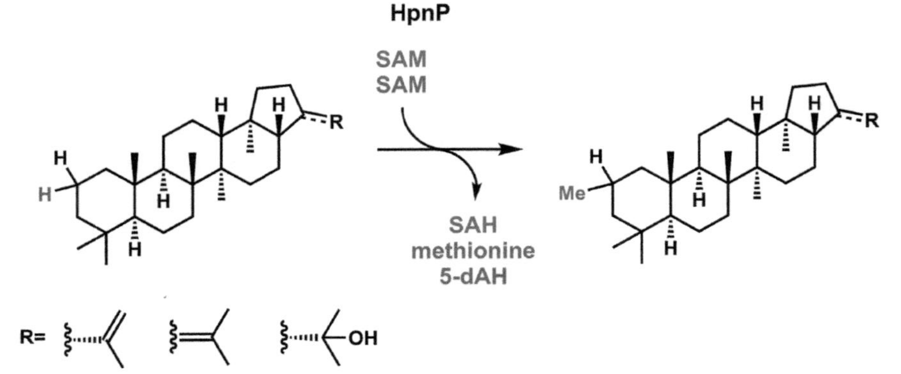

Figure 10.30 The methyl group at C_2 of this hopene derives from SAM by [$CH_3^•$] transfer. The other four methyl groups arise from the isopentenyl-PP building blocks.

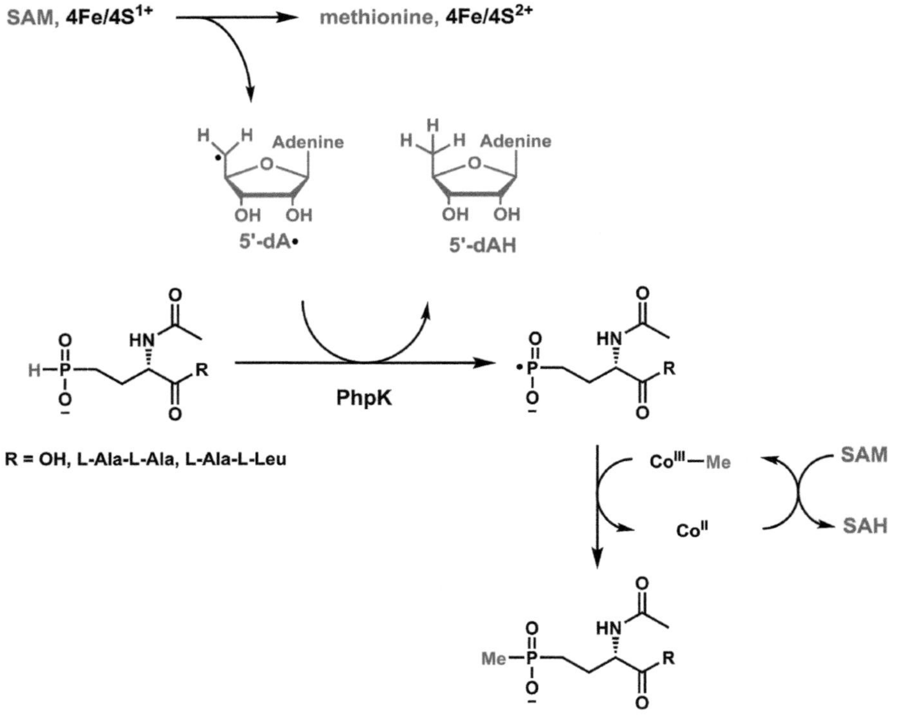

Figure 10.31 Phosphinothricin biogenesis. The CH_3–P bond arises from a deoxyadenosyl radical that makes the phosphorus radical species. The donor of the [$CH_3^•$] is the methyl cobalt atom in methyl coenzyme B12.

formations). The second P–C bond is put in by a variant radical SAM subclass that contains both SAM as a radical generator and coenzyme B12 (Ding, Li *et al.* 2016) (Figure 10.31). The first SAM is utilized to generate the phosphite radical. The second SAM transfers its methyl group to the cobalt in B12, which then executes the methyl radical transfer to produce the methyl phosphinate product.

Vignette 10.2 Wybutosine Maturation: The Versatility of *S*-Adenosyl-methionine in Six Steps.

Among the most remarkable of the natural perturbations of the guanosine nucleoside building block is the tricyclic core of the modified base wybutosine (Figure 10.V2) found at position 37 in eukaryotic phenylalanine tRNAs, specifically, next to the anticodon. It is one of the several modifications of canonical purine and pyrimidine bases involved in the maturation of tRNAs used for protein synthesis.

Although the details of every step in the conversion of guanosine to wybutosine are not fully worked out, it is clear from inspection that six molecules of *S*-adenosylmethionine (SAM) are consumed during the maturation process. The chemical versatility of SAM is demonstrated in this set of enzymatic conversions (Noma, Kirino *et al.* 2006) (Perche-Letuvee, Molle *et al.* 2014).

Two SAM molecules are methyl donors [CH_3^+], the most typical of the SAM transformations in biosynthesis. These include the *O*-methylation and *O*-methoxycarbonylation of the aminocarboxypropyl side chain by enzyme TYW (shown in yellow). Perche-Levetuvee *et al.* (2014) also argue that the methylations at the amide N_1 of G37-tRNA by the enzyme TRM5 and later the methylation at N_3 of the maturing tricyclic 7-aminocarboxypropyl-demethylwyosine by enzyme TYW3 are also likely to involve canonical type transfers of [CH_3^+] equivalents from *two* more molecules of SAM (yellow).

The *fifth* SAM is instead a source of the 5′-deoxyadenosyl radical equivalent during catalytic action of TYW1. As depicted in the boxed (dash) portion of Figure 10.V2, SAM is bound to one 4Fe/4S cluster while cosubstrate pyruvate is bound to a second such 4Fe/4S cluster. The 5′-dA$^\bullet$ radical from the SAM abstracts an H$^\bullet$ from the N_3-methyl group. This carbon radical is proposed to mediate a homolytic cleavage between C_{1*} and C_{2*} of the pyruvyl moiety with a net transfer of the acetyl portion to the $CH_2^\bullet$ equivalent, creating the *N*-CH_2–CO–CH_{3+} substituent transiently. Internal capture by the exocyclic amine builds the imidazoline ring and creates the unusual tricyclic framework in 4-demethylwyosine. Carbons 2* and 3* of cosubstrate pyruvate are depicted in pink.

The *last of the six* SAMs utilized reveals a third distinct mode of reactivity: fragmentation and transfer not of the C1-methyl unit but instead the C4-aminobutyric acid unit (brown) from the methionyl arm of SAM

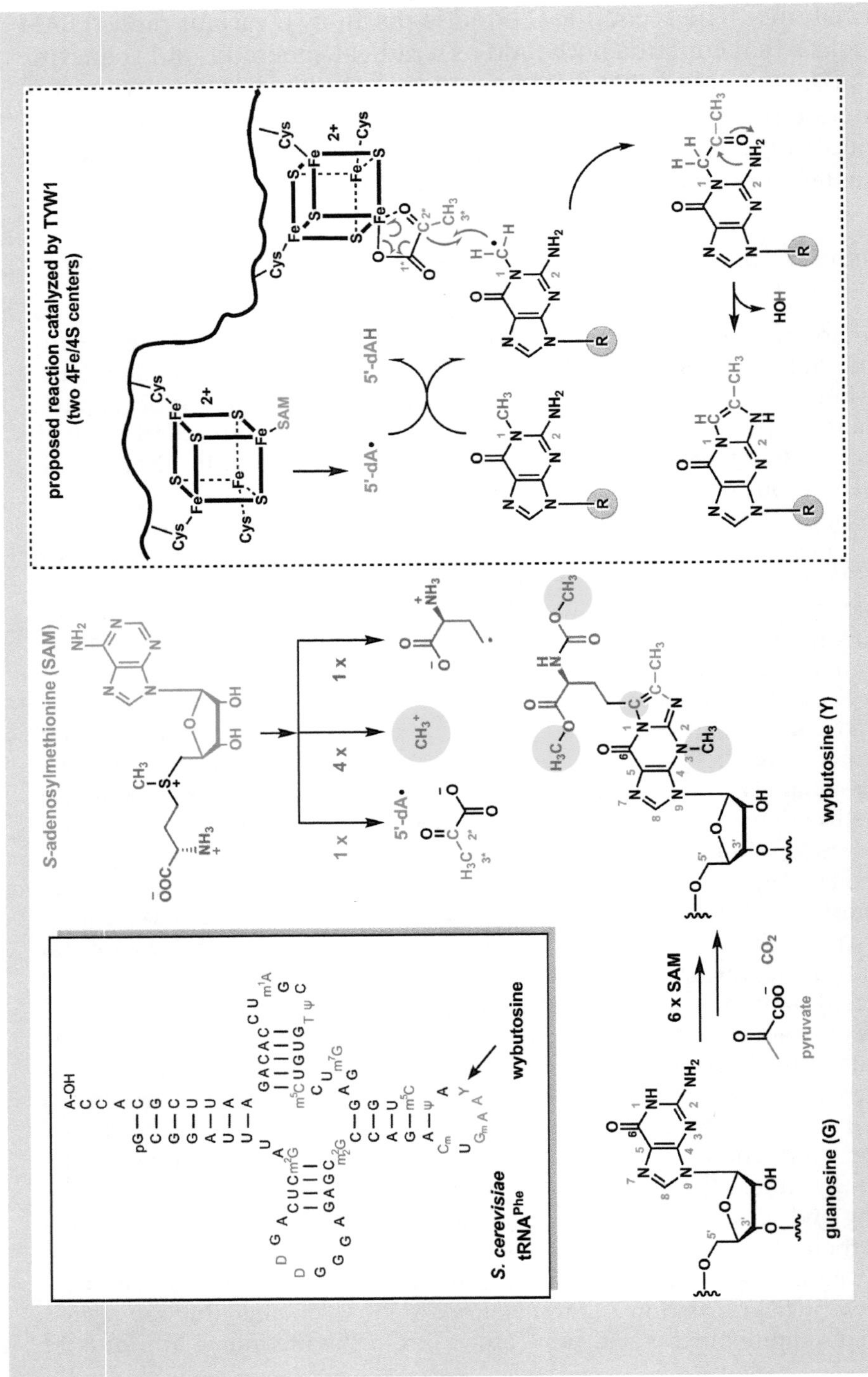

Figure 10.V2 The versatility of *S*-adenosylmethionine demonstrated through the synthesis of wybutosine.

(see Figure 10.14) (Umitsu, Nishimasu *et al.* 2009) to C_7 of the tricyclic core of 4-demethylwyosine, catalyzed by TYW2. It is not obvious that the imidazolyl carbon atom to which the aminobutyryl side chain becomes attached is a sufficient nucleophile to proceed *via* a two electron/carbanion pathway, although a tautomerization reaction could make that carbon a sufficient nucleophile. Perche-Letuvee *et al.* (2014) lean towards an ionic mechanism over a radical one but no conclusive data have been reported.

10.8 Summary on SAM reactivity and utility

S-Adenosylmethionine is probably an ancient coenzyme. It is certainly a chemically versatile one. SAM is the preeminent biological donor of C_1 fragments at the methyl oxidation state. Most of the methyl group flux from SAM in primary metabolism is transfer of [CH_3^+] equivalents to nucleophilic cosubstrates, mostly nitrogen and oxygen atoms. Carbons *ortho* to phenols are sufficiently nucleophilic to be alkylatable by this mechanism. In this chapter we have noted briefly that aminobutyryl and ribosyl transfers to cosubstrate nucleophiles can also occur.

The bulk of the chapter has dealt with the orthogonal mode of action of SAM. Bioinformatic predictions indicate tens to hundreds of thousands of enzymes that use SAM as a radical generator in the presence of a 4Fe/4S cluster with an open coordination site at one of the iron atoms. The *in situ* generation of the 5′-deoxyadenosyl radical sets off C–H bond homolysis in adjacently bound specific substrates. The resultant carbon radicals can undergo three types of fate: (1) rearrangement and then regeneration of SAM; (2) rearrangement and irreversible cleavage of SAM to 5′-deoxyadenosine and methionine; (3) capture of the substrate radical by a second molecule of SAM, leading to [$CH_3^\bullet$] transfer.

There are logical analogies to the generation of carbon-centered radicals in the iron oxygenases discussed in Chapter 9. A highly reactive oxidant is generated in the microenvironment of an enzyme active site: either a high valent oxo-iron or the 5′-dA•. The immediate job of each is to cleave a cosubstrate C–H bond homolytically, transferring the hydrogen atom with its one electron to the oxidant (and deactivating it). The carbon radicals in each instance can partition to alternative fates. In the oxygenases we have described several examples where rearrangements or radical coupling compete effectively with oxygen rebound, effectively decoupling the carbon radical generation for $OH^{\bullet}$ transfer. In the radical SAM enzymes, we have described similar partitions between rearrangements of radicals intramolecularly *vs.* capture by $CH_3^{\bullet}$ transfer. Perhaps most remarkably, Nature has evolved distinct mechanisms to carryout homolytic chemistry under two atmospheric extremes, anaerobically with SAM and in air with O_2. Both strategies are substantially in play in natural product biosyntheses.

References

Alexander, L. and D. Grierson (2002). "Ethylene biosynthesis and action in tomato: a model for climacteric fruit ripening". *J. Exp. Bot.* **53**(377): 2039–2055.

Booker, S. J. (2009). "Anaerobic functionalization of unactivated C-H bonds". *Curr. Opin. Chem. Biol.* **13**(1): 58–73.

Booker, S. J., R. M. Cicchillo and T. L. Grove (2007). "Self-sacrifice in radical S-adenosylmethionine proteins". *Curr. Opin. Chem. Biol.* **11**(5): 543–552.

Breckau, D., E. Mahlitz, A. Sauerwald, G. Layer and D. Jahn (2003). "Oxygen-dependent coproporphyrinogen III oxidase (HemF) from Escherichia coli is stimulated by manganese". *J. Biol. Chem.* **278**(47): 46625–46631.

Broderick, J. B., B. R. Duffus, K. S. Duschene and E. M. Shepard (2014). "Radical S-adenosylmethionine enzymes". *Chem. Rev.* **114**(8): 4229–4317.

Byers, R., D. Carbaugh and L. Combs (2005). "Ethylene Inhibitors Delay Fruit Drop, Maturity, and Increase Fruit Size of 'Arlet' Apples". *Hortic. Sci.* **40**(7): 2061–2065.

Charng, Y. Y., S. J. Chou, W. T. Jiaang, S. T. Chen and S. F. Yang (2001). "The catalytic mechanism of 1-aminocyclopropane-1-carboxylic acid oxidase". *Arch. Biochem. Biophys.* **385**(1): 179–185.

Cheek, J. and J. B. Broderick (2001). "Adenosylmethionine-dependent iron-sulfur enzymes: versatile clusters in a radical new role". *J. Biol. Inorg. Chem.* **6**(3): 209–226.

Chiang, P. K., R. K. Gordon, J. Tal, G. C. Zeng, B. P. Doctor, K. Pardhasaradhi and P. P. McCann (1996). "S-Adenosylmethionine and methylation". *FASEB J.* **10**(4): 471–480.

Cicchillo, R. M., D. F. Iwig, A. D. Jones, N. M. Nesbitt, C. Baleanu-Gogonea, M. G. Souder, L. Tu and S. J. Booker (2004). "Lipoyl synthase requires two equivalents of S-adenosyl-L-methionine to

synthesize one equivalent of lipoic acid". *Biochemistry* **43**(21): 6378–6386.
Ding, W., Q. Li, Y. Jia, X. Ji, H. Qianzhu and Q. Zhang (2016). "Emerging diversity of the cobalamin-dependent methyltransferases involving radical-based mechanisms". *ChemBioChem* **17**(13): 1191–1197.
Dowling, D. P., N. A. Bruender, A. P. Young, R. M. McCarty, V. Bandarian and C. L. Drennan (2014). "Radical SAM enzyme QueE defines a new minimal core fold and metal-dependent mechanism". *Nat. Chem. Biol.* **10**(2): 106–112.
Fontecave, M., M. Atta and E. Mulliez (2004). "S-adenosylmethionine: nothing goes to waste". *Trends Biochem. Sci.* **29**(5): 243–249.
Frey, P. A., A. D. Hegeman and F. J. Ruzicka (2008). "The Radical SAM Superfamily". *Crit. Rev. Biochem. Mol. Biol.* **43**(1): 63–88.
Gaston, M. A., L. Zhang, K. B. Green-Church and J. A. Krzycki (2011). "The complete biosynthesis of the genetically encoded amino acid pyrrolysine from lysine". *Nature* **471**(7340): 647–650.
Hiratsuka, T., K. Furihata, J. Ishikawa, H. Yamashita, N. Itoh, H. Seto and T. Dairi (2008). "An alternative menaquinone biosynthetic pathway operating in microorganisms". *Science* **321**(5896): 1670–1673.
Hutcheson, R. U. and J. B. Broderick (2012). "Radical SAM enzymes in methylation and methylthiolation". *Metallomics* **4**(11): 1149–1154.
Jarrett, J. T. (2003). "The generation of 5'-deoxyadenosyl radicals by adenosylmethionine-dependent radical enzymes". *Curr. Opin. Chem. Biol.* 7(2): 174–182.
Kim, H. J., R. M. McCarty, Y. Ogasawara, Y. N. Liu, S. O. Mansoorabadi, J. LeVieux and H. W. Liu (2013). "GenK-catalyzed C-6' methylation in the biosynthesis of gentamicin: isolation and characterization of a cobalamin-dependent radical SAM enzyme". *J. Am. Chem. Soc.* **135**(22): 8093–8096.
Komoto, J., T. Yamada, Y. Takata, G. D. Markham and F. Takusagawa (2004). "Crystal structure of the S-adenosylmethionine synthetase ternary complex: a novel catalytic mechanism of S-adenosylmethionine synthesis from ATP and Met". *Biochemistry* **43**(7): 1821–1831.
Kuchenreuther, J. M., W. K. Myers, T. A. Stich, S. J. George, Y. Nejatyjahromy, J. R. Swartz and R. D. Britt (2013). "A radical intermediate in tyrosine scission to the CO and CN- ligands of FeFe hydrogenase". *Science* **342**(6157): 472–475.
Lacey, R. F. and B. M. Binder (2014). "How plants sense ethylene gas–the ethylene receptors". *J. Inorg. Biochem.* **133**: 58–62.
Layer, G., E. Kervio, G. Morlock, D. W. Heinz, D. Jahn, J. Retey and W. D. Schubert (2005). "Structural and functional comparison of

HemN to other radical SAM enzymes". *Biol. Chem.* **386**(10): 971–980.
Layer, G., J. Moser, D. W. Heinz, D. Jahn and W. D. Schubert (2003). "Crystal structure of coproporphyrinogen III oxidase reveals cofactor geometry of Radical SAM enzymes". *EMBO J.* **22**(23): 6214–6224.
Mahanta, N., D. Fedoseyenko, T. Dairi and T. P. Begley (2013). "Menaquinone biosynthesis: formation of aminofutalosine requires a unique radical SAM enzyme". *J. Am. Chem. Soc.* **135**(41): 15318–15321.
McCarty, R. M. and V. Bandarian (2012). "Biosynthesis of pyrrolopyrimidines". *Bioorg. Chem.* **43**: 15–25.
McCarty, R. M., C. Krebs and V. Bandarian (2013). "Spectroscopic, steady-state kinetic, and mechanistic characterization of the radical SAM enzyme QueE, which catalyzes a complex cyclization reaction in the biosynthesis of 7-deazapurines". *Biochemistry* **52**(1): 188–198.
McCarty, R. M., A. Somogyi, G. Lin, N. E. Jacobsen and V. Bandarian (2009). "The deazapurine biosynthetic pathway revealed: in vitro enzymatic synthesis of PreQ(0) from guanosine 5'-triphosphate in four steps". *Biochemistry* **48**(18): 3847–3852.
Mehta, A. P., S. H. Abdelwahed, N. Mahanta, D. Fedoseyenko, B. Philmus, L. E. Cooper, Y. Liu, I. Jhulki, S. E. Ealick and T. P. Begley (2015). "Radical S-adenosylmethionine (SAM) enzymes in cofactor biosynthesis: a treasure trove of complex organic radical rearrangement reactions". *J. Biol. Chem.* **290**(7): 3980–3986.
Mulder, D. W., E. S. Boyd, R. Sarma, R. K. Lange, J. A. Endrizzi, J. B. Broderick and J. W. Peters (2010). "Stepwise [FeFe]-hydrogenase H-cluster assembly revealed in the structure of HydA(DeltaEFG)". *Nature* **465**(7295): 248–251.
Murphy, L. J., K. N. Robertson, S. G. Harroun, C. L. Brosseau, U. Werner-Zwanziger, J. Moilanen, H. M. Tuononen and J. A. Clyburne (2014). "A simple complex on the verge of breakdown: isolation of the elusive cyanoformate ion". *Science* **344**(6179): 75–78.
Murray, B., S. V. Antonyuk, A. Marina, S. C. Lu, J. M. Mato, S. S. Hasnain and A. L. Rojas (2016). "Crystallography captures catalytic steps in human methionine adenosyltransferase enzymes". *Proc. Natl. Acad. Sci. U. S. A.* **113**(8): 2104–2109.
Noma, A., Y. Kirino, Y. Ikeuchi and T. Suzuki (2006). "Biosynthesis of wybutosine, a hyper-modified nucleoside in eukaryotic phenylalanine tRNA". *EMBO J.* **25**(10): 2142–2154.
Peiser, G. D., T. T. Wang, N. E. Hoffman, S. F. Yang, H. W. Liu and C. T. Walsh (1984). "Formation of cyanide from carbon 1 of

1-aminocyclopropane-1-carboxylic acid during its conversion to ethylene". *Proc. Natl. Acad. Sci. U. S. A.* **81**(10): 3059–3063.

Perche-Letuvee, P., T. Molle, F. Forouhar, E. Mulliez and M. Atta (2014). "Wybutosine biosynthesis: structural and mechanistic overview". *RNA Biol.* **11**(12): 1508–1518.

Roje, S. (2006). "S-Adenosyl-L-methionine: beyond the universal methyl group donor". *Phytochemistry* **67**(15): 1686–1698.

Sofia, H. J., G. Chen, B. G. Hetzler, J. F. Reyes-Spindola and N. E. Miller (2001). "Radical SAM, a novel protein superfamily linking unresolved steps in familiar biosynthetic pathways with radical mechanisms: functional characterization using new analysis and information visualization methods". *Nucleic Acids Res.* **29**(5): 1097–1106.

Struck, A. W., M. L. Thompson, L. S. Wong and J. Micklefield (2012). "S-adenosyl-methionine-dependent methyltransferases: highly versatile enzymes in biocatalysis, biosynthesis and other biotechnological applications". *ChemBioChem* **13**(18): 2642–2655.

Trievel, R. C. (2004). "Structure and function of histone methyltransferases". *Crit. Rev. Eukaryotic Gene Expression* **14**(3): 147–169.

Umitsu, M., H. Nishimasu, A. Noma, T. Suzuki, R. Ishitani and O. Nureki (2009). "Structural basis of AdoMet-dependent aminocarboxypropyl transfer reaction catalyzed by tRNA-wybutosine synthesizing enzyme, TYW2". *Proc. Natl. Acad. Sci. U. S. A.* **106**(37): 15616–15621.

Wang, K. L., H. Li and J. R. Ecker (2002). "Ethylene biosynthesis and signaling networks". *Plant Cell* **14**: Suppl: S131–151.

Welander, P. V., M. L. Coleman, A. L. Sessions, R. E. Summons and D. K. Newman (2010). "Identification of a methylase required for 2-methylhopanoid production and implications for the interpretation of sedimentary hopanes". *Proc. Natl. Acad. Sci. U. S. A.* **107**(19): 8537–8542.

Yang, S. F. and N. E. Hoffman (1984). "Ethylene Biosynthesis and its Regulation in Higher Plants". *Annu. Rev. Plant Physiol.* **35**: 155–189.

Zhang, Q., W. A. van der Donk and W. Liu (2012). "Radical-mediated enzymatic methylation: a tale of two SAMS". *Acc. Chem. Res.* **45**(4): 555–564.

Zhang, X., H. Wen and X. Shi (2012). "Lysine methylation: beyond histones". *Acta Biochim. Biophys. Sin* **44**(1): 14–27.

Zhang, Y., X. Zhu, A. T. Torelli, M. Lee, B. Dzikovski, R. M. Koralewski, E. Wang, J. Freed, C. Krebs, S. E. Ealick and H. Lin (2010). "Diphthamide biosynthesis requires an organic radical generated by an iron-sulphur enzyme". *Nature* **465**(7300): 891–896.

Zhang, Z., J. S. Ren, I. J. Clifton and C. J. Schofield (2004). "Crystal structure and mechanistic implications of 1-aminocyclopropane-1-carboxylic acid oxidase–the ethylene-forming enzyme". *Chem. Biol.* **11**(10): 1383–1394.

Zhi, X. Y., J. C. Yao, S. K. Tang, Y. Huang, H. W. Li and W. J. Li (2014). "The futalosine pathway played an important role in menaquinone biosynthesis during early prokaryote evolution". *Genome Biol. Evol.* **6**(1): 149–160.

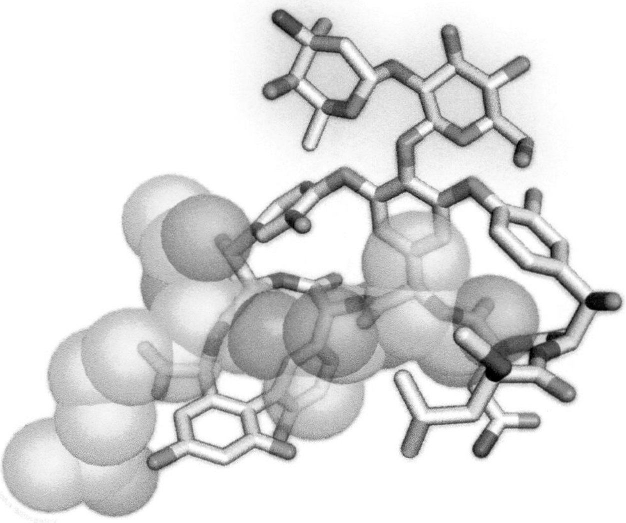

Crystal structure of vancomycin binding to Di-acetyl-L-Lys-D-Ala-D-Ala (PDB ID: 1FVM). The disaccharide (glucose–vancosamine) of vancomycin is shaded in blue. Vancomycin belongs to the larger family of glycopeptides. Copyright (2016) John Billingsley.

11 Natural Product Oligosaccharides and Glycosides

11.1 Introduction

Although polysaccharides such as cellulose, glucans such as starch, and the 1,3-glucans of fungal cell walls are abundant biopolymers, there are very few *oligosaccharides* that fall in the category of bioactive natural product scaffolds. Instead, sugar units are covalently combined with every major class of natural product in *glycoside linkages*, where C_1 of the sugar, usually a hexose derived ultimately from glucose, is attached to a nucleophilic atom in one or more parts of the aglycone scaffold. The sugar units in glycoside linkages can range commonly from monosaccharides through trisaccharides (anthracyclines, digoxins), up to heptasaccharide tails in angucycline polyketides.

The aminoglycoside antibiotics are trisaccharides, without any aglycone portion, that were isolated and characterized for structure and mechanism in the golden age of antibiotic natural products in the 1940s and 1950s (Figure 11.1) (Walsh and Wencewicz 2016). The moenomycins, potent inhibitors of bacterial cell wall transglycosylases, are glycolipidated pentasaccharides (Ostash and Walker 2010) (Figure 11.2). The orthosomycins range from hexasaccharide avilamycins, heptasaccharide everninomicins, to saccharomycins with 17 sugar building blocks (McCranie and Bachmann 2014) (Figure 11.3).

Natural Product Biosynthesis: Chemical Logic and Enzymatic Machinery
By Christopher T. Walsh and Yi Tang

Published by the Royal Society of Chemistry, www.rsc.org

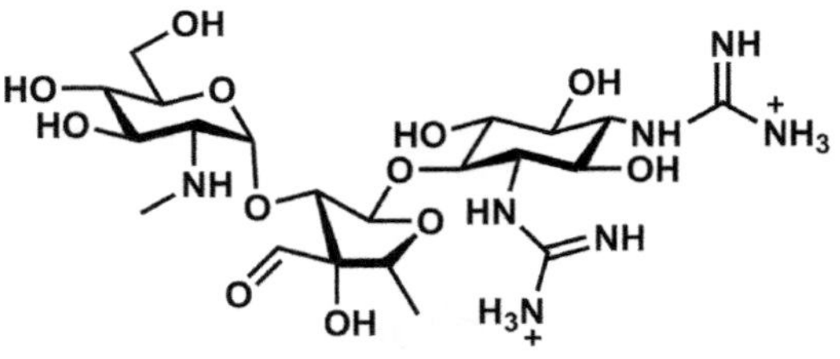

streptomycin (1944)

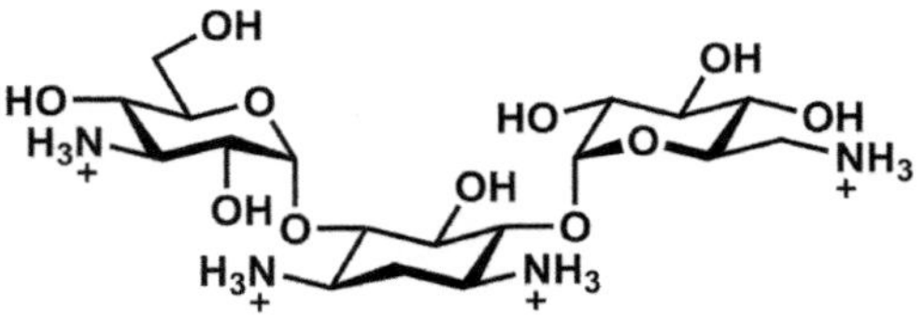

kanamycin (1957)

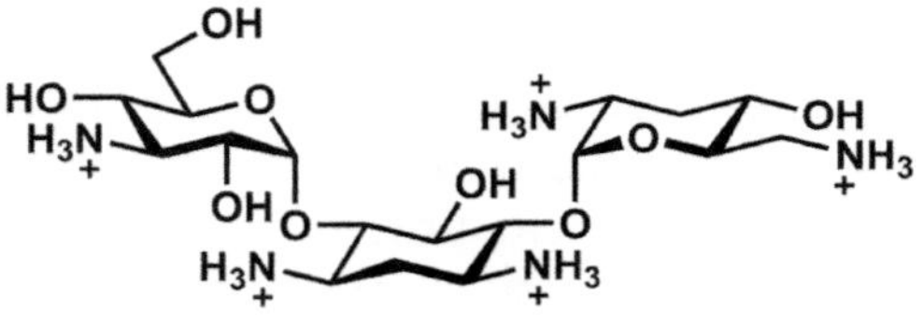

tobramycin (1967)

Figure 11.1 Three generations of tricyclic aminoglycoside antibiotics.

The orthosomycins target an rRNA site on the 50S bacterial ribosome. In avilamycin and everninomycin A there are two ortho ester linkages and one methylene dioxy bridge. We have noted the chemical logic for enzymatic formation of methylene dioxy bridges in prior chapters, including the conversion of pinoresinol to sesamin *via* iron-based oxygenation (Chapter 9). Analogously, the orthoester linkages appear to require a mononuclear nonheme iron oxygenase. Although the details of orthoester construction are not yet clear, this does expand the known chemistry of Fe(IV)=O enzyme catalysts (McCulloch, McCranie *et al.* 2015) (Figure 11.3).

The building blocks for the aminoglycosides and orthosomycins are hexose units but they are not the most abundant ones – glucose,

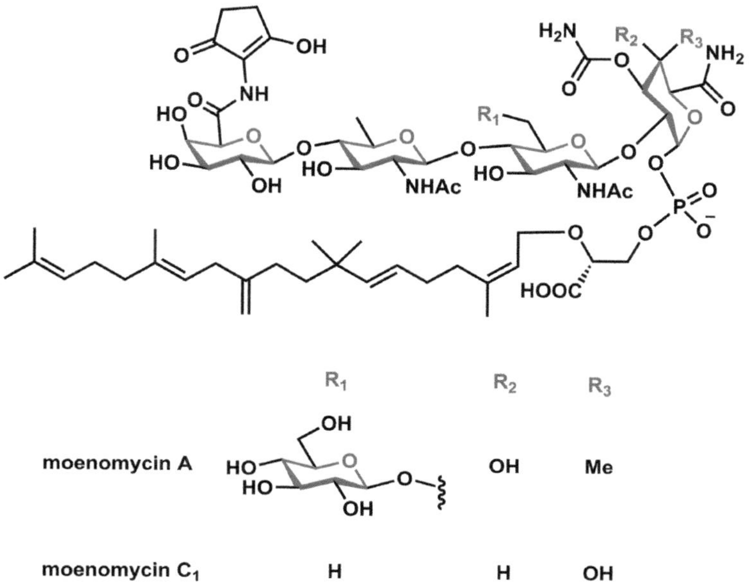

Figure 11.2 The moenomycin family of pentasaccharide antibiotics.

mannose, galactose – found in primary metabolism. Rather, they reflect dedicated biosynthetic routes from the central nucleoside diphosphosugar UDP-glucose or the closely related TDP-glucose (Figure 11.4) to deoxy sugars which have a balance of hydrophobic and hydrophilic surfaces, suggesting active roles in recognition of biological targets (Thibodeaux, Melancon *et al.* 2007, Thibodeaux, Melancon *et al.* 2008). The saccharomicins with 17 linked sugars encode 10 glycosyltransferases in the biosynthetic gene cluster, indicating that some are likely to be specific and some promiscuous in the stage at which they act (McCranie and Bachmann 2014) (Figure 11.5)

We will delve into NDP-glucose formation and reactivity, and its biosynthetic role as glucosyl donor of an electrophilic glucosyl moiety to cosubstrate nucleophiles. This will give insight into the chemical logic for enzymatic glycosylation of all the categories of natural products discussed in Chapters 2–8 and account for the connection of glycosyl moieties through C_1 to a variety of O, N, S, and even C atoms in the natural product aglycones. Then we will turn back to assembly of the trisaccharide and pentasaccharide frameworks at the end of this chapter.

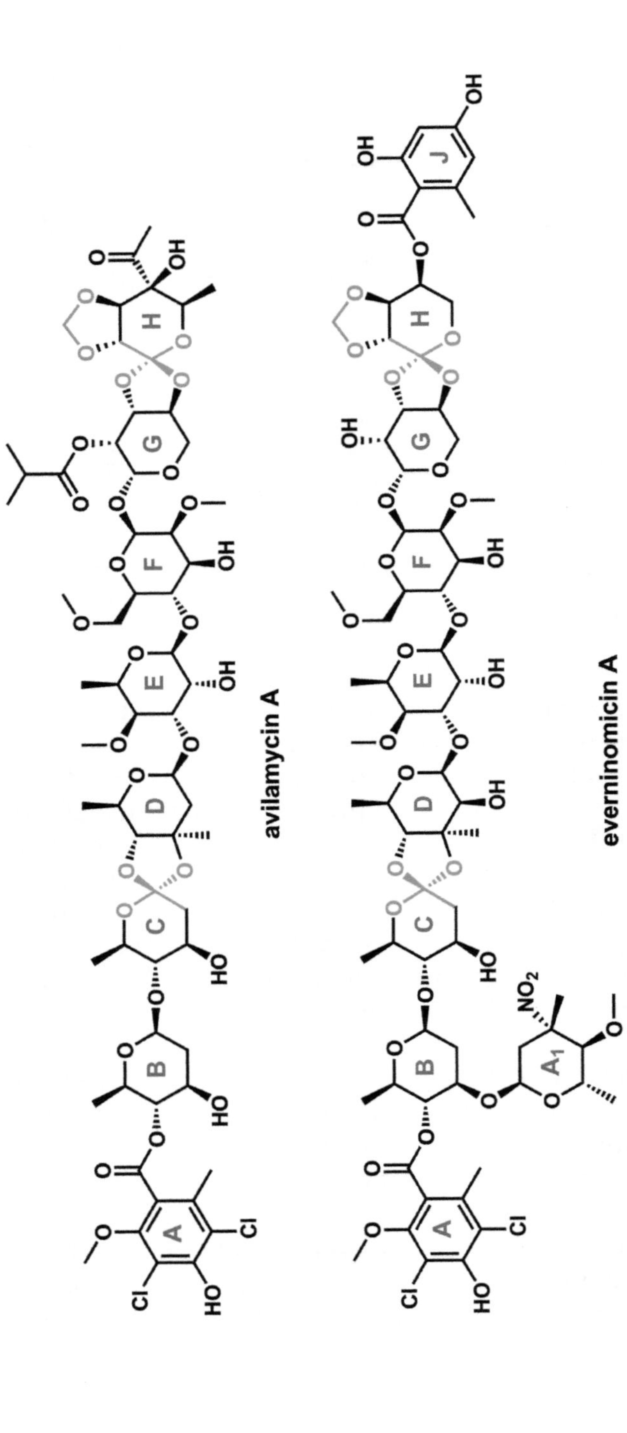

Figure 11.3 Avilamycin and everninomicin A and their gene clusters; ortho esters (red) rather than glycoside links.

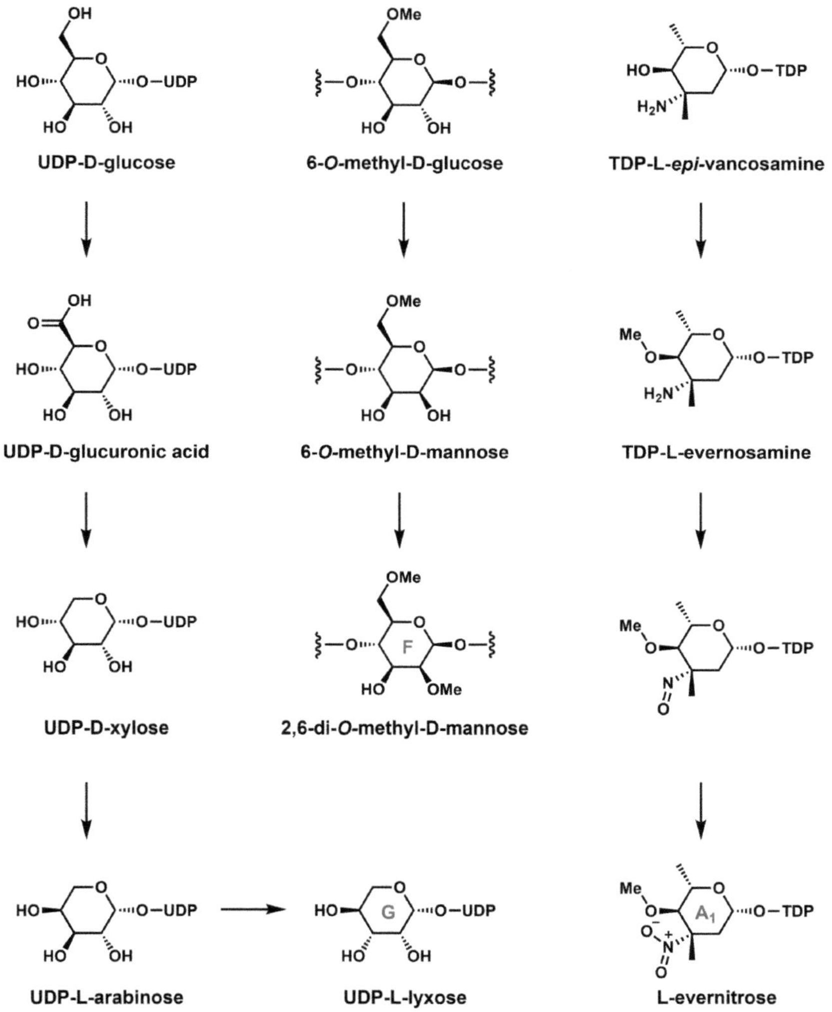

Figure 11.4 UDP-hexose building blocks for sugars F and G in avilamycin, and A1 in everninomicin, as indicated in Figure 11.3.

Figures 11.6–11.9 display a range of natural products, many of them noted largely for their aglycone moieties in previous chapters. Polyketides are represented by the macrolides erythromycin and ivermectin, with two deoxysugar monosaccharides and one modified disaccharide residue, respectively. Erythromycin has two highly modified hexoses, desosamine and cladinose, in *O*-glycoside linkages (Figure 11.6). We have commented in Chapter 2 that the monosaccharide in the polyene polyketide nystatin is essential for its

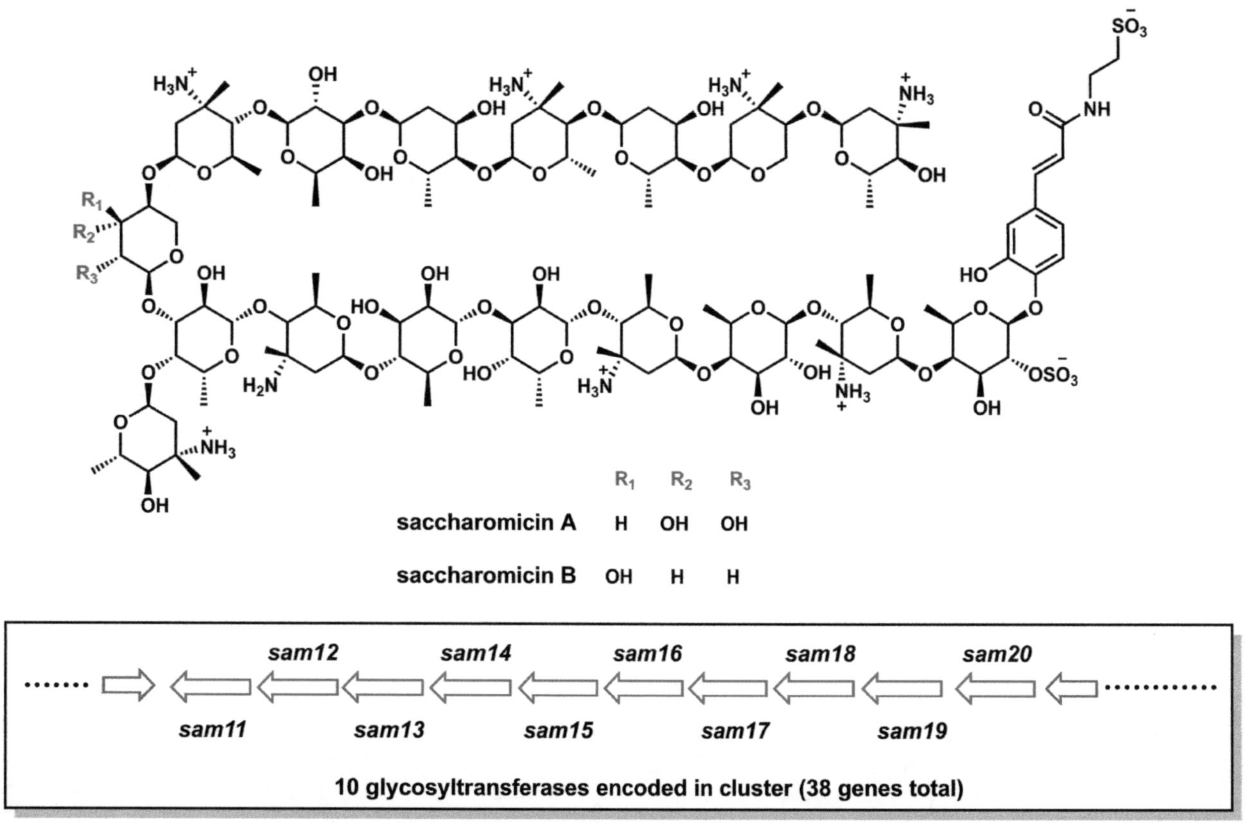

Figure 11.5 The saccharomicin biosynthetic gene cluster encodes 10 predicted glycosyltransferases for the assembly of the heptadecaoligosaccharide product.

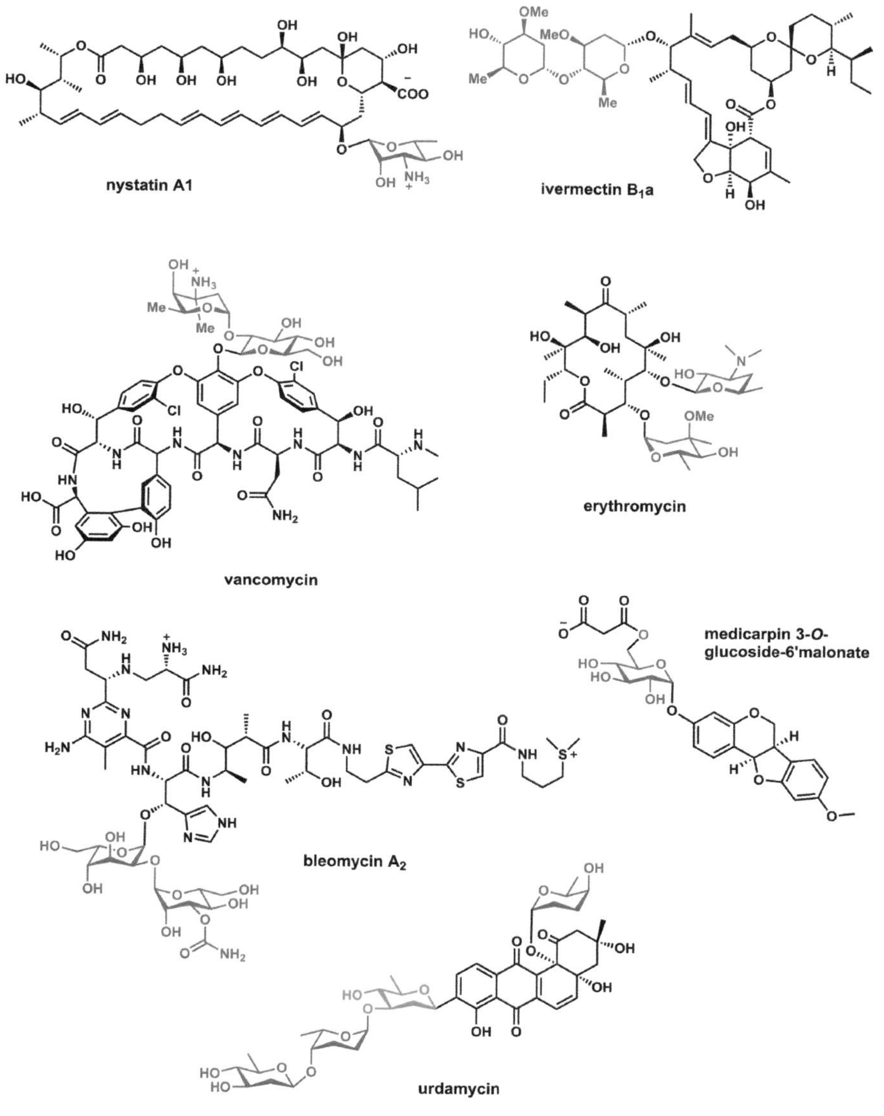

Figure 11.6 Seven natural products that are glycosylated in their mature forms.

antifungal activity. The fourth polyketide in Figure 11.6 is the angular urdamycin which has a *C*-glycoside link (red) in addition to the more traditional *O*-glycoside links.

The glycopeptide antibiotic vancomycin has a 1,2-glucosyl-L-vancosamine disaccharide attached as an *O*-glycoside to 4-hydroxyphenylglycyl

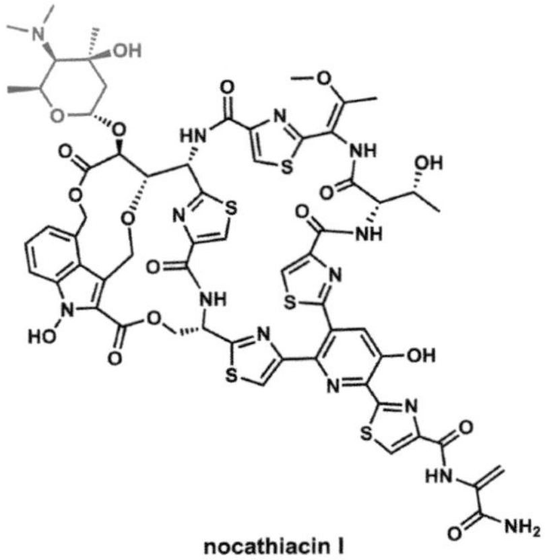

nocathiacin I

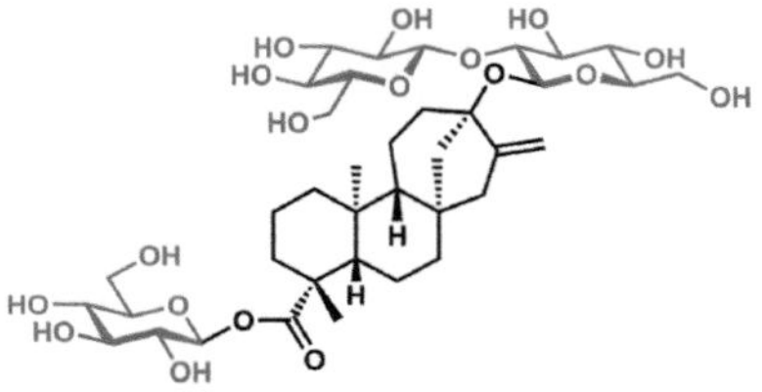

stevioside

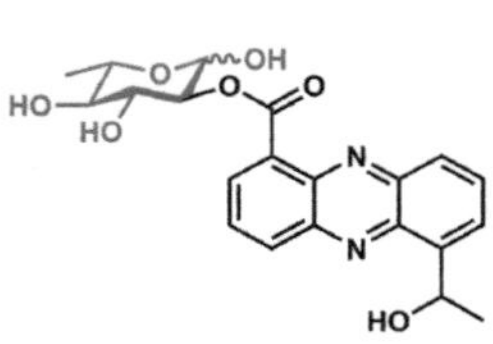

2'-L-quinovose saphenate

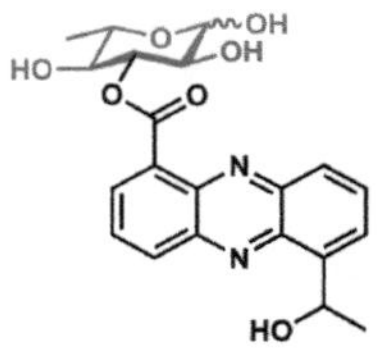

3'-L-quinovose saphenate

Figure 11.7 Glycosylation occurs with other families of natural products, including *O*-glycosylation of the thiazole peptide antibiotic nocathiacin I and the diterpene stevioside. The phenazine framework of saphenate (arising from dimerization of anthranilate) can be found as two related quinovose acetal isomers.

residue 4 of the heptapeptide backbone (Chapter 3). The NRP–PK hybrid bleomycin has an L-gulose-*O*-carbamoyl-D-mannose disaccharide attached to the β-OH-His residue. The last example is the isoflavone glucoside medicarpin-3-*O*-glucoside-6′-malonate. Here the glucose is added to increase the solubility and for storage of the phytoalexin.

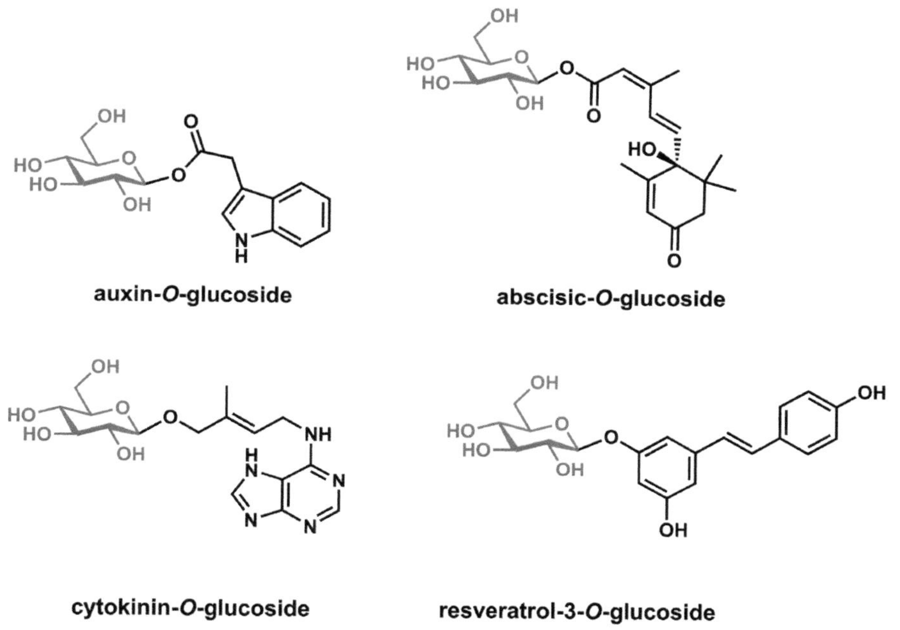

Figure 11.8 Plant hormones and the stilbene metabolite resveratrol are accumulated and stored as *O*-glycosides.

The ribosomally generated and extensively posttranslationally modified thiazolyl peptide antibiotic nocathiacin I is *O*-glycosylated with an aminodeoxyhexose unit dangling from the third loop of the antibiotic (Figure 11.7). The diterpene framework in steviol (related to the kaurene scaffold) has a diglucoside in acetal linkage and a monoglucoside in ester linkage. Stevioside is stored in vacuoles of the producing plants and is 30–300 times sweeter than sucrose; 2′- and 3′-*O*-quinovosyl-saphenate are storage forms of the phenazine metabolite (Figure 11.7).

Of the 14 hexosyl groups arrayed around the aglycone moieties of the seven natural products in Figure 11.6, only two are simple glucose residues (in vancomycin and the phytoalexin medicarpin). The other 12 reflect specialized sugar modification enzyme machinery to create the specific aglycone moiety to be coupled biosynthetically to the aglycone piece (Thibodeaux, Melancon *et al.* 2007). We will turn to the deoxyhexose biosynthetic inventory after dealing with NDP-glucose formation and its utilization by glycosyl transferases. Figure 11.8 shows that three of the major classes of plant hormones, auxin,

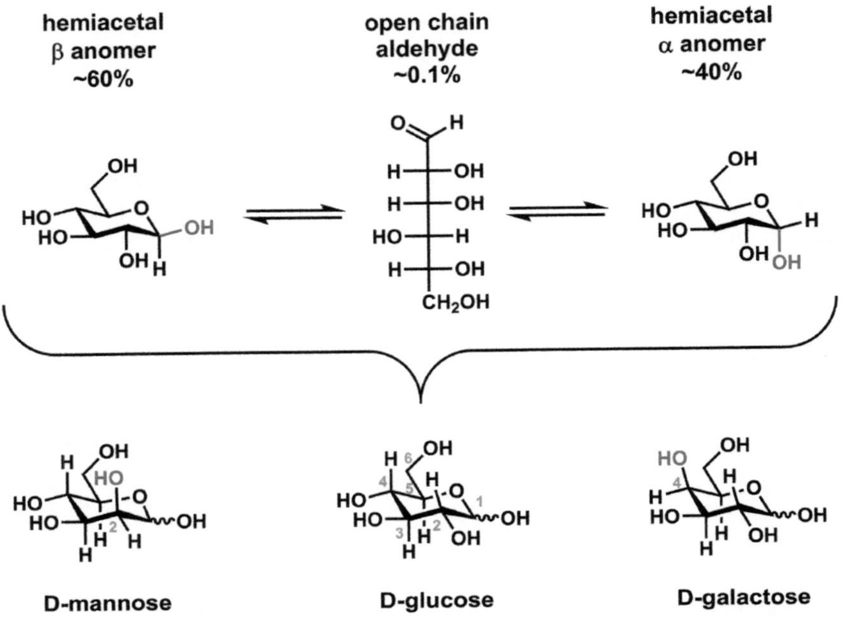

The three most common hexoses in primary metabolism

Figure 11.9 Glucose is the predominant biological hexose and exists largely as an equilibrating pair of α- and β-cyclic hemiacetals (pyranose ring forms) in equilibrium with the open chain aldehyde. D-Mannose is the C_2-axial epimer of D-glucose; D-galactose is the C_4-axial epimer.

abscisic acid, and cytokines, are stored as glycosides, as is the stilbene metabolite resveratrol.

11.2 Glucose is the Predominant Hexose in Primary Metabolism

Glucose, a central energy source for many cells, can exist as a rapidly equilibrating mixture of open chain aldehyde and a pair of cyclic six-membered pyranose isomers, designated α- and β-, based on whether the C_1–OH in the pyranose form is down or up, as shown in Figure 11.9. The pyranoses, strain free six-membered cyclic hemiacetals, represent greater than 99% of the bulk glucose population under physiologic conditions. This is the case despite the fact that all the flux of glucose down the glycolysis pathway goes through the aldehyde minor component (by action of the key enzyme fructose-1,6-diphosphate aldolase). In the chair pyranose forms of glucose all the C_2–C_6 substituents are equatorial, the lowest energy structures. The other two common hexoses in primary metabolism, galactose, the C_4-axial epimer, and mannose, the C_2-axial epimer, are thus higher in energy by virtue of those axial-OH groups (Figure 11.9).

11.2.1 Glucose-6-P and Glucose-1-P

Once glucose has diffused into cells, *via* a set of transmembrane transporter proteins, it is trapped inside by enzymatic phosphorylation. The responsible enzyme, hexokinase/glucokinase, is specific for phosphorylation of the primary, most reactive –OH group to give glucose-6-phosphate (G-6-P), still a mixture of pyranose anomers and the minor open chain C_1 aldehyde form. While G-6-P is the first and predominant phosphorylated regioisomer, the one that feeds into glycolysis, the citrate cycle, and the mitochondrial respiratory chain to liberate energy, it can be enzymatically converted to glucose-α-1-phosphate (G-α-1-P).

The relevant enzyme is phosphoglucomutase. The enzyme sits at rest as a covalent phosphoseryl-protein. When G-6-P binds in the active site, the Enz-PO_3^{2-} undergoes phosphoryl transfer to form glucose-1,6-di-phosphate and dephosphoenzyme. It stays tightly bound, can rotate and transfer the 6-PO_3 back to the active site Ser-OH side chain. At the end of the catalytic cycle the Enz-PO_3^{2-} has been regenerated and G-α-1-P is released. The enzyme can also catalyze the reverse reaction. Figure 11.10 shows that phosphoglucomutase turns out only the α anomer of G-1-P, reflecting a geometric constraint of orientation of substrate and Enz-PO_3^{2-} in the active site.

11.2.2 Glucose-1-P to UDP-Glucose

The glucose-α-1-P is the regioisomer that can react with the first committed enzyme in glucosyltransfer metabolic pathways, UTP-glucose pyrophosphorylase. This enzyme uses uridine triphosphate (UTP) as cosubstrate and catalyzes attack of one of the G-α-1-P phosphate oxygens on the electrophilic α phosphorus of UTP. As shown in Figure 11.11, when the pentacovalent phosphorane adduct

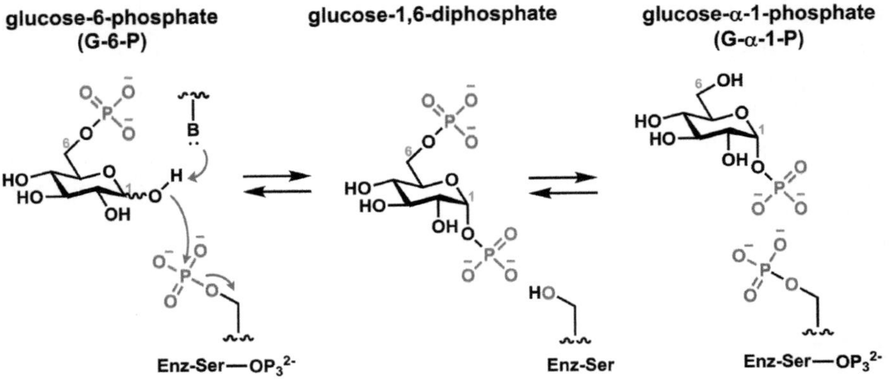

Figure 11.10 Phosphoglucomutase interconverts glucose-6-phosphate and glucose-α-1-phosphate *via* 1.6-diphospho-glucose intermediacy.

breaks down in the forward direction, the products are PP_i and UDP-glucose.

Some plant enzymes make ADP-glucose for hexosyl oligomerization to form starch granules. We will note that the microbial pathways to the deoxy and aminodeoxyhexoses, noted in Figure 11.4, use thymidylate triphosphate rather than UTP. This may be a way to segregate some of the G-α-1-P pool for secondary metabolism. There are related enzymes that can make UDP-galactose for subsequent galactosyl transfer. Mannose-1-P is converted to GDP-mannose rather than UDP-mannose, perhaps to provide specificity to the transferases; and perhaps to balance the nucleoside triphosphate pools so UTP is not excessively depleted by these routes. GDP-mannose then serves as the donor for both the L-gulose and carbamoyl-D-mannose residues in bleomycin (Shen, Du *et al.* 2002) (Figure 11.12).

11.2.3 Glycosyltransferases use NDP-Hexoses to Transfer Glycosyl Units to Cosubstrate Nucleophiles

Nucleoside diphosphosugars are the predominant biological donors of sugars in both primary and secondary metabolic pathways. As noted in Figure 11.6 the cosubstrates can span an enormous range of chemical space. They must each have a site that can function as the nucleophilic component. As Figure 11.13 delineates, *O*-, *N*-, and *C*-glycosides exist and arise by capture of the glycosyl moiety at C_1 of the nucleoside diphosphosugar substrate by an amine nucleophile, an oxygen nucleophile, or a carbon nucleophile.

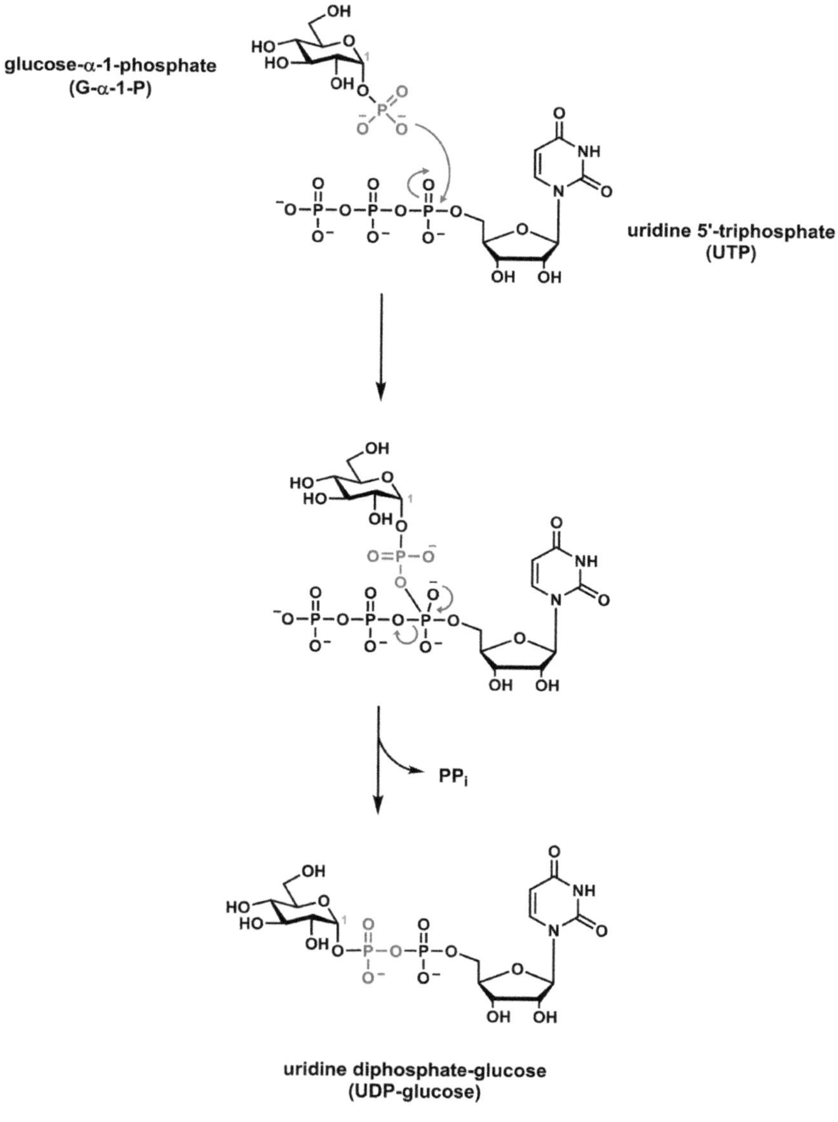

Figure 11.11 UDP-glucose formation from G-α-1-P and UTP.

Because the universal donors are NDP-sugars (almost always NDP-hexoses) that are potentially electrophilic at C_1 of the sugar, *all the natural product glycosides have a bond from the natural product nucleophile to C_1 of the sugar.* Subsequent elongation of the glycoside unit to oligosaccharides also always involves a C_1 connection of each glycosyl unit added.

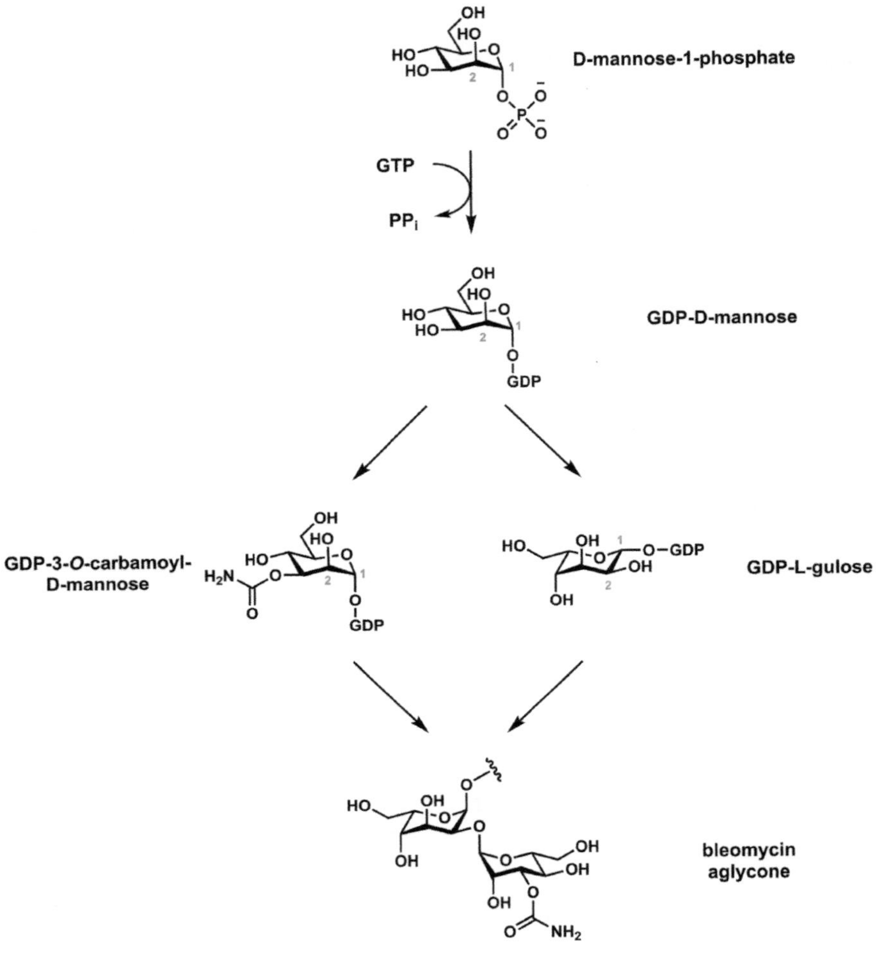

Figure 11.12 GDP-mannose is the precursor to both the L-glucose and O-carbamoyl-D-mannose residues on the bleomycin disaccharide.

The NDP-hexose substrates deliver the glycosyl unit as an electrophilic fragment at C_1. The strategy in converting glucose to G-1-P and then the G-1-UDP derivative is to lower the energy barrier for C_1–O bond cleavage. As featured in Figure 11.14, C_1–O bond cleavage is postulated to occur early in the reaction coordinate such that an oxocarbenium ion transition state is being attacked by the incoming nucleophile. This is the canonical mechanism for thousands of glycosyltransferases (Gtfs) in protein data bases (Lairson, Henrissat *et al.* 2008). While a direct S_N2-like attack would yield inversion of

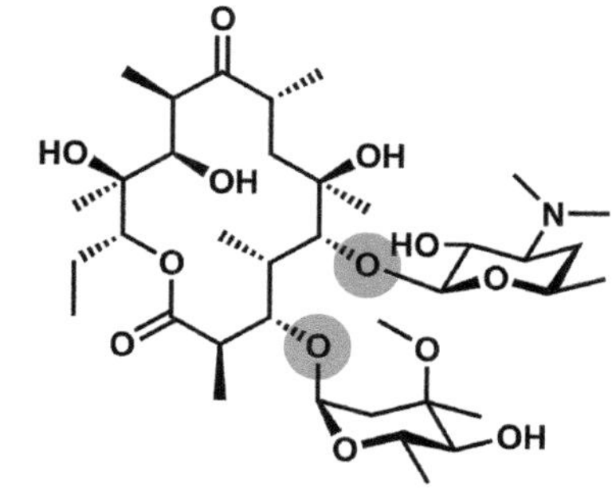

erythromycin (*O*-glycoside)

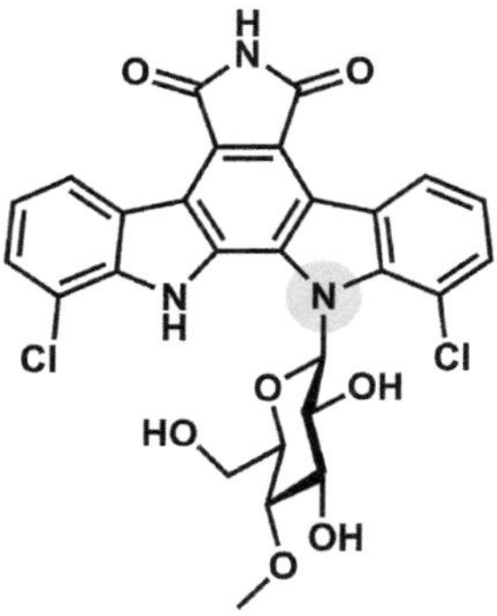

rebeccamycin (*N*-glycoside)

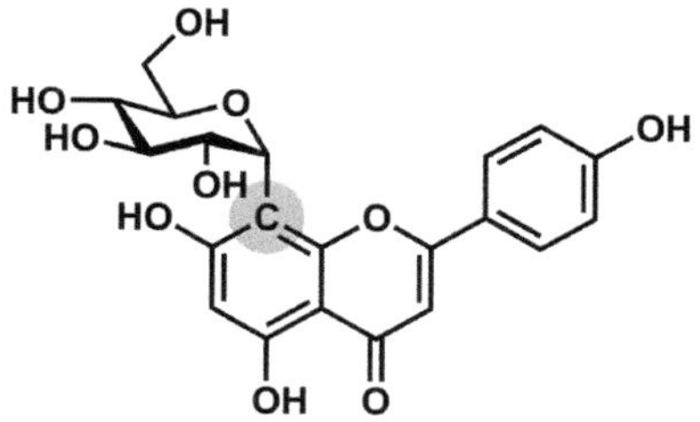

vitexin (*C*-glycoside)

Figure 11.13 Glycosyl groups attached to O, N, or C atoms arise from attack of nucleophilic amines, alcohols, and carbanionic transition states.

stereochemistry at C_1 of the glycosyl unit transferred, there are also many examples of retention of stereochemistry. In the absence of a double displacement, covalent Gtf enzyme mechanistic alternative, the early oxocarbenium ion formation would allow leaving group exit to allow the incoming nucleophile to come in on the same face.

There is a heavy commitment to glucosyltransferases in *Arabidopsis*, with about 125 *gtf* genes, and 165 in the barrel clover plant *Medicago truncatula.* A large fraction of the encoded Gtf enzymes are likely

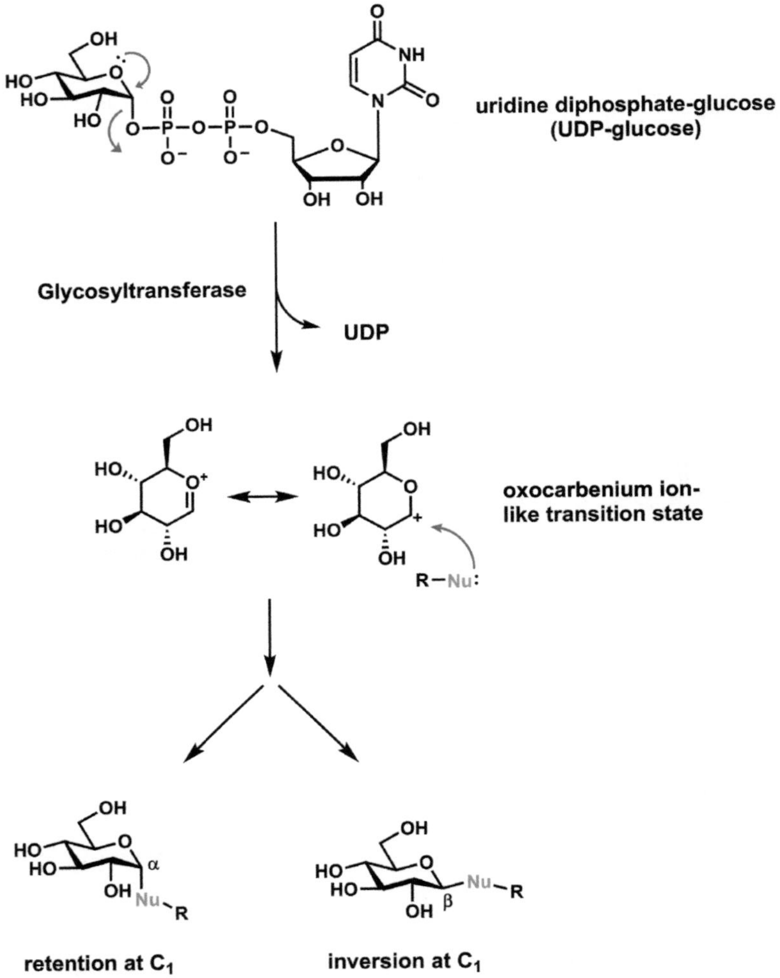

Figure 11.14 Glycosyltransferases number in the tens of thousands. They catalyze early C_1–O bond breakage *via* oxocarbenium ion transition states, and both retention and inversion of the stereochemistry at C_1 of the transferring glycosyl group are observed.

involved in glycosylation of secondary metabolites for solubilization, storage, transport, or metabolite compartmentalization (Keegstra and Raikhel 2001).

These include *phytoanticipins*, molecules made in plants in anticipation of a defense role (in contrast to phytoalexins, which are synthesized inducibly in real-time response to a perceived threat to the

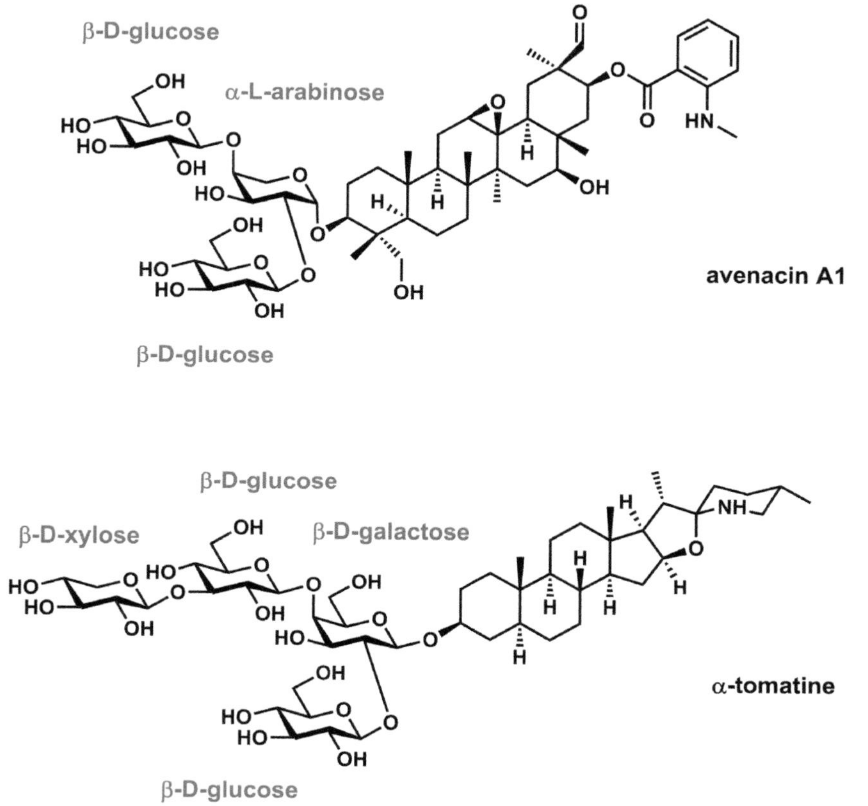

Figure 11.15 Phytoanticipins: avenacin A from oak roots and α-tomatine from tomatoes are steroidal alkaloids (see Chapter 5) that are stored as oligosaccharide acetals.

plant) (VanEtten, Mansfield *et al.* 1994). Two examples of glycosylated phytoanticipins (both steroidal alkaloid aglycones) are avenacin A from oak roots and α-tomatine in tomatoes (Figure 11.15). Glycosylated molecules can be divided into the glycosyl moiety and the aglycone. Both avenacin (Bowyer, Clarke *et al.* 1995, Osbourn, Bowyer *et al.* 1995) and α-tomatine contain triterpene aglycones with branched chain oligosaccharide chains in acetal linkage to the 3-OH of the aglycone (see Figure 5.36). As will be noted below, triterpene glycosides are designated as saponins because of their amphipathic behavior (hydrophobic and hydrophilic moieties joined together) and they localize to cell membranes in the producing plants in suitable locations to be defensive molecules.

The branched glycoside chains in avenacin and α-tomatine contain two pentoses, xylose and arabinose, as well as glucose and galactose. Both of the pentoses are available from UDP-glucose, as noted in Figure 11.16. Oxidation of the C_6 alcohol up to the

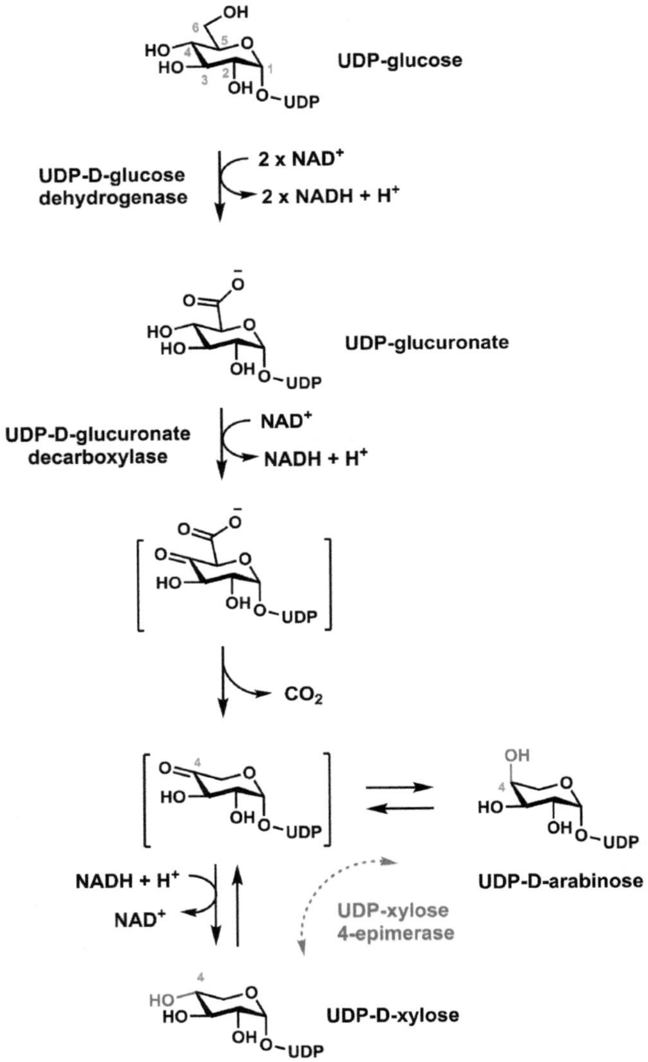

Figure 11.16 UDP-arabinose and UDP-xylose are formed by oxidation of UDP-glucose to UDP-glucuronate and subsequent decarboxylation.

acid generates UDP-glucuronate. Subsequent oxidation to the 4-keto sets up facile decarboxylation. Chiral reduction produces UDP-D-xylose. A UDP-xylose-4-epimerase utilizes NAD, remakes the 4-keto species and generates UDP-D-arabinose. UDP-xylose and UDP-arabinose are cosubstrates for particular Gtfs in oat roots and in tomato plants.

Other features of enzymatic glycosylation in plant metabolism include chemical stabilization of the anthocyanin flower pigments *via* 3-*O*-glycosylation. The balance between free hormones and hormone-*O*-glycosides was noted in Figure 11.8. The localization and transport of the glycosylated plant hormones differs from the distribution after glycosidase action.

In turn, the balance between the glycosylated forms and their aglycones is determined by the effective catalytic ratio of the biosynthetic glycosyltransferases to the hydrolyzing glycosidases (Bowles, Isayenkova *et al.* 2005, Bowles, Lim *et al.* 2006, Osbourn and Lanzotti 2009). This will be a multiparametric ratio, depending on tissue and subcompartment location of the synthetic and hydrolytic enzymes and their intrinsic catalytic efficiencies. It may be that plant metabolites modified with UDP-glucose rather than some modified UDP-hexose (more energy spent in its biosynthesis, see the next section) are the ones that are more prone to reversible glycosidase-mediated hydrolysis to liberate back the free aglycones (Figure 11.17). This is certainly relevant in the glucosinolates and cyanogenic glycosides noted in a later section of this chapter.

11.2.4 NDP-Hexoses *vs.* 5-Phosphoribose-1-PP

In Chapter 6 we noted that in the tripartite nucleotides – purine or pyrimidine base/sugar/5′-phosphates – the sugar is D-ribose not D-glucose in the building blocks for RNA and DNA. There are nucleosides with glucose moieties in some small molecule natural products noted in that chapter but they are rare.

The strategy for D-ribose activation parallels but is distinct from D-glucose activation. Both are pyrophosphorylated at C_1 to convert that carbon to an electrophilic site, and to ensure that cosubstrate nucleophiles attack there regiospecifically. However, ribose-5-P is directly pyrophosphorylated by a pyrophosphokinase, where the C_1–OH of the α-anomer of 5-P-ribose attacks the β-phosphorus of ATP (Eriksen, Kadziola *et al.* 2000). In the hexose manifold it is glucose-α-1-P that, more conventionally, is oriented to attack the α-P of UTP

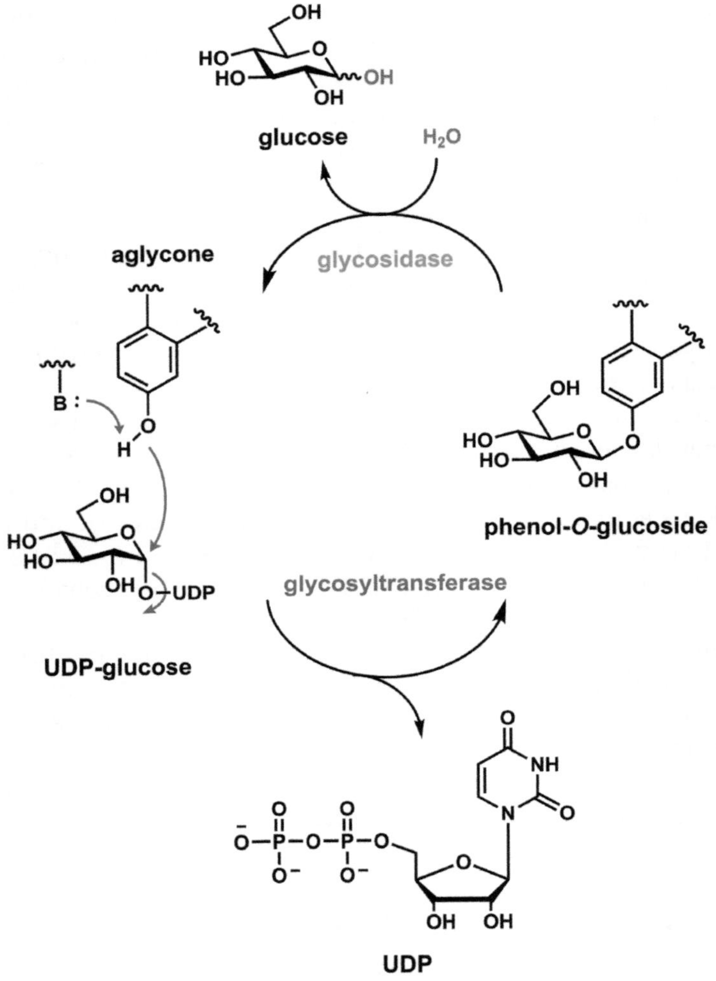

Figure 11.17 Opposing actions of glycosyltransferases and glycosidases balance the ratios of glycosides to aglycones.

rather than the β-P (Kleczkowski, Geisler *et al.* 2004). The result is a substituted C_1-pyrophosphosphoryl linkage in UDP-glucose.

In both the ribose and the glucose series, the C_1–OPP– linkages serve the purpose of setting up the C_1–O bond for cleavage to oxocarbonium ion transition states and ensuring the universal connection of nucleophiles to C_1 of the pentose and hexose in acetal linkages. UDP-ribose can be formed and used as ribosyl donor, as we have noted earlier in this chapter.

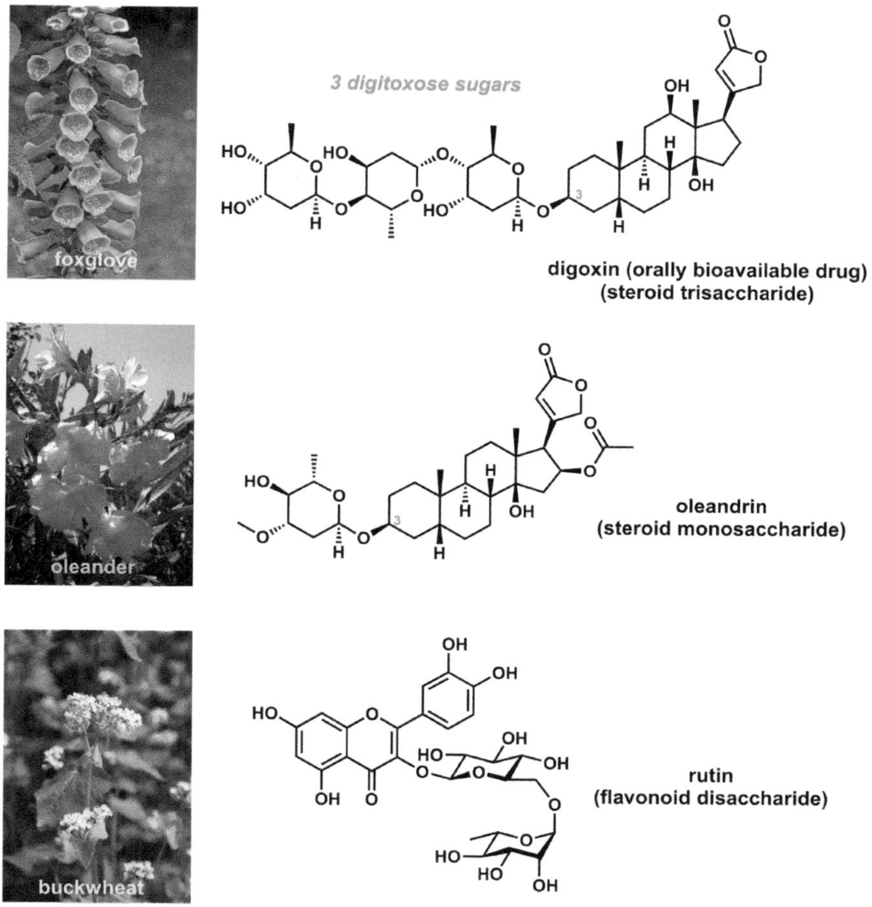

Figure 11.18 Steroidal and phenylpropanoid glycosides. The cardiotonic steroid glycosides in foxglove include digoxin.

11.3 A Gallery of Glycosylated Natural Products

Among the best known of the steroid glycosides that are not phyto-anticipins (*e.g.* Figure 11.15, above) are the cardiotonic metabolites from the foxglove plants. Figure 11.18 shows the structure of digoxin (Soldin 1986), noting the three digitose hexose units in a trisaccharide linkage, attached to the 3-hydroxyl group which derives biosynthetically from squalene 2,3-epoxide (see Chapter 4). Oleander contains the eponymous oleandrin, a steroid 3-*O*-monosaccharide (Kumar, De *et al.* 2013) (Figure 11.18). Rutin is the flavonol quercetin attached to rutinose (glucosyl-β-1,6-L-rhamnose) (also see Figure 7.21), and is found in abundance in many plants, including buckwheat. Adding the disaccharide improves aqueous solubility, allows transport to different organelles and tissues from the producing plant cells, and assists in oxygen scavenging defense.

Two other steroid glycosides, shown in Figure 11.19, fall into the saponin subclass of which avenacin and α-tomatine, noted above in Figure 11.15, are also members. The predominant saponin in fenugreek seeds (curry flavoring) is the indicated triterpene tetrasaccharide (Dewick 2009). The trisaccharide-containing triterpene solanine has insect antifeeding properties and sits as an amphiphile in potato and tomato plant cell membranes. The biosynthesis of the aglycone solanidine from cholesterol is described in Figure 5.35. These amphiphilic saponins have foaming characteristics that offer utility in soaps and shampoos (Hostettmann and Marston 1995).

In the polyketide tricyclic anthracycline family, glycosylation is typically at the 7-hydroxy group indicated for daunomycin and

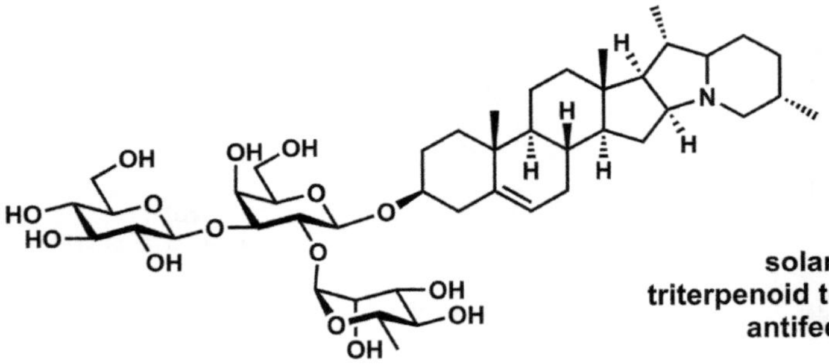

solanine
triterpenoid trisaccharide
antifeedant

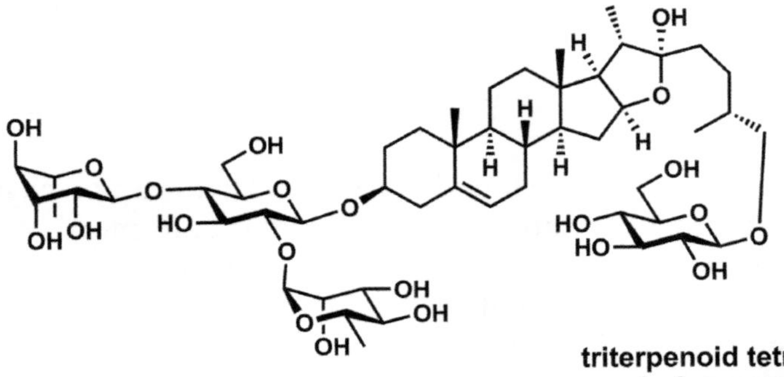

triterpenoid tetrasaccharide
in fenugreek seeds
("greek hay"/curries)

Figure 11.19 Structures of two saponins. Solanine, found in potato leaves, flowers and sprouts, is the trisaccharide of solanidine.

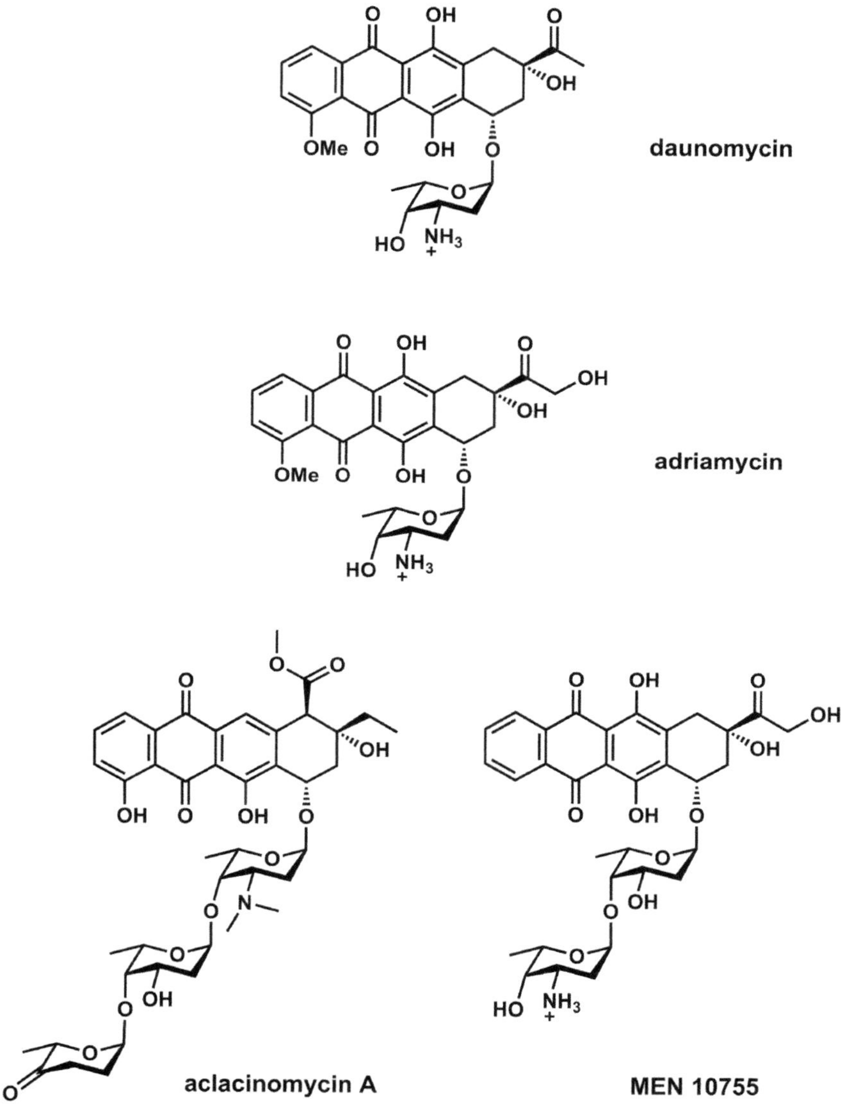

Figure 11.20 Common glycosylation patterns in anthracycline natural products.

aclacinomycin A, with the latter having an unusual trisaccharide of deoxysugars which help in binding to target DNA (Figure 11.20).

The indolecarbazole class of dimeric alkaloids features *N*-glycosylation at either one or both of the indole nitrogens (Figure 11.21). As shown for staurosporine maturation, both nitrogens have become

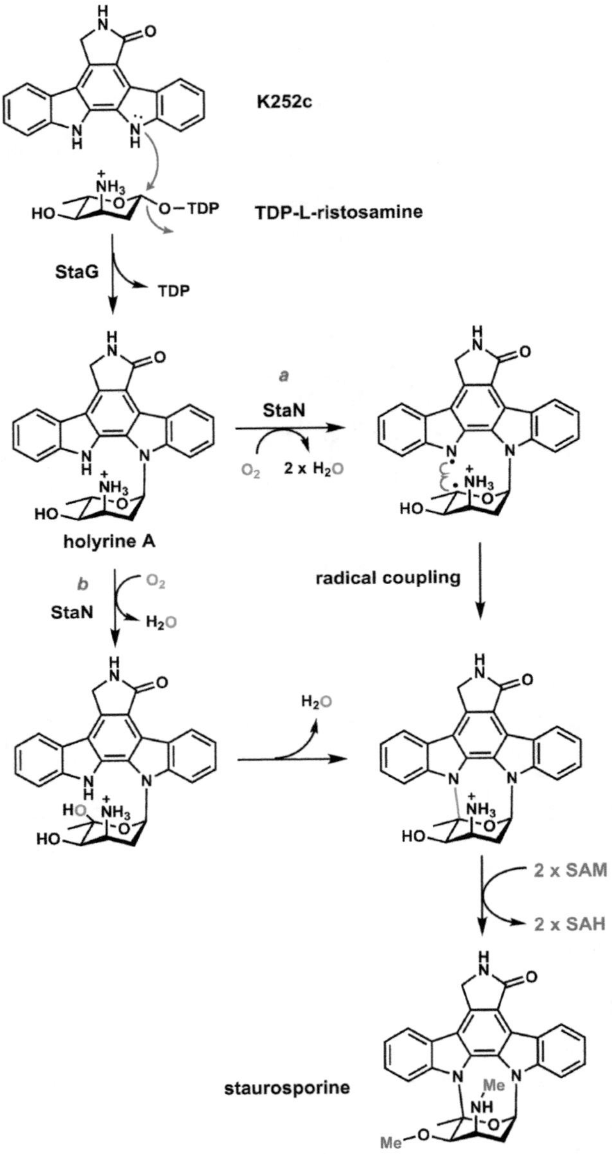

Figure 11.21 Staurosporine: Two *N*-glycoside bonds are formed to the L-ristosamine sugar at $C_{1'}$ and $C_{5'}$. The $C_{1'}$ link is *via* an *N*-glycosyltransferase. Two mechanistic proposals for forging the $N–C_{5'}$ link by the P450 type StaN enzyme are shown, a) *via* coupling of a diradical intermediate and b) *via* a transient $C_{5'}$ hydroxylation.

attached to the ristosamine sugar. Attachment at C_1 is mediated by a typical glycosyltransferase utilizing TDP-L-ristosamine (StaG) to yield holyrine A. On the other hand, the attachment of the other indole N to C_5 of the ristosamine requires a different kind of chemistry. StaN turns out to be a cytochrome P450 hemeprotein (Salas, Zhu *et al.* 2005, Sanchez, Zhu *et al.* 2005). It is of the subclass that makes the high valent Fe(V)=O oxidant as radical initiator but does not complete the [OH•] transfer (see Chapter 9). Instead, coupling of a nitrogen radical and the C_5 radical is proposed to create the N–C bond in the doubly tethered sugar–aglycone framework (route a). An alternative proposal (route b) is the hydroxylation of $C_{5'}$ of the ristosamine, followed by dehydration to yield the oxonium ion that can be attacked by the amine to furnish the C–N bond. Two SAM-dependent methylations would complete the maturation of staurosporine.

Four examples of glycosylation of different types of polyketide frameworks illustrate the range of glycosylation enzymology in this natural product class. The erythromycin system (see Figure 11.6) is the best characterized in terms of NDP sugar donors and timing of glycoside formation. TDP-mycarose is directed to the C_3–OH of the deoxyerythronolide aglycone by a dedicated Gtf. As always, the attachment is from the 3-O– to C_1 of the transferred mycarose. Then, the desosaminyl transferase utilizes TDP-desosamine to link the desosaminyl sugar *via* C_1 to O_5 of the 14-membered macrolactone. Nystatin, a polyene antifungal agent by virtue of membrane disruption, contains a single *O*-linked sugar D-mycosamine (Figure 11.6). This is added late in the biosynthetic pathway, from GDP-mycosamine by the Gtf Nysd1. The antiparasitic drug ivermectin (Figure 11.6), the first line therapy for treatment of river blindness, contains an oleandrose disaccharide at one of the two OH groups on the macrolactone.

The fourth polyketide glycoside exemplified is urdamycin A, where four Gtfs encoded in the biosynthetic gene cluster are specific for adding the four modified hexose residues (Figure 11.6). Three of the residues are *O*-linked, the fourth is an unusual C-linked glycoside (the first residue on the trisaccharide chain) (Trefzer, Hoffmeister *et al.* 2000). The carbon atom is adjacent to the phenolic oxygen in the A ring of the angular tetracyclic structure and so can function as a carbon nucleophile to attack C_1 of the NDP-hexose cosubstrate (Mittler, Bechthold *et al.* 2007). An additional C-glycoside is found in flavonoids (Falcone Ferreyra, Rodriguez *et al.* 2013, Nagatomo, Usui *et al.* 2014) and a mechanism for formation is shown in Figure 11.22.

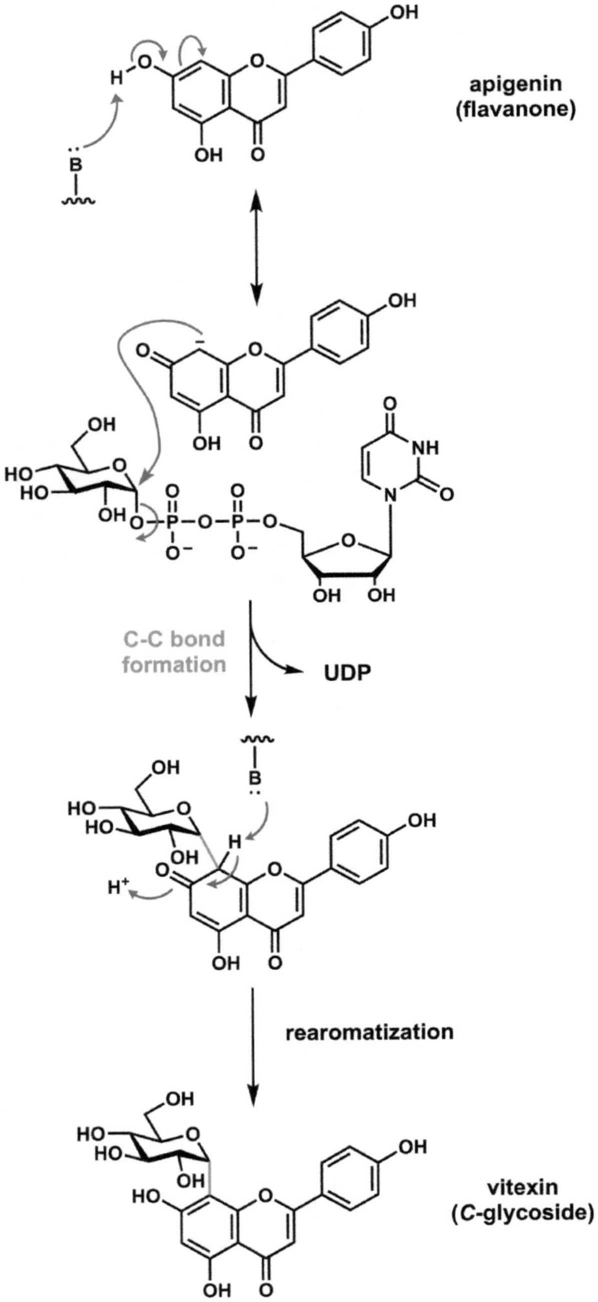

Figure 11.22 C-glucosylation of a flavonoid *ortho* to a phenol hydroxyl group.

Bleomycin is a mixed polyketide–nonribosomal peptide antitumor agent that also contains a disaccharide that is attached to the β-OH-histidine residue. The disaccharide contains L-gulose and an *O*-carbamoyl-D-mannose residue, both of which are available from GDP-mannose (see Figure 11.12). In the NRP class of natural products vancomycin and its congeners, the attachment of the glucosyl-1,2-L-vancosamine disaccharide has been dissected in detail (Figure 11.23). The coumarin antibiotics target the GyrB subunit of DNA gyrase. As depicted in Figure 11.24, they all contain at least one modified hexose. That sugar is L-noviose in novobiocin, clorobiocin, and the dimeric coumermycin. In the novobiocin system the NovM noviosyltransferase (Figure 11.24) has been examined in detail for timing, mechanism, and specificity (Freel Meyers, Oberthur *et al.* 2003) (Albermann, Soriano *et al.* 2003). Two subsequent modifications occur on the noviosyl ring, *O*-methylation and *O*-carbamylation to finish the mature antibiotic scaffold (Freel Meyers, Oberthur *et al.* 2004).

11.4 The Chemical Logic for Converting NDP-Glucose to NDP-Modified Hexoses

As noted in the several examples in the preceding figures of this chapter, many of the hexoses appended to microbial and plant aglycones are not the common glucose, *N*-acetylglucosamine, galactose, or mannose that are the workhorses of primary metabolism of hexoses. All these modified hexoses in glycoside biosynthesis derive from an NDP-glucose, typically TDP-glucose in bacteria and fungi and UDP-glucose in higher plants. The common early steps and the late branching pathways have been worked through extensively in bacterial systems (He and Liu 2002, He and Liu 2002).

All the diversification starts with an initial enzymatic conversion of TDP-glucose to TDP-4-keto-6-deoxyglucose (Thibodeaux, Melancon *et al.* 2007, Thibodeaux, Melancon *et al.* 2008). Establishment of the 4-keto is the critical functional group change that allows subsequent chemistry at C_2, C_3, C_4, C_5, and C_6. The first committed enzyme in the pathway, NDP-hexose-4,6-dehydratase, contains tightly bound NAD^+. Once TDP-glucose has docked in the enzyme active site, the C_4–OH is oxidized to the ketone while NAD^+ is reduced to NADH (Figure 11.25). The presence of the keto group allows abstraction of the C_5–H as a proton because the resultant C_5 carbanion can be stabilized as the enolate. The negative charge can assist in expulsion of the C_6–OH,

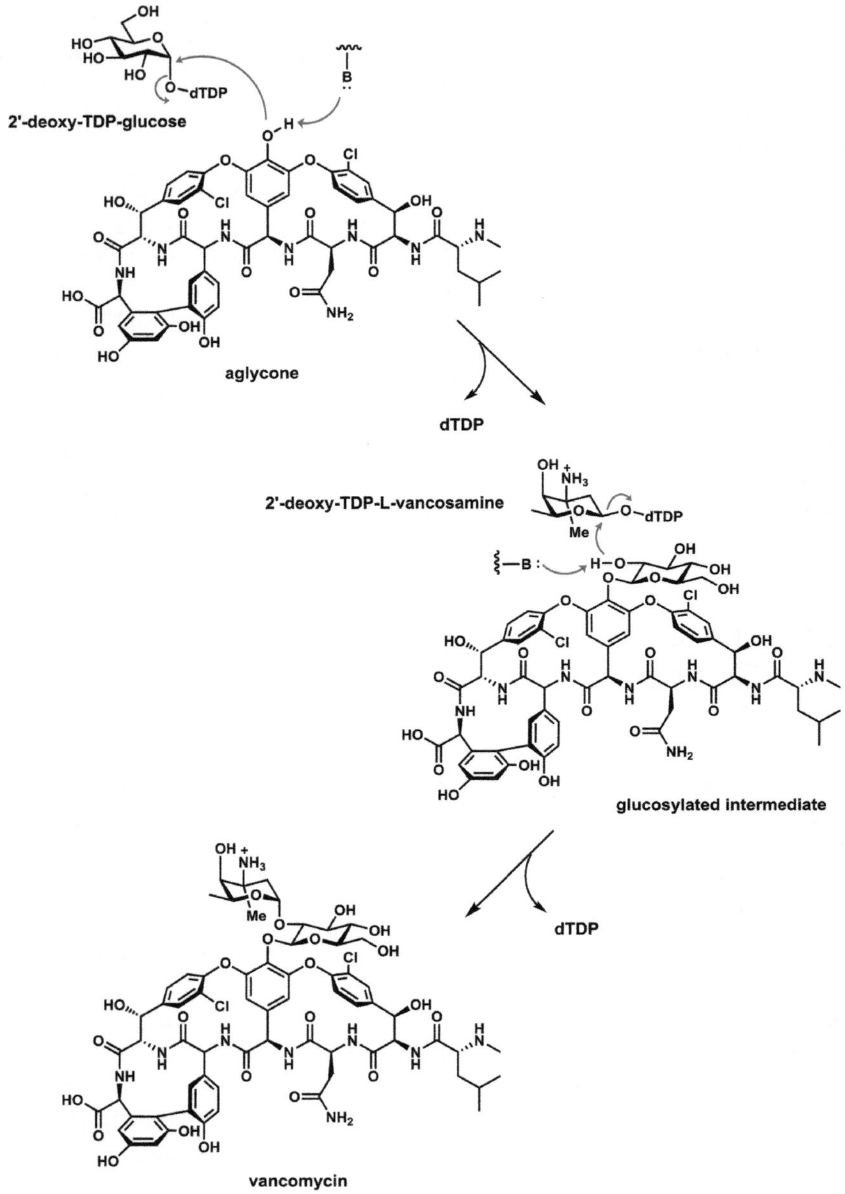

Figure 11.23 Post assembly line glycosylation of the vancomycin aglycone.

with C–O bond cleavage to generate the conjugated 6-ene-4-one. Now the enzyme transfers a hydride from NADH, not back to the original C_4 carbon of the ketone but to the olefinic terminus of the ene-one

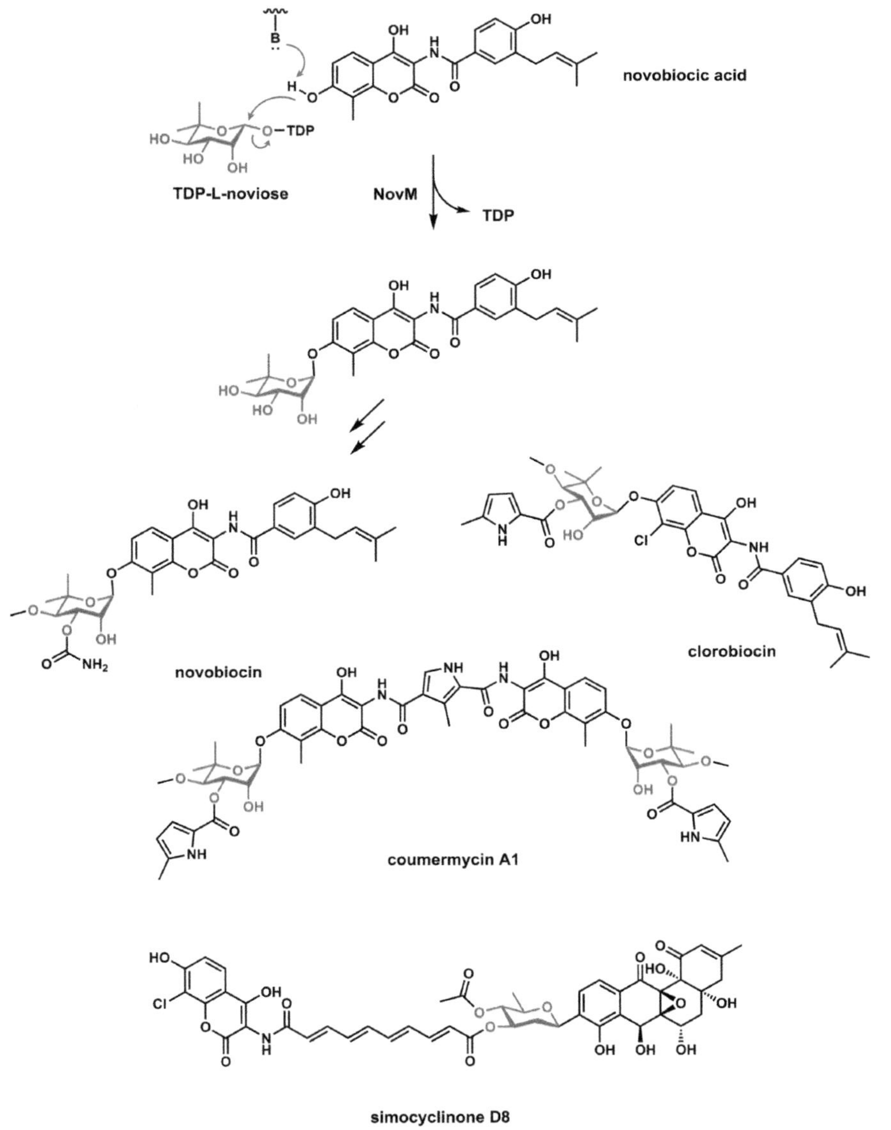

Figure 11.24 Glycosylated coumarin antibiotics contain an L-noviose deoxyhexose. The phenol hydroxyl of novobiocic acid is the nucleophile that captures the noviosyl group from dTDP-L-noviose in the third to last step of novobiocin biosynthesis.

at C_6. This completes the catalytic cycle and yields the TDP-4-keto-6-deoxyglucose for all downstream NDP-deoxy- and amino deoxyhexose variants.

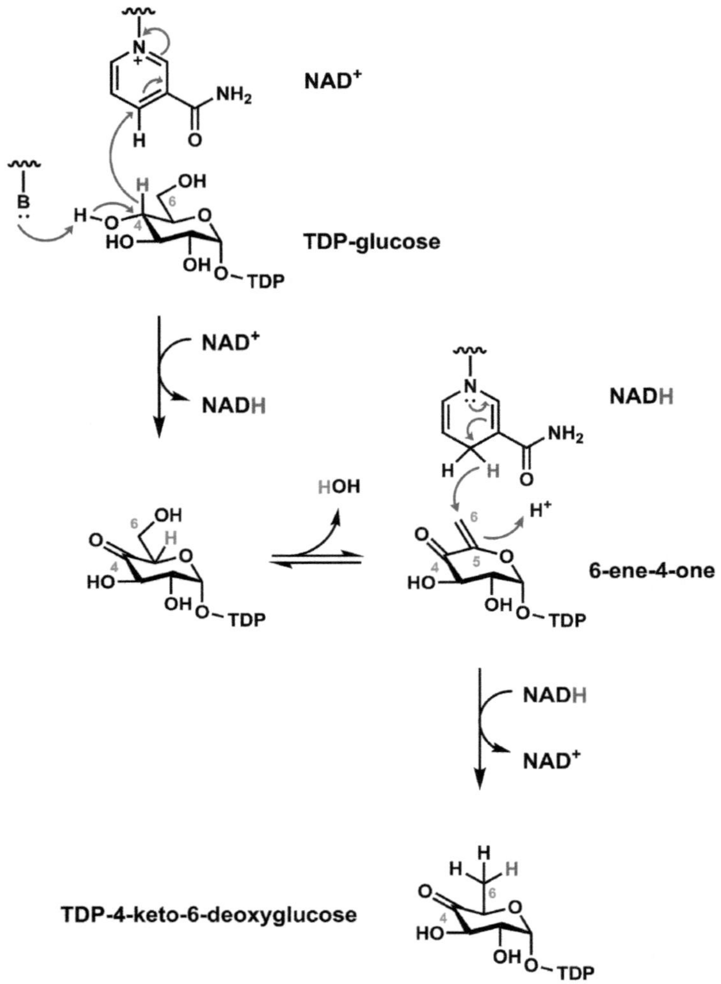

Figure 11.25 The first committed enzyme in the generation of TDP-deoxyhexoses converts TDP-glucose to TDP-4-keto-6-deoxyglucose.

These transformations are effected with only four enzyme types: oxidoreductases, dehydratases, epimerases, and C-methylases (Figure 11.26). In bacterial systems the genes encoding these enzymes are typically colocalized with the genes for the biosynthesis of the aglycone. So are the specific glycosyltransferase genes. *Inter alia*, these genetic colocalizations point out how important the glycosylations are for activity of the mature natural product.

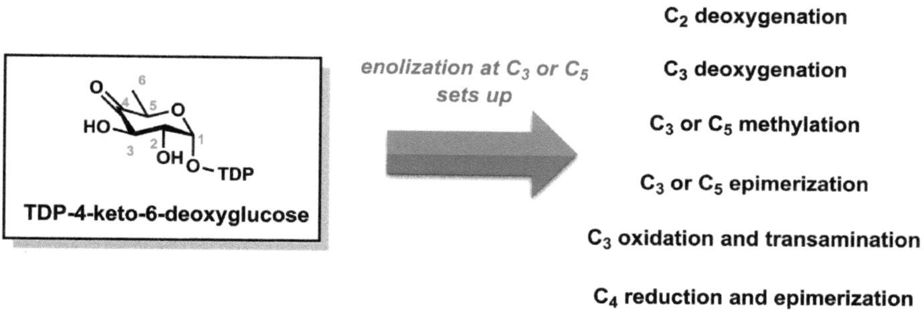

Figure 11.26 A limited inventory of enzymatic reactions to deoxygenate, epimerize, transaminate, and C-methylate NDP-hexoses in secondary metabolism.

As noted in Figure 11.27 for the vancomycin analog chloroeremomycin molecules, five genes, *evaA–E*, found in the cluster encode enzymes that act sequentially on the first committed intermediate TDP-4-keto-6-deoxyglucose. Figure 11.28 shows the progression for formation of TDP-L-epivancosamine (TDP-L-eremosamine) and also TDP-daunosamine (Thibodeaux, Melancon *et al.* 2008). The first step by EvaA is removal of the C_3–H as a proton and use of the stabilized C_3 carbanion to assist in elimination of the 2-OH (in strict analogy to formation of the 6-deoxy group above). The first product is the $C_{2,3}$-enol that will ketonize to TDP-3,4-diketo-2,6-dideoxy-D-glucose. The next enzyme, EvaB, carries out a reductive transamination of the 3-keto to TDP-3-β-amino-4-keto-2,3,6-trideoxy-D-glucose with stereocontrol. The next step is *C*-methylation from SAM by EvaC. Again the 4-keto group allows C_3 carbanion formation required for attack on SAM for $[CH_3^+]$ transfer, with the indicated chirality control. To this point the C_4 ketone has allowed both dehydrative elimination chemistry to the 2-deoxy and C_3-methylation. The next step is epimerization of the C_5 methyl to yield the L-sugar by EvaD. The final step in the pathway is reduction of the C_4 ketone in the L-deoxysugar to either the α-OH (in the L-vancosamine pathway) or β-OH (L-epivancosamine) stereochemistry, the latter catalyzed by EvaE. Figure 11.29 shows in tabular form the changes that occur at each of C_2 to C_6 as TDP-D-glucose is converted to the TDP-L-epivancosamine proximal sugar donor (Chen, Thomas *et al.* 2000).

The TDP-3-β-amino-4-keto-2,3,6-trideoxy hexose metabolite in the vancosamine pathway is epimerizable to the 3-α-amino isomer (Figure 11.28). Reduction of the 4-keto to the 4-α-alcohol provides

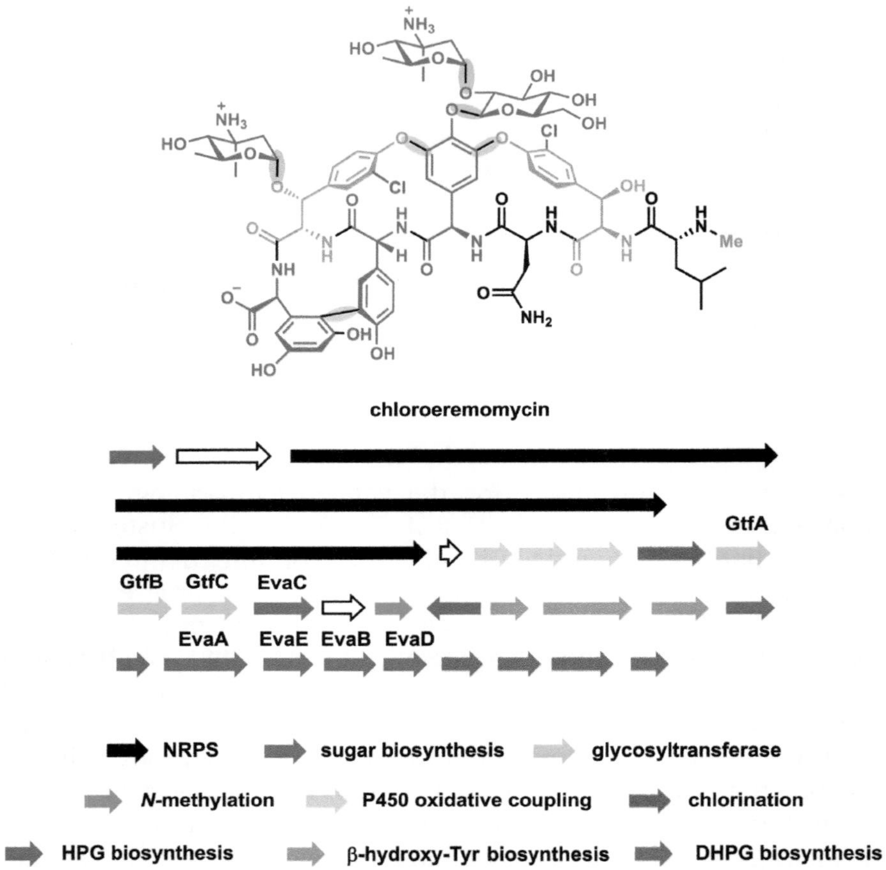

Figure 11.27 Genes encoding the three post-assembly glycosyltransferases and the five enzymes needed to convert TDP-4-keto-6-deoxyglucose to TDP-L-vancosamine are encoded in the bacterial chloroeremomycin biosynthetic gene cluster.

TDP-L-daunosamine. This is the proximal sugar donor to the 7-OH of the tetracyclic polyketide aglycone to complete biosynthesis of the antitumor agent daunomycin (see Figure 11.20). By analogous logic and enzymatic machinery in novobiocin scaffold maturation (Figure 11.30), the TDP-4-keto-6-D-deoxyglucose is epimerizable to the L-sugar series, methylatable at C_5 (*via* enolate anion chemistry) and then reduced at the C_4 ketone (Thibodeaux, Melancon *et al.* 2007, Thibodeaux, Melancon *et al.* 2008). The C_4-ketone reduction is almost always the last step in TDP-hexose modifications since its chemical effects on C_3 and C_5 carbanion formation are central to

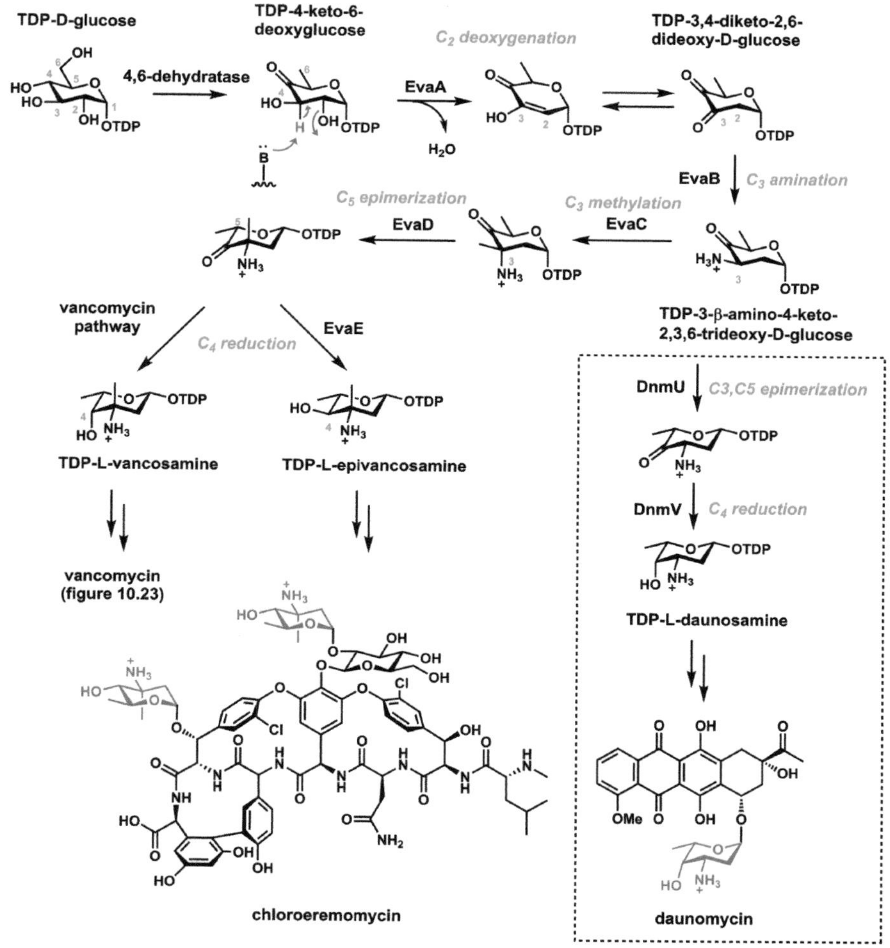

Figure 11.28 Stepwise generation of TDP-L-daunosamine, TDP-L-epivancosamine and TDP-L-vancosamine. A key step is the conversion of NDP-4-keto-6-deoxyhexose to NDP-4-keto-2,6-dideoxyhexose *via* the C_3 enolate and then the 3,4-diketo species before NADH-mediated reduction at C_3.

epimerizations and *C*-methylations. In this example the product is TDP-L-noviose, ready for processing by the NovM glycosyltransferase (see Figure 11.24).

One final example is conversion of TDP-glucose *via* the 4-keto-6-deoxy metabolite to the trideoxy-amino sugar TDP-L-desosamine that is the donor crucial for the antibiotic activity of erythromycin. The key enzyme is DesII, a member of the radical-SAM family of

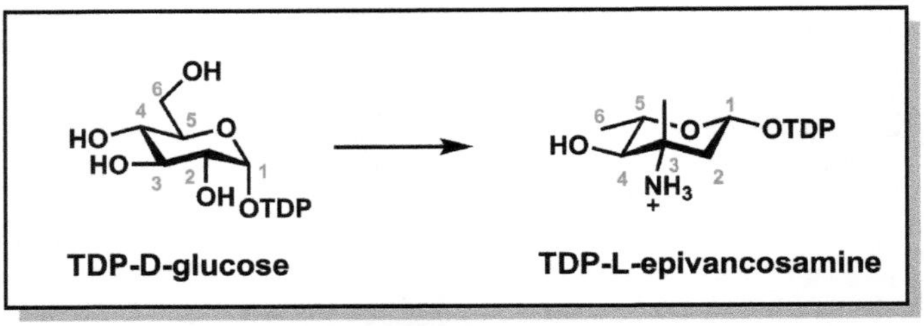

order of reactions

carbon number	modification	Eva gene
4	oxidation	4,6-dehydratase
6	deoxygenation	4,6-dehydratase
2	deoxygenation	EvaA
3	transamination	EvaB
3	methylation	EvaC
5	epimerization	EvaD
4	ketoreduction (epimerization)	EvaE

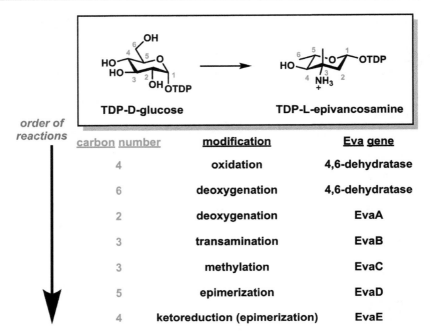

Figure 11.29 Enzymatic changes at C_2–C_6 of the hexosyl moiety in going from TDP-glucose to TDP-L-epivancosamine.

catalysts detailed in Chapter 10. DesII catalyzes a complex set of transformations, among them conversion of the C_4-amine to the C_4–CH_2 group, using radical chemistry to effect this unobvious transformation.

As depicted in Figure 11.31, the 5′-deoxy-adenosyl radical generated by SAM fragmentation can abstract the C_3–H as $H^\bullet$ to create a $C_3^\bullet$ radical (Szu, He *et al.* 2005, Szu, Ruszczycky *et al.* 2009), from which the adjacent C_4–NH_3 substituent can be lost. The resultant cation radical has resonance forms putting the radical at C_3 or C_4 as shown. Addition of the nascent NH_3 back to the C_3 cation would yield the C_3 carbinolamine along with radical density at neighboring C_4. Transfer of the $H^\bullet$ from 5-deoxyadenosine back to C_4 would be followed by decomposition of the carbinolamine to the 3-keto-4,6-deoxy sugar. This requires an enzymatic transamination of the 3-keto to the 3-amino group, and *N*-dimethylation to get to the end product TDP-L-desosamine.

The examples noted above are a small fraction of the modified hexoses produced from NDP hexoses. A recent review reported that

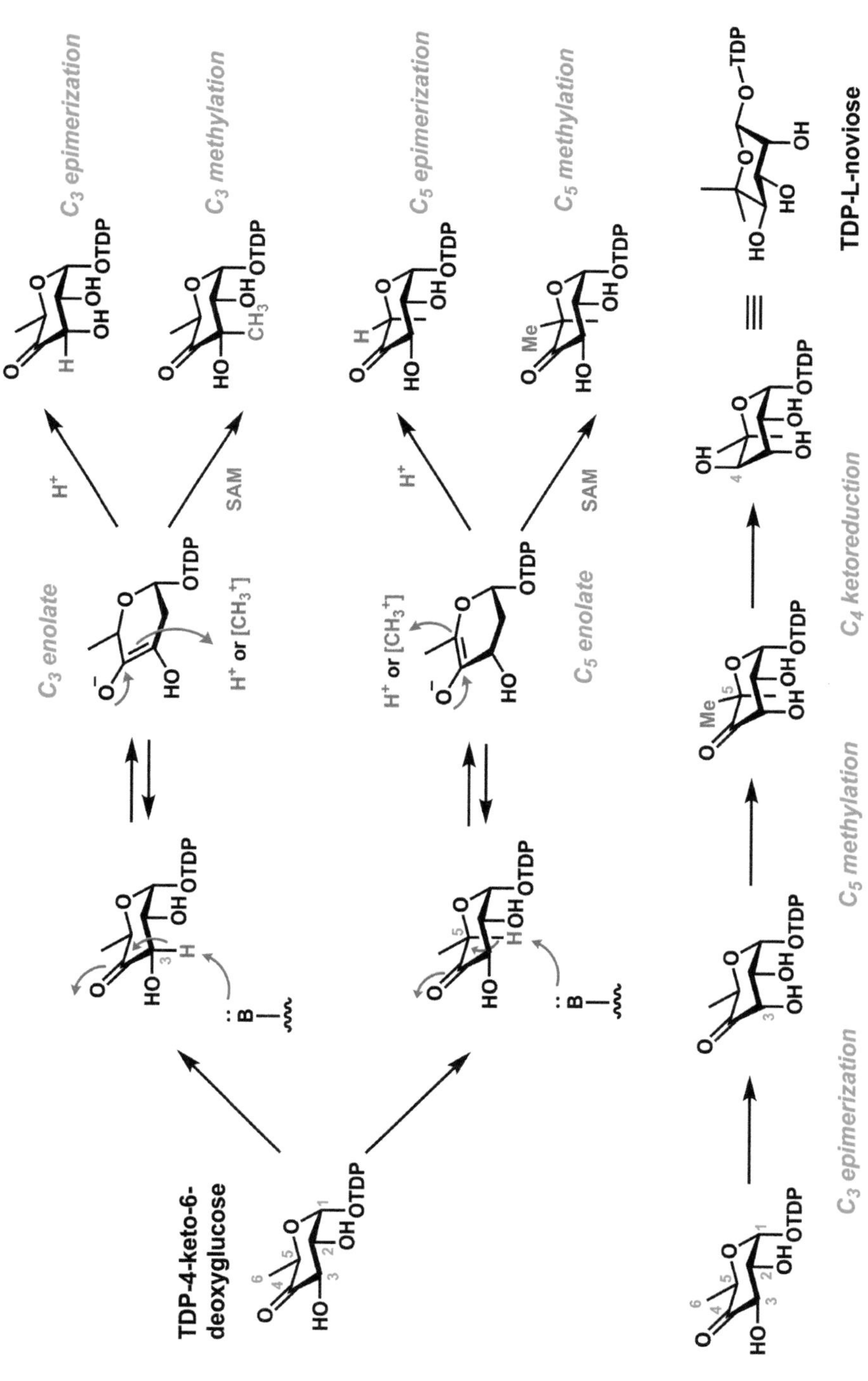

Figure 11.30 Enolization of the C_4-ketone in TDP-4-keto-6-deoxy hexoses enables *C*-methylation at either C_3 or C_5.

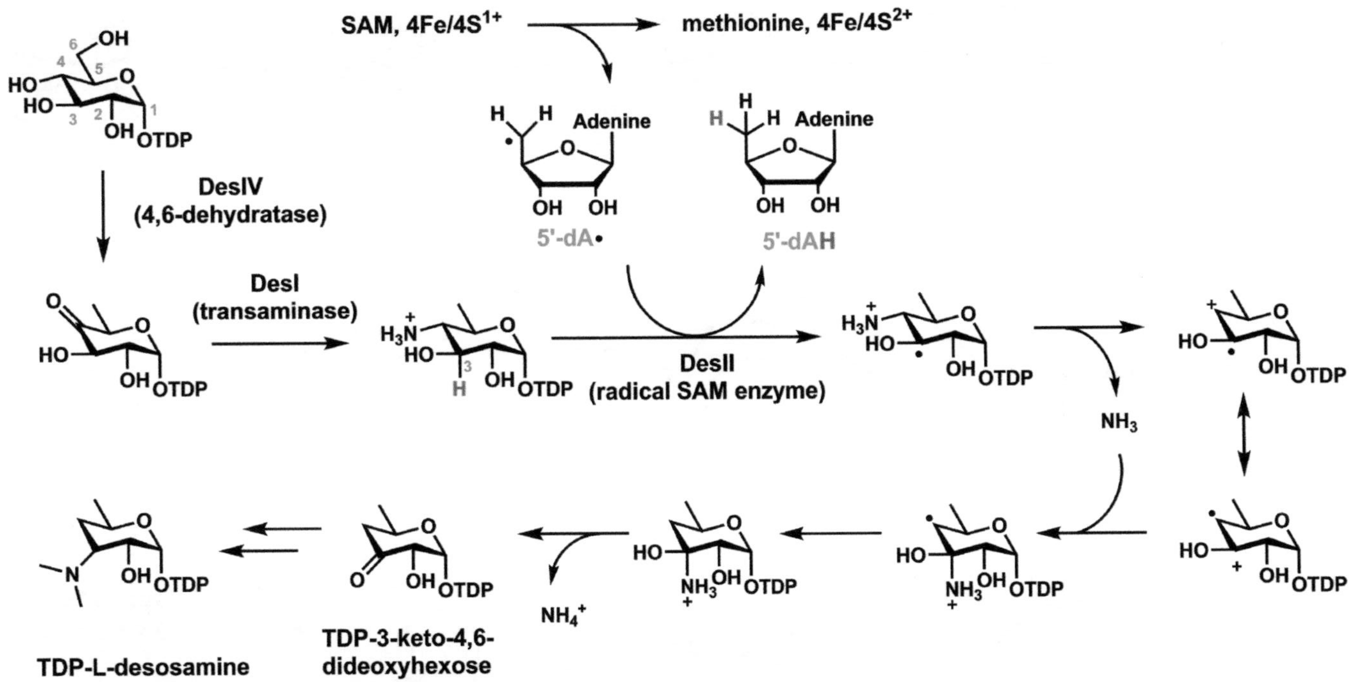

Figure 11.31 Steps in the conversion of TDP-glucose to TDP-D-desosamine for erythromycin maturation.

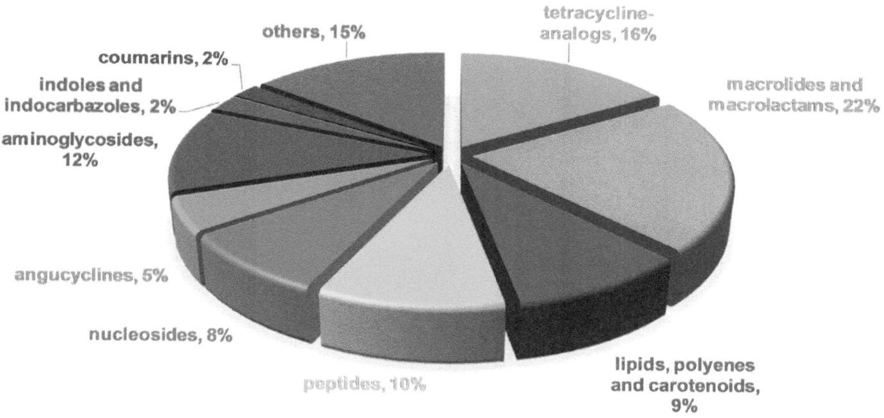

Figure 11.32 Distribution of types of natural products that are glycosylated. Adapted from Elshahawi, S. I., K. A. Shaaban, M. K. Kharel and J. S. Thorson (2015). "A comprehensive review of glycosylated bacterial natural products". *Chem. Soc. Rev.*, **44**: 7591–7697 with permission from The Royal Society of Chemistry.

"of the 15 940 known bacterial natural products, 3426 (21%) were glycosylated, with 344 distinct sugar moieties" (Elshahawi, Shaaban *et al.* 2015) (Figure 11.32).

11.5 Balance of Gtfs and Glycosidases: Cyanogenic Glycosides and Glucosinolates

In some, perhaps most, of the glycosylated natural products, the glycosyl units, monomers to heptasaccharides, are essential components of the mature scaffolds. They are not reversibly removed by enzymatic hydrolysis. This is the case for antibiotics such as erythromycin, vancomycin, and novobiocin, for the antiparasitic avermectin, the anticancer agent daunomycin, and the indolecarbazoles rebeccamycin and staurosporine, noted earlier in this chapter.

On the other hand, there are two classes of plant secondary metabolites, cyanogenic glycosides (Gleadow and Moller 2014) and glucosinolates (Halkier and Gershenzon 2006), Figure 11.33, where the glycosylated forms are storage depots for latent functional groups that become activated by glycosidase action. In these molecular

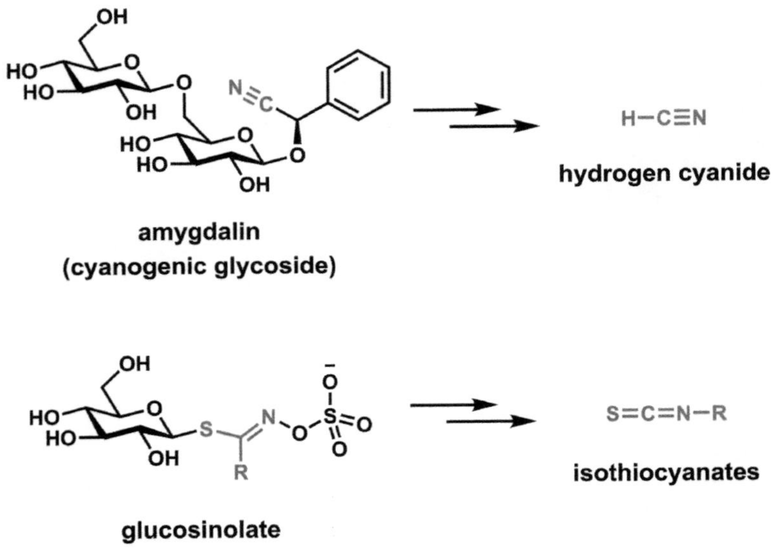

Figure 11.33 Glucosinolates and cyanogenic glycosides: glucosylation as a reversible protecting strategy for latently reactive toxins cyanide ion and isothiocyanate ion.

classes the balance of biosynthetic Gtfs and hydrolytic glycosidases reflect the utility of preformed toxins, either cyanide ion or isothiocyanate ions, to be released in a rapid defensive response by the plants against predators. As the name implies, cyanogenic glucosides give rise to hydrogen cyanide from protected cyanohydrin acetals. Glucosinolates are defined as "any of various bitter sulfur-containing glycosides found especially in cruciferous plants (as broccoli, cabbage, and mustard) that when hydrolyzed form bioactive compounds (as isothiocyanates) including some which are anticarcinogenic" (definition from the *Webster Medical Dictionary*).

Figure 11.34 shows the plant sources for some of the common cyanogenic glucosides. The figure also indicates the structures of amygdalin from almonds, dhurrin from sorghum, and linamarin from cassava. The aglycones are extremely simple cyanohydrin scaffolds. They arise from *N*-oxidation of amino acids as exemplified by the two cytochromes P450 which take tyrosine to *para*-hydroxymandelonitrile (Figure 11.35). The tetrahedral cyanohydrin can break down to release cyanide ion, a potent ligand to heme proteins, and the parent aldehyde. In cells generating these hydroxynitriles (cyanohydrins) they are trapped enzymatically by glucosyltransferases.

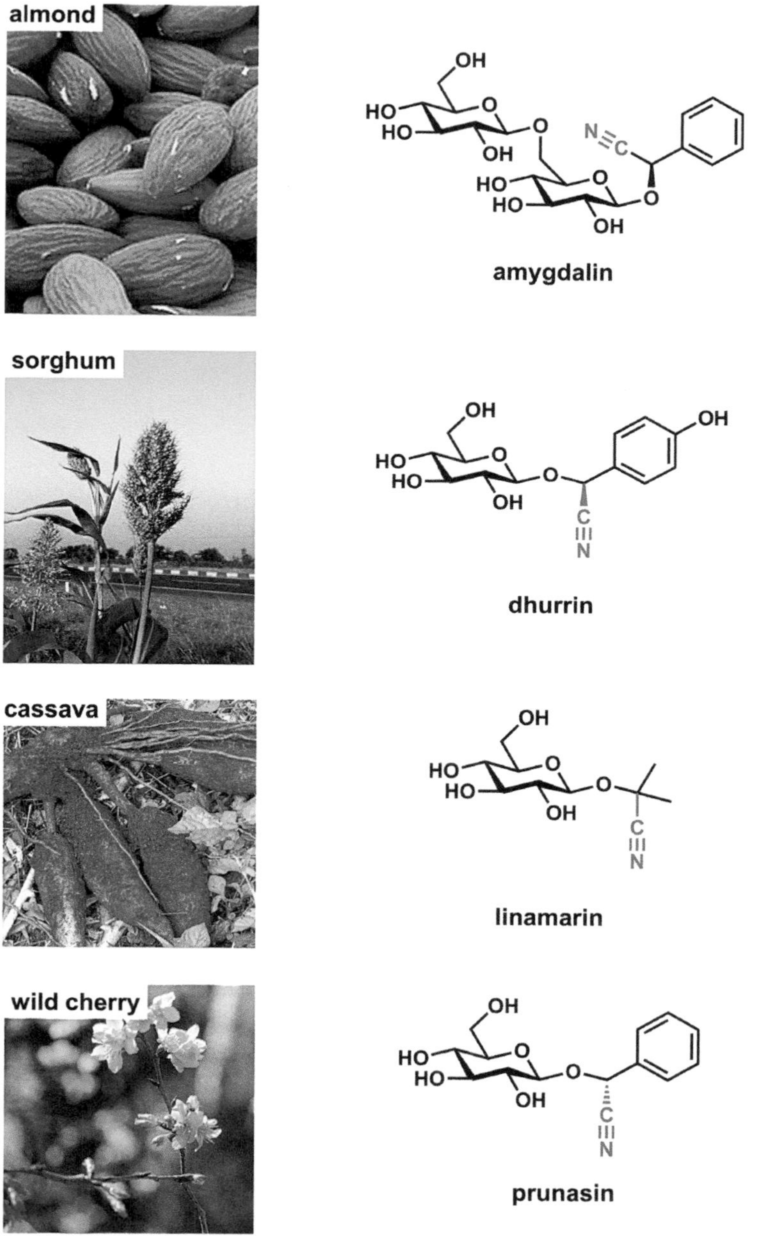

Figure 11.34 Plant sources and structures for some cyanogenic glycosides.

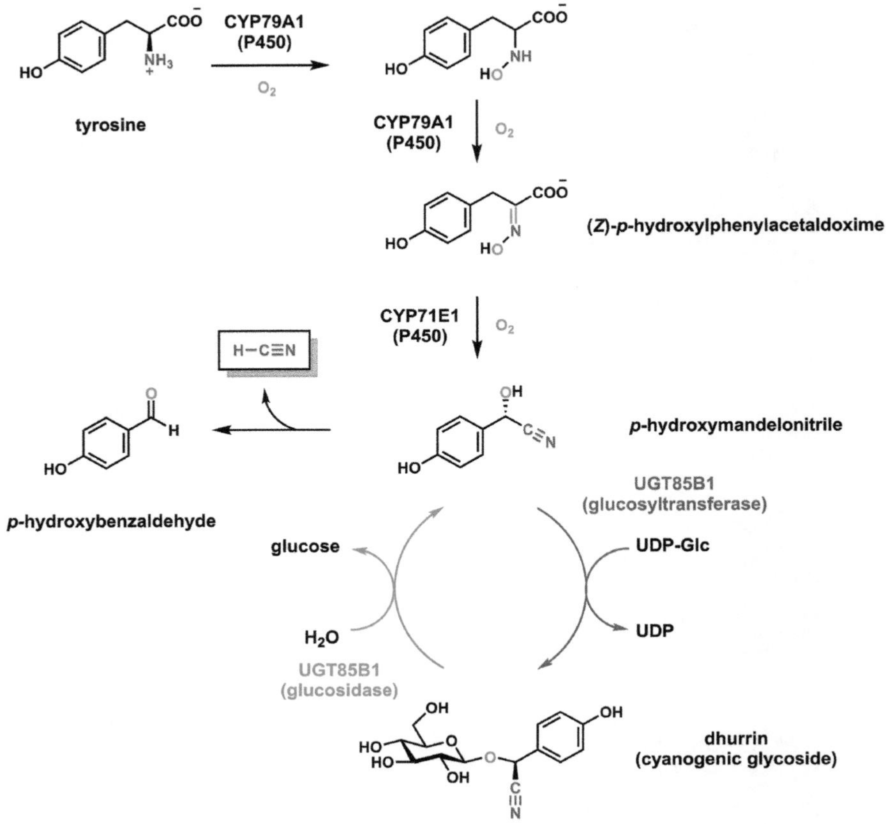

Figure 11.35 Tyrosine to dhurrin by two *N*-oxygenating P450s and a glucosyltransferase.

In this example the secondary hydroxyl group of the hydroxynitrile is the nucleophile that attacks C_1 of UDP-glucose. The resultant dhurrin (Halkier and Moller 1989, Halkier, Olsen *et al.* 1989) now has the cyanohydrin blocked, in acetal linkage to the glucose moiety. They are chemically stable and have no nonenzymatic tendency to release cyanide from these glycosides. The reversal to cyanohydrin that is the immediate progenitor of the respiratory poison cyanide ion is mediated by specific cyanogenic glycosidases. These hydrolytic enzymes can be regulated in terms of synthesis and/or localization in distinct subcellular/cellular compartments until appropriate signals are received.

Similarly, the metabolite prunasin (Nahrstedt and Rockenbach 1993), shown in Figure 11.36 (one of its sources, the Japanese

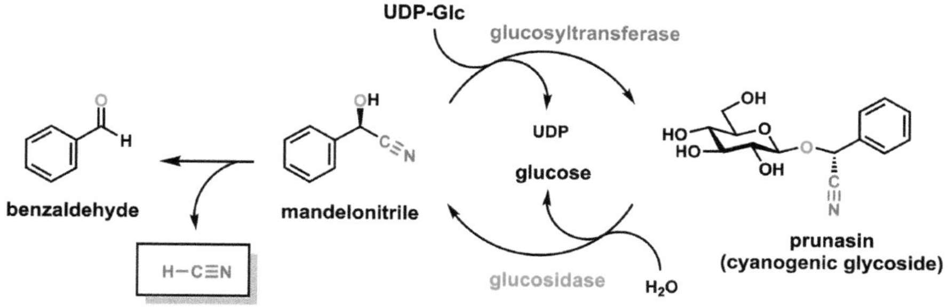

Figure 11.36 The Japanese cherry bush (see Figure 11.34) contains prunasins as a latent defense molecule. The glycosidic linkage in prunasin is hydrolyzed on contact with a glucosidase to yield benzaldehyde and cyanide ion.

cherry bush, is shown in Figure 11.34) can be acted on by a specific glucosidase. In turn the cyanohydrin decomposition can be accelerated by enzyme action.

Glucosinolate biosynthesis proceeds with parallel logic, using amino acids as starting units that undergo *N*-oxidation, in this case to the aldoximes after decarboxylation (Figure 11.37). The carbonyl of an aldoxime is electrophilic and can be attacked by thiolates, for example by cysteine or reduced glutathione, to generate *S*-alkylthiohydroximate adducts. Action of C–S lyases cleaves away the side chain(s) of the thiol donor to give the thiohydroximate. This is the species that is substrate for the UDP-glucose dependent glucosyltransferases. The thiolate anion is the strongest nucleophile and the result is an *S*-glucoside rather than a more typical *O*-glucoside. A final maturative step is sulfuryl group transfer from phosphoadenosine phosphosulfate (PAPS) to the oxygen of the hydroximate group. The glucosinolates are then stored until a need for a chemical defensive response is realized by the plant.

The reversal of *S*-glucosylation in plants is hydrolytic, typically catalyzed by thioglucosidases as exemplified by myrosinase (Figure 11.38). The reaction mechanism of myrosinase appears to involve a covalent glucosyl-enzyme intermediate (Burmeister, Cottaz *et al.* 1997). The released aglycones can eliminate sulfate (nonenzymatically) and generate electrophilic isothiocyanates as the proximal defense agent. The virtue of sulfation in the biosynthetic pathway becomes evident in this hydrolytic reversal mode to facilitate N–O bond cleavage in isothiocyanate formation. The glucosinolates are sequestered in plant cell vacuoles and kept away from

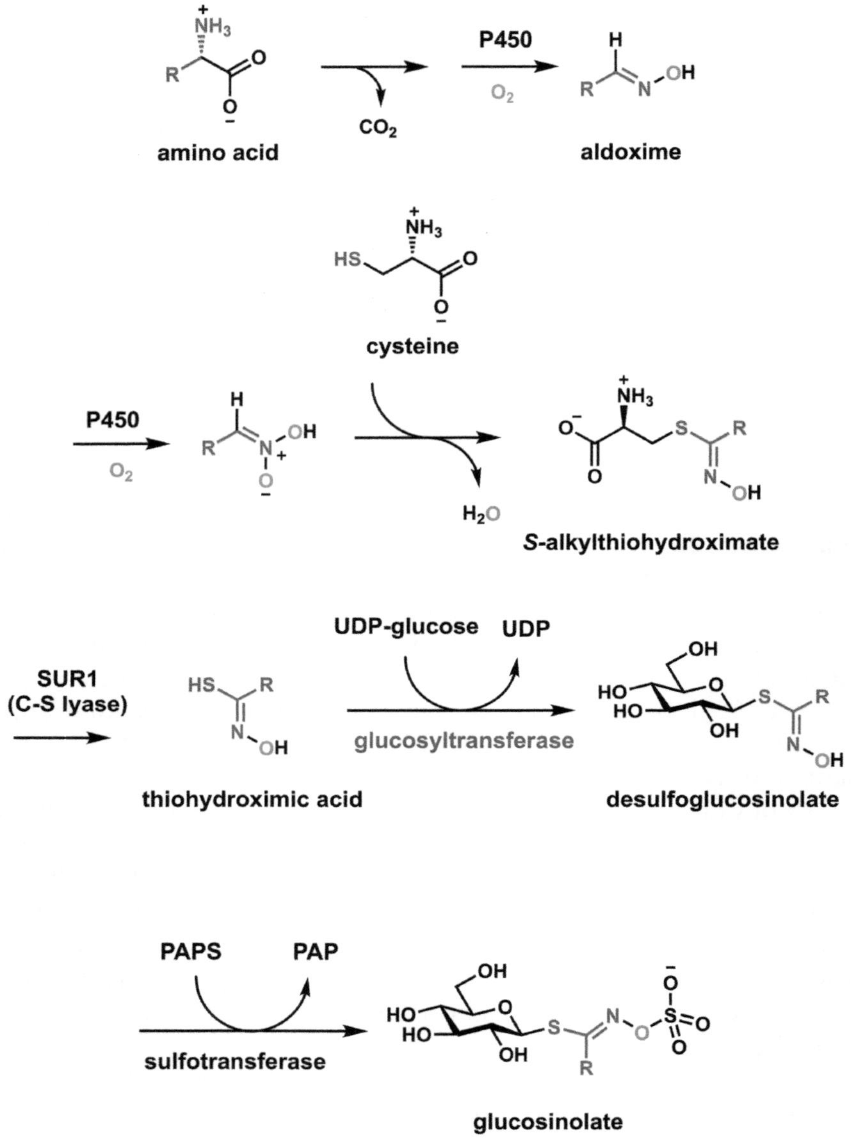

Figure 11.37 Glucosinolate biosynthesis. Amino acid oxidation to aldoximes, thiohydroximates, and *S*-glucosylation yields thioglycosides.

myrosinase in calm times. Under stress conditions, regulated fusion of vacuoles with other vesicles or cell membrane allows contact with intracellular or extracellular myrosinase and a controlled defensive response.

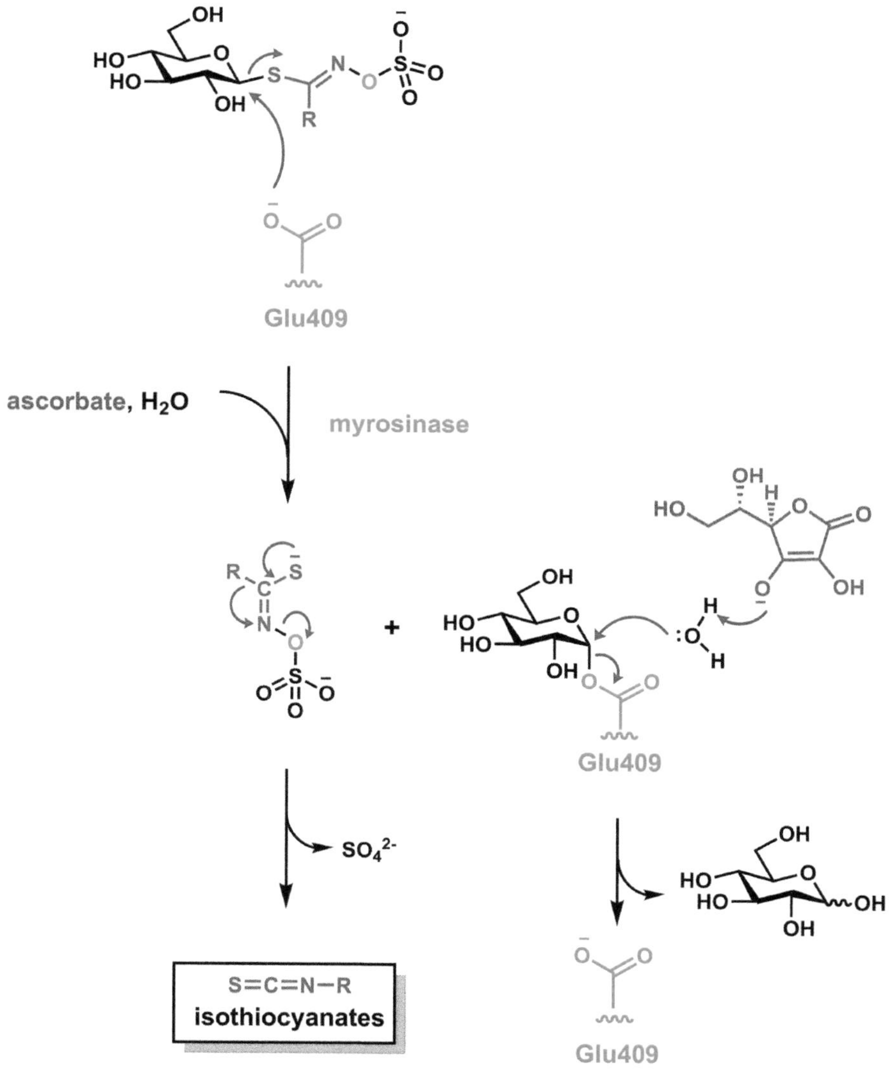

Figure 11.38 Myrosinases are thioglucosidases. Hydrolysis proceeds *via* a covalent glycosyl-enzyme intermediate. Isothiocyanate is the first released product.

A combinatorial approach to latent toxins is found in garlic mustard (Figure 11.39). Methionine is precursor to the glucosinolate sinigrin. The allyl side chain comes from β,γ-elimination on the methionine framework, most probably by pyridoxal-P dependent enzymology. The sulfoxidation is likely to be carried out by a

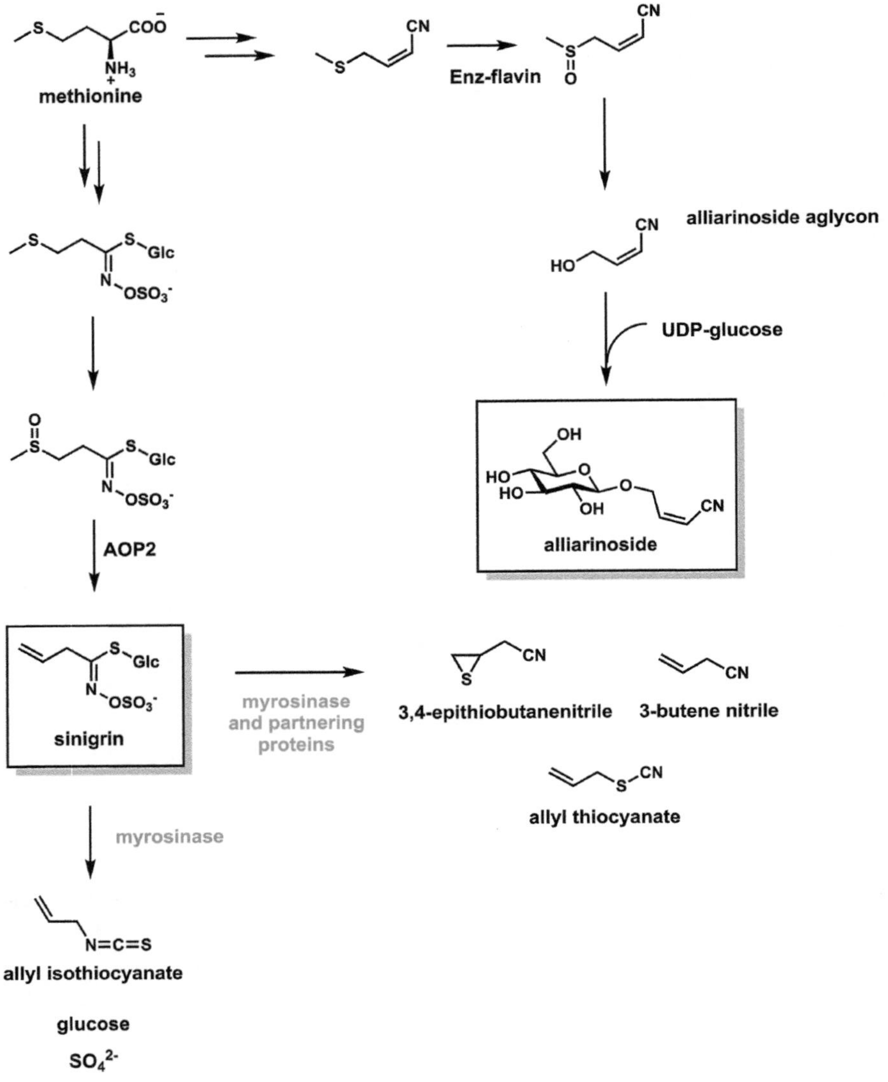

Figure 11.39 Methionine as precursor to two latently reactive chemotypes, alliarinoside and sinigrin, in garlic mustard.

flavin-dependent *S*,*N*-oxygenase but that is incompletely determined. Partner proteins in garlic mustard can direct myrosinase to allylthiocycanate and alkylnitriles. The plant can also generate alliarinoside as an *O*-glucoside in addition to the sinigrin-*S*-glycoside.

11.6 Aminoglycosides: Oligosaccharides without an Aglycone

The aminoglycoside antibiotics have a prominent place in the discovery of natural antibiotics. Streptomycin was the second class of commercially important antibiotics obtained from Nature, after the large scale fermentation of penicillin in the mid 1940s. Waksman and colleagues isolated streptomycin in 1947. Two additional family members include kanamycin in 1957 and tobramycin in 1967 (see Figure 11.1).

The aminoglycosides target specific regions of the 16S rRNA backbone in the 30S bacterial ribosome subunit, setting up miscoding of aminoacyl-tRNAs, incorporation of incorrect amino acids, and formation of defective proteins, leading to cell death. Central to the interaction is electrostatic attraction between the amino groups of the aminoglycosides and the negative charges on the phosphate backbones of rRNA, along with a network of hydrogen bonds. Thus, enzymatic introduction of the amino groups is a key strategic aspect of trisaccharide framework assembly (Walsh and Wencewicz 2016).

The central ring in these trisaccharide antibiotics is a cyclohexane, an inositol, rather than a pyranose. It arises enzymatically from glucose-6-P *via* the ring opened aldehyde form, catalyzed by inositol-3-phosphate synthase. Enzymatic oxidation of the C_5–OH (only available in the ring-opened aldehyde) to the ketone allows intramolecular aldol condensation from the C_6-enolate on the C_1 aldehyde (Figure 11.40). The C_5 ketone has done its job and is now re-reduced to the alcohol, yielding myoinositol-3-P, progenitor to inositol *via* phosphatase action.

11.7 Kanamycin, Tobramycin, Neomycin

Figure 11.41 schematizes enzymatic assembly of the kanamycin scaffold. The key steps are reductive transamination of keto groups to cationic amines and tandem action of two Gtfs. Glucose-6-P conversion to the keto-inositol, as above, is followed by reductive transamination from an amino acid cosubstrate donor. Then the alcohol to ketone to amine is iterated one more time to produce 2-deoxystreptamine (2-DOS). The 4-OH of 2-DOS is the nucleophile in a Gtf-mediated condensation with UDP-*N*-acetylglucosamine (UDP-GlcNAc). Deprotection of the *N*-acetyl moiety by deacetylase action produces the triamino

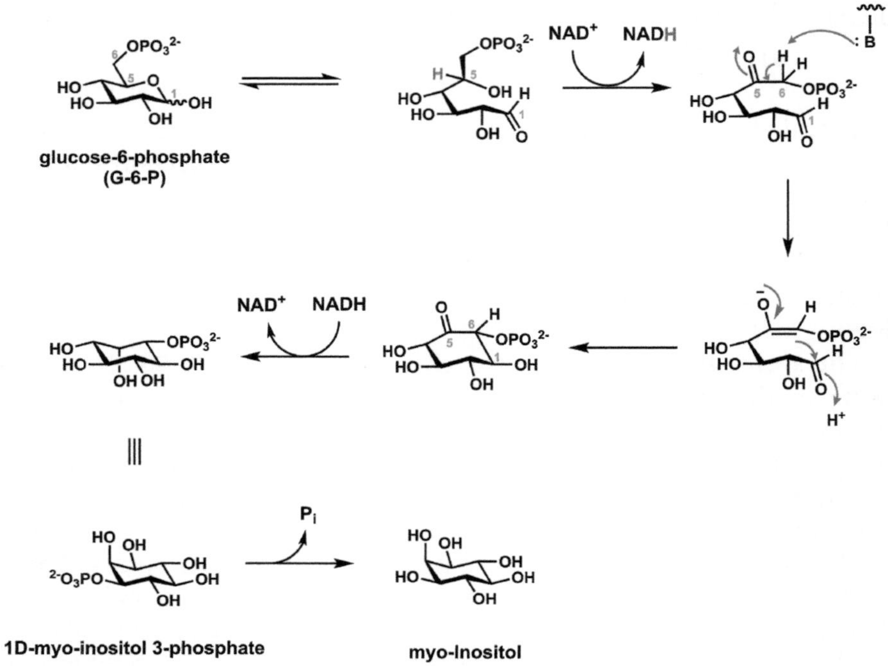

Figure 11.40 Aminoglycoside antibiotic biosynthesis: a common early step in streptomycin and kanamycin formation is conversion of the glucose-6-P pyranose ring to the hydroxycyclohexane ring of inositol. This reaction is catalyzed by inositol-3-phosphate synthase. Action by phosphatase forms myo-inositol.

paromamine. It was recently shown that paromamine is first reductively aminated at C_6 to give the triamine neamine (Park, Park *et al.* 2011). A second Gtf, specific for the C_3-aminated UDP-kanosamine that is reductively aminated from UDP-glucose, glycosylates neamine to complete the biosynthesis of kanamycin B. The final molecule has five amine groups, all cationic at physiologic pH.

Production of tobramycin and the ribose-containing aminoglycoside neomycin is shown in Figure 11.42. Kanamycin B is proposed to be regiospecifically deoxygenated in a radical SAM-dependent reduction (see Chapter 10) to give tobramycin (3′-deoxykanamycin B) in one enzymatic step. It is also possible that paromamine is first deoxygenated at the same position, then glycosylated in parallel fashion to that of kanamycin B to give tobramycin (pathway not shown) (Kudo and Eguchi 2016). Paromamine is precursor to ribostamycin by transfer of a ribosyl group (from UDP-ribose)

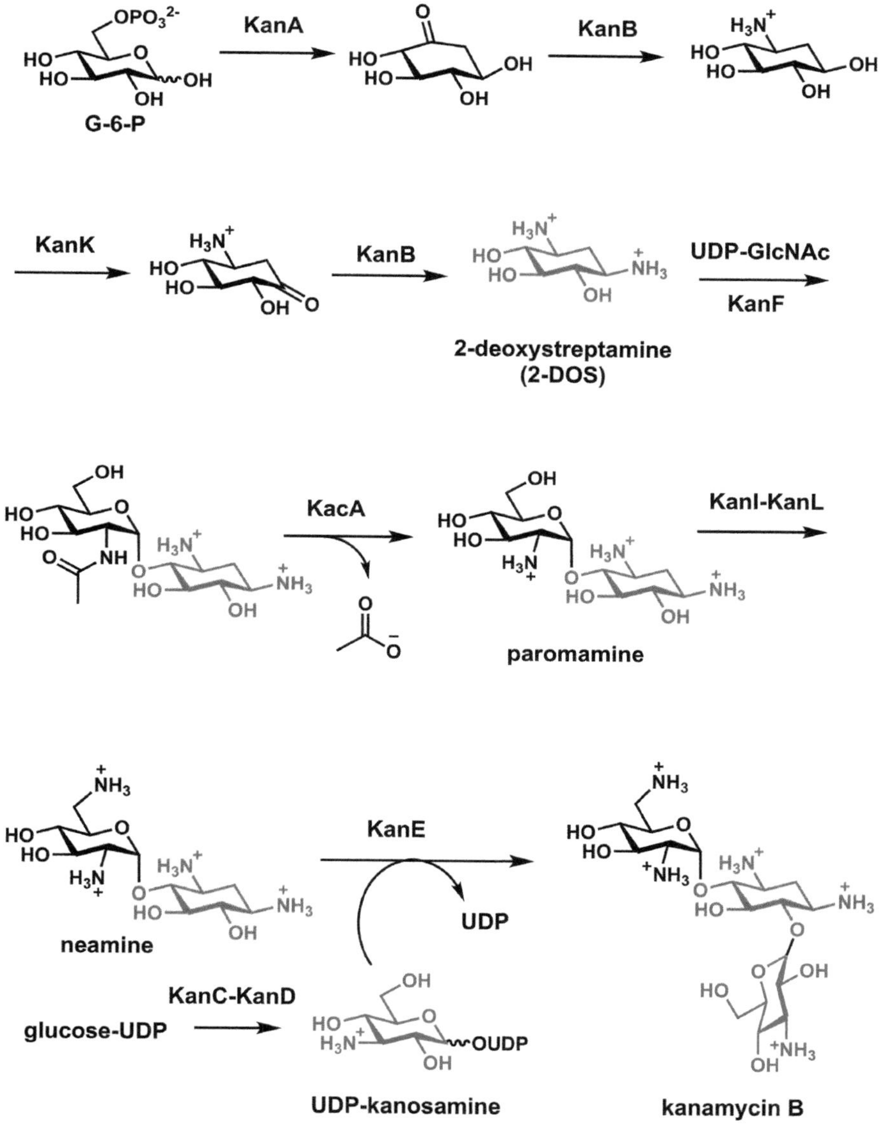

Figure 11.41 Enzymatic formation of kanamycin B from glucose-6-P production of 2-deoxystreptamine, UDP-GlcNAc, and UDP-glucose.

and then a glucosyltransfer and oxidation/reductive amination to reach neomycin, a component of the topical antibiotic ointment Neosporin.

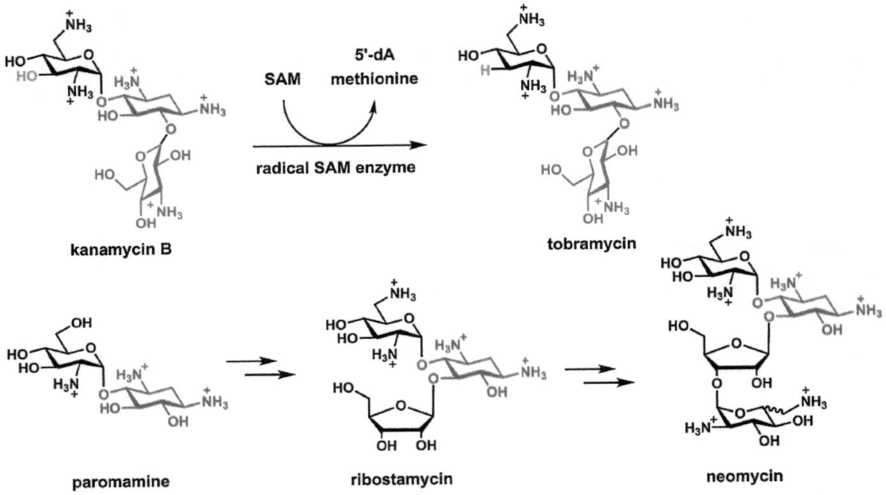

Figure 11.42 Tobramycin and neomycin biosynthesis.

11.8 Streptomycin

The provenance and assembly of the three six-carbon sugar units in streptomycin-6-P are depicted in Figure 11.43 (also see Walsh and Wencewicz 2016, chapter 17). Glucose-6-P is the common progenitor. The lefthand column diagrams its conversion to myoinositol *via* the internal aldol route of Figure 11.39. Then two sets of oxidation/ reductive aminations ensue, followed by guanidino group transfers and phosphorylation to give streptidine-6-P.

The second column notes formation of TDP-D-glucose by the standard route from G-1-P and its conversion on to TDP-L-rhamnose *via* 4-keto-6-deoxy glucose and epimerase action. Conversion from the pyranose to the five-membered furanose ring produces TDP-dihydrostreptose, which is coupled to streptidine-6-P. The third hexose arrives by way of CDP-glucose and CDP-*N*-methyl-L-glucosamine. The initial trisaccharide is the dihydro form of streptomycin-6-P and it is oxidized to the aldehyde in the penultimate biosynthetic step. As noted in Figure 11.44, streptomycin-6-P is shipped out of the producing cell, as is a 6-phosphatase. Cleavage of the phosphate protecting group, and removal of the negative charge, yields the active cationic streptomycin, but only outside of the producing streptomycete cells, a strategy of self-protection.

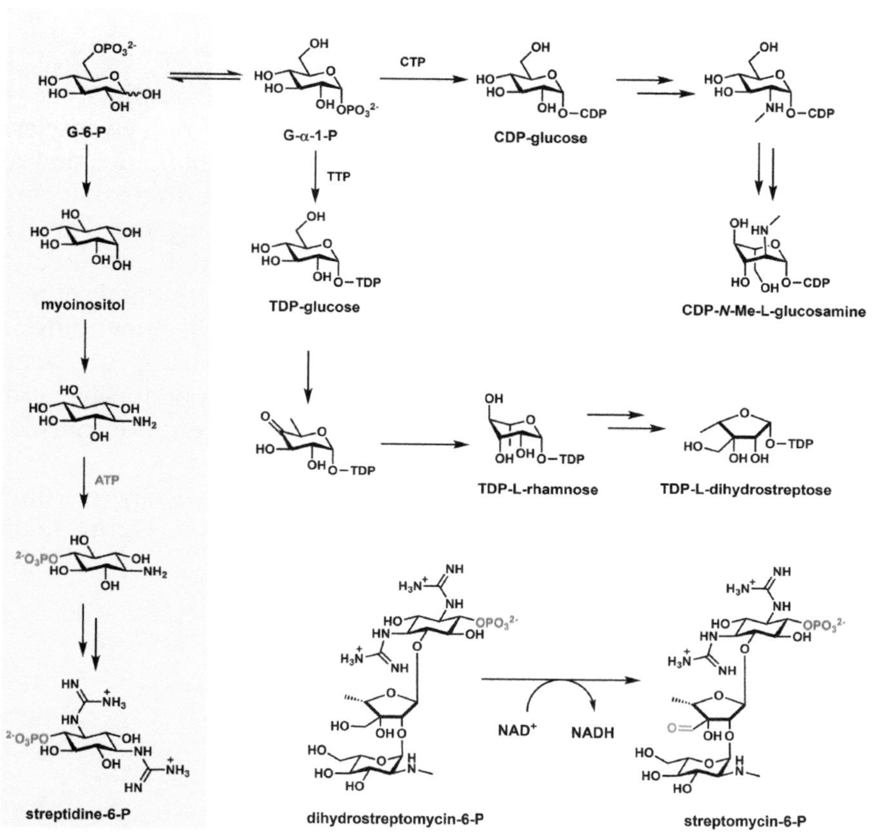

Figure 11.43 Streptomycin-6-P assembly.

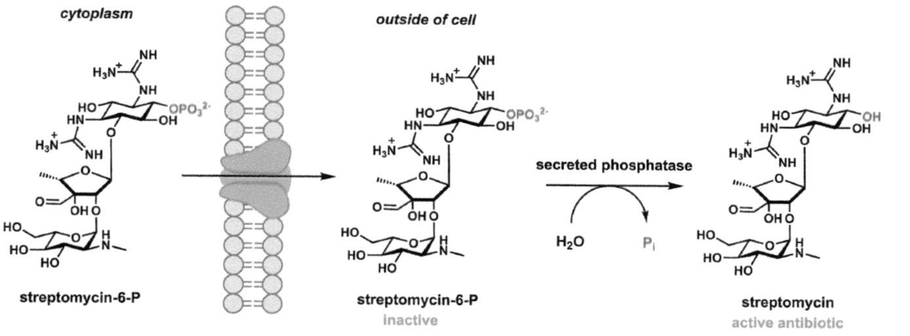

Figure 11.44 Dephosphorylation of streptomycin-6-P to the active antibiotic occurs enzymatically outside the producing cell.

11.9 Moenomycins

The moenomycins (see Figure 11.2), *Streptomyces* metabolites, are the most potent inhibitors known for bacterial cell wall transglycosylases (Ostash and Walker 2010). Because of poor pharmacokinetics they are too toxic for human use. They do point the way to scaffolds that could be of great clinical use in life-threatening Gram-negative bacterial infections if the therapeutic index could be optimized. Re-engineering of the biosynthetic pathway has been proposed as one route to intermediates amenable to medicinal chemistry. The moenomycins can be described as glycophospholipids, comprising a glycerol phosphate core to which a C_{25} isoprenoid lipid and a pentasaccharide are attached. Alternatively one could argue that the pentasaccharide is the key scaffolding element.

Biosynthesis begins with capture of the farnesyl C_1 allyl cation by the 1-OH group of the primary metabolite glycerol-3-P (Figure 11.45).

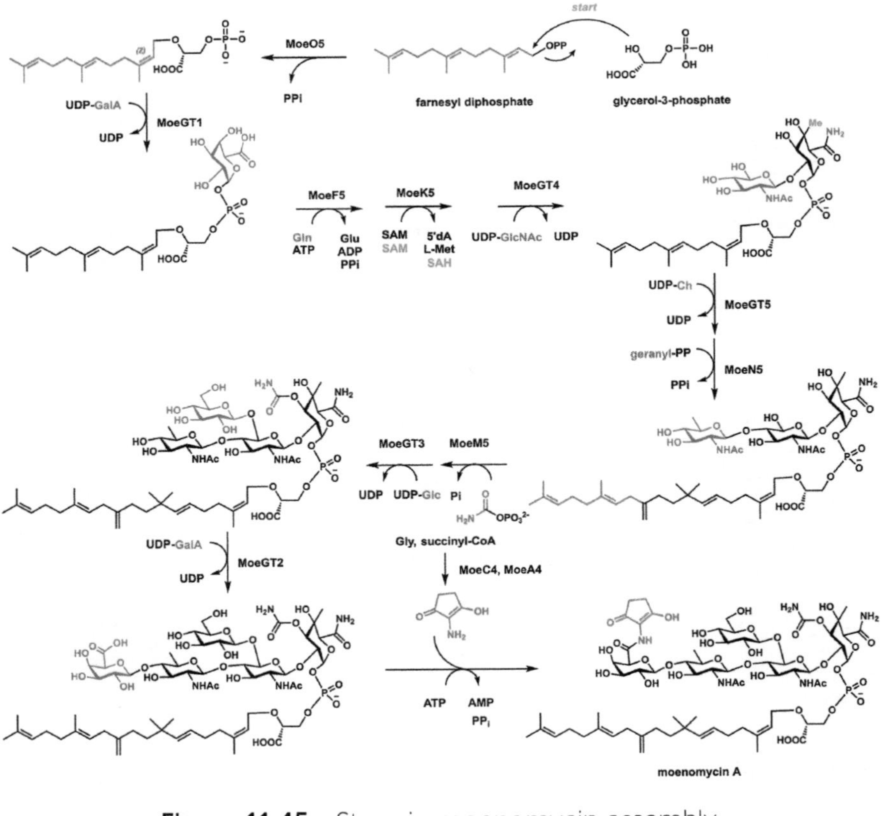

Figure 11.45 Steps in moenomycin assembly.

Then, consecutive addition of the hexosyl units by five Gtfs commences, with the phosphate oxygen as initiating nucleophile on the first UDP-hexose. Two more Gtfs act to build the trisaccharide before the second isoprenoid moiety, from the C_{10} geranyl-PP donor, occurs, along with as yet unproved rearrangements to yield the C_{25} moecinol lipid chain. Two additional Gtfs finish the pentasaccharide core and late addition of the aminocyclopentadione ring completes moenomycin assembly.

Vignette 11.1 Ascarosides – An Integrating Chemical Language in Nematodes.

The isomeric 3,6-dideoxyhexoses ascarylose and paratose are constituents of the outer chains of lipopolysaccharides of Gram-negative bacteria and can be antigenic determinants. They arise from NDP-glucose modification but as the CDP-glucose derivatives rather than the more typical UDP- or TDP-glucoses used for deoxysugar biosynthesis (Figure 11.V1). The unique step in maturation occurs as CDP-4-keto-6-deoxyglucose is converted to CDP-4-keto-3,6-dideoxyglucose by an enzyme that uses pyridoxal-phosphate, NADPH, and radical chemistry to effect the reductive loss of the C_3–OH (Thibodeaux, Melancon *et al.* 2007, Thibodeaux, Melancon *et al.* 2008).

In a separate biological context, ascarylose has been found linked to a long chain lipid as a structural component of the egg shell of *Ascaris* round worms (Figure 11.V2). More intriguingly, over 100 ascarides have been detected over the past decade as chemical communication signals in the nematode *Caenorhabditis elegans*. Ascarylose is at the core of molecules that can combine building blocks from amino acid metabolism, the Krebs cycle, and fatty acid biosynthesis (Ludewig and Schroeder 2012). Presumably the ascaride variants reflect metabolic states and building block availability in the organism.

These ascarides were originally identified as molecules that accumulated in nutrient limiting conditions and drove the *C. elegans* worms into a dauer state, where they limited metabolism and development. More recently the distinct forms of ascarides, presumptively acting as ligands for G-protein coupled receptors in distinct populations of neuronal cells, drive complex sets of behaviors, including sexual mating, development, metabolic choices, and lifespan (Figure 11.V3). Different species of nematodes also make related ascarides; some of them, such as npar#1, contain the isomeric paratose core rather than ascarylose, presumably for species selectivity.

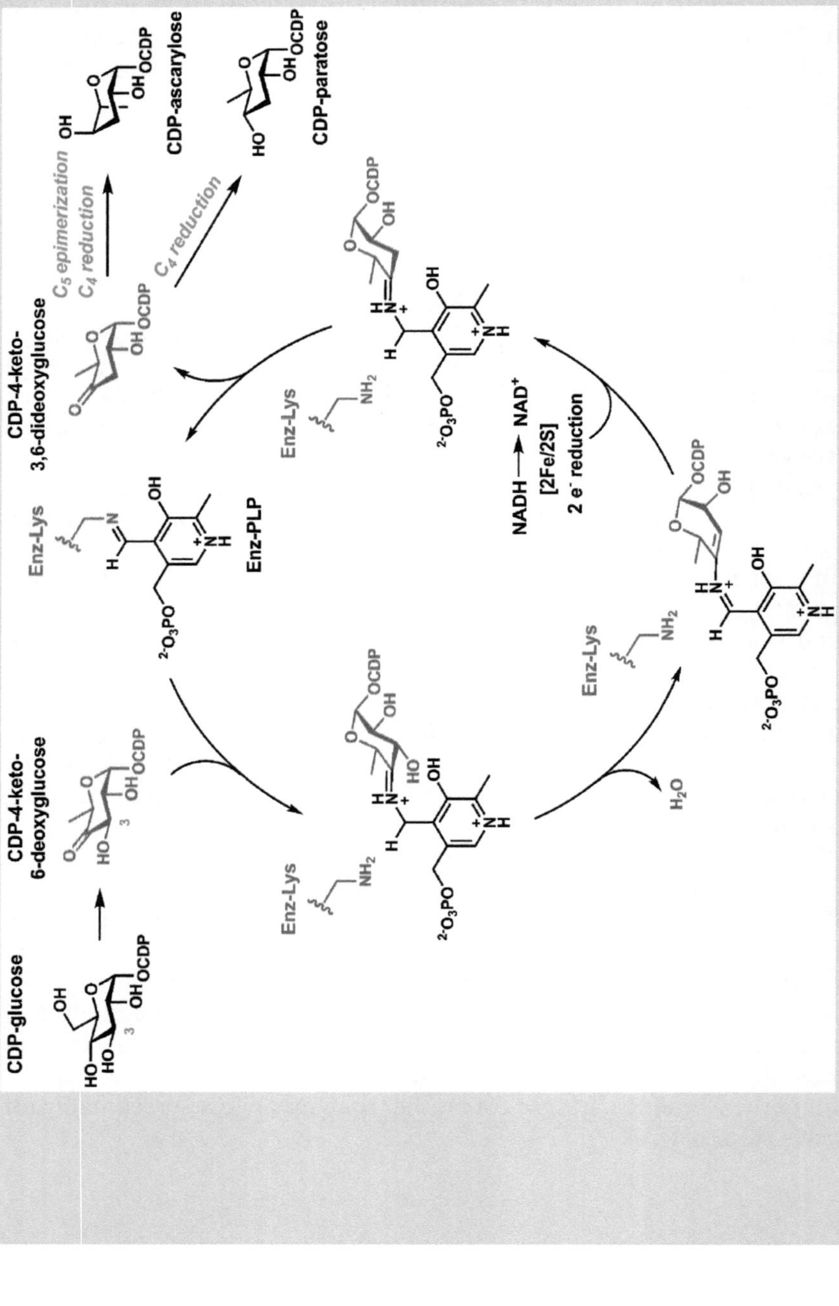

Figure 11.V1 Biosynthesis of ascarylose and paratose from glucose.

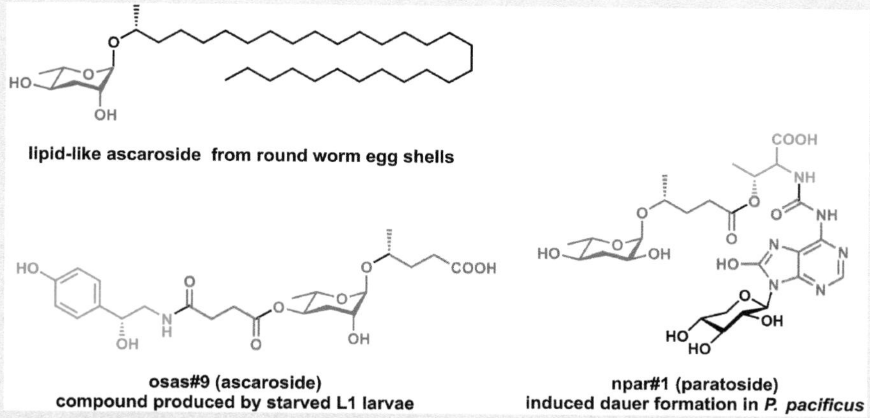

Figure 11.V2 Ascaroside and paratoside from roundworms.

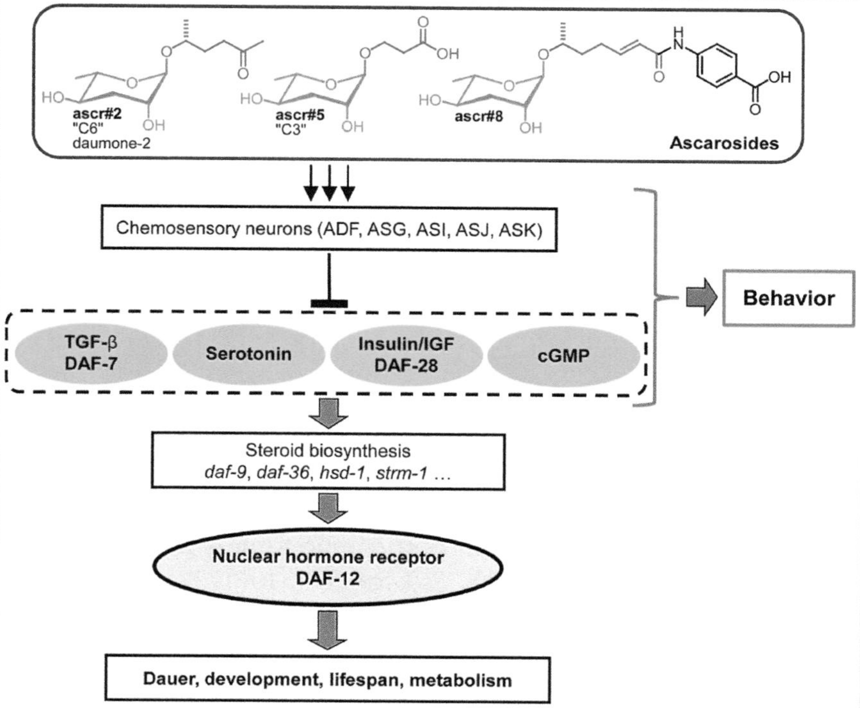

Figure 11.V3 Ascarides, involved in neuronal signaling of *C. elegans*. Image from Ludwig, A. H. and F. C. Schroeder (2013). "Ascaroside signaling in C. elegans". *Wormbook*, DOI: 10.1895/wormbook.1.155.1. Image under Creative Commons License (https://creativecommons.org/licenses/by/2.5/legalcode).

11.10 Summary

Glycosyltransferases comprise key tailoring enzyme classes for essentially every type of natural product. Many of the Gtf genes are encoded adjacent to the core biosynthetic genes, indicating their important roles in generating functional pathway end products. The sugar residues can vary from mono- to hepta-saccharides, linear and branched, and specialized hexose residues play important roles in hydrophobic/hydrophilic balance in the mature natural product scaffolds. They can determine cellular, subcellular, and tissue location as exemplified for plant hormones. While most natural product glycosylations are functionally irreversible, the cyanogenic glucosides and the glucosinolates are preformed latent defensive chemical weaponry. Controlled temporal and spatial access of these latently reactive metabolites to hydrolytic glycosidases to release the reactive warheads allows rapid responses by plants to immediate predation.

References

Albermann, C., A. Soriano, J. Jiang, H. Vollmer, J. B. Biggins, W. A. Barton, J. Lesniak, D. B. Nikolov and J. S. Thorson (2003). "Substrate specificity of NovM: implications for novobiocin biosynthesis and glycorandomization". *Org. Lett.* **5**(6): 933–936.

Bowles, D., J. Isayenkova, E. K. Lim and B. Poppenberger (2005). "Glycosyltransferases: managers of small molecules". *Curr. Opin. Plant Biol.* **8**(3): 254–263.

Bowles, D., E. K. Lim, B. Poppenberger and F. E. Vaistij (2006). "Glycosyltransferases of lipophilic small molecules". *Annu. Rev. Plant Biol.* **57**: 567–597.

Bowyer, P., B. R. Clarke, P. Lunness, M. J. Daniels and A. E. Osbourn (1995). "Host range of a plant pathogenic fungus determined by a saponin detoxifying enzyme". *Science* **267**(5196): 371–374.

Burmeister, W. P., S. Cottaz, H. Driguez, R. Iori, S. Palmieri and B. Henrissat (1997). "The crystal structures of Sinapis alba myrosinase and a covalent glycosyl-enzyme intermediate provide insights into the substrate recognition and active-site machinery of an S-glycosidase". *Structure* **5**(5): 663–675.

Chen, H., M. G. Thomas, B. K. Hubbard, H. C. Losey, C. T. Walsh and M. D. Burkart (2000). "Deoxysugars in glycopeptide antibiotics: enzymatic synthesis of TDP-L-epivancosamine in chloroeremomycin biosynthesis". *Proc. Natl. Acad. Sci. U. S. A.* **97**(22): 11942–11947.

Dewick, P. (2009). *Medicinal Natural Products, a Biosynthetic Approach.* UK, Wiley.

Elshahawi, S. I., K. A. Shaaban, M. K. Kharel and J. S. Thorson (2015). "A comprehensive review of glycosylated bacterial natural products". *Chem. Soc. Rev.* **44**(21): 7591–7697.

Eriksen, T. A., A. Kadziola, A. K. Bentsen, K. W. Harlow and S. Larsen (2000). "Structural basis for the function of Bacillus subtilis phosphoribosyl-pyrophosphate synthetase". *Nat. Struct. Biol.* 7(4): 303–308.

Falcone Ferreyra, M. L., E. Rodriguez, M. I. Casas, G. Labadie, E. Grotewold and P. Casati (2013). "Identification of a bifunctional maize C- and O-glucosyltransferase". *J. Biol. Chem.* **288**(44): 31678–31688.

Freel Meyers, C. L., M. Oberthur, J. W. Anderson, D. Kahne and C. T. Walsh (2003). "Initial characterization of novobiocic acid noviosyl transferase activity of NovM in biosynthesis of the antibiotic novobiocin". *Biochemistry* **42**(14): 4179–4189.

Freel Meyers, C. L., M. Oberthur, H. Xu, L. Heide, D. Kahne and C. T. Walsh (2004). "Characterization of NovP and NovN: completion of novobiocin biosynthesis by sequential tailoring of the noviosyl ring". *Angew. Chem., Int. Ed.* **43**(1): 67–70.

Gleadow, R. M. and B. L. Moller (2014). "Cyanogenic glycosides: synthesis, physiology, and phenotypic plasticity". *Annu. Rev. Plant Biol.* **65**: 155–185.

Halkier, B. A. and J. Gershenzon (2006). "Biology and biochemistry of glucosinolates". *Annu. Rev. Plant Biol.* **57**: 303–333.

Halkier, B. A. and B. L. Moller (1989). "Biosynthesis of the Cyanogenic Glucoside Dhurrin in Seedlings of Sorghum bicolor (L.) Moench and Partial Purification of the Enzyme System Involved". *Plant Physiol.* **90**(4): 1552–1559.

Halkier, B. A., C. E. Olsen and B. L. Moller (1989). "The biosynthesis of cyanogenic glucosides in higher plants. The (E)- and (Z)-isomers of p-hydroxyphenylacetaldehyde oxime as intermediates in the biosynthesis of dhurrin in Sorghum bicolor (L.) Moench". *J. Biol. Chem.* **264**(33): 19487–19494.

He, X. and H. W. Liu (2002). "Mechanisms of enzymatic C-O bond cleavages in deoxyhexose biosynthesis". *Curr. Opin. Chem. Biol.* **6**(5): 590–597.

He, X. M. and H. W. Liu (2002). "Formation of unusual sugars: mechanistic studies and biosynthetic applications". *Annu. Rev. Biochem.* **71**: 701–754.

Hostettmann, K. and A. Marston (1995). *Saponins.* Cambridge, UK, Cambridge University Press.

Keegstra, K. and N. Raikhel (2001). "Plant glycosyltransferases". *Curr. Opin. Plant Biol.* **4**(3): 219–224.

Kleczkowski, L. A., M. Geisler, I. Ciereszko and H. Johansson (2004). "UDP-glucose pyrophosphorylase. An old protein with new tricks". *Plant Physiol.* **134**(3): 912–918.

Kudo, F. and T. Eguchi (2016). "Aminoglycoside Antibiotics: New Insights into the Biosynthetic Machinery of Old Drugs". *Chem. Rec.* **16**(1): 4–18.

Kumar, A., T. De, A. Mishra and A. K. Mishra (2013). "Oleandrin: A cardiac glycosides with potent cytotoxicity". *Pharmacogn. Rev.* 7(14): 131–139.

Lairson, L. L., B. Henrissat, G. J. Davies and S. G. Withers (2008). "Glycosyltransferases: structures, functions, and mechanisms". *Annu. Rev. Biochem.* 77: 521–555.

Ludewig, A. H. and F. C. Schroeder (2012). "Ascaroside Signaling in C. elegans". In *WormBook*, ed. P. Kuwabara. The C. elegans Research Community, WormBook, doi/10.1895/wormbook.1.155.1, http://www.wormbook.org.

McCranie, E. K. and B. O. Bachmann (2014). "Bioactive oligosaccharide natural products". *Nat. Prod. Rep.* **31**(8): 1026–1042.

McCulloch, K. M., E. K. McCranie, J. A. Smith, M. Sarwar, J. L. Mathieu, B. L. Gitschlag, Y. Du, B. O. Bachmann and T. M. Iverson (2015). "Oxidative cyclizations in orthosomycin biosynthesis expand the known chemistry of an oxygenase superfamily". *Proc. Natl. Acad. Sci. U. S. A.* **112**(37): 11547–11552.

Mittler, M., A. Bechthold and G. E. Schulz (2007). "Structure and action of the C-C bond-forming glycosyltransferase UrdGT2 involved in the biosynthesis of the antibiotic urdamycin". *J. Mol. Biol.* **372**(1): 67–76.

Nagatomo, Y., S. Usui, T. Ito, A. Kato, M. Shimosaka and G. Taguchi (2014). "Purification, molecular cloning and functional characterization of flavonoid C-glucosyltransferases from Fagopyrum esculentum M. (buckwheat) cotyledon". *Plant J.* **80**(3): 437–448.

Nahrstedt, A. and J. Rockenbach (1993). "Occurrence of the cyanogenic glucoside prunasin and II corresponding mandelic acid amide glucoside in Olinia species (oliniaceae)". *Phytochemistry* **34**(2): 433–436.

Osbourn, A., P. Bowyer, P. Lunness, B. Clarke and M. Daniels (1995). "Fungal pathogens of oat roots and tomato leaves employ closely related enzymes to detoxify different host plant saponins". *Mol. Plant-Microbe Interact.* **8**(6): 971–978.

Osbourn, A. and V. Lanzotti, eds. (2009). *Plant Derived Natural Products: Synthesis, Function, Application*. Springer Science and Business Media.

Ostash, B. and S. Walker (2010). "Moenomycin family antibiotics: chemical synthesis, biosynthesis, and biological activity". *Nat. Prod. Rep.* **27**(11): 1594–1617.

Park, J. W., S. R. Park, K. K. Nepal, A. R. Han, Y. H. Ban, Y. J. Yoo, E. J. Kim, E. M. Kim, D. Kim, J. K. Sohng and Y. J. Yoon (2011). "Discovery of parallel pathways of kanamycin biosynthesis allows antibiotic manipulation". *Nat. Chem. Biol.* **7**(11): 843–852.

Salas, A. P., L. Zhu, C. Sanchez, A. F. Brana, J. Rohr, C. Mendez and J. A. Salas (2005). "Deciphering the late steps in the biosynthesis of the anti-tumour indolocarbazole staurosporine: sugar donor substrate flexibility of the StaG glycosyltransferase". *Mol. Microbiol.* **58**(1): 17–27.

Sanchez, C., L. Zhu, A. F. Brana, A. P. Salas, J. Rohr, C. Mendez and J. A. Salas (2005). "Combinatorial biosynthesis of antitumor indolocarbazole compounds". *Proc. Natl. Acad. Sci. U. S. A.* **102**(2): 461–466.

Shen, B., L. Du, C. Sanchez, D. J. Edwards, M. Chen and J. M. Murrell (2002). "Cloning and characterization of the bleomycin biosynthetic gene cluster from Streptomyces verticillus ATCC15003". *J. Nat. Prod.* **65**(3): 422–431.

Soldin, S. J. (1986). "Digoxin–issues and controversies". *Clin. Chem.* **32**(1 Pt 1): 5–12.

Szu, P. H., X. He, L. Zhao and H. W. Liu (2005). "Biosynthesis of TDP-D-desosamine: identification of a strategy for C4 deoxygenation". *Angew. Chem., Int. Ed.* **44**(41): 6742–6746.

Szu, P. H., M. W. Ruszczycky, S. H. Choi, F. Yan and H. W. Liu (2009). "Characterization and mechanistic studies of DesII: a radical S-adenosyl-L-methionine enzyme involved in the biosynthesis of TDP-D-desosamine". *J. Am. Chem. Soc.* **131**(39): 14030–14042.

Thibodeaux, C. J., C. E. Melancon 3rd and H. W. Liu (2008). "Natural-product sugar biosynthesis and enzymatic glycodiversification". *Angew. Chem., Int. Ed.* **47**(51): 9814–9859.

Thibodeaux, C. J., C. E. Melancon and H. W. Liu (2007). "Unusual sugar biosynthesis and natural product glycodiversification". *Nature* **446**(7139): 1008–1016.

Trefzer, A., D. Hoffmeister, E. Kunzel, S. Stockert, G. Weitnauer, L. Westrich, U. Rix, J. Fuchser, K. U. Bindseil, J. Rohr and

A. Bechthold (2000). "Function of glycosyltransferase genes involved in urdamycin A biosynthesis". *Chem. Biol.* 7(2): 133–142.

VanEtten, H. D., J. W. Mansfield, J. A. Bailey and E. E. Farmer (1994). "Two Classes of Plant Antibiotics: Phytoalexins versus "Phytoanticipins"". *Plant Cell* **6**(9): 1191–1192.

Walsh, C. T. and T. Wencewicz (2016). *Antibiotics Challeneges, Mechanisms, Opportunities*. Washington DC, ASM Press.

Walsh, C. T. and T. Wencewicz (2016). *Antibiotics Challeneges, Mechanisms, Opportunities*. Washington DC, ASM Press, ch 17.

Section IV

Genome-independent and Genome-dependent Detection of Natural Products

The two chapters in this final section change focus from a detailed consideration of reactions in the prior sections to some general strategies used historically and concurrently to isolate natural products.

Chapter 12 starts with the isolation of the first half dozen alkaloids as pure natural products, decades before their structures were determined, reflecting the long ethnopharmacologic history of recognition of usefully active agents in nature. Short summaries for production and isolation of some medicinally active natural products are presented to illustrate the different challenges posed by products of plant *vs.* microbial origin, especially for production at useful scale. These include five plant natural products: quinine, morphine, digoxin, taxol, and the pair of vinca alkaloids vinblastine and vincristine. Three microbial metabolites of note are the peptide-based penicillins and cephalosporins, vancomycin, and cyclosporine.

Expanding the inventory of natural products by gene independent approaches continues the theme of Chapter 12, including searching in new biological niches and culturing the microbial unculturables. The *o*ne *s*train *ma*ny *c*ompounds (OSMAC) approach is highlighted as prelude to its genomic molecular underpinnings in Chapter 13.

No single chapter could do justice to the changes that the post-genomic era has wrought to natural products research. Indeed the burgeoning literature across plants, fungi, and bacteria biosynthetic gene clusters shows steep increases in activity in this interdisciplinary

Natural Product Biosynthesis: Chemical Logic and Enzymatic Machinery
By Christopher T. Walsh and Yi Tang

Published by the Royal Society of Chemistry, www.rsc.org

area of science. In particular, the rapid pace of methodology changes is likely to make detailed methodologic descriptions almost instantly out of date.

To that end we have not delved significantly into the many particular approaches to gene cluster computation, the myriad cloning and expression approaches, or the tool chest of synthetic biology methods brought to bear on heterologous expression constructs. Instead we have raised some questions about strategies to find "unknown unknown" scaffolds amidst the "known unknowns". We note the limited class of phosphonate natural products and the prospect that one could approach a saturation limit at 40 000 actinomycete genomes. This may be an illustrative example of how to determine the biological universe for larger classes of natural product families.

The chapter ends with some recent successful case studies on reconstitution of plant biosynthetic pathways for alkaloids in tobacco plants, but most commonly in yeast. Also, the engineering of commercial levels of artemisinic acid in yeast represents a clear metabolic engineering successful trajectory. Chief among the questions going forward is how to increase the efficiency of finding and expressing novel natural product scaffolds among the noise of the known scaffolds, on the premise that completely new scaffolds may be most useful in discovery.

Laboratory growth, isolation, and characterization of natural product structures.

12 Natural Products Isolation and Characterization: Gene Independent Approaches

12.1 Introduction

In this chapter we take up gene independent traditional and contemporary approaches to isolation of some notable natural products and to some selected elements and challenges of their production. We also summarize efforts to increase metabolite numbers specifically from fungal and bacterial cultures.

In the next chapter we turn to the approaches in the genomic era when biosynthetic genes are often known or suspected. In those contexts, gene-based informatics and molecular biology approaches guide efforts to turn on cryptic biosynthetic pathways and to find new ones based on genome mining.

12.2 Historic and Contemporary Isolation Protocols for Natural Products

Natural products are generated from three major biological sources: plants, fungi, and bacteria. Historically, over the millennia and more recently the past two centuries, the macrobiotic plants were the primary sources of natural products. Prior to the 19th century, hundreds of plants were collected for medicinal remedies, often as mixtures of

Natural Product Biosynthesis: Chemical Logic and Enzymatic Machinery
By Christopher T. Walsh and Yi Tang

Published by the Royal Society of Chemistry, www.rsc.org

plants or crude infusions for pharmacologic purposes. From the time of Pliny and Galen in the Roman Empire, through the Arab medical efforts from the 7th to 12th centuries, and in parallel efforts in China and Indian Ayurvedic traditions, plant therapeutic sources were dominant. In Mesopotamia in 2600 BC there were more than 1000 plant based medical recipes. In Egypt, by 1500 BC the Ebers Papyrus noted >700 plant therapeutic sources (Atanasov, Waltenberger *et al.* 2015).

It was in the late 18th and early 19th centuries, starting with W. Withering's treatment of patients with congestive heart failure ("dropsy") in England and in the isolation of morphine from poppy seeds by Serturner in 1817 (Serturner 1817), that the transition from plant mixtures, extracts, and infusions to isolation of single active molecules began.

Microbes, fungi and bacteria, became more serious sources of natural products after the discovery of penicillin in 1928 and the intense follow up by Chain and Florey in the late 1930s. During the two decades of the golden age of natural antibiotic isolation from the 1940s to the 1960s, aminoglycosides, tetracyclines, the erythromycin class of macrolactones, and the glycopeptides represented by vancomycin were all isolated (Walsh and Wencewicz 2016).

Figure 12.1 lists the time line, starting in the early 19th century, for isolation of pure compounds in the alkaloid class of natural products. The isolation of half a dozen pure alkaloids from plant sources in the decade 1816–1826 reflected in part the ability of these basic nitrogen-containing scaffolds to be partitioned between aqueous and organic solvents with pH control. They also were targets in substantial part based on ethnopharmacology traditions that biologically active substances were present in the plant materials. A third parameter was the availability of source material and the ability of isolation chemists to obtain crystals of the natural product as they approached high concentration and purity in organic solvents. This set of isolations changed millennial long studies of complex mixtures of natural products to the subsequent and current chemical tradition of pure compound isolation, structural characterization, and independent validation of structural assignment by total synthesis.

Figure 12.1 also illustrates that the ability to detect, isolate, and even crystallize this first suite of pharmacologically varied natural products ran far ahead of the ability to determine structures. Coniine, a relatively simple alkaloid, yielded to structural characterization in

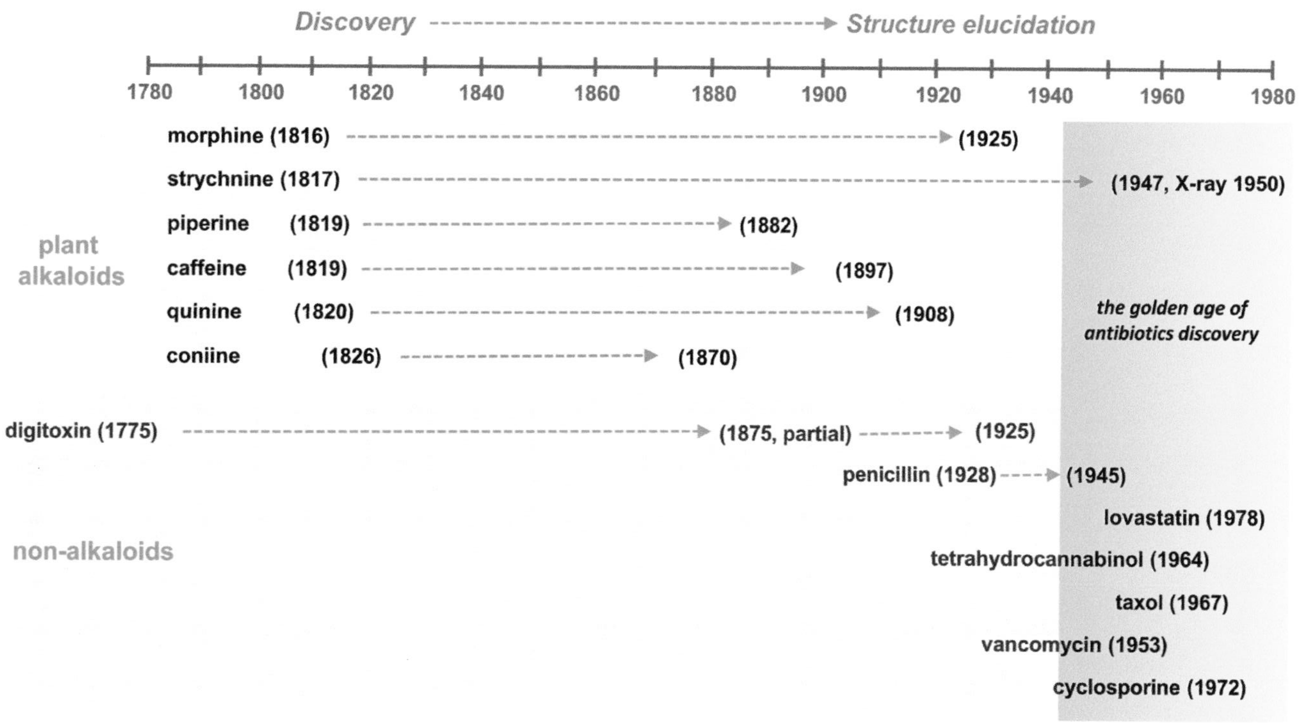

Figure 12.1 The top half lists alkaloids as signature natural products from higher plants. The bottom half lists seven non-alkaloid natural products that may be "molecules that changed the world".

1870 but quinine, morphine, and strychnine progressed in some cases far into the 20th century before their complex molecular scaffolds were fully determined.

A substance in the bark of the cinchona trees in South America, later determined to be quinine, was appreciated as an agent to combat fever in the 1600s and was listed in the *London Pharmacopeia* in 1677 (Atanasov, Waltenberger *et al.* 2015). Analogously, the economic value to European nations of spices from the Far East in the 15th century was a significant factor in the funding of Columbus's visits to the West Indies to find a new route to India. During his second trip in 1493, Columbus and crew found both black pepper in Haiti (piperine as the key spice molecule) and also chili peppers, and thereby set off the eventual identification of capsaicin (Chapter 7).

The bottom half of Figure 12.1 lists a subset of seven non-alkaloidal natural products and the dates on which they were isolated and/or structures determined. With the exception of the cardiotonic steroid glycoside digitoxin, from wooly foxglove flowers, the other six molecular discoveries are from the mid 20th century. While penicillin was discovered at the end of the 1920s, and was in large scale fermentation by the early 1940s, the structure was conclusively established only on completion of X-ray analysis by Dorothy Crowfoot Hodgkin in 1945.

There have been books and articles written in recent times about molecules that changed the world (Le Couteur and Burreson 2004, Nicolaou and Montagnon 2008). The various candidates that have been advanced span biological and therapeutic roles into material sciences. Those lists include natural products, semisynthetic ones (aspirin as the acetyl derivative of the natural product salicylate), and synthetic ones (nylon). In the more bounded universe of natural products the 13 molecules of Figure 12.1 represent a subset of natural products that have had significant worldwide impact. Many are on the WHO list of essential medicines, as are others we have noted in prior chapters.

An orthogonal approach to the evaluation of natural product importance, at least in the therapeutic sphere, would be a list of molecules approved for human clinical use over the past 30 years. A subset of these, noted in Figure 12.2, would include the antimalarial natural product artemisinin (1987), capsaicin for pain in 1999, galatamine in 2001 for Alzheimer's patients, tetrahydrocannabinol for pain in 2005, and perhaps most notably taxol in 1993 for ovarian cancer.

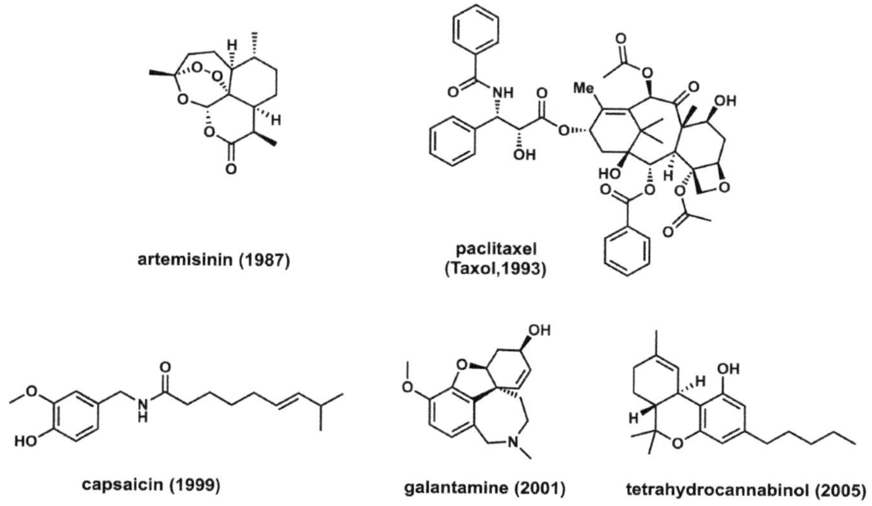

Figure 12.2 Five natural products approved as therapeutic agents since 1987.

12.3 Isolation and Characterizations of Specific Natural Products

The detailed strategies and steps involved in contemporary isolation of natural products, both from plants and from microbial sources, are beyond the scope of this volume, which is not a technical manual on either purification or structure determination. Two comprehensive review articles in *Natural Product Reports*, one by Sticher in 2008 (Sticher 2008) and the more recent one by Bucar *et al.* (Bucar, Wube *et al.* 2013), lay out strategies and methodologies for isolation of secondary metabolites from both plant and microbial sources. They review the current armamentarium of various extraction methods, including supercritical fluids, microwave assisted extractions, and protocols for going from nonpolar to polar solvent partitioning with aqueous layers. They also describe various high throughput and high resolution chromatographic methods to obtain pure natural products. These include liquid–solid chromatographic partition columns, hydrophilic interaction chromatography, and chiral stationary phase resins.

Small quantities, from hundred of nanograms to micrograms, of material suffice for the major structural methods of NMR and mass spectrometry to yield exact mass, scaffold connectivity, and often complete stereochemical assignments. X-ray analysis of natural product single crystals can be used as a final method for structure determination/confirmation.

Some 20–30 years ago, NMR and mass spectrometry (MS, MS^2 and MS^3) methodologies were not as well developed as currently, when essentially any complex small molecule structure can be solved by this pair of solution methods. In those days single crystal X-ray structure determination resolved the uncertainties of the structures of such architecturally complex molecules as saxitoxin (Schantz, Ghazarossian *et al.* 1975), bryostatin (Petit, Herald *et al.* 1982), brevetoxin (Lin, Risk *et al.* 1981, Shimizu, Bando *et al.* 1986), okadaic acid (Tachibana, Scheuer *et al.* 1981), and the enediyne dynemicin (Konishi, Ohkuma *et al.* 1990) (Figure 12.3).

X-ray analysis is no longer one of the equal legs of a three-legged stool of natural product structure determination. Instead, dramatic developments in both NMR and mass spectrometry (Figure 12.4) over the five decades 1960–2010 have made them the dominant, and generally sufficient, methodologies for structure determination of any organic compound (Gaudencio and Pereira 2015). Readers interested in both MS and NMR technologies will find this review article a useful entry into the structure determination literature.

For example, by 2007, NMR methodologies had become so advanced that it was possible to determine the structure of the *Bacillus subtilis* polyketide/nonribosomal peptide bacillaene by two-dimensional (2D) NMR methods, walking along the carbon backbone for complete assignment (Butcher, Schroeder *et al.* 2007). This includes five *E*- and three *Z*-olefins, as noted in Figure 12.5. Also shown in the figure is the 10-protein hybrid NRP–PK assembly line for bacillaene biosynthesis. The logic of hybrid NRPS/PKS assembly lines was presented in Chapter 3.

Features of note in operation of the assembly line include the unusual initiation module: the first monomer loaded by the adenylation domain of PksK is the branched chain α-ketoacid rather than its branched chain amino acid congener L-valine (Calderone, Bumpus *et al.* 2008). The α-ketone is surprisingly reduced to the α-hydroxyl group by the ketoreductase of the third module. Second, the five proteins PksCFGHI build an unusual β-methyl enoyl-CoA that gets inserted into the growing chain by PksL (Calderone 2008).

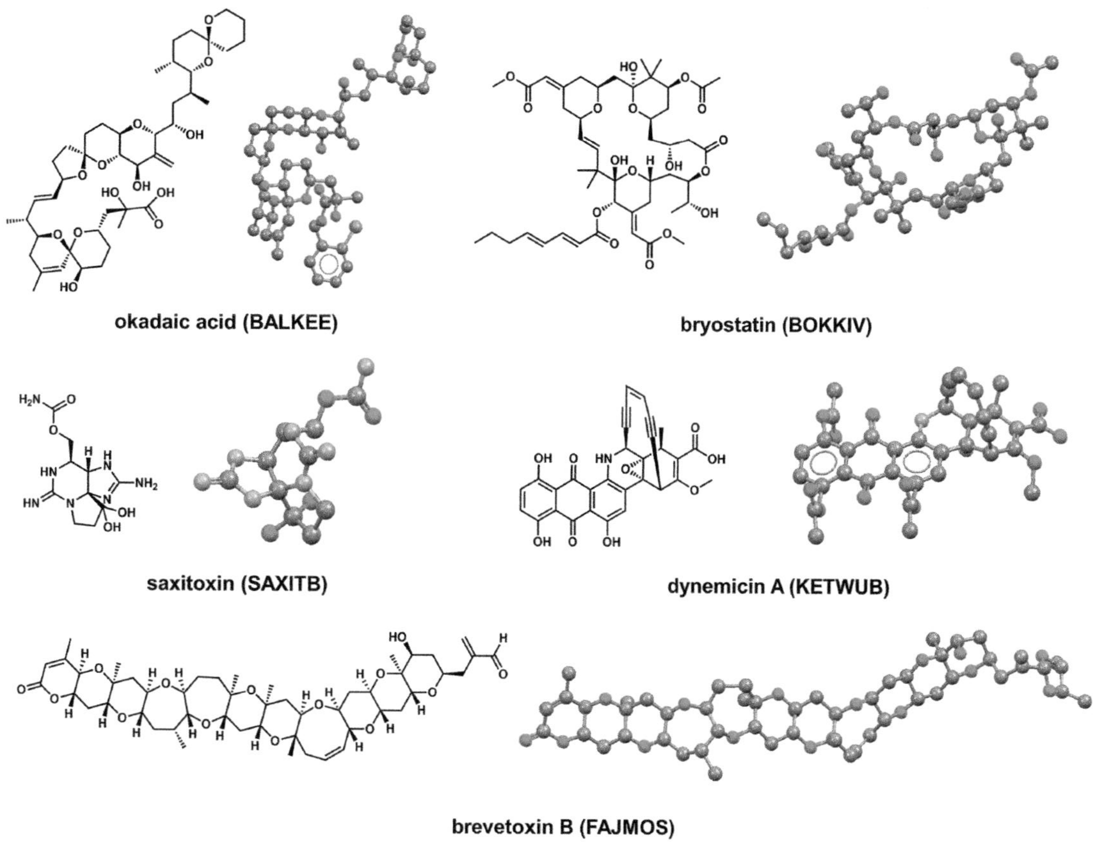

Figure 12.3 Five representative natural product structures determined by X-ray analysis 1975–1985. The X-ray crystal structures of the natural products or derivatized compounds are shown adjacent to the two-dimensional structures. Hydrogen atoms are omitted in the three-dimensional structures for a clearer view. The six letter CIF codes from The Cambridge Crystallographic Data Centre (CCDC) are shown in parenthesis.

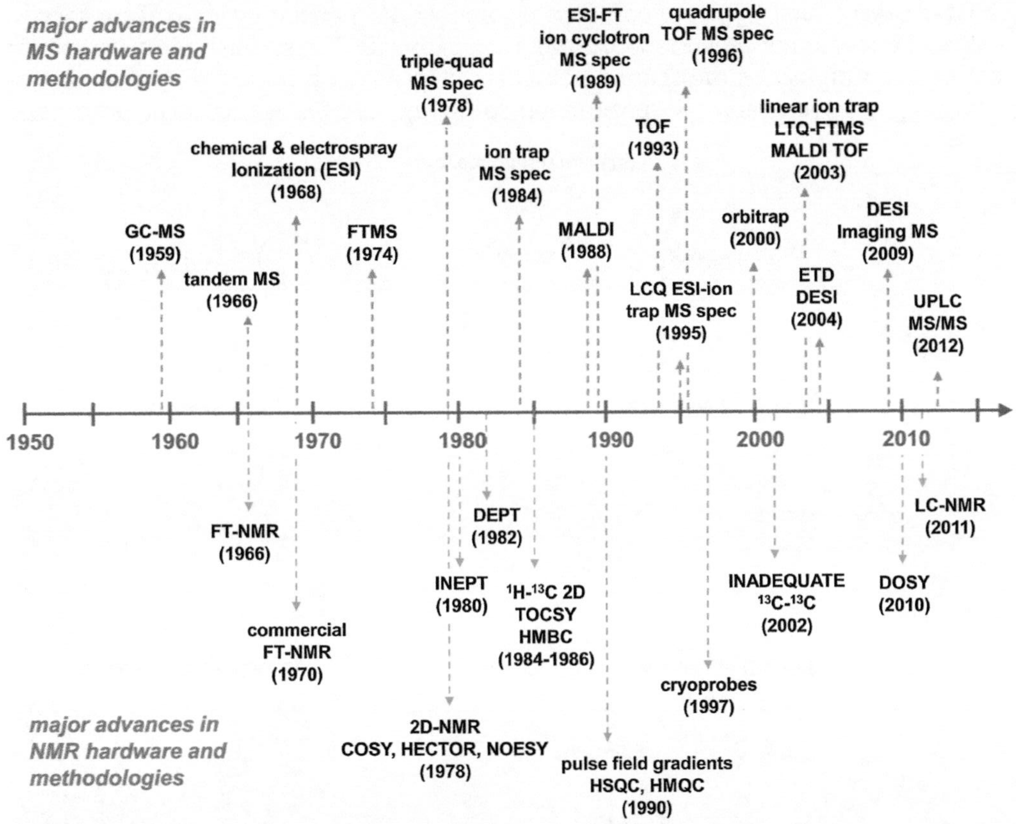

Figure 12.4 Time lines for advances in NMR and mass spectrometry (MS) hardware and methods relevant to natural product structure determination, 1960–2010.
Adapted from Gaudêncio, S. P. and F. Pereira (2015). "Dereplication: racing to speed up the natural products discovery process". *Nat. Prod. Rep.* **32**: 779 with permission from The Royal Society of Chemistry.

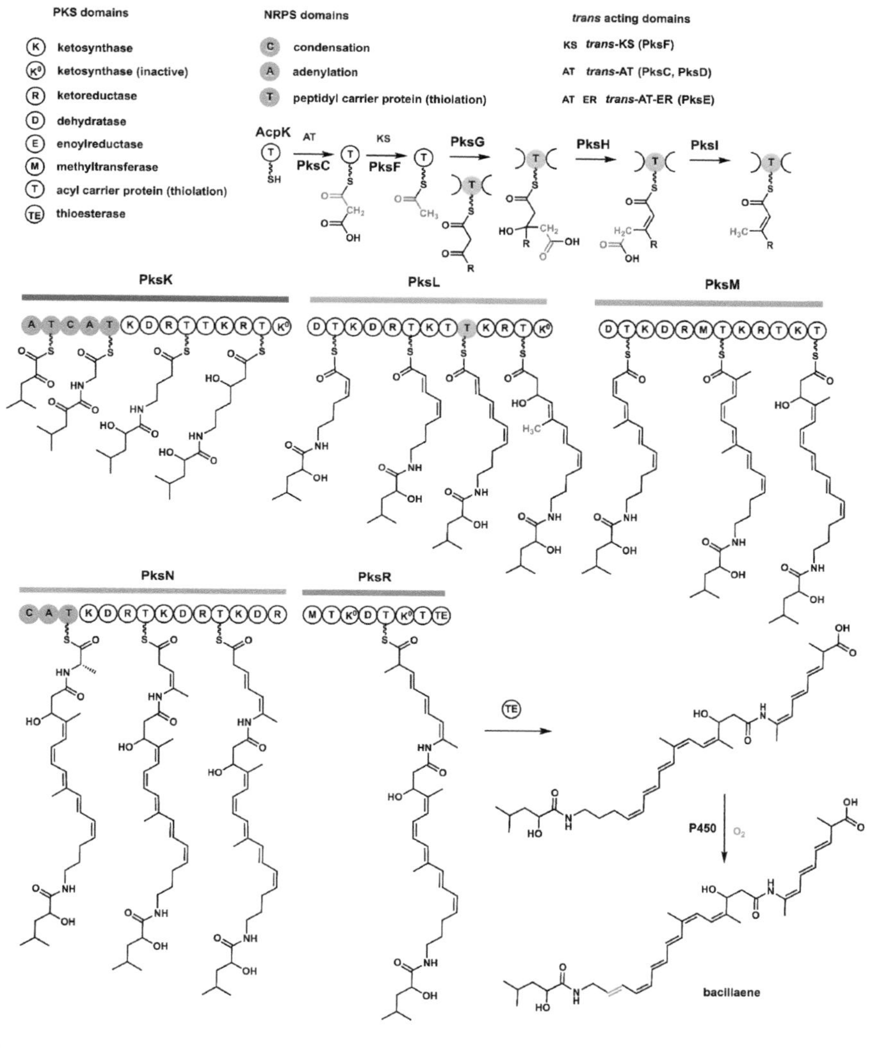

Figure 12.5 Bacillaene and its hybrid NRPS–PKS biosynthetic assembly line.

12.4 Case Studies: Historical and Current

Figure 12.6 shows a general schematic for isolation of a pure compound from a crude biological sample, from plant, fungal, or microbial extracts (Sarker, Latif *et al.* 2005). Early assessment by MS and NMR methods on crude mixtures allows determination of any novel compounds. A positive finding will then lead to further purification and characterization of novel molecules structurally and for biologic

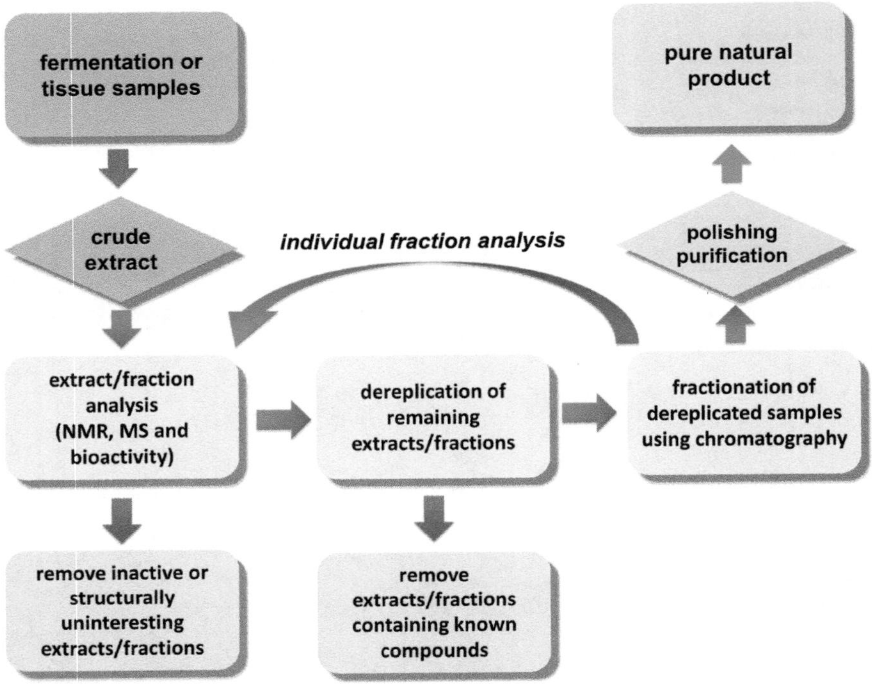

Figure 12.6 Generic scheme for isolation and characterization of natural products.
Adapted from Sarker, S. D., Z. Latif and A. I. Gray (2005). *Methods in Biotechnology, Natural Product Isolation*. Totowa New Jersey, Humana Press, vol. 20.

activities. Each class of natural product has its own lessons and requirements for purification at a useful scale; we take up selected examples of molecules below.

The first set of examples includes molecules from plant sources, whose initial isolations probably span four centuries, from quinine in the 16th century to vincristine and taxol in the second half of the 20th century. The second set of examples comprises microbial metabolites, all isolated during the 20th century and amenable to fermentations in large tanks for commercial production.

12.5 Five Plant Derived Natural Products

12.5.1 Quinine

This alkaloid, found in the bark of cinchona trees endemic to South America, was an herbal remedy for fevers used by indigenous South

American Indian tribes well before the Spanish landed in the West Indies and South America. By the mid 17th century the Jesuits in Peru had appropriated the treatment and it had become a frontline treatment for malaria among Europeans by the end of the 17th century (C&E News Top Pharmaceuticals, 2005, (Kaufman and Ruveda 2005)).

The active molecule quinine was isolated from bark in 1820 by Pelletier and Caventou, reviewed in Kaufman and Ruveda, 2005. The formal synthesis of quinine by von Doering and Woodward in the 1940s was viewed as a *tour de force* strategically and tactically and ushered in the golden age of total synthesis as a high art form, especially of complex natural products, for the next two generations of organic chemists. The first stereocontrolled total synthesis of quinine was by Stork in 2001 (Stork, Niu *et al.* 2001). Quinine as a frontline antimalarial was superseded in the mid 20th century by synthetic analogs including chloroquine and mefloquine, until widespread resistance made artemisinin the agent of choice for first line combination therapies.

The starting material for quinine isolation is cinchona bark (Figure 12.7). Quicklime is added for alkaline maceration of the bark (Roth and Streiler 2013). One route of workup is extraction with supercritical CO_2 at elevated temperature and pressure. Alternatively, the alkaline crude mixture is extracted into hot paraffin oil, filtered, and mixed with sulfuric acid. Neutralization with sodium carbonate yields the insoluble quinine sulfate on cooling. Free base quinine is obtained by titrating the sulfate salt with ammonia.

A parallel route is to treat the macerated bark-treated alkaline solutions with an aromatic hydrocarbon–alcohol mixture, and then

Figure 12.7 Cinchona bar is the starting material for quinine isolation.

extract with 5% HCl. The aqueous layer is back titrated to pH 4–5 and crude quinine HCl salt precipitated This can be redissolved, treated with activated charcoal to decolorize, and then crystallized at ambient temperature to give the hydrochloride salt of quinine (Patent Appl PCT/IN2011/000404).

12.5.2 Morphine

The alkaloid morphine is one of about 2500 distinct benzylisoquinoline molecules generated by the many species of poppy plants around the world. The opium poppy *Papaver somniferum* itself makes about 80 alkaloids, of which morphine is the principal constituent of that molecular class (Chemical & Engineering News, Special Issue on Top Pharmaceuticals that Changed the World, Volume 83, Issue 25, 2005). The poppy plants have been cultivated for an estimated 4600 years, starting with the Sumerians in 3000 BC (Blakemore and White 2002). Paracelsus reportedly introduced laudanum, an alcoholic solution of opium alkaloids plus essential oils, to Europe in the 16th century. Estimates of licit and illicit global production of opium straw that contains morphine are in the range of 30 000 to 50 000 tons per year. About 10% of the opium content is morphine.

As noted in Chapter 5, the pure compound morphine, named after the Greek god of sleep, Morpheus, was isolated by Serturner in the early 1800s. Morphine is used for analgesia and pain control (Kapoor 1995). Its *O,O*-diacetyl derivative was marketed, starting in 1898, in Germany under the trade name heroin (Blakemore and White 2002). Heroin is highly addictive and, in undergoing metabolic deacetylation to morphine, produces the addictive "rush" so characteristic of this addictive agent. The third molecule of note in the series is the morphine biosynthetic precursor codeine which does not have the same pharmacology as heroin. Instead codeine is used as an anti-cough medicine with mild analgesic effect. Figure 12.8 shows the structures of morphine, codeine, and heroin.

Morphine is an exogenous mimic of a series of small peptides – endorphins (endogenous morphines) – that bind to subtypes of opioid receptors. Morphine is a selective agonist of the μ-receptor subclass while semisynthetic ligands such as naloxone (Figure 12.8) are antagonists of all the opioid receptor subtypes (Kapoor 1995).

Some 100 years after its isolation as a pure compound, the structure of morphine was finally correctly deduced by Robinson in 1925, in part from degradative experiments and in part from integration of 60 years of prior experimental work by others. The total synthesis was

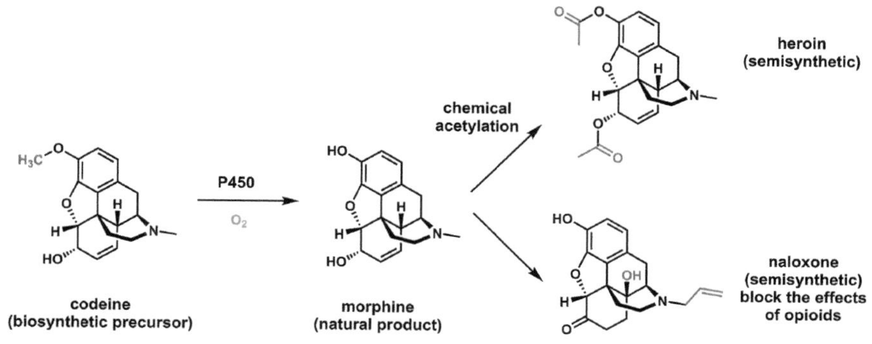

Figure 12.8 Structures of codeine, morphine, heroin, and naloxone

first achieved in 1952 but is not competitive with the isolation from natural sources (Blakemore and White 2002).

It is likely that isolation methods have varied only slightly over the millennia. Latex from the scored poppy seed capsule is air dried to a black resin, known as Indian opium. The alkaloid content of the resin is typically 10–15% morphine, 3–4% codeine, small amounts of papaverine, and up to 2% thebaine. The morphine is purified 6–10 fold to purity by partition back and forth between organic and aqueous phases with appropriate pH control, to go from the free base to the amine salt and back.

Chemical methylation at the phenolic –OH converts morphine back to its *O*-methylated biosynthetic precursor codeine. Codeine is presumably acted on biosynthetically by a cytochrome P450 oxygenase at that CH_3 to generate a CH_2OH group that unravels spontaneously to formaldehyde and morphine. The chemical bis acetylation of morphine yields heroin. The 3-monoacetyl intermediate is weakly pharmacologically active as a heroin mimetic, while the 6-*O*-acetyl metabolite is a strong heroin mimetic.

12.5.3 Digoxin and Related Cardiotonic Steroidal Glycosides From the Wooly Foxglove

The best source of the cardiotonic steroidal glycosides digitoxin and digoxin is the wooly foxglove, *Digitalis lanata* (Fonin and Khorlin 2003) (Figure 12.9). The dried aerial portions of the plants, including the flowers, are ground into a crude digitalis powder. Lipophilic material is removed from the amphiphilic glycosides by petroleum ether extraction. A wash with water and aqueous ethanol partitions the glycosides into the aqueous phase. Tannins can be precipitated with lead hydroxide and the cardiac glycosides purified by alumina gel filtration and elution with polar organic solvents such as alcohols.

The primary cardiotonic glycosides in *D. lanata* are actually lanatosides A–C, which have a tetrasaccharide tail: three digitose residues and a terminal glucose. The other difference between the lanatosides and the desired digoxin and digitoxin is the presence of an acetyl group on the third digitose residue (Figure 12.9). Alkaline treatment selectively removes the *O*-acetyl group while glucosidase activity in the crude extract removes the glucose. The digoxin and digitoxin can then

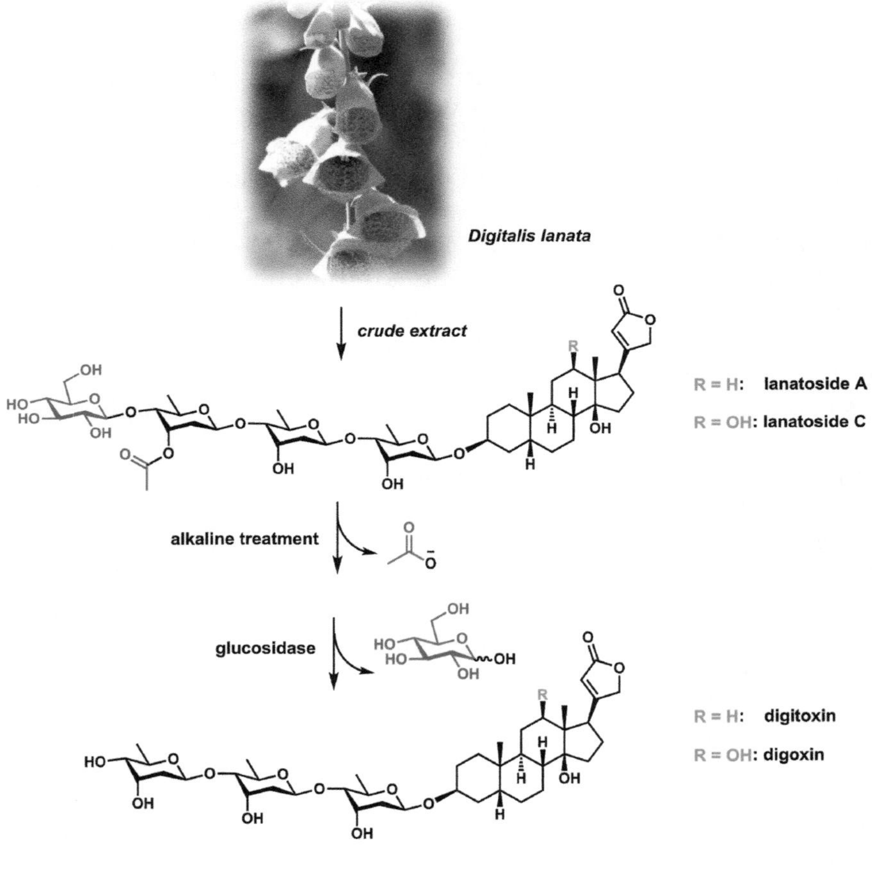

Figure 12.9 *Digitalis lanata*, the wooly foxglove, a source of lanatosides that are precursors to the secondary glycosides digoxin and digitoxin that can be obtained from deacetylation and deglucosylation.

be purified by the above schemes. This is a case of chemoenzymatic conversion of the primary digitalis glycosides to the desired secondary glycosides as an integral part of the isolation scheme.

Vignette 12.1 Van Gogh and Digitalis.

In a 1994 report in *Clinical Chemistry*, Paul Wolf (Wolf 1994) speculated on diagnoses and treatments that might have been made if clinical analytic infrastructure had existed in the lifetimes of Beethoven, Mozart, Chopin, George III, and Vincent Van Gogh. His musings on the painter Van Gogh (1853–1890) noted the manic composition in his last painting, *Wheat Fields*. Wolf noted that the various post mortem speculations on Van Gogh's illnesses included "epilepsy, psychosis, syphilis, cerebral tumor and sunstroke.... turpentine poisoning... epilepsy... acute mania with delirium".

Van Gogh's treatment for epilepsy included digitalis extracts. Subsequent studies in 1925 noted the "dominance of yellow and green visual disturbance in digitalis poisoning". Van Gogh's house was totally yellow and his late palette was dominated by yellow colors. The portrait of his final physician, Paul Ferdinand Gachet, holding the foxglove flowers has been argued as a focus of Van Gogh on foxglove and digitalis (Figure 12.V1).

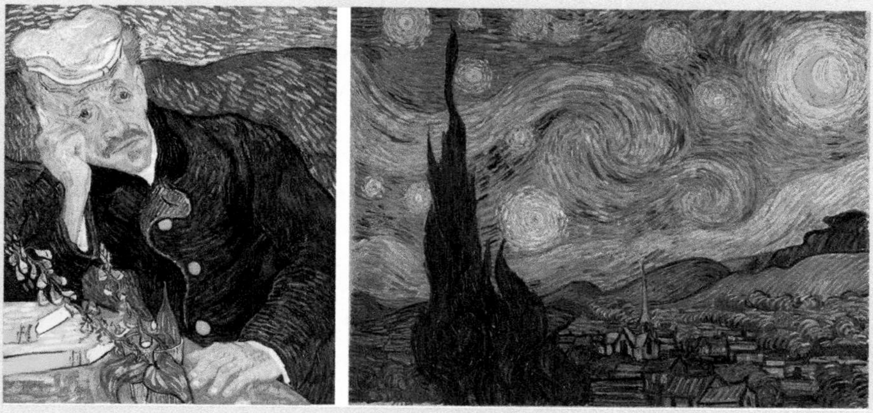

Figure 12.V1 Van Gogh and *Digitalis* (left: *Digitalis purpurae* held by Paul-Ferdinand Gachet in van Gogh's 1890 portrait of him, reprinted with permission from Musée d'Orsay, Paris; right: Gogh, Vincent van (1853–1890): *The Starry Night*, 1889. New York, Museum of Modern Art (MoMA). Oil on canvas, 29×36 OE (73.7×92.1 cm). Acquired through the Lillie P. Bliss Bequest. Acc. n.:472.1941.1 2016. Digital image, The Museum of Modern Art, New York/Scala, Florence).

Anisocoria, pupils of unequal size due to dilatation and contraction, is a side effect of digitalis and Wolfe noted that Van Gogh did a self portrait showing him with anisocoria. Other theories have been raised including terpene poisoning from camphor, which Van Gogh used in his mattress and pillow, and also from drinking turpentine. But the suggestion that overmedication with digitalis "resulted in the yellow coronas and halos surrounding the stars in *The Starry Night*", which he painted in 1889 one year before his death may be most compelling.

12.5.4 Taxol: from Yew Trees to Plant Cell Culture

Taxol was detected in extracts of the pacific yew tree (*Taxus brevifolia*) in 1962, purified by 1971, and found to have potent antitumor activity, due to its action on tubulin (Chapter 4). In the English/European yew *Taxus baccata* the main taxane, 1 part per thousand in extracts of the needles, is the 10-deacetylbaccatin III (Figure 12.10). This serves as an advanced intermediate for synthetic acetylation at C_{10} and acylation at C_{13} with either the *N*-benzoyl-α-hydroxy-β-Phe chain to yield taxol (paclitaxel) or with the *N*-*t*-butanoyl-α-hydroxy-β-Phe to the equivalently acting taxotere (docetaxel), as practiced by the company Indena among others.

From the outset it was clear that supply was going to be a serious obstacle to widespread use of taxol. The yield from Pacific yew was 1 kg of taxol from 3000 trees, or about six trees per course of therapy for one patient. The Chinese company Yewcare cultivates 30 km^2 of a Chinese yew tree (*T. chinensis*) as an initial step in taxol production in Hunan province (Malik, Mirjalili *et al.* 2011). A competitive production approach is to grow yew cells in culture under conditions of taxane production. By this route Phyton Biotech has achieved production levels of 500 kg of taxanes per year. This involved selection of *Taxus* cell lines that are producers, with optimization of nutrients and culture conditions, including elicitor molecules, phytohormones, and inducing factors to get to yields of taxol of ~20 mg L^{-1}. Two-phase cultures are utilized to sequester taxol out of the producing plant cells and prevent feedback inhibition. The crude taxane is purified by column chromatography and then crystallization. At a commercial plant cell fermentation bioreactor scale of 75 000 L each run would give ~1.5 kg of taxanes *via* the deacetylbaccatin III metabolite.

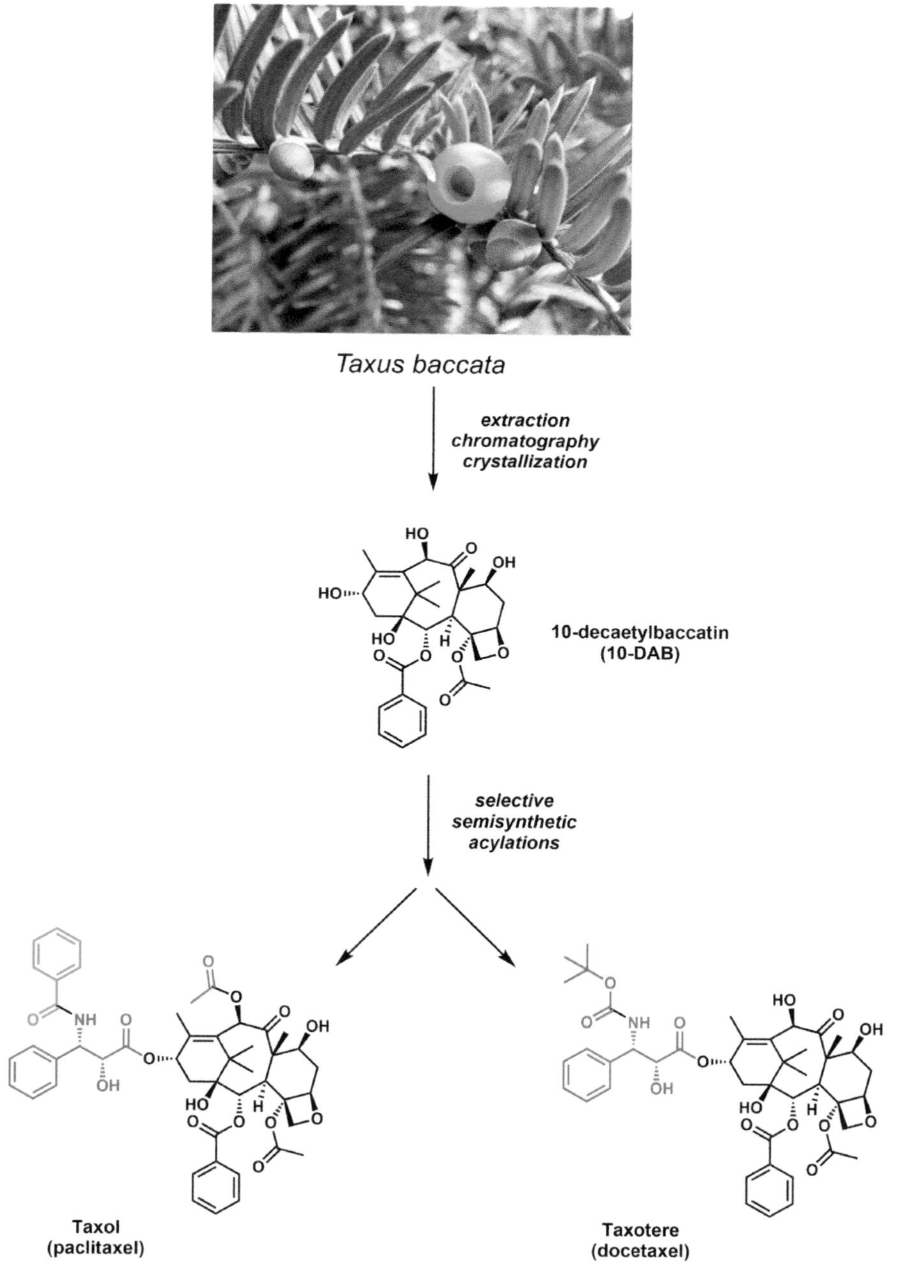

Figure 12.10 From *Taxus baccata* to purified taxol by way of 10-deacetylbaccatin, which is also a precursor to the semisynthetic taxotere.

12.5.5 Vinblastine and Vincristine from *Catharanthus roseus*, the Madagascar Periwinkle

The vinca alkaloids vinblastine (trade name Velban), discovered in 1953, and vincristine (tradename Oncovorin), discovered in 1961, may be among the most complex natural products that have been commercialized and approved as human therapeutic agents (FDA approval in 1963). We noted in Chapter 8 how catharanthine and vindoline in the Madagascar periwinkle plant are coupled biosynthetically to give vinblastine. The rarer metabolite vincristine is two oxidations further down the pathway as the *N*-methyl of vinblastine is converted to the indole aldehyde functionality in vincristine (Figure 12.11).

Vincristine, like taxol, targets tubulin, and is an agent commonly used in combination chemotherapy regimes for a variety of cancers including leukemias and lymphomas. Vinblastine has been part of regimens for testicular cancer and Hodgkin's disease (Moudi, Go *et al.* 2013). The history of use of the plant as an active agent in Africa stretches back into the distant past as a folk medicine, particularly for diabetes, although none of the purified compounds from the plant was found to be active for that therapeutic indication.

As with taxol, noted above, supply from the periwinkle plant has been a challenge due to low content of these dimeric alkaloids ($\sim$1 μg g^{-1} plant tissue of vincristine), interspersed with about 80 other alkaloids. As it happens Oncovorin is sufficiently potent that it has been estimated that 5–6 kg of the pure alkaloid is sufficient to treat all the indicated cancer patients in the USA in a given year. In China it has been estimated that 2 million pounds of the fresh periwinkle plants give about 250 000 pounds of dried leaves as starting material for the annual supply of vincristine needed for in-country treatments. Traditional schemes for alkaloid isolation are employed, including solvent extraction of the macerated leaves, and acid and base adjustment of aqueous phases to move the free base and salt forms of alkaloids back and forth between phases and thereby to effect separation from other material classes. Also, in parallel with taxol plant cell culture, alternatives have been explored but none has risen to commercial levels to date.

12.6 Three Microbial Metabolites

Microbial fermentations at scale became common for commodity chemicals including citric acid and acetone in the USA during

Madagascar periwinkle *(Catharanthus roseus)*

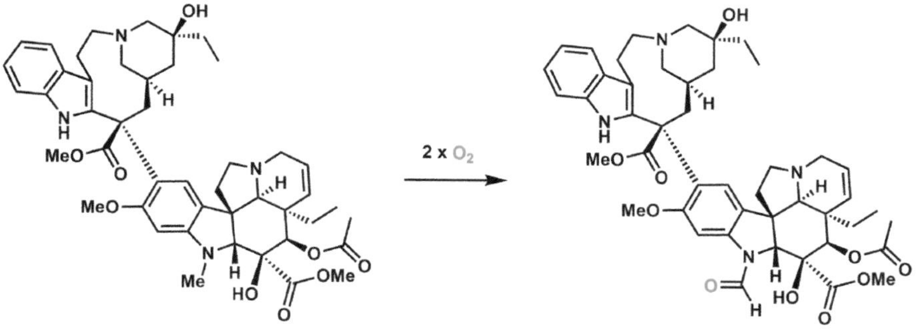

vinblastine **vincristine**

Figure 12.11 Metabolic conversion of vinblastine to vincristine. Left: the producing plant, the Madagascar periwinkle.

World War I. Because both bacteria and fungi have proved amenable to large scale high density fermentations, the starting material availability has generally been less of an issue than with the low yield from medicinal plant sources noted above.

12.6.1 β-Lactam Antibiotics

In the 75 years of penicillin fermentations and the 60 years of the ring expanded cephalosporins, many of the production parameters were incrementally optimized. These included strain improvement by undirected mutagenesis and selection for increased production, development of feed stocks and fermentation parameters, and isolation of the largely secreted β-lactam antibiotics, with solvent partitions (penicillin G dissolves in butyl acetate), column chromatography steps, and/or repeated crystallizations to high final purities. The lactams were the best selling antibiotic class for decades, and close to a million tons have been produced in aggregate since commercial production began (Figure 12.12) (Gordon, Grenfell *et al.* 1947).

Because of widespread deployment of penicillins and cephalosporins, waves of resistant bacterial pathogens were selected, leading to the need for successive generations of semisynthetic penicillins and cephalosporins to stay ahead of resistance development. The bicyclic 4,5-penam core and the corresponding fused 4,6-cepehem core are the warheads and are kept intact. Variations of the acyl chains are the most common synthetic changes wrought by medicinal chemists. To that end it became crucial to be able to deacylate naturally occurring acyl chains in metabolites such as penicillin G and cephalosporin C. Specific penicillin deacylases and cephalosporin deacylases yield the 6-aminopenicillanic acid (6-APA) and 7-aminocephalosporonic acid (7-ACA), respectively (Figure 12.13) on multi-ton scales. The 6-APA and 7-ACA are the substrates for chemical acylation, *e.g.* to produce contemporary lactam antibiotics such as piperacillin and ceftobiprole (which has synthetic groups at both the amine and the 3′ site of the cephem ring) (Walsh and Wencewicz 2016).

12.6.2 Vancomycin

In the 1950s, Eli Lilly and Co and other pharmaceutical organizations were conducting large scale screens for soil organisms that could produce novel antibiotics. One soil sample was collected in the interior jungles of Borneo. This sample contained a bacterium initially termed *Streptomyces orientalis* and later *Amycolocaptosis orientalis* that produced a molecule, compound 05865, that was bactericidal against pathogenic *Streptococcus* culture. It was renamed vancomycin as a substance that could "vanquish" streptococcal infections. As an index of purification procedures at that time, the vancomycin-containing solutions from fermentations were precipitated with picric acid to give a material that was colloquially described as "Mississippi mud" (Figure 12.14) (Griffith 1981). The liquor sold as Mississippi Mud may

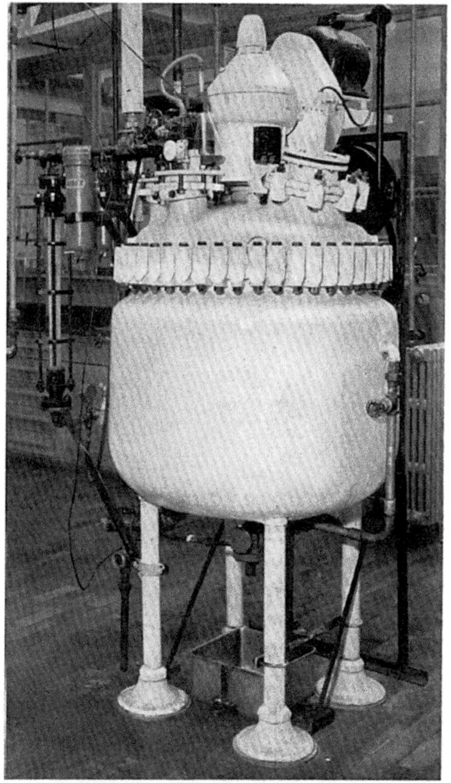

50 gall. pilot-plant penicillin fermenter.

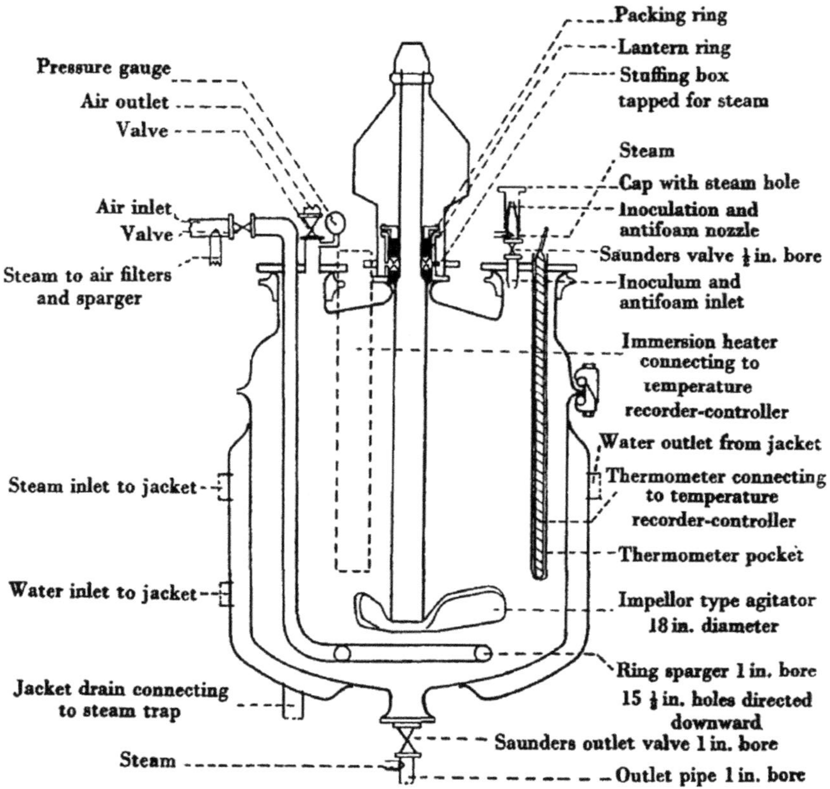

Figure 12.12 The 1947 50-gallon pilot plant penicillin fermenter and its design are reprinted with permission from Gordon, J. J., E. Grenfeel, E. Knowles, B. J. Legge, R. C. A. Mcallister and T. White (1947). "Methods of Penicillin Production in Submerged Culture on a Pilot-Plant Scale". *Microbiology* **1**: 187–202. Copyright 1947 Microbiology Society.

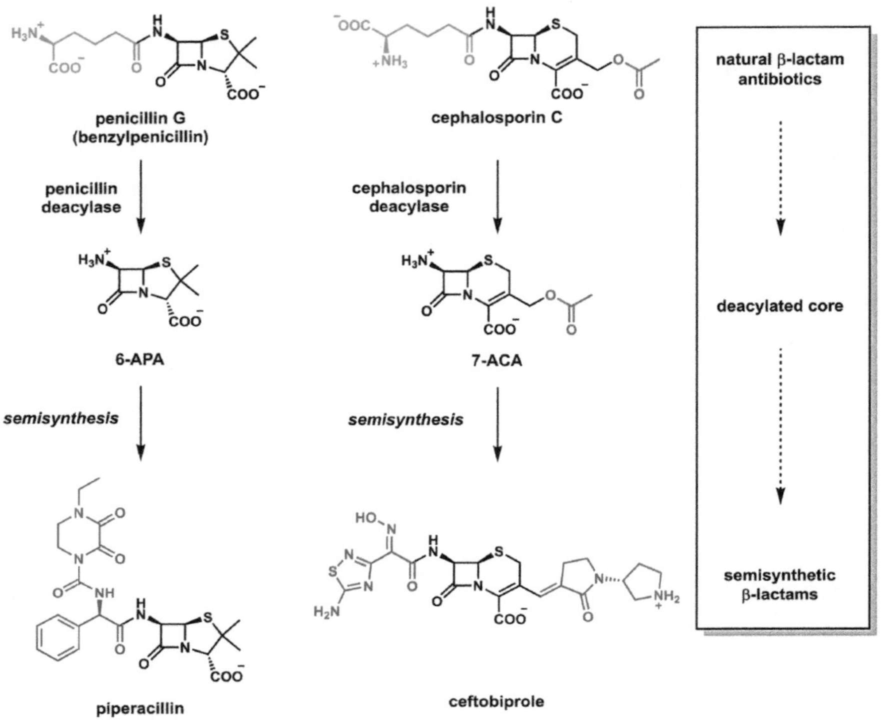

Figure 12.13 Enzymatic deacylation and synthetic reacylation of penicillin G and cephalosporin C: 6-APA and 7-ACA as intermediates for chemical reacylation to semisynthetic β-lactams.

have spawned the comparison and is shown on the right of the figure. This purification protocol was superseded by ion exchange chromatography, formation of a crystalline copper–vancomycin complex, and finally crystallization as vancomycin hydrochloride, a white powder (Griffith 1984).

A more evolved production and purification protocol from a 1993 US patent involves an 8 L growth on glycerol–soy flower and calcium gluconate to yield 3.7 kg of vancomycin which is adsorbed from the culture filtrate on Amberlite IRC 50 (H + phase), washed and eluted with 0.5 N NH_4OH. After pH adjustment and clarification with activated charcoal, the glycopeptide is adsorbed onto and then eluted from cation-exchange columns and finally precipitated with 11 volumes of acetone and dried to yield 2.9 kg of purified vancomycin (US Patent 5223413).

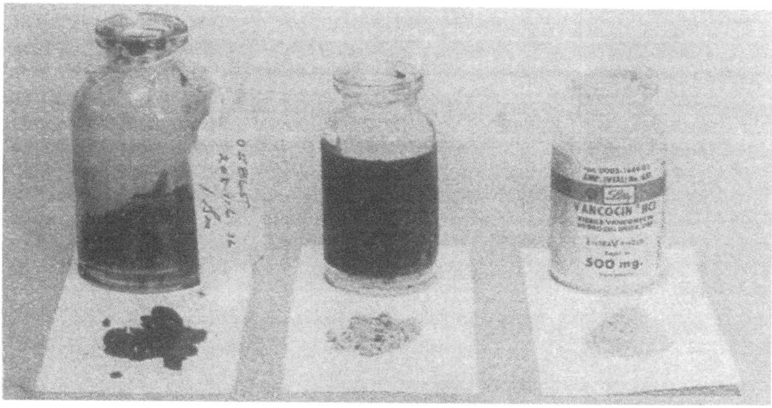

Figure 12.14 Increased purity of vancomycin from the initial crude sample "Mississippi mud" (left), following ion-exchange chromatography (center), and final formulation ready vancomycin hydrochloride (right).
Reprinted from Griffith, R. (1984). "Vancomycin use – an historical perspective." *J. Antimicrob. Chemother.* **14**, by permission of Oxford University Press.

12.6.3 Cyclosporine A

The second of the fungal metabolites to be discussed is cyclosporine A, again a nonribosomal peptide, as are the beta lactams and the vancomycin glycopeptides. Cyclosporine A, as noted in Chapter 3 (Figure 3.23), is not an antibiotic but rather the first of the immunosuppressive drugs, which blocks T cell development and allows organ transplant medicine, to grow up and become a major medical specialty. This *N*-methylated undecapeptide was discovered in fungi from two independent soil samples, from Wisconsin and Norway, in 1970 from *Cylindrocaprim lucidum* and *Tolypocladium inflatum.* It is typical that nonribosomal peptide synthetase adenylation domains lack editing functions but the promiscuity of cyclosporine synthetase is particularly high. There are 32 closely related analogs produced by native cultures, with some conservative replacements of hydrophobic amino acid building blocks and the failure to *N*-methylate at one or more of the eight residues in cyclosporine A (Survase, Kagliwal *et al.* 2011).

Strains of *T. inflatum* have been the production organism, with classical strain improvement by mutation and selection for high production and fermentation parameters. Effects of added precursor amino acids and aeration and agitation rates have been thoroughly explored during the 35 years of production at scale (Survase, Kagliwal *et al.* 2011). A typical downstream purification protocol involves extraction of cyclosporine A from the solid fermentation mass with butyl acetate, gel filtration chromatography, and silica gel and/or alumina columns with final crystallization to get cyclosporine A of pharmaceutical purity.

12.7 Expanding the Inventory of Natural Products

New natural product structures continue to be reported from plant, fungal, and bacterial sources, although the latter two groups have become major sources over the past three decades. Not surprisingly, most new isolates are congeners of existing natural product scaffolds and architectures. On the one hand, this could suggest an approach to saturation of Nature's inventory and biosynthetic capacity to make a limited set of molecular scaffolds and functional group arrays from simple primary metabolic building blocks. On the other hand, this may reflect limitations in the traditional approaches to natural product isolation protocols that start with organic solvent extraction of starting materials, particularly plant solids. One measure of under-examined metabolite complexity can be estimated by Liquid chromatography–mass spectrometry (LC-MS) analyses of crude extracts with a substantial number of unidentified peaks of exact mass.

The input parameter that gives a molecular estimate of natural products yet to be characterized comes from genome analyses, as noted in the next chapter. To date there are few completed genomes from plants and the lack of biosynthetic gene clustering has made predictions of novel molecules difficult. By contrast, the clustering of biosynthetic genes in fungi and bacteria has made bioinformatic predictive analyses of biosynthetic capacity almost routine. It is estimated there may be millions of such microbial gene clusters recorded within the next decade, the great majority of them coding for as yet unidentified natural products.

How to choose biological starting materials as promising sources for novel natural products, especially ones with novel connectivities, scaffold architectures, and rare functional groups (*e.g.* diazo groups or *N*–*N* linkages), in a gene-independent manner will continue to be a challenge for innovative natural product characterization.

One approach is to look in new or underexplored biological niches to seek specialized molecules.

12.7.1 Marine Actinomycetes

Thus, marine organisms, compared to terrestrial organisms, have been a source of novel natural product scaffolds for much of the past half century. These include sponge consortia communities where associated bacterial populations may be most often the responsible chemists. Two cases in point are the biosyntheses of polytheonamides (Hamada, Matsunaga *et al.* 2005) (with 18 D-amino acid residues that shape the molecule into a membrane-spanning ion channel, see Figure 3.17) and patellamides, which contain thiazole and oxazoline heterocycles wrapped in a macrocyclic framework (see Figure 3.7) (Figure 12.15). Both patellamides are enzymatically processed from the single protein precursor PatE (Schmidt, Nelson *et al.* 2005). Analogously, polytheonamide A, a 48-residue highly modified peptide, derives from a ribosomally generated precursor protein that *inter alia* has undergone 22 radical SAM-mediated *C*-methylations (see Chapter 10) of side chain residues.

Marine actinomycetes of the *Salinospora* genus have been prolific producers of novel molecules, exemplified by the isolation and characterization efforts of Fenical and colleagues (Jensen, Williams *et al.* 2007). A small selection of such *Salinospora*-derived metabolites in Figure 12.16 shows known compounds such as rifamycin and staurosporine but also some unusual scaffolds in salinosporamide (a proteasome inhibitor, see Figure 9.47), sporalide A, and cyanosporaside.

12.7.2 Endophytic Fungi

Plants harbor internal communities of microorganisms. The endophytic fungi are well known, especially in association with particular grasses, but there are also many endophytic bacteria (Hardoim, van Overbeek *et al.* 2015) that functionally synergistically with plants. Some fungi make auxins, some make taxol, and there are many other molecules that have now been isolated from such endophytes (Strobel, Daisy *et al.* 2004). Figure 12.17 displays four natural products made by different endophytic fungi which have been found to have diverse activity, from antifungal activities (cryptocin and ambuic acid), to anticancer (torreyanic acid), to immunosuppressant (subglutinol) functions.

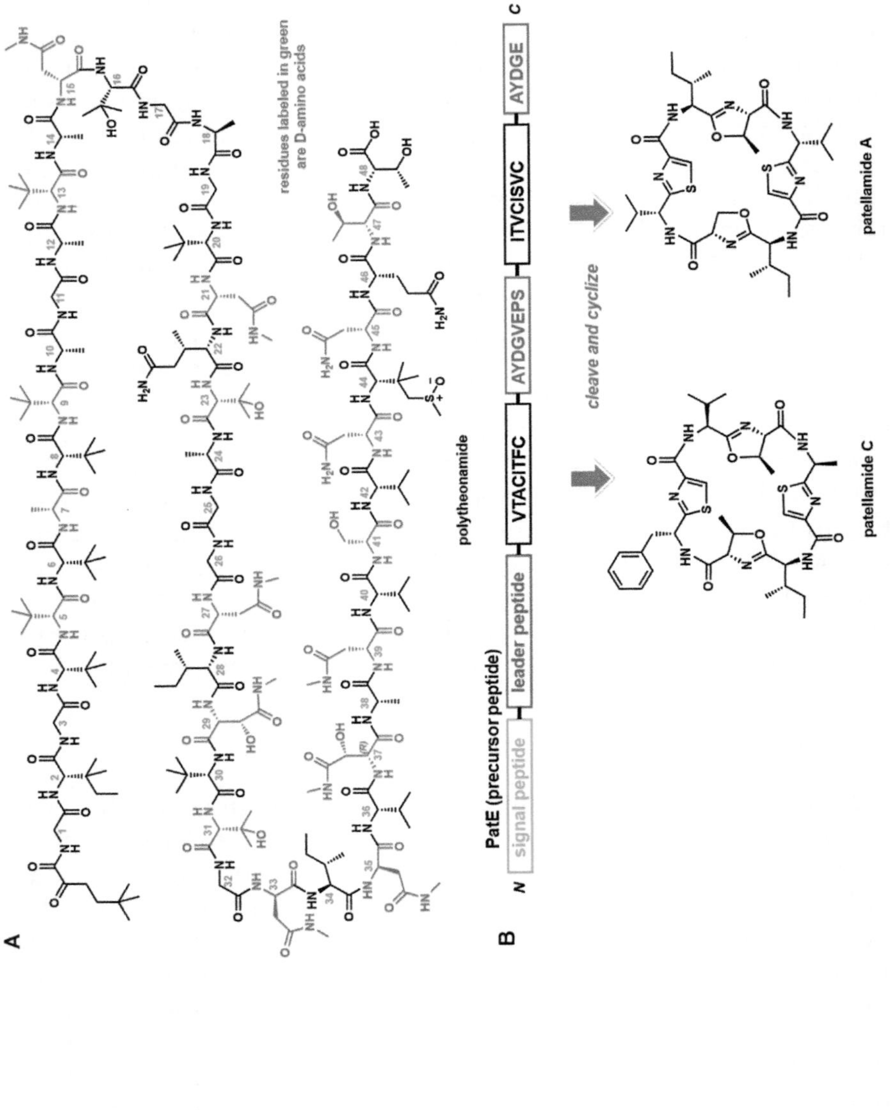

Figure 12.15 (A) polytheonamide A is a helical peptide that inserts into target cell membranes; (B) patellamides A and C from the same precursor protein, PatE.

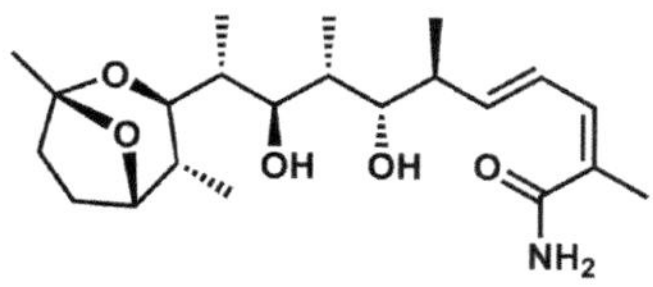

saliniketal A

Salinispora arenicola

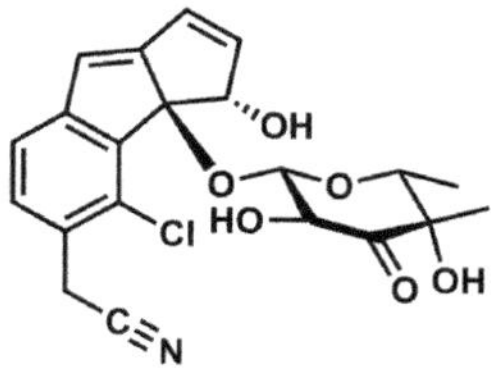

cyanosporaside A

Salinispora pacifica

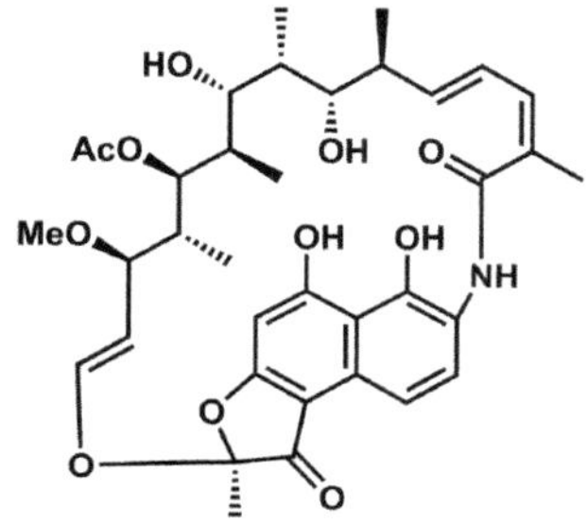

rifamycin

Salinispora arenicola

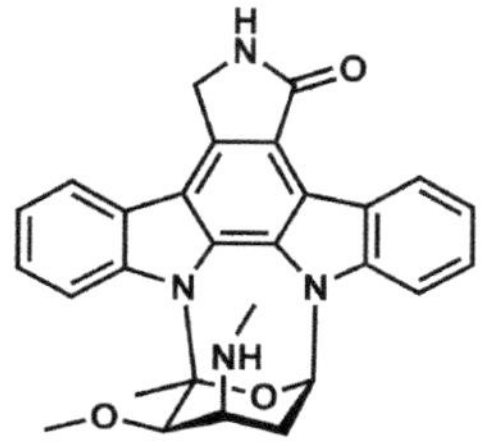

staurosporine

Salinispora arenicola

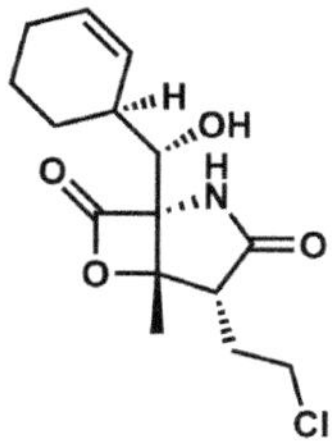

salinosporamide A

Salinispora tropica

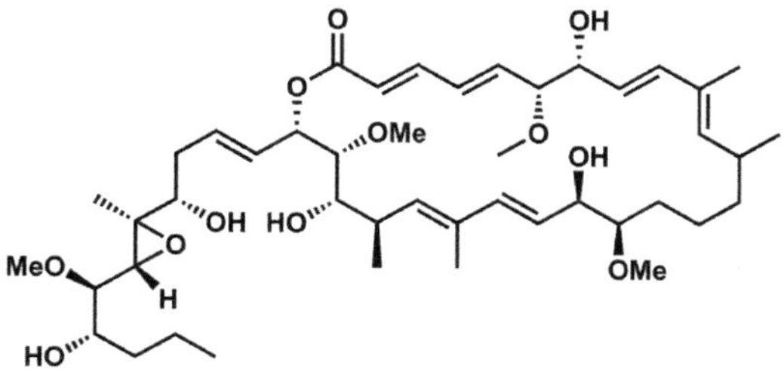

arenicolide A

Salinispora arenicola

Figure 12.16 A selection of metabolites from *Salinospora* strains.

12.7.3 Insect–Fungi–Bacteria Symbioses

There are many examples of mutualistic symbiosis between insects, fungi, and actinomycetes for protection against potential microbial pathogens (Poulsen, Erhardt *et al.* 2007, Poulsen, Oh *et al.* 2011). Chemical protection is afforded by the actinomycetes' production of antifungal compounds. One such molecule is the cyclic depsipeptide dentigerumycin produced by the *Pseudonocardia* bacterium that

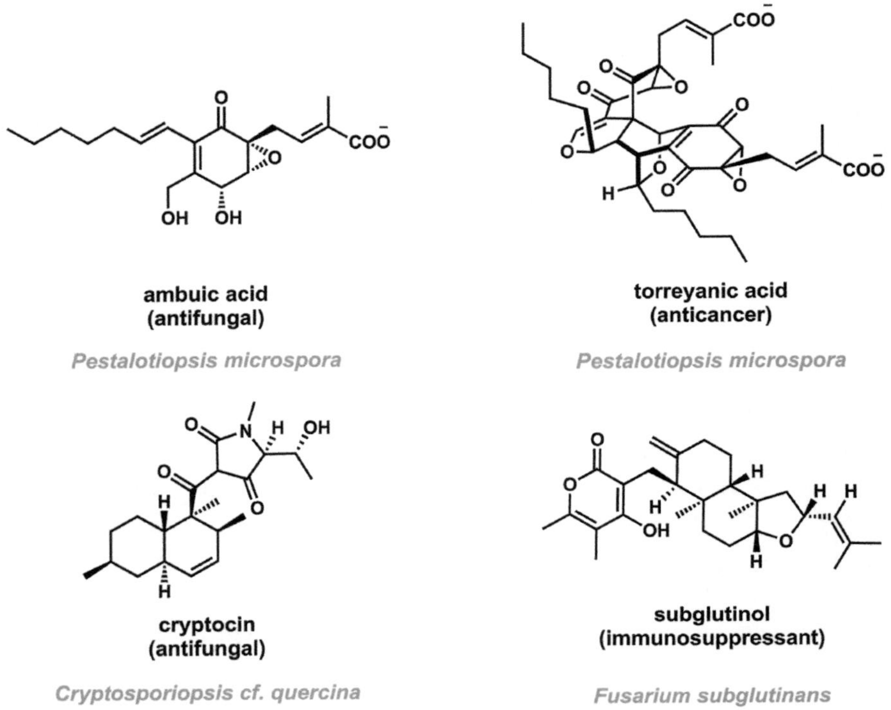

Figure 12.17 Four natural products from endophytic fungi.

associates with the fungus-growing ant *Apterostigma dentigerum* (Figure 12.18) (Oh, Poulsen *et al.* 2009). Analogously, southern pine beetles maintain a beneficial symbiosis with the fungus *Entomocorticum* sp. which produces mycangimycin to fight off the fungal pathogen *Ophiostoma minus* (Figure 12.18) (Oh, Scott *et al.* 2009).

Each of these dedicated antifungal metabolites has unusual chemical functionalities. Mycangimycin is a polyene endoperoxide and could be a light-activated radical generator. Dentigerumycin is a depsipeptide that belongs to a family of nonribosomal peptide–polyketide hybrid macrolactones that contain the unusual piperazate building block. Piperazate is an analog of pipecolate but has the unusual *N–N* single bond in the six-membered ring; its biosynthetic origin is incompletely understood.

12.7.4 From the Depths of the Abyss

One last example showing that exploration of new biological niches can increase the probability of new natural product scaffolds

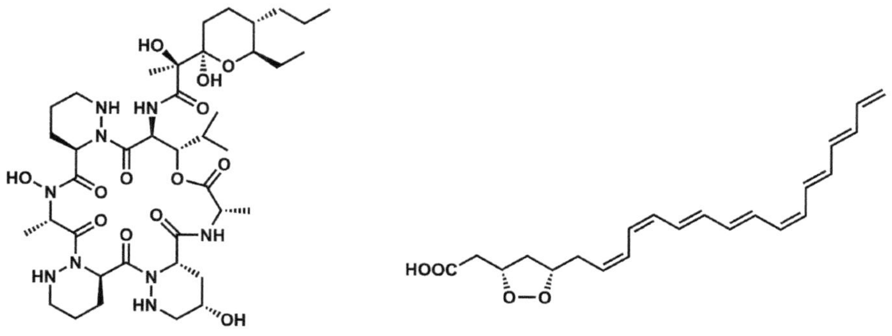

dentigerumycin
(3 piperazates)

Pseudonocardia

mycangimycin
(polyene endoperoxide)

Entomocorticum sp.

Figure 12.18 Molecules from mutualistic symbioses between microbes and insects.

comes from a soil sample gathered in the depths of the Sea of Japan (Bister, Bischoff *et al.* 2004, Riedlinger, Reicke *et al.* 2004). The recovered actinomycetes make the compound abyssomycin (Figure 12.19) and its *atrop*-isomer. The two are separable owing to restricted rotation in the macrocyclic ring. Both are inactivators of 4-amino-4-deoxychorismate synthase (Keller, Schadt *et al.* 2007), a key building block in folate biosynthesis and in turn for DNA biosynthesis, but the *atrop*-abyssomycin is more potent at capturing an active site cysteine side chain. The scaffold is novel and the restricted rotation adds another functional feature to this natural product pair, emphasizing that new natural chemistry is still to be found.

12.7.5 Intersection of More than Two Biosynthetic Pathways

A set of natural product scaffolds that may be of more than routine interest going forward would be the convergence of more than two kinds of biosynthetic pathways, beyond the by now familiar indole–terpene or NRP–PK intersections. To that point, leupyrrin A (an unsymmetrical antifungal macrolide from *Sorangium cellulosum*) and kibdelomycin (Bode, Irschik *et al.* 2003, Phillips, Goetz *et al.* 2011) (Figure 12.20) represent three or more convergent biosynthetic logics and have created some unusual scaffolds. It is probable that the bioinformatic approaches noted in the next chapter would

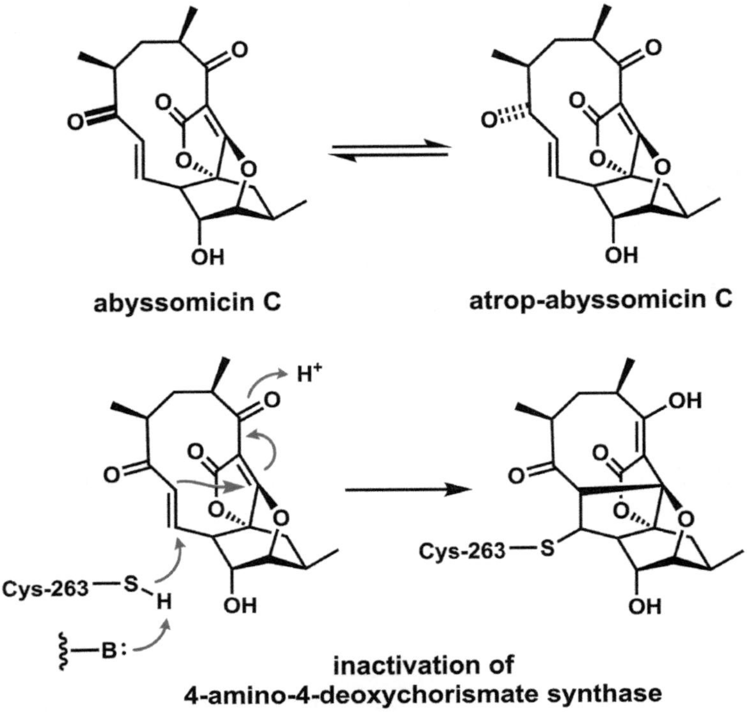

Figure 12.19 Abyssomycin C and its *atrop*-isomer are found in the depths of the Sea of Japan. The mechanism of action is through covalent inactivation of 4-amino-4-deoxychorismate synthase.

allow focus on such multiple pathway convergences in microbial genomes.

12.7.6 Culturing the Unculturables

It is likely that the microbial world of fungi and bacteria holds much as yet undiscovered natural product biosynthetic capacity. A minor fraction of global microbial diversity has been sampled over the 70–80 years of cultivation. Estimates are that 100 000 fungal strains have been examined out of perhaps a million that are anticipated (Schueffler and Anke 2014). It is not clear what fraction of fungal strains is captured by contemporary culture methods but as little as 1% of bacterial diversity has been thought to be susceptible to common laboratory culture conditions. This is one of the driving forces for microbial genome sequencing, detailed in the next chapter. Even

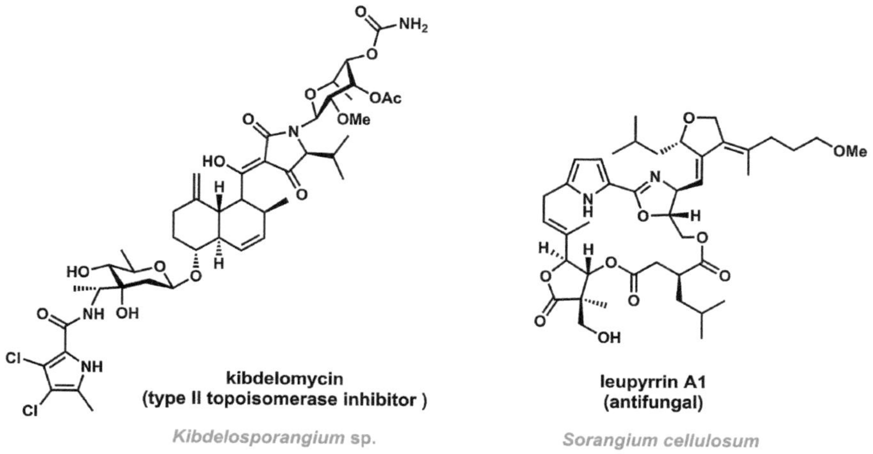

Figure 12.20 Kibdelomycin and leupyrrin A: metabolic intersections of several elements of natural product chemical logic and convergent biosynthetic machinery.

independent of gene-based indices of microbial habitat complexity, it is clear that expanding the range of culturable microbes is likely to pay dividends in terms of both previously under-catalogued organisms and perhaps novel natural products.

To that end, recent advances in bacterial culture methods have allowed screens of these rare, previously unattainable, organisms and led to isolation of the undecapeptidolactone teixobactin (Ling, Schneider *et al.* 2015) (Figure 12.21). Four of the 11 residues have the D-configuration, consistent with, but not proof, of a non-ribosomal origin (recall the ribosomal origin of polytheonamide noted earlier in this chapter). The target of teixobactin is lipid II, a key carrier molecule for the growing peptidoglycan chain in bacterial cell wall assembly. If advanced culture methods increase the number of culturable bacteria even from well explored niches, it is likely that new molecular frameworks will emerge.

12.7.7 Redox Enzymes as Predictors of New Functionality in Natural Products?

Between the encoding genes and the structures of the secondary metabolites, both pathway intermediates and final end products, are the biosynthetic enzymes that assemble the bonds and functional groups, and create the architectural scaffolds. As we have noted in

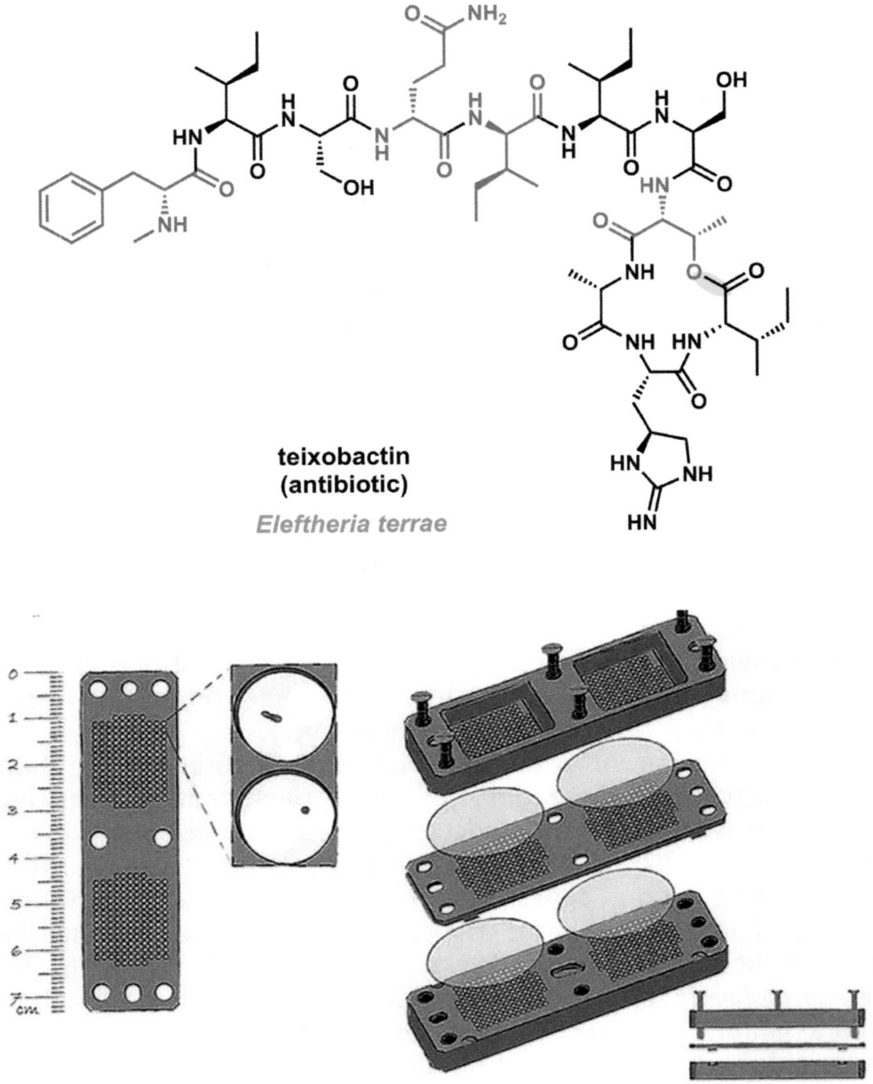

Figure 12.21 Teixobactin obtained by culturing the "unculturables": the iChip used to culture the bacteria directly in soil.
Reprinted by permission from Macmillan Publishers Ltd: Nature (Ling, L. L., T. Schneider, A. J. Peoples, A. L. Spoering, I. Engels, B. P. Conlon, A. Mueller, T. F. Schäberle, D. E. Hughes, S. Epstein, M. Jones, L. Lazarides, V. A. Steadman, D. R. Cohen, C. R. Felix, K. Ashley Fetterman, W. P. Millett, A. G. Nitti, A. M. Zullo, C. Chen and K. Lewis (2015). "A New Antibiotic Kills Pathogens Without Detectable Resistance". *Nature* **517**: 455–459), copyright (2015).

earlier chapters, some of the enzymes utilize chemistry that is commonly applied in primary metabolic pathways.

Others, though, are more prevalent in conditional metabolic pathways. We have noted the putative Diels–Alderases in polyketide assemblies (Chapter 2) and a possible aza-Diels–Alderase in construction of the pyridine ring in the trithiazolyl pyridine core of thiazolyl peptides of the thiocillin class (Chapter 3). We have emphasized the plethora of oxygenative enzymes dispersed at critical biosynthetic junctures in all the major classes of natural products in Chapter 9 and earlier chapters.

The ability of oxygen-reducing enzymes to generate substrate radicals also opens up scaffold recombination pathways without incorporation of oxygen. Analogously, in Chapter 10 we noted that hundreds of thousands of ORFs with the signature of radical SAM catalysis exist in databases. Together these radical SAM enzymes and the iron-based oxygenases, both of the cytochrome P450 class and the nonheme mononuclear iron class (*e.g.* isopenicillin N synthase), may be the catalysts most likely to carry out novel chemistry in natural product formation.

Following closely may be the flavin-dependent oxygenases, whose known members have shown unexpected chemical virtuosity in the scope of reactions catalyzed. These include Baeyer–Villigerases and also the enzyme EntM in enterocin biosynthesis, with its intermediacy of novel FAD-N_5-oxide oxygen transfer agent active for a favorskiiase transformation (Teufel, Miyanaga *et al.* 2013) (Figure 12.22). The flavoenzymes with epoxidase activity on indole rings may also be signposts for unusual oxidative rearrangement chemistry (*e.g.* in fumitremorgin to spirotryprostatin frameworks, noted in Chapter 8). Looking under these kinds of catalytic lampposts may increase the chances of finding pathways that generate unanticipatedly rearranged molecular frameworks.

The gene-based approaches of Chapter 13 are in principle ideal for scanning genomes for the above kinds of oxidative enzymes, especially if they are in the vicinity of what look to be biosynthetic gene clusters for known natural product classes. However, they are also worth investigation even in the absence of any gene based anchors. We also have argued in Chapter 11 that some of the dedicated glycosyltransferases are worth special focus, especially the ones likely to add unusual pentose and hexose units to aglycones as a route to increase the probability of novel and composite natural product frameworks.

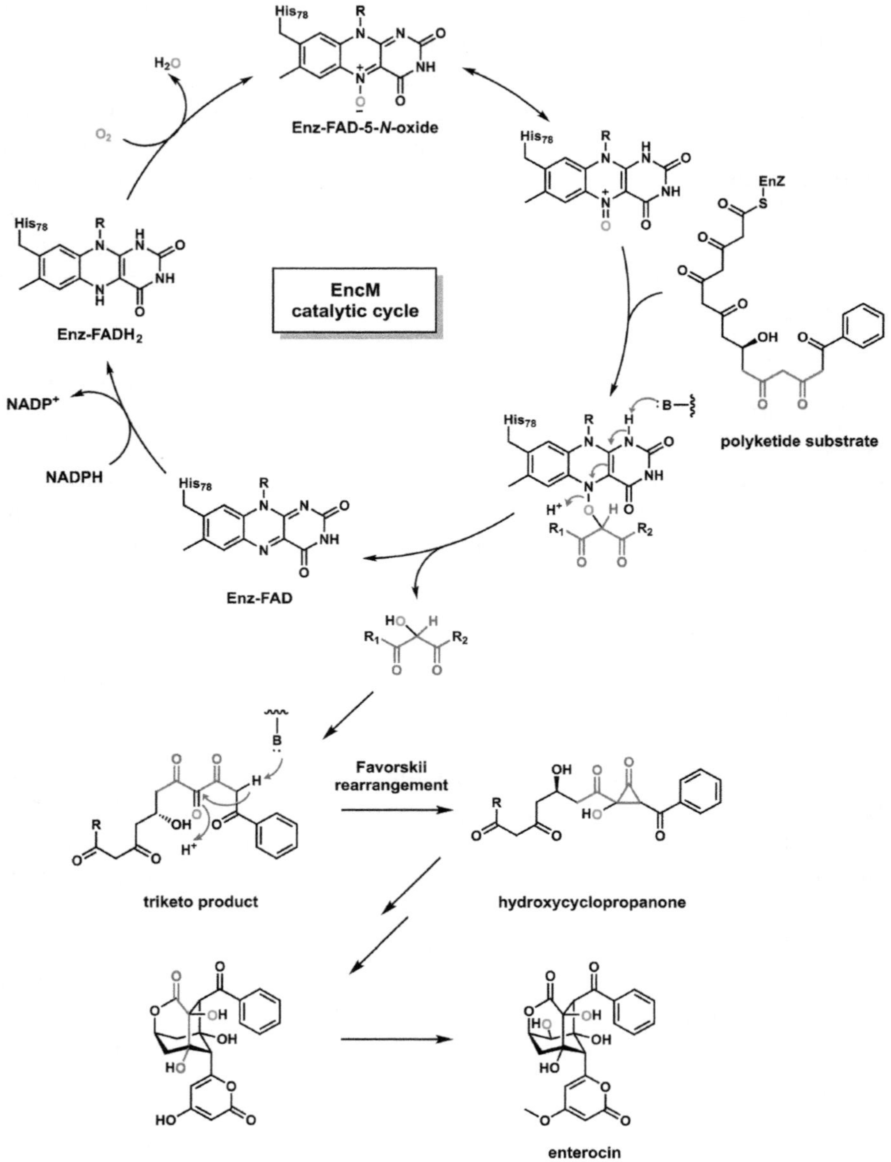

Figure 12.22 EntM is a flavoenzyme in the enterocin pathway that catalyzes a Favorskii-type rearrangement from a triketo intermediate.

12.7.8 Overlooked Building Blocks: NRPS Assembly Lines

One of the potential strategic flexibilities of building natural product scaffolds derived from peptide units from NRPS assembly lines, rather than posttranslational maturation of nascent ribosomal proteins, is the freedom from the restraints that the inventory of cellular aminoacyl-tRNA synthetases imposes on proteinogenic amino acid building blocks. It is certainly the case that all the effort in reengineering of the genetic code in both prokaryotes and eukaryotes is rapidly advancing the day when one could routinely insert several nonproteinogenic amino acids at will in specific sites of any desired protein.

However, in complementarity to those approaches, NRPS assembly line adenylation domains can select and activate up to hundreds of nonproteinogenic amino acids. Furthermore, some of them work selectively on the corresponding α-keto acids and α-hydroxy acids (Magarvey, Ehling-Schulz *et al.* 2006). We have already noted that convergence of PKS and NRPS machinery can build some of these nonproteinogenic amino acid monomers, such as statine and isostatine (see Figure 3.24), and vinyl tyrosine. It is worth calling out two additional nonproteinogenic amino acids from primary metabolism that in the past few years have come into focus as NRPS building blocks, with concomitant appreciation of novel natural product scaffolds that ensue.

These are the *ortho* and *para* isomers of aminobenzoates (Figure 12.23), each of them key metabolites in primary metabolism in microbes. Anthranilate, as we have detailed in Chapter 8, is the precursor to the central amino acid tryptophan and PABA is a key building block in construction of the folate scaffold. They also are now known to serve as both chain initiator and chain extender molecules for fungal and bacterial NPRS assembly lines, as noted in Chapter 5 (Walsh, Haynes *et al.* 2013). Six residues of *para*-aminobenzoate are built into the linear heptapeptide Cystobactamid 919-2 (from the myxobacterial *Cystobacteria*), which is a novel antibiotic by virtue of inhibition of bacterial DNA gyrase (Baumann, Herrmann *et al.* 2014). One or two residues of *ortho*-aminobenzoate (anthranilate) are utilized in fungal production of fumiquinazoline A and asperlicin, respectively (Chapter 5).

12.7.9 OSMAC (*O*ne *S*train *M*any *C*ompounds)

Prior to the availability of bacterial and fungal genomes it had been observed in many studies that cultivation of some fungal and bacterial strains under different conditions could lead to different amounts and even different compositions of natural products from the fermentations. The molecular bases of these observations were subsequently revealed when multiple biosynthetic gene clusters were detected in those microbial genomes. The expectation that there were silent secondary pathways that could be turned on by different inputs led Zeeck

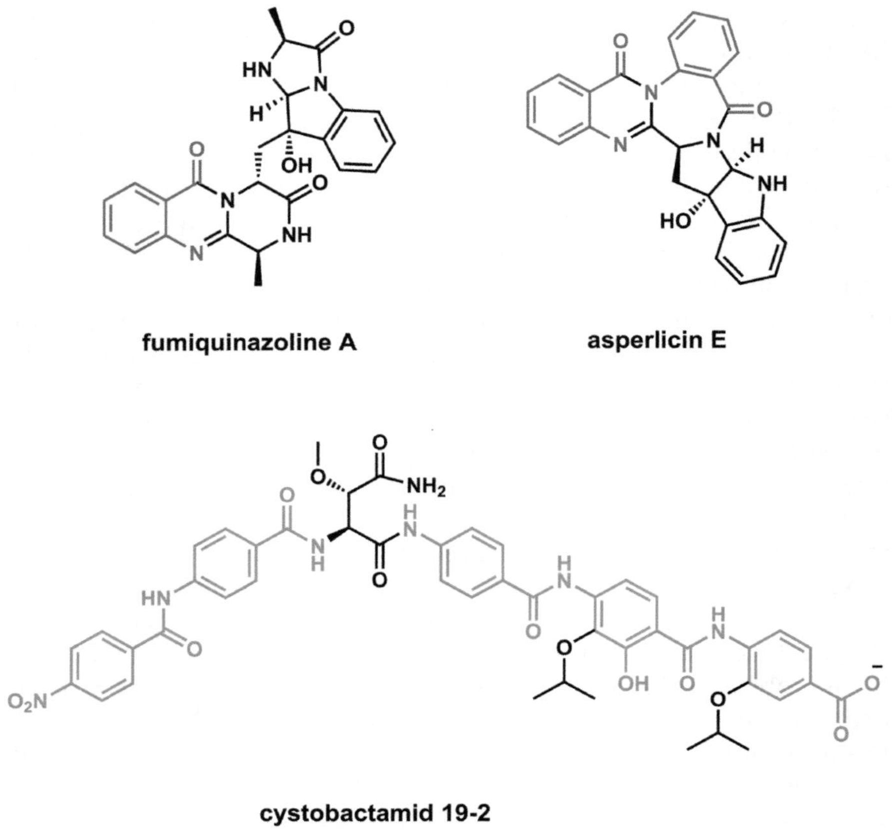

Figure 12.23 Three natural products utilizing either *ortho*-aminobenzoate (blue) or *para*-aminobenzoate (red) as NRPS assembly line building blocks.

and colleagues (Bode, Bethe *et al.* 2002) to coin the term OSMAC (*o*ne *s*train *ma*ny *c*ompounds) to formalize the hypotheses that one could see distinct natural product profiles by altering readily manipulatable parameters during culture. Variations could be as simple as change in the culture vessels, altered agitation (differential dissolution of O_2), pH control, change from liquid to solid media, temperature variation, the use of specific enzyme inhibitors, *etc.* For example, changing the nature of the liquid and solid substratum on which the fungus *Glomerella acutata* is grown not only changes the macroscopic morphology of the cultures but also the chemical inventory of natural product pathways activated (Figure 12.24) (Vandermolen, Raja *et al.* 2013). In sum, any set of conditions that might affect the response of a fungus or bacterium to external conditions might be expected to alter the transcriptome and thus the proteome and finally be read out in an altered secondary metabolome.

Over the decade and a half since the 2002 paper, many investigators have validated the OSMAC concept and plumbed the biosynthetic capacity of bacteria and fungi. Three examples are noted below. The first comes from the minireview of Bode Bethe *et al.*, 2002 (Bode, Bethe *et al.* 2002). They noted that, with two fungi and four actinomycete bacterial strains, simple alteration in culture conditions led to >100 new metabolites in 25 structural subclasses, where previously one compound had been dominant in each fermentation. Figure 12.25 shows the set of 11 new metabolites observed in cultures of *Aspergillus ochraceus* DSM 742 on varying culture conditions. These represent the activation of numerous biosynthetic pathways previously silent under the original cultivation condition. The authors noted that the practice of OSMAC by varying one or more culture condition inputs was essentially a random walk for selective desilencing of secondary pathways. When a microbial genome is known, as noted in the next chapter, it is possible to do a more targeted genetic approach to relieve repression or carry out selective activation on the genes of a given pathway.

The endolichenic fungus *Myxotrichium* sp., resolved from its lichen niche in Yunnan province in China, was known to make three polyketides when cultured on typical potato dextrose broth (PDB) liquid culture, as shown in Figure 12.26. When the endolichenic fungus was instead cultured on rice for 45 days, four new polyketides were produced (Yuan, Guo *et al.* 2016). The genome of this microbe is not yet available so it is not known what fraction of its biosynthetic capacity is being revealed by these two growth substrates. A third example, of many that could be elaborated, involves the endophytic fungus *Dothideomycete* sp. Again, a switch from the potato dextrose broth liquid medium to malt as a solid substrate, more likely to represent a mimic of the normal medium in which the fungus grows, led to a number of new molecular scaffolds, as depicted in Figure 12.26 (Hewage, Aree *et al.* 2014).

Given the accruing evidence that actinomycetes and fungi examined to date express 10–15% of their biosynthetic capacity when grown under standard laboratory conditions, OSMAC will continue to be valuable as a systems approach to eliciting a larger fraction of the biosynthetic potential of bacteria and fungi, both before and after genome information is acquired. The perturbations to culture conditions can be extended in many directions. For fungi, eukaryotes with histones and chromatin remodeling, modulators of histone acetylation and methylation status can alter the activation of secondary metabolic pathways (Netzker, Fischer *et al.* 2015).

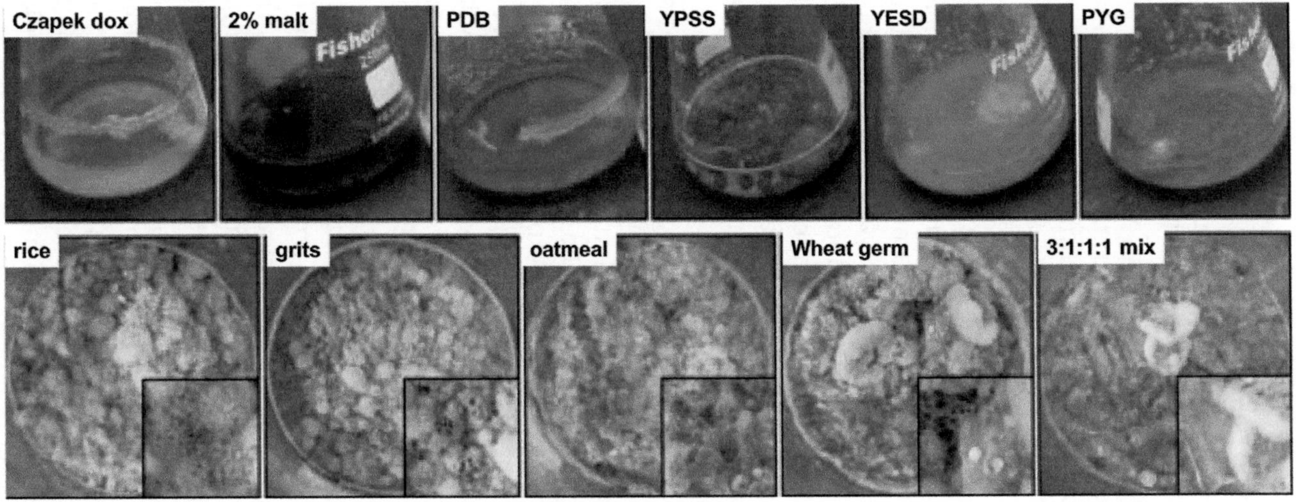

Figure 12.24 Different solid substrata induce different fungal growth and development morphologies and different expression of natural product pathways.
Reprinted and modified from VanderMolen, K. M., H. A. Raja, T. El-Elimat and N. H. Oberlies (2016). "Evaluation of Culture Media for the Production of Secondary Metabolites in a Natural Products Screening Program". *AMB Express* **3**: 71. Original article distributed under CC-BY 2.0 license (https://creativecommons.org/licenses/by/2.0/).

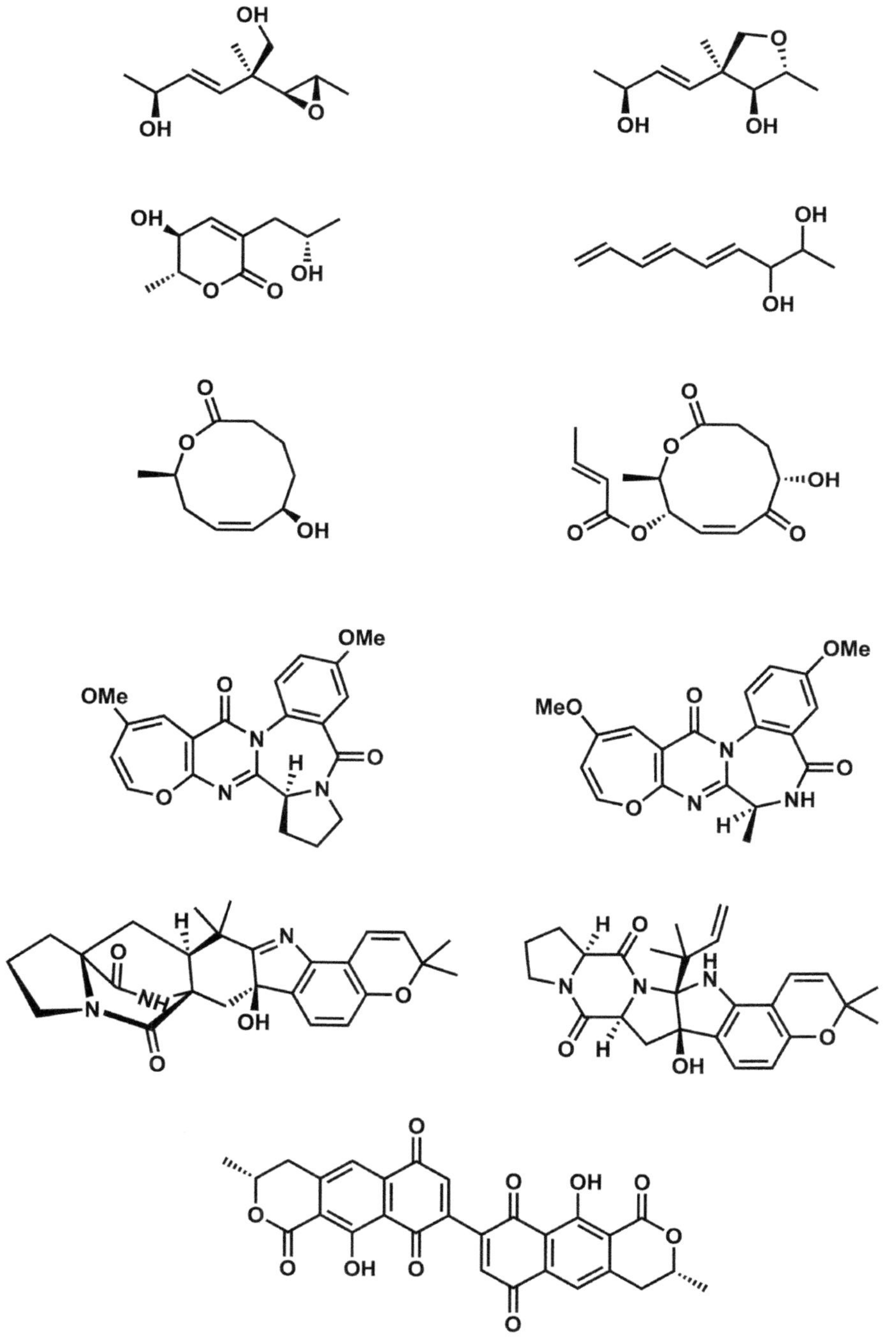

Figure 12.25 New metabolites from *Aspergillus ochraceus* DSM 7428 by alteration of culture conditions.

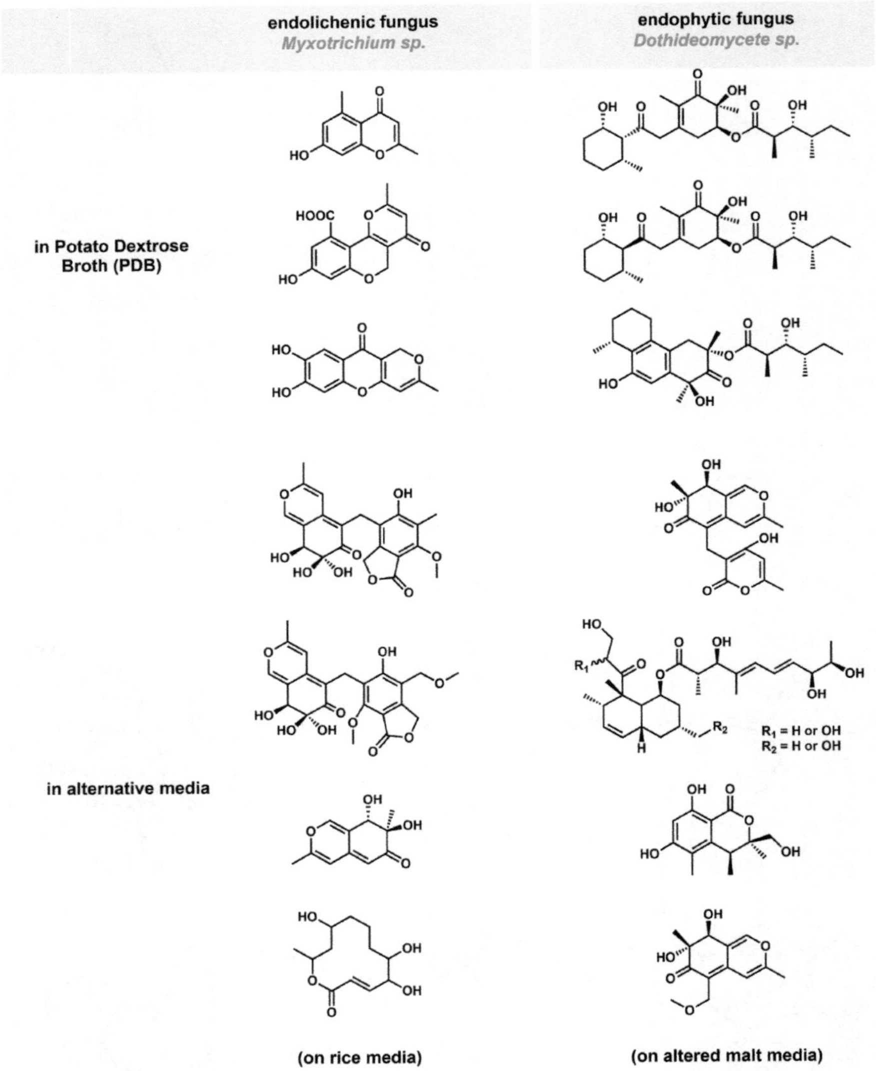

Figure 12.26 The effect of using different media on metabolite profile, using the endolichenic fungus *Myxotrichium* sp. and the endophytic fungus *Dothideomycete* sp. as examples.

In that connection, one of the ways to elicit activation of bacterial or fungal natural product pathways is to cocultivate an organism with another microbe. Figure 12.27 shows that exposure of *Aspergillus nidulans* to *Streptomyces rapamycincus* leads to activation of polyketide pathways to generate the indicated four molecules. Histone H3

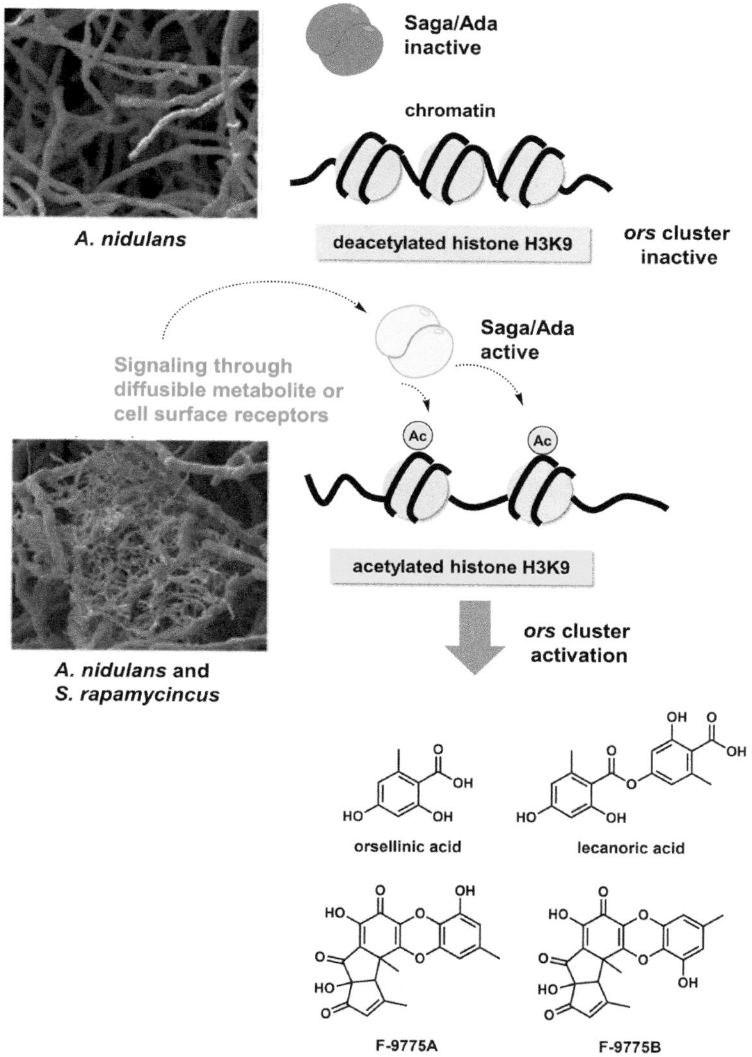

Figure 12.27 Exposure of *Aspergillus fumigatus* to *Streptomyces rapamycincus* leads to four new metabolites that reflect pathway(s) turned on by action of histone acetyltransferase on side chains of lysines in N-terminal tails of histones in chromatin. The scanning electron microscopy (SEM) images of the fungi and fungi with bacteria are reprinted with permission from Nützmann, H.-W., Y. Reyes-Dominguez, K. Scherlach, V. Schroeckh, F. Horn, A. Gacek, J. Schümann, C. Hertweck, J. Strauss and A. A. Brakhage (2011). "Bacteria-induced natural product formation in the fungus *Aspergillus nidulans* requires Saga/Ada-mediated histone acetylation". *Proc. Natl. Acad. Sci. U. S. A.* **108**(34): 14282.

lysine side chain acetylation by the Saga–Ada complex was a requisite signal for the natural product biosynthetic response (Nutzmann, Reyes-Dominguez *et al.* 2011) (Figure 12.27). More generally, stimulator strains are likely, in two-way chemical communication with the responder strains (Figure 12.28), to make and modulate signaling molecules that go back and forth and mutually perturb the chemical inventory of both stimulator and responder strains (Scherlach and Hertweck 2009).

Many examples of new metabolites seen in cocultures have been reported, including acremostatin lipopeptides and aspergicins in mixed fungal cultures (see citations in (Netzker, Fischer *et al.* 2015)) (Figure 12.29). Cultivation of the bacterium *Thalassopia* sp. with the marine fungus *Pestalotia* led to production of the chlorinated polyketide aldehyde pestalone (Figure 12.29) (Cueto, Jensen *et al.* 2001). Analogously, coculture of *Streptomyces bulli* with *A. fumigatus* turned on the fungal production of the diketopiperazine alkaloids noted in Chapter 8, brevianamide F, spirotryprostatin A, 6-methoxytryprostatin B, fumitremorgin B, verruculogen, and other molecules (Rateb, Hallyburton *et al.* 2013).

Vignette 12.2 Peptide Antibiotic Warfare between Staphylococcal Strains in the Human Nasal Cavity.

About 30% of humans carry *Staphylococcus aureus* colonies in the anterior nares, and *S. aureus* is one of the most problematic human pathogens in disease causation (Walsh and Wencewicz, 2016). The anterior nares is typically a nutrient-poor ecological niche, where it is likely that various bacterial species are in competition. To that end, screens of a collection of nasal staphylococcal isolates revealed that *Staphylococcus lugdunensis* IVK28 strain had strong *S. aureus* antigrowth activity. Isolation of a noninhibitory *S. lugdunensis* mutant revealed insertion in a putative nonribosomal peptide synthetase-encoding gene, one of four (lugD, A, B, C) ORFs (Figure 12.V2) (Zipperer, Konnerth *et al.* 2016), implicating the pathway end product as the active *S. aureus* antibiotic.

The LugDABC four ORF assembly line has only five adenylation domains but turns out to encode a heptapeptide termed lugdunin (molecular weight = 783), implicating triple use of the adenylation domain in LugC for activating three L-valine residues, the second one of which is epimerized by the epimerization (E) domain. There are also multiple carrier protein (thiolation) domains (T) in LugC, indicating a noncanonical assembly line organization. LugC ends in a reductase domain, consistent with the expectation that a linear heptapeptidyl aldehyde may be the nascent product (see Chapter 3).

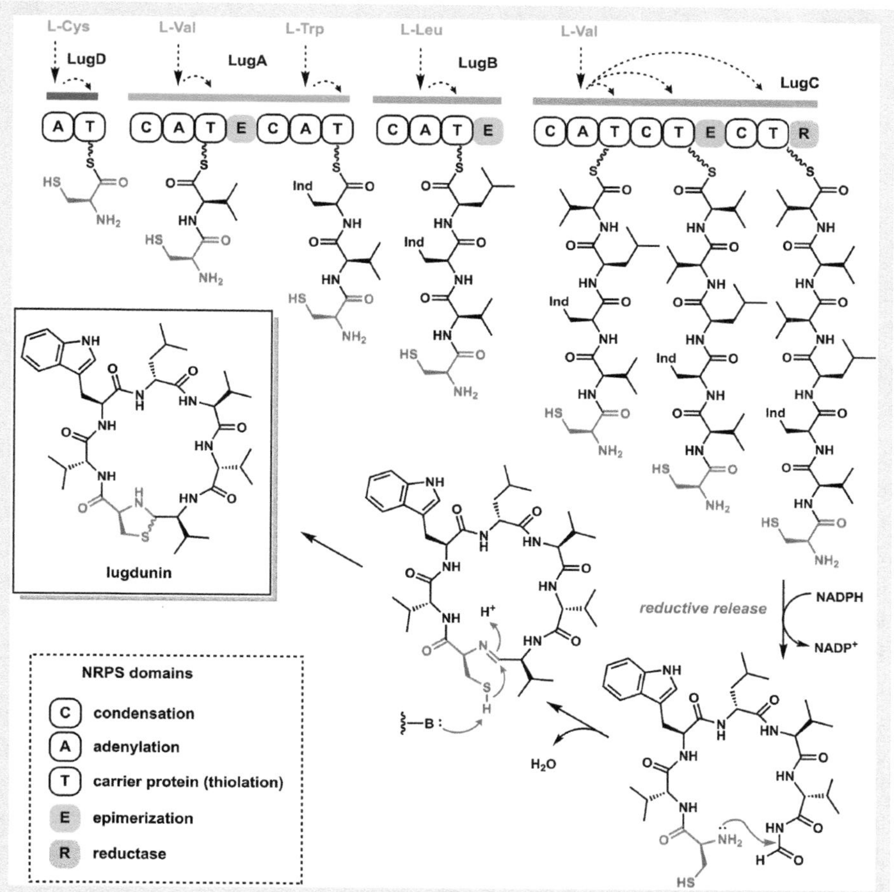

Figure 12.V2 Biosynthesis of lugdunin by *Staphylococcus lugdunensis* IVK28.

On isolation, lugdunin is a cyclic peptide in which the *N*-terminal free amine of Cys_1 is proposed to react with the L-Val_7-aldehyde, followed by addition of the thiolate side chain into the resulting imine to produce the cyclic five-member thiazolidine heterocycle.

Lugdunin was bactericidal against a range of Gram-positive bacteria including pathogenic methicillin resistant *Staphylococcus aureus* (MRSA) strains. No resistance was developed in multiple passages of susceptible *S. aureus* strains. Lugdunin was curative in a mouse skin infection model and lugdunin-producing *S. lugdunensis* cultures outcompeted and killed *S. aureus* cells in cocultures. Analysis of 181 humans indicated that *S. lugdunensis* was present in about 9%, *vs.* 32% who were *S. aureus* positive. In the individuals with *S. lugdunensis* cultures the *S. aureus* carriage rate was reduced five-fold compared to the negative controls.

The killing mode of action of lugdunin remains to be determined (the low resistance rate could mean a step involved in membrane assembly or maintenance and may be reminiscent of action of teixobactin on lipid II) (Ling, Schneider *et al.* 2015). The results raise the prospect that *S. lugdunensis* should be viewed as a probiotic bacterium for protection against *S. aureus* infections. Zipper *et al.* (Zipperer, Konnerth *et al.* 2016) argue that lugdunin is the first in a class of macrocyclic thiazolidine peptide antibiotics and suggest that the human microbiota may be a new source of antibiotics produced under mixed culture challenges.

12.7.10 Real-time Mass Spectrometry of Microbes on Plates and Platforms

One of the new methodologies of particular use for metabolites in cocultures is imaging mass spectrometry by desorption electrospray ionization (DESI) (Fang and Dorrestein 2014) (Hsu and Dorrestein 2015), which allows examination of molecules in transit between organisms by analysis of their spatial distribution in media on culture plates (Figure 12.30) (Esquenazi, Yang *et al.* 2009, Moree, Phelan *et al.* 2012). The large amount of information collected in mass spectrometric analyses allows for network connectivity and a data base that facilitates both dereplication of molecular signals in crude extracts and the use of fragmentation patterns to confirm known molecular structures and related ones, *e.g.* different fatty acyl chains in lipopeptides by nested sets of *m/z* separated by 14 mass units, and for amino acid variants in peptide frameworks by fragment sequencing.

Another recent methodologic advance is an open surface micro-metabolomics platform (Barkal, Theberge *et al.* 2016) that lowers culture volumes, and allows fungal and/or bacterial cultures to be arrayed and metabolites assessed by microfluidic-based LC-MS (Figure 12.31). The micro-metabolomic device was calibrated with *A. fumigatus* and *A. nidulans* strains, allowing rapid detection of 33 natural products. This approach should allow facile characterization of thousands of fungal samples, including systematic approaches to OSMAC studies. In one example, cultures were extracted not only with chloroform or 1-pentanol but also with γ-caprolactone, a nontraditional solvent in natural product isolation. The metabolite asperfuranone was detected only in the caprolactone extractions, a reminder of the limitations in metabolite detection that can be imposed from the very first step of choice of extracting solvent. The device is also

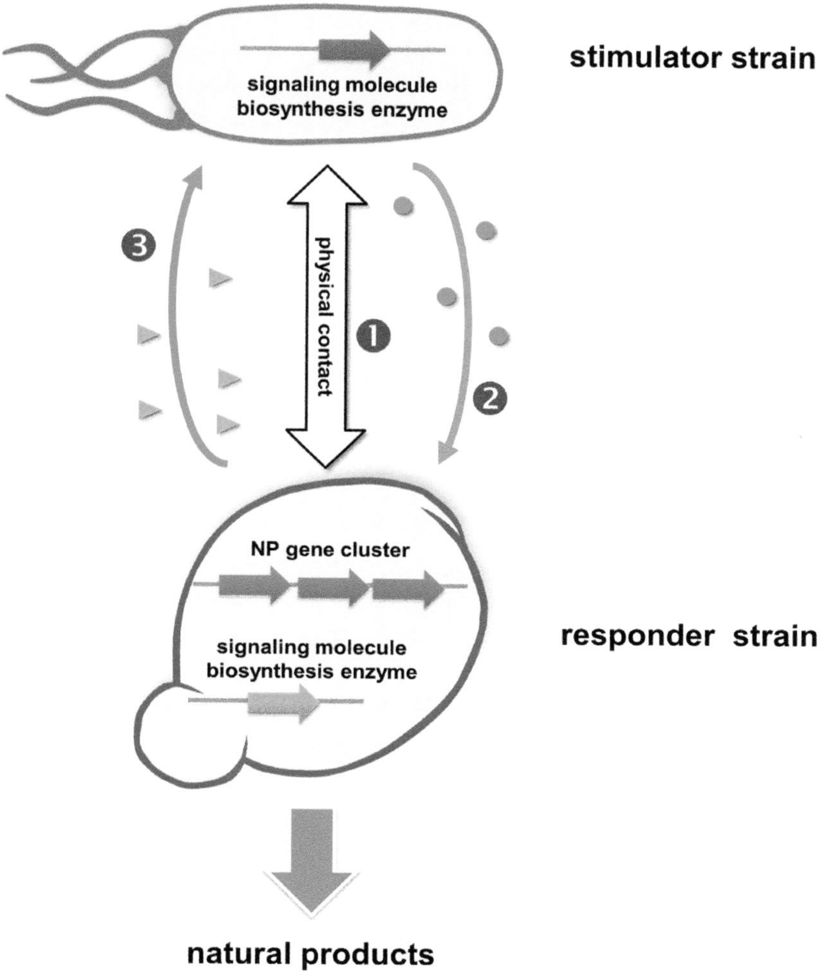

Figure 12.28 Proposed two-way chemical communication between stimulator strains and producer strains. The first route is direct physical contact that leads to intracellular signaling. The second route is the stimulator strain producing a diffusible molecule that can lead to activation of a gene cluster in the responder strain. The third route is a two-way communication in which the responder strain can produce its own signaling molecule that leads to triggering of signaling molecule production from the stimulator strain.
Adapted from Scherlach, K. and C. Hertweck (2009). "Triggering Cryptic Natural Product Biosynthesis in Microorganisms". *Org. Biomol. Chem.* **7**: 1753 with permission from The Royal Society of Chemistry.

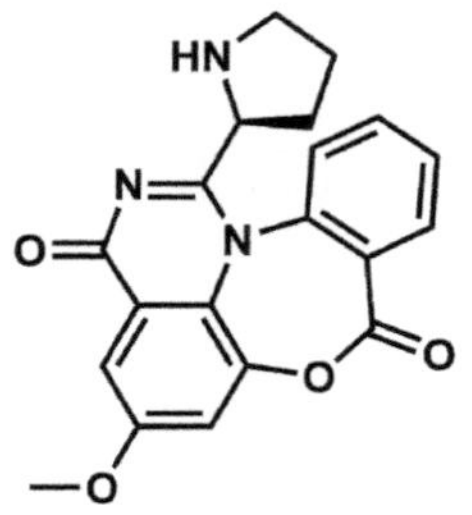

aspergicin

from mixed fermentation of two marine *Aspergilli*

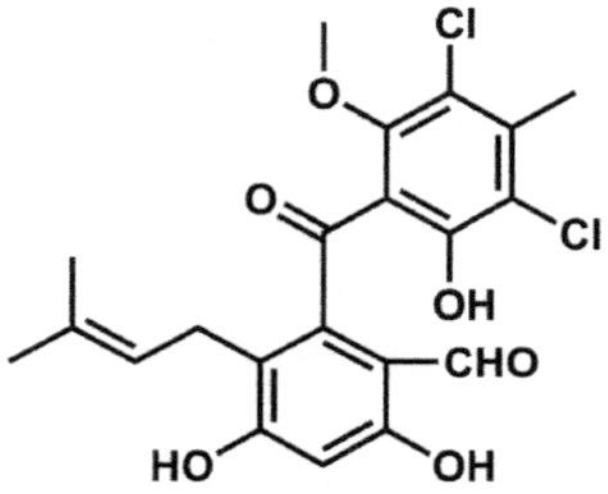

pestalone

produced by *Pestalotia* under bacterial challenge

Figure 12.29 Aspergicin and pestalone: fungal metabolites elicited during coculture.

easily configured to study interkingdom interactions, both in physical contact, as in the *Streptomyces rapamycincus–Aspergillus nidulans* pair noted above, but also where the two organisms are next to each other and only diffusible chemicals interact.

12.7.11 Target Screen Improvements

As a last subsection to this overview chapter on genetic- and molecular biology-independent approaches to natural product production, isolation, and characterization that have defined historical and much of contemporary efforts in the natural products arena, we note a recent phenotypic approach to provide short cuts to identification of host targets of both existing natural products and novel molecules as well. This high image phenotypic screen

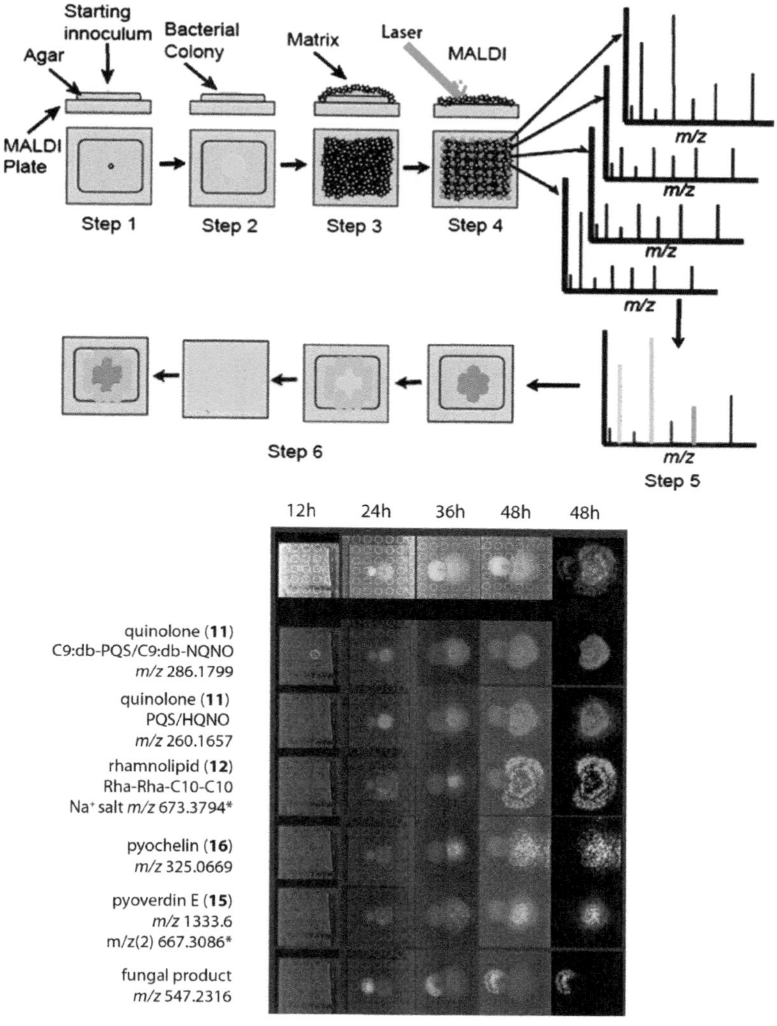

Figure 12.30 Interkingdom metabolic transformations detected by microbial imaging mass spectrometry. On the top is the work flow of the imaging MS technique. Reproduced from Esquenazi, E., Y.-L. Yang, J. Watrous, W. H. Gerwick, P. C. Dorrestein (2009). "Imaging Mass Spectrometry of Natural Products". *Nat. Prod. Rep.* **26**: 1521 with permission from The Royal Society of Chemistry. On the bottom is the MS-based imaging of metabolites produced through the interspecies interaction between *Pseudomonas aeruginosa* and *A. fumigatus*. The partial image is reprinted with permission from Moree, W. J., V. V. Phelan, C.-H. Wu, N. Bandeira, D. S. Cornett, B. M. Duggan and P. C. Dorrestein (2012). "Interkingdom metabolic transformations captured by microbial imaging mass spectrometry". *Proc. Natl. Acad. Sci. U. S. A.* **109**(34), 13811.

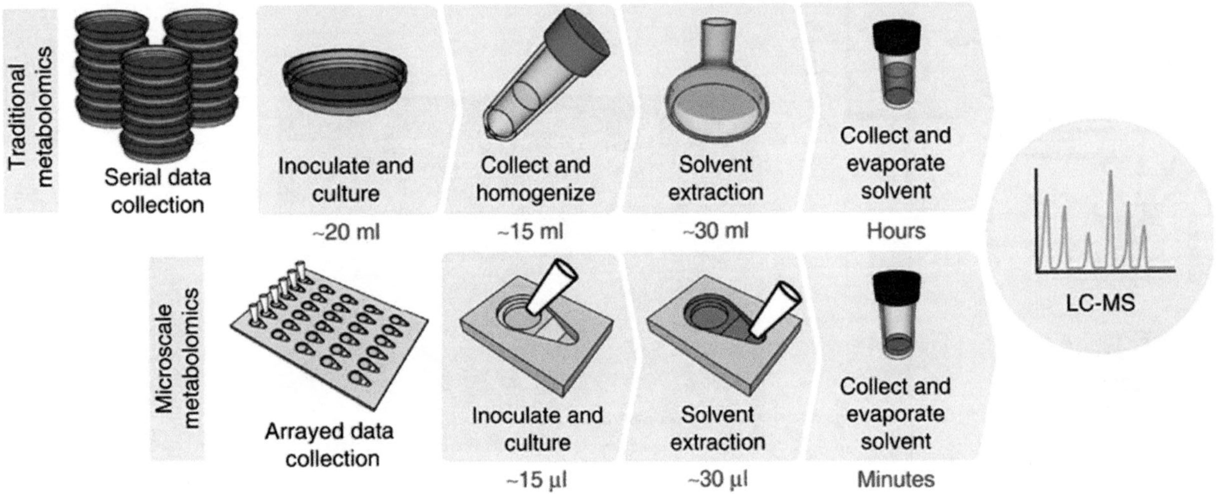

Figure 12.31 Schematic comparison of traditional and microscale metabolomics.
Reprinted from Barkal, L. J., A. B. Theberge, C.-J. Guo, J. Spraker, L. Rappert, J. Berthier, K. A. Brakke, C. C. C. Wang, D. J. Bebbe, N. P. Keller and E. Berthier (2016). "Microbial Metabolomics in Open Microscale Platforms". *Nat. Commun.* **7**: 10610. Original article distributed under CC-BY 4.0 license (https://creative-commons.org/licenses/by/4.0/).

complements an siRNA methodology to compare the effects on knockdown of a given gene expression with a matching phenotype from a natural small molecule on cells. This latter technique is squarely in the gene/messenger RNA-dependent arena of the next chapter.

In an initial study a high content image screen in a cytological platform was conducted for the specific endpoint of mitotic arrest (Ochoa, Bray *et al.* 2015). The approach is to use automated image analysis for alterations in cell morphology that are characteristic of specific target blockade and correlatable with known mechanism of action of a 480-compound reference set of molecules/natural product extracts on cultured Hela cells. In 5300 prefractionated extracts from marine *Actinobacteria*, two compounds were identified, one acting to mediate mitotic arrest, the other as a calcium channel blocker.

Figure 12.32 shows that the mitotic arrest molecule XR334 has a doubly dehydro diketopiperazine scaffold. We have noted two routes to such DKP frameworks in Chapter 9, both NRPS assembly line-directed release of dipeptides and aminoacyl-tRNA cyclodipeptide synthases. In this case it is likely that Phe and Tyr are the building blocks, with *O*-methylation and two dehydrogenation events most likely after DKP formation as the reaction sequence. The second molecule, identified as the calcium channel blocker, is identical to the known molecule nocapyrone L, of polyketide origin. High content imaging of cell-based phenotypes is likely to be ever more central to mechanism of action studies when a molecule, natural or synthetic, has shown sufficient cell-based potency and selectivity to warrant rapid characterization.

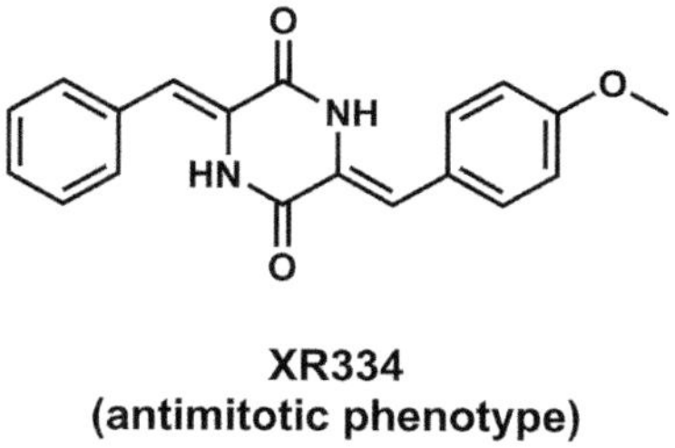

XR334
(antimitotic phenotype)

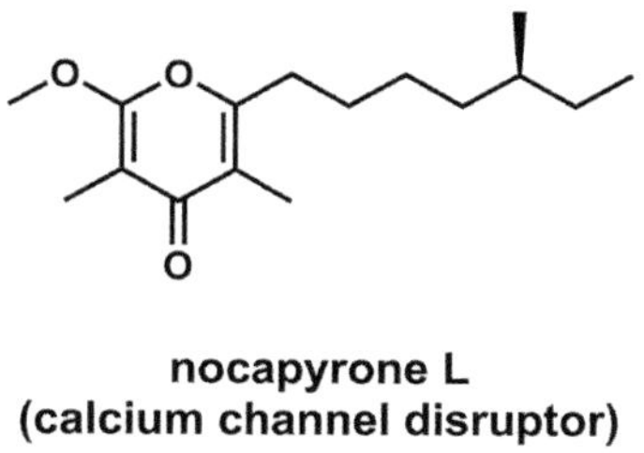

nocapyrone L
(calcium channel disruptor)

Figure 12.32 Two actinobacterial metabolites identified in phenotypic high content imaging screens in HeLa cells.

References

Atanasov, A. G., B. Waltenberger, E. M. Pferschy-Wenzig, T. Linder, C. Wawrosch, P. Uhrin, V. Temml, L. Wang, S. Schwaiger, E. H. Heiss, J. M. Rollinger, D. Schuster, J. M. Breuss, V. Bochkov, M. D. Mihovilovic, B. Kopp, R. Bauer, V. M. Dirsch and H. Stuppner (2015). "Discovery and resupply of pharmacologically active plant-derived natural products: A review". *Biotechnol. Adv.* **33**(8): 1582–1614.

Barkal, L. J., A. B. Theberge, C. J. Guo, J. Spraker, L. Rappert, J. Berthier, K. A. Brakke, C. C. Wang, D. J. Beebe, N. P. Keller and E. Berthier (2016). "Microbial metabolomics in open microscale platforms". *Nat. Commun.* **7**: 10610.

Baumann, S., J. Herrmann, R. Raju, H. Steinmetz, K. I. Mohr, S. Huttel, K. Harmrolfs, M. Stadler and R. Muller (2014). "Cystobactamids: myxobacterial topoisomerase inhibitors exhibiting potent antibacterial activity". *Angew. Chem., Int. Ed.* **53**(52): 14605–14609.

Bister, B., D. Bischoff, M. Strobele, J. Riedlinger, A. Reicke, F. Wolter, A. T. Bull, H. Zahner, H. P. Fiedler and R. D. Sussmuth (2004). "Abyssomicin C-A polycyclic antibiotic from a marine Verrucosispora strain as an inhibitor of the p-aminobenzoic acid/tetrahydrofolate biosynthesis pathway". *Angew. Chem., Int. Ed.* **43**(19): 2574–2576.

Blakemore, P. R. and J. D. White (2002). "Morphine, the Proteus of organic molecules". *Chem. Commun.* **11**: 1159–1168.

Bode, H. B., B. Bethe, R. Hofs and A. Zeeck (2002). "Big effects from small changes: possible ways to explore nature's chemical diversity". *ChemBioChem* **3**(7): 619–627.

Bode, H. B., H. Irschik, S. C. Wenzel, H. Reichenbach, R. Muller and G. Hofle (2003). "The leupyrrins: a structurally unique family of secondary metabolites from the myxobacterium Sorangium cellulosum". *J. Nat. Prod.* **66**(9): 1203–1206.

Bucar, F., A. Wube and M. Schmid (2013). "Natural product isolation-how to get from biological material to pure compounds". *Nat. Prod. Rep.* **30**: 525–545.

Butcher, R. A., F. C. Schroeder, M. A. Fischbach, P. D. Straight, R. Kolter, C. T. Walsh and J. Clardy (2007). "The identification of bacillaene, the product of the PksX megacomplex in Bacillus subtilis". *Proc. Natl. Acad. Sci. U. S. A.* **104**(5): 1506–1509.

Calderone, C. T. (2008). "Isoprenoid-like alkylations in polyketide biosynthesis". *Nat. Prod. Rep.* **25**(5): 845–853.

Calderone, C. T., S. B. Bumpus, N. L. Kelleher, C. T. Walsh and N. A. Magarvey (2008). "A ketoreductase domain in the PksJ protein of the bacillaene assembly line carries out both alpha- and beta-ketone reduction during chain growth". *Proc. Natl. Acad. Sci. U. S. A.* **105**(35): 12809–12814.
Cueto, M., P. R. Jensen, C. Kauffman, W. Fenical, E. Lobkovsky and J. Clardy (2001). "Pestalone, a new antibiotic produced by a marine fungus in response to bacterial challenge". *J. Nat. Prod.* **64**(11): 1444–1446.
Esquenazi, E., Y. L. Yang, J. Watrous, W. H. Gerwick and P. C. Dorrestein (2009). "Imaging mass spectrometry of natural products". *Nat. Prod. Rep.* **26**(12): 1521–1534.
Fang, J. and P. C. Dorrestein (2014). "Emerging mass spectrometry techniques for the direct analysis of microbial colonies". *Curr. Opin. Microbiol.* **19**: 120–129.
Fonin, V. S. and Y. Khorlin (2003). "Preparation of Biologically Transformed Raw Material of Woolly Foxglove (Digitalis lanata Ehrh.) and Isolation of Digoxin Therefrom". *Appl. Biochem. Microbiol.* **39**: 519–523.
Gaudencio, S. P. and F. Pereira (2015). "Dereplication: racing to speed up the natural products discovery process". *Nat. Prod. Rep.* **32**(6): 779–810.
Gordon, J. J., E. Grenfell, E. Knowles, B. J. Legge, R. C. A. Mcallister and T. White (1947). "Methods of Penicillin Production in Submerged Culture on a Pilot-Plant Scale". *J. Gen. Microbiol.* **1**(2): 187–202.
Griffith, R. S. (1981). "Introduction to Vancomycin". *Rev. Infect. Dis.* **3**(Supplement 3): S200–S204.
Griffith, R. S. (1984). "Vancomycin use–an historical review". *J. Antimicrob. Chemother.* **14**(Suppl D): 1–5.
Hamada, T., S. Matsunaga, G. Yano and N. Fusetani (2005). "Polytheonamides A and B, highly cytotoxic, linear polypeptides with unprecedented structural features, from the marine sponge, Theonella swinhoei". *J. Am. Chem. Soc.* **127**(1): 110–118.
Hardoim, P. R., L. S. van Overbeek, G. Berg, A. M. Pirttila, S. Compant, A. Campisano, M. Doring and A. Sessitsch (2015). "The Hidden World within Plants: Ecological and Evolutionary Considerations for Defining Functioning of Microbial Endophytes". *Microbiol. Mol. Biol. Rev.* **79**(3): 293–320.
Hewage, R. T., T. Aree, C. Mahidol, S. Ruchirawat and P. Kittakoop (2014). "One strain-many compounds (OSMAC) method for production of polyketides, azaphilones, and an isochromanone using

the endophytic fungus Dothideomycete sp''. *Phytochemistry* **108:** 87–94.

Hsu, C. C. and P. C. Dorrestein (2015). ''Visualizing life with ambient mass spectrometry''. *Curr. Opin. Biotechnol.* **31:** 24–34.

Jensen, P. R., P. G. Williams, D. C. Oh, L. Zeigler and W. Fenical (2007). ''Species-specific secondary metabolite production in marine actinomycetes of the genus Salinispora''. *Appl. Environ. Microbiol.* **73**(4): 1146–1152.

Kapoor, L. (1995). *Opium Poppy: Botany, Chemistry, and Pharmacology*. United States, CRC Press.

Kaufman, T. S. and E. A. Ruveda (2005). ''The quest for quinine: those who won the battles and those who won the war''. *Angew. Chem., Int. Ed.* **44**(6): 854–885.

Keller, S., H. S. Schadt, I. Ortel and R. D. Sussmuth (2007). ''Action of atrop-abyssomicin C as an inhibitor of 4-amino-4-deoxychorismate synthase PabB''. *Angew. Chem., Int. Ed.* **46**(43): 8284–8286.

Konishi, M., H. Ohkuma, T. Tsuno, T. Oki, G. Van Duyne and J. Clardy (1990). ''Crystal and molecular structure of dynemicin A: a novel 1,5-diyn-3-ene antitumor antibiotic''. *J. Am. Chem. Soc.* **112:** 3715–3716.

Le Couteur, P. and J. Burreson (2004). *Napoleon's Buttons*. Tarcher Perigree/Penguin Group.

Lin, Y. Y., M. Risk, S. M. Ray, D. Van Engen, J. Clardy, J. Golik and K. Nakanishi (1981). ''Isolation and structure of brevetoxin B from the ''red tide'' dinoflagellate Ptychodiscus brevis (Gymnodinium breve)''. *J. Am. Chem. Soc.* **103:** 6773–6775.

Ling, L. L., T. Schneider, A. J. Peoples, A. L. Spoering, I. Engels, B. P. Conlon, A. Mueller, T. F. Schaberle, D. E. Hughes, S. Epstein, M. Jones, L. Lazarides, V. A. Steadman, D. R. Cohen, C. R. Felix, K. A. Fetterman, W. P. Millett, A. G. Nitti, A. M. Zullo, C. Chen and K. Lewis (2015). ''A new antibiotic kills pathogens without detectable resistance''. *Nature* **517**(7535): 455–459.

Magarvey, N. A., M. Ehling-Schulz and C. T. Walsh (2006). ''Characterization of the cereulide NRPS alpha-hydroxy acid specifying modules: activation of alpha-keto acids and chiral reduction on the assembly line''. *J. Am. Chem. Soc.* **128**(33): 10698–10699.

Malik, S., H. M. Mirjalili, E. Moyano, J. Palazon and M. Nonfil (2011). ''Production of the anticancer drug taxol in Taxus baccata suspension cultures: A review''. *Process Biochem.* **46:** 23–34.

Moree, W. J., V. V. Phelan, C. H. Wu, N. Bandeira, D. S. Cornett, B. M. Duggan and P. C. Dorrestein (2012). ''Interkingdom metabolic transformations captured by microbial imaging mass spectrometry". *Proc. Natl. Acad. Sci. U. S. A.* **109**(34): 13811–13816.

Moudi, M., R. Go, C. Y. Yien and M. Nazre (2013). "Vinca alkaloids". *Int. J. Prev. Med.* **4**(11): 1231–1235.

Netzker, T., J. Fischer, J. Weber, D. J. Mattern, C. C. Konig, V. Valiante, V. Schroeckh and A. A. Brakhage (2015). "Microbial communication leading to the activation of silent fungal secondary metabolite gene clusters". *Front Microbiol.* **6**: 299.

Nicolaou, K. C. and T. Montagnon (2008). *Molecules that Changed the World*. Wiley.

Nutzmann, H. W., Y. Reyes-Dominguez, K. Scherlach, V. Schroeckh, F. Horn, A. Gacek, J. Schumann, C. Hertweck, J. Strauss and A. A. Brakhage (2011). "Bacteria-induced natural product formation in the fungus Aspergillus nidulans requires Saga/Ada-mediated histone acetylation". *Proc. Natl. Acad. Sci. U. S. A.* **108**(34): 14282–14287.

Ochoa, J. L., W. M. Bray, R. S. Lokey and R. G. Linington (2015). "Phenotype-Guided Natural Products Discovery Using Cytological Profiling". *J. Nat. Prod.* **78**(9): 2242–2248.

Oh, D. C., M. Poulsen, C. R. Currie and J. Clardy (2009). "Dentigerumycin: a bacterial mediator of an ant-fungus symbiosis". *Nat. Chem. Biol.* **5**(6): 391–393.

Oh, D. C., J. J. Scott, C. R. Currie and J. Clardy (2009). "Mycangimycin, a polyene peroxide from a mutualist Streptomyces sp". *Org. Lett.* **11**(3): 633–636.

Petit, G. R., C. L. Herald, D. L. Doubek, D. L. Herald, E. Arnold and J. Clardy (1982). "Isolation and Structure of Bryostatin 1". *J. Am. Chem. Soc.* **104**: 6846–6848.

Phillips, J. W., M. A. Goetz, S. K. Smith, D. L. Zink, J. Polishook, R. Onishi, S. Salowe, J. Wiltsie, J. Allocco, J. Sigmund, K. Dorso, S. Lee, S. Skwish, M. de la Cruz, J. Martin, F. Vicente, O. Genilloud, J. Lu, R. E. Painter, K. Young, K. Overbye, R. G. Donald and S. B. Singh (2011). "Discovery of kibdelomycin, a potent new class of bacterial type II topoisomerase inhibitor by chemical-genetic profiling in Staphylococcus aureus". *Chem. Biol.* **18**(8): 955–965.

Poulsen, M., D. P. Erhardt, D. J. Molinaro, T. L. Lin and C. R. Currie (2007). "Antagonistic bacterial interactions help shape host-symbiont dynamics within the fungus-growing ant-microbe mutualism". *PLoS One* **2**(9): e960.

Poulsen, M., D. C. Oh, J. Clardy and C. R. Currie (2011). "Chemical analyses of wasp-associated streptomyces bacteria reveal a prolific potential for natural products discovery". *PLoS One* **6**(2): e16763.

Rateb, M., I. Hallyburton, W. Houssen, A. Bull, M. Goodfellow, R. Santhanam, M. Jaspars and R. Ebel (2013). "Induction of diverse secondary metabolites in Aspergillus fumigatus by microbial coculture". *RSC Adv.* **3**: 14444–14450.

Riedlinger, J., A. Reicke, H. Zahner, B. Krismer, A. T. Bull, L. A. Maldonado, A. C. Ward, M. Goodfellow, B. Bister, D. Bischoff, R. D. Sussmuth and H. P. Fiedler (2004). "Abyssomicins, inhibitors of the para-aminobenzoic acid pathway produced by the marine Verrucosispora strain AB-18-032". *J. Antibiot.* **57**(4): 271–279.

Roth, K. and S. Streiler (2013). From Pharmacy to the PUb: A Bark Conquers the World: Part 1. Chem Views, May 7, 2013, Chemie in unserer Zeit/Wiley-VCH.

Sarker, S. D., Z. Latif and A. I. Gray, eds. (2005). *Methods in Biotechnology, Natural Product Isolation.* Totowa, NJ, Humana Press.

Schantz, E. J., V. E. Ghazarossian, H. K. Schnoes, F. M. Strong, J. P. Springer, J. O. Pezzanite and J. Clardy (1975). "Letter: The structure of saxitoxin". *J. Am. Chem. Soc.* **97**(5): 1238.

Scherlach, K. and C. Hertweck (2009). "Triggering cryptic natural product biosynthesis in microorganisms". *Org. Biomol. Chem.* 7(9): 1753–1760.

Schmidt, E. W., J. T. Nelson, D. A. Rasko, S. Sudek, J. A. Eisen, M. G. Haygood and J. Ravel (2005). "Patellamide A and C biosynthesis by a microcin-like pathway in Prochloron didemni, the cyanobacterial symbiont of Lissoclinum patella". *Proc. Natl. Acad. Sci. U. S. A.* **102**(20): 7315–7320.

Schueffler, A. and T. Anke (2014). "Fungal natural products in research and development". *Nat. Prod. Rep.* **31**(10): 1425–1448.

Serturner, F. (1817). "Über das Morphium, eine neue salzfähige Grundlage, und die Mekonsäure, als Hauptbestandteile des Opiums". *Ann. Phys.* **25**: 56–90.

Shimizu, Y., H. Bando, H. N. Chou, G. Van Duyne and J. Clardy (1986). "Absolute configuration of BT-B". *J. Chem. Soc. Chem. Commun.* 1656.

Sticher, O. (2008). "Natural product isolation". *Nat. Prod. Rep.* **25**(3): 517–554.

Stork, G., D. Niu, R. A. Fujimoto, E. R. Koft, J. M. Balkovec, J. R. Tata and G. R. Dake (2001). "The first stereoselective total synthesis of quinine". *J. Am. Chem. Soc.* **123**(14): 3239–3242.

Strobel, G., B. Daisy, U. Castillo and J. Harper (2004). "Natural products from endophytic microorganisms". *J. Nat. Prod.* **67**(2): 257–268.

Survase, S. A., L. D. Kagliwal, U. S. Annapure and R. S. Singhal (2011). "Cyclosporin A–a review on fermentative production, downstream processing and pharmacological applications". *Biotechnol. Adv.* **29**(4): 418–435.

Tachibana, K., P. Scheuer, Y. Tsukitani, H. Kikuchi, D. Vanengen, J. Clardy, Y. Gopchand and F. Schmitz (1981). "Okadaic acid, a cytotoxic polyether from2 marine sponges of the genus halichondra". *J. Am. Chem. Soc.* **103**: 2469–2471.

Teufel, R., A. Miyanaga, Q. Michaudel, F. Stull, G. Louie, J. P. Noel, P. S. Baran, B. Palfey and B. S. Moore (2013). "Flavin-mediated dual oxidation controls an enzymatic Favorskii-type rearrangement". *Nature* **503**(7477): 552–556.

Vandermolen, K. M., H. A. Raja, T. El-Elimat and N. H. Oberlies (2013). "Evaluation of culture media for the production of secondary metabolites in a natural products screening program". *AMB Express* **3**(1): 71.

Walsh, C. T., S. W. Haynes, B. D. Ames, X. Gao and Y. Tang (2013). "Short pathways to complexity generation: fungal peptidyl alkaloid multicyclic scaffolds from anthranilate building blocks". *ACS Chem. Biol.* **8**(7): 1366–1382.

Walsh, C. T. and T. Wencewicz (2016). *Antibiotics Challeneges, Mechanisms, Opportunities.* Washington DC, ASM Press.

Wolf, P. L. (1994). "If clinical chemistry had existed then". *Clin. Chem.* **40**(2): 328–335.

Yuan, C., Y. H. Guo, H. Y. Wang, X. J. Ma, T. Jiang, J. L. Zhao, Z. M. Zou and G. Ding (2016). "Allelopathic Polyketides from an Endolichenic Fungus Myxotrichum SP. by Using OSMAC Strategy". *Sci. Rep.* **6**: 19350.

Zipperer, A., M. C. Konnerth, C. Laux, A. Berscheid, D. Janek, C. Weidenmaier, M. Burian, N. A. Schilling, C. Slavetinsky, M. Marschal, M. Willmann, H. Kalbacher, B. Schittek, H. Brotz-Oesterhelt, S. Grond, A. Peschel and B. Krismer (2016). "Human commensals producing a novel antibiotic impair pathogen colonization". *Nature* **535**(7613): 511–516.

Bioinformatic analysis plays an important role in the post-genomic era of natural product biosynthesis.

13 Natural Products in the Post Genomic Era

13.1 Introduction

It has been more than 30 years since the 1984 publication of the first cloning of the genes encoding a biosynthetic pathway for a natural product with antibiotic activity - actinorhodin from *Streptomyces coelicolor* (Malpartida and Hopwood 1984). Malpartida and Hopwood isolated a contiguous segment of DNA from *S. coelicolor*, and moved it *via* a plasmid into *Streptomyces parvulus*, which now produced the dimeric polyketide antibiotic with its characteristic deep blue–purple color (Figure 13.1). The responsible PKS encoded in Act II-ORF2 is a type II enzyme that makes a octaketonic chain tethered to the ACP. Cyclizing and aromatizing release creates the tricyclic monomer framework in dihydrokalafungin. Redox adjustment precedes phenoxy radical formations and regiospecific dimerization to actinorhodin. Among other conclusions, this study established that the necessary biosynthetic genes were clustered on a contiguous stretch of *S. coelicolor* DNA, and could be packaged in a plasmid and transferred to a heterologous, albeit related, host.

In 1990 and 1991, a pair of papers in *Nature* from the groups of Leonard Katz at Abbott (Donadio, Staver *et al.* 1991) and Peter Leadlay (Cortes, Haydock *et al.* 1990) at the University of Cambridge were revelatory in reporting that the genes required to assemble the 14-member macrolactone core of erythromycin were contiguous in *Saccharopolyspora erythraea* and encoded a set of three proteins containing six fatty acid synthase-like modules. This was the original

Natural Product Biosynthesis: Chemical Logic and Enzymatic Machinery
By Christopher T. Walsh and Yi Tang

Published by the Royal Society of Chemistry, www.rsc.org

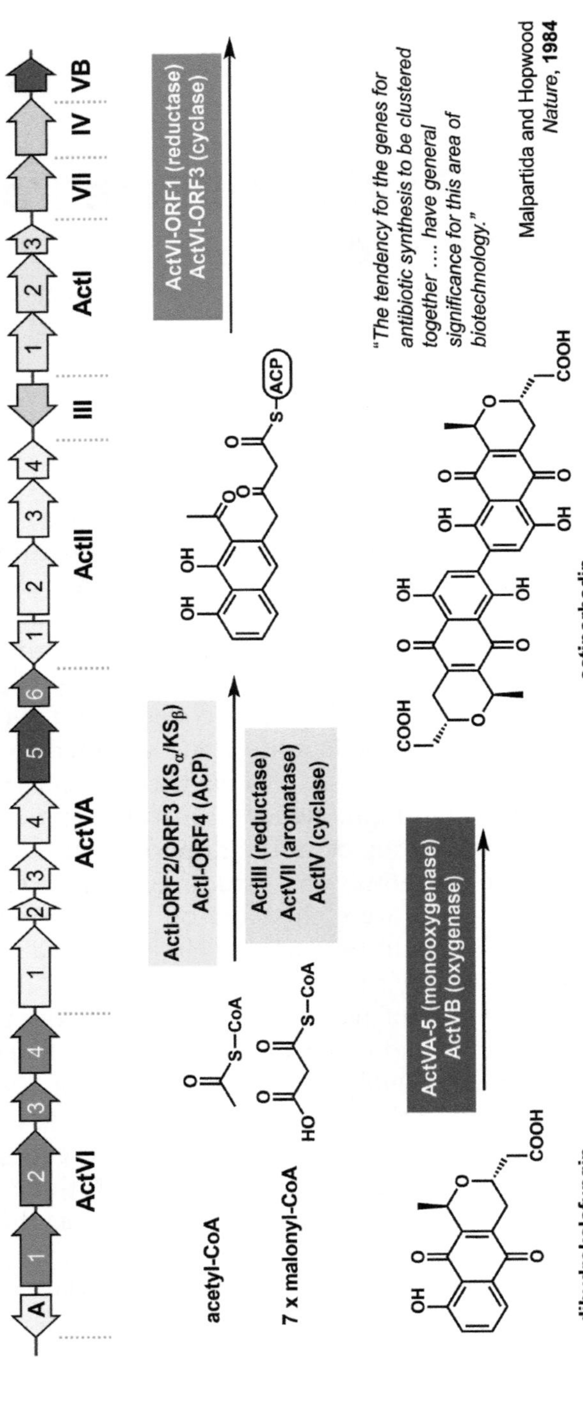

Figure 13.1 The actinorhodin gene cluster published in 1984 represents the first clustered biosynthetic pathway discovered.

example of a Type I polyketide synthase (Figure 13.2) with colinearity of the modules for predicting the events in chain growth and processing within each module. In the intervening 25 years the flood gates of genome sequencing of both prokaryotic and eukaryotic organisms have opened and led to thousands of full and/or draft genome sequences.

One interesting organizational twist is found on chromosome VIII of the human pathogen *Aspergillus fumigatus* strain Af293, where a fungal secondary metabolite supercluster is found (Wiemann, Guo *et al.* 2013). The genes for three natural products, fumitremorgin, fumagillin, and pseurotin (Figure 13.3), are intertwined. These are, respectively, indole alkaloid, polyketide, and nonribosomal peptide–polyketide hybrid pathways and may reflect an economy of regulated simultaneous production.

Bacterial and fungal biosynthetic genes for complete pathways from primary metabolic building blocks to secondary metabolic end products are generally tightly clustered, perhaps reflecting both efficient regulation for turning on a compete pathway and a history of horizontal gene transfers. It had been proposed that the comparable genomic organization in plants is less tight and that at least some of the genes required for full pathway maturation of plant secondary metabolites are dispersed throughout the genomes. To the extent that is the general case it makes natural product biosynthetic pathway definition and reconstruction much more difficult in plants.

However, Nutzmann and Osbourn (Nutzmann and Osbourn 2014) have compiled a list of several examples that indicate substantial gene clustering in specialized metabolic pathways. As shown in Figure 13.4 these include terpenes such as thalianol, cyanogenic glycosides, the steroidal glycosides α-tomatine and avenacin, and the phenylpropanoid noscapine, spanning a set of distinct natural product scaffolds. A more detailed analysis of triterpene gene clustering can be found in the original paper (Thimmappa, Geisler *et al.* 2014). Such clustering allows investigators to zero in on the core of a biosynthetic pathway but some of the necessary genes are typically dispersed in other chromosomal locations. In contrast to the clustering of some of the core genes for triterpenes and benzylisoquinoline alkaloids in plant genomes, the flavonoid, carotenoid, and glucosinolate biosynthetic genes are not clustered (Nutzmann, Huang *et al.* 2016). The patterns of gene clusters and possible utility to the plants of such organizations are discussed in a recent review (Nutzmann, Huang *et al.* 2016).

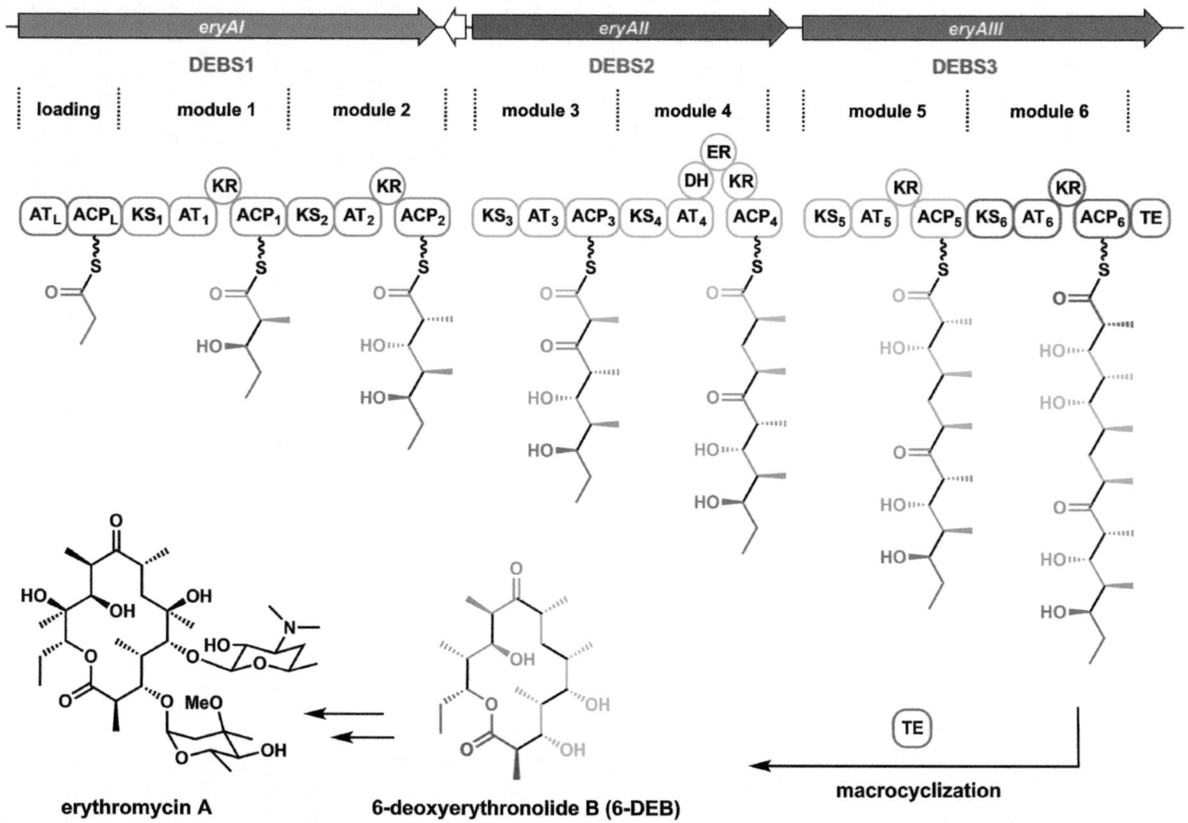

Figure 13.2 The three modular polyketide synthases are found in the *ery* cluster in the genome of *Saccharopolyspora erythraea*.

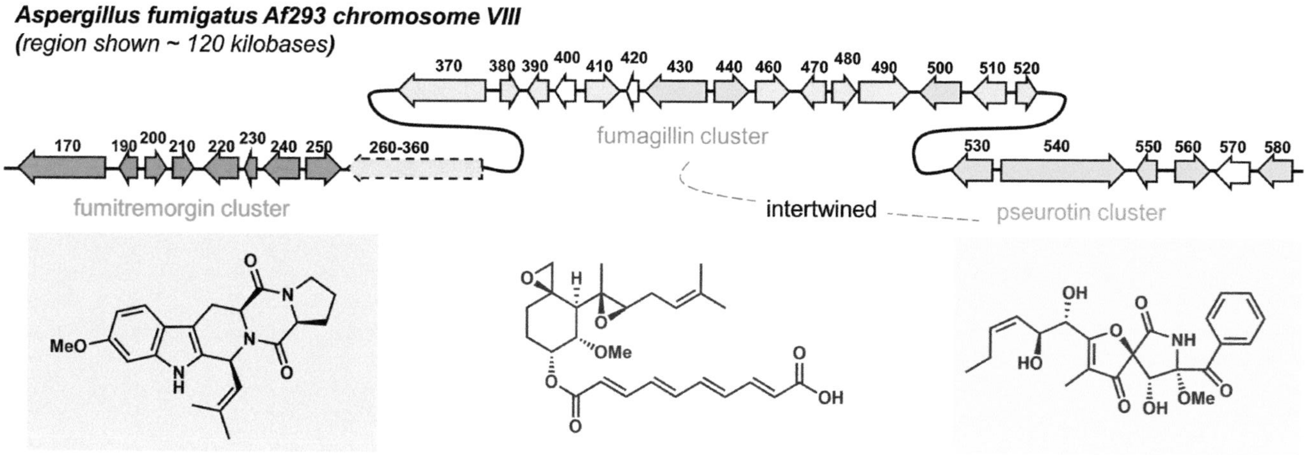

Figure 13.3 The genes for three natural products, fumitremorgin, fumagillin, and pseurotin, are intertwined on the chromosome of *Aspergillus fumigatus* Af293.

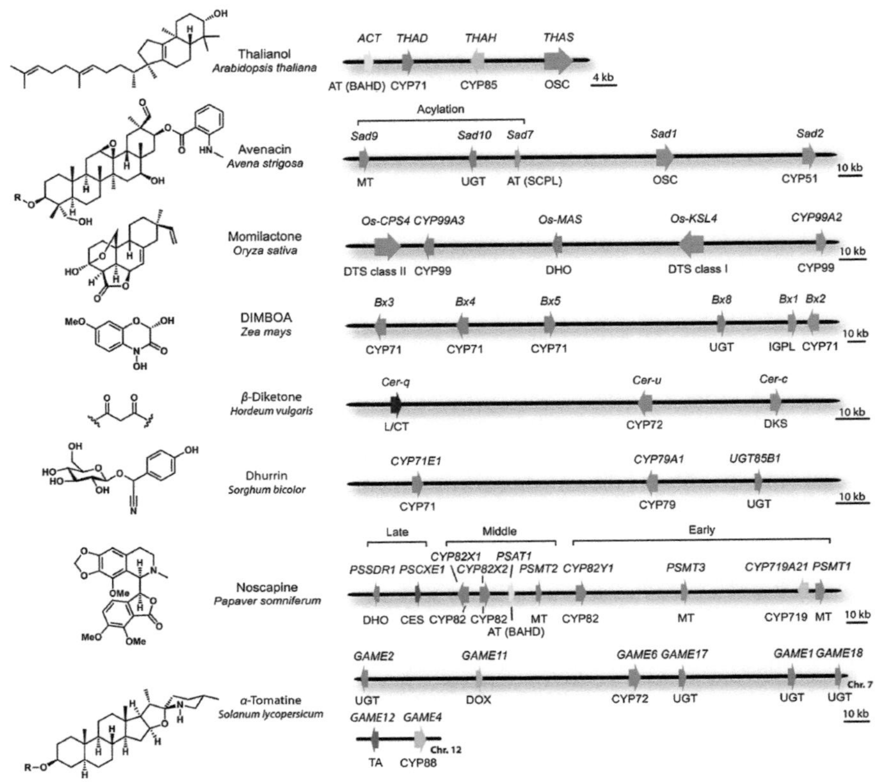

Figure 13.4 Clustered biosynthetic pathways in plants. Reprinted with permission from John Wiley and Sons. Nützmann, H.-W., A. Huang and A. Osbourn (2016). "Plant metabolic clusters – from genetics to genomics". *New Phytol.* **211**. © 2016 The Authors. New Phytologist © 2016 New Phytologist Trust.

13.2 Bioinformatic and Computational Predictions of Biosynthetic Gene Clusters

One of the persistent surprises from the early genome sequences of fungi and bacteria with large genomes, such as actinomycetes and myxobacteria, but not small ones, such as *E. coli*, was the prediction of many PKs and NRPs and hybrid NRPS–PKS clusters that outstripped the known biosynthetic capacity of the producer strains. As one example, the industrial strain of *Streptomyces avermitilis*, the producer of avermectins, was known to make three to five natural products. The genome sequence indicated 22 nonribosomal peptide synthetase clusters and 30 polyketide synthase clusters (Omura, Ikeda *et al.* 2001). The situation was comparable when the *Aspergillus fumigatus* and *Aspergillus nidulans* genomes were sequenced, with a predicted PKS, NRPS, NRPS–PKS capacity of 30–50 natural products (Galagan, Calvo *et al.* 2005, Nierman, Pain *et al.* 2005, Hoffmeister and Keller 2007, Inglis, Binkley *et al.* 2013).

The general expectation currently is that, under standard culture conditions, bacteria and fungi may express 10% of their biosynthetic capacity. This is the genetic underpinning of the OSMAC protocols discussed in Chapter 12. Tweaking conditions of growth, signal input, chromatin modifications, *etc.* has a reasonable chance of turning on silent pathways. Figure 13.5 shows in diagrammatic form several inputs to fungal cells and their intracellular signaling protein partners that can affect transcription of silent PKS and NRPS gene clusters (Bode, Bethe *et al.* 2002, Brakhage 2013). Figure 13.6 elaborates on the probes that are used in OSMAC and related efforts to activate silent biosynthetic gene clusters, including overexpression of transcription factors in and around the gene clusters, promoter exchange for specific induction, exchange or modulation of global regulators, chromatin modifiers that affect histone methylation and acetylation

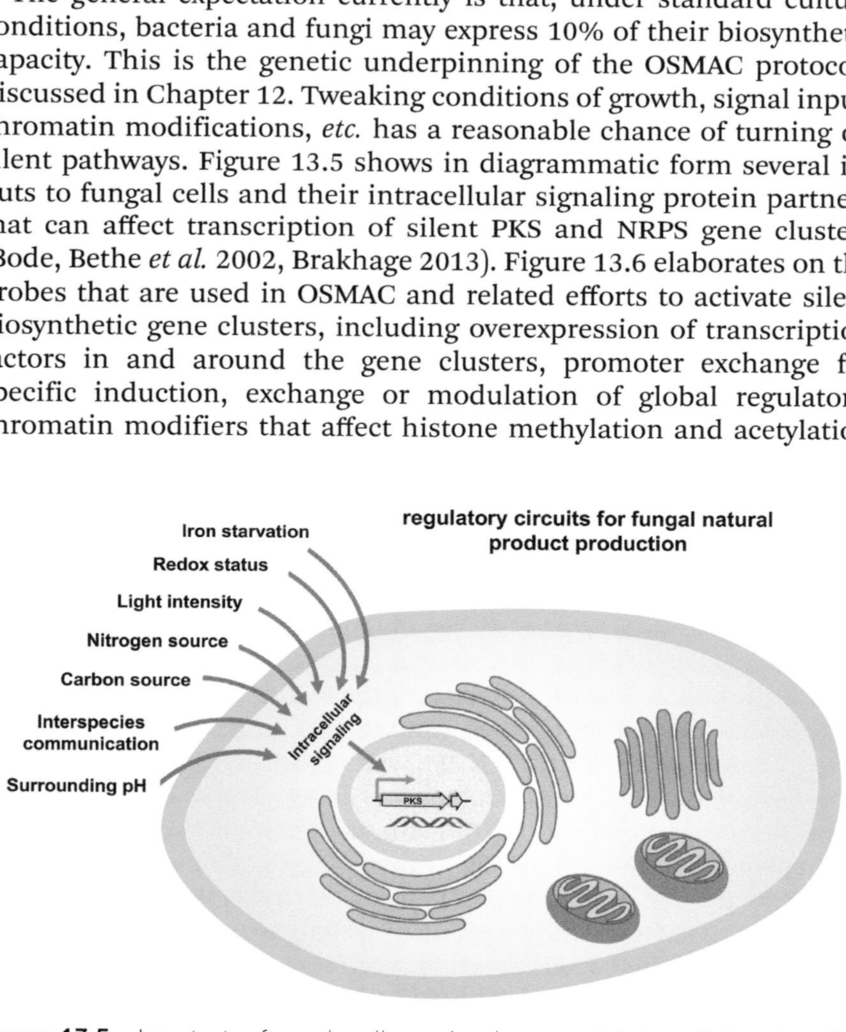

Figure 13.5 Inputs to fungal cells and subsequent intracellular signaling can affect transcription of natural product gene clusters. Adapted from Brakhage, A. A. (2013). "Regulation of fungal secondary metabolism". *Nat. Rev. Microbiol.* **11**: 21–32.

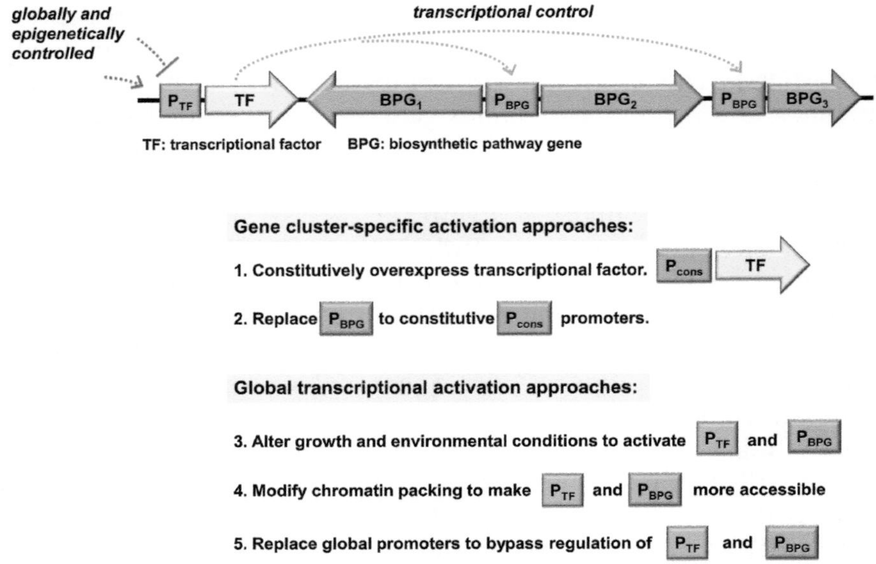

Figure 13.6 Methods that can be used to activate silent genes in microorganisms; strategies for fungi outlined.
Adapted from Brakhage. A. A. (2013). "Regulation of fungal secondary metabolism". *Nat. Rev. Microbiol.* **11**: 21–32.

states, and the array of physiologic conditions, noted in passing in the Chapter 12 discussion of OSMAC (Bode, Bethe *et al.* 2002, Brakhage 2013).

A recent computational analysis of 2700 microbial genomes, using algorithms such as cluster finder and antiSMASH 1, 2 (Blin, Medema *et al.* 2013) and 3 (Weber, Blin *et al.* 2015) which look for NRPS, PKS, and terpene biosynthetic genes, among others, turned up 3300 predicted gene clusters for PKS, NRPS, and hybrid NRPS–PKS metabolites, essentially all (>90%) unknown (Wang, Fewer *et al.* 2014). A more recent compilation of 581 fungal genomes computes 4984 PKS clusters, 2983 NRPS, 550 dimethylallyltransferases, and 336 geranylgeranyl-PP synthases (Li, Tsai *et al.* 2016) (Figure 13.7). At the anticipated near-term future point of 20 000 sequenced genomes of fungi and potential bacterial natural product producers, there may be up to 1 million predicted biosynthetic gene clusters that encode >99% unknown secondary metabolites.

With so many predicted clusters to activate and sample, several questions arise. One set involves ways of prioritizing clusters to investigate. One criterion for prioritization could be predicted structural

~97% of Fungal biosynthetic gene clusters are uncharacterized		
Type of pathway	Characterized	Total
Polyketides	127	4984
Nonribosomal peptides	81	2983
Alkaloids	44	550
Diterpenes	25	336
Total	277 (3.1%)	8853

Figure 13.7 Predicted number of PKS, NRPS, and PKS–NRPS hybrid gene clusters from different kingdoms of life.

novelty. Another might be to isolate all predicted glycopeptides of the vancomycin and teicoplanin class, to see which scaffold variations gave optimal properties for a given therapeutic application, or as starting point for semisynthesis of next generation antibiotics against newly emerging pathogens. Many other criteria could be applied, including structural class choices, *e.g.* more terpenes, or more fungal peptidyl alkaloids utilizing *ortho*-aminobenzoate as building block, and so on, limited only by the creativity and objectives of investigators.

Pursuit of structural novelty as an enrichment criterion might devolve into the categories of "known unknowns" *vs.* "unknown unknowns". Known unknowns would fall into the class of predicted structural types, for example polyketides, or isoprenoids, or indole terpene natural product classes. The core open reading frames (ORFs) for these classes are known and form the bases for the predictive bioinformatic algorithms. Expression of such silent gene clusters is likely to give known structural framework cores. The separation from known molecules may come in the suite of dedicated tailoring enzymes, and that is an opportunity for metabolic engineering.

Finding the "unknown unknown" molecule categories is both more difficult and potentially an opening to new regions of natural product chemical space. In such searches one may need to ignore/avoid the categories of ORFs that define the core regions of the known major classes of natural products and search instead for clusters of ORFs that are not making recognizable primary or secondary molecules. The kinds of redox enzymes we noted in Chapters 9 and 10 and most recently in Chapter 12 may be clues to unusual chemical transformations, but whether these become predictive rules will depend on experimental findings. This latter category is subsumed in the question, *a priori* unknowable, of how many kinds of new small molecule natural product frameworks are yet to be discovered. They may lie at the intersection of novel enzyme transformations and nonenzymatic chemistry (*e.g.* spontaneous Diels–Alder or Claisen or Cope rearrangements) that occur once a molecular architecture has been established.

13.3 The Phosphonate Class of Natural Products: Can Genomics Define the Complete Set?

One of the potential powers of high throughput microbial genome sequencing could be the exploration of the complete set of natural product structural variants in a given molecular class. The class would have to be small enough such that a few tens of thousands of genome acquisitions would allow *approach to saturation* with regard to new compounds, allowing calculation of essentially the complete array of molecules that contemporary microbes make.

The natural products with direct carbon–phosphorus linkages, phosphonates and phosphinates (Figure 13.8), are bacterial metabolites where the $-PO_3$ or $-PO_2$ groups act as carboxylate mimics and serve as antimetabolites that soil actinomycetes produce and secrete against their neighbors. The great preponderance of phosphorus-containing metabolites have C–O–P bonds (most commonly phosphate ester linkages), not the direct C–P bonds in this natural product class (Metcalf and van der Donk 2009, Peck and van der Donk 2013, Walsh and Wencewicz 2016).

The gate keeper enzyme moving flux from abundant phosphate esters to rare phosphonates is PepM, interconverting the primary

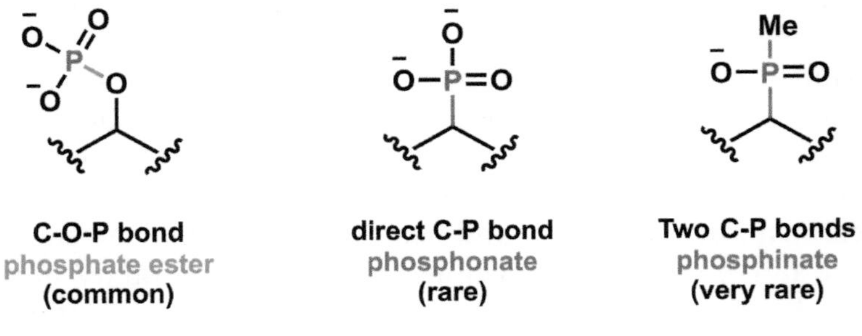

Figure 13.8 Naturally occurring C–P containing molecules. Approximately 80 phosphonates and phosphinates are known.

glycolytic intermediate phosphoenolpyruvate (PEP) and phosphonopyruvate (Figure 13.9). The equilibrium favors PEP, so flux into the C–P pathways is enabled by the next enzyme which decarboxylates phosphonopyruvate nonoxidatively to the corresponding aldehyde in a typical thiamin diphosphate-dependent reaction. The aldehyde can be transaminated to aminoethylphosphonate, found as a phospholipid head group in organisms such as tetrahymena, or reduced to the hydroxyethylphosphonate (HEP).

As also depicted in Figure 13.9, HEP can be subsequently methylated, by a radical SAM- and cobalt-dependent enzyme, and then closed to the epoxide-bearing antibiotic fosfomycin (a clinically approved inhibitor of the first committed step in bacterial cell wall peptidoglycan assembly). Both of these enzymatic steps reveal rare biological chemistry. The *C*-methylation involves both a 5′-deoxyadenosine radical initiator and a subsequent methyl cobalt intermediate for the $C–CH_3$ bond-forming step. The last enzyme in the fosfomycin pathway is an Fe(II)-utilizing enzyme that reduces hydrogen peroxide (rather than molecular oxygen) to water as it makes a high valent Fe(IV)=O. This potent oxidant is capable of carrying out two one-electron oxidations on the C–H bond where the new O–C bond will form. The epoxide closure step is featured as capture of a cation (Figure 13.10). Thus, this hidden corner of natural product biosynthesis reveals some novel and unusual chemistry.

William Metcalf at Illinois has led a consortium of investigators over two decades to explore the universe of biologically generated phosphonates (Metcalf and van der Donk 2009). They have utilized the *pepM* gene as a marker for the first committed step from phosphate ester to carbon–phosphorus metabolites and have reported screening of 10 000 actinomycetes (Ju, Gao *et al.* 2015). The result has been the detection of 278 strains with the *pepM* gene, signifying the potential to make phosphonate and phosphinate natural product metabolites (about a 3% abundance). Bioinformatic clustering suggested 64 subgroups and ORFs likely to make some 55 new C–P products. Investigation of five of those strains revealed 11 new phosphonate metabolites. Figure 13.11 shows an extrapolation from the current 278 strains, by gene cluster (red) or amino acid content (blue) of unique phosphonate natural product gene clusters. The authors estimate that actinomycetes have the capacity to utilize 125 distinct pathways to phosphonates and phosphinates. The extrapolation is from the 10 000 actinomycetes screened to a projected set of 40 000 such strains that would be needed to capture all the structural variation in this relatively compact natural product family.

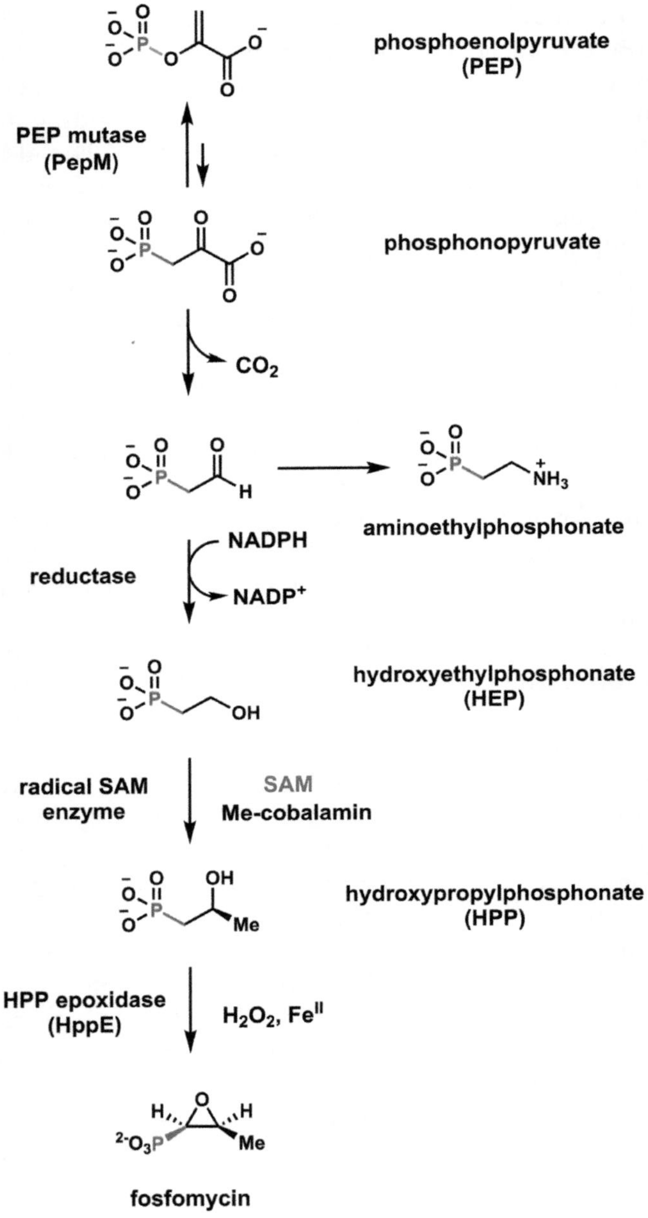

Figure 13.9 PepM as gatekeeper to phosphonate metabolites including the antibiotic fosfomycin.

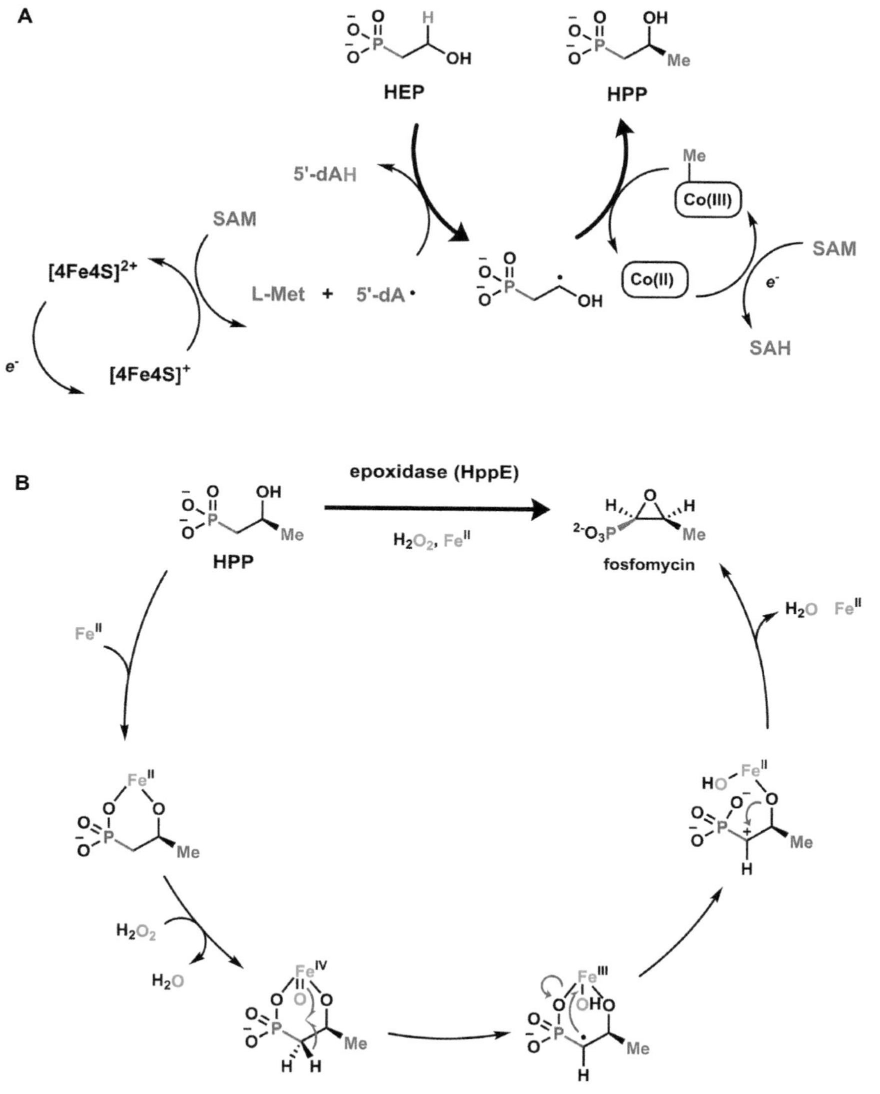

Figure 13.10 The last two enzymes in the fosfomycin pathway catalyze unusual chemical transformations. (A) The methyltransferase that converts HEP to HPP is a radical SAM family cobalamin-dependent enzyme; (B) the epoxidase HppE is a noncofactor iron-dependent enzyme that uses H_2O_2 as the oxidant in converting HPP to fosfomycin.

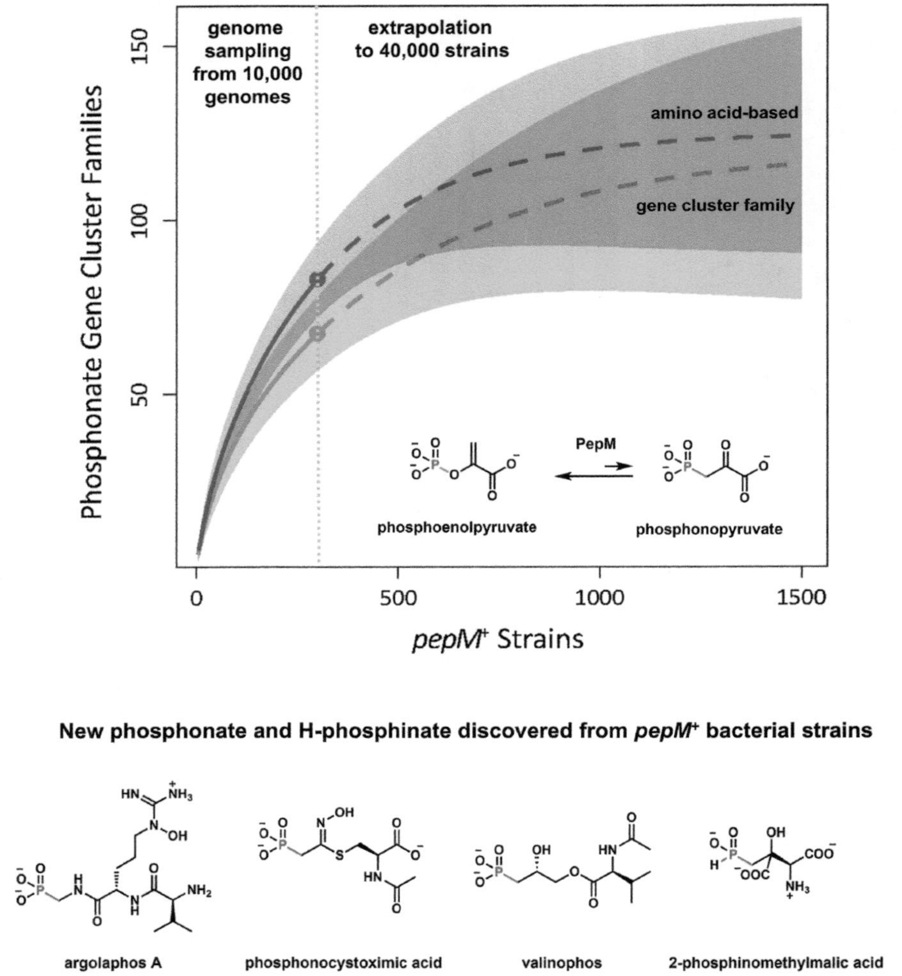

Figure 13.11 The potential of discovering new phosphonate and phosphinate natural products from bacterial species using the phosphoenolpyruvate mutase (PepM) gene sequence as an indicator. Four new metabolites were isolated from *pepM*+ strains.
The plot on the left is adapted with permission from Ju, K.-S., J. Gao, J. R. Doroghazi, K.-K. A. Wang, C. J. Thibodeaux, S. Li, E. Metzger, J. Fudala, J. Su, J. K. Zhang, J. Lee, J. P. Cioni, B. S. Evans, R. Hirota, D. P. Labeda, W. A. van der Donk and W. W. Metcalf (2015). "Discovery of phosphonic acid natural products by mining the genomes of 10,000 actinomycetes". *Proc. Natl. Acad. Sci. U. S. A.* **112**(39): 12175–12180.

This is an intriguing application of genomic information and algorithms based on the logic of a single gate keeper gene/enzyme that builds the crucial direct C–P bond that defines this natural product class. It gives a perspective on how one might analogously predict the extant biological universes of other natural product subclasses. The larger the natural product subclass the correspondingly greater the amount of genome sequence data required to extrapolate an asymptotic limit for the diversity in a natural product family.

13.4 Overview of Heterologous Expression Systems

Figure 13.12 shows a simple logic diagram for two options for expression of biosynthetic gene clusters identified by bioinformatic predictions from genomic mining *via* DNA sequencing (Kim, Charusanti *et al.* 2016). Sometimes the native host is available, sometimes not, as in metagenomic and/or environmental DNA samples. If the native host, for example a fungal or bacterial strain, can be reproducibly cultivated and the biosynthetic genes expressed under achievable growth conditions, then one may decide it is worthwhile to bring to bear the systems and synthetic biology tools to optimize production of the encoded natural product in that context. Alternatively, at several points in the decision tree one may opt instead for biosynthetic gene cluster (and tailoring enzymes) expression in one or more heterologous hosts. If successful detection of product ensues, then the systems metabolic engineering efforts can be undertaken for optimization of production.

The next sections will illustrate various choices for specific gene clusters, in large part driven by the origin of the gene clusters (Figure 13.13). One notable plant example involves the identification and collection of six genes from the mayapple lignan pathway for the conversion of pinoresinol into the advanced phenylpropanoid scaffold of 4-deoxy-epipodophyllotoxin. The heterologous system chosen was the tobacco plant *Nicotiana benthamiana*, using *Agrobacterium* as a transient expression vector (Lau and Sattely 2015). In three other examples, genes for significant parts of alkaloid pathways, for thebaine and hydrocodone, for strictosidine, and for tabersonine to vindoline, *Saccharomyces cerevisiae* was the preferred heterologous host. This choice reflects the vast knowledge of

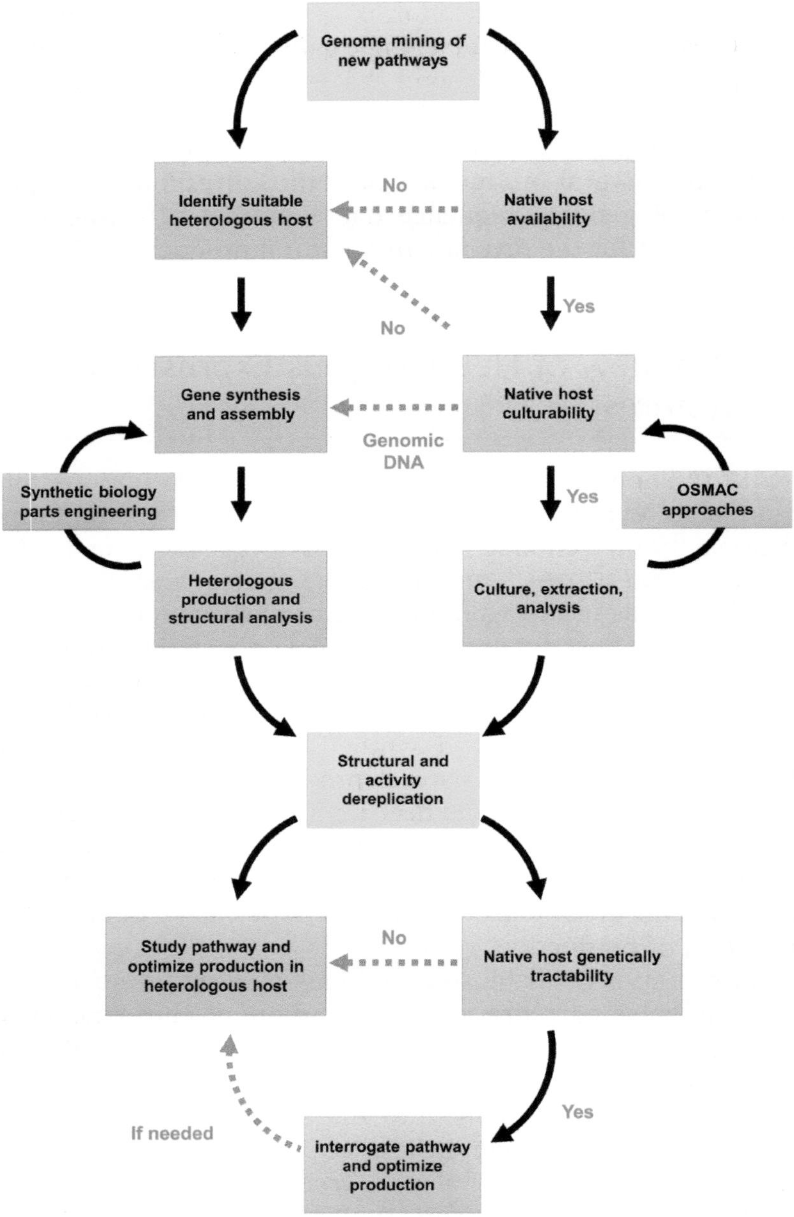

Figure 13.12 A logic diagram for host choice options in expressing biosynthetic gene clusters identified by bioinformatic predictions from genomic mining *via* DNA sequencing.
Adapted from Kim, H. U., P. Charusanti, S. Y. Lee and T. Weber (2016). "Metabolic engineering with systems biology tools to optimize production of prokaryotic secondary metabolites". *Nat. Prod. Rep.* **33**: 933 with permission from The Royal Society of Chemistry.

Native Hosts (Biosynthetic origin)	Related heterologous hosts	Model heterologous hosts
Plants	Tobacco plant (*Nicotiana benthamiana*)	*Saccharomyces cerevisiae*
Fungi	*Aspergillus nidulans* *Aspergillus oryzae*	*Saccharomyces cerevisiae*
Bacteria	*Streptomyces lividans* *Streptomyces albus*	*Escherichia coli*

Figure 13.13 The various choices for specific gene clusters are in large part driven by the origin of the gene clusters.

genetics, vectors, and potential for scale up of fermentations offered by yeast.

For fungal gene cluster expressions, yeast has also been a popular workhorse for heterologous expression, but strains of *A. nidulans* with several large gene clusters deleted offer the prospect of simplified analytical patterns in metabolite detection in crude extracts and less opportunity for competition for PKS and NRPS building blocks during heterologous biosynthetic gene expression (Chiang, Oakley *et al.* 2013). For bacterial expression of many streptomycete and other actinomycete biosynthetic gene clusters, strains of streptomycetes have been useful for compatible codon usage, promoters, and the likely availability of building block metabolites. On the other hand the deep knowledge of *E. coli* genetics and the wealth of molecular biology reagents have made *E. coli* a system that has been tried early and often.

13.5 Selected Examples of Pathway Reconstitutions

13.5.1 Phenylpropanoid: Pinoresinol to 4-Deoxyepipodophyllotoxin in *Nicotiana benthamiana*

We noted the biosynthesis of the natural product podophyllotoxin from the mayapple *Podophyllum hexandrum* in Chapter 7 as an example of the metabolic progression of the lignan pinoresinol into this pentacyclic end product. The scaffold has been of special interest because the antitumor drug etoposide derives semisynthetically from the podophyllotoxin framework.

The recent elucidation of the full six genes that encode the enzymes that catalyze the six steps from pinoresinol to 4′-desmethyl-epipodophyllotoxin was established in the absence of physical contiguity of these mayapple genes (Lau and Sattely 2015). Two lines of reasoning were productive. One was the chemical expectation that oxidative changes as the scaffold matured might be catalyzed by iron-based oxygenases while methylations would be mediated by SAM-dependent methyltransferases. The second was that transcriptome analysis during the period when mayapple produced podophyllotoxin would yield candidates for the missing genes. Thus, P450-encoding genes, nonheme iron oxygenase encoding genes, and methyltransferases that were transcriptionally active in the right time period were chosen.

Mayapple is not a well developed host for molecular biology and genetic manipulations so candidate genes were transferred combinatorially by *Agrobacterium* infections of *N. benthamiana* plants. The plant extracts were subsequently interrogated by high resolution, high sensitivity mass spectroscopic (MS) analysis to look for intermediates along the podophyllotoxin pathway. In this manner, four *O*-methyltransferase, 12 cytochromes P450, and two nonheme iron oxygenase candidates were coexpressed with one or more of the pathway intermediates infiltrated as substrate. As shown in Figure 13.14, 10 genes allow 4′-desmethylepipodophyllotoxin formation from pinoresinol and six from the more advanced intermediate 5′-desmethoxy-yatein.

Thus, this strategy identifies the missing genes for a medicinally important scaffold, offers promise of dispensing with mayapple plants as ultimate starting material, and offers the chance for scaffold analoging by feeding alternate intermediates and/or protein engineering for altered specificity. It is worth explicit note that etoposide has the epipodophyllotoxin configuration, so the tobacco expression results yield the desired framework for ready semisynthetic finishing. The pairing of transcriptionally active genes with insights into what kinds of tailoring enzymes are likely to be involved in biosynthetic pathway steps looks to be a powerful method for identification of genes in plant pathways that are not clustered all together.

13.5.2 Alkaloid Expression in Yeast: Tyrosine to Thebaine and Hydrocodone in Yeast

Several examples of reconstitution of parts of plant alkaloid pathways in yeast have been published over the last five years. A particularly

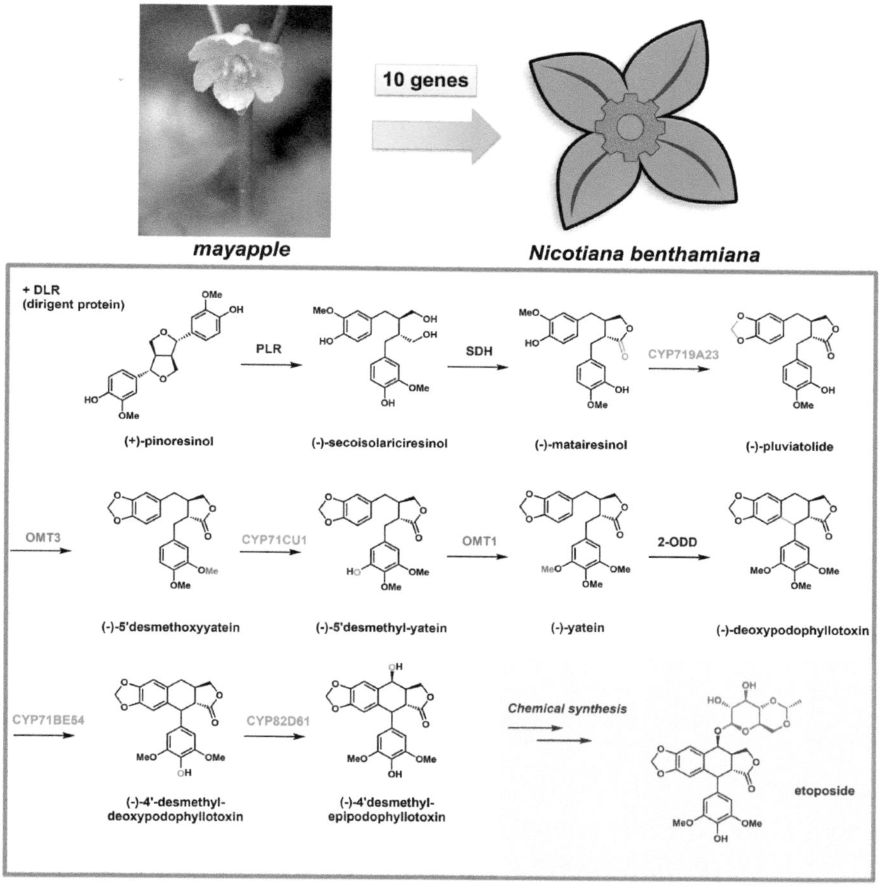

Figure 13.14 Introduction of 10 mayapple genes allows 4′-desmethyl-epipodophyllotoxin formation from pinoresinol. The discovery of six new genes was needed to start from the more advanced intermediate 5′-desmethoxyyatein.

notable example is the reconstitution of the morphine alkaloid pathway up to thebaine and hydrocodone, which is three steps away from morphine (Figure 13.15). To get to thebaine required 21 enzymes collected from a diverse set of biological sources, including the discovery of an isomerase for equilibration of *S*- and *R*-reticuline, as the *R*-isomer but not the *S*-isomer is on pathway (Galanie, Thodey *et al.* 2015). To proceed on to hydrocodone required another two genes, for a total of 23 enzymes encoded functionally in the hydrocodone-producing yeast. Several technical problems had to be

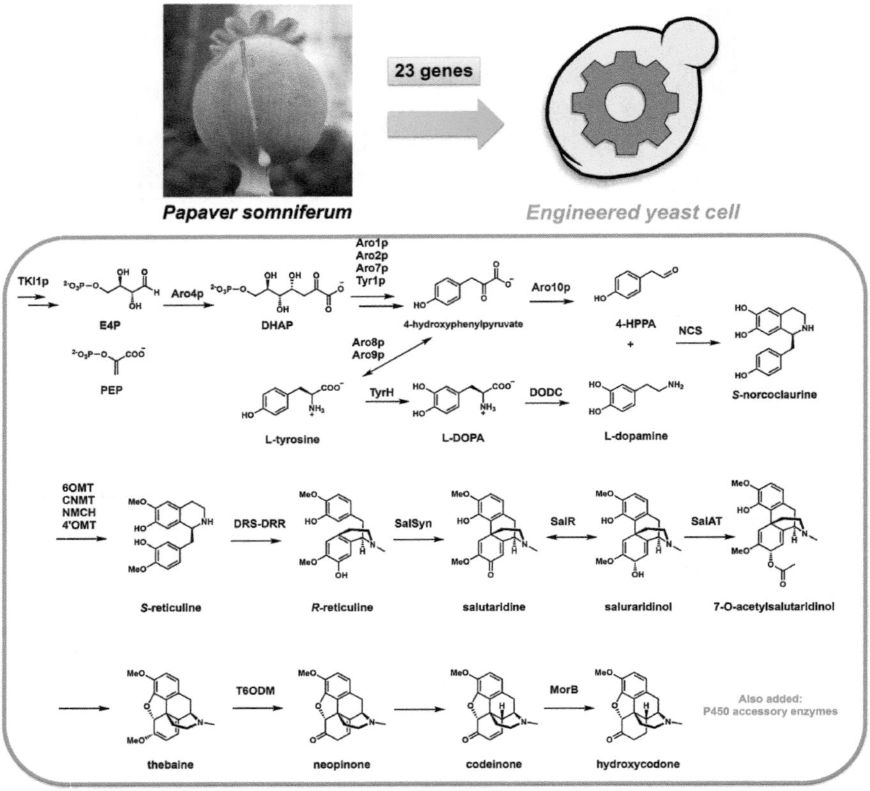

Figure 13.15 The reconstitution of the morphine alkaloid pathway in yeast up to the baine and hydrocodone, which is three steps away from morphine.

surmounted including integration of the genes into yeast chromosome without deleterious effects on growth or alkaloid production, recycling of cofactor machinery, and the protein folding capacity of a chimeric protein.

In the end the yield was very low, ~6 μg of thebaine or 0.3 μg of hydrocodone per liter. The authors note this yield would need to be raised almost 10^7 fold to get to a commercially viable level. So at present it stands as a proof of principle that a dramatically complex 21- to 23-gene pathway can be reconstituted in yeast and sets the stage for multifactorial approaches to yield enhancements to see how robust production of the opioid alkaloids in *S. cerevisiae* could become.

The thebaine/hydrocodone yeast production example builds on extensive infrastructural experimental work by the Smolke group and

Tools for controlling gene expression in *Saccharomyces cerevisiae*			
Introducing & maintaining DNA	plasmid copy number	chromosomal integration	artificial yeast chromosome
Transcriptional regulation	constitutive promoters	inducible promoter	promoter library
Posttranscriptional & posttranslational regulation	RNA degradation rates	bicistronic expression	protein degradation rates
Spatial regulation	protein fusion	scaffold anchoring	organelle localization

Figure 13.16 Tools for controlling gene expression and protein localization in yeast.
Adapted from Siddiqui, M. S., K. Thodey, I. Trenchard and C. D. Smolke (2012). "Advancing secondary metabolite biosynthesis in yeast with synthetic biology tools". *FEMS Yeast Res.* **12**(2): 144–1710. © 2012 Federation of European Microbiological Societies. Published by Elsevier B.V.

others in vector design, chromosome integration methods, tools for regulating specific gene transcription, posttranscriptional and posttranslational control of RNA stability, posttranslational modification of proteins, and finally spatial regulation to send particular proteins to cell compartments where they can function in their intended pathway roles (Figure 13.16).

13.5.3 Alkaloid Expression in Yeast: 10 *Catharanthus roseus* Genes Produce Strictosidine in Yeast

A second and a third example of alkaloid pathway reconstitution in yeast, in these instances partial reconstitution, are at the front end and back end of vinca alkaloid metabolism. Figure 13.17 reminds us that strictosidine is a key common intermediate to >1000 complex downstream alkaloid scaffolds including strychnine, quinine, and vincristine.

The O'Connor group at John Innes Center have reported the assembly of 14 known terpene and indole alkaloid pathway genes from *Catharanthus roseus* (the vinca alkaloid producing plant) encoding the requisite enzymes to produce strictosidine from the primary isoprenyl-PP building block isomers (Brown, Clastre *et al.* 2015). Strategic gene deletions were also constructed in the yeast strain to increase on-pathway availability of SAM and NADPH. Partner

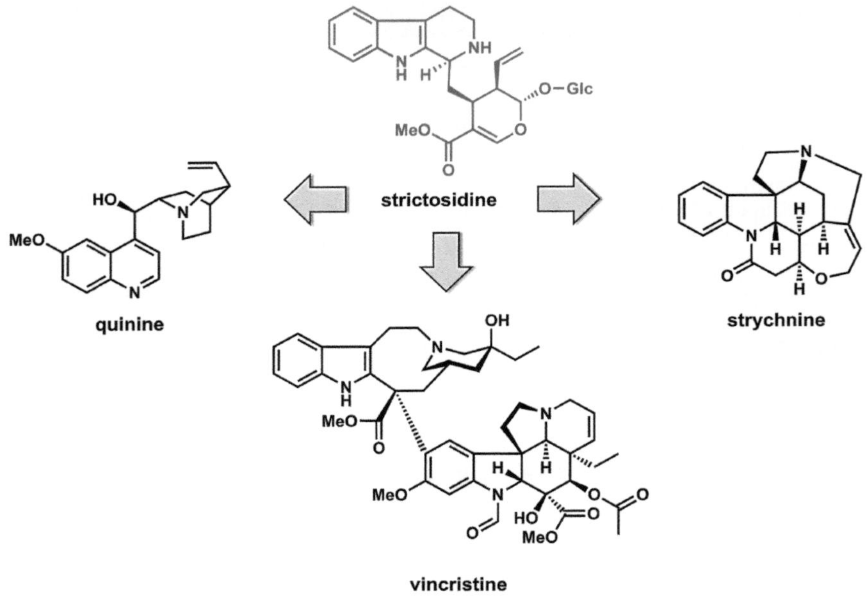

Figure 13.17 Strictosidine is a key common intermediate to >1000 complex downstream alkaloid scaffolds including strychnine, quinine, and vincristine.

proteins for cytochrome P450 reduction were also engineered to increase flux through the pathway. Monitoring of the buildup of pathway intermediates gave insights into what downstream enzymatic steps may be limiting and these were addressed *ad seriatim*. In the end the best producing strain, with 15 plant-derived genes, three yeast gene deletions, one animal-derived gene, and five enhanced yeast gene copies (Figure 13.18), gave strictosidine at levels of ~0.5 mg L^{-1}. These are low but real titers and represent a starting platform for optimization.

Given the central role of strictosidine for literally thousands of downstream indole alkaloids this is an achievement worth continuing development. We noted in Chapter 5 that the conversion from geranyl-PP to the secologanin framework involves reactive dialdehydes and then is awash with multiple oxygenative transformations. The condensation of secologanin with tryptamine is the Pictet Spengler condensation catalyzed by strictosidine synthase. This is intriguing chemistry to have reconstituted in genetically manipulatable yeast.

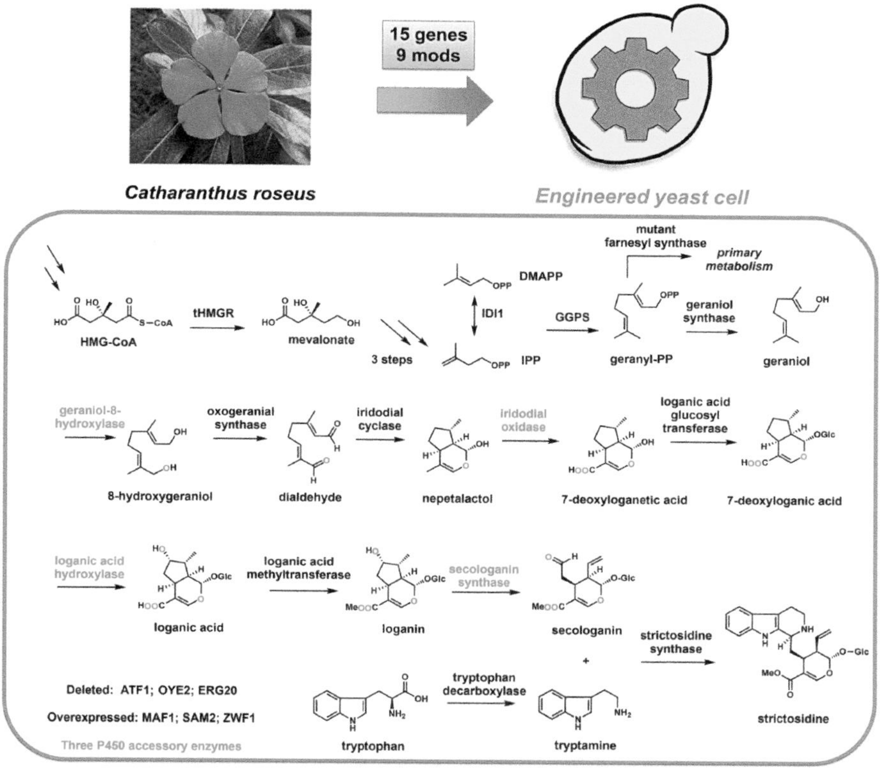

Figure 13.18 Complete reconstitution of strictosidine in yeast from 15 genes and nine other gene modifications in yeast.

13.5.4 Alkaloid Expression in Yeast: Tabersonine to Vindoline

Near the back end of the vinca alkaloid pathway tabersonine undergoes a set of seven enzymatic tailorings at N_1, C_3, C_4, and C_{16} of the tetracyclic periphery to yield vindoline, one of the two condensation substrates taken on to dimeric vinblastine and vincristine. The modifications involve three oxygenations (C_3, C_4, and C_{16}, and then covering up the C_{16}–OH as the methoxy derivative and the C_4–OH as the acetoxy group) (Figure 13.19). Additionally, N_1 of the indole moiety is methylated. The hydroxylation/acetylation at C_4 occurs once the metabolites have transited to the lactifer/idioblast layer, while the prior conversions occur in the *Catharanthus roseus* epidermis. These seven steps have been reconstituted in yeast (Qu, Easson *et al.* 2015), closing in on the remaining, not fully charted, middle of the vinca alkaloid pathway (strictosidine to tabersonine).

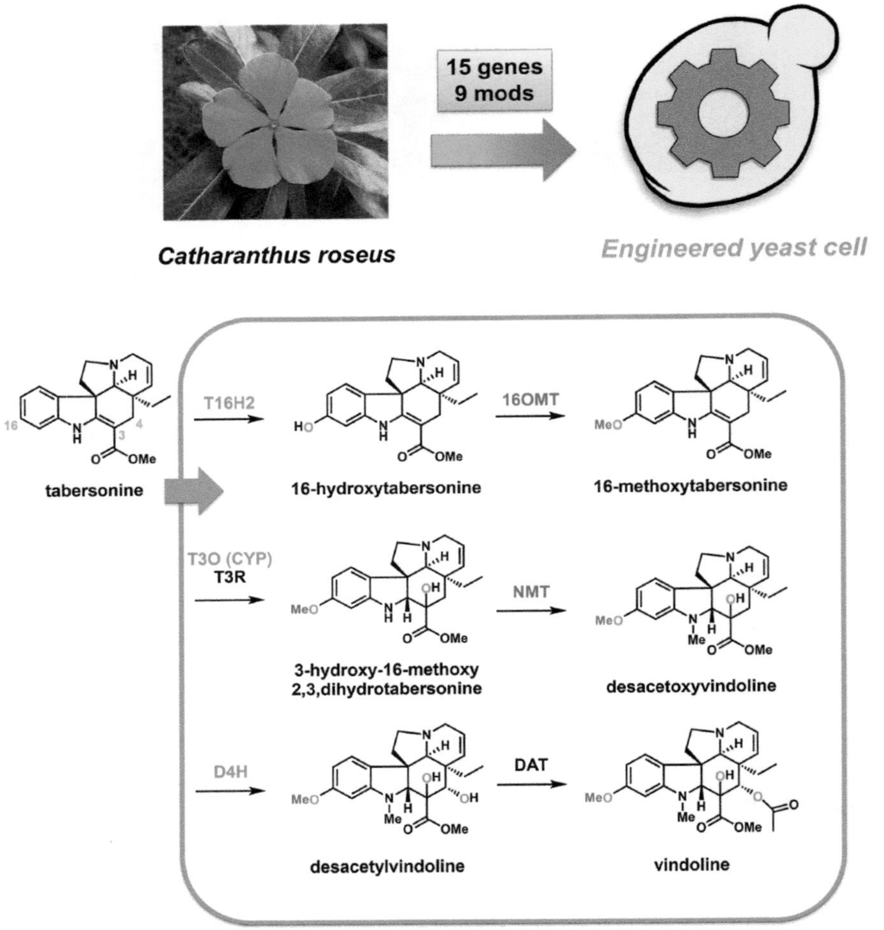

Figure 13.19 The seven-enzyme pathway that transforms tabersonine to vindoline was reconstituted in yeast.

13.5.5 Artemisinic Acid Production in Yeast and Chemical Conversion to Artemisinin

In earlier chapters we have noted that the isoprenoid cyclic endoperoxide metabolite artemisinin from *Artemisia annua* is a key component of frontline therapy for treatment of uncomplicated malarial infections (>200 million cases globally). In an early publication, in 2006, Keasling and team (Ro, Paradise *et al.* 2006) reported the production of the advanced precursor artemisinic acid (Figure 13.20) in *S. cerevisiae* by engineering a modified mevalonate pathway to the isoprenyl-PP isomeric building blocks, adding a plant

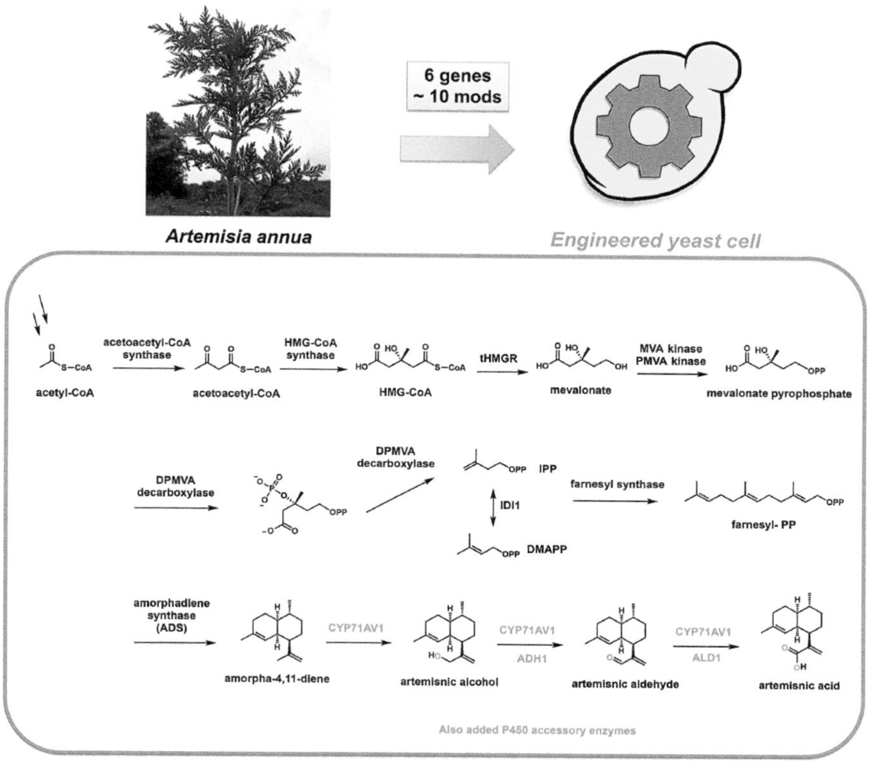

Figure 13.20 Production of artemisinic acid in *S. cerevisiae* by engineering a modified mevalonate pathway.

amorphadiene synthase (Chapter 4) and a cytochrome P450 that converts one of the amorphadiene methyl groups to the carboxylate of artemisinic acid by three successive oxygenations. These engineering manipulations led to yields of artemisinic acid of 1.7 g L^{-1}. In a follow up effort in 2013, the commercial team at Amyris obtained much higher yields with strain improvements (Paddon, Westfall *et al.* 2013), most notably using a cytochrome P450 for only the first hydroxylation step and importing genes for an alcohol dehydrogenase and then an aldehyde dehydrogenase to carryout the final two redox steps to artemisinic acid. To avoid problematic crystallization of the artemisinic acid at these production levels, fermentations were carried out in the presence of isopropyl myristate oil which solubilized the accumulating extracellular product and gave seven-fold higher yield in fed batch mode. Titers of 25 g L^{-1} were obtained. In this publication the authors also reported a chemical conversion of artemisinic acid to the cyclic

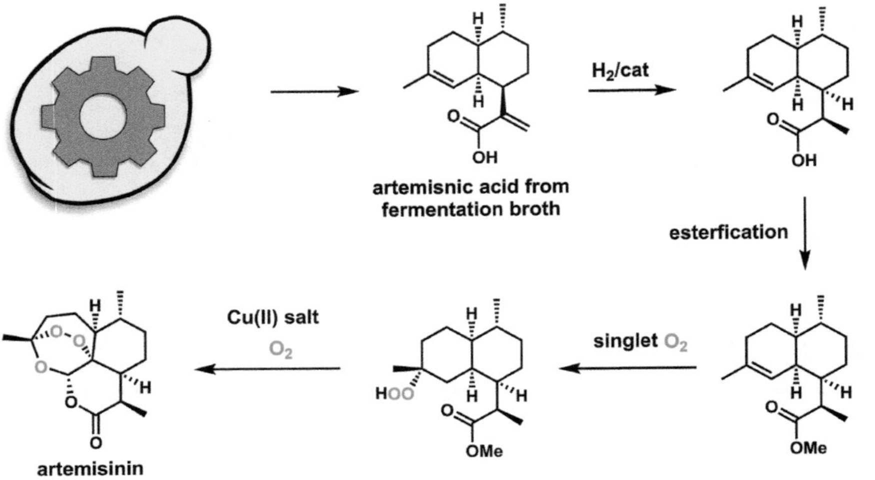

Figure 13.21 The chemical conversion of artemisinic acid isolated from yeast fermentation to the cyclic endoperoxide artemisinin is a four step, scalable procedure in 40–45% yield.

endoperoxide artemisinin in a four step, scalable procedure in 40–45% yield (Figure 13.21) at 99.6% purity after recrystallization. The authors conclude: "The chemical conversion procedure is notable for its simplicity, scalability, economy of reagents, and the high yield obtained".

13.5.6 Yeast as a Heterologous Host for Unknown Fungal Gene Clusters: Further Improvements and Tools

The above examples indicate that yeast is a suitable eukaryotic single-cell host for expression of plant derived alkaloid and isoprenoid and indole–isoprenoid biosynthetic pathway enzymes. Yeast also serves for polyketide (such as lovastatin and 6-methylsalicylic acid) and NRP gene cluster expression (such as fumiquinazoline F). The vast reservoir of knowledge in yeast genetics and molecular biology allows choice of several promoters of different strengths, plasmids, and vectors for introduction and maintenance of foreign DNA sequences.

Technology development includes identifying and refactoring biosynthetic gene clusters, for example so that they can be dispersed into vectors of appropriate size. This involves design and assembly of the desired biosynthetic pathway as prelude to expression in engineered yeast strains and LC-MS analysis for natural products even at minute levels (Siddiqui, Thodey *et al.* 2012, Billingsley, DeNicola *et al.* 2016) (Figure 13.22). As schematized in the right side of

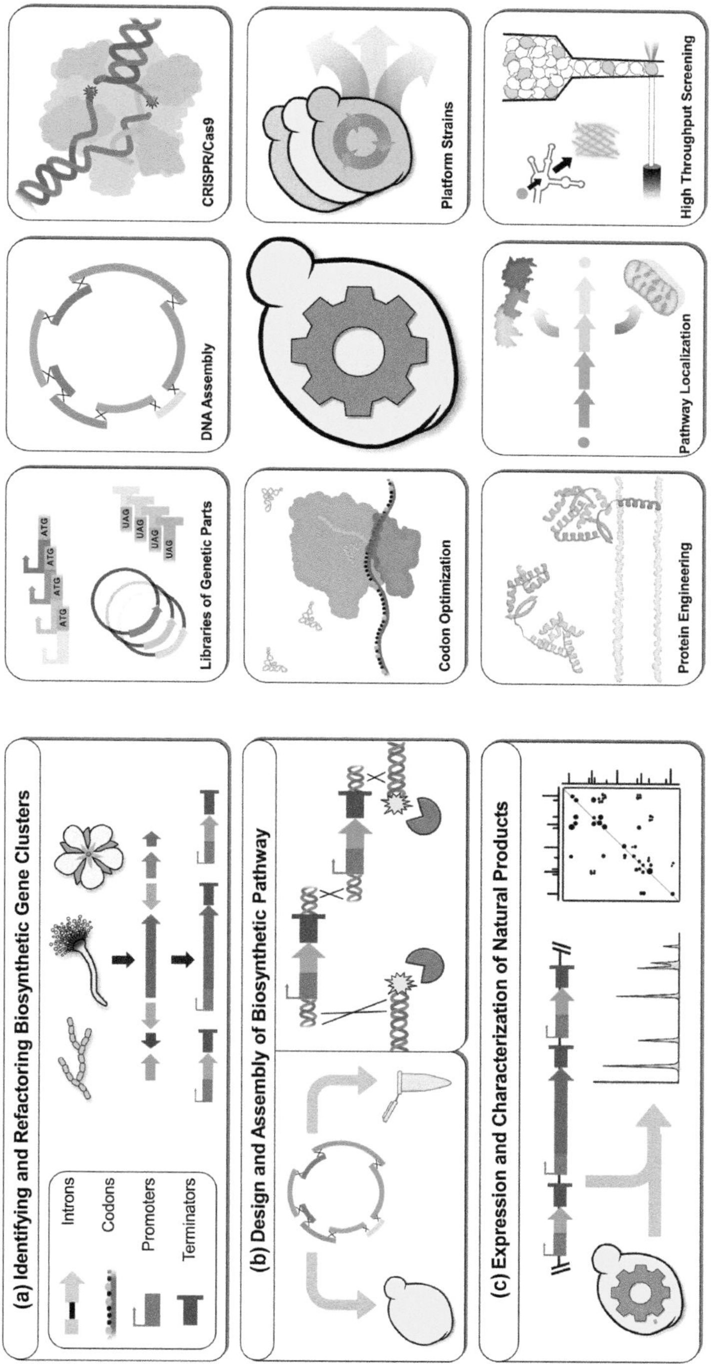

Figure 13.22 Technology development for producing heterologous natural products in *Saccharomyces cerevisiae*.

Figure 13.22, the inventory of tools will include libraries of genetic parts (often from synthetic DNA), DNA assembly methods that are efficient and essentially error free, codon optimization programs for synthetic DNA, a set of yeast platform strains for DNA assembly and expression, and protein engineering capability, including sequences to send protein to desired subcellular yeast compartments. Finally, there are the robotic, parallel culture growths, and analytical LC-MS needs to conduct high throughput screening of strains for natural product detection.

Pathway refactoring of gene clusters identified from eukaryotic species requires the removal of non-coding intron sequences. For example, fungal genes involved in natural product pathways frequently contain as many as five to nine introns. Although there are several methods for successful *in silico* intron prediction from genome sequences, even one improperly annotated intron can derail pathway characterization efforts owing to incorrect protein translation. Intron-less complementary (c)DNA can be generated from transcriptively active gene clusters, but silent gene clusters cannot be interrogated by this approach. The native spliceosome of *S. cerevisiae* does not have the capability to remove introns from distant fungi and plants, and has been shown to reject foreign introns that differ by as little as a single base from consensus yeast introns. With the continued accumulation of genomic data and the use of yeast as a platform host, it will become increasingly important to develop new computational and experimental tools to solve the intron problem. One promising approach is to engineer the yeast splicing machinery towards recognition of foreign intron sequences, either through the introduction of heterologous spliceosome components, or *via* directed evolution of incompatible components.

Vignette 13.1 Generation of Unnatural Molecular Complexity using Synthetic Genes for Terpene Cyclases Paired with Cytochromes P450.

Based on a new class of membrane bound, as opposed to soluble, cytoplasmic terpene cyclases, genome mining was performed to identify this subclass of cyclases of potentially unknown functions as well as associated P450 enzymes that oxidatively modify the hydrocarbon scaffolds.

To demonstrate the potential to generate chemical diversity, synthetic DNA encoding two different terpene cyclases and two different P450s (AF510 and Tv86) identified from fungal genomes were prepared. Using gene assembly and artificial promoters, different pairwise combinations of the cyclases and P450 were constructed and expressed from *Saccharomyces cerevisiae*. The two terpene cyclases produced isomeric sesquiterpenes α- and β-trans-bergamotene (Figure 13.V1). In the presence of the two distinct P450s, the olefinic terpenes can be converted to a large collection of oxygenated terpene products, some of which are new to Nature. The P450 enzymes were especially impressive in performing C–H activation and C–C rearrangements at different sites in the molecules and may have value in driving oxygenative transformations and rearrangements on other scaffolds. This study illustrates the power of using synthetic gene combinations in the generation of natural product structural diversity and complexity, and indicates that yeast should be a suitable host for such combinatorial synthetic biology operations.

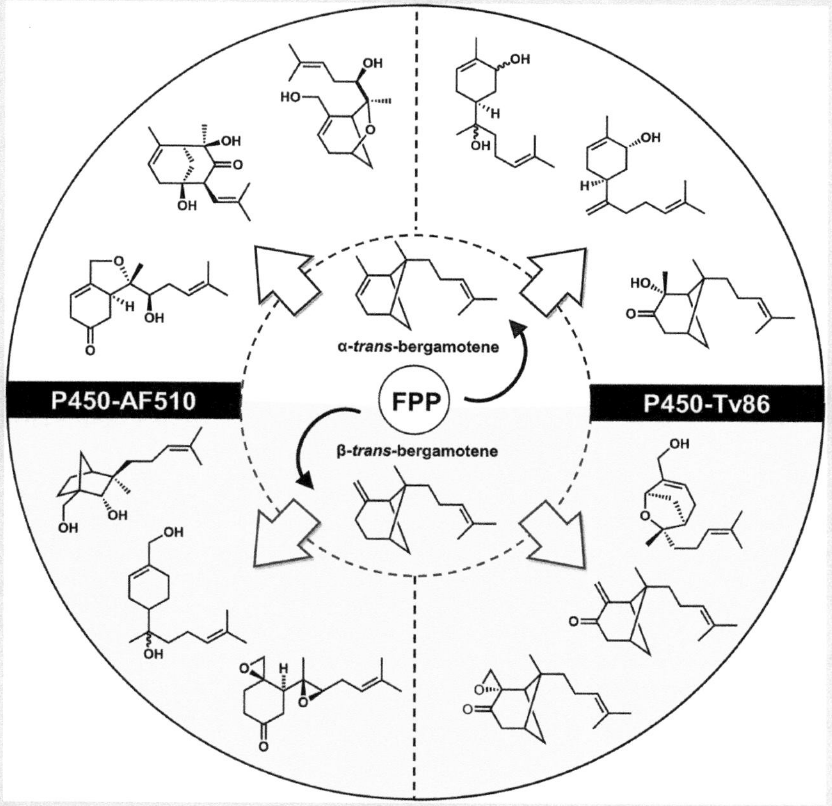

Figure 13.V1 Generation of unnatural molecular complexity from genome mined enzymes.

13.6 Bioinformatics-based Natural Product Prospecting in Bacterial Genomes

In parallel with the efforts to activate silent biosynthetic gene clusters in fungi, many investigations have gone into routes to selectively turn on unutilized bacterial gene clusters in organisms grown under laboratory conditions. A recent review (Zarins-Tutt, Barberi *et al.* 2016) provides an overview of approaches for upregulation of bacterial cryptic natural product pathways. Two examples of successes are depicted in Figure 13.23. The top part of the figure shows four structures, among many that could be cited (Zarins-Tutt *et al.* 2016), that are induced in different bacteria by quorum sensor molecules. (Quorum sensing is a measure of local bacterial cell density and correlates with exhaustion of sufficient nutrients for growth.) These are known ligands for DNA-binding proteins that control transcription of specific gene sets. The four product molecules shown, pyrocycanin, pyrrolnitrin, carbapenems, and fosfomycin, span a wide range of different natural product pathways, emphasizing that understanding the (often hierarchical) regulatory circuitry of bacterial development will inform conditions of cultivation for turning on many types of natural product production.

A second approach is exemplified in the bottom half of Figure 13.23, showing two molecules that are expressed in heterologous bacterial hosts. Luminmycin A is one of a series of related molecules encoded by a silent *Photorhabdus luminescens* cluster that is turned on when expressed in *E. coli* (Bian, Plaza *et al.* 2012). Many examples involve moving gene clusters from one streptomycete, often genetically intractable owing to absence of developed molecular biology reagents, to an alternate streptomycete with useful molecular biology tools for expression of heterologous genes. One such case is provided by alteramide encoded by cryptic *S. griseus* genes. The complex natural product is detectably produced in an engineered *Streptomyces lividans* strain (Olano, Garcia *et al.* 2014).

Additional strains have been used as expression hosts by Sean Brady and colleagues to test expression of gene cluster retrieved from metagenomic bacterial DNA samples. These strains serve as a rapid way to assess whether alternative host cells have the requisite machinery for transcription and translation of the foreign DNA and any ancillary metabolites and coenzymes that may be necessary for specific pathway requirements (Charlop-Powers, Milshteyn *et al.* 2014, Milshteyn, Schneider *et al.* 2014).

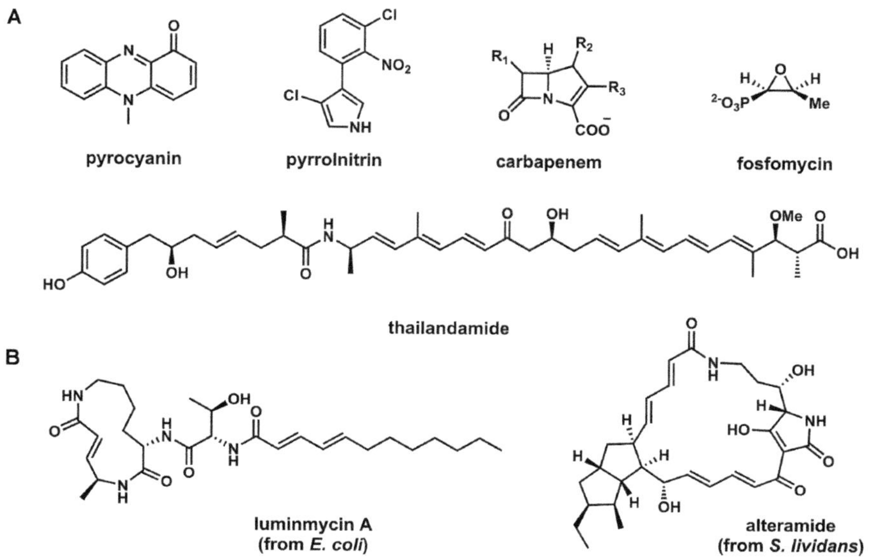

Figure 13.23 Representative compounds identified from engineered bacterial strains. (A) Five compounds identified through manipulation of quorum sensing systems of bacteria. (B) Two molecules that are expressed in heterologous bacterial hosts.

13.7 Heterologous expression in *E. coli*

13.7.1 Deoxyerythronolide B

Given the 70 years of genetic and molecular biology reagent development associated with *E. coli* strains, this bacterium is often tested as a default strain for introduction of heterologous bacterial genes and evaluation of novel natural product production (Ahmadi and Pfeifer 2016). One of the known limitations is that *E. coli* cells may not produce requisite building blocks at appropriate levels for the flux needed to power a natural product pathway.

This is illustrated and solved in the case of expression of the three deoxyerythronolide synthase genes from *S. erythraea* in *E. coli* (Figure 13.24) (Pfeifer, Admiraal *et al.* 2001). In particular, 2*S*-methylmalonyl-CoA is not available in *E. coli* so it would not be able to produce any of the polyketides that use this building block. The three subunit, six module DEB synthase uses 2*S*-methylmalonyl-CoA in all six elongation steps. Thus, the overproducing candidate strain was equipped with genes that encode the enzymes that

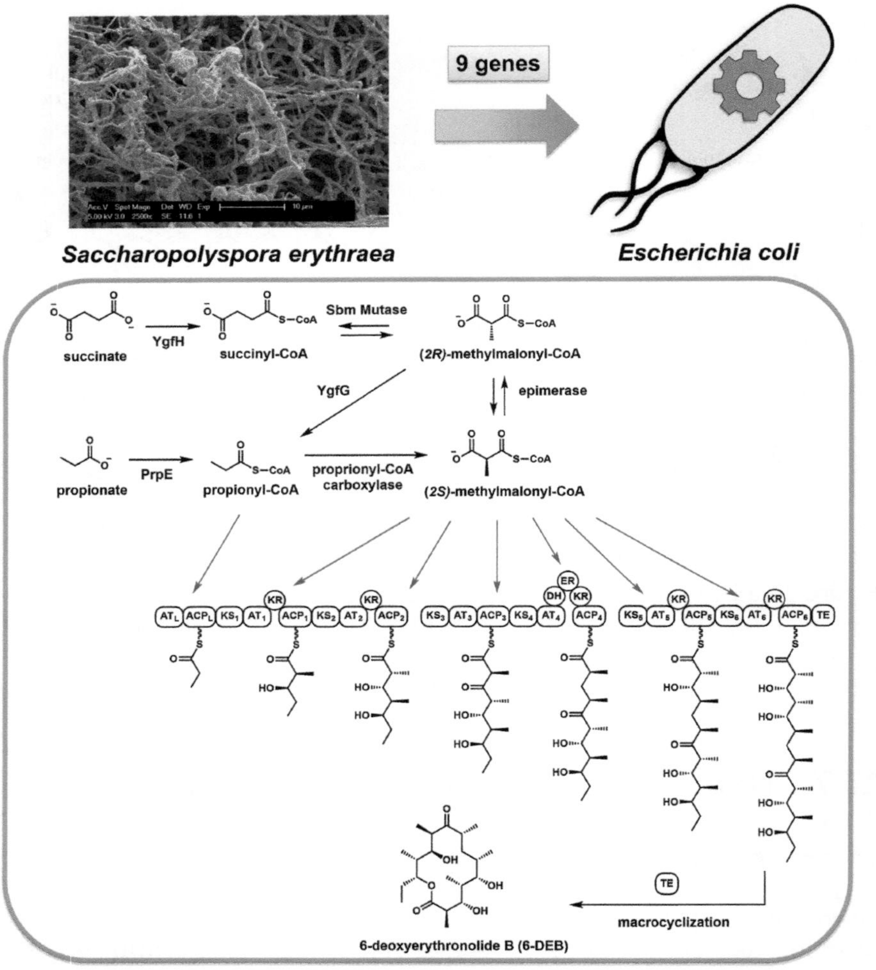

Figure 13.24 Engineered biosynthesis of 6-deoxyerythronolide B in *E. coli*. The electron micrograph of *S. erythraea* mycelium is courtesy of Jeremy Skepper, University of Cambridge.

activate propionate to propionyl-CoA and then the multisubunit biotin-containing carboxylase that converts propionyl-CoA to 2*S*-methylmalonyl-CoA. The epimerase that mediates the conversion of 2*R*- to 2*S*-methylmalonyl-CoA was also provided. In the event the 14-membered macrocyclic deoxyerythronolide B was produced, setting up both optimization prospects and facile manipulations of the genes for protein engineering, *e.g.* to make analogs (Jiang, Fang *et al.* 2013).

13.7.2 Phenylpropanoid Production in *E. coli* Cells

Phenylpropanoid metabolites encoded by plant genes have also been successfully generated in *E. coli* cells (Santos, Koffas *et al.* 2011, Trantas, Koffas *et al.* 2015). Simple flavonoids, stilbenoids, and curcuminoids are generated *via* the action of chalcone synthases, stilbene synthases, and curcumin synthases, respectively (Chapter 7). A more detailed schematic for the production of the flavonoid 2*S*-pinocembrin in *E. coli* cells is shown in Figure 13.25 in which eight

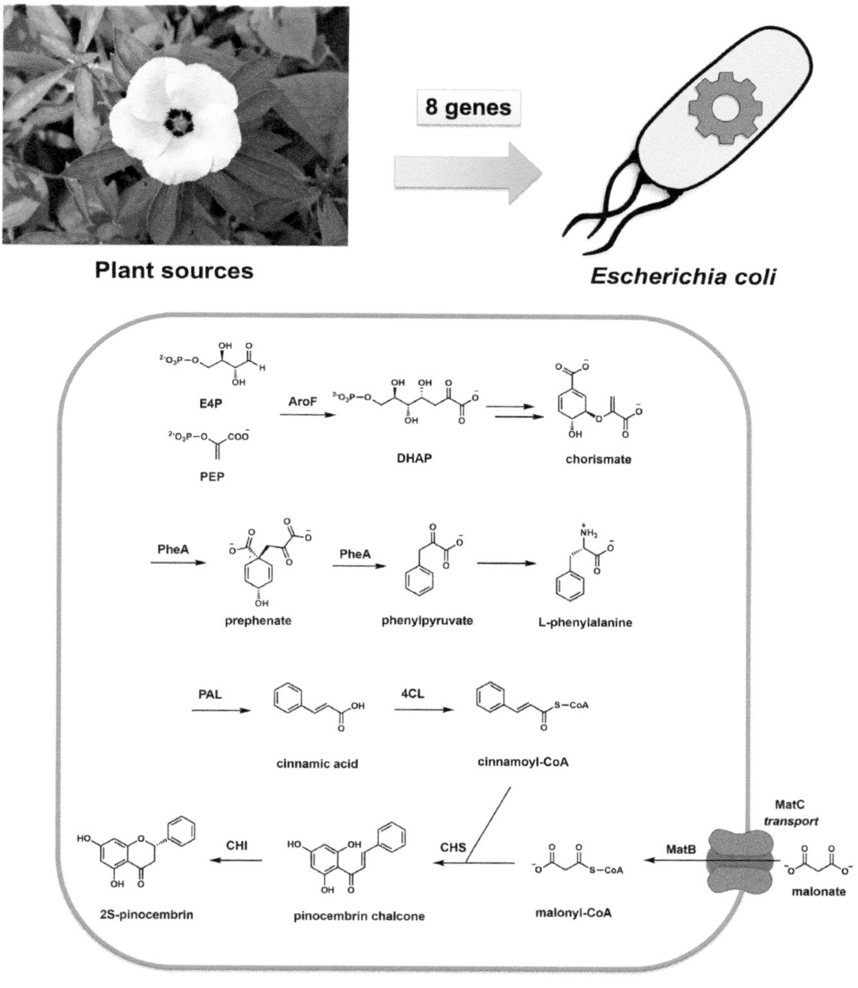

Figure 13.25 Engineered biosynthesis of the flavonoid 2*S*-pinocembrin in *E. coli*. MatC is expressed to increase transport of malonate into the cell.

relevant genes were distributed pairwise into four vectors (Wu, Du *et al.* 2013, Guo, Chen *et al.* 2016).

13.7.3 Monacolin J to Simvastatin in *E. coli* cells

In a variant use of *E. coli* as a production host, 30 g L^{-1} of monacolin J was converted to simvastatin in a 24-hour period (Figure 13.26), using a mutagenized LovD esterification enzyme to accept a low molecular weight surrogate 2,2-dimethylbutyryl thioester as the electrophilic partner in the LovD condensation that is the final step in assembly of this commercial statin (Jimenez-Oses, Osuna *et al.* 2014).

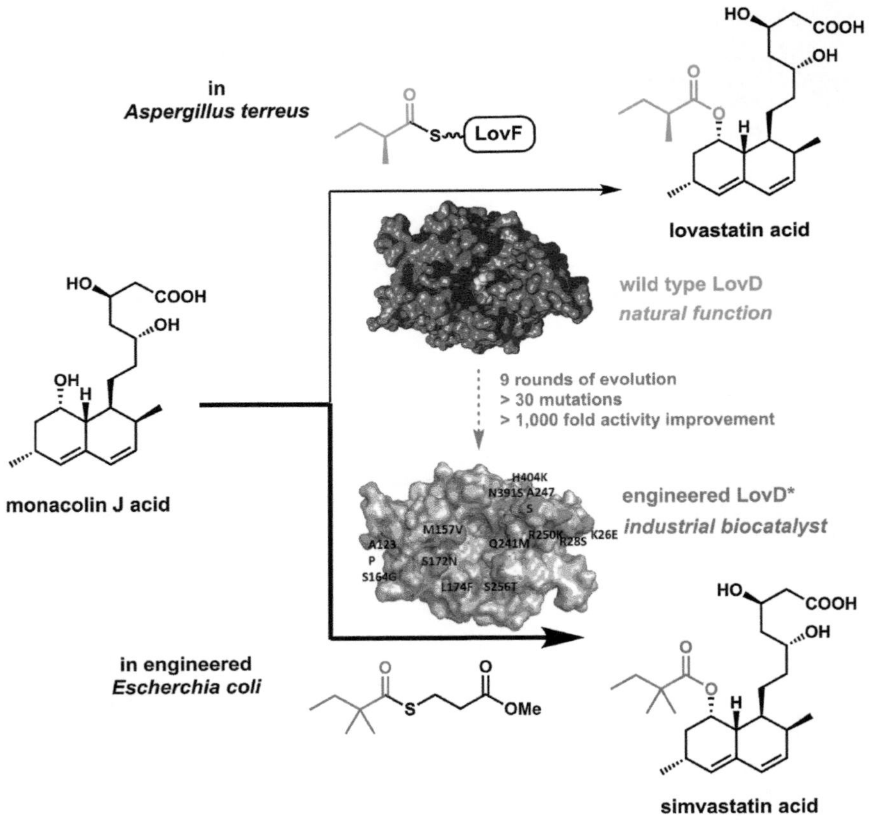

Figure 13.26 Using an evolved LovD mutant for the production of the cholesterol lowering drug simvastatin in *E. coli*.

13.8 Metabolic Engineering for Diversity Generation

Metabolic engineering of natural products has been practiced for decades before the advent of full genome sequence information, and afterwards (Pickens, Tang *et al.* 2011). Classical examples of modifications not dependent on genome information include the conversions of natural penicillins and cephalosporins to 6-aminopenicillanic acid and 7-aminocephalosporonic acid, respectively, by enzymatic deacylation (see Chapter 12). The chemical reacylations with literally hundreds of different side chains have been successful, several at the multi-ton scale. A second commercially useful example has been to add decanoic acid to fed batch fermentations of *Streptomyces roseosporus*, the daptomycin producer, to ensure that the lipopeptide product has only the saturated C_{10} chain and not intermediates from the endogenous fatty acid biosynthetic pathways.

Three examples, out of a plethora of possibilities, in the post-genomic era include the construction of a 380-member library of antimycin derivatives that contain modified alkyl groups at C_7 or alternative acyl groups at C_8 (Yan, Chen *et al.* 2013). The C_7 groups arise from alternative alkylmalonyl-CoAs generated by reductive carboxylation of the precursor enoyl-CoAs by AntE. These alkylmalonyl-CoA variants were then accepted as starter units and processed down the *Streptomyces* sp. NRRL 2288 PKS assembly line. The *ca.* 25 acyl chain variants at C_8 represent permissive processing of acyl-CoA variants by the post assembly line dedicated 8-*O*-acyltransferase AntB encoded in the antimycin biosynthetic cluster (Yan, Chen *et al.* 2013) (Figure 13.27). The second example is also a case of exchange of tailoring genes, in the pathways to the novobiocin and clorobiocin antibiotics that target the GyrB subunit of DNA gyrase. Methyltransferase and aromatic ring halogenase genes were swapped between the two pathways and gave the morphed antibiotic scaffolds (Figure 13.28A) (Heide 2009).

The third example deals with the multiple posttranslational conversion steps of the nascent 52 residue TclE protein into the trithiazolylpyridine core of the thiocillin antibiotics (Chapter 3). A *ΔtclE* strain of the producer *Bacillus cereus* could be complemented with mutant forms of the 156 bp (base pairs) *tclE* gene to evaluate several structure–activity features of the maturation process. Among the notable findings was that changing the distance between the two

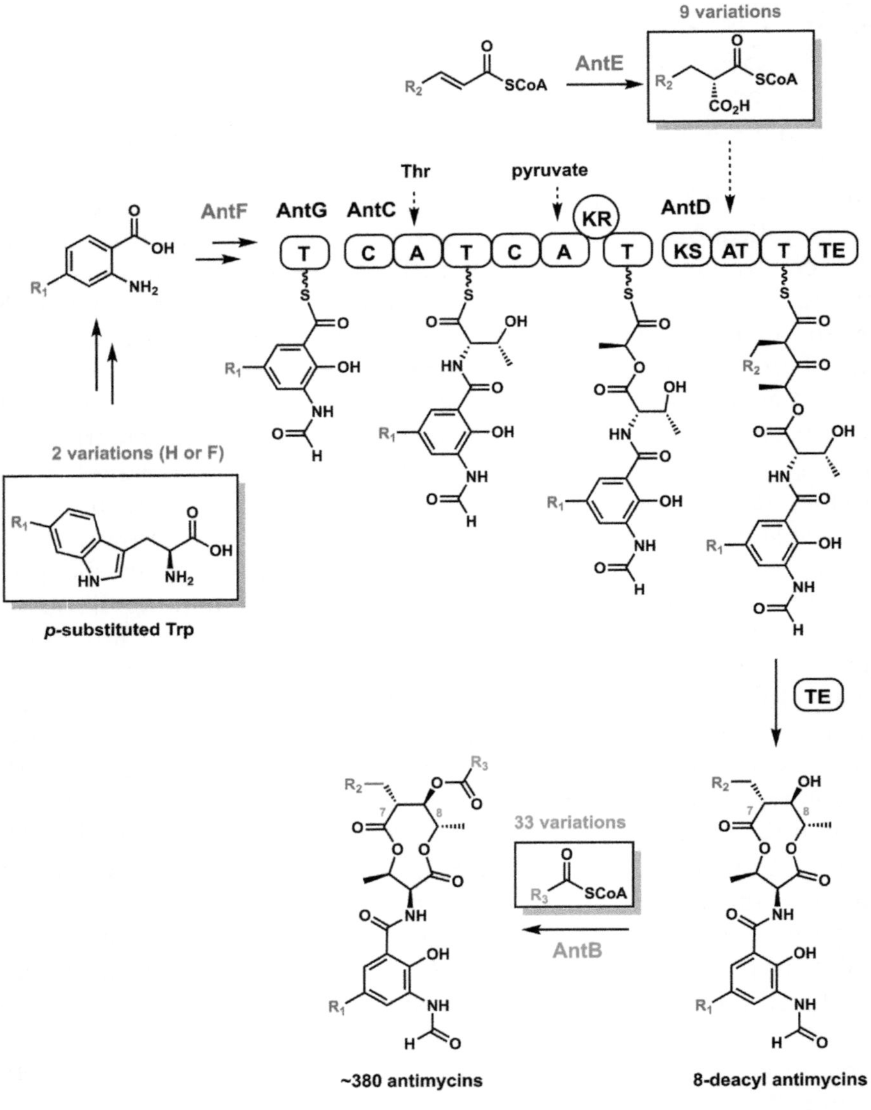

Figure 13.27 The construction of a 380-member library of antimycin derivatives that contain modified alkyl groups at C_7, alternative acyl groups at C_8, and substituted phenyl ring. The substrate promiscuities of gatekeeping enzymes were explored combinatorially.

serine residues (at relative positions 1 and 10 to yield a 26 atom macrocycle) that give rise to the central pyridine in the last step to positions 1 and 9 or 1 and 11 allowed formation of macrocyclic

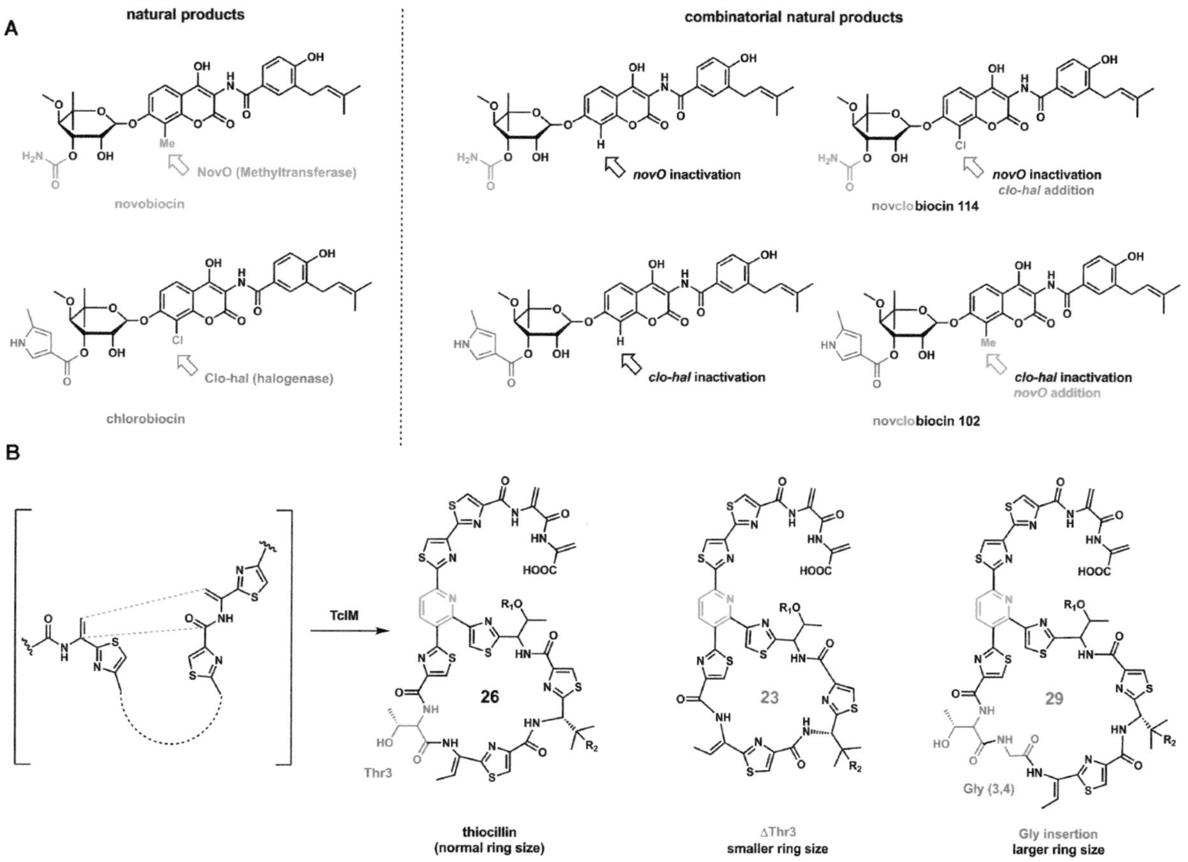

Figure 13.28 Engineering pathways for diversity generation. (A) Mix and match of one gene in the novobiocin and chlorobiocin pathways leads to production of hybrid compounds; (B) rational alteration of peptide length of thiocillin precursor leads to formation of thiopeptides of varying ring sizes.

lactams of altered rings sizes, 23 and 29 atoms, with the characteristic substituted thiazolyl pyridine cores (Figure 13.28B) (Bowers, Acker *et al.* 2010, Bowers, Acker *et al.* 2012). These gene-based manipulations on an encoded ribosomal protein give direct access to structure–activity relationships (SAR) studies on scaffolds forbiddingly complex to explore by total synthesis (Walsh, Acker *et al.* 2010). This strategy also applies to variants of patellamides and other cyanobactins.

13.9 Plug and Play Approach to Cloning Transcriptionally Silent Gene Clusters of Unknown Function from Bacteria

The intersection of synthetic biology methods and microbial genomic sequencing has resulted in the assembly of full molecular toolboxes to go from DNA to RNA to protein regulation in efforts to facilitate the identification of unknown natural products predicted to be encoded in transcriptionally silent gene clusters. This can involve refactoring of orphan gene clusters, at one end from fully synthetic DNA (cost likely to approach as low as $1000 per 20 kb DNA) to the other end where natural DNA sequences are revamped for promoter optimization, transcriptional regulation, installation of (improved) ribosome binding sites, and even codon usage changes to fit the preference of candidate heterologous hosts. It is also likely that the tools in the synthetic biology tool box will continue to evolve and improve over the near term.

One example gives some insight into lowering the barriers for capturing and moving an unknown gene cluster into a streptomycete strain, *S. coelicolor* M1146, engineered for biosynthetic pathway expression. The specific goal was to capture a silent 67 kb gene cluster with homology to the highly successful lipopeptide antibiotic daptomycin from a marine actinomycete, *Saccharomonospora* sp. CNQ-490 (Yamanaka, Reynolds *et al.* 2014). The cloning approach has been termed transformation associated recombination (TAR) and in this case involved a vector with elements that allowed (1) homologous recombination in yeast, constituting the foreign DNA capture step with 73 kb of entrained CNQ-490 DNA. Potential repressor genes among the 20 in the captured cluster could be replaced *in vivo* in *S. cerevisiae* by marker replacement. (2) The circular construct could be shuttled into *E. coli* for any additional recombination plans, *e.g.* by the well-established λ-RED system. (3) Finally, the tailored DNA construct could be conjugally transferred from *E. coli* to the *S. coelicolor* host and examined for expression. As shown in Figure 13.29, a lipopeptide homologous to daptomycin, termed taromycin, was produced. It is chlorinated at two residues, Trp1 and kyneurinine14, and has calcium-dependent antibiotic activity against Gram-positive bacteria, consonant with the action of daptomycin.

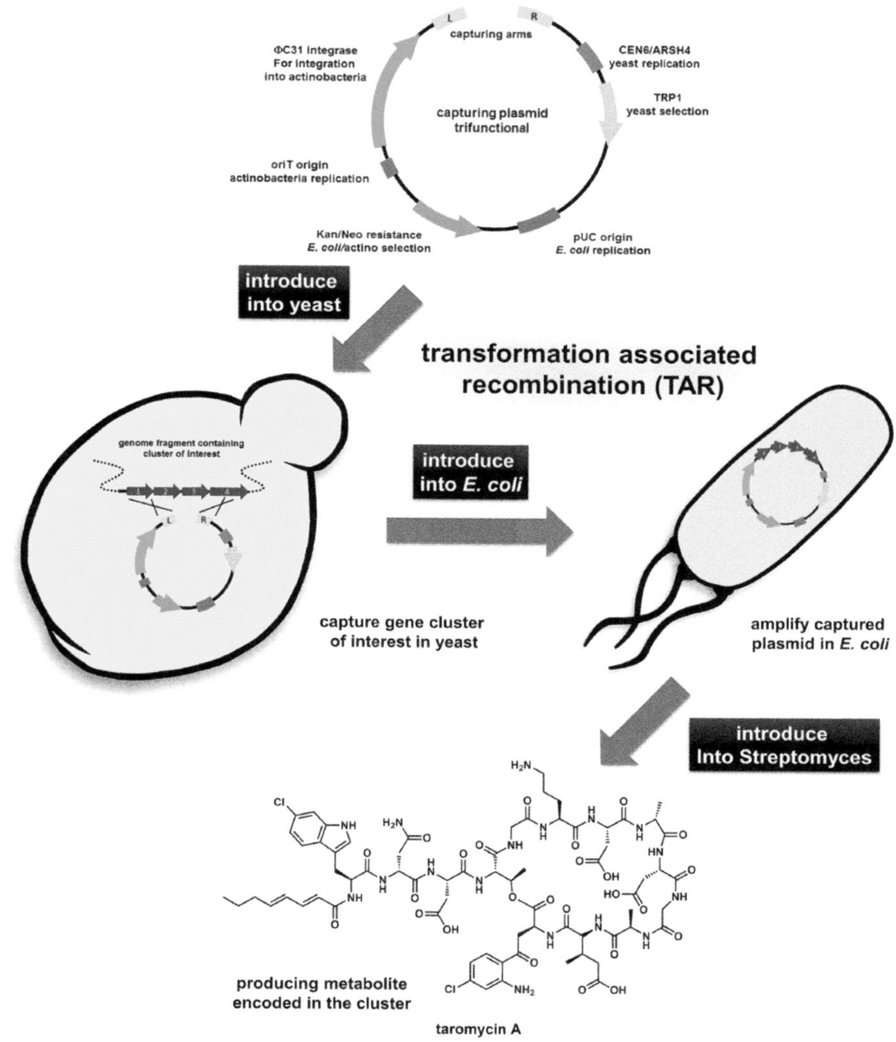

Figure 13.29 Transformation associated recombination (TAR) can be used to rapidly capture large gene clusters from genomic DNA for heterologous production of natural product, as demonstrated in the production of taromycin A here.

The ability to capture foreign DNA of a size (70 kb) likely to encode full biosynthetic pathways for the great majority of polyketides, nonribosomal peptides, and terpenes, and to passage from yeast to *E. coli* to a designated bacterial host strain lowers many technical barriers for examination of silent biosynthetic gene clusters. It is illustrative of how genome mining and synthetic biology have converged to enable investigators to take on almost any silent, unknown microbial gene cluster and give it a high probability of heterologous expression.

13.10 Comparative Metabolomics in the Post Genomic Era

Among the many contemporary advances in characterization of metabolomes under different physiologic conditions, we conclude this chapter with a short summary of the deployment of a sensitive two-dimensional NMR technique termed DANS (*d*ifferential *a*nalysis of 2D *N*MR *s*pectra) pioneered by the research group of Frank Schroeder for natural product samples (Schroeder 2015). The sensitivity is appropriate for small amounts of material in crude extracts and one application is the structural assignment of molecules that may be too unstable to isolate in pure form without decomposition and/or enormous scale up efforts. Figure 13.30 shows four molecules whose structures were determined in different biological crude extract fluids: myrmicarin 430A (a heptacyclic alkaloid of novel connectivity from myrmicine ants), a 2′,5′-bis-sulfated nucleoside from spider venom, the PK-NRP bacillaene noted in Chapter 12, and a xantheurenyl glucoside from human urine (Robinette, Bruschweiler *et al.* 2012).

The comparative power of DANS comes into play with NMR analysis of small molecule content in matched extracts of wild type and one or more defined genetic mutants. This is illustrated in Figure 13.31 for metabolite-containing extracts from wild type *Caenorhabditis elegans* and the *daf 22* mutant that is inactive for onset of the dauer phenotype (see the prior discussion in Chapter 11). The overlay spectrum allowed assignment of peaks that were present in wild type but not in mutant and/or in molecular species accumulating in the *daf* 22 mutant. With a knowledge of the active ascaroside end product, *e.g.* asc#8, it was possible to assign what was missing in the daf22

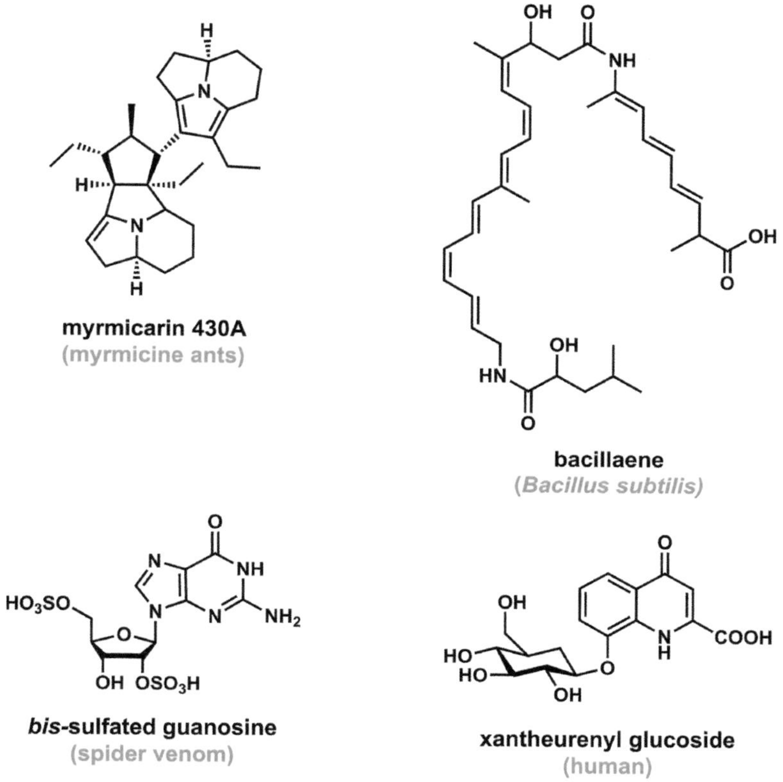

Figure 13.30 Four molecules whose structures were determined in different biological crude extract fluids.
Adapted from Robinette, S. L., R. Brüschweiler, F. C. Schroeder and A. S. Edison (2012). "NMR in metabolomics and natural products research: two sides of the same coin". *Acc. Chem. Res.* **45**(2): 288. http://pubs.acs.org/doi/full/10.1021/ar2001606.

extract and thereby the structure of the intermediate asc#7 (Ludewig and Schroeder 2012). With a series of mutant metabolome comparisons combined with genetics (identifying which genes are upstream or downstream of each other in a pathway) the NMR spectra can serially deconvolute a biosynthetic pathway. DANS analyses have also been used by the Schroeder and Keller groups in a collaboration on intermediates in biosynthesis of gliotoxin, a diketopiperazine alkaloid that is among the longest known toxic metabolites of *Aspergillus* strains (Forseth, Fox *et al.* 2011).

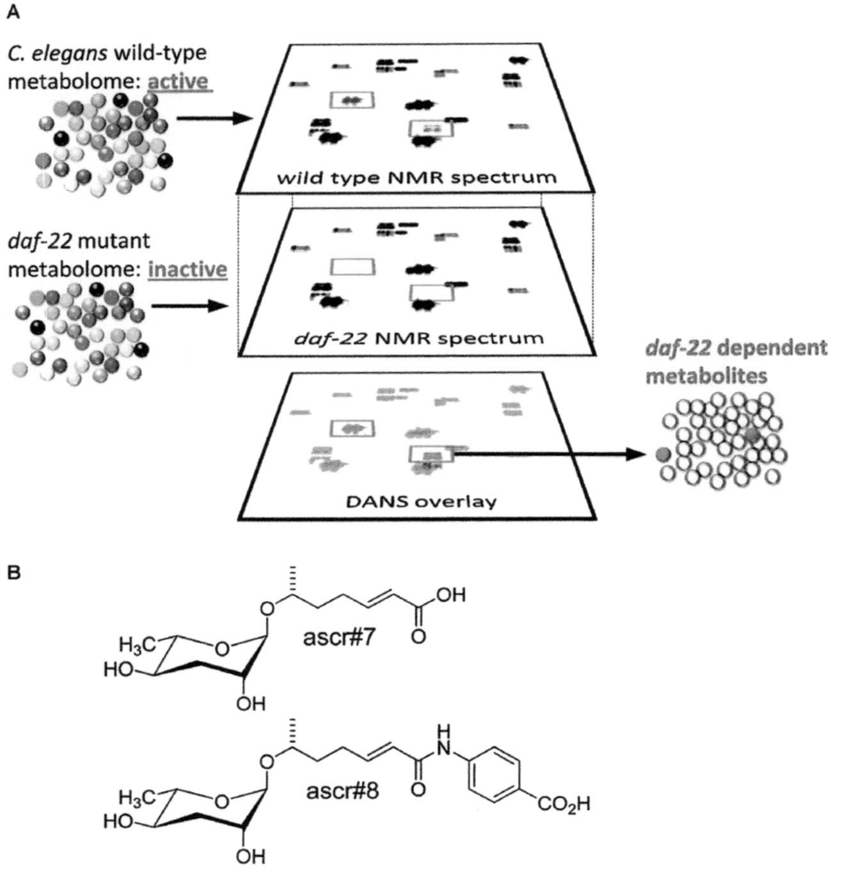

Figure 13.31 Using DANS (*d*ifferential *a*nalysis of 2D *N*MR *s*pectra) to identify metabolites from *Caenorhabditis elegans*.
Reprinted with permission from Robinette, S. L., R. Brüschweiler, F. C. Schroeder and A. S. Edison (2012). "NMR in metabolomics and natural products research: two sides of the same coin". *Acc. Chem. Res.* **45**(2): 288. http://pubs.acs.org/doi/full/10.1021/ar2001606.

Vignette 13.2 Pattern Matching of Natural Product and siRNAs/microRNAs for Cell-Based Target Identification.

One of the applied goals of natural product research is to define the target of action of biologically active natural products. This has been part and parcel historically of the conversion of ethnopharmacologic information about the utility of herbal extracts into identification of both specific human targets and targets in microbes causing infectious diseases. The artemisinin case stands out.

In recent years, investigators have examined the action of purified natural products in cell culture with great success. Some of these have become major therapeutic agents such as cyclosporine, rapamycin, and FK506. Others do not have the requisite specificity or therapeutic ratio to function safely or efficaciously in humans but have been inspirational for semisynthetic optimization, as exemplified by the indolecarbazole framework of staurosporine as a core structure for protein kinase inhibitor design and the utilization of the podophyllotoxin core in etoposide.

As discussed in earlier chapters, some molecules, such as the alkaloid cyclopamine, were identified as toxins and mechanism of action studies have led to deeper understanding of the hedgehog signaling pathways. Analogously, the fungal steroid metabolite wortmannin has proven to be a useful probe of phosphatidylinositol-3-kinase cell biology (see Vignette 4.2). One of the most successful roles for natural products as tool molecules for cell biology has been in the secretory pathways of mammalian cells. As noted in Figure 13.V2 the polyketide brefeldin A has become the default inhibitor for blockade of secretory cargo moving from the endoplasmic reticulum to the Golgi organelles. Similarly, thapsigargin is a potent blocker of sarcoplasmic Ca^{++}-dependent ATPase (as is cyclopiazonate; see Figure 8.24). The epoxytetrapeptide lactam trapoxin is a covalent inhibitor of histone deacetylases.

A recent effort to use genomic data, read out in microRNAs and small interfering RNA reagents, to accelerate the identification of natural product targets is worth noting (Potts, Kim *et al.* 2013). The action of natural products is compared to readouts of a set of microRNAs and siRNAs that had been designed and calibrated against hundreds of messenger RNAs in cells. The goal was pattern matching of gene expression signatures from the specific RNA treatments to gene expression signatures with natural products in semipurified extracts. When a closely matching pattern was observed, the natural products were further purified and their mode of action confirmed in specific target assays. The authors demonstrated matches in autophagy assays, in receptor-mediated chemotaxis assays, and in assay of the AKT protein kinase.

As shown in the bottom part of Figure 13.V2, the known microbial metabolite bafilomycin A1 was purified as the agent causing inhibition of a vacuolar ATPase required for autophagy. In a separate set of assays the compound discoipyrrole A, with unusual scaffold connectivity, was purified based on its matching pattern to siRNAs that inhibited the DDR2 receptor which controls aspects of cell migration. It may be that this approach can be converted into high throughput formats to

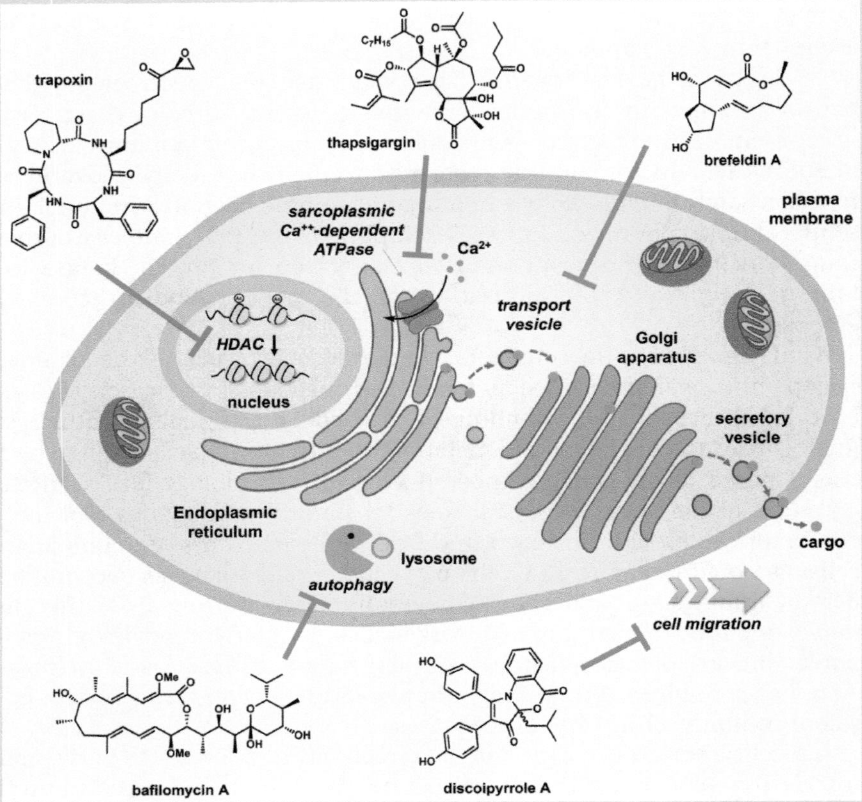

Figure 13.V2 Cell-based natural product target identification.

interrogate natural product extracts that have been fractionated to low complexity (two to six compounds) to determine quickly whether molecules with promising activity are worth purification and structure determination.

References

Ahmadi, M. K. and B. A. Pfeifer (2016). "Recent progress in therapeutic natural product biosynthesis using Escherichia coli". *Curr. Opin. Biotechnol.* **42**: 7–12.

Bian, X., A. Plaza, Y. Zhang and R. Muller (2012). "Luminmycins A-C, cryptic natural products from Photorhabdus luminescens identified by heterologous expression in Escherichia coli". *J. Nat. Prod.* 75(9): 1652–1655.

Billingsley, J. M., A. B. DeNicola and Y. Tang (2016). "Technology development for natural product biosynthesis in Saccharomyces cerevisiae". *Curr. Opin. Biotechnol.* **42**: 74–83.

Blin, K., M. H. Medema, D. Kazempour, M. A. Fischbach, R. Breitling, E. Takano and T. Weber (2013). "antiSMASH 2.0–a versatile platform for genome mining of secondary metabolite producers". *Nucleic Acids Res.* **41**(Web Server issue): W204–212.

Bode, H. B., B. Bethe, R. Hofs and A. Zeeck (2002). "Big effects from small changes: possible ways to explore nature's chemical diversity". *ChemBioChem* **3**(7): 619–627.

Bowers, A. A., M. G. Acker, A. Koglin and C. T. Walsh (2010). "Manipulation of thiocillin variants by prepeptide gene replacement: structure, conformation, and activity of heterocycle substitution mutants". *J. Am. Chem. Soc.* **132**(21): 7519–7527.

Bowers, A. A., M. G. Acker, T. S. Young and C. T. Walsh (2012). "Generation of thiocillin ring size variants by prepeptide gene replacement and in vivo processing by Bacillus cereus". *J. Am. Chem. Soc.* **134**(25): 10313–10316.

Brakhage, A. A. (2013). "Regulation of fungal secondary metabolism". *Nat. Rev. Microbiol.* **11**(1): 21–32.

Brown, S., M. Clastre, V. Courdavault and S. E. O'Connor (2015). "De novo production of the plant-derived alkaloid strictosidine in yeast". *Proc. Natl. Acad. Sci. U. S. A.* **112**(11): 3205–3210.

Charlop-Powers, Z., A. Milshteyn and S. F. Brady (2014). "Metagenomic small molecule discovery methods". *Curr. Opin. Microbiol.* **19**: 70–75.

Chiang, Y. M., C. E. Oakley, M. Ahuja, R. Entwistle, A. Schultz, S. L. Chang, C. T. Sung, C. C. Wang and B. R. Oakley (2013). "An efficient system for heterologous expression of secondary metabolite genes in Aspergillus nidulans". *J. Am. Chem. Soc.* **135**(20): 7720–7731.

Cortes, J., S. F. Haydock, G. A. Roberts, D. J. Bevitt and P. F. Leadlay (1990). "An unusually large multifunctional polypeptide in the erythromycin-producing polyketide synthase of Saccharopolyspora erythraea". *Nature* **348**(6297): 176–178.

Donadio, S., M. J. Staver, J. B. McAlpine, S. J. Swanson and L. Katz (1991). "Modular organization of genes required for complex polyketide biosynthesis". *Science* **252**(5006): 675–679.

Forseth, R. R., E. M. Fox, D. Chung, B. J. Howlett, N. P. Keller and F. C. Schroeder (2011). "Identification of cryptic products of the gliotoxin gene cluster using NMR-based comparative metabolomics and a model for gliotoxin biosynthesis". *J. Am. Chem. Soc.* **133**(25): 9678–9681.

Galagan, J. E., S. E. Calvo, C. Cuomo, L. J. Ma, J. R. Wortman, S. Batzoglou, S. I. Lee, M. Basturkmen, C. C. Spevak, J. Clutterbuck, V. Kapitonov, J. Jurka, C. Scazzocchio, M. Farman, J. Butler, S. Purcell, S. Harris, G. H. Braus, O. Draht, S. Busch, C. D'Enfert, C. Bouchier, G. H. Goldman, D. Bell-Pedersen, S. Griffiths-Jones, J. H. Doonan, J. Yu, K. Vienken, A. Pain, M. Freitag, E. U. Selker, D. B. Archer, M. A. Penalva, B. R. Oakley, M. Momany, T. Tanaka, T. Kumagai, K. Asai, M. Machida, W. C. Nierman, D. W. Denning, M. Caddick, M. Hynes, M. Paoletti, R. Fischer, B. Miller, P. Dyer, M. S. Sachs, S. A. Osmani and B. W. Birren (2005). "Sequencing of Aspergillus nidulans and comparative analysis with A. fumigatus and A. oryzae". *Nature* **438**(7071): 1105–1115.

Galanie, S., K. Thodey, I. J. Trenchard, M. Filsinger Interrante and C. D. Smolke (2015). "Complete biosynthesis of opioids in yeast". *Science* **349**(6252): 1095–1100.

Guo, L., X. Chen, L. N. Li, W. Tang, Y. T. Pan and J. Q. Kong (2016). "Transcriptome-enabled discovery and functional characterization of enzymes related to (2S)-pinocembrin biosynthesis from Ornithogalum caudatum and their application for metabolic engineering". *Microb Cell Fact.* **15**: 27.

Heide, L. (2009). "Aminocoumarins mutasynthesis, chemoenzymatic synthesis, and metabolic engineering". *Methods Enzymol.* **459**: 437–455.

Hoffmeister, D. and N. P. Keller (2007). "Natural products of filamentous fungi: enzymes, genes, and their regulation". *Nat. Prod. Rep.* **24**(2): 393–416.

Inglis, D. O., J. Binkley, M. S. Skrzypek, M. B. Arnaud, G. C. Cerqueira, P. Shah, F. Wymore, J. R. Wortman and G. Sherlock (2013). "Comprehensive annotation of secondary metabolite biosynthetic genes and gene clusters of Aspergillus nidulans, A. fumigatus, A. niger and A. oryzae". *BMC Microbiol.* **13**: 91.

Jiang, M., L. Fang and B. A. Pfeifer (2013). "Improved heterologous erythromycin A production through expression plasmid re-design". *Biotechnol. Prog.* **29**(4): 862–869.

Jimenez-Oses, G., S. Osuna, X. Gao, M. R. Sawaya, L. Gilson, S. J. Collier, G. W. Huisman, T. O. Yeates, Y. Tang and K. N. Houk (2014). "The role of distant mutations and allosteric regulation on LovD active site dynamics". *Nat. Chem. Biol.* **10**(6): 431–436.

Ju, K. S., J. Gao, J. R. Doroghazi, K. K. Wang, C. J. Thibodeaux, S. Li, E. Metzger, J. Fudala, J. Su, J. K. Zhang, J. Lee, J. P. Cioni, B. S. Evans, R. Hirota, D. P. Labeda, W. A. van der Donk and W. W. Metcalf (2015). "Discovery of phosphonic acid natural

products by mining the genomes of 10,000 actinomycetes". *Proc. Natl. Acad. Sci. U. S. A.* **112**(39): 12175–12180.
Kim, H. U., P. Charusanti, S. Y. Lee and T. Weber (2016). "Metabolic engineering with systems biology tools to optimize production of prokaryotic secondary metabolites". *Nat. Prod. Rep.* **33**(8): 933–941.
Lau, W. and E. Sattely (2015). "Six enzymes from mayapple that complete the biosynthetic pathway to the etoposide aglycone". *Science* **349**: 1224–1228.
Li, Y. F., K. J. Tsai, C. J. Harvey, J. J. Li, B. E. Ary, E. E. Berlew, B. L. Boehman, D. M. Findley, A. G. Friant, C. A. Gardner, M. P. Gould, J. H. Ha, B. K. Lilley, E. L. McKinstry, S. Nawal, R. C. Parry, K. W. Rothchild, S. D. Silbert, M. D. Tentilucci, A. M. Thurston, R. B. Wai, Y. Yoon, R. S. Aiyar, M. H. Medema, M. E. Hillenmeyer and L. K. Charkoudian (2016). "Comprehensive curation and analysis of fungal biosynthetic gene clusters of published natural products". *Fungal Genet. Biol.* **89**: 18–28.
Ludewig, A. H. and F. C. Schroeder (2012). "Ascaroside Signaling in C. elegans". In *WormBook*, ed. P. Kuwabara. The C. elegans Research Community, WormBook, doi/10.1895/wormbook.1.155.1, http://www.wormbook.org.
Malpartida, F. and D. A. Hopwood (1984). "Molecular cloning of the whole biosynthetic pathway of a Streptomyces antibiotic and its expression in a heterologous host". *Nature* **309**(5967): 462–464.
Metcalf, W. and W. A. van der Donk (2009). "Biosynthesis of phosphonic and phosphinic acid natural products". *Annu. Rev. Biochem.* **78**: 65–94.
Milshteyn, A., J. S. Schneider and S. F. Brady (2014). "Mining the metabiome: identifying novel natural products from microbial communities". *Chem. Biol.* **21**(9): 1211–1223.
Nierman, W. C., A. Pain, M. J. Anderson, J. R. Wortman, H. S. Kim, J. Arroyo, M. Berriman, K. Abe, D. B. Archer, C. Bermejo, J. Bennett, P. Bowyer, D. Chen, M. Collins, R. Coulsen, R. Davies, P. S. Dyer, M. Farman, N. Fedorova, N. Fedorova, T. V. Feldblyum, R. Fischer, N. Fosker, A. Fraser, J. L. Garcia, M. J. Garcia, A. Goble, G. H. Goldman, K. Gomi, S. Griffith-Jones, R. Gwilliam, B. Haas, H. Haas, D. Harris, H. Horiuchi, J. Huang, S. Humphray, J. Jimenez, N. Keller, H. Khouri, K. Kitamoto, T. Kobayashi, S. Konzack, R. Kulkarni, T. Kumagai, A. Lafon, J. P. Latge, W. Li, A. Lord, C. Lu, W. H. Majoros, G. S. May, B. L. Miller, Y. Mohamoud, M. Molina, M. Monod, I. Mouyna, S. Mulligan, L. Murphy, S. O'Neil, I. Paulsen, M. A. Penalva, M. Pertea, C. Price, B. L. Pritchard, M. A. Quail, E. Rabbinowitsch, N. Rawlins, M. A. Rajandream, U. Reichard,

H. Renauld, G. D. Robson, S. Rodriguez de Cordoba, J. M. Rodriguez-Pena, C. M. Ronning, S. Rutter, S. L. Salzberg, M. Sanchez, J. C. Sanchez-Ferrero, D. Saunders, K. Seeger, R. Squares, S. Squares, M. Takeuchi, F. Tekaia, G. Turner, C. R. Vazquez de Aldana, J. Weidman, O. White, J. Woodward, J. H. Yu, C. Fraser, J. E. Galagan, K. Asai, M. Machida, N. Hall, B. Barrell and D. W. Denning (2005). "Genomic sequence of the pathogenic and allergenic filamentous fungus Aspergillus fumigatus". *Nature* **438**(7071): 1151–1156.

Nutzmann, H. W., A. Huang and A. Osbourn (2016). "Plant metabolic clusters - from genetics to genomics". *New Phytol.* **211**(3): 771–789.

Nutzmann, H. W. and A. Osbourn (2014). "Gene clustering in plant specialized metabolism". *Curr. Opin. Biotechnol.* **26**: 91–99.

Olano, C., I. Garcia, A. Gonzalez, M. Rodriguez, D. Rozas, J. Rubio, M. Sanchez-Hidalgo, A. F. Brana, C. Mendez and J. A. Salas (2014). "Activation and identification of five clusters for secondary metabolites in Streptomyces albus J1074". *Microb. Biotechnol.* **7**(3): 242–256.

Omura, S., H. Ikeda, J. Ishikawa, A. Hanamoto, C. Takahashi, M. Shinose, Y. Takahashi, H. Horikawa, H. Nakazawa, T. Osonoe, H. Kikuchi, T. Shiba, Y. Sakaki and M. Hattori (2001). "Genome sequence of an industrial microorganism Streptomyces avermitilis: deducing the ability of producing secondary metabolites". *Proc. Natl. Acad. Sci. U. S. A.* **98**(21): 12215–12220.

Paddon, C. J., P. J. Westfall, D. J. Pitera, K. Benjamin, K. Fisher, D. McPhee, M. D. Leavell, A. Tai, A. Main, D. Eng, D. R. Polichuk, K. H. Teoh, D. W. Reed, T. Treynor, J. Lenihan, M. Fleck, S. Bajad, G. Dang, D. Dengrove, D. Diola, G. Dorin, K. W. Ellens, S. Fickes, J. Galazzo, S. P. Gaucher, T. Geistlinger, R. Henry, M. Hepp, T. Horning, T. Iqbal, H. Jiang, L. Kizer, B. Lieu, D. Melis, N. Moss, R. Regentin, S. Secrest, H. Tsuruta, R. Vazquez, L. F. Westblade, L. Xu, M. Yu, Y. Zhang, L. Zhao, J. Lievense, P. S. Covello, J. D. Keasling, K. K. Reiling, N. S. Renninger and J. D. Newman (2013). "High-level semi-synthetic production of the potent antimalarial artemisinin". *Nature* **496**(7446): 528–532.

Peck, S. and W. A. van der Donk (2013). "Phosphonate biosynthesis and catabolism: a treasure trove of unusual enzymology". *Curr. Opin. Chem. Biol.* **17**: 580–588.

Pfeifer, B. A., S. J. Admiraal, H. Gramajo, D. E. Cane and C. Khosla (2001). "Biosynthesis of complex polyketides in a metabolically engineered strain of E. coli". *Science* **291**(5509): 1790–1792.

Pickens, L. B., Y. Tang and Y. H. Chooi (2011). "Metabolic engineering for the production of natural products". *Annu. Rev. Chem. Biomol. Eng.* **2**: 211–236.

Potts, M. B., H. S. Kim, K. W. Fisher, Y. Hu, Y. P. Carrasco, G. B. Bulut, Y. H. Ou, M. L. Herrera-Herrera, F. Cubillos, S. Mendiratta, G. Xiao, M. Hofree, T. Ideker, Y. Xie, L. J. Huang, R. E. Lewis, J. B. MacMillan and M. A. White (2013). "Using functional signature ontology (FUSION) to identify mechanisms of action for natural products". *Sci. Signaling* **6**(297) ra90.

Qu, Y., M. L. Easson, J. Froese, R. Simionescu, T. Hudlicky and V. De Luca (2015). "Completion of the seven-step pathway from tabersonine to the anticancer drug precursor vindoline and its assembly in yeast". *Proc. Natl. Acad. Sci. U. S. A.* **112**(19): 6224–6229.

Ro, D. K., E. M. Paradise, M. Ouellet, K. J. Fisher, K. L. Newman, J. M. Ndungu, K. A. Ho, R. A. Eachus, T. S. Ham, J. Kirby, M. C. Chang, S. T. Withers, Y. Shiba, R. Sarpong and J. D. Keasling (2006). "Production of the antimalarial drug precursor artemisinic acid in engineered yeast". *Nature* **440**(7086): 940–943.

Robinette, S. L., R. Bruschweiler, F. C. Schroeder and A. S. Edison (2012). "NMR in metabolomics and natural products research: two sides of the same coin". *Acc. Chem. Res.* **45**(2): 288–297.

Santos, C. N., M. Koffas and G. Stephanopoulos (2011). "Optimization of a heterologous pathway for the production of flavonoids from glucose". *Metab. Eng.* **13**(4): 392–400.

Schroeder, F. C. (2015). "Modular assembly of primary metabolic building blocks: a chemical language in C. elegans". *Chem. Biol.* **22**(1): 7–16.

Siddiqui, M. S., K. Thodey, I. Trenchard and C. D. Smolke (2012). "Advancing secondary metabolite biosynthesis in yeast with synthetic biology tools". *FEMS Yeast Res.* **12**(2): 144–170.

Thimmappa, R., K. Geisler, T. Louveau, P. O'Maille and A. Osbourn (2014). "Triterpene biosynthesis in plants". *Annu. Rev. Plant Biol.* **65**: 225–257.

Trantas, E. A., M. A. Koffas, P. Xu and F. Ververidis (2015). "When plants produce not enough or at all: metabolic engineering of flavonoids in microbial hosts". *Front. Plant Sci.* **6**: 7.

Walsh, C. T., M. G. Acker and A. A. Bowers (2010). "Thiazolyl peptide antibiotic biosynthesis: a cascade of post-translational modifications on ribosomal nascent proteins". *J. Biol. Chem.* **285**(36): 27525–27531.

Walsh, C. T. and T. Wencewicz (2016). *Antibiotics Challeneges, Mechanisms, Opportunities*. Wshington DC, ASM Press.

Wang, H., D. P. Fewer, L. Holm, L. Rouhiainen and K. Sivonen (2014). "Atlas of nonribosomal peptide and polyketide biosynthetic pathways reveals common occurrence of nonmodular enzymes". *Proc. Natl. Acad. Sci. U. S. A.* **111**(25): 9259–9264.

Weber, T., K. Blin, S. Duddela, D. Krug, H. U. Kim, R. Bruccoleri, S. Y. Lee, M. A. Fischbach, R. Muller, W. Wohlleben, R. Breitling, E. Takano and M. H. Medema (2015). "antiSMASH 3.0-a comprehensive resource for the genome mining of biosynthetic gene clusters". *Nucleic Acids Res.* **43**(W1): W237–243.

Wiemann, P., C. J. Guo, J. M. Palmer, R. Sekonyela, C. C. Wang and N. P. Keller (2013). "Prototype of an intertwined secondary-metabolite supercluster". *Proc. Natl. Acad. Sci. U. S. A.* **110**(42): 17065–17070.

Wu, J., G. Du, J. Zhou and J. Chen (2013). "Metabolic engineering of Escherichia coli for (2S)-pinocembrin production from glucose by a modular metabolic strategy". *Metab. Eng.* **16**: 48–55.

Yamanaka, K., K. A. Reynolds, R. D. Kersten, K. S. Ryan, D. J. Gonzalez, V. Nizet, P. C. Dorrestein and B. S. Moore (2014). "Direct cloning and refactoring of a silent lipopeptide biosynthetic gene cluster yields the antibiotic taromycin A". *Proc. Natl. Acad. Sci. U. S. A.* **111**(5): 1957–1962.

Yan, Y., J. Chen, L. Zhang, Q. Zheng, Y. Han, H. Zhang, D. Zhang, T. Awakawa, I. Abe and W. Liu (2013). "Multiplexing of combinatorial chemistry in antimycin biosynthesis: expansion of molecular diversity and utility". *Angew. Chem., Int. Ed.* **52**(47): 12308–12312.

Zarins-Tutt, J. S., T. T. Barberi, H. Gao, A. Mearns-Spragg, L. Zhang, D. J. Newman and R. J. Goss (2016). "Prospecting for new bacterial metabolites: a glossary of approaches for inducing, activating and upregulating the biosynthesis of bacterial cryptic or silent natural products". *Nat. Prod. Rep.* **33**(1): 54–72.

Subject Index